STUDENT STUDY GUIDE

FOR

BIOLOGY

CAMPBELL • REECE

MARTHA R. TAYLOR
Ithaca, New York

SIXTH EDITION

Benjamin Cummings

San Francisco Boston New York
Capetown Hong Kong London Madrid Mexico City
Montreal Munich Paris Singapore Sidney Tokyo Toronto

Senior Editor: Beth Wilbur
Project Editor: Evelyn Dahlgren
Publishing Assistant: David DeRouen
Production Editor: Steven Anderson
Production Service: The Left Coast Group
Copy Editor: Anna Reynolds Trabucco
Cover Designer: Roy Neuhaus

Cover Photo:

ISBN 0-8053-6634-2

Copyright © 2002 Pearson Education, Inc., publishing as Benjamin Cummings, 1301 Sansome St., San Francisco, CA 94111. All rights reserved. Manufactured in the United States of America. This publication is protected by Copyright and permission should be obtained from the publisher prior to any prohibited reproduction, storage in a retrieval system, or transmission in any form or by any means, electronic, mechanical, photocopying, recording, or likewise. To obtain permission(s) to use material from this work, please submit a written request to Pearson Education, Inc., Permissions Department, 1900 E. Lake Ave., Glenview, IL 60025. For information regarding permissions, call 847/486/2635.

Many of the designations used by manufacturers and sellers to distinguish their products are claimed as trademarks. Where those designations appear in this book, and the publisher was aware of a trademark claim, the designations have been printed in initial caps or all caps.

Credits

Figures 5.AS.1, 5.AS.2, 5.IQ.2, 5.IQ.10, 11.IQ.2, 11.IQ.3, 12.IQ.4, 16.IQ.2, 21.IQ.10, 29.UN.1, 30.IQ.2, 30.IQ.4, 36.IQ.2, 36.IQ.6, 39.IQ.3, 39.UN.1, 41.IQ.3, 43.IQ.4, 46.IQ.7, 47.UN.1, 50.IQ.5: Adapted from Neil Campbell, Jane Reece, and Larry Mitchell, Biology, 5th ed. (Menlo Park, CA: Benjamin/Cummings, 1999). ©1999 The Benjamin/Cummings Publishing Company, Inc.
Photo 12.UN.2: © Carolina Biological Supply/Phototake.
Photo 27.IQ.6a: © H. S. Pankratz, Michigan State University/BPS.
Photo 27.IQ.6b: © P. W. Johnson and Jon M. Sieburth, University of Rhode Island/BPS.
Figure 47.IQ.4: Adapted from Neil Campbell, Larry Mitchell, and Jane Reece, Biology: Concepts and Connections, 2nd ed. (Menlo Park, CA: Benjamin/Cummings, 1997). ©1997 The Benjamin/Cummings Publishing Company, Inc.
Figure 48.IQ.1: Adapted from Elaine Marieb, Human Anatomy and Physiology, 4th ed. (Menlo Park, CA: Benjamin/Cummings, 1998). ©1998 The Benjamin/Cummings Publishing Company, Inc.

Benjamin
Cummings

7 8 9 10—CRS—04
www.aw.com/bc

CONTENTS

Another name for the *Student Study Guide* for Campbell/Reece, *Biology,* Sixth Edition, could be "A Student Structuring Guide." The purpose of this guide is to help you to structure and organize your developing knowledge of biology and to create your own personal understanding of the topics covered in the text.

Biology is a visual science, and this *Student Study Guide* includes many diagrams and pictures taken from *Biology,* Sixth Edition, to help you see and understand the relationships among the parts of structures and processes. Interactive Questions, which are interspersed throughout each chapter's narrative, are designed to help you develop an active learning approach to your study of biology. The numerous multiple choice questions in each chapter will help you test your growing understanding of biological facts and concepts. (See the end of this Preface for tips on taking objective tests.) New to this sixth edition is a Word Roots section in most of the chapters, intended to help you learn the large new vocabulary that is a necessary part of a biology course.

The five divisions of each study guide chapter are as follows:

- The *Framework* identifies the overall picture; it provides a conceptual framework into which the chapter information fits.
- The *Chapter Review* is a condensation of each textbook chapter, with page references given for each concept heading within the chapter. All text bold terms are also shown in boldface in this section. Interspersed in this summary are **Interactive Questions** that help you stop and synthesize the material just covered. You are asked to complete tables, label diagrams, respond to short answer or essay questions, and complete or construct concept maps.
- The *Word Roots* section presents the derivation of key biological prefixes, suffixes, and word roots. Examples of bold-faced terms from the chapter are then defined. Breaking a complicated term down to identifiable components will help you to recognize and learn many new biological terms.
- The *Structure Your Knowledge* section directs you to organize and relate the main concepts of the chapter. It helps you to piece together the key ideas into a bigger picture.

- In the *Test Your Knowledge* section, you are provided with objective questions to test your understanding. The multiple choice questions presented in each chapter ask you to choose the best answer. Some answers may be partly correct; almost all choices have been written to test your ability to think and discriminate among alternatives. Make sure you understand why the other choices are incorrect as well as why the correct answer is correct.
- Suggested answers to the Interactive Questions, the Structure Your Knowledge sections, and the Test Your Knowledge questions are provided in the *Answer Section* at the end of the book.

Using Concept Maps: What are the **concept maps** that appear throughout this study guide? A concept map is a diagram that shows how ideas are organized and related. The *structure* of a concept map is a hierarchically organized cluster of concepts, enclosed in boxes and connected with labeled lines, that explicitly shows the relationships among the concepts. The *function* of a concept map is to help you structure your understanding of a topic and create meaning. The *value* of a concept map is in the thinking and organizing required to create a map.

Developing a concept map for a group of concepts requires you to evaluate the relative importance of the concepts (Which are most inclusive and important? Which are less important and subordinate to other concepts?), arrange the concepts in a meaningful cluster, and draw connections between them that help you make their meanings explicit.

This book uses concept maps in several ways. A map of a chapter may be presented in the Framework section to show the organization of the key concepts in that chapter. An Interactive Question may provide a skeleton concept map, with some concepts provided and boxes for you to complete. This technique is intended to help you become more familiar with concept maps and to illustrate one possible approach to organizing the concepts of a particular section.

You will also be asked to develop your own concept maps on certain subsets of ideas. In these cases, the Answer Section will present a suggested concept map. A concept map is an individual picture of your understanding at the time you make the map. As your understanding of an area develops, your concept map will evolve—sometimes becoming more complex and interrelated, sometimes becoming simplified and streamlined. Do not look to the Answer Section for the "right" concept map. After you have organized your own thoughts, look at the answer map to make sure you have included the key concepts (although you may have added more), to check that the connections you have made are reasonable, and perhaps to see another way to organize the information.

Tips for Using this Study Guide: This guide certainly is not a replacement for your textbook or biology class. But it should help support and even streamline your learning process. Some students read the assigned text chapters before lecture and then the study guide chapters after class to reinforce and review. Others reverse that order, skimming the study guide chapter before lecture and then carefully dissecting the text after listening to the professor's presentation. This study guide is especially helpful in preparation for tests. With the text open beside it so you can refer to the essential textbook diagrams, reread the relevant study guide chapters for a quick review. Return to the text for a more complete description of any sections that you don't fully understand.

Because this book is intended to help you learn, most incorrect choices given for multiple choice questions are written to be educational: to review other concepts covered in the chapter, to help you distinguish between closely related ideas, to point out common misconceptions of a particular concept. Get your money's worth out of this book. Even when you can quickly identify the correct answer to a question, take the time to read the other choices to see what they can teach you. Page

back through the Chapter Summary section and the textbook to review material that you have not yet fully understood. Take your time with the Test Your Knowledge section—don't rush through it right before an exam. Treat it as an important learning opportunity. And use it to practice good test-taking skills.

Tips for Taking Multiple Choice Tests: Don't do all of your thinking in your head. Write in the margins and blank spaces of your test. Interact with each question. Read the stem of the question carefully, underlining or even using a highlighter to identify the key concept. Read each answer slowly. Cross out the ones you know are wrong. Circle the key idea that you think identifies the correct answer. Now read the question and the answer you chose together, making sure your choice really does answer what is asked. If you aren't sure about a question, try rereading the question and each choice individually. Draw yourself diagrams and pictures. Write down what you do know, and it may jog your memory. If you still are not sure, mark the question to come back to later. As you work through related questions, you may find information that helps you figure out that question. And remember, there is no substitute for good preparation, proper rest and nutrition, and a positive attitude.

Biology is a fascinating, broad, and exciting subject. Campbell/Reece, *Biology* is filled with terminology and facts organized in a manner that will help you build a conceptual framework of the major themes of modern biology. This *Student Study Guide* is intended to help you learn and recall information and, most importantly, to encourage and guide you as you develop your own understanding of and appreciation for biology.

Martha R. Taylor

CHAPTER 1

INTRODUCTION:
TEN THEMES IN THE STUDY OF LIFE

FRAMEWORK

This chapter outlines broad themes that unify the study of biology and describe the scientific construction of biological knowledge. A course in biology is neither a vocabulary course nor a classification exercise for the diverse forms of life. Biology is a collection of facts and concepts structured within theories and organizing principles. Recognizing the common themes within biology will help you to structure your knowledge of this fascinating and challenging study of life.

CHAPTER REVIEW

Biology, the scientific study of life, is an extension of our innate interest in life in its diverse forms. The scope of biology is immense, spanning the submicroscopic level to the complex web of ecosystems from the present back through nearly 4 billion years of evolutionary history. Recent advances in research methods have aided the efforts of a large number of contemporary biologists working throughout the many subfields of biology to achieve an explosion of information. A beginning student can make sense of this expanding body of knowledge by focusing on a few enduring themes that unify the study of biology.

Exploring Life on Its Many Levels

Each level of biological organization has emergent properties (2–4)

A Hierarchy of Organization   The hierarchy of biological structure includes atoms, biological molecules, organelles, cells, tissues, organs, organ systems, organisms, populations, communities, and ecosystems. The study of biology encompasses these various levels of structure and the interactions among them.

Emergent Properties   Interactions among components at each level of biological organization lead to

the emergence of novel properties at the next level: The whole is greater than the sum of its parts. Structural arrangement is central to these emergent properties.

The properties of life include order, reproduction, growth and development directed by heritable programs, energy utilization, responsiveness to the environment, homeostasis, and evolutionary adaptation.

Reductionism in Biology   Biology combines the powerful and pragmatic reductionist strategy, which breaks down complex systems to simpler components, with the study of higher organizational levels of life.

Cells are an organism's basic units of structure and function (4–6)

The cell is the simplest structural level capable of performing all the activities of life.

The Cell Theory   Hooke first described and named cells in 1665 when he observed a slice of cork with a simple microscope. Leeuwenhoek, a contemporary, developed lenses that permitted him to view the world of microscopic organisms. In 1839, Schleiden and Schwann concluded that all living things consist of cells. This cell theory now includes the idea that all cells come from other cells.

The Two Main Cell Types   Two major types of cells are recognized. The simpler prokaryotic cell, unique to bacteria and archaea, lacks both a nucleus to enclose its DNA and most cytoplasmic organelles. The eukaryotic cell, with its nucleus, DNA organized into chromosomes, and numerous membrane-bound compartments, is typical of all other living organisms.

The continuity of life is based on heritable information in the form of DNA (6–7)

The biological instructions for the development and functioning of organisms are coded in the arrangement of the four kinds of nucleotides in DNA molecules. A precise mechanism for replicating the DNA double helix is essential for cell division and for the transmission from parent to offspring of the units of inheritance called genes. All forms of life use essentially the same genetic code.

The first sequencing of the human genome, or complete set of genetic instructions, was published in 2001.

Structure and function are correlated at all levels of biological organization (7–8)

The form of a biological structure gives information about its function, and a study of function provides insight into structural organization. The principle that form fits function is illustrated at all levels of biological organization.

Organisms are open systems that interact continuously with their environments (8)

An organism exchanges materials and energy with its surroundings. Organisms affect and are affected by the physical and biological environments with which they interact.

Ecosystem Dynamics Within an ecosystem, nutrients cycle between the abiotic and biotic components, and energy flows from sunlight to photosynthetic organisms (producers) to consumers and exits in the form of heat.

Energy Conversion Photosynthetic plants convert solar energy to chemical energy. The work of cells relies on chemical energy, which may be converted to kinetic energy and which eventually dissipates as heat.

Regulatory mechanisms ensure a dynamic balance in living systems (8–9)

Enzymes are organic catalysts produced by a cell that speed up its chemical reactions. Precise regulation of its enzymes allows a cell to respond to changing conditions or needs. Organisms maintain an internal balance (homeostasis) through positive or negative feedback systems that either speed up or slow down body processes.

Evolution, Unity, and Diversity

Diversity and unity are the dual faces of life on Earth (9–12)

About 1.5 million species, out of an estimated total of 5–30 million, have been identified and named.

Grouping Species: The Basic Concept Taxonomy is the branch of biology that names organisms and classifies species into hierarchical groups.

The Three Domains of Life Traditionally, life forms have been organized into five kingdoms, although schemes of six, eight, or more kingdoms have recently been proposed. These varying numbers of kingdoms can be grouped into three domains. The prokaryotes are split into the domains Archaea and Bacteria, and the eukaryotes are placed in the domain Eukarya. Within the Eukarya, the traditional kingdom Protista contains mostly unicellular or simple multicellular forms. The other three kingdoms contain multicellular organisms characterized to a large extent by their mode of nutrition. Plants are photosynthetic, fungi absorb their nutrients from decomposing organic material, and animals ingest other organisms.

Unity in the Diversity of Life Within this diversity, living forms share a universal genetic language of DNA and similarities in cell structure.

Evolution is the core theme of biology (12–15)

Evolution connects all of life by common ancestry. The history of living forms extends back over 3.5 billion years to the ancient prokaryotes, and the incredible diversity of life is the result of evolution.

Darwin and Natural Selection In *The Origin of Species*, published in 1859, Charles Darwin presented his case for evolution, or "descent with modification," that present forms evolved from a succession of ancestral forms. Darwin synthesized the theory of natural selection as the mechanism of evolution by drawing an inference from two observations: Individuals vary in many heritable traits, and the overproduction of offspring sets up a struggle for existence. Individuals with traits best suited for an environment leave a larger proportion of offspring than do less fit individuals. This natural selection, or differential reproductive success within a population, results in the gradual accumulation of favorable adaptations to the challenges of an environment.

Natural Selection and the Diversity of Life According to Darwin, new species originate when isolated populations diversify over time in response to different environmental selective pressures. Evolution makes sense of both the unity and diversity of life.

The Process of Science

Science is a process of inquiry that includes repeatable observations and testable hypotheses (16–20)

Science is a way of knowing that involves asking and endeavoring to answer questions about nature.

Discovery Science and Induction   Careful and verifiable observation and description are the basis of discovery science. Using inductive reasoning, a generalized conclusion can often be drawn from collections of observations.

Hypothetico-Deductive Science   The scientific process is based on hypothetico-deductive reasoning. A hypothesis is a tentative explanation for an observation or question. Using "if . . . then" logic, deductive reasoning proceeds from the general to the specific, from a general hypothesis to specific predictions of results if the general premise is true. A hypothesis is usually tested by performing experiments or making observations to see whether predicted results occur.

The field and laboratory experiments of Reznick and Endler illustrate the hypothetico-deductive approach. By maintaining control populations and following generations of guppies transplanted from sites with pike-cichlids to sites with killifish for 11 years, they were able to conclude that natural selection due to differential predation was the most likely explanation for differences in life history characteristics between guppy populations.

The Reznick and Endler experimental design also illustrates the scientific use of controlled experiments in which subjects are divided into an experimental group and a control group. Both groups are treated alike except for the one variable that the experiment is trying to test.

Theories in Science   Facts, in the form of observations and experimental results, are prerequisites of science, but it is the new ways of organizing and relating those facts that advance science. A theory is broader in scope than a hypothesis and is supported by a large body of evidence.

Science as a Social Process   Most scientists work in teams and share their results with a broader research community in journal articles. Science is characterized as progressive and self-correcting. Scientists build on the work done by others, refining or refuting their ideas, both cooperating and competing with each other.

The Cultural Context of Science   The political and cultural environment influences the ways in which scientists approach their work. But the adherence to the criteria of hypothesis testing and verifiable observations sets science apart from other ways of "knowing nature."

Science and technology are functions of society (21)

Science and technology are interwoven as the information generated by science is used in the development of goods and services, and as technological advances are used to extend scientific knowledge. The fields of molecular biology and genetic engineering illustrate this relationship. Technology contributes to our standard of living and has made it possible for the human population to expand rapidly but at a price of severe environmental consequences. Solutions to environmental problems and future uses of scientific knowledge and technologies will involve politics, economics, and cultural values as well as science and technology.

Review: Using Themes to Connect the Concepts of Biology

Biology is a demanding science—partly because living systems are so complex and partly because biology incorporates concepts from chemistry, physics, and math. This book presents a wealth of information. The basic themes of biology will help you understand, appreciate, and structure your growing knowledge of biology.

STRUCTURE YOUR KNOWLEDGE

1. This chapter presents ten unifying themes of biology. Briefly describe each of these in your own words:
    a. emergent properties
    b. the cell
    c. heritable information
    d. correlation of structure and function
    e. interaction with the environment
    f. regulation
    g. unity and diversity
    h. evolution
    i. scientific inquiry
    j. science, technology, and society

THE CHEMISTRY OF LIFE

CHAPTER 2

THE CHEMICAL CONTEXT OF LIFE

FRAMEWORK

This chapter considers the basic principles of chemistry that explain the behavior of atoms and molecules and that form the basis for our modern understanding of biology. Emergent properties are associated with each new level of structural organization as the subatomic particles—protons, neutrons, and electrons—are organized into atoms and atoms are combined by covalent or ionic bonds into molecules.

CHAPTER REVIEW

Chemical Elements and Compounds

Matter consists of chemical elements in pure form and in combinations called compounds (26–28)

Matter is anything that takes up space and has mass. The basic forms of matter are **elements**, substances that cannot be chemically broken down to other types of matter. A **compound** is made up of two or more elements combined in a fixed ratio. A compound usually has characteristics quite different from its constituent elements, an example of the emergence of novel properties in higher levels of organization.

■ **INTERACTIVE QUESTION 2.1**

On Earth, mass and weight may be considered synonymous.

a. Define mass.

b. Define weight.

Life requires about 25 chemical elements (27–28)

Carbon (C), oxygen (O), hydrogen (H), and nitrogen (N) make up 96% of living matter. The remaining 4% is composed of the seven elements listed on the following page. Some elements, like iron (Fe) and iodine (I), may be required in very minute quantities and are called **trace elements**.

Atoms and Molecules

Atomic structure determines the behavior of an element (28–33)

An **atom** is the smallest unit of an element retaining the physical and chemical properties of that element. Each element has its own unique type of atom.

■ INTERACTIVE QUESTION 2.2

Fill in the names beside the symbols of the following elements commonly found in living matter.

Symbol	Element
Ca	
P	
K	
S	
Na	
Cl	
Mg	

Subatomic Particles   Three stable subatomic particles are important to our understanding of atoms. Uncharged **neutrons** and positively charged **protons** are packed tightly together to form the **atomic nucleus** of an atom. Negatively charged **electrons** orbit rapidly about the nucleus.

Protons and neutrons have a similar mass of about 1.7×10^{-24} g or 1 **dalton** each. A dalton is the measurement unit for atomic mass. Electrons have negligible mass.

Atomic Number and Atomic Weight   Each element has a characteristic **atomic number,** or number of protons in the nucleus of its atom. Unless otherwise indicated, the number of protons in an atom is equal to the number of electrons, and the atom has a neutral electrical charge. A subscript to the left of the symbol for an element indicates its atomic number; a superscript indicates mass number. The **mass number** is equal to the number of protons and neutrons in the nucleus and approximates the mass of an atom of that element in daltons. The term **atomic weight** is often used to refer to the mass of an atom.

■ INTERACTIVE QUESTION 2.3

The difference between the mass number and the atomic number of an atom is equal to the number of _____. An atom of phosphorus, $^{31}_{15}$ P, contains _____ protons, _____ electrons, and _____ neutrons. The atomic weight of phosphorus is _____.

Isotopes   Although the number of protons is constant, the number of neutrons can vary among the atoms of an element, creating different **isotopes** that have slighlty different masses but the same chemical behavior. Some isotopes are unstable, or radioactive; their nuclei spontaneously decay, giving off particles and energy.

Radioactive isotopes are important tools in biological research and medicine. Techniques using scintillation counters or autoradiography can determine the quantity and location of radioactively labeled molecules within a cell or tissue. Chemical processes can be located and monitored within an organism using radioactive tracers and PET (positron-emission tomography). Too great an exposure to radiation from decaying isotopes poses a significant health hazard.

The Energy Levels of Electrons   **Energy** is defined as the ability to do work. **Potential energy** is energy stored in matter as a consequence of the relative position of masses. Matter naturally tends to move toward a more stable lower level of potential energy and requires the input of energy to return to a higher potential energy.

The potential energy of electrons increases as their distance from the positively charged nucleus increases. Electrons can orbit in several different potential energy states, called **energy levels** or **electron shells,** surrounding the nucleus.

■ INTERACTIVE QUESTION 2.4

To move to a shell farther from the nucleus, an electron must (absorb/release) energy; energy is (absorbed/released) when an electron moves to a closer shell. (Circle correct terms.)

Electron Configuration and Chemical Properties   The chemical behavior of an atom is a function of its electron configuration—in particular, the number of **valence electrons** in its outermost electron shell, or **valence shell.** A valence shell of eight electrons is complete, resulting in an unreactive or inert atom. Atoms with incomplete valence shells are chemically reactive because of their unpaired electrons. The periodic table of elements is arranged in order of the sequential addition of electrons to orbitals in the electron shells.

Electron Orbitals   An **orbital** is the three-dimensional space or volume within which an electron is most likely to be found. No more than two electrons can occupy the same orbital. The first electron shell can contain two electrons in a single spherical orbital, called the 1*s* orbital. The second electron shell can hold a maximum of eight electrons in its four orbitals, which are a 2*s* spherical orbital and three dumbbell-shaped *p* orbitals located along the *x, y,* and *z* axes.

■ INTERACTIVE QUESTION 2.5

Draw the electron configuration for these atoms.

a. $_7$N   **c.** $_{12}$Mg

b. $_8$O   **d.** $_{17}$Cl

Atoms combine by chemical bonding to form molecules (33–36)

Atoms with incomplete valence shells can either share electrons with or completely transfer electrons to or from other atoms such that each atom is able to complete its valence shell. These interactions usually result in attractions, called **chemical bonds,** that hold the atoms together.

■ INTERACTIVE QUESTION 2.6

Fill in the blanks in the following concept map to help you review the atomic structure of atoms.

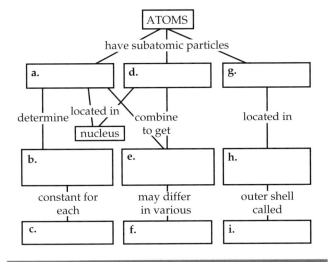

Covalent Bonds   When two atoms share a pair of valence electrons, a **covalent bond** is formed. A **molecule** consists of two or more atoms held together by covalent bonds. A **structural formula,** such as H—H, indicates both the number and type of atoms and also the bonding within a molecule. A **molecular formula,** such as O_2, indicates only the kinds and numbers of atoms in a molecule. In an oxygen molecule, two pairs of valence electrons are shared between oxygen atoms, forming a **double covalent bond.**

The **valence,** or bonding capacity, of an atom equals the number of unpaired electrons in its valence shell.

■ INTERACTIVE QUESTION 2.7

What are the valences of the four most common elements of living matter?

a. hydrogen ＿＿＿   **c.** nitrogen ＿＿＿

b. oxygen ＿＿＿   **d.** carbon ＿＿＿

Electronegativity is the attraction of an atom for shared electrons. If the atoms in a molecule have similar electronegativities, the electrons remain equally shared between the two nuclei, and the covalent bond is said to be **nonpolar.** If one element is more electronegative, it pulls the shared electrons closer to itself, creating a **polar covalent bond.** This unequal sharing of electrons results in a slight negative charge associated with the more electronegative atom and a slight positive charge associated with the atom from which the electrons are pulled.

■ INTERACTIVE QUESTION 2.8

Explain whether the following molecules contain nonpolar or polar covalent bonds. (Hint: N and O both have high electronegativities.)

a. nitrogen molecule N≡N   **b.** methane H–C–H (with H above and H below)

c. ammonia H–N(–H)(–H)   **d.** formaldehyde (H)(H)C=O

Ionic Bonds   If two atoms are very different in their attraction for the shared electrons, the more electronegative atom may completely transfer an electron from another atom, resulting in the formation of charged atoms called **ions.** The atom that lost the electron is a positively charged **cation.** The negatively charged atom that gained the electron is called an **anion.** An **ionic bond** may hold these atoms together because of the attraction of their opposite charges. **Ionic compounds,** called **salts,** often exist as three-dimensional crystalline lattice arrangements held together by electrical attraction. The number of ions present in a salt crystal is not fixed, but the atoms are present in specific ratios. Salts have strong ionic bonds when dry, but the crystal dissolves in water.

Ion also refers to entire covalent molecules that are electrically charged. Ammonium (NH_4^+) is a cation; this covalently bonded molecule is missing one electron.

■ **INTERACTIVE QUESTION   2.9**

Calcium ($_{20}$Ca) and chlorine ($_{17}$Cl) can combine to form the salt calcium chloride. Based on the number of electrons in their valence shells and their bonding capacities, what would the molecular formula for this salt be?
a. _____ Which atom becomes the cation? **b.** _____

Weak chemical bonds play important roles in the chemistry of life (36–37)

Weak bonds, such as ionic bonds in water, form temporary interactions between molecules and are involved in many biological signals and processes. Weak bonds within large molecules such as proteins help to create the three-dimensional shape and resulting activity of these molecules.

Hydrogen Bonds   When a hydrogen atom is covalently bonded with an electronegative atom and thus has a partial positive charge, it can be attracted to another electronegative atom and form a **hydrogen bond.**

Van der Waals Interactions   All atoms and molecules are attracted to each other when in close contact by **van der Waals interactions.** Momentary uneven electron distributions produce changing positive and negative regions that create these weak attractions.

■ **INTERACTIVE QUESTION   2.10**

Sketch a water molecule, showing oxygen's electron shells and the covalently shared electrons. Indicate the areas with slight negative and positive charges that enable a water molecule to form hydrogen bonds with other polar molecules.

A molecule's biological function is related to its shape (37–38)

A molecule's characteristic size and shape affect how it interacts with other molecules. When atoms form covalent bonds, their *s* and three *p* orbitals hybridize to form four teardrop-shaped orbitals in a tetrahedral arrangement. These hybrid orbitals dictate the specific shapes of different molecules.

Chemical reactions make and break chemical bonds (38–39)

Chemical reactions involve the making or breaking of chemical bonds in the transformation of matter into different forms. Matter is conserved in chemical reactions; the same number and kinds of atoms are present in both **reactants** and **products,** although the rearrangement of electrons and atoms causes the properties of these molecules to be different.

■ **INTERACTIVE QUESTION   2.11**

Fill in the missing coefficients for respiration, the conversion of glucose and oxygen to carbon dioxide and water, so that all atoms are conserved in the chemical reaction.

$$C_6H_{12}O_6 + \underline{} O_2 \longrightarrow \underline{} CO_2 + \underline{} H_2O$$

Most reactions are reversible—the products of the forward reaction can become reactants in the reverse reaction. Increasing the concentrations of reactants can speed up the rate of a reaction. **Chemical equilibrium** may be reached when the forward and reverse reactions proceed at the same rate, and the relative concentrations of reactants and products no longer change.

WORD ROOTS

an- = not (*anion:* a negatively charged ion)

co- = together; **-valent** = strength (*covalent bond:* an attraction between atoms that share one or more pairs of outer-shell electrons)

electro- = electricity (*electronegativity:* the tendency for an atom to pull electrons towards itself)

iso- = equal (*isotope:* an element having the same number of protons and electrons but a different number of neutrons)

neutr- = neither (*neutron:* a subatomic particle with a neutral electrical charge)

pro- = before (*proton:* a subatomic particle with a single positive electrical charge)

STRUCTURE YOUR KNOWLEDGE

Take the time to write out or discuss your answers to the following questions. Then refer to the suggested answers at the end of the book.

1. Fill in the following chart for the major subatomic particles of an atom.

Particle	Charge	Mass	Location

2. Atoms can have various numbers associated with them.
   a. Define the following and show where each of them is placed relative to the symbol of an element such as C: atomic number, mass number, atomic weight.
   b. Define valence.
   c. Which of these four numbers is most related to the chemical behavior of an atom? Explain.

3. Explain what is meant by saying that the sharing of electrons between atoms falls on a continuum from covalent bonds to ionic bonds.

TEST YOUR KNOWLEDGE

MULTIPLE CHOICE: *Choose the one best answer.*

1. Each element has its own characteristic atom in which
   a. the atomic weight is constant.
   b. the atomic number is constant.
   c. the mass number is constant.
   d. two of the above are correct.
   e. all of the above are correct.

2. Radioactive isotopes can be used in studies of metabolic pathways because
   a. their half-life allows a researcher to time an experiment.
   b. they are more reactive.
   c. the cell does not recognize the extra protons in the nucleus, so isotopes are readily used in metabolism.
   d. their location or quantity can be experimentally determined because of their radioactivity.
   e. their extra neutrons produce different colors that can be traced through the body.

3. In a reaction in chemical equilibrium,
   a. the forward and reverse reactions are occurring at the same rate.
   b. the reactants and products are in equal concentration.
   c. the forward reaction has gone further than the reverse reaction.
   d. there are equal numbers of atoms on both sides of the equation.
   e. a, b, and d are correct.

4. Oxygen has eight electrons. You would expect the arrangement of these electrons to be:
   a. eight in the second energy shell, creating an inert element.
   b. two in the first energy shell and six in the second, creating a valence of six.
   c. two in the 1*s* orbital and two each in the three 2*p* orbitals, creating a valence of zero.
   d. two in the 1*s* orbital, one each in the 2*s* and three 2*p* orbitals, and two in the 3*s* orbital, creating a valence of two.
   e. two in the 1*s* orbital, two in both the 2*s* and 2*px* orbitals, and one each in the 2*py* and 2*pz* orbitals, creating a valence of two.

5. A covalent bond between two atoms is likely to be polar if
   a. one of the atoms is much more electronegative than the other.
   b. the two atoms are equally electronegative.
   c. the two atoms are of the same element.
   d. the bond is part of a tetrahedrally shaped molecule.
   e. one atom is an anion.

6. A triple covalent bond would
   a. be very polar.
   b. involve the bonding of three atoms.
   c. involve the bonding of six atoms.
   d. produce a triangularly shaped molecule.
   e. involve the sharing of six electrons.

7. A cation
   a. has gained an electron.
   b. can easily form hydrogen bonds.
   c. is more likely to form from an atom with six or seven electrons in its valence shell.
   d. has a positive charge.
   e. Both c and d are correct.

8. What types of bonds are identified in the following illustration of a water molecule interacting with an ammonia molecule?

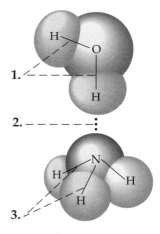

   a. Bonds 1 are polar covalent bonds, bond 2 is a hydrogen bond, and bonds 3 are nonpolar covalent bonds.
   b. Bonds 1 and 3 are polar covalent bonds and bond 2 is a hydrogen bond.
   c. Bonds 1 and 3 are polar covalent bonds and bond 2 is an ionic bond.
   d. Bonds 1 and 3 are nonpolar covalent bonds and bond 2 is a hydrogen bond.
   e. Bonds 1 and 3 are polar covalent bonds and bond 2 is a nonpolar covalent bond.

9. Which of the following weak bonds may form between any closely aligned molecules?
   a. nonpolar covalent
   b. polar covalent
   c. ionic
   d. hydrogen
   e. van der Waals interactions

10. The ability of morphine to mimic the effects of the body's endorphins is due to
   a. a chemical equilibrium developing between morphine and endorphins.
   b. the one-way conversion of morphine into endorphin.
   c. molecular shape similarities that allow morphine to bind to endorphin receptors.
   d. the similarities between morphine and heroin.
   e. hydrogen bonding and other weak bonds forming between morphine and endorphins.

The six elements most common in living organisms are:

$$^{12}_{6}C \qquad ^{16}_{8}O \qquad ^{1}_{1}H \qquad ^{14}_{7}N \qquad ^{32}_{16}S \qquad ^{31}_{15}P$$

Use this information to answer questions 11 through 16.

11. How many electrons does phosphorus have in its valence shell?
   a. 5
   b. 7
   c. 8
   d. 15
   e. 16

12. What is the atomic weight of phosphorus?
   a. 15
   b. 16
   c. 31
   d. 46
   e. 62

13. A radioactive isotope of carbon has the mass number 14. How many neutrons does this isotope have?
   a. 6
   b. 7
   c. 8
   d. 12
   e. 14

14. How many covalent bonds is a sulfur atom most likely to form?
   a. 1
   b. 2
   c. 3
   d. 4
   e. 5

15. Based on electron configuration, which of these elements would have chemical behavior most like that of oxygen?
    a. C
    b. H
    c. N
    d. P
    e. S

16. How many of these elements are found next to each other (side by side) on the periodic table?
    a. one group of two
    b. two groups of two
    c. one group of two and one group of three
    d. one group of three
    e. all of them

17. Taking into account the bonding capacities or valences of carbon (C) and oxygen (O), how many hydrogen (H) must be added to complete the structural diagram of this molecule?

$$O \atop O \diagdown C-C-C-C=C-C-C$$

    a. 9
    b. 10
    c. 11
    d. 12
    e. 13

18. A sodium ion (Na^+) contains 10 electrons, 11 protons, and 12 neutrons. What is the atomic number of sodium?
    a. 10
    b. 11
    c. 12
    d. 23
    e. 33

19. What type of bond would you expect potassium ($^{39}_{19}K$) to form?
    a. ionic; it would donate one electron and carry a positive charge
    b. ionic; it would donate one electron and carry a negative charge
    c. covalent; it would share one electron and make one covalent bond
    d. covalent; it would share two electrons and form two bonds
    e. none; potassium is an inert element

20. What is the molecular shape of methane (CH_4)?
    a. planar or flat, with the H arranged around the C
    b. pentagonal, or a flat five-sided arrangement
    c. tetrahedral, due to the hybridization of the s and three *p* orbits of the C
    d. circular, with the four H attached in a ring around the C
    e. linear, since all the bonds are nonpolar covalent

21. Which of the following is a molecule capable of forming hydrogen bonds?
    a. CH_4
    b. H_2O
    c. NaCl
    d. H_2
    e. a, b, and d can form hydrogen bonds.

22. Chlorine has an atomic number of 17 and a mass number of 35. How many electrons would a chloride ion have?
    a. 16
    b. 17
    c. 18
    d. 33
    e. 34

CHAPTER 3

WATER AND THE FITNESS
OF THE ENVIRONMENT

FRAMEWORK

Water makes up 70% to 95% of the cell content of living organisms and covers 75% of the Earth's surface. Its unique properties make the external environment fit for living organisms and the internal environment of organisms fit for the chemical and physical processes of life.

Hydrogen bonding between polar water molecules creates a cohesive liquid with a high specific heat and high heat of vaporization, both of which help to regulate environmental temperature. Ice floats and protects oceans and lakes from freezing. The polarity of water makes it a versatile solvent. An organism's pH may be regulated by buffers. Acid precipitation poses a serious environmental threat.

CHAPTER REVIEW

The Effects of Water's Polarity

**The polarity of water molecules
results in hydrogen bonding (41–42)**

A water molecule consists of two hydrogen atoms each covalently bonded to a more electronegative oxygen atom. This **polar molecule** has a V shape with a slight positive charge on each hydrogen atom and a slight negative charge associated with the oxygen. Hydrogen bonds form between the hydrogen atoms of one water molecule and the oxygen atoms of other water molecules, creating a higher level of structural organization and leading to the emergent properties of water.

**Organisms depend on the cohesion
of water molecules (42)**

Liquid water is unusually cohesive due to the constant forming and re-forming of hydrogen bonds that hold the molecules together. This **cohesion** creates a

■ **INTERACTIVE QUESTION   3.1**

Draw the four water molecules that can hydrogen-bond to this water molecule. Indicate the slight negative and positive charges that account for the formation of hydrogen bonds.

more structurally organized liquid and enables water to move against gravity in plants. The **adhesion** of water molecules to the walls of plant vessels also contributes to water transport. Hydrogen bonding between water molecules produces a high **surface tension** at the interface between water and air.

**Water moderates temperatures
on Earth (42–44)**

Heat and Temperature   In a body of matter, **heat** is a measure of the total quantity of **kinetic energy**, the energy associated with the movement of atoms and molecules. **Temperature** measures the average kinetic energy of the molecules in a substance.

Temperature is measured using a **Celsius scale.** Water at sea level freezes at 0°C and boils at 100°C. Heat may be measured by the **calorie (cal).** A calorie is the amount of heat energy it takes to raise 1 g of water 1°C. A **kilocalorie (kcal)** is 1,000 calories, the amount of heat required to raise 1 kg of water 1°C. A **joule (J)** equals 0.239 cal; a calorie is 4.184 J.

Water's High Specific Heat   **Specific heat** is the amount of heat absorbed or lost when 1 g of a substance changes its temperature by 1°C. Water's specific heat of 1 cal/g/°C is unusually high compared with that of other common substances; water must absorb or release a relatively large quantity of heat in order for its temperature to change. Heat must be absorbed to break hydrogen bonds before water molecules can move faster and the temperature can rise, and conversely, heat is released when hydrogen bonds form as the temperature of water drops. The ability of large bodies of water to stabilize air temperature is due to the high specific heat of water. The high proportion of water in the environment and within organisms keeps temperature fluctuations within limits that permit life.

Evaporative Cooling   The transformation from a liquid to a gas is called vaporization or evaporation and happens when molecules with sufficient kinetic energy overcome their attraction to other molecules and escape into the air as gas. The **heat of vaporization** is the quantity of heat that must be absorbed for 1 g of a liquid to be converted to a gas. Water has a high heat of vaporization (580 cal/g) because a large amount of heat is needed to break the hydrogen bonds holding water molecules together. This property of water helps moderate the climate on Earth as solar heat is dissipated from tropical seas during evaporation and heat is released when moist tropical air condenses to form rain.

As a substance vaporizes, the liquid left behind loses the kinetic energy of the escaping molecules and cools down. **Evaporative cooling** helps to protect terrestrial organisms from overheating and contributes to the stability of temperatures in lakes and ponds.

Oceans and lakes don't freeze solid because ice floats (44–45)

As water cools below 4°C, it expands. By 0°C, each water molecule becomes hydrogen-bonded to four other molecules, creating a crystalline lattice that spaces the molecules apart. Ice is less dense than liquid water, and therefore, it floats. The floating ice insulates the liquid water below.

■ INTERACTIVE QUESTION 3.2

The following concept map is one way to show how the breaking and forming of hydrogen bonds is related to temperature regulation. Fill in the blanks and compare your choice of concepts to those given in the answer section. Or, better still, create your own map to help you understand how water stabilizes temperature.

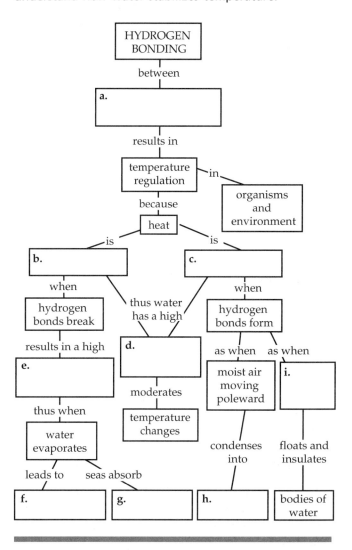

Water is the solvent of life (45–47)

A **solution** is a liquid homogeneous mixture of two or more substances; the dissolving agent is called the **solvent** and the substance that is dissolved is the **solute.** An **aqueous solution** is one in which water is the solvent. The positive and negative regions of water molecules are attracted to oppositely charged ions or partially charged regions of polar molecules. Thus, solute molecules become surrounded by water molecules (a **hydration shell**) and dissolve into solution.

Hydrophilic and Hydrophobic Substances Ionic and polar substances are **hydrophilic;** they have an affinity for water due to electrical attractions and hydrogen bonding. Nonpolar and non-ionic compounds are **hydrophobic;** they will not mix with or dissolve in water.

■ INTERACTIVE QUESTION 3.3

Indicate whether the following are hydrophilic or hydrophobic. Do these substances contain ionic, polar, or nonpolar bonds?

a. olive oil

b. sugar

c. salt

d. candle wax

Solute Concentration in Aqueous Solutions Most of the chemical reactions of life take place in water. A **mole (mol)** is the amount of a substance that has a mass in grams numerically equivalent to its **molecular weight** (sum of the weight of all atoms in the molecule) in daltons. A mole of any substance has exactly the same number of molecules—6.02×10^{23}, called Avogadro's number. The **molarity** of a solution (abbreviated *M*) refers to the number of moles of a solute dissolved in 1 liter of solution.

■ INTERACTIVE QUESTION 3.4

a. How many grams of lactic acid ($C_3H_6O_3$) are in a 0.5 *M* solution of lactic acid? (^{12}C, ^1H, ^{16}O)

b. How many grams of salt (NaCl) must be dissolved in water to make 2 liters of a 2 *M* salt solution? (^{23}Na, ^{34}Cl)

The Dissociation of Water Molecules

A water molecule can dissociate into a **hydrogen ion,** H^+ (which binds to another water molecule to form a hydronium ion, H_3O^+), and a **hydroxide ion,** OH^-. Although reversible and statistically rare, this dissociation into the highly reactive hydrogen and hydroxide ions has important biological consequences. In pure water, the concentrations of H^+ and OH^- ions are the same; both are equal to 10^{-7} *M*.

Organisms are sensitive to changes in pH (47–49)

Acids and Bases When acids or bases dissolve in water, the H^+ and OH^- balance shifts. An **acid** adds H^+ to a solution, whereas a **base** reduces H^+ in a solution by accepting hydrogen ions or by adding hydroxide ions (which then combine with H^+ and thus remove hydrogen ions). A strong acid or strong base dissociates completely when mixed with water. A weak acid or base reversibly dissociates, releasing or binding H^+.

The pH Scale In any solution, the product of the $[H^+]$ and $[OH^-]$ is constant at 10^{-14} *M^2*. Brackets, [ ], indicate molar concentration. If the $[H^+]$ is higher, then the $[OH^-]$ is lower, due to the tendency of excess hydrogen ions to combine with the hydroxide ions in solution and form water. Likewise, an increase in $[OH^-]$ causes an equivalent decrease in $[H^+]$. If $[OH^-]$ is equal to 10^{-10} *M*, then $[H^-]$ will equal 10^{-4} *M*.

The logarithmic pH scale compresses the range of hydrogen and hydroxide ion concentrations, which can vary in different solutions by many orders of magnitude. The **pH** of a solution is defined as the negative log (base 10) of the $[H^+]$: pH $= -\log [H^+]$. For a neutral solution, $[H^+]$ is 10^{-7} M, and the pH equals 7. As the $[H^+]$ increases in an acidic solution, the pH value decreases. The difference between each unit of the pH scale represents a tenfold difference in the concentration of $[H^+]$ and $[OH^-]$.

■ INTERACTIVE QUESTION 3.5

Complete the following table to review your understanding of pH.

$[H^+]$	$[OH^-]$	pH	Acidic, Basic, or Neutral?
	10^{-11}	3	acidic
10^{-8}			
	10^{-7}		
		1	

Buffers   Most cells have an internal pH close to 7. **Buffers** within the cell maintain a constant pH by accepting excess H^+ ions or donating H^+ ions when H^+ concentration decreases. Weak acid-base pairs that reversibly bind hydrogen ions are typical of most buffering systems.

■ **INTERACTIVE QUESTION 3.6**

The carbonic acid/bicarbonate system is an important biological buffer. Label the molecules and ions in this equation and indicate which is the H^+ donor and acceptor.

$$H_2CO_3 \rightleftharpoons HCO_3^- + H^+$$

In which direction will this reaction proceed

a. when the pH of a solution begins to fall?

b. when the pH rises above normal level?

Acid precipitation threatens the fitness of the environment (49–50)

Acid precipitation, with a pH lower than normal pH 5.6, is due to the reaction of water in the atmosphere with the sulfur oxides and nitrogen oxides released by the combustion of fossil fuels. Lowering the pH of the soil solution affects the solubility of minerals needed by plants. In lakes and ponds, a lowered pH and the accumulation of minerals leached from the soil by acid rain harm many species of fishes, amphibians, and aquatic invertebrates.

WORD ROOTS

kilo- = a thousand (*kilocalorie:* a thousand calories)
hydro- = water; **-philos** = loving; **-phobos** = fearing (*hydrophilic:* having an affinity for water; *hydrophobic:* having an aversion to water)

STRUCTURE YOUR KNOWLEDGE

1. Fill in the table below that summarizes the properties of water that contribute to the fitness of the environment for life.

2. To become proficient in the use of the concepts relating to pH, develop a concept map to organize your understanding of the following terms: pH, $[H^+]$, $[OH^-]$, acidic, basic, neutral, buffer, 1–14, acid-base pair. Remember to label connecting lines and add additional concepts as you need them. *A suggested concept map is given in the answer section, but remember that your concept map should represent your own understanding. The value of this exercise is in organizing these concepts for yourself.*

Property	Explanation of Property	Example of Benefit to Life
a.	Hydrogen bonds hold molecules together and adhere them to hydrophilic surfaces.	**b.**
High specific heat	**c.**	Temperature changes in environment and organisms are moderated.
d.	Hydrogen bonds must be broken for water to evaporate.	**e.**
f.	Water molecules with high kinetic energy evaporate; remaining molecules are cooler.	**g.**
Ice floats	**h.**	**i.**
j.	**k.**	Most chemical reactions in life involve solutes dissolved in water.

TEST YOUR KNOWLEDGE

MULTIPLE CHOICE: *Choose the one best answer.*

1. Each water molecule is capable of forming
   a. one hydrogen bond.
   b. three hydrogen bonds.
   c. four hydrogen bonds.
   d. one covalent bond and two hydrogen bonds.
   e. one covalent bond and three hydrogen bonds.

2. The polarity of water molecules
   a. promotes the formation of hydrogen bonds.
   b. helps water to dissolve nonpolar solutes.
   c. lowers the heat of vaporization and leads to evaporative cooling.
   d. creates a crystalline structure in liquid water.
   e. does all of the above.

3. What accounts for the movement of water up xylem vessels in a plant?
   a. cohesion
   b. hydrogen bonding
   c. adhesion
   d. hydrophilic vessel walls
   e. all of the above

4. Climates tend to be moderate near large bodies of water because
   a. a large amount of solar heat is absorbed by the gradual rise in temperature of the water.
   b. the gradual cooling of the water releases heat to the environment.
   c. the high specific heat of water helps to regulate air temperatures.
   d. a great deal of heat is absorbed and released by the breaking and forming of hydrogen bonds.
   e. of all of the above.

5. Temperature is a measure of
   a. specific heat.
   b. average kinetic energy of molecules.
   c. total kinetic energy of molecules.
   d. Celsius degrees.
   e. joules.

6. Evaporative cooling is a result of
   a. a low heat of vaporization.
   b. a high heat of melting.
   c. a high specific heat.
   d. a reduction in the average kinetic energy of the liquid remaining after molecules enter the gaseous state.
   e. release of heat caused by the breaking of hydrogen bonds when water molecules escape.

7. Ice floats because
   a. air is trapped in the crystalline lattice.
   b. the formation of hydrogen bonds releases heat; warmer objects float.
   c. it has a smaller surface area than liquid water.
   d. it insulates bodies of water so they do not freeze from the bottom up.
   e. hydrogen bonding spaces the molecules farther apart, creating a less dense structure.

8. The molarity of a solution is equal to
   a. Avogadro's number of molecules in 1 liter of solvent.
   b. the number of moles of a solute in 1 liter of solution.
   c. the molecular weight of a solute in 1 liter of solution.
   d. the number of solute particles in 1 liter of solvent.
   e. 342 g if the solute is sucrose.

9. Some archaea are able to live in lakes with pH values of 11. How does pH 11 compare with the pH 7 typical of your body cells?
   a. It is four times more acidic than pH 7.
   b. It is four times more basic than pH 7.
   c. It is a thousand times more acidic than pH 7.
   d. It is a thousand times more basic than pH 7.
   e. It is ten thousand times more basic than pH 7.

10. A buffer
    a. changes pH by a magnitude of 10.
    b. absorbs excess OH^-.
    c. releases excess H^+.
    d. is often a weak acid-base pair.
    e. always maintains a neutral pH.

11. Which of the following is least soluble in water?
    a. polar molecules
    b. nonpolar compounds
    c. ionic compounds
    d. hydrophilic molecules
    e. anions

12. Which would be the best method for reducing acid precipitation?
    a. Raise the height of smokestacks so that exhaust enters the upper atmosphere.
    b. Add buffers and bases to bodies of water whose pH has dropped.
    c. Use coal-burning generators rather than nuclear power to produce electricity.
    d. Tighten emission control standards for factories and automobiles.
    e. Reduce the concentration of heavy metals in industrial exhaust.

13. What bonds must be broken for water to vaporize?
    a. polar covalent bonds
    b. nonpolar covalent bonds
    c. hydrogen bonds
    d. ionic bonds
    e. polar covalent and hydrogen bonds

14. How would you make a 0.1 *M* solution of glucose ($C_6H_{12}O_6$)? The mass numbers for these elements are C = 12, O = 16, H = 1.
    a. Mix 6 g C, 12 g H, and 6 g O in 1 liter of water.
    b. Mix 72 g C, 12 g H, and 96 g O in 1 liter of water.
    c. Mix 18 g of glucose with enough water to make 1 liter of solution.
    d. Mix 29 g of glucose with enough water to yield 1 liter of solution.
    e. Mix 180 g of glucose with enough water to yield 1 liter of solution.

15. How many molecules of glucose would be in the solution in question 14?
    a. 0.1
    b. 6
    c. 60
    d. 6×10^23
    e. 6×10^22

16. Why is water such an excellent solvent?
    a. As a polar molecule, it can surround and dissolve ionic and other polar molecules.
    b. It forms ionic bonds with ions, hydrogen bonds with polar molecules, and hydrophobic interactions with nonpolar molecules.
    c. It forms hydrogen bonds with itself.
    d. It has a high specific heat and high heat of vaporization.
    e. It is wet and has a great deal of surface tension.

17. Adding a base to a solution would
    a. raise the pH.
    b. lower the pH.
    c. decrease [H^+].
    d. do both a and c.
    e. do both b and c.

18. The following are pH values: cola–2; orange juice–3; beer–4; coffee–5; human blood–7.4. Which of these liquids has the *highest* molar concentration of OH^-?
    a. cola
    b. orange juice
    c. beer
    d. coffee
    e. human blood

19. Comparing the [H^+] of orange juice and coffee,
    a. the [H^+] of coffee is two times higher.
    b. the [H^+] of coffee is 100 times higher.
    c. the [H^+] of orange juice is 10 times higher.
    d. the [H^+] of orange juice is 100 times higher.
    e. the [H^+] of orange juice is 1,000 times higher.

20. The ability of water molecules to form hydrogen bonds accounts for water's
    a. high specific heat.
    b. evaporative cooling.
    c. high heat of vaporization.
    d. cohesiveness and surface tension.
    e. All of the above result from water's hydrogen-bonding capacity.

CHAPTER 4

CARBON AND THE MOLECULAR DIVERSITY OF LIFE

FRAMEWORK

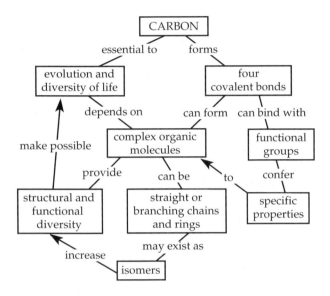

CHAPTER REVIEW

The Importance of Carbon

Organic chemistry is the study of carbon compounds (52–53)

Organic chemistry is the study of carbon-containing molecules. Early organic chemists could not synthesize the complex molecules found in living organisms and therefore attributed the existence of life and the formation of these molecules to a life force independent of physical and chemical laws, a belief known as vitalism. Mechanism, the philosophy underlying modern organic chemistry, holds that physical and chemical laws and explanations are sufficient to account for all natural phenomena, even the evolution of life.

Carbon atoms are the most versatile building blocks of molecules (53–55)

Carbon has six electrons. To complete its valence shell, carbon forms four covalent bonds with other atoms. This tetravalence is at the center of carbon's ability to form large and complex molecules with characteristic three-dimensional shapes and properties. When carbon forms four single covalent bonds, its hybrid orbitals create a tetrahedral shape. When two carbons are joined by a double bond, the other carbon bonds are in the same plane, forming a flat molecule.

Variation in carbon skeletons contributes to the diversity of organic molecules (55–57)

Carbon atoms readily bond with each other, producing chains or rings of carbon atoms. These molecular backbones can vary in length, branching, placement of double bonds, and location of atoms of other elements. The simplest organic molecules are **hydrocarbons,** consisting of only carbon and hydrogen. The nonpolar C—H bonds in hydrocarbon chains account for their hydrophobic behavior.

■ INTERACTIVE QUESTION   4.1

From what you know about the valences of C, H, and O, sketch the structural formulas for the following molecules: C_3H_8O (propanol) and C_2H_4 (ethene). (Alcohols such as propanol always have an —OH group.)

Isomers Isomers are compounds with the same molecular formula but different structural arrangements and, thus, different properties. **Structural isomers** differ in the arrangement of atoms and often in the location of double bonds. **Geometric isomers** have the same sequence of covalently bonded atoms but differ in spatial arrangement due to the inflexibility of double bonds. **Enantiomers** are left- and right-handed versions of each other and can differ greatly in their biological activity. An asymmetric carbon is one that is covalently bonded to four different kinds of atoms or groups of atoms. Due to the tetrahedral shape of the asymmetric carbon, the four groups can be attached in spatial arrangements that are not superimposable on each other.

■ **INTERACTIVE QUESTION   4.2**

Identify the structural isomers, geometric isomers, and enantiomers from the following compounds.

l-lactic acid

ethanol

d-lactic acid

fumaric acid

dimethyl ether

maleic acid

Functional Groups

Functional groups also contribute to the molecular diversity of life (57–59)

The properties of organic molecules are determined by groups of atoms, known as **functional groups,** that bond to the carbon skeleton and behave consistently from one carbon-based molecule to another. Because the functional groups considered here are hydrophilic, they increase the solubility of organic compounds in water.

The **hydroxyl group** consists of an oxygen and hydrogen (—OH) covalently bonded to the carbon skeleton. Organic molecules with hydroxyl groups are called **alcohols,** and their names often end in *-ol.*

Carbonyl groups consist of a carbon double-bonded to an oxygen ($>$CO). If the carbonyl group is at the end of the carbon skeleton, the compound is called an **aldehyde.** Otherwise, the compound is called a **ketone.**

A **carboxyl group** consists of a carbon double-bonded to an oxygen and also attached to a hydroxyl group (—COOH). Compounds with a carboxyl group are called **carboxylic acids** or organic acids because they tend to dissociate to release H^+.

An **amino group** consists of a nitrogen atom bonded to two hydrogens (—NH$_2$). Compounds with an amino group, called **amines,** can act as bases. The nitrogen, with its pair of unshared electrons, can attract a hydrogen ion, becoming —NH$_3^+$.

The **sulfhydryl group** consists of a sulfur atom bonded to a hydrogen (—SH). **Thiols** are compounds containing sulfhydryl groups.

A **phosphate group** is bonded to the carbon skeleton by its oxygen attached to the phosphorus atom that is bonded to three other oxygen atoms (—OPO$_3^{-2}$). The group is an anion due to the dissociation of hydrogen ions.

The chemical elements of life: a review (59)

Carbon, oxygen, hydrogen, nitrogen, and smaller quantities of sulfur and phosphorus, all capable of forming strong covalent bonds, are combined into the complex organic molecules of living matter. The versatility of carbon in forming four covalent bonds, linking readily with itself to produce chains and rings, and binding with other elements and functional groups makes possible the incredible diversity of organic molecules.

WORD ROOTS

hydro- = water (*hydrocarbon:* an organic molecule consisting only of carbon and hydrogen)

iso- = equal (*isomer:* one of several organic compounds with the same molecular formula but different structures and, therefore, different properties)

enanti- = opposite (*enantiomer:* molecules that are mirror images of each other)

carb- = coal (*carboxyl group:* a functional group present in organic acids, consisting of a carbon atom double-bonded to an oxygen atom)

sulf- = sulfur (*sulfhydryl group:* a functional group that consists of a sulfur atom bonded to an atom of hydrogen)

thio- = sulfur (*thiol:* organic compounds containing sulfhydryl groups)

STRUCTURE YOUR KNOWLEDGE

1. Construct a concept map that illustrates your understanding of the characteristics and significance of the three types of isomers. *A suggested map is in the answer section. Comparing and discussing your map with that of a study partner would be most helpful.*

2. Fill in the following table on the functional groups.

Functional Group	Molecular Formula	Names and Characteristics of Organic Compounds Containing Functional Group
	—OH	
		Aldehyde or ketone; polar group
Carboxyl		/
	—NH₂	
		Thiols; cross-links stabilize protein structure
Phosphate		

TEST YOUR KNOWLEDGE

MULTIPLE CHOICE: *Choose the one best answer.*

1. The tetravalence of carbon most directly results from
   a. its tetrahedral shape.
   b. its very slight electronegativity.
   c. its four electrons in the valence shell that can form four covalent bonds.
   d. its ability to form single, double, and triple bonds.
   e. its ability to form chains and rings of carbon atoms.

2. Hydrocarbons are not soluble in water because
   a. they are hydrophilic.
   b. the C—H bond is very nonpolar.
   c. they do not ionize.
   d. they store energy in the many C—H bonds along the carbon backbone.
   e. they are lighter than water.

3. Which of the following is not true of an asymmetric carbon atom?
   a. It is attached to four different atoms or groups.
   b. It results in right- and left-handed versions of a molecule.
   c. It can create enantiomers.
   d. Its configuration is in the shape of a tetrahedron.
   e. It can create geometric isomers.

4. A reductionist approach to considering the structure and function of organic molecules would be based on
   a. mechanism.
   b. holism.
   c. determinism.
   d. vitalism.
   e. evolution.

5. The functional group that can cause an organic molecule to act as a base is
   a. —COOH.   c. —SH.   e. —OPO322.
   b. —OH.   d. —NH2.

6. The functional group that confers acidic properties to organic molecules is
   a. —COOH.   c. —SH.   e. $>C = O$.
   b. —OH.   d. —NH₂.

7. Which is not true about structural isomers?
   a. They have different chemical properties.
   b. They have the same molecular formula.
   c. Their atoms and bonds are arranged in different sequences.
   d. They are a result of restricted movement around a carbon double bond.
   e. Their possible numbers increase as carbon skeletons increase in size.

8. How many asymmetric carbons are there in the sugar ribose?
   a. 1   c. 3   e. 5
   b. 2   d. 4

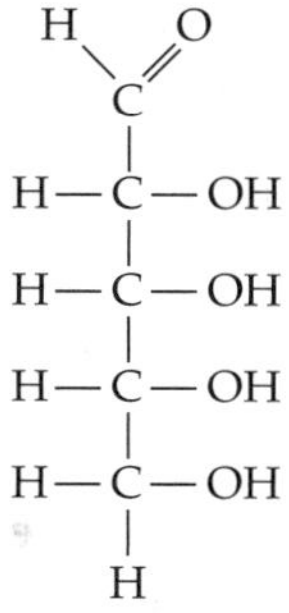

MATCHING: *Match the formulas (a–f) to the terms at the right. Choices may be used more than once; more than one right choice may be available.*

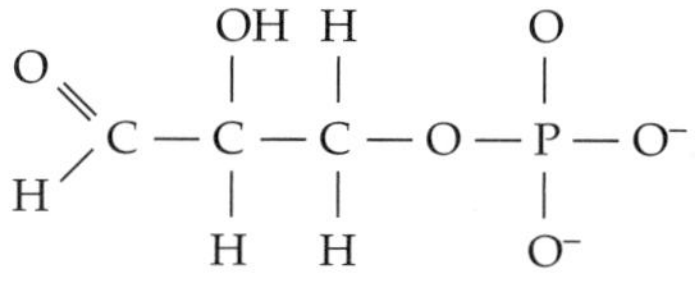

a.

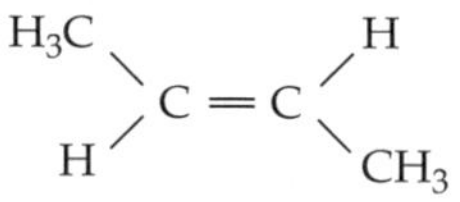

b.

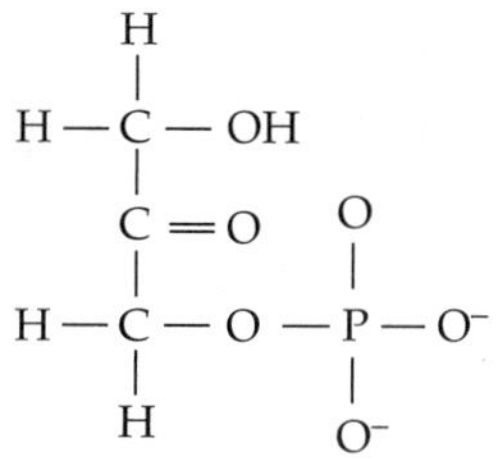

c.

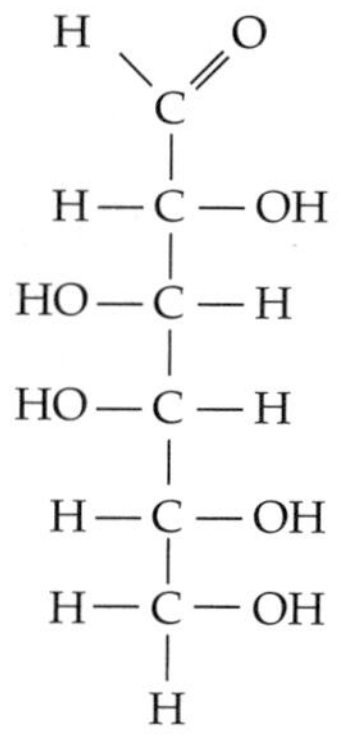

d.

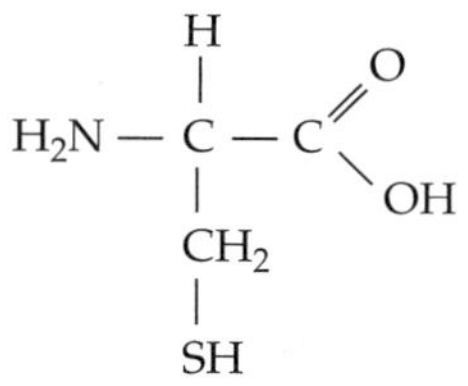

e.

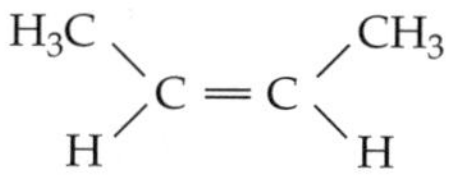

f.

_____ 1. structural isomers

_____ 2. geometric isomers

_____ 3. can have enantiomers

_____ 4. carboxylic acid

_____ 5. can make cross-link in protein

_____ 6. hydrophilic

_____ 7. hydrocarbon

_____ 8. amino acid

_____ 9. organic phosphate

_____ 10. aldehyde

_____ 11. amine

_____ 12. ketone

CHAPTER 5

THE STRUCTURE AND FUNCTION OF MACROMOLECULES

FRAMEWORK

The central ideas of this chapter are that molecular function relates to molecular structure and that the diversity of molecular structure is the basis for the diversity of life. Combining a small number of monomers or subunits into unique sequences and three-dimensional structures creates a huge variety of macromolecules. The table below briefly summarizes the major characteristics of the four classes of macromolecules.

CHAPTER REVIEW

Smaller organic molecules are joined together to form carbohydrates, lipids, proteins, and nucleic acids. These giant molecules, called **macromolecules,** represent another level in the hierarchy of biological organization, and their functions derive from their complex and unique architectures.

Polymer Principles

Most macromolecules are polymers (62–63)

Polymers are chainlike molecules formed from the linking together of many similar or identical small molecules, called **monomers.** Monomers are joined by **condensation reactions** (or **dehydration reactions**), in which one monomer provides a hydroxyl (–OH) and the other contributes a hydrogen (–H) to release a water molecule, and a covalent bond between the monomers is formed. Energy is required to join monomers, and the process is facilitated by enzymes.

Hydrolysis is the breaking of bonds between monomers through the addition of water molecules. A hydroxyl is joined to one monomer while a hydrogen is bonded with the other. Enzymes also control hydrolysis.

An immense variety of polymers can be built from a small set of monomers (63–64)

Macromolecules are constructed from about 40 to 50 common monomers and a few rarer molecules. The seemingly endless variety of polymers arises from the essentially infinite number of possibilities in the sequencing and arrangement of these basic subunits.

Carbohydrates—Fuel and Building Material

Carbohydrates include sugars and their polymers.

Sugars, the smallest carbohydrates, serve as fuel and carbon sources (64–65)

Monosaccharides have the general formula of $(CH_2O)_n$. The number of these units forming a sugar varies from three to seven, with hexoses ($C_6H_{12}O_6$), trioses, and pentoses found most commonly. Sugar

Class	Monomers or subunits	Functions
Carbohydrates	Monosaccharides	Energy, raw materials, energy storage, structural compounds
Lipids	Glycerol and fatty acids → fats	Energy storage, membranes, steroids, hormones
Proteins	Amino acids	Enzymes, transport, movement, receptors, defense, structure
Nucleic acids	Nucleotides	Heredity, code for amino acid sequence

molecules may be enantiomers due to the spatial arrangement of parts around asymmetric carbons.

Glucose is broken down to yield energy in cellular respiration. Monosaccharides serve also as the raw materials for synthesis of other organic molecules and as monomers that are synthesized into disaccharides or polysaccharides.

■ INTERACTIVE QUESTION 5.1

Fill in the blanks to review the structure of monosaccharides.

You can recognize a monosaccharide by its multiple **(a)** _____ groups and its one **(b)** _____ group, whose location determines whether the sugar is an **(c)** _____ or a **(d)** _____. In aqueous solutions, most monosaccharides form **(e)** _____ .

Sucrose, or table sugar, is a **disaccharide** consisting of a glucose and a fructose molecule. A **glycosidic linkage** is a covalent bond formed by a dehydration reaction between two monosaccharides.

Polysaccharides, the polymers of sugars, have storage and structural roles (66–68)

Polysaccharides are storage or structural macromolecules made from a few hundred to a few thousand monosaccharides. **Starch,** a storage molecule in plants, is a polymer made of glucose molecules joined by 1–4 linkages that give starch a helical shape. Most animals have enzymes to hydrolyze plant starch into glucose. Animals produce **glycogen,** a highly branched polymer of glucose, as their energy storage form.

Cellulose, the major component of plant cell walls, is the most abundant organic compound on Earth. It differs from starch by the configuration of the ring form of glucose and the resulting geometry of the glycosidic bonds. In a plant cell wall, hydrogen bonds between hydroxyl groups hold parallel cellulose molecules together to form strong microfibrils.

Enzymes that digest starch are unable to hydrolyze the β linkages of cellulose. Only a few organisms (some bacteria, microorganisms, and fungi) have enzymes that can digest cellulose.

Chitin is a structural polysaccharide formed from glucose monomers with nitrogen-containing groups and found in the exoskeleton of arthropods and the cell walls of many fungi.

■ INTERACTIVE QUESTION 5.2

Circle the atoms of these two glucose molecules that will be removed by a dehydration reaction. Then draw the resulting maltose molecule with its 1–4 glycosidic linkage (between the number 1 carbon of the first glucose and the number 4 carbon of the second).

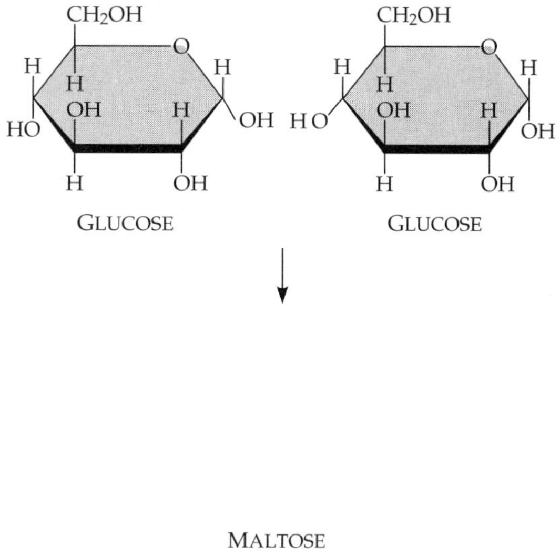

MALTOSE

■ INTERACTIVE QUESTION   5.3

Fill in the following concept map that summarizes this section on carbohydrates.

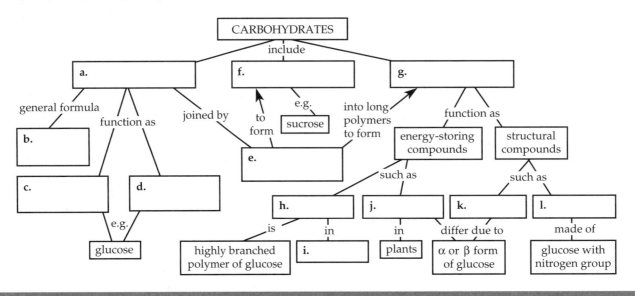

Lipids—Diverse Hydrophobic Molecules

Fats, phospholipids, and steroids are a diverse assemblage of macromolecules that are classed together as **lipids** based on their hydrophobic behavior. Lipids do not form polymers.

Fats store large amounts of energy (69–70)

Fats are composed of fatty acids attached to the three-carbon alcohol, glycerol. A **fatty acid** consists of a long hydrocarbon "tail" with a carboxyl group at the "head" end. The nonpolar hydrocarbons make a fat hydrophobic.

A **triacylglycerol,** or fat, consists of three fatty acids, each linked to glycerol by an ester linkage, a bond that forms between a hydroxyl and a carboxyl group. Triglyceride is another name for fats.

Fatty acids with double bonds in their carbon skeletons are called **unsaturated fatty acids.** The double bonds create a kink in the shape of the molecule and prevent the fat molecules from packing closely together and becoming solidified at room temperature. **Saturated fatty acids** have no double bonds in their carbon skeletons. Most animal fats are saturated and solid at room temperature. The fats of plants and fishes are generally unsaturated and are called oils. Diets rich in saturated fats have been linked to cardiovascular disease.

Fats are excellent energy storage molecules, containing twice the energy reserves of carbohydrates such as starch. Adipose tissue, made of fat storage cells, also cushions organs and insulates the body.

Phospholipids are major components of cell membranes (70–71)

Phospholipids consist of a glycerol linked to two fatty acids and a negatively charged phosphate group, to which other small molecules may be attached. The phosphate head of this molecule is hydrophilic and water soluble, whereas the two fatty acid chains are hydrophobic.

The unique structure of phospholipids makes them ideal constituents of cell membranes. Arranged in a bilayer, the hydrophilic heads face toward the aqueous solutions inside and outside the cell, and the hydrophobic tails mingle in the center of the membrane, forming a boundary between the cell and its external environment.

■ INTERACTIVE QUESTION   5.4

Sketch a section of a phospholipid bilayer of a membrane, and label the hydrophilic head and hydrophobic tail of one of the phospholipids.

■ INTERACTIVE QUESTION 5.5

Fill in this concept map to help you organize your understanding of lipids

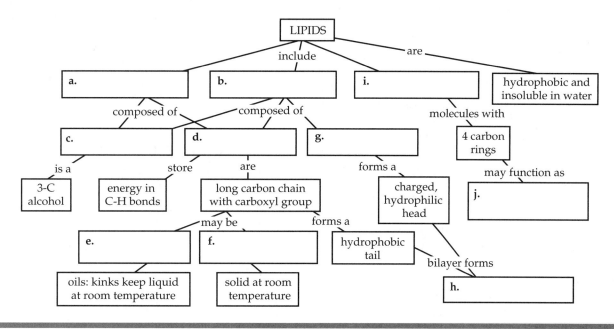

Steroids include cholesterol and certain hormones (71)

Steroids are a class of lipids distinguished by four connected carbon rings with various functional groups attached. **Cholesterol** is an important steroid that is a common component of animal cell membranes and a precursor for other steroids, including many hormones.

Proteins—Many Structures, Many Functions

Proteins are central to almost every function of life. A **polypeptide** is a polymer of amino acids. Proteins consist of one or more polypeptide chains folded into a unique three-dimensional shape or conformation.

A polypeptide is a polymer of amino acids connected in a specific sequence (71–74)

Amino acids are composed of an asymmetric carbon, (called the α carbon) bonded to a hydrogen, a carboxyl group, an amino group, and a variable side chain called the R group. The R group confers the unique physical and chemical properties of each amino acid. Side chains may be either nonpolar and hydrophobic, or polar or charged (acidic or basic) and thus hydrophilic.

A **peptide bond** links the amino group of one amino acid with the carboxyl group of another. A polypeptide chain has a free carboxyl group at one end and a free amino group at the other. Polypeptides vary in length from a few to a thousand or more amino acids.

■ INTERACTIVE QUESTION 5.6

a. Draw the amino acids alanine (R group—CH_3) and serine (R group—CH_2OH) and then show how a dehydration reaction will form a peptide bond between them.

b. Which of these amino acids has a polar R group? _____ a nonpolar R group? _____

c. What does this molecule segment represent? Note the N-C-C-N-C-C sequence.

```
       H      H      H
       |      |      |
    -N-C-C-N-C-C-N-C-C-
       |  ‖   |  ‖   |  ‖
       H  O   H  O   H  O
```

**A protein's function depends on
its specific conformation (74–80)**

Proteins have unique three-dimensional shapes created by the twisting or folding of one or more polypeptide chains. The unique conformation of a protein, which results from its sequence of amino acids, enables it to recognize and bind to other molecules.

Four Levels of Protein Structure   **Primary structure** is the unique, genetically coded sequence of amino acids within a protein. Even a slight deviation from the sequence of amino acids can severely affect a protein's function by altering the protein's conformation. In the early 1950s, Sanger determined the primary structure of insulin through the laborious process of hydrolyzing the protein into small peptide chains, determining their amino acid sequences, and then overlapping the sequences of small fragments created with different agents to reconstruct the whole polypeptide. Most of these steps are now automated.

Secondary structure involves the coiling or folding of the polypeptide backbone, stabilized by hydrogen bonds between the electronegative oxygen of one peptide bond and the weakly positive hydrogen attached to the nitrogen of another peptide bond. An **alpha (α) helix** is a coil produced by hydrogen bonding between every fourth amino acid. A β **pleated sheet** is also held by repeated hydrogen bonds along the polypeptide backbone. This secondary structure forms when regions of the polypeptide chain lie parallel to each other.

Interactions between the various side chains of the constituent amino acids produce a protein's **tertiary structure. Hydrophobic interactions** between nonpolar side groups in the center of the molecule, van der Waals interactions, hydrogen bonds, and ionic bonds between negatively and positively charged side chains produce the stable and unique shape of the protein. Strong covalent bonds, called **disulfide bridges,** may occur between the sulfhydryl side groups of cysteine monomers that have been brought close together by the folding of the polypeptide.

Quaternary structure occurs in proteins that are composed of more than one polypeptide chain. The individual polypeptide subunits are held together in a precise structural arrangement.

What Determines Protein Conformation?   Protein conformation is dependent upon the interactions among the amino acids making up the polypeptide chain and usually arises spontaneously as soon as the protein is synthesized in the cell. These interactions can be disrupted by changes in pH, salt concentration, temperature, or other aspects of the environment, and the protein may **denature,** losing its native conformation and thus its function.

The Protein-Folding Problem   The amino acid sequences of thousands of proteins have been determined. Using the technique of **X-ray crystallography,** coupled with computer modeling and graphics, biochemists have established the three-dimensional shape of many of these molecules. Researchers have developed methods for following a protein through its intermediate states on the way to its final form and have discovered **chaperonins,** chaperone proteins that assist the folding of other proteins.

■ **INTERACTIVE QUESTION   5.7**

In the following diagram of a portion of a polypeptide, label the types of interactions that are shown. What level of structure are these interactions producing?

■ **INTERACTIVE QUESTION   5.8**

a. Why would a change in pH cause a protein to denature?

b. Why would transfer to a nonpolar organic solvent (such as ether) cause denaturation?

c. A denatured protein may re-form to its functional shape when returned to its normal environment. What does that indicate about a protein's conformation?

■ INTERACTIVE QUESTION 5.9

Now that you have gained more experience with concept maps, create your own map to help you organize the key concepts you have learned about proteins. Try to include the concepts of structure and function and look for cross-links on your map. One version of a protein map is included in the answer section, but remember that the real value is in the thinking process you must go through to create your own map.

Nucleic Acids—Informational Polymers

Genes are the units of inheritance that determine the primary structure of proteins. **Nucleic acids** are macromolecules that carry and transmit this code.

Nucleic acids store and transmit hereditary information (80–81)

DNA, deoxyribonucleic acid, is the genetic material that is inherited from one generation to the next and is reproduced in each cell of an organism. The instructions coded in DNA are transcribed to **RNA, ribonucleic acid,** which directs the synthesis of proteins, the ultimate enactors of the genetic program. In a eukaryotic cell, DNA resides in the nucleus and messenger RNA carries the instructions for protein synthesis to ribosomes located in the cytoplasm.

■ INTERACTIVE QUESTION 5.10

Show the flow of genetic information in a cell.

_____ → _____ → _____

A nucleic acid strand is a polymer of nucleotides (82)

Nucleic acids are polymers of **nucleotides,** monomers that consist of a pentose (five-carbon sugar) covalently bonded to a phosphate group and a nitrogenous base. In DNA, the pentose is **deoxyribose;** in RNA it is **ribose.** There are two families of nitrogenous bases: **Pyrimidines,** including cytosine (C), thymine (T), and uracil (U), are characterized by six-membered rings of carbon and nitrogen atoms.

Purines, adenine (A) and guanine (G), add a five-membered ring to the pyrimidine ring. Thymine is only in DNA; uracil is only in RNA.

Nucleotides are linked together into a DNA polymer, or **polynucleotide,** by phosphodiester linkages, which join the phosphate of one nucleotide with the sugar of the next. The nitrogenous bases extend from this backbone of repeating sugar-phosphate units. The unique sequence of bases in a gene codes for the specific amino acid sequence of a protein.

■ INTERACTIVE QUESTION 5.11

a. Label the three parts of this nucleotide. Indicate with an arrow where the phosphate group of the next nucleotide would attach to build a polynucleotide. Number the carbons of the pentose sugar.

b. Is this a purine or a pyrimidine?

c. Is this a DNA or RNA nucleotide?

Inheritance is based on replication of the DNA double helix (82–83)

DNA molecules consist of two chains of polynucleotides spiraling around an imaginary axis in a **double helix.** In 1953, Watson and Crick first proposed this double-helix arrangement, which consists of two sugar-phosphate backbones on the outside of the helix with their nitrogenous bases pairing and hydrogen-bonding together in the inside. Adenine pairs only with thymine; guanine always pairs with cytosine. Thus, the sequences of nitrogenous bases on the two strands of DNA are complementary. Because of this specific base-pairing property, DNA can replicate itself and precisely copy the genes of inheritance.

■ INTERACTIVE QUESTION   5.12

Take the time to create a concept map that summarizes what you have just reviewed about nucleic acids. Compare your map with that of a study partner or explain it to a friend. One version of a map on nucleic acids is included in the answer section. Refer to Figures 5.29 and 5.30 in your textbook to help you visualize polynucleotides and the double helix of DNA.

We can use DNA and proteins as tape measures of evolution (84)

Genes form the hereditary link between generations. Closely related members of the same species share many common DNA sequences and proteins. More closely related species have a larger proportion of their DNA and proteins in common. This "molecular genealogy" provides evidence of evolutionary relationships.

WORD ROOTS

con- = together (*condensation reaction:* a reaction in which two molecules become covalently bonded to each other through the loss of a small molecule, usually water)

di- = two (*disaccharide:* two monosaccharides joined together)

glyco- = sweet (*glycogen:* a polysaccharide sugar used to store energy in animals)

hydro- = water; **-lyse** = break (*hydrolysis:* breaking chemical bonds by adding water)

macro- = large (*macromolecule:* a large molecule)

meros- = part (*polymer:* a chain made from smaller organic molecules)

mono- = single; **-sacchar** = sugar (*monosaccharide:* simplest type of sugar)

poly- = many (*polysaccharide:* many monosaccharides joined together)

tri- = three (*triacylglycerol:* three fatty acids linked to one glycerol molecule)

STRUCTURE YOUR KNOWLEDGE

1. Describe the four structural levels in the conformation of a protein.

2. Identify the type of monomer or group shown by these formulae. Then match the chemical formulae with their description. Answers may be used more than once.

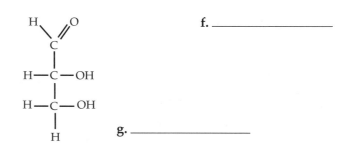

_____ 1. molecules that would combine to form a fat

_____ 2. molecule that would be attached to other monomers by a peptide bond

_____ 3. molecules or groups that would combine to form a nucleotide

_____ 4. molecules that are carbohydrates

_____ 5. molecule that is a purine

_____ 6. monomer of a protein

_____ 7. groups that would be joined by phosphodiester bonds

_____ 8. most nonpolar (hydrophobic) molecule

TEST YOUR KNOWLEDGE

MATCHING: *Match the molecule with its class of macromolecules.*

_____ **1.** glycogen

_____ **2.** cholesterol

_____ **3.** RNA

_____ **4.** collagen

_____ **5.** hemoglobin

_____ **6.** a gene

_____ **7.** triacylglycerol

_____ **8.** enzyme

_____ **9.** cellulose

_____**10.** chitin

A. carbohydrate

B. lipid

C. protein

D. nucleic acid

MULTIPLE CHOICE: *Choose the one best answer.*

1. Polymerization is a process that
   a. creates bonds between amino acids in the formation of a peptide chain.
   b. involves the removal of a water molecule.
   c. links the sugar of one nucleotide with the phosphate of the next.
   d. requires a condensation or dehydration reaction.
   e. involves all of the above.

2. Which of the following is *not* true of a pentose?
   a. It can be found in nucleic acids.
   b. It can occur in a ring structure.
   c. It has the formula $C_5H_{12}O_5$.
   d. It has one carbonyl and four hydroxyl groups.
   e. It may be an aldose or a ketose.

3. Disaccharides can differ from each other in all of the following ways *except*
   a. in the number of their monosaccharides.
   b. as enantiomers.
   c. in the monomers involved.
   d. in the location of their glycosidic linkage.
   e. in their structural formulae.

4. Which of the following is not true of cellulose?
   a. It is the most abundant organic compound on Earth.
   b. It differs from starch because of the configuration of glucose and the geometry of the glycosidic linkage.
   c. It may be hydrogen-bonded to neighboring cellulose molecules to form microfibrils.

   d. Few organisms have enzymes that hydrolyze its glycosidic linkages.
   e. Its monomers are glucose with nitrogen-containing appendages.

5. Plants store most of their energy as
   a. glucose.
   b. glycogen.
   c. starch.
   d. sucrose.
   e. cellulose.

6. What happens when a protein denatures?
   a. It loses its primary structure.
   b. It loses its secondary and tertiary structure.
   c. It becomes irreversibly insoluble and precipitates.
   d. It hydrolyzes into component amino acids.
   e. Its hydrogen bonds, ionic bonds, hydrophobic interactions, disulfide bridges, and peptide bonds are disrupted.

7. The alpha helix of proteins is
   a. part of the tertiary structure and is stabilized by disulfide bridges.
   b. a double helix.
   c. stabilized by hydrogen bonds and commonly found in fibrous proteins.
   d. found in some regions of globular proteins and stabilized by hydrophobic interactions.
   e. a complementary sequence to messenger RNA.

8. A fatty acid that has the formula $C_{16}H_{32}O_2$ is
   a. saturated.
   b. unsaturated.
   c. branched.
   d. hydrophilic.
   e. part of a steroid molecule.

9. Three molecules of the fatty acid in question 8 are joined to a molecule of glycerol ($C_3H_8O_3$). The resulting molecule has the formula
   a. $C_{48}H_{96}O_6$.
   b. $C_{48}H_{98}O_9$.
   c. $C_{51}H_{102}O_8$.
   d. $C_{51}H_{98}O_6$.
   e. $C_{51}H_{104}O_9$.

10. β pleated sheets are characterized by
   a. disulfide bridges between cysteine amino acids.
   b. parallel regions of the polypeptide chain held together by hydrophobic interactions.
   c. folds stabilized by hydrogen bonds between segments of the polypeptide backbone.
   d. membrane sheets composed of phospholipids.
   e. hydrogen bonds between adjacent cellulose molecules.

11. Cows can derive nutrients from cellulose because
    a. they can produce the enzymes that break the b linkages between glucose molecules.
    b. they chew and rechew their cud so that cellulose fibers are finally broken down.
    c. one of their stomachs contains bacteria that can hydrolyze the bonds of cellulose.
    d. their intestinal tract contains termites, which produce enzymes to hydrolyze cellulose.
    e. they can convert cellulose to starch and then hydrolyze starch to glucose.

12. Which of these molecules would provide the most energy (kcal/g) when eaten?
    a. glucose
    b. starch
    c. glycogen
    d. fat
    e. protein

13. What *determines* the sequence of the amino acids in a particular protein?
    a. its primary structure
    b. the sequence of nucleotides in RNA, which was determined by the sequence of nucleotides in the gene for that protein
    c. the sequence of nucleotides in DNA, which was determined by the sequence of nucleotides in RNA
    d. the sequence of RNA nucleotides making up the ribosome
    e. the three-dimensional shape of the protein

14. Sucrose is made from joining a glucose and a fructose molecule in a dehydration reaction. What is the molecular formula for this disaccharide?
    a. $C_6H_{12}O_6$
    b. $C_{12}H_{24}O_{12}$
    c. $C_{12}H_{24}O_{13}$
    d. $C_{12}H_{22}O_{11}$
    e. $C_{10}H_{20}O_{10}$

15. How are the nucleotide monomers connected to form a polynucleotide?
    a. hydrogen bonds between complementary nitrogenous base pairs
    b. ionic attractions between phosphate groups
    c. disulfide bridges between cysteine amino acids
    d. covalent bonds between the sugar of one nucleotide and the phosphate of the next
    e. ester linkages between the carboxyl group of one nucleotide and the hydroxyl group on the ribose of the next

16. Which of the following would be a hydrophobic molecule?
    a. cholesterol
    b. nucleotide
    c. amino acid
    d. chitin
    e. glucose

17. What is the best description of this molecule?

    a. chitin   d. nucleotide
    b. amino acid   e. protein
    c. polypeptide
       (tripeptide)

18. Which number(s) in the molecule in question 17 refer(s) to a peptide bond?
    a. 1   c. 3   e. both 2 and 4
    b. 2   d. 4

19. If the nucleotide sequence of one strand of a DNA helix is GCCTAA, what would be the sequence on the complementary strand?
    a. GCCTAA
    b. CGGAUU
    c. CGGATT
    d. ATTCGG
    e. TAAGCC

20. Monkeys and humans share many of the same DNA sequences and have similar proteins, indicating that
    a. the two groups belong to the same species.
    b. the two groups share a relatively recent common ancestor.
    c. humans evolved from monkeys.
    d. monkeys evolved from humans.
    e. the two groups first appeared on Earth at about the same time.

FILL IN THE BLANKS

1. The man who determined the amino acid sequence of insulin was _____.

2. Cytosine always pairs with _____.

3. Adenine and guanine are _____.

4. A pentose joined to a nitrogenous base and a phosphate group is called a _____.

5. The conformation of a protein is determined by its _____.

6. Proteins with more than one polypeptide chain have _____ structure.

7. The carbohydrate energy storage molecule of animals is _____.

8. Membranes are composed of a bilayer of _____.

9. The insoluble fiber listed on food packages consists primarily of _____.

10. Proteins that assist the proper folding of newly synthesized proteins are called _____.

CHAPTER 6

AN INTRODUCTION TO METABOLISM

FRAMEWORK

This chapter considers metabolism, the totality of the chemical reactions that take place in living organisms. Two key topics are emphasized: the energy transformations that underlie all chemical reactions and the role of enzymes in the "cold chemistry" of the cell. Enzymes are biological catalysts that lower the activation energy of a reaction and thus greatly speed up metabolic processes. An enzyme is a three-dimensional protein molecule with an active site specific for its substrate. Intricate control and feedback mechanisms produce the metabolic integration necessary for life.

The reactions in a cell either release or consume energy. The following illustration summarizes some of the components of the energy changes within a cell.

CHAPTER REVIEW

Metabolism, Energy, and Life

The chemistry of life is organized into metabolic pathways (87–88)

Metabolism includes the thousands of precisely coordinated chemical reactions in an organism. These reactions are ordered into metabolic pathways —sequenced and branching routes controlled by enzymes. Through these pathways the cell transforms and creates the organic molecules that provide the energy and material for life.

Catabolic pathways release the energy stored in complex molecules through the breaking down of these molecules into simpler compounds. **Anabolic**

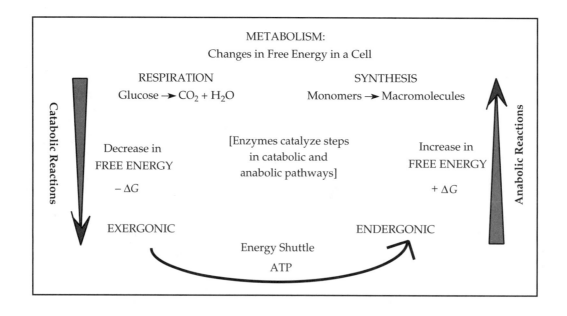

METABOLISM:
Changes in Free Energy in a Cell

RESPIRATION
Glucose → CO_2 + H_2O

SYNTHESIS
Monomers → Macromolecules

Catabolic Reactions

Anabolic Reactions

Decrease in
FREE ENERGY

$-\Delta G$

[Enzymes catalyze steps
in catabolic and
anabolic pathways]

Increase in
FREE ENERGY

$+\Delta G$

EXERGONIC

ENDERGONIC

Energy Shuttle
ATP

pathways require energy to combine simpler molecules into more complicated ones. This energy is supplied through energy coupling, the linking of catabolic and anabolic pathways in a cell. Energy is involved in all metabolic processes. Thus, the study of how organisms transform energy, called **bioenergetics,** is central to the understanding of metabolism.

Organisms transform energy (88–89)

Energy has been defined as the capacity to do work, to move matter against an opposing force, to rearrange matter. **Kinetic energy** is the energy of motion, of matter that is moving. This matter does its work by transferring its motion to other matter. **Potential energy** is the capacity of matter to do work as a consequence of its location or arrangement. **Chemical energy** is a form of potential energy stored in the arrangement of atoms in molecules.

Energy can be converted from one form to another. Plants convert light energy (a form of kinetic energy) to the chemical energy in sugar, and cells release this potential energy to drive cellular processes.

The energy transformations of life are subject to two laws of thermodynamics (89–90)

Thermodynamics is the study of energy transformations. In an open system, energy (and matter) may be exchanged between the system and the surroundings.

The First Law of Thermodynamics The **first law of thermodynamics** states that energy can be neither created nor destroyed. According to this principle of the *conservation of energy,* energy can be transferred between matter and transformed from one kind to another, but the total energy of the universe is constant.

The Second Law of Thermodynamics The **second law of thermodynamics** states that every energy transformation or transfer results in an increasing disorder within the universe. **Entropy** is used as a measure of disorder or randomness. No energy transfer or transformation is 100% efficient; some of the energy is converted to heat, the least ordered form of energy.

An open system, such as a cell, may become more ordered, but it does so with an attendant increase in the entropy of its surroundings. An organism may take in and use highly ordered organic molecules as a source of the energy needed to create and maintain its own organized structure, but it returns heat and the simple molecules of carbon dioxide and water to the environment.

Fill in the concept map in Interactive Question 6.1 to review the key concepts of energy.

Organisms live at the expense of free energy (91–94)

A spontaneous process is a change that occurs without the input of external energy and that results in greater stability, reflected as an increase in entropy.

Free Energy: A Criterion for Spontaneous Change **Free energy** can be defined as the portion of a system's energy available to perform work when the system's temperature is uniform. Free energy (G) depends on the total energy of a system (H) and its

■ **INTERACTIVE QUESTION 6.1**

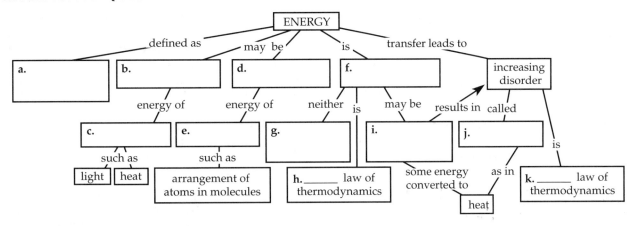

entropy (*S*), such that $G = H - TS$. Absolute temperature (°Kelvin, or 0°C + 273) is a factor because higher temperature increases random molecular motion, disrupting order and increasing entropy. The change in free energy during a reaction is represented by $\Delta G = \Delta H - T\Delta S$. For a reaction to be spontaneous, the free energy of the system must decrease ($-\Delta G$): the system must lose energy (decrease *H*), become more disordered (increase *S*), or both. An unstable system is rich in free energy and has a tendency to change spontaneously to a more stable stage, potentially performing work in the process.

Free Energy and Equilibrium At equilibrium in a chemical reaction, the forward and backward reactions are proceeding at the same rate, and $\Delta G = 0$ because there is no net free energy change. Moving toward equilibrium is spontaneous; the ΔG of the reaction is negative. To move away from equilibrium is nonspontaneous; energy must be added to the system.

■ INTERACTIVE QUESTION 6.2

Complete the table to show how the free energy of a system or reaction relates to its stability, tendency for spontaneous change, equilibrium, and capacity to do work.

	System with high free energy	System with low free energy
Stability		
Spontaneous		
Equilibrium		
Work capacity		

Free Energy and Metabolism An **exergonic reaction** ($-\Delta G$) proceeds with a net release of free energy and is spontaneous. The magnitude of ΔG indicates the maximum amount of work the reaction can do. **Endergonic reactions** ($+\Delta G$) are nonspontaneous; they must absorb free energy from the surroundings. The energy released by an exergonic reaction ($-\Delta G$) is equal to the energy required by the reverse reaction ($+\Delta G$).

Metabolic disequilibrium is essential to life. Respiration and other cellular chemical reactions are reversible and could reach equilibrium if the cell did not maintain a steady supply of reactants and siphon off the products (as reactants for new processes or as waste products to be expelled).

Central to a cell's bioenergetics is **energy coupling**, using exergonic processes to power endergonic ones.

■ INTERACTIVE QUESTION 6.3

Develop a concept map on free energy and ΔG. The value in this exercise is for you to wrestle with and organize these concepts for yourself. Do not turn to the suggested concept map until you have worked on your own understanding. Remember that the concept map in the answer section is only one way of structuring these ideas—have confidence in your own organization.

ATP powers cellular work by coupling exergonic to endergonic reactions (94–96)

A cell usually uses ATP as the immediate source of energy for its mechanical, transport, and chemical work.

The Structure and Hydrolysis of ATP **ATP (adenosine triphosphate)** consists of the nitrogenous base adenine bonded to the sugar ribose, which is connected to a chain of three phosphate groups. This triphosphate tail is unstable; ATP can be hydrolyzed to ADP (adenosine diphosphate) and an inorganic phosphate molecule ($Ⓟ_i$), releasing 7.3 kcal (31 kJ) of energy per mole of ATP when measured under standard conditions. The ΔG of the reaction in the cell is estimated to be closer to -13 kcal/mol.

How ATP Performs Work In a cell, the free energy released from the hydrolysis of ATP is used to transfer the phosphate group to another molecule, producing a **phosphorylated intermediate** that is more reactive. The phosphorylation of other molecules by ATP forms the basis for almost all cellular work.

The Regeneration of ATP A cell regenerates ATP at a phenomenal rate. The formation of ATP from ADP and $Ⓟ_i$ is endergonic, with a ΔG of $+7.3$ kcal/mol (standard conditions). Cellular respiration (the catabolic processing of glucose and other organic molecules) provides the energy for the regeneration of ATP. Plants can also produce ATP using light energy.

Label the three components (**a** through **c**) of the ATP molecule shown below.

a. _____

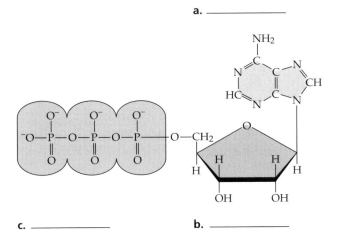

c. _____   b. _____

d. Circle the unstable region in this molecule, and show which bond is likely to break.

e. Explain why that part of the molecule is unstable. By what chemical mechanism is the bond broken?

f. Why is energy released?

Enzymes

Enzymes speed up metabolic reactions by lowering energy barriers (96–97)

Enzymes are biological **catalysts**—agents that speed the rate of a reaction but are unchanged by the reaction.

Chemical reactions rearrange atoms by breaking and forming chemical bonds. Energy must be absorbed to break bonds and is released when bonds form. **Activation energy,** or the **free energy of activation** (E_A), is the energy that must be absorbed by reactants to reach the unstable *transition state,* in which bonds are more fragile and likely to break, and from which the reaction can proceed.

The E_A barrier is essential to life because it prevents the energy-rich macromolecules of the cell from decomposing spontaneously. For metabolism to proceed in a cell, however, E_A must be reached. Heat, a possible source of activation energy in reactions, would be harmful to the cell and would also speed metabolic reactions indiscriminately. Enzymes are able to lower E_A for specific reactions so that metabolism can proceed at cellular temperatures. Enzymes do not change the ΔG for a reaction.

In this graph of an exergonic reaction with and without an enzyme catalyst, label the parts **a** through **e**.

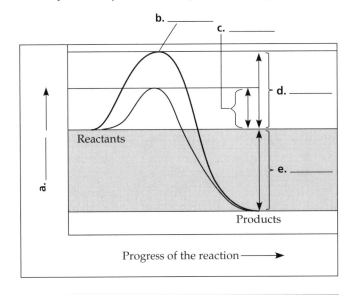

Enzymes are substrate specific (97–98)

Most enzymes are proteins, macromolecules with characteristic three-dimensional shapes. The specificity of an enzyme for a particular **substrate** is determined by this shape. The substrate is temporarily bound to its enzyme at the **active site,** a pocket or groove found on the surface of the enzyme that has a shape complementary to the substrate. When a substrate molecule enters the active site, the enzyme changes shape slightly, creating what is called an **induced fit** that enhances the ability of the enzyme to catalyze the chemical reaction.

The active site is an enzyme's catalytic center (98–99)

The substrate is often held in the active site by hydrogen or ionic bonds, creating an enzyme-substrate complex. The side chains (R groups) of some of the surrounding amino acids in the active site facilitate the conversion of substrate to product. The product then leaves the active site and the enzyme can bind with another substrate molecule.

Enzymes can catalyze reactions involving the joining of two reactants by binding the substrates closely together and properly oriented. An induced fit can stretch critical bonds in the substrate molecule and make them easier to break. An active site may provide a microenvironment, such as a lower pH, that is necessary for a particular reaction. Enzymes may also actually participate in a reaction by forming brief covalent bonds with the substrate.

The rate of a reaction will increase with increasing substrate concentration up to the point at which all enzyme molecules are saturated with substrate molecules and working at full speed.

■ INTERACTIVE QUESTION   6.6

Outline a catalytic cycle using the diagrammatic enzyme below. Sketch the appropriate substrate and products, and identify the key steps of the cycle.

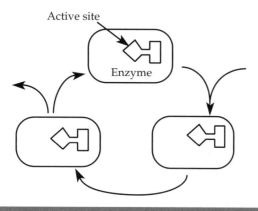

Active site

Enzyme

A cell's physical and chemical environment affects enzyme activity (99–101)

Effects of Temperature and pH   The velocity of an enzyme-catalyzed reaction may increase with rising temperature up to the point at which increased thermal agitation begins to disrupt the hydrogen and ionic bonds and other interactions that stabilize protein conformation. A change in pH may denature an enzyme by disrupting the hydrogen bonding of the molecule. Each enzyme has a temperature and pH optimum at which it is most active.

Cofactors   Cofactors are small molecules that bind either permanently or reversibly with enzymes and are necessary for enzyme catalytic function. They may

be inorganic, such as various metal atoms, or organic molecules called **coenzymes.** Most vitamins are coenzymes or precursors of coenzymes.

Enzyme Inhibitors   Enzyme inhibitors selectively disrupt the action of enzymes, either reversibly by binding with the enzyme with weak bonds or irreversibly by attaching with covalent bonds. **Competitive inhibitors** compete with the substrate for the active site of the enzyme. Increasing the concentration of substrate molecules may overcome this type of inhibition. **Noncompetitive inhibitors** bind to a part of the enzyme separate from the active site and change the conformation of the enzyme, thus impeding enzyme action.

■ INTERACTIVE QUESTION   6.7

Return to your diagram in 6.6. Draw a competitive and a noncompetitive inhibitor, and indicate where they would bind to the enzyme molecule.

The Control of Metabolism

Metabolic control often depends on allosteric regulation (101–103)

Allosteric Regulation   Molecules that inhibit or activate enzyme activity may bind to an **allosteric site,** a receptor site separate from the active site. Complex enzymes made of two or more polypeptide chains, each with its own active site, may have allosteric sites located where subunits join. The entire unit may oscillate between two conformational states, and the binding of an activator (or inhibitor) stabilizes the catalytically active (or inactive) conformation. Allosteric enzymes may be critical regulators of metabolic pathways.

Feedback Inhibition   Metabolic pathways are commonly regulated by **feedback inhibition,** in which the product of a pathway acts as an allosteric inhibitor of an enzyme early in the pathway.

Cooperativity   Through a phenomenon called **cooperativity,** the induced-fit binding of a substrate molecule to one polypeptide subunit can change the conformation such that the active sites of all subunits are more active.

**The localization of enzymes within
a cell helps order metabolism (103)**

Specialized cellular compartments may contain high concentrations of the enzymes and substrates needed for a particular pathway. Enzymes also may be incorporated into the membranes of cellular compartments. The complex internal structures of the cell facilitate metabolic order.

**The theme of emergent properties
is manifest in the chemistry
of life:** *a review* **(103)**

The chemistry of life involves a hierarchy of structural levels, from atoms to macromolecules to metabolism, with the emergence of new properties accompanying each increase in organizational level.

WORD ROOTS

cata- = down (*catabolic pathway:* a metabolic pathway that releases energy by breaking down complex molecules into simpler ones)

ana- = up (*anabolic pathway:* a metabolic pathway that consumes energy to build complex molecules from simpler ones)

bio- = life (*bioenergetics:* the study of how organisms manage their energy resources)

kinet- = movement (*kinetic energy:* the energy of motion)

therm- = heat (*thermodynamics:* the study of the energy transformations that occur in a collection of matter)

ex- = out (*exergonic reaction:* a reaction that proceeds with a net release of free energy)

endo- = within (*endergonic reaction:* a reaction that absorbs free energy from its surroundings)

allo- = different (*allosteric site:* a specific receptor site on some part of an enzyme molecule remote from the active site)

STRUCTURE YOUR KNOWLEDGE

This chapter introduced many new and complex concepts. See if you can step back from the details and answer the following general questions.

1. Relate the concept of free energy to metabolism.

2. What role do enzymes play in metabolism?

TEST YOUR KNOWLEDGE

MULTIPLE CHOICE: *Choose the one best answer.*

1. Catabolic and anabolic pathways are often coupled in a cell because
   a. the intermediates of a catabolic pathway are used in the anabolic pathway.
   b. both pathways use the same enzymes.
   c. the free energy released from one pathway is used to drive the other.
   d. the activation energy of the catabolic pathway can be used in the anabolic pathway.
   e. their enzymes are controlled by the same activators and inhibitors.

2. When glucose is converted to CO_2 and H_2O, changes in total energy, entropy, and free energy are as follows:
   a. $-\Delta H, -\Delta S, -\Delta G$
   b. $-\Delta H, +\Delta S, -\Delta G$
   c. $-\Delta H, +\Delta S, +\Delta G$
   d. $+\Delta H, +\Delta S, +\Delta G$
   e. $+\Delta H, -\Delta S, +\Delta G$

3. When a protein forms from amino acids, the following changes apply:
   a. $+\Delta H, -\Delta S, +\Delta G$
   b. $+\Delta H, +\Delta S, -\Delta G$
   c. $+\Delta H, +\Delta S, +\Delta G$
   d. $-\Delta H, +\Delta S, +\Delta G$
   e. $-\Delta H, -\Delta S, +\Delta G$

4. A negative ΔG means that
   a. the quantity G of energy is available to do work.
   b. the reaction is spontaneous.
   c. the reactants have more free energy than the products.
   d. the reaction is exergonic.
   e. all of the above are true.

5. According to the first law of thermodynamics,
   a. for every action there is an equal and opposite reaction.
   b. every energy transfer results in an increase in disorder or entropy.
   c. the total amount of energy in the universe is conserved or constant.
   d. energy can be transferred or transformed, but disorder always increases.
   e. potential energy is converted to kinetic energy, and kinetic energy is converted to heat.

6. What is meant by an induced fit?
   a. The binding of the substrate is an energy-requiring process.
   b. A competitive inhibitor can outcompete the substrate for the active site.
   c. The binding of the substrate changes the shape of the active site, which can stress or bend substrate bonds.
   d. The active site creates a microenvironment ideal for the reaction.
   e. Substrates are held in the active site by hydrogen and ionic bonds.

7. One way in which a cell maintains metabolic disequilibrium is to
   a. siphon products of a reaction off to the next step in a metabolic pathway.
   b. provide a constant supply of enzymes for critical reactions.
   c. use feedback inhibition to turn off pathways.
   d. use allosteric enzymes that can bind to activators or inhibitors.
   e. use the energy from anabolic pathways to drive catabolic pathways.

8. In an experiment, changing the pH from 7 to 6 resulted in an increase in product formation. From this we could conclude that
   a. the enzyme became saturated at pH 6.
   b. the enzyme's optimal pH is 6.
   c. this enzyme works best in a neutral pH.
   d. the temperature must have increased when the pH was changed to 6.
   e. the enzyme was in a more active conformation at pH 6.

9. When substance A was added to an enzyme reaction, product formation decreased. The addition of more substrate did not increase product formation. From this we conclude that substance A could be
   a. product molecules.
   b. a cofactor.
   c. an allosteric enzyme.
   d. a competitive inhibitor.
   e. a noncompetitive inhibitor.

10. The formation of ATP from ADP and inorganic phosphate
    a. is an exergonic process.
    b. transfers the phosphate to another intermediate that becomes more reactive.
    c. produces an unstable energy compound that can drive cellular work.
    d. has a ΔG of -7.3 kcal/mol under standard conditions.
    e. involves the hydrolysis of a phosphate bond.

11. At equilibrium,
    a. no enzymes are functioning.
    b. $\Delta G = 0$.
    c. the forward and backward reactions have stopped.
    d. the products and reactants have equal values of H.
    e. a reaction has a $+ \Delta G$.

12. In cooperativity,
    a. a cellular organelle contains all the enzymes needed for a metabolic pathway.
    b. a product of a pathway serves as a competitive inhibitor of an early enzyme in the pathway.
    c. a molecule bound to the active site of one subunit of an enzyme affects the active site of other subunits.
    d. the allosteric site is filled with an activator molecule.
    e. the product of one reaction serves as the substrate for the next in intricately ordered metabolic pathways.

13. When a cell breaks down glucose, only about 40% of the energy is captured in ATP molecules. The remaining 60% of the energy is
    a. used to increase the order necessary for life to exist.
    b. lost as heat because of the second law of thermodynamics.
    c. used to increase the entropy of the system by converting kinetic energy into potential energy.
    d. stored in starch or glycogen for use later by the cell.
    e. released when the ATP molecules are hydrolyzed.

14. An endergonic reaction could be described as one that will
    a. proceed spontaneously with the addition of activation energy.
    b. produce products with more free energy than the reactants.
    c. not be able to be catalyzed by enzymes.
    d. release energy.
    e. produce ATP for energy coupling.

15. What is most directly responsible for the specificity of a protein enzyme?
    a. its primary structure
    b. its secondary and tertiary structure
    c. the conformation of its allosteric site
    d. its cofactors
    e. the R groups of the amino acids in its active site

16. Which of the following parameters does an enzyme raise?
    a. ΔG
    b. ΔH
    c. equilibrium of a reaction
    d. speed of a reaction
    e. free energy of activation

17. Zinc, an essential trace element, may be found bound to the active site of some enzymes. What would be the most likely function of such zinc ions?
    a. coenzyme derived from a vitamin
    b. a cofactor necessary for catalysis
    c. a substrate of the enzyme
    d. a competitive inhibitor of the enzyme
    e. an allosteric activator of the enzyme

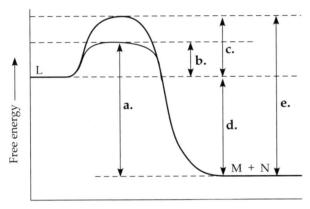

18. Which lines in the diagram above indicate the ΔG of the enzyme-catalyzed reaction of L → M + N?

19. Which of the following terms would best describe this reaction?
    a. nonspontaneous
    b. $+\Delta G$
    c. exergonic
    d. coupled reaction
    e. anabolic reaction

20. In the metabolic pathway, A → B → C → D → E, what effect would molecule E likely have on the enzyme that catalyzes A → B?
    a. allosteric inhibitor
    b. allosteric activator
    c. competitive inhibitor
    d. feedback activator
    e. coenzyme

FILL IN THE BLANKS

_____ 1. the totality of an organism's chemical processes

_____ 2. pathways that require energy to combine molecules together

_____ 3. the energy of motion

_____ 4. enzymes that change between two conformations, depending on whether an activator or inhibitor is bound to them

_____ 5. term for the measure of disorder or randomness

_____ 6. the energy that must be absorbed by molecules to reach the transition state

_____ 7. inhibitors that decrease an enzyme's activity by binding to the active site

_____ 8. organic molecules that bind to enzymes and are necessary for their functioning

_____ 9. regulatory device in which the product of a pathway binds to an enzyme early in the pathway

_____ 10. more reactive molecules created by the transfer of a phosphate group from ATP

THE CELL

CHAPTER 7

A TOUR OF THE CELL

FRAMEWORK

This chapter deals with the fundamental unit of life—the cell. The complexities in the processes of life are reflected in the complexities of the structure of the cell. It is easy to become overwhelmed by the number of new vocabulary terms for this array of cell organelles and membranes. The following concept map provides an organizational framework for the wealth of detail found in "a tour of the cell."

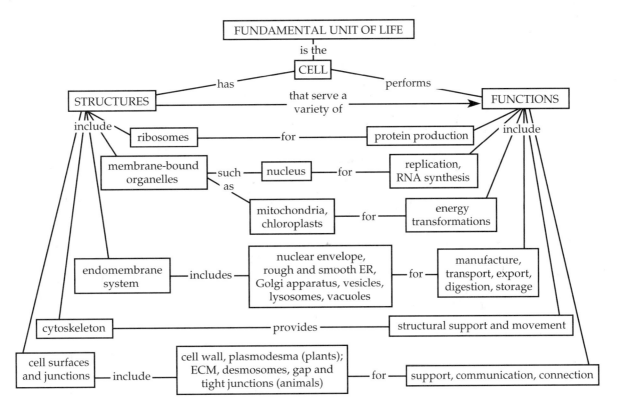

CHAPTER REVIEW

The cell is the basic structural and functional unit of all living organisms. In the hierarchy of biological organization, the capacity for life emerges from the structural order of the cell. The cell senses and responds to its environment and exchanges materials and energy with its surroundings. All cells are related through common descent, but evolution has shaped diverse adaptations.

How We Study Cells

Microscopes provide windows to the world of the cell (109–111)

The glass lenses of **light microscopes (LMs)** refract (bend) the visible light passing through a specimen such that the projected image is magnified. **Resolving power** is a measure of the clarity of an image—determined by the minimum distance two points must be separated to be distinguished. The resolving power of the light microscope is limited by the shortest wavelength of visible light, so that details finer than 0.2 μm (micrometers = 10^{-3} mm) cannot be resolved. Staining of specimens and using techniques such as fluorescence, phase-contrast, and confocal microscopy improve visibility by increasing contrast between structures that are large enough to be resolved.

Most subcellular structures, or **organelles,** cannot be resolved by the light microscope. Cells were discovered by Robert Hooke in 1665, but their ultrastructure was largely unknown until the development of the **electron microscope (EM)** in the 1950s. The electron microscope focuses a beam of electrons through the specimen. The short wavelength of the electron beam allows a resolution of about 2 nm (nanometers), a hundred times greater than that of the light microscope.

In a **transmission electron microscope (TEM),** a beam of electrons is passed through a thin section of a specimen, and electromagnets, acting as lenses, focus and magnify the image. Contrast is increased by staining preserved cells with atoms of heavy metals.

In a **scanning electron microscope (SEM),** the electron beam scans the surface of a specimen coated with a thin gold film, exciting electrons from the specimen and collecting and focusing them onto a screen. The resulting image appears three dimensional.

Modern cell biology integrates cytology with biochemistry to understand relationships between cellular structure and function.

■ INTERACTIVE QUESTION 7.1

a. Define cytology.

b. What do cell biologists use a TEM to study?

c. What does an SEM show best?

d. What advantages does light microscopy have over TEM and SEM?

Cell biologists can isolate organelles to study their functions (111)

Cell fractionation is a technique that separates major organelles of a cell so that their functions can be studied. Cells are homogenized and the resulting cellular soup is separated into component fractions by differential centrifugation. The homogenate is first spun slowly, and nuclei settle to form a pellet. The remaining supernatant is centrifuged at increasing speeds, each time isolating smaller and smaller cellular components in the pellet. **Ultracentrifuges** can spin at speeds up to 130,000 rpm.

A Panoramic View of the Cell

Prokaryotic and eukaryotic cells differ in size and complexity (112–114)

The bacteria and archaea have **prokaryotic cells,** which are cells with no nucleus or membrane-enclosed organelles. The DNA of prokaryotic cells is concentrated in a region called the **nucleoid. Eukaryotic cells,** which are much more structurally complex, are found in protists, plants, fungi, and animals. They have a true nucleus enclosed in a nuclear membrane and numerous organelles suspended in a semifluid medium called **cytosol.** The **cytoplasm** is the entire region between the nucleus and the membrane enclosing the cell.

Most bacterial cells range from 1 to 10 μm in diameter, whereas eukaryotic cells are ten times larger, ranging from 10 to 100 μm.

The small size of cells is dictated by geometry and the requirements of metabolism. Area is proportional to the square of linear dimension, while volume is proportional to its cube. The **plasma membrane** surrounding every cell must provide sufficient surface area for exchange of oxygen, nutrients, and wastes relative to the volume of the cell.

■ **INTERACTIVE QUESTION 7.2**

a. If a eukaryotic cell has a diameter that is 10 times that of a bacterial cell, proportionally how much more surface area would the eukarotic cell have?

b. Proportionally how much more volume would it have?

Internal membranes compartmentalize the functions of a eukaryotic cell (114–116)

Membranes compartmentalize the eukaryotic cell, providing local environments for specific metabolic functions and participating in metabolism through membrane-bound enzymes. Membranes are composed of a bilayer of phospholipid molecules associated with diverse proteins.

The Nucleus and Ribosomes

The nucleus contains a eukaryotic cell's genetic library (117)

The **nucleus** is surrounded by the nuclear envelope, a double membrane perforated by pores that regulate the movement of large macromolecules between the nucleus and the cytoplasm. The inner membrane is lined by the **nuclear lamina,** a layer of protein filaments that helps to maintain the shape of the nucleus. There is evidence of a framework of fibers, called a nuclear matrix, extending through the nucleus.

Most of the cell's DNA is located in the nucleus, where, along with associated proteins, it appears as a diffuse mass called **chromatin.** Each eukaryotic species has a characteristic chromosomal number. Individual **chromosomes** are visible only when coiled and condensed in a dividing cell.

The **nucleolus,** a dense structure visible in the nondividing nucleus, synthesizes ribosomal RNA and combines it with protein to assemble ribosomal subunits that pass through nuclear pores to the cytoplasm.

■ **INTERACTIVE QUESTION 7.3**

How does the nucleus control protein synthesis in the cytoplasm?

Ribosomes build a cell's proteins (117–118)

Most of the proteins produced by free **ribosomes** are used within the cytosol. Bound ribosomes, attached to the endoplasmic reticulum, usually make proteins that will be included within membranes, packaged into organelles, or exported from the cell.

The Endomembrane System

The **endomembrane system** of a cell consists of the nuclear envelope, endoplasmic reticulum, Golgi apparatus, lysosomes, vacuoles, and the plasma membrane. These membranes are all related either through direct contact or by the transfer of membrane segments by membrane-bound sacs called **vesicles.**

The endoplasmic reticulum manufactures membranes and performs many other biosynthetic functions (118–119)

The **endoplasmic reticulum (ER)** is the most extensive portion of the endomembrane system. It is continuous with the nuclear envelope and encloses a network of interconnected tubules or compartments called cisternae. Ribosomes are attached to the cytoplasmic surface of **rough ER; smooth ER** lacks ribosomes.

Functions of Smooth ER   Smooth ER serves diverse functions in different cells: Its enzymes are involved in phospholipid, steroid, and sex hormone synthesis; carbohydrate metabolism; and detoxification of drugs and poisons. Barbiturates, alcohol, and other drugs increase a liver cell's production of smooth ER, thus leading to an increased tolerance (and thus reduced effectiveness) for these and other drugs. Smooth ER also functions in storage and release of calcium ions during muscle contraction.

Rough ER and Synthesis of Secretory Proteins   Proteins intended for secretion are manufactured by membrane-bound ribosomes and then threaded into the cisternal space of the rough ER, where they fold into their native conformation. Many are covalently bonded to small carbohydrates to form **glycoproteins.**

The secretory proteins are transported from the rough ER in membrane-bound **transport vesicles.**

Rough ER and Membrane Production   Rough ER manufactures membranes by inserting proteins formed by the attached ribosomes into the membrane. Membrane phospholipids are assembled with the aid of enzymes built into the membrane. The newly formed membrane can be transferred to other parts of the endomembrane system by transport vesicles.

The Golgi apparatus finishes, sorts, and ships cell products (119–121)

The **Golgi apparatus** consists of a stack of flattened membranous sacs. Vesicles that bud from the ER join to the cis face of a Golgi stack, adding to it their contents and membrane. Products that travel through the Golgi apparatus are usually modified or refined as they move from one cisterna to the next. Some polysaccharides are manufactured by the Golgi. Golgi products are sorted into vesicles, which pinch off from the trans face of the Golgi apparatus. These vesicles may have surface molecules that help direct them to the plasma membrane or to other organelles.

Lysosomes are digestive compartments (121–122)

Lysosomes are membrane-enclosed sacs of hydrolytic enzymes used by the cell to digest macromolecules. The cell can maintain an acidic pH for these enzymes and also protect itself from unwanted digestion by containing hydrolytic enzymes within lysosomes.

In macrophages and some protists, lysosomes fuse with food vacuoles to digest material ingested by **phagocytosis.** They also recycle the cell's own macromolecules by engulfing organelles or small bits of cytosol, a process known as autophagy. During development or metamorphosis, lysosomes may be programmed to destroy their cells to create a specific body form.

Storage diseases are inherited defects in which a lysosomal enzyme is missing or defective, and lysosomes become packed with indigestible substances.

Vacuoles have diverse functions in cell maintenance (122–123)

Vacuoles are membrane-enclosed sacs that are larger than vesicles. **Food vacuoles** are formed as a result of phagocytosis. **Contractile vacuoles** pump excess water out of freshwater protists.

A large **central vacuole** is found in mature plant cells, surrounded by a membrane called the **tonoplast.** This vacuole stores organic compounds and inorganic ions for the cell. Poisonous or unpalatable compounds, which may protect the plant from

predators, and dangerous metabolic by-products may also be contained in the vacuole. Plant cells increase in size with a minimal addition of new cytoplasm as their vacuoles absorb water and expand.

■ INTERACTIVE QUESTION   7.4

Name the components of the endomembrane system shown in this diagram and review the functions of each of these membranes.

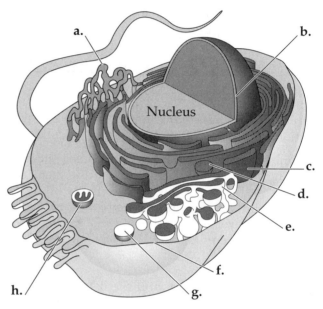

a.

b.

c.

d.

e.

f.

g.

h.

Other Membranous Organelles

Mitochondria and chloroplasts are the main energy transformers of cells (123–125)

Cellular respiration, the catabolic processing of fuels to produce ATP, occurs within the **mitochondria** of eukaryotic cells. Photosynthesis occurs in the **chloroplasts** of plants and eukaryotic algae, which produce organic compounds from carbon dioxide and water by absorbing solar energy. The membrane proteins of mitochondria and chloroplasts are made by ribosomes either free in the cytosol or contained within these organelles. They also contain a small amount of DNA that directs the synthesis of some of their proteins.

Mitochondria Two membranes, each a phospholipid bilayer with unique embedded proteins, enclose a mitochondrion. A narrow intermembrane space exists between the smooth outer membrane and the convoluted inner membrane. The folds of the inner membrane, called **cristae,** create a large surface area and enclose the **mitochondrial matrix.** Many respiratory enzymes, mitochondrial DNA, and ribosomes are housed in this matrix. Other respiratory enzymes and proteins are built into the inner membrane.

Chloroplasts **Plastids** are plant organelles that include amyloplasts, which store starch; chromoplasts, which contain pigments; and chloroplasts, which contain the green pigment chlorophyll and function in photosynthesis.

Chloroplasts are bounded by two membranes that enclose a fluid called the **stroma** and a membranous system of flattened sacs, called **thylakoids,** which, in turn, enclose the thylakoid space. Thylakoids may be stacked together to form structures called **grana.**

■ **INTERACTIVE QUESTION 7.5**

Sketch a mitochondrion and a chloroplast and label their membranes and compartments.

Peroxisomes generate and degrade H₂O₂ in performing various metabolic functions (125–126)

Peroxisomes are membrane-enclosed compartments filled with enzymes that function in a variety of metabolic pathways, such as breaking down fatty acids for energy or detoxifying alcohol and other poisons. An enzyme that converts hydrogen peroxide (H_2O_2), a toxic by-product of these pathways, is also packaged into peroxisomes.

■ **INTERACTIVE QUESTION 7.6**

Why are peroxisomes not considered part of the endomembrane system?

The Cytoskeleton

Providing structural support to the cell, the cytoskeleton also functions in cell motility and regulation (126–132)

The **cytoskeleton** is a network of fibers that function to give mechanical support; maintain or change cell shape; anchor or direct the movement of organelles and cytoplasm; control movement of cilia, pseudopods, and even contraction of muscle cells; and mechanically transmit signals from the cell's surface to its interior. The cytoskeleton interacts with special proteins called motor molecules that change shape to produce cellular movements. The three main types of fibers involved in the cytoskeleton are **microtubules, microfilaments,** and **intermediate filaments.**

Microtubules All eukaryotic cells have microtubules, which are hollow rods constructed of columns of a globular protein called tubulin. In addition to providing the supporting framework of the cell, microtubules serve as tracks along which organelles move with the aid of motor molecules.

In many cells, microtubules radiate out from a region near the nucleus called a **centrosome.** In animal cells, a pair of **centrioles,** each composed of nine sets of triplet microtubules arranged in a ring, may help to organize microtubule assembly for cell division.

Cilia and **flagella** are extensions of some eukaryotic cells. Cilia are numerous and short; flagella occur one or two to a cell and are longer. Many protists use cilia or flagella to move through aqueous media. Cilia or flagella attached to stationary cells of a tissue move fluid past the cell.

Both cilia and flagella are composed of two single microtubules surrounded by a ring of nine doublets of microtubules (a nearly universal "9 + 2" arrangement), all of which are enclosed in an extension of the plasma membrane. A **basal body,** structurally identical to a centriole, anchors the tubules in the cell. The sliding of the microtubule doublets past each other occurs as arms, composed of the motor protein **dynein,** alternately attach to adjacent doublets, pull

■ **INTERACTIVE QUESTION 7.7**

Fill in the following table to organize what you have learned about the components of the cytoskeleton. You may wish to refer to the textbook for additional details.

Cytoskeleton	Structure and Monomers	Functions
Microtubules	a.	b.
Microfilaments (actin filaments)	c.	d.
Intermediate filaments	e.	f.

down, release, and reattach. In conjunction with anchoring cross-links and radial spokes, this action —driven by ATP—causes the bending of the flagellum or cilium.

Microfilaments (Actin Filaments) Microfilaments, probably present in all eukaryotic cells, are solid rods consisting of a twisted double chain of molecules of the globular protein **actin.** They function in support, forming a network just inside the plasma membrane and the core of small cytoplasmic extensions called microvilli.

In muscle cells, thousands of actin filaments interdigitate with thicker filaments made of the protein **myosin.** The sliding of actin and myosin filaments past each other causes the contraction of muscles.

Actin and myosin also interact in localized contractions such as cleavage furrows in animal cell division and amoeboid movements. Actin subunits reversibly assemble into microfilaments and then networks, driving the conversion of cytoplasm from sol to gel during the extension and retraction of **pseudopodia. Cytoplasmic streaming** in plant cells appears to involve both actin–myosin interactions and sol–gel conversions.

Intermediate Filaments Intermediate filaments are intermediate in size between microtubules and microfilaments and are more diverse in their composition. Intermediate fibers appear to be important in maintaining cell shape. The nucleus is securely held in a web of intermediate filaments, and the nuclear lamina lining the inside of the nuclear envelope is composed of intermediate filaments.

Cell Surfaces and Junctions

Plant cells are encased by cell walls (132)

A **cell wall** is a diagnostic feature of plant cells. Plant cell walls are composed of microfibrils of cellulose embedded in a matrix of polysaccharides and protein.

The **primary cell wall** secreted by a young plant cell is relatively thin and flexible. Adjacent cells are connected by the **middle lamella,** a thin layer of polysaccharides (called pectins) that glue the cells together. When they stop growing, some cells secrete a thicker and stronger **secondary cell wall** between the plasma membrane and the primary cell wall.

■ **INTERACTIVE QUESTION 7.8**

Sketch two adjacent plant cells, and show the location of the primary and secondary cell walls and the middle lamella.

The extracellular matrix (ECM) of animal cells functions in support, adhesion, movement, and regulation (133)

Animal cells secrete an **extracellular matrix (ECM)** composed primarily of glycoproteins. **Collagen,** the most abundant glycoprotein of the ECM, forms strong fibers that are embedded in a network of **proteoglycans.** Cells may be attached to the ECM by **fibronectins** that bind to **integrins,** receptor proteins that span the plasma membrane and bind to microfilaments of the cytoskeleton. Thus, information about changes inside and outside the cell can be exchanged through a mechanical signaling pathway involving fibronectins, integrins, and the cytoskeleton. Signals from the ECM appear to influence the activity of genes in the nucleus.

Intercellular junctions help integrate cells into higher levels of structure and function (133–134)

Plasmodesmata are channels in plant cell walls through which the plasma membranes of bordering cells connect, thus linking most cells of a plant into a living continuum. Water, small solutes, and even some proteins and RNA molecules can move through these channels.

There are three main types of intercellular junctions between animal cells. **Tight junctions** are fusions between adjacent cell membranes that create an impermeable seal across a layer of epithelial cells. **Desmosomes,** reinforced by intermediate filaments, are anchoring junctions between adjacent cells. **Gap junctions** (also called communicating junctions) are cytoplasmic connections that allow for the exchange of ions and small molecules between cells through protein-surrounded pores.

■ **INTERACTIVE QUESTION   7.9**

Return to your sketch of plant cells in 7.8 and draw in a plasmodesma.

The cell is a living unit greater than the sum of its parts (135)

The compartmentalization and the number of specialized organelles typical of cells exemplify the principle that structure correlates with function. The intricate functioning of a living cell emerges from the complex interactions of its many parts.

WORD ROOTS

centro- = the center; **-soma** = a body (*centrosome:* material present in the cytoplasm of all eukaryotic cells and important during cell division)

chloro- = green (*chloroplast:* the site of photosynthesis in plants and eukaryotic algae)

cili- = hair (*cilium:* a short hair-like cellular appendage with a microtubule core)

cyto- = cell (*cytosol:* a semifluid medium in a cell in which are located organelles)

-ell = small (*organelle:* a small formed body with a specialized function found in the cytoplasm of eukaryotic cells)

endo- = inner (*endomembrane system:* the system of membranes within a cell that includes the nuclear envelope, endoplasmic reticulum, Golgi apparatus, lysosomes, vacuoles, and the plasma membrane)

eu- = true (*eukaryotic cell:* a cell that has a true nucleus)

extra- = outside (*extracellular matrix:* the substance in which animal tissue cells are embedded)

flagell- = whip (*flagellum:* a long whip-like cellular appendage that moves cells)

glyco- = sweet (*glycoprotein:* a protein covalently bonded to a carbohydrate)

lamin- = sheet/layer (*nuclear lamina:* a netlike array of protein filaments that maintains the shape of the nucleus)

lyso- = loosen (*lysosome:* a membrane-bounded sac of hydrolytic enzymes that a cell uses to digest macromolecules)

micro- = small; **-tubul** = a little pipe (*microtubule:* a hollow rod of tubulin protein in the cytoplasm of almost all eukaryotic cells)

nucle- = nucleus; **-oid** = like (*nucleoid:* the region where the genetic material is concentrated in prokaryotic cells)

phago- = to eat; **-kytos** = vessel (*phagocytosis:* a form of cell eating in which a cell engulfs a smaller organism or food particle)

plasm- = molded; **-desma** = a band or bond (*plasmodesmata:* an open channel in a plant cell wall)

pro- = before; **karyo-** = nucleus (*prokaryotic cell:* a cell that has no nucleus)

pseudo- = false; **-pod** = foot (*pseudopodium:* a cellular extension of amoeboid cells used in moving and feeding)

thylaco- = sac or pouch (*thylakoid:* a series of flattened sacs within chloroplasts)

tono- = stretched; **-plast** = molded (*tonoplast:* the membrane that encloses a large central vacuole in a mature plant cell) trans- = across; -port = a harbor (*transport vesicle:* a membranous compartment used to enclose and transport materials from one part of a cell to another)

ultra- = beyond (*ultracentrifuge:* a machine that spins test tubes at the fastest speeds to separate liquids and particles of different densities)

vacu- = empty (*vacuole:* sac that buds from the ER, Golgi, or plasma membrane)

STRUCTURE YOUR KNOWLEDGE

1. The table below lists the general functions performed by an animal cell. List the cellular structures associated with each of these functions.

Functions	Associated Organelles and Structures
Cell division	a.
Information storage and transferal	b.
Energy conversions	c.
Manufacture of membranes and products	d.
Lipid synthesis, drug detoxification	e.
Digestion, recycling	f.
Conversion of H2O2 to water	g.
Structural integrity	h.
Movement	i.
Exchange with environment	j.
Cell to cell connections	k.

2. This table lists structures that are unique to plant cells. Fill in the functions of these structures.

Plant Cell Structures	Functions
Cell wall	a.
Central vacuole	b.
Chloroplast	c.
Amyloplast	d.
Plasmodesmata	e.

3. Label the indicated structures in this diagram of an animal cell.

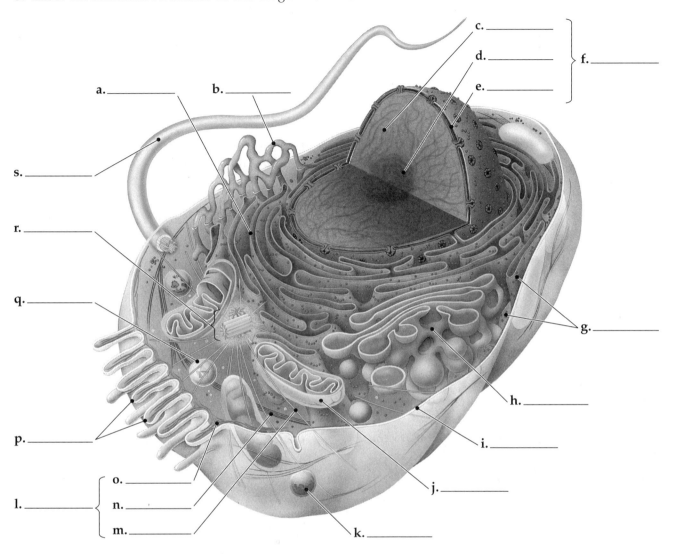

a. _____

b. _____

c. _____

d. _____

e. _____

f. _____

g. _____

h. _____

i. _____

j. _____

k. _____

l. _____

m. _____

n. _____

o. _____

p. _____

q. _____

r. _____

s. _____

4. Create a diagram or flow chart in the space below to trace the development of a secretory product (such as a digestive enzyme) from the DNA code to its export from the cell.

TEST YOUR KNOWLEDGE

MULTIPLE CHOICE: *Choose the one best answer.*

1. Which of the following is/are not found in a prokaryotic cell?
   a. ribosomes
   b. plasma membrane
   c. mitochondria
   d. a and c
   e. a, b, and c

2. Resolving power of a microscope is
   a. the distance between two separate points.
   b. the sharpness or clarity of an image.
   c. the degree of magnification of an image.
   d. the depth of focus on a specimen's surface.
   e. the wavelength of light.

3. Which of the following is *not* a similarity among the nucleus, chloroplasts, and mitochondria?
   a. They contain DNA.
   b. They are bounded by two phospholipid bilayer membranes.
   c. They can divide to reproduce themselves.
   d. They are derived from the endoplasmic reticulum system.
   e. Their membranes are associated with specific proteins.

4. The pores in the nuclear envelope provide for the movement of
   a. proteins into the nucleus.
   b. ribosomal subunits out of the nucleus.
   c. mRNA out of the nucleus.
   d. enzymes into the nucleus.
   e. all of the above.

5. The ultrastructure of a chloroplast could be seen best using
   a. transmission electron microscopy.
   b. scanning electron microscopy.
   c. phase-contrast light microscopy.
   d. cell fractionation.
   e. fluorescence microscopy.

6. Which of the following is *incorrectly* paired with its function?
   a. peroxisome—contains enzymes that break down H_2O_2
   b. nucleolus—produces ribosomal RNA, assembles ribosome subunits
   c. Golgi apparatus—processes, tags, and ships cellular products
   d. lysosome—contains pigments in plant cells
   e. ECM (extracellular matrix)—supports and anchors cells, communicates between inside and outside of cell

7. The cytoskeleton is composed of which type of molecule?
   a. protein
   b. cellulose
   c. chitin
   d. phospholipid
   e. calcium phosphate

8. A growing plant cell elongates primarily by
   a. increasing the number of vacuoles.
   b. synthesizing more cytoplasm.
   c. taking up water into its central vacuole.
   d. synthesizing more cellulose.
   e. producing a secondary cell wall.

9. The innermost portion of a mature plant cell wall is the
   a. primary cell wall.
   b. secondary cell wall.
   c. middle lamella.
   d. plasma membrane.
   e. plasmodesmata.

10. Contractile elements of muscle cells are
    a. intermediate filaments.
    b. centrioles.
    c. microtubules.
    d. actin filaments (microfilaments).
    e. fibronectins.

11. Microtubules are components of all of the following *except*
    a. centrioles.
    b. the spindle apparatus for separating chromosomes in cell division.
    c. tracks along which organelles can move using motor molecules.
    d. flagella and cilia.
    e. the pinching apart of the cytoplasm in animal cell division.

12. Of the following, which is probably the most common route for membrane flow in the endomembrane system?
    a. rough ER → Golgi → lysosomes → nuclear membrane → plasma membrane
    b. rough ER → transport vesicles→ Golgi → vesicles → plasma membrane
    c. nuclear envelope→ rough ER → Golgi → smooth ER → lysosomes
    d. rough ER → vesicles → Golgi → smooth ER → plasma membrane
    e. smooth ER → vesicles → Golgi → vesicles → peroxisomes

13. Proteins to be used within the cytosol are generally synthesized
    a. by ribosomes bound to rough ER.
    b. by free ribosomes.
    c. by the nucleolus.
    d. within the Golgi apparatus.
    e. by mitochondria and chloroplasts.

14. Plasmodesmata in plant cells are similar in function to
    a. desmosomes.
    b. tight junctions.
    c. gap junctions.
    d. the extracellular matrix.
    e. integrins.

15. In a cell fractionation procedure, the first pellet formed would most likely contain
    a. the extracellular matrix.
    b. ribosomes.
    c. mitochondria.
    d. lysosomes.
    e. nuclei.

Use the cells described as follows to answer questions 16–20.
    a. muscle cell in the thigh muscle of a long-distance runner
    b. pancreatic cell that manufactures digestive enzymes
    c. macrophage (white blood cell) that engulfs bacteria
    d. epithelial cell lining digestive tract
    e. ovarian cell that produces estrogen (a steroid hormone)

16. In which cell would you expect to find the most tight junctions?

17. In which cell would you expect to find the most lysosomes?

18. In which cell would you expect to find the most smooth endoplasmic reticulum?

19. In which cell would you expect to find the most bound ribosomes?

20. In which cell would you expect to find the most mitochondria?

FILL IN THE BLANKS *with the appropriate cellular organelle or structure.*

_____ 1. transports membranes and products to various locations

_____ 2. infoldings of mitochondrial membrane with attached enzymes

_____ 3. consists of collagen, proteoglycans, and fibronectins

_____ 4. small sacs with specific enzymes for a particular metabolic pathway

_____ 5. stacks of flattened sacs inside chloroplasts

_____ 6. anchoring structure for cilia and flagella

_____ 7. semifluid medium between nucleus and plasma membrane

_____ 8. system of fibers that maintains cell shape, anchors organelles

_____ 9. connection between animal cells that creates impermeable layer

_____10. membrane surrounding central vacuole of plant cells

CHAPTER 8

MEMBRANE STRUCTURE AND FUNCTION

FRAMEWORK

This chapter presents the fluid mosaic model of membrane structure, relating the molecular structure of biological membranes to their function of regulating the passage of substances into and out of the cell.

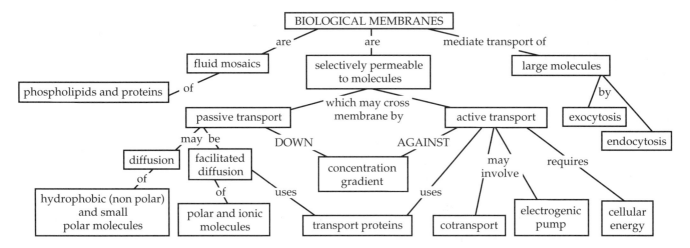

CHAPTER REVIEW

The plasma membrane is the boundary of life; this **selectively permeable membrane** allows the cell to maintain a unique internal environment and to control the movement of materials into and out of the cell.

Membrane Structure

According to the currently accepted **fluid mosaic model,** the structure of biological membranes consists of various proteins that attach to or float in a bilayer of **amphipathic** phospholipids.

Membrane models have evolved to fit new data: *the process of science* (138–141)

In 1925, two Dutch scientists suggested that cell membranes must be a phospholipid bilayer, with the hydrophobic hydrocarbon tails in the center and the hydrophilic heads facing the aqueous solution on both sides of the membrane. In 1935, H. Davson and J. Danielli proposed a model in which a double layer of phospholipids is covered with a coat of globular proteins. This sandwich model was consistent with the triple-layer staining results seen with electron microscopy, and by the 1960s, the Davson-Danielli model was widely accepted for all cellular membranes.

In 1972, S. J. Singer and G. Nicolson proposed their fluid mosaic model in which amphipathic membrane proteins are embedded in the phospholipid bilayer with their hydrophilic regions extending out into the aqueous environment. The phospholipid bilayer is envisioned as fluid, with a mosaic of protein molecules floating in it.

The fluid mosaic model is supported by evidence from freeze-fracture electron microscopy. This technique involves freezing a specimen, fracturing it with a cold knife, and coating the fractured surface with platinum and then carbon. The resulting replica of the surface is examined with the electron microscope.

The fluid mosaic model is currently the most accepted and useful model for organizing existing knowledge and extending further research on membrane structure.

■ INTERACTIVE QUESTION 8.1

Label the components in this diagram of the fluid mosaic model of membrane structure. Indicate the regions that are hydrophobic and those that are hydrophilic.

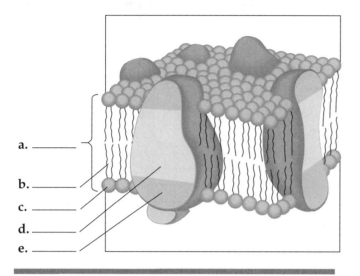

a. _____

b. _____

c. _____

d. _____

e. _____

Membranes are fluid (141–142)

Membranes are held together primarily by weak hydrophobic interactions that allow the lipids and some of the proteins to drift laterally. Some membrane proteins seem to be held rigid by attachments to the cytoskeleton; others appear to be directed in their movements by cytoskeletal fibers.

Phospholipids with unsaturated hydrocarbon tails maintain membrane fluidity at lower temperatures. The steroid cholesterol, common in plasma membranes of animals, restricts movement of phospholipids and thus reduces fluidity at warmer temperatures. Cholesterol also prevents the close packing of lipids and thus enhances fluidity at lower temperatures.

■ INTERACTIVE QUESTION 8.2

a. Cite some experimental evidence that shows that membrane proteins drift.

b. How might the plasma membrane of a plant cell change in response to the cold temperatures of winter?

Membranes are mosaics of structure and function (142–143)

Each membrane has its own unique complement of membrane proteins, which determine most of the specific functions of that membrane. **Integral proteins** generally extend through the membrane, with two hydrophilic ends and a hydrophobic midsection consisting of one or more α-helical stretches of hydrophobic amino acids. **Peripheral proteins** are attached to the surface of the membrane, often to integral proteins. Membranes are asymmetric; they have distinct inner and outer faces related to the directional orientation of their proteins, the composition of the lipid bilayers, and, in the plasma membrane, the attachment of carbohydrates to the exterior surface.

Membrane carbohydrates are important for cell-cell recognition (143–144)

The ability of a cell to distinguish other cells is based on recognition of membrane carbohydrates. The glycolipids and glycoproteins attached to the outside of plasma membranes vary from species to species, from individual to individual, and even among cell types.

■ INTERACTIVE QUESTION 8.3

List the six major kinds of functions that membrane proteins may perform.

Traffic Across Membranes

A membrane's molecular organization results in selective permeability (144–145)

The plasma membrane permits a regular exchange of nutrients, waste products, oxygen, and inorganic ions. Biological membranes are selectively permeable; the ease and rate at which small molecules pass through them differ.

Permeability of the Lipid Bilayer   Hydrophobic molecules, such as hydrocarbons, CO_2, and O_2, can dissolve in and cross through a membrane. Very small polar molecules, including H_2O, can cross a plasma membrane easily.

■ INTERACTIVE QUESTION 8.4

What types of molecules have difficulty crossing the plasma membrane? Why?

Transport Proteins   Ions and polar molecules may move across the plasma membrane with the aid of **transport proteins,** which may provide a hydrophilic channel or may physically bind and transport a specific molecule.

Passive transport is diffusion across a membrane (145–146)

Diffusion is the movement of a substance down its **concentration gradient** due to random motion. This spontaneous process decreases free energy and increases entropy (disorder, randomness) in a system. The cell does not expend energy when substances diffuse across membranes down their concentration gradient; therefore, the process is called **passive transport.**

Osmosis is the passive transport of water (146)

Osmosis is the diffusion of water across a selectively permeable membrane. Water diffuses down its own concentration gradient, which is affected by the binding of water molecules to solute particles, thus lowering the proportion of unbound water that is free to cross the membrane. Water will move from a **hypotonic** solution across a membrane into a **hypertonic** solution—in other words, from a region with a lesser concentration of solutes (and greater concentration of unbound water) into a region with a greater concentration of solutes. **Isotonic** solutions have equal solute concentrations, and there is no net movement of water across a membrane separating them. The direction of osmosis is determined by the difference in total solute concentration, regardless of the kinds of solute molecules involved.

■ INTERACTIVE QUESTION 8.5

A solution of 1 *M* glucose is separated by a selectively permeable membrane from a solution of 0.2 *M* fructose and 0.7 *M* sucrose. The membrane is not permeable to the sugar molecules. Indicate which side is initially hypertonic and which is hypotonic. Show the direction of osmosis.

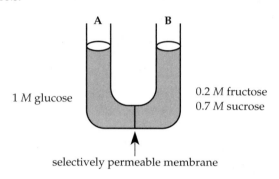

selectively permeable membrane

Cell survival depends on balancing water uptake and loss (146–147)

Water Balance of Cells Without Walls   An animal cell placed in a hypertonic environment will lose water and shrivel. If placed in a hypotonic environment, the cell will gain water, swell, and possibly lyse (burst). Cells without rigid walls must either live in an isotonic environment, such as salt water or isotonic body fluids, or have adaptations for **osmoregulation.**

■ INTERACTIVE QUESTION 8.6

a. What osmotic problems do freshwater protists face?

b. What adaptations may help them osmoregulate?

Water Balance of Cells With Walls   The cell walls of plants, fungi, prokaryotes, and some protists play a role in water balance within hypotonic environments. Water moving into the cell causes the cell to swell against its cell wall, creating a **turgid** cell. Turgid cells provide mechanical support for nonwoody plants.

Plant cells in an isotonic surrounding are **flaccid.** In a hypertonic medium, a plant cell undergoes **plasmolysis,** the pulling away of the plasma membrane from the cell wall as water leaves and the cell shrivels.

■ INTERACTIVE QUESTION 8.7

a. The ideal osmotic environment for animal cells is _____.

b. The ideal osmotic environment for plant cells is _____.

Specific proteins facilitate the passive transport of selected solutes (147–148)

Facilitated diffusion involves the diffusion of polar molecules and ions across a membrane with the aid of transport proteins. Transport proteins are similar to enzymes in several ways: They are specialized and may have a specific binding site for the solute they transport; they can reach a maximum rate of transport when all transport proteins are saturated by a high concentration of the solute; and they can be inhibited by molecules that resemble their normal solute.

The binding of the solute to the transport protein may cause a conformational change that serves to translocate the binding site and the attached solute. *Channel proteins* may provide hydrophilic passageways for specific ions ans other solutes. Channel proteins called **aquaporins** facilitate the rapid diffusion of water through a membrane. **Gated channels** that open in response to electrical or chemical stimulation.

■ INTERACTIVE QUESTION 8.8

Why is facilitated diffusion considered passive transport?

Active transport is the pumping of solutes against their gradients (148–149)

Active transport, requiring the expenditure of energy by the cell, is essential for a cell to maintain internal concentrations of small molecules that differ from environmental concentrations. The terminal phosphate group of ATP may be transferred to a transport protein, inducing it to change its conformation and translocate the bound solute across the membrane. The **sodium-potassium pump** allows the cell to exchange Na^+ and K^+ across animal cell membranes, creating a greater concentration of potassium ions and a lesser concentration of sodium ions within the cell.

Some ion pumps generate voltage across membranes (149–150)

Cells have a **membrane potential,** a voltage across their plasma membrane. This electrical potential energy results from the separation of opposite charges: The cytoplasm of a cell is negatively charged compared to extracellular fluid. The membrane potential favors the diffusion of cations into the cell and anions out of the cell. Diffusion of an ion is affected by both the membrane potential and its own concentration gradient; thus, an ion diffuses down its **electrochemical gradient.**

Electrogenic pumps are membrane proteins that generate voltage across a membrane by the active transport of ions. A **proton pump** that transports H^+ out of the cell generates membrane potentials in plants, fungi, and bacteria.

■ INTERACTIVE QUESTION 8.9

The Na^+-K^+ pump, the major electrogenic pump in animal cells, exchanges sodium ions for potassium ions, both of which are cations. How does this exchange generate a membrane potential?

In cotransport, a membrane protein couples the transport of one solute to another (150–151)

Cotransport is a mechanism through which the active transport of a solute is indirectly driven by an ATP-powered pump that transports another substance against its gradient. As that substance diffuses back down its concentration gradient through specialized transport proteins, the solute may be cotransported against its concentration gradient across the membrane.

Exocytosis and endocytosis transport large molecules (151–152)

In **exocytosis,** the cell secretes macromolecules by the fusion of vesicles with the plasma membrane. In **endocytosis,** a region of the plasma membrane sinks inward and pinches off to form a vesicle containing material that had been outside the cell. **Phagocytosis** is a form of endocytosis in which pseudopodia wrap around a food particle, creating a vacuole that then fuses with a lysosome containing hydrolytic enzymes. In **pinocytosis,** droplets of extracellular fluid are taken into the cell in small vesicles. **Receptor-mediated endocytosis** allows a cell to acquire specific substances from extracellular fluid. **Ligands,** molecules that bind specifically to receptor sites, attach to proteins usually clustered in coated pits on the cell surface and are carried into the cell when a vesicle forms.

■ INTERACTIVE QUESTION   8.10

a. How is cholesterol, which is used for the synthesis of other steroids and membranes, transported into human cells?

b. Explain why cholesterol accumulates in the blood of individuals with the disease familial hypercholesteremia.

WORD ROOTS

amphi- = dual (*amphipathic molecule:* a molecule that has both a hydrophobic and a hydrophilic region)

aqua- = water; **-pori** = a small opening (*aquaporin:* a transport protein in the plasma membrane of a plant or animal cell that specifically facilitates the diffusion of water across the membrane)

co- = together; **trans-** = across (*cotransport:* the coupling of the "downhill" diffusion of one substance to the "uphill" transport of another against its own concentration gradient)

electro- = electricity; **-genic** = producing (*electrogenic pump:* an ion transport protein generating voltage across a membrane)

endo- = inner; **cyto-** = cell (*endocytosis:* the movement of materials into a cell. Cell-eating)

exo- = outer (*exocytosis:* the movement of materials out of a cell)

hyper- = exceeding; **-tonus** = tension (*hypertonic:* a solution with a higher concentration of solutes)

hypo- = lower (*hypotonic:* a solution with a lower concentration of solutes)

iso- = same; (*isotonic:* solutions with equal concentrations of solutes)

phago- = eat (*phagocytosis:* cell eating)

pino- = drink (*pinocytosis:* cell drinking)

plasm- = molded; **-lyso** = loosen (*plasmolysis:* a phenomenon in walled cells in which the cytoplasm shrivels and the plasma membrane pulls away from the cell wall when the cell loses water to a hypertonic environment)

STRUCTURE YOUR KNOWLEDGE

1. Create a concept map to illustrate your understanding of osmosis. This exercise will help you practice using the words *hypotonic, isotonic,* and *hypertonic,* and it will help you focus on the effect of these osmotic environments on plant and animal cells. Explain your map to a friend.

2. The following diagram illustrates passive and active transport across a plasma membrane. Use it to answer questions a–d.

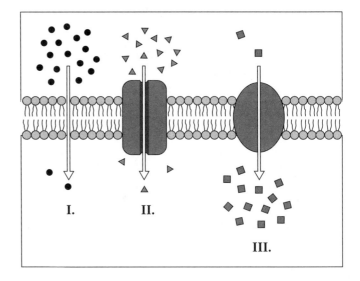

a. Which section represents facilitated diffusion?

How can you tell?

Does the cell expend energy in this transport?

Why or why not?

What types of solute molecules may be moved by this type of transport?

b. Which section shows active transport?

How can you tell?

Does the cell expend energy in this transport?

Why or why not?

c. Which section shows diffusion?

What types of solute molecules may be moved by this type of transport?

d. Which of these sections are considered passive transport?

TEST YOUR KNOWLEDGE

MULTIPLE CHOICE: *Choose the one best answer.*

1. Glycoproteins and glycolipids are important for
   a. facilitated diffusion.
   b. active transport.
   c. cell-cell recognition.
   d. cotransport.
   e. signal-transduction pathways.

2. A single layer of phospholipid molecules coats the water in a beaker. Which part of the molecules will face the air?
   a. the phosphate groups
   b. the hydrocarbon tails
   c. both head and tail because the molecules are amphipathic and will lie sideways
   d. the phospholipids would dissolve in the water and not form a membrane coat
   e. the glycolipid regions

3. Which of the following is *not* true about osmosis?
   a. It increases free energy in a system.
   b. Water moves from a hypotonic to a hypertonic solution.
   c. Solute molecules bind to water and decrease the water available to move.
   d. It increases the entropy in a system.
   e. There is no net osmosis between isotonic solutions.

4. Support for the fluid mosaic model of membrane structure comes from
   a. the freeze-fracture technique of electron microscopy.
   b. the movement of proteins in hybrid cells.
   c. the amphipathic nature of membrane proteins.
   d. both a and c.
   e. all of the above.

5. A freshwater *Paramecium* is placed into salt water. Which of the following events would occur?
   a. an increase in the action of its contractile vacuole
   b. swelling of the cell until it becomes turgid
   c. swelling of the cell until it lyses
   d. shriveling of the cell
   e. diffusion of salt ions out of the cell

6. Ions diffuse across membranes down their
   a. electrochemical gradient.
   b. electrogenic gradient.
   c. electrical gradient.
   d. concentration gradient.
   e. osmotic gradient.

7. The fluidity of membranes in a plant in cold weather may be maintained by
   a. increasing the number of phospholipids with saturated hydrocarbon tails.
   b. activating an H^+ pump.
   c. increasing the concentration of cholesterol in the membrane.
   d. increasing the proportion of peripheral proteins.
   e. increasing the number of phospholipids with unsaturated hydrocarbon tails.

8. A plant cell placed in a hypotonic environment will
   a. plasmolyze.
   b. shrivel.
   c. become turgid.
   d. become flaccid.
   e. lyse.

9. Which of the following is *not* true of the carrier molecules involved in facilitated diffusion?
   a. They increase the speed of transport across a membrane.
   b. They can concentrate solute molecules on one side of the membrane.
   c. They may have specific binding sites for the molecules they transport.
   d. They may undergo a conformational change upon binding of solute.
   e. They may be inhibited by molecules that resemble the solute to which they normally bind.

10. The membrane potential of a cell favors the
    a. movement of cations into the cell.
    b. movement of anions into the cell.
    c. action of an electrogenic pump.
    d. movement of sodium out of the cell.
    e. action of a proton pump.

11. Cotransport may involve
    a. active transport of two solutes through a transport protein.
    b. passive transport of two solutes through a transport protein.
    c. ion diffusion against the electrochemical gradient created by an electrogenic pump.
    d. a pump such as the Na⁺-K⁺ pump that moves ions in two different directions.
    e. transport of one solute against its concentration gradient in tandem with another that is diffusing down its concentration gradient.

12. Exocytosis involves all of the following *except*
    a. ligands and coated pits.
    b. the fusion of a vesicle with the plasma membrane.
    c. a mechanism to transport carbohydrates to the outside of plant cells during the formation of cell walls.
    d. a mechanism to rejuvenate the plasma membrane.
    e. a means of exporting large molecules.

13. The proton pump in plant cells is the functional equivalent of an animal cell's
    a. cotransport mechanism.
    b. sodium-potassium pump.
    c. contractile vacuole for osmoregulation.
    d. receptor-mediated endocytosis of cholesterol.
    e. ATP pump.

14. Pinocytosis involves
    a. the fusion of a newly formed food vacuole with a lysosome.
    b. receptor-mediated endocytosis and the formation of vesicles.
    c. the pinching in of the plasma membrane around small droplets of external fluid.
    d. pseudopod extension as vesicles move along the cytoskeleton and fuse with the plasma membrane.
    e. the accumulation of specific large molecules in a cell.

15. Watering a houseplant with too concentrated a solution of fertilizer can result in wilting because
    a. the uptake of ions into plant cells makes the cells hypertonic.
    b. the soil solution becomes hypertonic, causing the cells to lose water.
    c. the plant will grow faster than it can transport water and maintain proper water balance.
    d. diffusion down the electrochemical gradient will cause a disruption of membrane potential and accompanying loss of water.
    e. the plant will suffer fertilizer burn due to a caustic soil solution.

16. A cell is manufacturing receptor proteins for cholesterol. How would those proteins be oriented in the following membranes before they reach the plasma membrane?
    a. facing inside the ER lumen but outside the transport vesicle membrane
    b. facing inside the ER lumen and inside the transport vesicle
    c. attached outside the ER and outside the transport vesicle
    d. attached outside the ER but facing inside the transport vesicle
    e. completely embedded in the hydrophobic center of both the ER and transport vesicle membranes

17. Which of the following is the most probable description of an integral, transmembrane protein?
    a. amphipathic with a hydrophilic head and a hydrophobic tail region
    b. a globular protein with hydrophobic amino acids in the interior and hydrophilic amino acids arranged around the outside
    c. a fibrous protein coated with hydrophobic sugar residues
    d. a glycoprotein with oligosaccharides attached to the portion of the protein facing the exterior of the cell and cytoskeletal elements facing inside the cell
    e. a middle region composed of α-helical stretches of hydrophobic amino acids, with hydrophilic regions at both ends of the protein

Use the U-tube setup to answer questions 18 through 20.

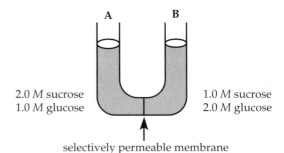

2.0 *M* sucrose
1.0 *M* glucose

1.0 *M* sucrose
2.0 *M* glucose

selectively permeable membrane

The solutions in the two arms of this U-tube are separated by a membrane that is permeable to water and glucose but not to sucrose. Side A is filled with a solution of 2.0 *M* sucrose and 1.0 *M* glucose. Side B is filled with 1.0 *M* sucrose and 2.0 *M* glucose.

18. *Initially,* the solution in side A, with respect to that in side B, is
    a. hypotonic.
    b. hypertonic.
    c. isotonic.
    d. lower.
    e. higher.

19. *After* the system reaches equilibrium, what changes are observed?
    a. The water level is higher in side A than in side B.
    b. The water level is higher in side B than in side A.
    c. The molarity of glucose is higher in side A than in side B.
    d. The molarity of sucrose has increased in side A.
    e. Both a and c have occurred.

20. During the period *before* equilibrium is reached, which molecule(s) will show net movement through the membrane?
    a. water
    b. glucose
    c. sucrose
    d. water and sucrose
    e. water and glucose

21. Facilitated diffusion across a cellular membrane requires _____ and moves a solute _____ its concentration gradients.
    a. energy and transport proteins . . . down
    b. energy and transport proteins . . . up (against)
    c. energy . . . up
    d. transport proteins . . . down
    e. transport proteins . . . up

22. The extracellular fluids that surround the cells of a multicellular animal must be _____ to the cells.
    a. buffers
    b. isotonic
    c. hypotonic
    d. hypertonic
    e. homeotonic

23. LDLs (low-density lipoproteins) enter animal cells by
    a. diffusion through the lipid bilayer.
    b. pinocytosis.
    c. exocytosis.
    d. receptor-mediated endocytosis.
    e. diffusion through transport proteins

24. The fluid mosaic model describes biological membranes as consisting of
    a. a phospholipid bilayer with proteins sandwiched between the layers.
    b. a lipid bilayer with proteins coating the outside of this hydrophobic structure.
    c. a phospholipid bilayer with proteins embedded in and attached to it.
    d. a protein bilayer with phospholipids embedded in it.
    e. a cholesterol bilayer with proteins embedded in the hydrophobic center.

25. You observe plant cells under a microscope that have just been placed in an unknown solution. First the cells plasmolyze; after a few minutes, the plasmolysis reverses and the cells appear normal. What would you conclude about the unknown solute.
    a. It is hypertonic to the plant cells, and its solute cannot cross the plant cell membranes.
    b. It is hypotonic to the plant cells, and its solute cannot cross the plant cell membranes.
    c. It is isotonic to the plant cells, but its solute can cross the plant cell membranes.
    d. It is hypertonic to the plant cells, but its solute can cross the plant cell membranes.
    e. It is hypotonic to the plant cells, but its solute can cross the plant cell membranes.

CHAPTER 9

CELLULAR RESPIRATION: HARVESTING CHEMICAL ENERGY

FRAMEWORK

The catabolic pathways of glycolysis and respiration capture the chemical energy in glucose and other fuels and store it in ATP. Glycolysis, occurring in the cytosol, produces ATP, pyruvate, and NADH; the latter two may then enter the mitochondria for respiration. A mitochondrion consists of a matrix in which the enzymes of the Krebs cycle are localized, a highly folded inner membrane in which enzymes and the molecules of the electron transport chain are embedded, and an intermembrane space between the two membranes to temporarily house H^+ that has been pumped across the inner membrane during the redox reactions of the electron transport chain. Through a chemiosmotic mechanism, a proton motive force drives oxidative phosphorylation as protons move back through ATP synthases located in the membrane.

CHAPTER REVIEW

Principles of Energy Harvest

Cellular respiration and fermentation are catabolic, energy-yielding pathways (155–156)

Fermentation, which occurs without oxygen, is the partial degradation of sugars to release energy. **Cellular respiration,** the most common catabolic pathway, uses oxygen in the breakdown of glucose (or other energy-rich organic compounds) to obtain energy in the usable form of ATP. This exergonic process has a free energy change of -686 kcal/mol of glucose.

■ **INTERACTIVE QUESTION 9.1**

Fill in this summary equation for cellular respiration.

$$\underline{\hspace{2cm}} + 6\,O_2 \longrightarrow \underline{\hspace{2cm}} + 6\,H_2O + \underline{\hspace{2cm}}$$

Cells recycle the ATP they use for work (156)

ATP, adenosine triphosphate, is an energy source because of the instability of the bonds between its three negatively charged phosphate groups. When enzymes shift the terminal phosphate group to another molecule, this newly phosphorylated molecule can perform work. A cell regenerates its supply of ATP from ADP and inorganic phosphate, using energy from cellular respiration.

Redox reactions release energy when electrons move closer to electronegative atoms (156–158)

Oxidation-reduction or **redox reactions** involve the partial or complete transfer of one or more electrons from one reactant to another. The substance that loses electrons is **oxidized** and acts as a **reducing agent** (electron donor) to the substance that gains electrons. By gaining electrons, a substance is **reduced** and acts as an **oxidizing agent** (electron acceptor).

Oxygen strongly attracts electrons and is one of the most powerful oxidizing agents. As electrons shift toward a more electronegative atom, they give up potential energy. Thus, chemical energy is released in a redox reaction that relocates electrons closer to oxygen.

■ **INTERACTIVE QUESTION 9.2**

Fill in the appropriate terms in this equation.

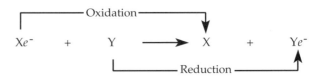

X is the reducing agent; it becomes **a.** _____.

Y is the **b.** _____ ; it becomes **c.** _____.

Electrons "fall" from organic molecules to oxygen during cellular respiration (158)

Organic molecules with an abundance of hydrogen are rich in "hilltop" electrons that release their potential energy when they "fall" closer to oxygen.

■ INTERACTIVE QUESTION 9.3

a. In the conversion of glucose and oxygen to carbon dioxide and water, which molecule is reduced?

b. Which molecule is oxidized?

c. What happens to the energy that is released in this redox reaction?

The "fall" of electrons during respiration is stepwise, via NAD⁺ and an electron transport chain (158–159)

At certain steps in the oxidation of glucose, two hydrogen atoms are removed by enzymes called dehydrogenases and the two electrons and one proton are passed to a coenzyme, **NAD⁺** (nicotinamide adenine dinucleotide).

Energy from respiration is slowly released in a series of small steps as electrons are passed from NADH down an **electron transport chain,** a group of carrier molecules located in the inner mitochondrial membrane, to a stable location close to a highly electronegative oxygen atom.

■ INTERACTIVE QUESTION 9.4

a. NAD⁺ is called an _____ .

b. Its reduced form is _____ .

The Process of Cellular Respiration

Respiration involves glycolysis, the Krebs cycle, and electron transport: *an overview* (160)

Glycolysis, occurring in the cytosol, breaks glucose into two molecules of pyruvate. The **Krebs cycle,** located in the mitochondrial matrix, converts a derivative of pyruvate into carbon dioxide. In some of the steps of glycolysis and the Krebs cycle, dehydrogenase enzymes transfer electrons to NAD⁺. NADH passes electrons to the electron transport chain, from which they eventually combine with hydrogen ions and oxygen to form water. The energy released in each step of the chain is used to synthesize ATP by **oxidative phosphorylation.**

About 10% of the ATP generated for each molecule of glucose oxidized to carbon dioxide is produced by **substrate-level phosphorylation,** in which an enzyme transfers a phosphate group from a substrate to ADP.

■ INTERACTIVE QUESTION 9.5

Fill in the three stages of respiration (**a–c**). Indicate whether ATP is produced by substrate-level or oxidative phosphorylation (**d–f**). Label the arrows indicating electrons carried by NADH.

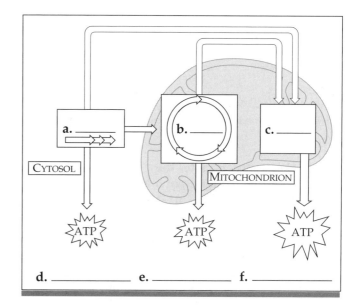

Glycolysis harvests chemical energy by oxidizing glucose to pyruvate: *a closer look* (161)

Glycolysis, a ten-step process occurring in the cytosol, has an energy-investment phase and an energy-payoff phase. Two molecules of ATP are consumed as glucose is split into two molecules of glyceraldehyde phosphate. The conversion of these molecules to pyruvate produces two NADH and four ATP by substrate-level phosphorylation. For each molecule of glucose, glycolysis yields a net gain of two ATP and two NADH.

Enzymes catalyze each step in glycosis. Kinases transfer phosphate groups; a dehydrogenase oxidizes glyceraldehyde phosphate and reduces NAD^+; and other enzymes rearrange atoms in substrate molecules.

■ INTERACTIVE QUESTION   9.6

Fill in the blanks in this summary diagram of glycolysis.

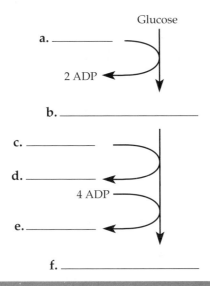

The Krebs cycle completes the energy-yielding oxidation of organic molecules: *a closer look* (161–164)

Before the Krebs cycle begins, a series of steps occurs within a multienzyme complex in the mitochondria: a carboxyl group is removed from the three-carbon pyruvate and released as CO_2; the remaining two-carbon group is oxidized to form acetate with the accompanying reduction of NAD^+ to NADH; and coenzyme A is attached to the acetate by an unstable bond, forming **acetyl CoA.**

In the Krebs cycle, the acetyl fragment of acetyl CoA is added to oxaloacetate to form citrate, which is progressively decomposed back to oxaloacetate. For each turn of the Krebs cycle, two carbons enter in the reduced form from acetyl CoA; two carbons exit completely oxidized as CO_2; three NADH and one $FADH_2$ are formed; and one ATP is made by substrate-level phosphorylation. There are two turns of the Krebs cycle for each glucose molecule oxidized.

■ INTERACTIVE QUESTION   9.7

Fill in the blanks in this summary figure of the Krebs cycle. Balls represent carbon atoms.

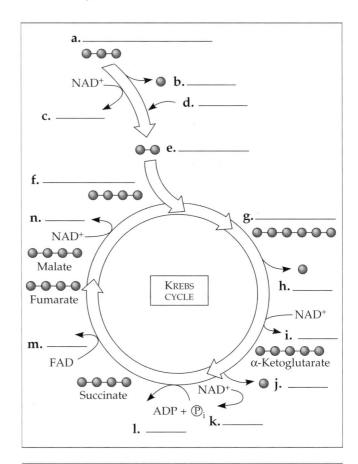

The inner mitochondrial membrane couples electron transport to ATP synthesis: *a closer look* (164–168)

Pathway of Electron Transport   Most components of the electron transport chain are proteins with tightly bound, nonprotein prosthetic groups that shift between reduced and oxidized states as they accept and donate electrons. The transfer of electrons goes from NADH to a flavoprotein to an iron–sulfur protein to ubiquinone (Q) to a series of molecules called cytochromes, proteins with an iron-containing heme group. The last **cytochrome,** cyt a_3, passes its electrons to oxygen, which picks up a pair of H^+ and forms water.

$FADH_2$ adds its electrons to the chain at a lower energy level; thus, one-third less energy is provided for ATP synthesis by $FADH_2$ as compared to NADH.

Chemiosmosis: The Energy-Coupling Mechanism

The electron carrier molecules are organized into three complexes. Electrons are transferred between the complexes by the mobile carriers, Q and cytochrome *c*. Some members of the chain also accept and release protons, which are pumped into the intermembrane space at three points. The resulting proton gradient stores potential energy, referred to as the **proton-motive force.** The exergonic passage of protons back through the **ATP synthase** complexes that span the membrane drives ATP synthesis. The flow of H^+ ions down their gradient through the cylinder part of the ATP synthase complex causes the cylinder and attached rod to rotate, activating catalytic sites on the knob portion where ADP and inorganic phosphate join to make ATP.

In 1961, P. Mitchell first proposed that an energy-coupling mechanism, called **chemiosmosis,** uses the energy of an H^+ gradient to drive cellular work. In mitochondria, exergonic redox reactions produce the H^+ gradient that drives the production of ATP. Chloroplasts use light energy to create the proton-motive force used to make ATP by chemiosmosis. Prokaryotes use H^+ gradients generated across their plasma membranes to transport molecules, make ATP, and rotate flagella.

Cellular respiration generates many ATP molecules for each sugar molecule it oxidizes: *a review* (169–170)

The efficiency of respiration in its energy conversions is approximately 40%. The rest of the energy is released as heat. (See Interactive Question 9.9, p. 64)

Related Metabolic Processes

Fermentation enables some cells to produce ATP without the help of oxygen (170–172)

Glycolysis produces a net of 2 ATP per glucose molecule, under both **aerobic** and **anaerobic** conditions. Without oxygen, glycolysis is part of fermentation—the anaerobic catabolism of organic nutrients to generate ATP and a waste product that regenerates NAD^+, the oxidizing agent for glycolysis.

In **alcohol fermentation,** pyruvate is converted into acetaldehyde, and CO_2 is released. Acetaldehyde is then reduced by NADH to form ethanol (ethyl alcohol), and NAD^+ is regenerated. In **lactic acid fermentation,** pyruvate is reduced directly by NADH to form lactate and recycle NAD^+. Muscle cells make ATP by lactic acid fermentation when energy demand is high and oxygen supply is low.

■ INTERACTIVE QUESTION 9.8

Label this diagram of the electron transport chain and chemiosmosis in a mitochondrial membrane.

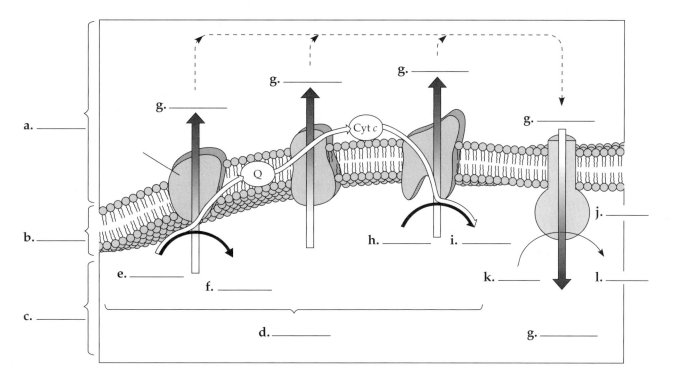

■ INTERACTIVE QUESTION 9.9

Fill in the tally for maximum ATP yield from the oxidation of one molecule of glucose to six molecules of carbon dioxide.

Process	# ATP
Initial phosphorylation of glucose:	a. _____
Substrate-level phosphorylation: in glycolysis	b. _____
in c. _____	__2__
Oxidative phosphorylation:*	d. _____
Maximum Total	e. _____

*approximately 3 ATP for each of the **f.** _____ NADH from pyruvate ⟶ acetyl CoA and the **g.** _____ NADH from Krebs; **h.** _____ ATP for each of the **i.** _____ FADH₂ from Krebs; 2 or 3 ATP for the **j.** _____ NADH from glycolysis, depending on whether the shuttle passes electrons across the membrane to NAD⁺ or FAD.

Fermentation and Respiration Compared   Both fermentation and respiration use glycolysis with NAD^+ as the oxidizing agent to convert glucose and other organic fuels to pyruvate. To oxidize NADH back to NAD^+, fermentation uses pyruvate or acetaldehyde as the final electron acceptor, whereas respiration uses oxygen, via the electron transport chain.

Facultative anaerobes, such as yeasts and many bacteria, can make ATP by fermentation or respiration, depending upon whether oxygen is available.

Evolutionary Significance of Glycolysis   Glycolysis is common to fermentation and respiration. This most widespread of all metabolic processes probably evolved in ancient prokaryotes before oxygen was available. The cytosolic location of glycolysis is also evidence of its antiquity.

■ INTERACTIVE QUESTION 9.10

How much more ATP can be generated by respiration than by fermentation? Explain why.

Glycolysis and the Krebs cycle connect to many other metabolic pathways (172–173)

The Versatility of Catabolism   Fats, proteins, and carbohydrates can all be used by cellular respiration to make ATP. Proteins are digested into amino acids, which are then deaminated (amino group removed) and can enter into respiration at several sites. The digestion of fats yields glycerol, which is converted to an intermediate of glycolysis, and fatty acids, which are broken down by **beta oxidation** to two-carbon fragments that enter the Krebs cycle as acetyl CoA.

Biosynthesis (Anabolic Pathways)   The organic molecules of food also provide carbon skeletons for biosynthesis. Some monomers, such as amino acids, can be directly incorporated into the cell's macromolecules. Intermediates of glycolysis and the Krebs cycle serve as precursors for anabolic pathways. The molecules of carbohydrates, fats, and proteins can all be interconverted to provide for a cell's needs.

Feedback mechanisms control cellular respiration (173)

Through feedback inhibition, the end product of a pathway inhibits an enzyme early in the pathway, thus preventing a cell from producing an excess of a particular substance. The supply of ATP in the cell regulates respiration. The allosteric enzyme that catalyzes the third step of glycolysis, phosphofructokinase, is inhibited by ATP and activated by AMP (derived from ADP). Phosphofructokinase is also inhibited by citrate transported from the mitochondria into the cytosol, thus synchronizing the rates of glycolysis and the Krebs cycle. Other enzymes located at key intersections help to maintain metabolic balance.

WORD ROOTS

aero- = air (*aerobic:* chemical reaction using oxygen)
an- = not (*anaerobic:* chemical reaction not using oxygen)
chemi- = chemical (*chemiosmosis:* the production of ATP using the energy of hydrogen ion gradients across membranes to phosphorylate ADP)
glyco- = sweet; *-lysis* = split (*glycolysis:* the splitting of glucose into pyruvate)

STRUCTURE YOUR KNOWLEDGE

1. This chapter describes how the catabolic pathways of glycolysis and respiration release chemical energy and store it in ATP. One of the best ways to learn the three main components of cellular respiration is to explain them to someone. Find two study partners and have each person be responsible for learning and explaining the important concepts and steps of either glycolysis, the Krebs cycle, or the electron transport chain and chemiosmosis. You may want to involve a fourth person to teach fermentation, or learn that together as a group. Use diagrams and sketches to help you explain your process to your partners.

2. Fill in the table below to summarize the major inputs and outputs of glycolysis, the Krebs cycle, the electron transport chain and chemiosmosis, and fermentation. Base inputs and outputs on one glucose molecule.

3. Create a concept map to organize your understanding of oxidative phosphorylation and chemiosmosis.

TEST YOUR KNOWLEDGE

MULTIPLE CHOICE: *Choose the one best answer.*

1. When electrons move closer to a more electronegative atom,
   a. energy is released.
   b. energy is consumed.
   c. a proton gradient is established.
   d. water is produced.
   e. ATP is synthesized.

2. In the reaction $C_6H_{12}O_6 + 6\ O_2 \rightarrow 6\ CO_2 + 6\ H_2O$,
   a. oxygen becomes reduced.
   b. glucose becomes reduced.
   c. oxygen becomes oxidized.
   d. water is a reducing agent.
   e. oxygen is a reducing agent.

3. A substrate that is phosphorylated
   a. has a stable phosphate bond.
   b. has been formed by the reaction ADP + P_i → ATP.
   c. has an increased reactivity; it is primed to do work.
   d. has been oxidized.
   e. will pass its electrons to the electron transport chain.

Process	Main Function	Inputs	Outputs
Glycolysis			
Pyruvate to acetyl CoA			
Krebs cycle			
Electron transport chain and chemiosmosis			
Fermentation			

4. Which of the following is *not* true of oxidative phosphorylation?
   a. It produces approximately three ATP for every NADH that is oxidized.
   b. It involves the redox reactions of the electron transport chain.
   c. It involves an ATP synthase located in the inner mitochondrial membrane.
   d. It uses oxygen as the initial electron donor.
   e. It depends on chemiosmosis.

5. Substrate-level phosphorylation
   a. involves the shifting of a phosphate group from ATP to a substrate.
   b. can use NADH or $FADH_2$.
   c. takes place only in the cytosol.
   d. accounts for 10% of the ATP formed by fermentation.
   e. is the energy source for facultative anaerobes under anaerobic conditions.

6. The *major* reason that glycolysis is not as energy-productive as respiration is that
   a. NAD^+ is regenerated by alcohol or lactate production, without the high-energy electrons passing through the electron transport chain.
   b. it is the pathway common to fermentation and respiration.
   c. it does not take place in a specialized membrane-bound organelle.
   d. pyruvate is more reduced than CO_2; it still contains much of the energy from glucose.
   e. substrate-level phosphorylation is not as energy efficient as oxidative phosphorylation.

7. The products of glycolysis are
   a. 2 ATP, 2 CO_2, 2 ethanol.
   b. 2 ATP, 2 NAD^+, 2 acetate.
   c. 2 ATP, 2 NADH, 2 pyruvate.
   d. 38 ATP, 6 CO_2, 6 H_2O.
   e. 4 ATP, 2 $FADH_2$, 2 pyruvate.

8. The electron carrier molecules Q and cytochrome *c*
   a. are reduced as they pass electrons on to the next molecule.
   b. contain heme prosthetic groups.
   c. shuttle protons to ATP synthase.
   d. transport H^+ into the mitochondrial matrix, establishing the proton-motive force.
   e. are mobile carriers that transfer electrons between the electron carrier complexes.

9. When pyruvate is converted to acetyl CoA,
   a. CO_2 and ATP are released.
   b. a multienzyme complex removes a carboxyl group, transfers electrons to NAD^+, and attaches a coenzyme.

   c. one turn of the Krebs cycle is completed.
   d. NAD^+ is regenerated so that glycolysis can continue to produce ATP by substrate-level phosphorylation.
   e. phosphofructokinase is activated and glycolysis continues.

10. How many molecules of CO_2 are generated for each molecule of acetyl CoA introduced into the Krebs cycle?
    a. 1   c. 3   e. 6
    b. 2   d. 4

11. In the chemiosmotic mechanism,
    a. ATP production is linked to the proton gradient established by the electron transport chain.
    b. the difference in pH between the intermembrane space and the cytosol drives the formation of ATP.
    c. the flow of H^+ through ATP synthases from the matrix to the intermembrane space drives the phosphorylation of ADP.
    d. the energy released by the reduction and subsequent oxidation of components of the electron transport chain is transferred as a phosphate to ADP.
    e. the production of water in the matrix by the reduction of oxygen leads to a net flow of water out of a mitochondrion.

12. Which of the following reactions is *incorrectly* paired with its location?
    a. ATP synthesis—inner membrane of the mitochondrion (produced in matrix)
    b. fermentation—cell cytosol
    c. glycolysis—cell cytosol
    d. substrate-level phosphorylation—cytosol and matrix
    e. Krebs cycle—cristae of mitochondrion

13. When glucose is oxidized to CO_2 and water, approximately 40% of its energy is transferred to
    a. heat.
    b. ATP.
    c. acetyl CoA.
    d. water.
    e. the Krebs cycle.

14. From an energetic viewpoint, what do muscle cells in oxygen deprivation gain from the reduction of pyruvate?
    a. ATP and lactate
    b. ATP and recycled NAD^+
    c. CO_2 and lactate
    d. ATP, alcohol, and NAD^+
    e. ATP, lactate, and CO_2

15. Glucose, made from six radioactively labeled carbon atoms, is fed to yeast cells in the absence of oxygen. How many molecules of radioactive alcohol (C_2H_5OH) are formed from each molecule of glucose?
    a. 0   **c.** 2   **e.** 6
    b. 1   **d.** 3

16. Which of the following produces the most ATP per gram?
    a. glucose, because it is the starting place for glycolysis
    b. glycogen or starch, because they are polymers of glucose
    c. fats, because they are highly reduced compounds
    d. proteins, because of the energy stored in their tertiary structure
    e. amino acids, because they can be fed directly into the Krebs cycle

17. Fats and proteins can be used as fuel in the cell because they
    a. can be converted to glucose by enzymes.
    b. can be converted to intermediates of glycolysis or the Krebs cycle.
    c. can pass through the mitochondrial membrane to enter the Krebs cycle.
    d. contain unstable phosphate bonds.
    e. contain more energy than glucose.

18. Which is *not* true of the enzyme phosphofructokinase? It is
    a. an allosteric enzyme.
    b. inhibited by citrate.
    c. the pacemaker of glycolysis and respiration.
    d. inhibited by ADP.
    e. an early enzyme in the glycolytic pathway.

19. Substrate-level phosphorylation accounts for approximately what percentage of ATP formation when glucose is oxidized to CO_2 and water?
    a. 0%   **c.** 10%   **e.** 20%
    b. 4%   **d.** 15%

20. Cyanide is a poison that blocks the passage of electrons along the electron transport chain. Which of the following is a metabolic effect of this poison?
    a. The pH of the intermembrane space is much lower than normal.
    b. Electrons are passed directly to oxygen, causing cells to explode.
    c. Alcohol would build up in the cells.
    d. NADH supplies would be exhausted, and ATP synthesis would cease.
    e. No proton gradient would be produced, and ATP synthesis would cease.

21. Which enzyme would use NAD^+ as a coenzyme?
    a. phosphofructokinase
    b. phosphoglucoisomerase
    c. triose phosphate dehydrogenase
    d. hexokinase
    e. phosphoglyceromutase

22. List the order of the following compounds as they occur in cellular respiration.
    1. glyceraldehyde phosphate
    2. pyruvate
    3. glucose
    4. acetyl CoA
    5. fructose bisphosphate
    6. CO_2

    a. 5, 3, 1, 2, 6, 4
    b. 3, 5, 1, 2, 6, 4
    c. 3, 5, 2, 1, 4, 6
    d. 6, 4, 1, 2, 3, 5
    e. 5, 3, 2, 1, 6, 4

23. Which compound has the highest free energy (will produce the most ATP when oxidized)?
    a. acetyl CoA
    b. glucose
    c. pyruvate
    d. fructose bisphosphate
    e. glyceraldehyde phosphate

24. Why is glycolysis considered one of the first metabolic pathways to have evolved?
    a. It relies on fermentation, which is characteristic of the archaea and bacteria.
    b. It is found only in prokaryotes, whereas eukaryotes use their mitochondria to produce ATP.
    c. It produces much less ATP than does the electron transport chain and chemiosmosis.
    d. It relies totally on enzymes that are produced by free ribosomes, and bacteria have only free ribosomes and no bound ribosomes.
    e. It is nearly universal, is located in the cytosol, and does not involve O_2.

25. The metabolic function of fermentation is to
    a. oxidize NADH to NAD^+ so that glycolysis can continue in the absence of oxygen.
    b. reduce NADH so that more ATP can be produced by the electron transport chain.
    c. produce lactate during aerobic exercise.
    d. oxidize pyruvate in order to release more energy.
    e. make beer.

26. Which of the following conversions represents a reduction reaction?
    a. pyruvate $\rightarrow$ acetyl CoA
    b. $C_6H_{12}O_6 \rightarrow 6\ CO_2$
    c. $NADH + H^+ \rightarrow NAD^+ + 2\ H$
    d. glyceraldehyde phosphate $\rightarrow$ pyruvate
    e. acetaldehyde $(C_2H_4O) \rightarrow$ ethanol (C_2H_6O)

27. The oxidation of a molecule of $FADH_2$ yields less ATP than a molecule of NADH yields because $FADH_2$
    a. carries fewer electrons.
    b. is formed in the cytosol and energy is lost when it shuttles its electrons across the mitochondrial membrane.
    c. passes its electrons to a transport molecule later in the chain and at a lower energy level.
    d. is the last molecule produced by the Krebs cycle, and little energy is left to be captured.
    e. has a much lower energy conformation than does NADH.

28. What is the role of oxygen in cellular respiration?
    a. It is reduced in glycolysis as glucose is oxidized.
    b. It provides electrons to the electron transport chain.
    c. It provides the activation energy needed for oxidation to occur.
    d. It is the final electron acceptor for the electron transport chain.
    e. It combines with the carbon removed during the Krebs cycle to form CO_2.

CHAPTER 10

PHOTOSYNTHESIS

FRAMEWORK

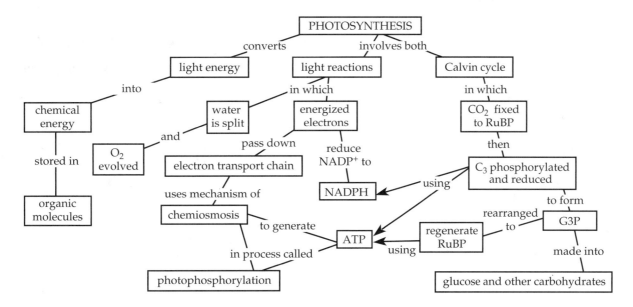

CHAPTER REVIEW

Photosynthesis in Nature

Plants and other autotrophs are the producers of the biosphere (176–177)

In **photosynthesis,** the light energy of the sun is converted into chemical energy stored in organic molecules. Organisms obtain the organic molecules they require for energy and carbon skeletons by autotrophic or heterotrophic nutrition. **Autotrophs** "feed themselves" in the sense that they make their own organic molecules from inorganic raw materials. Plants, algae, and some protists and prokaryotes are photoautotrophs. Some bacteria are chemoautotrophs, which use energy from oxidizing inorganic substances to produce organic compounds.

Heterotrophs are consumers. They may eat plants or animals or decompose organic litter, but almost all are ultimately dependent on photoautotrophs for food and oxygen.

Chloroplasts are the sites of photosynthesis in plants (178)

Chloroplasts, found mainly in the **mesophyll** tissue of the leaf, contain **chlorophyll,** the green pigment that absorbs the light energy that drives photosynthesis. CO_2 enters and O_2 exits the leaf through **stomata.** Veins carry water from the roots to leaves and distribute sugar to nonphotosynthetic tissue.

A chloroplast consists of a double membrane surrounding a dense fluid called the stroma and an elaborate thylakoid membrane system enclosing the thylakoid space. Thylakoid sacs may be stacked to form grana. Chlorophyll is contained in the thylakoid membrane.

■ INTERACTIVE QUESTION   10.1

Label the indicated parts in this diagram of a chloroplast.

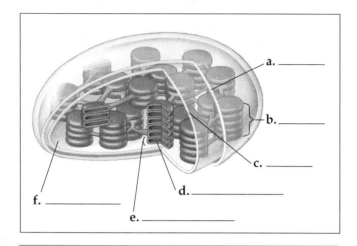

The Pathways of Photosynthesis

**Evidence that chloroplasts split
water molecules enabled researchers
to track atoms through photosynthesis:**
the process of science **(179–180)**

If only the net consumption of water is considered, the equation for photosynthesis is the reverse of respiration:

$$6\ CO_2 + 6\ H_2O + \text{Light energy} \rightarrow C_6H_{12}O_6 + 6\ O_2$$

The Splitting of Water   Using evidence from bacteria that utilize hydrogen sulfide (H_2S) for photosynthesis, C. B. van Niel hypothesized that all photosynthetic organisms need a hydrogen source and that plants split water for their hydrogen source, releasing oxygen. Scientists confirmed this hypothesis by using a heavy isotope of oxygen (^{18}O). Labeled O_2 was produced in photosynthesis only when water, rather than carbon dioxide, contained the labeled oxygen.

Photosynthesis as a Redox Process   Photosynthesis is a redox process like respiration, but differs in the direction of electron flow. The electrons increase their potential energy when they travel from water to reduce CO_2 into sugar, and light provides this energy.

**The light reactions and the Calvin
cycle cooperate in converting
light energy to the chemical energy
of food:** *an overview* **(180–181)**

Solar energy is converted into chemical energy in the **light reactions.** Light energy absorbed by

chlorophyll drives the transfer of electrons and hydrogen from water to the electron acceptor **NADP$^+$**, which is reduced to NADPH and temporarily stores energized electrons. Oxygen is released when water is split. ATP is formed during the light reactions by **photophosphorylation.**

In the **Calvin cycle,** carbon dioxide is incorporated into existing organic compounds by **carbon fixation,** and these compounds are then reduced to form carbohydrate. NADPH and ATP from the light reactions supply the reducing power and chemical energy needed for the Calvin cycle.

■ INTERACTIVE QUESTION   10.2

Fill in the blanks in this overview of photosynthesis in a chloroplast.

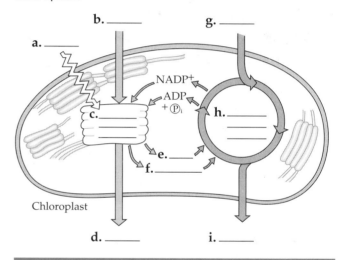

Chloroplast

**The light reactions convert solar
energy to the chemical energy of ATP
and NADPH:** *a closer look* **(181–189)**

The Nature of Sunlight   Electromagnetic energy, also called radiation, travels as rhythmic wave disturbances of electrical and magnetic fields. The distance between the crests of the electromagnetic waves, their **wavelength,** ranges across the **electromagnetic spectrum.** The small band of radiation from about 380 to 750 nm is called **visible light.**

Light also behaves as if it consists of discrete particles called **photons,** which have a fixed quantity of energy. The amount of energy in a photon is inversely related to its wavelength.

■ **INTERACTIVE QUESTION 10.3**

A photon of which color of light would contain more energy: orange (620 nm) or blue (480 nm)?

Photosynthetic Pigments: The Light Receptors A **spectrophotometer** measures the amounts of light of different wavelengths absorbed by a pigment. The **absorption spectrum** of **chlorophyll *a***, the pigment that participates directly in the light reactions, shows that it absorbs blue and red light best. Accessory pigments such as **chlorophyll *b*** and some **carotenoids** absorb light of different wavelengths and transfer energy to chlorophyll *a*. Some carotenoids function in photoprotection by absorbing excessive light energy.

■ **INTERACTIVE QUESTION 10.4**

An **action spectrum** shows the relative rates of photosynthesis under different wavelengths of light. Label the absorption and action spectra on this graph. Why are these lines different?

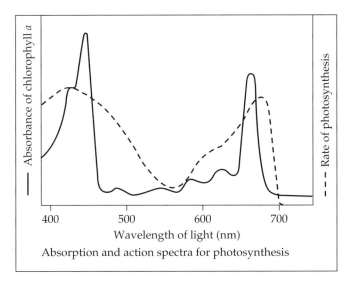

Absorption and action spectra for photosynthesis

Excitation of Chlorophyll by Light When a pigment molecule absorbs energy from a photon, one of the molecule's electrons is elevated to an orbital where it has more potential energy. Only photons whose energy is equal to the difference between the ground state and the excited state for that molecule are absorbed.

The excited state is unstable. Energy is released as heat as the electron drops back to its ground-state orbital. Some pigments, including chlorophyll when it is isolated from thylakoids, may also emit photons of light, called fluorescence, as their electrons return to ground state.

Photosystems: Light-Harvesting Complexes of the Thylakoid Membrane Chlorophyll, accessory pigments, proteins, and other small molecules are organized in the thylakoid membrane into **photosystems,** which contain an antenna complex of a few hundred pigment molecules, a **reaction-center** chlorophyll a, and its **primary electron acceptor.** Pigment molecules in the antenna complex absorb photons and pass the energy from molecule to molecule until it reaches the reaction center. In a redox reaction, an excited electron of the reaction-center chlorophyll is trapped by the primary electron acceptor before it can return to the ground state.

There are two types of photosystems in the thylakoid membrane. The chlorophyll *a* molecule at the reaction center of **photosystem I** is called P700, after the wavelength of light (700 nm) it absorbs best. At the reaction center of **photosystem II** is a chlorophyll *a* molecule called P680.

■ **INTERACTIVE QUESTION 10.5**

List the components of a photosystem.

Noncyclic Electron Flow In **noncyclic electron flow,** electrons pass continuously from water to $NADP^+$. Excited electrons of P680 in photosystem II are trapped by the primary electron acceptor. Electrons are restored to oxidized P680 by an enzyme that removes electrons from water, splitting it into two H^+ and an oxygen atom that immediately combines with another oxygen to form O_2.

The primary electron acceptor passes the photoexcited electrons to an electron transport chain consisting of plastoquinone (Pq), two cytochromes, and plastocyanin (Pc). The energy released as electrons "fall" through the electron transport chain is used to produce ATP using a chemiosmotic mechanism similar to that found in mitochondria. ATP synthesis during noncyclic electron flow is called **noncyclic photophosphorylation.**

■ INTERACTIVE QUESTION 10.6

Fill in the steps of noncyclic electron flow in the diagram below. Circle the important products that will be used to provide chemical energy and reducing power to the Calvin cycle.

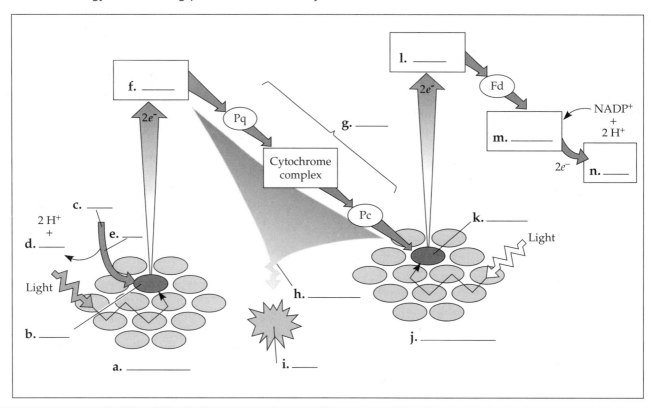

Electrons finally pass to P700 in photosystem I to replace the photo-excited electrons passed to its primary electron acceptor. This primary electron acceptor passes electrons through a second electron transport chain to ferredoxin (Fd), from which the enzyme $NADP^+$ reductase transfers them to $NADP^+$.

Cyclic Electron Flow In **cyclic electron flow**, electrons excited from P700 are passed from Fd to the cytochrome complex and back to P700. Neither NADPH nor O_2 is generated. The Calvin cycle requires more ATP than NADPH, and the additional ATP may be supplied by **cyclic photophosphorylation**, perhaps in response to a buildup of NADPH.

A Comparison of Chemiosmosis in Chloroplasts and Mitochondria Chemiosmosis in mitochondria and in chloroplasts is very similar. The key difference is that in respiration chemical energy from food is converted into ATP, whereas in chloroplasts light energy is converted to chemical energy stored in ATP.

■ INTERACTIVE QUESTION 10.7

a. On the diagram above, sketch the path that electrons from P700 take during cyclic electron flow.

b. Why is neither oxygen nor NADPH generated by cyclic electron flow?

c. How, then, is ATP produced by cyclic electron flow?

■ INTERACTIVE QUESTION 10.8

a. In the light, the proton gradient across the thylakoid membrane is as great as 3 pH units. On which side is the pH lowest?

b. What three factors contribute to the formation of this large difference in H^+ concentration between the thylakoid space and the stroma?

In chloroplasts, the electron transport chain pumps protons from the stroma into the thylakoid space. As H^+ diffuses back through ATP synthase, ATP is formed on the stroma side, where it is available for the Calvin cycle.

The Calvin cycle uses ATP and NADPH to convert CO_2 to sugar: *a closer look* (189–191)

The Calvin cycle turns three times to fix three molecules of CO_2 and produce one molecule of **glyceraldehyde 3-phosphate (G3P)**. The cycle can be divided into three phases:

(1) *Carbon fixation:* CO_2 is fixed by being added to a five-carbon sugar, ribulose bisphosphate (RuBP), in a reaction catalyzed by the enzyme RuBP carboxylase (**rubisco**). The unstable six-carbon intermediate that is formed splits into two molecules of 3-phosphoglycerate.

(2) *Reduction:* Each molecule of 3-phosphoglycerate is then phosphorylated by ATP to form 1,3-bisphosphoglycerate. Two electrons from NADPH reduce this compound to create G3P. The cycle must turn three times to create a net gain of one molecule of G3P.

(3) *Regeneration of CO_2 acceptor (RuBP):* The rearrangement of five molecules of G3P into the three molecules of RuBP requires three more ATP.

Nine molecules of ATP and six of NADPH are required to synthesize one G3P. (Complete the summary diagram of the Calvin cycle in Interactive Question 10.11 on page 74.)

Alternative mechanisms of carbon fixation have evolved in hot, arid climates (191–193)

Photorespiration: An Evolutionary Relic? In most plants, CO_2 enters the Calvin cycle and the first product of carbon fixation is 3-phosphoglycerate. When these **C_3 plants** close their stomata on hot, dry days to limit water loss, CO_2 concentration in the leaf air spaces falls, slowing the Calvin cycle. As more O_2 than CO_2 accumulates, rubisco adds O_2 in place of CO_2 to RuBP. The product splits and a two-carbon compound leaves the chloroplast and is broken down to release CO_2. This seemingly wasteful process is called **photorespiration.**

■ INTERACTIVE QUESTION 10.9

What possible explanation is there for photorespiration, a process that can result in the loss of as much as 50% of the carbon fixed in the Calvin cycle?

C_4 Plants In **C_4 plants,** CO_2 is first added to a 3-carbon compound, PEP, with the aid of an enzyme **(PEP carboxylase)** that has a high affinity for CO_2. The resulting four-carbon compound formed in the **mesophyll cells** of the leaf is transported to **bundle-sheath** cells tightly packed around the veins of the leaf. The compound is broken down to release CO_2, creating concentrations high enough that rubisco will accept CO_2 and initiate the Calvin cycle.

■ INTERACTIVE QUESTION 10.10

a. Where does the Calvin cycle take place in C_4 plants?

b. How can C_4 plants successfully perform the Calvin cycle in hot, dry conditions when C_3 plants would be undergoing photorespiration?

■ INTERACTIVE QUESTION 10.11

Label the three phases (**a** through **c**) and key molecules (**d** through **o**) in this diagram of the Calvin cycle.

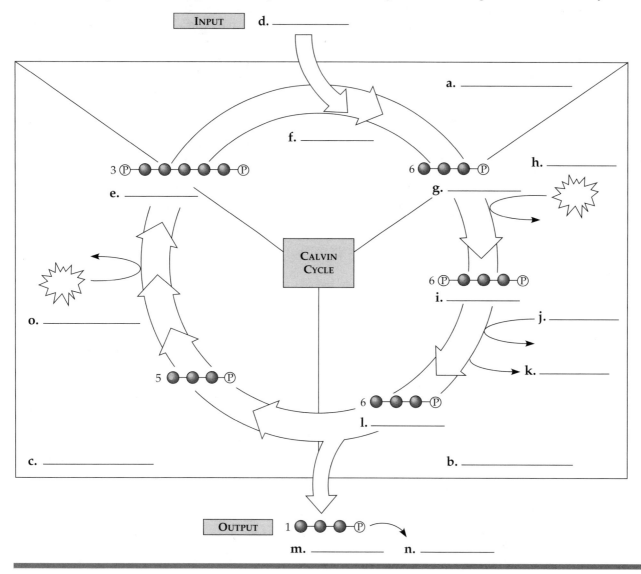

CAM Plants Many succulent plants close their stomata during the day to prevent water loss, but open them at night to take up CO₂ and incorporate it into a variety of organic acids. These compounds are broken down to release CO₂ during daylight. Unlike the C₄ pathway, the **crassulacean acid metabolism (CAM)** pathway does not structurally separate carbon fixation from the Calvin cycle; instead, the two processes are separated in time.

Photosynthesis is the biosphere's metabolic foundation: *a review* (193–194)

About 50% of the organic material produced by photosynthesis is used as fuel for cellular respiration in the mitochondria of plant cells; the rest is used as carbon skeletons for synthesis of organic molecules (proteins, lipids, and a great deal of cellulose), stored as starch, or lost through photorespiration. About 160 billion metric tons of carbohydrate per year are produced by photosynthesis.

WORD ROOTS

auto- = self; **-troph** = food (*autotroph:* an organism that obtains organic food molecules without eating other organisms)

chloro- = green; **-phyll** = leaf (*chlorophyll:* photosynthetic pigment in chloroplasts)

electro- = electricity; **magnet-** = magnetic (*electromagnetic spectrum:* the entire spectrum of radiation)

hetero- = other (*heterotroph:* an organism that obtains organic food molecules by eating other organisms or their by-products)

meso- = middle (*mesophyll:* the green tissue in the middle, inside of a leaf)

photo- = light (*photosystem:* cluster of pigment molecules)

STRUCTURE YOUR KNOWLEDGE

1. You have already filled in the blanks in several diagrams of photosynthesis. To really understand this process, however, you should create your own representation. In a diagrammatic form, outline the key events of photosynthesis. Trace the flow of electrons through photosystem II and I, the transfer of ATP and NADPH formed by the light reactions into the Calvin cycle, and the major steps in the production of G3P. Note where these reactions occur in the chloroplast. You can compare your creation to the sketch in the answer section. Then talk a friend through the steps in your diagram. Perhaps your study group can combine several representations into a clear, concise summary of this chapter.

2. Create a concept map to confirm your understanding of the chemiosmotic synthesis of ATP in photophosphorylation.

TEST YOUR KNOWLEDGE

MULTIPLE CHOICE: *Choose the one best answer.*

1. Which of the following is mismatched with its location?
   a. light reactions—grana
   b. electron transport chain—thylakoid membrane
   c. Calvin cycle—stroma
   d. ATP synthase—double membrane surrounding chloroplast
   e. splitting of water—thylakoid space

2. Photosynthesis is a redox process in which
   a. CO_2 is reduced and water is oxidized.
   b. $NADP^+$ is reduced and RuBP is oxidized.
   c. CO_2, $NADP^+$, and water are reduced.
   d. O_2 acts as an oxidizing agent and water acts as a reducing agent.
   e. G3P is reduced and the electron transport chain is oxidized.

3. Blue light has more energy than red light. Therefore, blue light
   a. has a longer wavelength than red light.
   b. has a shorter wavelength than red light.
   c. contains more photons than red light.
   d. has a broader electromagnetic spectrum than red light.
   e. is absorbed faster by chlorophyll *a*.

4. A spectrophotometer can be used to measure
   a. the absorption spectrum of a substance.
   b. the action spectrum of a substance.
   c. the amount of energy in a photon.
   d. the wavelength of visible light.
   e. the efficiency of photosynthesis.

5. Accessory pigments within chloroplasts are responsible for
   a. driving the splitting of water molecules.
   b. absorbing photons of different wavelengths of light and passing that energy to P680 or P700.
   c. providing electrons to the reaction-center chlorophyll after photo-excited electrons pass to $NADP^+$.
   d. pumping H^+ across the thylakoid membrane to create a proton-motive force.
   e. anchoring chlorophyll *a* within the reaction center.

6. Below is an absorption spectrum for an unknown pigment molecule. What color would this pigment appear to you?

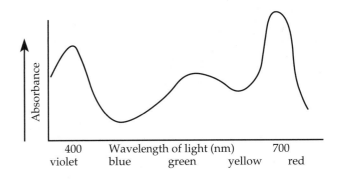

   a. violet
   b. blue
   c. green
   d. yellow
   e. red

7. Noncyclic electron flow in the chloroplast results in the production of
   a. ATP only.
   b. ATP and NADPH.
   c. ATP and G3P.
   d. ATP and O_2.
   e. ATP, NADPH, and O_2.

8. The chlorophyll known as P680 has its electron "holes" filled by electrons from
   a. photosystem I.
   b. photosystem II.
   c. water.
   d. NADPH.
   e. accessory pigments.

9. CAM plants avoid photorespiration by
   a. fixing CO_2 into organic acids during the night; these acids then release CO_2 during the day.
   b. performing the Calvin cycle at night.
   c. fixing CO_2 into four-carbon compounds in the mesophyll, which release CO_2 in the bundle-sheath cells.
   d. using PEP carboxylate to fix CO_2 to ribulose bisphosphate (RuBP).
   e. keeping their stomata closed during the day.

10. Electrons that flow through the two photosystems have their highest potential energy in
    a. water.
    b. P680.
    c. NADPH.
    d. the electron transport chain.
    e. photo-excited P700.

11. Chloroplasts can make carbohydrate in the dark if provided with
    a. ATP and NADPH and CO_2.
    b. an artificially induced proton gradient.
    c. organic acids or four-carbon compounds.
    d. a source of hydrogen.
    e. photons and CO_2.

12. In the chemiosmotic synthesis of ATP, H^+ diffuses through the ATP synthase
    a. from the stroma into the thylakoid space.
    b. from the thylakoid space into the stroma.
    c. from the cytoplasm into the matrix.
    d. from the cytoplasm into the stroma.
    e. from the matrix into the stroma.

13. In C_4 plants, the Calvin cycle
    a. takes place at night.
    b. only occurs when the stomata are closed.
    c. takes place in the mesophyll cells.
    d. takes place in the bundle-sheath cells.
    e. uses PEP carboxylase instead of rubisco because of its greater affinity for CO_2.

14. How many "turns" of the Calvin cycle are required to produce one molecule of glucose?
    a. 1   c. 3   e. 12
    b. 2   d. 6

15. In green plants, most of the ATP for synthesis of proteins, cytoplasmic streaming, and other cellular activities comes directly from
    a. photosystem I.
    b. the Calvin cycle.
    c. oxidative phosphorylation.
    d. noncyclic photophosphorylation.
    e. cyclic photophosphorylation.

16. The six molecules of G3P formed from three turns of the Calvin cycle are converted into
    a. three molecules of glucose.
    b. three molecules of RuBP and one G3P.
    c. one molecule of glucose and four molecules of 3-phosphoglycerate.
    d. one G3P and three four-carbon intermediates.
    e. none of the above, since three molecules of G3P result from three turns of the Calvin cycle.

17. A difference between chemiosmosis in photosynthesis and respiration is that in photophosphorylation
    a. NADPH rather than NADH passes electrons to the electron transport chain.
    b. ATP synthase releases ATP into the stroma rather than into the cytosol.
    c. light provides the energy to push electrons to the top of the electron chain, rather than energy from the oxidation of food molecules.
    d. an H^+ concentration gradient rather than a proton-motive force drives the phosphorylation of ATP.
    e. both a and c are correct.

18. NADPH and ATP from the light reactions are both needed
    a. in the carbon fixation stage to provide energy and reducing power to rubisco.
    b. to regenerate three RuBP from five G3P (glyceraldehyde 3-phosphate).
    c. to combine two molecules of G3P to produce glucose.
    d. to convert 3-phosphoglycerate to G3P.
    e. to reduce the H^+ concentration in the stroma and contribute to the proton-motive force.

19. What portion of an illuminated plant cell would you expect to have the lowest pH?
    a. nucleus
    b. vacuole
    c. chloroplast
    d. stroma of chloroplast
    e. thylakoid space

20. How does cyclic electron flow differ from non-cyclic electron flow?
    a. No NADPH is produced by cyclic electron flow.
    b. No O_2 is produced by cyclic electron flow.
    c. The cytochrome complex in the electron transport chain is not involved in cyclic electron flow.
    d. Both a and b are correct.
    e. a, b, and c are correct.

21 . What does rubisco do?
    a. reduces CO_2 to G3P
    b. regenerates RuBP with the aid of ATP
    c. combines electrons and H^+ to reduce $NADP^+$ to NADPH
    d. adds CO_2 to RuBP in the carbon fixation stage
    e. transfers electrons from NADPH to 1,3-bisphosphoglycerate to produce G3P

22. What are the final electron acceptors for the electron transport chains in the light reactions of photosynthesis and in cellular respiration?
    a. O_2 in both
    b. CO_2 in both
    c. H_2O in the light reactions and O_2 in respiration
    d. $NADP^+$ in the light reactions and NAD^+ or FAD in respiration
    e. $NADP^+$ in the light reactions and O_2 in respiration

23–28.
Indicate if the following events occur during
    a. respiration
    b. photosynthesis
    c. both respiration and photosynthesis
    d. neither respiration nor photosynthesis

_____ 23. Chemiosmotic synthesis of ATP

_____ 24. Reduction of oxygen

_____ 25. Reduction of CO_2

_____ 26. Reduction of NAD^+

_____ 27. Oxidation of $NADP^+$

_____ 28. Oxidative phosphorylation

CHAPTER 11

CELL COMMUNICATION

FRAMEWORK

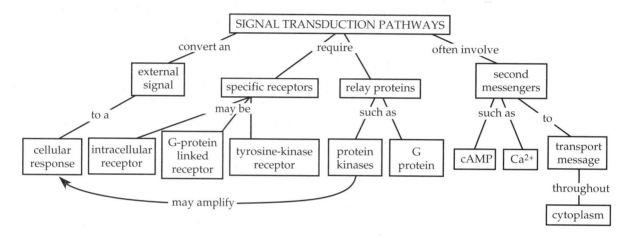

CHAPTER REVIEW

An Overview of Cell Signaling

Cell signaling evolved early in the history of life (197–198)

The series of steps involved in the conversion of a cell surface signal to a cellular response is called a **signal-transduction pathway.** Similarities among these pathways in bacteria, yeast, plants, and animals suggest an early evolution of cell-signaling mechanisms.

Communicating cells may be close together or far apart (199–200)

In paracrine signaling in animals, a secreting cell releases chemical signals into the extracellular fluid, and these **local regulators** influence nearby cells. Another type of local signaling is synaptic signaling, in which a nerve cell releases neurotransmitter molecules into the narrow synapse separating it from its target cell.

    Hormones are chemical signals that travel to more distant cells. In endocrine signaling in animals, the circulatory system transports hormones throughout the body to reach target cells with appropriate receptors.

Signals may also be communicated through direct cytoplasmic connections between cells or through contact of surface molecules.

■ INTERACTIVE QUESTION 11.1

a. Do plant cells communicate using hormones?

b. If so, how do those hormones travel between secreting cells and target cells?

The three stages of cell signaling are reception, transduction, and response: *the process of science* (200–201)

E. W. Sutherland's work studying epinephrine's effect on the hydrolysis of glycogen in liver cells established that cell signaling involves three stages: **reception** of a chemical signal by binding to a cell surface protein; **transduction** of the signal, often by a sequence of changes in cellular molecules; and the final **response** of the cell.

Signal Reception and the Initiation of Transduction

A signal molecule binds to a receptor protein, causing the protein to change shape (201)

A signal molecule acts as a **ligand,** which specifically binds to a receptor protein and usually induces a change in the receptor's conformation.

Most signal receptors are plasma membrane proteins (201–204)

G-Protein-Linked Receptors The various receptors that work with the aid of a G protein, called **G-protein-linked receptors,** are structurally similar, with seven α helices spanning the plasma membrane. Binding of the appropriate extracellular signal to a G-protein-linked receptor activates the receptor, which binds to and activates a specific **G protein** located on the cytoplasmic side of the membrane. This activation occurs when a GTP nucleotide replaces the GDP bound to the G protein. The G protein then activates a membrane-bound enzyme, after which it hydrolyzes its GTP and becomes inactive again. The activated enzyme triggers the next step in the pathway to the cell's response.

   G-protein pathways are used by many hormones and neurotransmitters and are involved in embryological development and sensory reception.

■ INTERACTIVE QUESTION 11.2

Explain why G-protein-regulated pathways shut down rapidly in the absence of a signal molecule.

Tyrosine-Kinase Receptors A **tyrosine kinase** is an enzyme that transfers phosphate groups from ATP to the amino acid tyrosine on a protein. **Tyrosine-kinase receptors** exist as single transmembrane polypeptides with a signal binding site, a transmembrane α helix, and a cytoplasmic tail with a series of tyrosine amino acids. Ligand binding causes two receptors to form a dimer, activating the tyrosine kinase portions of each polypeptide, which phosphorylate the tyrosines on each other's cytoplasmic tails. Different relay proteins now bind to specific phosphorylated tyrosines and become activated, triggering many different transduction pathways in response to one type of signal.

■ INTERACTIVE QUESTION 11.3

Label the parts in this diagram of a dimer of activated tyrosine-kinase receptors.

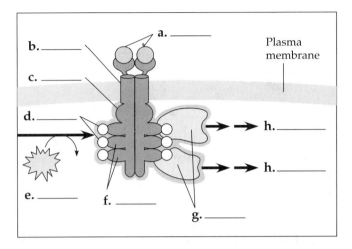

Ion-Channel Receptors The binding of a chemical signal to a **ligand-gated ion channel** opens or closes the protein pore, thus allowing or blocking the flow of specific ions through the membrane.

Intracellular Receptors Hydrophobic chemical messengers may cross a cell's plasma membrane and bind to receptors in the cytoplasm or nucleus. Steroid hormones activate receptors in target cells that function as transcription factors to regulate gene expression.

■ INTERACTIVE QUESTION 11.4

a. What determines whether a cell is a target cell for a particular signal molecule?

b. What determines whether a signal molecule binds to a membrane-surface receptor or an intracellular receptor?

Signal-Transduction Pathways

Multistep signal pathways allow a small number of extracellular signal molecules to be amplified to produce a large cellular response.

Pathways relay signals from receptors to cellular responses (205)

The relay molecules in a signal-transduction pathway are usually proteins, which interact as they pass the message from the extracellular signal to the protein that produces the cellular response.

Protein phosphorylation, a common mode of regulation in cells, is a major mechanism of signal transduction (205–206)

Protein kinases are enzymes that transfer phosphate groups from ATP to proteins, often to the amino acid serine or threonine. Relay molecules in signal-transduction pathways are often protein kinases, which sequentially phosphorylate each other, producing a conformational change that activates each enzyme. Hundreds of different kinds of protein kinases regulate the activity of a cell's proteins.

    Protein phosphatases are enzymes that remove phosphate groups from proteins. They effectively shut down signaling pathways when the extracellular signal is no longer present.

■ INTERACTIVE QUESTION 11.5

a. What does a protein kinase do?

b. What does a protein phosphatase do?

c. What is a phosphorylation cascade?

Certain small molecules and ions are key components of signaling pathways (second messengers) (206–209)

Small, water-soluble molecules or ions often function as **second messengers** that rapidly relay the signal from the membrane-receptor-bound "first messenger" into a cell's interior.

Cyclic AMP (and the process of science) Binding of an extracellular signal to a G-protein-linked receptor activates a G protein that may activate **adenylyl cyclase,** a membrane protein that converts ATP to cyclic adenosine monophosphate (**cyclic AMP** or **cAMP**). The cAMP often activates protein kinase A,

which phosphorylates other proteins. A cytoplasmic enzyme, phosphodiesterase, converts cAMP to inactive AMP, thus removing the second messenger.

    Some signal molecules may activate an inhibitory G protein that inhibits adenylyl cyclase.

■ INTERACTIVE QUESTION 11.6

Label the components in this diagram of the steps in a signal-transduction pathway that uses cAMP as a second messenger.

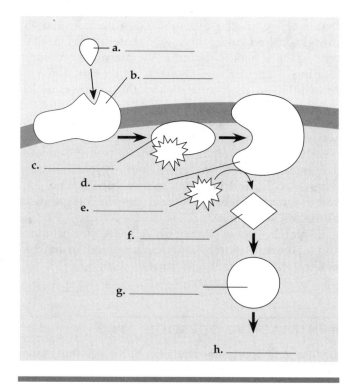

a. _____

b. _____

c. _____

d. _____

e. _____

f. _____

g. _____

h. _____

Calcium Ions and Inositol Triphosphate Calcium ions are widely used as a second messenger in many G-protein-linked and tyrosine-kinase receptor pathways. Cytosolic concentration of Ca^{2+} is usually kept very low by active transport of calcium out of the cell and into the endoplasmic reticulum. The release of Ca^{2+} from the ER in response to a signal involves **diacylglycerol (DAG)** and **inositol triphosphate (IP$_3$)**. Both of these messengers are produced when the enzyme phospholipase C cleaves a membrane phospholipid called PIP$_2$. IP$_3$ binds to and opens ligand-gated calcium channels in the ER. The calcium ions most often bind to the ubiquitous protein, **calmodulin,** which then activates or inactivates other relay proteins such as protein kinases and phosphatases.

■ **INTERACTIVE QUESTION 11.7**

Fill in the blanks to review the steps in a signal transduction pathway involving a tyrosine-kinase receptor and Ca^{2+} as a second messenger.

A **a.**_____ binds to a tyrosine-kinase receptor. Two such activated receptors aggregate to form a **b.**_____. The **c.**_____ on the cytoplasmic side of the receptor activate the enzyme **d.**_____, which cleaves a membrane phospholipid into DAG and **e.**_____. **e.**_____ opens a ligand-gated channel, releasing **f.**_____ from the **g.**_____. Ca^{2+} usually binds to **h.**_____, which regulates other proteins in the pathway to cellular responses.

■ **INTERACTIVE QUESTION 11.8**

How do the following mechanisms or molecules maintain a cell's ability to respond to fresh signals?

a. reversible binding of signal molecules

b. GTPase activity of G protein

c. phosphodiesterase

d. protein phosphatases

Cellular Responses to Signals

In response to a signal, a cell may regulate activities in the cytoplasm or transcription in the nucleus (209–210)

Signal-transduction pathways may lead to the activation of cytoplasmic enzymes or other proteins, or may lead to the synthesis of such proteins by affecting gene expression. Growth factors and certain animal and plant hormones may initiate pathways that ultimately activate transcription factors, which regulate the transcription of mRNA from specific genes.

Elaborate pathways amplify and specify the cell's response to signals (210–212)

Signal Amplification A signal-transduction pathway can amplify a signal in an enzyme cascade, as each successive enzyme in the pathway can activate multiple enzymes of the next step.

The Specificity of Cell Signaling As a result of their particular set of receptor proteins, relay proteins, and effector proteins, different cells can respond to different signals or can exhibit different responses to the same molecular signal. Pathways may branch to produce multiple responses, or two pathways may interact ("cross-talk") to mediate a single response.

Scaffolding proteins are large relay proteins that bind other relay proteins in a pathway and thus increase the efficiency of signal transduction.

Inactivation mechanisms that discontinue a cell's response to a signal are essential in keeping a cell responsive to regulation.

WORD ROOTS

liga- = bound or tied (*ligand:* a small molecule that specifically binds to a larger one)

trans- = across (*signal-transduction pathway:* the process by which a signal on a cell's surface is converted into a specific cellular response inside the cell)

-yl = substance or matter (*adenylyl cyclase:* an enzyme built into the plasma membrane that converts ATP to cAMP)

STRUCTURE YOUR KNOWLEDGE

1. Why is cell signaling such an important component of a cell's life?

2. Briefly describe the three stages of cell signaling.

3. Some signal pathways alter a protein's activity; others may result in the production of new proteins. Explain the mechanisms for these two different responses.

4. How does an enzyme cascade produce an amplified response to a signal molecule?

TEST YOUR KNOWLEDGE

1. When epinephrine binds to cardiac (heart) muscle cells, it speeds their contraction. When it binds to muscle cells of the small intestine, it inhibits their contraction. How can the same hormone have different effects on muscle cells?
   a. Cardiac cells have more receptors for epinephrine than do intestinal cells.
   b. Epinephrine circulates to the heart first and thus is in higher concentration around cardiac cells.
   c. The two types of muscle cells have different signal-transduction pathways for epinephrine and thus have different cellular responses.
   d. Cardiac muscle is stronger than intestinal muscle and thus has a stronger response to epinephrine.
   e. Epinephrine binds to G-protein-linked receptors in cardiac cells, and these receptors always increase a response to the signal. Epinephrine binds to tyrosine-kinase receptors in intestinal cells, and these receptors always inhibit a response to the signal.

2. Which of the following would be used in the type of local signaling called paracrine signaling in animals?
   a. the neurotransmitter acetylcholine
   b. the hormone epinephrine
   c. the neurotransmitter norepinephrine
   d. a local regulator such as a growth factor
   e. Both a and c are correct.

3. A signal molecule that binds to a plasma-membrane protein functions as a
   a. ligand.
   b. second messenger.
   c. protein phosphatase.
   d. protein kinase.
   e. receptor protein.

4. What is a G protein?
   a. a specific type of membrane-receptor protein
   b. a protein bound to the cytoplasmic side of a membrane that becomes activated when it binds GTP
   c. a membrane-bound enzyme that converts ATP to cAMP
   d. a tyrosine-kinase relay protein
   e. a guanine nucleotide that converts between GDP and GTP to activate and inactivate relay proteins

5. How do tyrosine-kinase receptors transduce a signal?
   a. They transport the signal molecule into the cell, where it binds to and activates a transcription factor. The transcription factor then alters gene expression.
   b. Signal binding causes a conformational change that activates membrane-bound tyrosine-kinase relay proteins that phosphorylate serine and threonine amino acids.
   c. Their activated tyrosine kinases convert ATP to cAMP; cAMP acts as a second messenger to activate other protein kinases.
   d. When activated, they cleave a membrane phospholipid into two second-messenger molecules. One of the molecules binds with calmodulin, which sets off an enzyme cascade.
   e. They form a dimer; they phosphorylate each other's tyrosines; specific relay or response proteins bind to and are activated by specific phosphorylated tyrosines.

6. Which of the following can activate a protein by transferring a phosphate group to it?
   a. cAMP
   b. G protein
   c. phosphodiesterase
   d. protein kinase
   e. protein phosphatase

7. Many signal-transduction pathways use second messengers to
   a. transport a signal through the lipid bilayer portion of the plasma membrane.
   b. relay a signal from the outside to the inside of the cell.
   c. relay the message from the inside of the membrane throughout the cytoplasm.
   d. amplify the message by phosphorylating proteins.
   e. dampen the message once the signal molecule has left the receptor.

8. What is the function of the second messenger IP_3?
   a. bind to and activate protein kinase A
   b. bind to calmodulin
   c. activate other membrane-bound relay molecules
   d. convert ATP to cAMP
   e. bind to and open ligand-gated calcium channels on the ER

9. The function of calmodulin is to
   a. begin an enzyme cascade by attaching Ca^{2+} to a protein kinase.
   b. bind with Ca^{2+} and regulate the activity of specific proteins.
   c. pump Ca^{2+} out of the cell to lower cytosolic calcium levels and stop a cell's response to a signal.
   d. release Ca^{2+} from the ER to initiate an enzyme cascade.
   e. dampen a cell's response to a signal by activating protein phosphatases.

10. Signal amplification is most often achieved by
    a. an enzyme cascade involving multiple protein kinases.
    b. the binding of multiple signal molecules.
    c. branching pathways that produce multiple cellular responses.
    d. activating transcription factors that affect gene expression.
    e. the action of adenylyl cyclase in converting ATP to cAMP.

11. From studying the effects of epinephrine on liver cells, Sutherland concluded that
    a. there is a one-to-one correlation between the number of epinephrine molecules bound to receptors and the number of glucose molecules released from glycogen.
    b. epinephrine enters liver cells and binds to receptors that function as transcription factors to turn on the gene for glycogen phosphorylase.
    c. there is a "second messenger" that transmits the signal of epinephrine binding on the plasma membrane to the enzymes involved in glycogen breakdown inside the cell.
    d. the signal-transduction pathway through which epinephrine signals glycogen breakdown involves tyrosine-kinase receptors and the release of calcium ions that activate glycogen phosphorylase.
    e. epinephrine functions as a ligand to open ion channels in the plasma membrane that allow calcium ions to enter and bind with calmodulin.

12. Which of the following is a similarity between G-protein-linked receptors and tyrosine-kinase receptors?
    a. signal-binding sites specific for steroid hormones
    b. formation of a dimer following binding of a signal molecule

   c. activation that results from binding of GTP
   d. phosphorylation of specific amino acids in response to signal binding
   e. α-helix regions of the receptor that span the plasma membrane

13. Which of the following is *incorrectly* matched with its description?
    a. scaffolding protein—large relay protein that may bind with several other relay proteins to increase the efficiency of a signaling pathway
    b. protein phosphatase—enzyme that transfers a phosphate group from ATP to a protein, causing a conformational change that usually activates that protein
    c. adenylyl cyclase—enzyme attached to plasma membrane that converts ATP to cAMP in response to an extracellular signal
    d. phospholipase C—enzyme that may be activated by a G protein or tyrosine-kinase receptor and cleaves a plasma-membrane phospholipid into the second messengers IP_3 and DAG
    e. G protein—relay protein attached to the inside of plasma membrane that, when activated by an activated G-protein-linked receptor, binds GTP and then usually activates another membrane-attached protein

14. Which of the following signal molecules pass through the plasma membrane, bind to intracellular receptors that move into the nucleus, and function as transcription factors to regulate gene expression?
    a. epinephrine
    b. growth factors
    c. yeast mating factors α and a
    d. testosterone, a steroid hormone
    e. neurotransmitter released into synapse between nerve cells

15. Many human diseases, including bacterial infections, and also many medicines used to treat these diseases produce their effects by influencing which of the following?
    a. cAMP concentrations in the cell
    b. Ca^{2+} concentrations in the cell
    c. G-protein pathways
    d. gene expression
    e. tyrosine-kinase receptors

CHAPTER 12

THE CELL CYCLE

FRAMEWORK

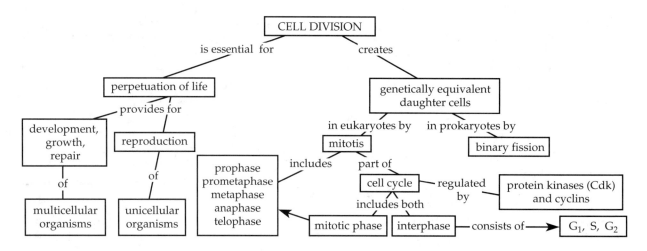

CHAPTER REVIEW

Cell division is the basis for the continuity of all life. The **cell cycle** extends from the creation of a new cell by the division of its parent cell to its own division into two genetically identical daughter cells.

The Key Roles of Cell Division

Cell division functions in reproduction, growth, and repair (215)

Cell division creates duplicate offspring in unicellular organisms and provides for growth, development, and repair in multicellular organisms. The process of re-creating a structure as intricate as a cell necessitates the exact duplication and equal division of the DNA containing the cell's genetic program.

■ INTERACTIVE QUESTION   12.1

A cell's complete complement of DNA is called its

_____.

Cell division distributes identical sets of chromosomes to daughter cells (216–217)

Each eukaryotic species has a characteristic number of chromosomes in each **somatic cell;** reproductive cells, or **gametes** (egg and sperm), have half that number of chromosomes.

Each **chromosome** is a very long DNA molecule with associated proteins that help to structure the chromosome and control the activity of the genes. This DNA–protein complex is called **chromatin.** Prior to cell division, a cell copies its DNA and each chromosome densely coils and shortens. Replicated chromosomes consist of two identical **sister chromatids,** joined along their length and at a specialized region called a **centromere.** The two sister chromatids separate during **mitosis** (the division of the nucleus), and then the cytoplasm divides during **cytokinesis,** producing two separate, genetically equivalent daughter cells.

A type of cell division called **meiosis** produces daughter cells that have half the number of chromosomes of the parent cell. With the fertilization of egg and sperm, which were formed by meiosis, the chromosome number is restored in the somatic cells of the new offspring.

■ **INTERACTIVE QUESTION 12.2**

a. How many chromosomes do you have in your somatic cells?

b. How many chromosomes in your gametes?

The Mitotic Cell Cycle

The mitotic phase alternates with interphase in the cell cycle: *an overview* (217–219)

The cell cycle consists of the **mitotic (M) phase,** which includes mitosis and cytokinesis, and **interphase,** during which the cell grows and replicates its chromosomes. Interphase, usually lasting 90% of the cell cycle, includes the G_1 **phase,** the **S phase,** and the G_2 **phase.** Mitosis is conventionally described in five subphases: **prophase, prometaphase, metaphase, anaphase,** and **telophase.**

■ **INTERACTIVE QUESTION 12.3**

a. How are the three subphases of interphase alike?

b. What key event happens during the S phase?

The mitotic spindle distributes chromosomes to daughter cells: *a closer look* (220–221)

The **mitotic spindle** consists of fibers made of microtubules and associated proteins. The assembly of the spindle begins in the **centrosome,** or microtubule-organizing center. A pair of centrioles is centered in each centrosome of an animal cell but is not required for normal spindle operation. Radial arrays of microtubules, called asters, extend from the centrosomes in animal cells.

Having duplicated during interphase, the two centrosomes move from near the nucleus toward the poles of the cell during prophase and prometaphase, as spindle microtubules elongate between them. During prophase, the nucleoli disappear and the chromatin fibers coil and fold into visible chromosomes, consisting of sister chromatids joined at the centromere. During prometaphase, some of the spindle microtubules attach to each chromatid's **kinetochore,** a structure of protein and DNA located at the centromere region. Nonkinetochore microtubules extend out from each centrosome and overlap at the midline. Alternate tugging on the chromosome by opposite kinetochore microtubules moves the chromosome to the midline of the cell. At metaphase, the chromosomes are aligned at the **metaphase plate,** across the midline of the spindle.

The proteins joining sister chromatids are inactivated in anaphase, and the now separate chromosomes move toward the poles. Motor proteins "walk" a chromosome along the kinetochore microtubules as these shorten by depolymerizing at their kinetochore end. The extension of the spindle poles away from each other as an animal cell elongates is probably due to the sliding of nonkinetochore microtubules past each other, also using motor proteins.

During telophase, equivalent sets of chromosomes gather at the two poles of the cell. Nuclear envelopes form, nucleoli reappear, and cytokinesis begins.

Cytokinesis divides the cytoplasm: *a closer look* (221–223)

Cleavage is the process that separates the two daughter cells in animals. A **cleavage furrow** on the cell surface forms, as a ring of actin microfilaments, interacting with myosin proteins, begins to contract on the cytoplasmic side of the membrane. The cleavage furrow deepens until the dividing cell is pinched in two.

■ INTERACTIVE QUESTION   12.4

The following diagrams depict interphase and the five subphases of mitosis in an animal cell. Assuming that this cell has four chromosomes, sketch the chromosomes as they would appear in each subphase. Identify the stages and label the indicated structures.

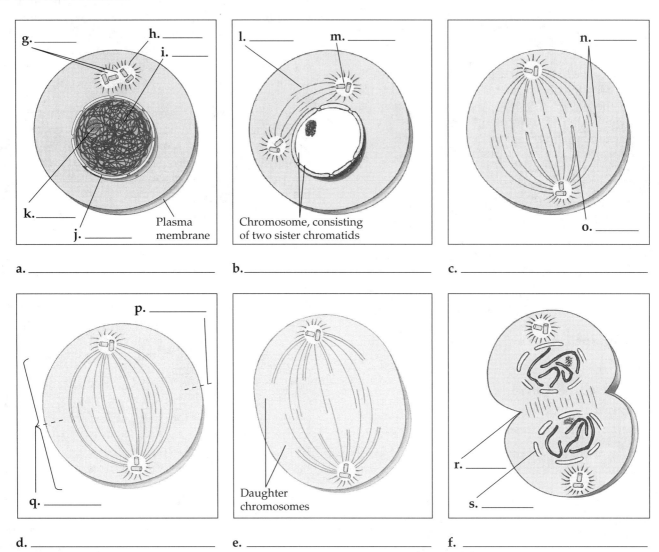

In plant cells, a **cell plate** forms from the fusion of membrane vesicles derived from the Golgi apparatus. The membrane of the enlarging cell plate joins with the plasma membrane, separating the two daughter cells. A new cell wall develops between the cells.

Mitosis in eukaryotes may have evolved from binary fission in bacteria (223–224)

Prokaryotes reproduce by a process known as **binary fission.** The single circular DNA molecule begins to replicate at the **origin of replication** and these regions move apart by an unknown mechanism. The cell doubles in size and the plasma membrane grows inward to divide the two identical daughter cells.

Possible intermediate stages in the evolution of mitosis from the binary fission of bacteria are seen in some groups of unicellular algae in which chromosomal division takes place within an intact nuclear envelope.

Regulation of the Cell Cycle

Normal growth, development, and maintenance depend on proper control of the timing and rate of cell division.

A molecular control system drives the cell cycle (224–227)

Experiments that fuse two cells at different phases of the cell cycle indicate that cytoplasmic chemical signals drive the cell cycle. A **cell cycle control system**, consisting of a set of molecules that function cyclically, coordinates the events of the cell cycle.

Cell Cycle Checkpoints   Important internal and external signals are monitored to determine whether the cell cycle will proceed past the three main **checkpoints** in the G_1, G_2, and M phases. If a mammalian cell does not receive a go-ahead signal at the G_1 checkpoint, called the "restriction point," the cell will usually exit the cell cycle to a nondividing state called the **G_0 phase.**

The Cell Cycle Clock: Cyclins and Cyclin-Dependent Kinases   Protein kinases are enzymes that activate or inactivate other proteins by phosphorylating them. The changing concentrations of **cyclins,** regulatory proteins that attach to these kinases, affect the activity of **cyclin-dependent kinases, or Cdks.**

A cyclin–Cdk complex called **MPF,** for maturation or M-phase-promoting factor, triggers passage past the G_2 checkpoint into M phase. In addition to phosphorylating proteins and other kinases that initiate mitotic events, MPF activates a protein breakdown process that destroys its cyclin and thus MPF activity. The Cdk portion of the complex remains to associate with new cyclin synthesized during interphase of the next cycle.

Other Cdk proteins and cyclins probably control the movement of a cell past the G_1 checkpoint.

■ **INTERACTIVE QUESTION   12.5**

a. What is MPF?

b. Describe the relative concentrations of MPF and its constituent molecules throughout the cell cycle:

MPF

Cdk

cyclin

Internal and external cues help regulate the cell cycle (227–228)

Internal Signals: Messages from the Kinetochores   To move past the M-phase checkpoint into anaphase, an anaphase-promoting complex (APC) must be activated. Kinetochores that are not yet attached to spindle microtubules set off a signaling pathway that keeps the APS inactive until all chromosomes are attached to the spindle at the metaphase plate.

External Signals: Growth Factors   Growing cells in cell culture has allowed researchers to identify chemical and physical factors that affect cell division. Certain nutrients and regulatory proteins called **growth factors** have been found to be essential for normal cell division. Mammalian fibroblast cells have receptors on their plasma membranes for platelet-derived growth factor (PDGF), which is released from blood platelets at the site of an injury. Binding of PDGF initiates a signal-transduction pathway that stimulates cell division.

Density-dependent inhibition of cell division is related to diminishing supplies of growth factors and essential nutrients. Most animal cells also show **anchorage dependence** and must be attached to a substratum in order to divide.

Cancer cells have escaped from cell cycle controls (228–229)

Cancer cells escape from the body's normal control mechanisms. When grown in cell culture, cancer cells do not exhibit density-dependent inhibition and may continue to divide indefinitely instead of stopping after the typical 20 to 50 divisions of normal mammalian cells.

When a normal cell is **transformed** or converted to a cancer cell, the body's immune system usually destroys it. If it proliferates to form a **tumor,** a mass of abnormal cells develops within a tissue. **Benign tumors** remain at their original site and can be removed by surgery. **Malignant tumors** cause cancer as they invade and disrupt the functions of one or more organs. Malignant tumor cells may have abnormal metabolism and unusual numbers of chromosomes. They lose their attachments to other cells and may **metastasize,** entering the blood and lymph systems and spreading to other sites. Radiation and chemicals are used to treat tumors that metastasize. Much remains to be learned about control of the cell division processes of both normal and cancerous cells.

WORD ROOTS

ana- = up, throughout, again (*anaphase:* the mitotic stage in which the chromatids of each chromosome have separated and the daughter chromosomes are moving to the poles of the cell)

bi- = two (*binary fission:* a type of cell division in which a cell divides in half)

centro- = the center; **-mere** = a part (*centromere:* the narrow "waist" of a condensed chromosome)

chroma- = colored (*chromatin:* DNA and the various associated proteins that form eukaryotic chromosomes)

cyclo- = a circle (*cyclin:* a regulatory protein whose concentration fluctuates cyclically)

cyto- = cell; **-kinet** = move (*cytokinesis:* division of the cytoplasm)

gamet- = a wife or husband (*gamete:* a haploid egg or sperm cell)

gen- = produce (*genome:* a cell's endowment of DNA)

inter- = between (*interphase:* time when a cell metabolizes and performs its various functions)

mal- = bad or evil (*malignant tumor:* a cancerous tumor that is invasive enough to impair functions of one or more organs)

meio- = less (*meiosis:* a variation of cell division that yields daughter cells with half as many chromosomes as the parent cell)

meta- = between (*metaphase:* the mitotic stage in which the chromosomes are aligned in the middle of the cell, at the metaphase plate)

mito- = a thread (*mitosis:* the division of the nucleus)

pro- = before (*prophase:* the first mitotic stage in which the chromatin is condensing)

soma- = body (*centrosome:* a nonmembranous organelle that functions throughout the cell cycle to organize the cell's microtubules)

telos- = an end (*telophase:* the final stage of mitosis in which daughter nuclei are forming and cytokinesis has typically begun)

trans- = across; **-form** = shape (*transformation:* the process that converts a normal cell into a cancer cell)

STRUCTURE YOUR KNOWLEDGE

1. Describe the life of one chromosome as it proceeds through an entire cell cycle, starting with interphase and ending with telophase of mitosis.

2. Draw a sketch of one half of a mitotic spindle. Identify and list the functions of the components.

3. In this photomicrograph of cells in an onion root tip, identify the cell cycle phases for the indicated cells.

a. _____

b. _____

c. _____

d. _____

TEST YOUR KNOWLEDGE

FILL IN THE BLANK: *Identify the appropriate phase of the cell cycle.*

_____ 1. most cells that will no longer divide are in this phase

_____ 2. sister chromatids separate and chromosomes move apart

_____ 3. mitotic spindle begins to form

_____ 4. cell plate forms or cleavage furrow pinches cells apart

_____ 5. chromosomes replicate

_____ 6. chromosomes line up at equatorial plane

_____ 7. nuclear membranes form around separated chromosomes

_____ 8. chromosomes become visible

_____ 9. kinetochore–microtubule interactions move chromosomes to midline

_____ 10. restriction point occurs in this phase

MULTIPLE CHOICE: *Choose the one best answer.*

1. One of the major differences in the cell division of prokaryotic cells compared to eukaryotic cells is that
   a. cytokinesis does not occur in prokaryotic cells.
   b. genes are not replicated on chromosomes in prokaryotic cells.
   c. the duplicated chromosomes are attached to the nuclear membrane in prokaryotic cells and are separated from each other as the membrane grows.
   d. the chromosomes do not separate along a mitotic spindle in prokaryotic cells.
   e. the chromosome number is reduced by half in eukaryotic cells but not prokaryotic cells.

2. A plant cell has 12 chromosomes at the end of mitosis. How many *chromosomes* would it have in the G_2 phase of its next cell cycle?
   a. 6
   b. 9
   c. 12
   d. 24
   e. It depends on whether it is undergoing mitosis or meiosis.

3. How many *chromatids* would this plant cell have in the G_2 phase of its cell cycle?
   a. 6
   b. 9
   c. 12
   d. 24
   e. 48

4. The longest part of the cell cycle is
   a. prophase.
   b. G_1 phase.
   c. G_2 phase.
   d. mitosis.
   e. interphase.

5. In animal cells, cytokinesis involves
   a. the separation of sister chromatids.
   b. the contraction of the contractile ring of microfilaments.
   c. depolymerization of kinetochore microtubules.
   d. a protein kinase that phosphorylates other enzymes.
   e. sliding of nonkinetochore microtubules past each other.

6. Humans have 46 chromosomes. That number of chromosomes will be found in
   a. cells in anaphase.
   b. the egg and sperm cells.
   c. the somatic cells.
   d. all the cells of the body.
   e. only cells in G_1 of interphase.

7. Sister chromatids
   a. have one-half the amount of genetic material as does the original chromosome.
   b. start to move along kinetochore microtubules toward opposite poles during telophase.
   c. each have their own kinetochore.
   d. are formed during prophase.
   e. slide past each other along nonkinetochore microtubules.

8. Which of the following would *not* be exhibited by cancer cells?
   a. changing levels of MPF concentration
   b. passage through the restriction point
   c. density-dependent inhibition
   d. metastasis
   e. response to growth factors

9. Which of the following is *not* true of a cell plate?
   a. It forms at the site of the metaphase plate.
   b. It results from the fusion of microtubules.
   c. It fuses with the plasma membrane.
   d. A cell wall is laid down between its membranes.
   e. It forms during telophase in plant cells.

10. A cell that passes the restriction point in G_1 will most likely
    a. undergo chromosome duplication.
    b. have just completed cytokinesis.
    c. continue to divide only if it is a cancer cell.
    d. show a drop in MPF concentration.
    e. move into the G_0 phase.

11. The rhythmic changes in cyclin concentration in a cell cycle are due to
    a. its increased production once the restriction point is passed.
    b. the cascade of increased production once its enzyme is phosphorylated by MPF.
    c. its degradation, which is initiated by active MPF.
    d. the correlation of its production with the production of Cdk.
    e. the binding of the growth factor PDGF.

12. In a plant cell, a centrosome functions in the formation of
    a. the cell plate.
    b. kinetochores.
    c. duplicate chromosomes.
    d. centromeres.
    e. microtubules of the spindle apparatus.

13. A cell in which of the following phases would have the *least* amount of DNA?
    a. G_0
    b. G_2
    c. prophase
    d. metaphase
    e. anaphase

14. What causes the anaphase-promoting complex (APC) to become active and initiate the separation of sister chromatids?
 a. the drop in MPF concentration
 b. a rapid rise in Cdk concentration
 c. movement past the G_2 checkpoint
 d. a signal pathway initiated by the binding of a growth factor
 e. the cessation of delay signals received from unattached kinetochores

15. Cells growing in cell culture that divide and pile up on top of each other are lacking
 a. anchorage dependence.
 b. density independence.
 c. PDGF.
 d. MPF.
 e. nutrients and growth factors.

16. Knowledge of the cell cycle control system will be most beneficial to the area of
 a. human reproduction.
 b. plant genetics.
 c. prokaryotic growth and development.
 d. cancer prevention and treatment.
 e. prevention and treatment of cardiovascular disease.

GENETICS

CHAPTER 13

MEIOSIS AND SEXUAL LIFE CYCLES

FRAMEWORK

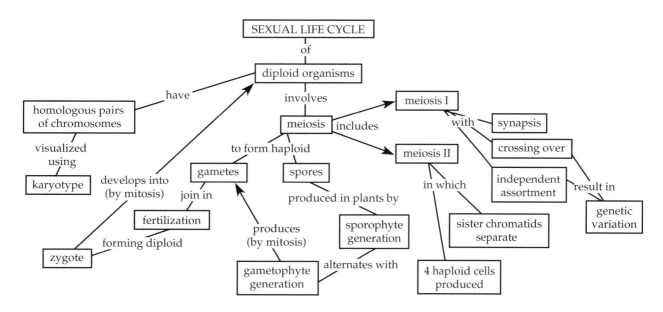

CHAPTER REVIEW

Genetics is the scientific study of the transmission of traits from parents to offspring **(heredity)** and the **variation** between and within generations.

An Introduction to Heredity

Offspring acquire genes from parents by inheriting chromosomes (234–235)

The inheritance of traits from parents to offspring involves the transmission of discrete units of information coded in segments of DNA known as **genes.** Specific sequences of the four nucleotides that comprise DNA translate into instructions for synthesizing proteins, such as enzymes, that then guide the development of inherited traits.

Precise copies of an organism's genes are packaged into sperm or ova. Upon fertilization, genes from both parents are present in the fertilized egg. The DNA of a eukaryotic cell is packaged into a species-specific number of chromosomes, each of which contains hundreds or thousands of genes. A gene's **locus** is its location on a chromosome.

**Like begets like, more or less: a comparison
of asexual and sexual reproduction (235)**

In **asexual reproduction,** a single parent passes copies
of all its genes on to its offspring. A **clone** is a group
of genetically identical offspring of an asexually reproducing individual.

In **sexual reproduction,** an individual receives a
unique combination of genes inherited from two
parents.

The Role of Meiosis in Sexual Life Cycles

An organism's **life cycle** is the sequence of stages
from conception to production of its own offspring.

**Fertilization and meiosis alternate
in sexual life cycles (236–239)**

The Human Life Cycle In **somatic cells,** there are
two chromosomes of each type, known as **homologous chromosomes** or homologs. A gene controlling
a particular trait is found at the same locus on each
chromosome of a homologous pair.

A **karyotype** is an ordered display of an individual's chromosomes. It is usually made by cutting the
individual chromosomes out of a photograph taken of
white blood cells that were stimulated to undergo mitosis, arrested in metaphase, and stained. The chromosomes are arranged into homologous pairs by size,
shape, and banding pattern. Karyotyping may be
used to identify chromosomal abnormalities associated with inherited disorders.

Sex chromosomes determine the sex of a person:
females have two homologous X chromosomes; males
have nonhomologous X and Y chromosomes. Chromosomes other than the sex chromosomes are called
autosomes.

Somatic cells contain a set of chromosomes from
each parent. **Gametes,** ova and sperm, are **haploid
cells** and contain a single set of chromosomes. The
haploid number (*n*) of chromosomes for humans is 23.
Fusion of sperm and ovum into a **zygote,** called **fertilization** or **syngamy,** combines the paternal and
maternal sets of chromosomes. The **diploid** zygote
then divides by mitosis to produce the somatic cells
of the body, all of which contain the diploid number
(2*n*) of chromosomes.

Meiosis is a special type of cell division that halves
the chromosome number and provides a haploid set

of chromosomes to each gamete. An alternation between diploid and haploid conditions, involving the
processes of fertilization and meiosis, is characteristic
of the life cycle of all sexually reproducing organisms.

■ INTERACTIVE QUESTION   13.1

If 2*n* = 14, how many chromosomes will be present in
somatic cells? **a.** _____ How many chromosomes will be
found in gametes? **b.** _____ If *n* = 14, how many chromosomes will be found in diploid somatic cells? **c.** _____
How many sets of homologous chromosomes will be
found in gametes? **d.** _____

The Variety of Sexual Life Cycles In most animals,
meiosis occurs in the formation of gametes, which are
the only haploid cells in the life cycle. In many fungi
and some protists, the only diploid stage is the zygote. Meiosis occurs after the gametes fuse, producing haploid cells that divide by mitosis to create a
multicellular haploid organism. Gametes are produced by mitosis in these organisms.

Plants and some species of algae have a type of life
cycle called **alternation of generations** that includes
both diploid and haploid multicellular stages. The
multicellular diploid **sporophyte** stage produces haploid **spores** by meiosis. These spores undergo mitosis and develop into a multicellular haploid plant, the
gametophyte, which produces gametes by mitosis.
Gametes fuse to form a diploid zygote that develops
into the next sporophyte generation. (See Interactive
Question 13.2, p. 93.)

**Meiosis reduces chromosome number from
diploid to haploid:** *a closer look* **(239–243)**

In meiosis, chromosome replication is followed by
two consecutive cell divisions: **meiosis I** and **meiosis
II,** producing four haploid daughter cells.

In interphase I, each chromosome replicates, producing two genetically identical sister chromatids that
remain attached. During prophase I, homologous chromosomes synapse and are initially held together by a
protein "zipper" called the synaptonemal complex.

■ INTERACTIVE QUESTION 13.2

Complete these three diagrams of sexual life cycles with the names of processes or cells. Shade in the portion of each life cycle that is diploid.

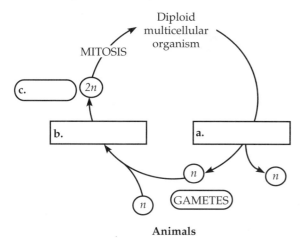

Animals

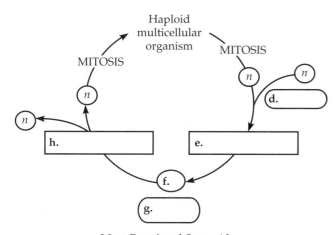

Most Fungi and Some Algae

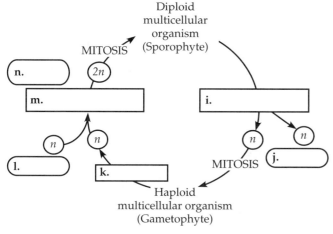

Plants and Some Algae

In metaphase I, the synapsed chromosomes line up on the metaphase plate with their kinetochores attached to spindle fibers from opposite poles. The homologous pairs separate in anaphase I, with one homolog moving toward each pole. Sister chromatids remain attached. In telophase I, a haploid set of chromosomes, each composed of two chromatids, reaches each pole. Cytokinesis usually occurs during telophase I. There is no replication of genetic material prior to the second division of meiosis.

Meiosis II looks like a regular mitotic division, in which chromosomes line up individually on the metaphase plate, and sister chromatids separate and move apart in anaphase II. At the end of telophase II, there are four haploid daughter cells. (See Interactive Question 13.4, p. 94.)

Mitosis and Meiosis Compared Mitosis produces daughter cells that are genetically identical to the parent cell. Meiosis produces haploid cells that differ genetically from their parent cell and from each other.

The three unique events that produce this result occur during meiosis I: In prophase I, when homologous chromosomes **synapse** to form a **tetrad,** genetic material is rearranged by crossing over between non-sister chromatids, which is visible during this stage by the appearance of X-shaped regions called **chiasmata.** In metaphase I, chromosomes line up in pairs, not as individuals, on the metaphase plate. During anaphase I, the homologous pairs separate and one homolog goes to each pole.

The sister chromatids of each homolog do not separate until meiosis II.

Origins of Genetic Variation

Sexual life cycles produce genetic variation among offspring (243–245)

Independent Assortment of Chromosomes The first meiotic division results in an assortment of maternal and paternal chromosomes in the two daughter cells. Each homologous pair lines up independently at the metaphase plate; the orientation of the maternal and paternal chromosomes is random. The number of possible combinations of maternal and paternal chromosomes is 2^n, where n is the haploid number.

■ INTERACTIVE QUESTION 13.3

How many assortments of maternal and paternal chromosomes are possible in human gametes?

■ INTERACTIVE QUESTION   13.4

The following diagrams represent some of the stages of meiosis (not in the right order). Label these stages.

a. _____   b. _____   c. _____   d. _____   e. _____   f. _____

Place these stages in the proper sequence.

_____   _____   _____   _____   _____   _____

Crossing Over   In prophase I, homologous segments of nonsister chromatids **cross over,** resulting in new genetic combinations of maternal and paternal genes on the same chromosome. The genetic variability of gametes is greatly increased as the no longer equivalent sister chromatids of **recombinant chromosomes** assort independently during meiosis II.

Random Fertilization   The random nature of fertilization adds to the genetic variability established in meiosis. Human parents can produce a zygote with any of about 64 trillion (8 million × 8 million) diploid combinations.

Evolutionary adaptation depends on a population's genetic variation (245)

In Darwin's theory of evolution by natural selection, genetic variations present in a population result in adaptation as the individuals best suited to an environment produce the most offspring. The process of sexual reproduction and mutation are the sources of this variation.

WORD ROOTS

-apsis = juncture (*synapsis:* the paring of replicated homologous chromosomes during prophase I of meiosis)

a- = not or without (*asexual:* type of reproduction not involving fertilization)

auto- = self (*autosome:* the chromosomes that do not determine gender)

chiasm- = marked crosswise (*chiasma:* the X-shaped microscopically visible region representing homologous chromosomes that have exchanged genetic material through crossing over during meiosis)

di- = two (*diploid:* cells that contain two homologous sets of chromosomes)

fertil- = fruitful (*fertilization:* process of fusion of a haploid sperm and a haploid egg cell)

haplo- = single (*haploid:* cells that contain only one chromosome of each homologous pair)

homo- = like (*homologous:* like chromosomes that form a pair)

karyo- = nucleus (*karyotype:* a display of the chromosomes of a cell)

meio- = less (*meiosis:* a variation of cell division that yields daughter cells with half as many chromosomes as the parent cell)

soma- = body (*somatic:* body cells with 46 chromosomes in humans)

sporo- = a seed; **-phyte** = a plant (*sporophyte:* the multicellular diploid form in organisms undergoing alternation of generations that results from a union of gametes and that meiotically produces haploid spores that grow into the gametophyte generation)

syn- = together; **gam-** = marriage (*syngamy:* the process of cellular union during fertilization)

tetra- = four (*tetrad:* the four closely associated chromatids of a homologous pair of chromosomes.

STRUCTURE YOUR KNOWLEDGE

1. Describe the key events of these stages of meiosis.

a. Interphase I
b. Prophase I
c. Metaphase I
d. Anaphase I
e. Metaphase II
f. Anaphase II

2. Create a concept map to help you organize your understanding of the similarities and differences between mitosis and meiosis. Compare your map with those of some classmates to see different ways of organizing this material.

TEST YOUR KNOWLEDGE

MULTIPLE CHOICE: *Choose the one best answer.*

1. The restoration of the diploid chromosome number after halving in meiosis is due to
   a. synapsis.
   b. fertilization.
   c. mitosis.
   d. DNA replication.
   e. chiasmata.

2. What is a karyotype?
   a. a genotype of an individual
   b. a unique combination of chromosomes found in a gamete
   c. a blood type determination of an individual
   d. a pictorial display of an individual's chromosomes
   e. a species-specific diploid number of chromosomes

3. What are autosomes?
   a. sex chromosomes
   b. chromosomes that occur singly
   c. chromosomal abnormalities that result in genetic defects
   d. chromosomes found in mitochondria and chloroplasts
   e. none of the above

4. A synaptonemal complex would be found during
   a. prophase I of meiosis.
   b. fertilization or syngamy of gametes.
   c. metaphase II of meiosis.
   d. prophase of mitosis.
   e. anaphase I of meiosis.

5. During the first meiotic division (meiosis I),
   a. homologous chromosomes separate.
   b. the chromosome number becomes haploid.
   c. crossing over between nonsister chromatids occurs.
   d. paternal and maternal chromosomes assort randomly.
   e. all of the above occur.

6. A cell with a diploid number of 6 could produce gametes with how many different combinations of maternal and paternal chromosomes?
   a. 6
   b. 8
   c. 12
   d. 64
   e. 128

7. The DNA content of a cell is measured in the G_2 phase. After meiosis I, the DNA content of one of the two cells produced would be
   a. equal to that of the G_2 cell.
   b. twice that of the G_2 cell.
   c. one-half that of the G_2 cell.
   d. one-fourth that of the G_2 cell.
   e. impossible to estimate due to independent assortment of homologous chromosomes.

8. In most fungi and some protists,
   a. the zygote is the only haploid stage.
   b. gametes are formed by meiosis.
   c. the multicellular organism is haploid.
   d. the gametophyte generation produces gametes by mitosis.
   e. reproduction is exclusively asexual.

9. In the alternation of generations found in plants,
   a. the sporophyte generation produces spores by mitosis.
   b. the gametophyte generation produces gametes by mitosis.
   c. the zygote will develop into a sporophyte generation by meiosis.
   d. spores develop into the haploid sporophyte generation.
   e. the gametophyte generation produces spores by meiosis.

10. Which of the following is *not* a source of genetic variation in sexually reproducing organisms?
    a. crossing over
    b. replication of DNA during S phase before meiosis I
    c. independent assortment of chromosomes
    d. random fertilization of gametes
    e. mutation

11. Meiosis II is similar to mitosis because
    a. sister chromatids separate.
    b. homologous chromosomes separate.
    c. DNA replication precedes the division.
    d. they both take the same amount of time.
    e. haploid cells are produced.

12. Homologous chromosomes
    a. have identical genes.
    b. have genes for the same traits at the same loci.
    c. are found in gametes.
    d. separate in meiosis II.
    e. have all of the above characteristics.

13. Asexual reproduction of a diploid organism would
    a. be impossible.
    b. involve meiosis.
    c. produce identical offspring.
    d. show variation among sibling offspring.
    e. involve spores produced by meiosis.

14. In a sexually reproducing species with a diploid number of 8, how many different combinations of paternal and maternal chromosomes would be possible in the *offspring*?
    a. 8
    b. 16
    c. 64
    d. 256
    e. 512

15. The calculation of offspring in question 14 includes only variation resulting from
    a. crossing over.
    b. random fertilization.
    c. independent assortment of chromosomes.
    d. a, b, and c.
    e. only b and c.

16. How many *chromatids* are present in metaphase II in a cell undergoing meiosis from an organism in which $2n = 24$?
    a. 12
    b. 24
    c. 36
    d. 48
    e. 96

17. Which of the following would *not* be considered a haploid cell?
    a. daughter cell after meiosis II
    b. gamete
    c. daughter cell after mitosis in gametophyte generation of a plant
    d. cell in prophase I
    e. cell in prophase II

18. Which of the following is *not* true of homologous chromosomes?
    a. They behave independently in mitosis.
    b. They synapse during the S phase of meiosis.
    c. They travel together to the metaphase plate in prometaphase of meiosis I.
    d. Each parent contributes one set of homologous chromosomes to an offspring.
    e. Crossing over between nonsister chromatids of homologous chromosomes is indicated by the presence of chiasmata.

19. Which of the following describes why or how recombinant chromosomes add to genetic variability?
    a. They are formed as a result of random fertilization when two sets of chromosomes combine in a zygote.
    b. They are the result of mutations that change alleles.
    c. They randomly orient during metaphase II and the nonequivalent sister chromatids separate in anaphase II.
    d. Genetic material from two parents is combined on the same chromosome.
    e. Both c and d are true.

20. A cell in G_2 before meiosis compared with one of the four cells produced by that meiotic division has
    a. twice as much DNA and twice as many chromosomes.
    b. four times as much DNA and twice as many chromosomes.
    c. four times as much DNA and four times as many chromosomes.
    d. half as much DNA but the same number of chromosomes.
    e. half as much DNA and half as many chromosomes.

21. A cell that has 22 autosomes and a Y chromosome would be
    a. a sperm.
    b. a male somatic cell.
    c. a male ovum.
    d. either a or b.
    e. either a or c.

CHAPTER 14

MENDEL AND THE GENE IDEA

FRAMEWORK

Through his work with garden peas in the 1860s, Mendel developed the fundamental principles of inheritance and the laws of segregation and independent assortment. This chapter describes the basic monohybrid and dihybrid crosses that Mendel performed to establish that inheritance involves particulate genetic factors (genes) that segregate independently in the formation of gametes and recombine to form offspring. The laws of probability can be applied to predict the outcome of genetic crosses.

The phenotypic expression of genotype may be affected by such factors as incomplete dominance, multiple alleles, pleiotropy, epistasis, and polygenic inheritance, as well as the environment. Genetic screening and counseling for recessively inherited disorders use new technologies and Mendelian principles to analyze human pedigrees.

CHAPTER REVIEW

What is the genetic basis for the variation evident in the individuals of a population and how is this variation transmitted from parents to offspring? The genetic material of two parents is neither blended nor irreversibly mixed in offspring but is passed on to future generations as discrete heritable units, or genes.

Gregor Mendel's Discoveries

Mendel brought an experimental and quantitative approach to genetics: *the process of science* (247–249)

Mendel worked with garden peas, a good choice of study organism because they are available in many varieties, their fertilization is easily controlled, and the characteristics of their offspring can be quantified.

Mendel studied seven **characters,** or heritable features, that occurred in alternative forms called **traits** He used **true-breeding** varieties of pea plants, which means that self-fertilizing parents always produce

offspring with the parental form of the character. To follow the transmission of these well-defined traits, Mendel performed **hybridizations** in which he cross-pollinated contrasting true-breeding varieties, and then allowed the next generation to self-pollinate. In a **monohybrid** cross, the inheritance of a single character is followed. The true-breeding parental plants are the **P generation** (parental); the results of the first cross are the F_1 **generation** (first filial); and the next generation, from the self-cross of the F_1, is known as the F_2 **generation.**

By the law of segregation, the two alleles for a character are packaged into separate gametes (249–252)

Mendel found that the F_1 offspring did not show a blending of the parental traits. Instead, only one of the parental forms of the character was found in the hybrid offspring. In the F_2 generation, however, the missing parental form reappeared in the ratio of 3 : 1—three offspring with the dominant trait shown by the F_1 to one offspring with the reappearing recessive trait.

Mendel's explanation for this phenomenon contains four parts: (1) alternate forms of genes, now called **alleles,** account for variations in characters; (2) an organism has two alleles for each inherited trait, one received from each parent; (3) when two different alleles occur together, one of them, called the **dominant allele,** may be completely expressed while the other, the **recessive allele,** has no observable effect on the organism's appearance; and (4) allele pairs separate (segregate) during the formation of gametes, so an egg or sperm carries only one allele for each inherited trait. This explanation is consistent with the behavior of chromosomes during meiosis. Mendel's **law of segregation** refers to this separation of alleles in the formation of gametes.

Mendel's law of segregation explains the 3 : 1 ratio observed in the F_2 plants. During the segregation of allele pairs in the formation of F_1 gametes, half the gametes receive one allele while the other half receive the alternate allele. Random fertilization of gametes results in one-fourth of the plants having two dominant alleles, half having one dominant and one recessive allele, and one-fourth receiving two recessive alleles, producing a ratio of plants showing the dominant to recessive trait of 3 : 1. A **Punnett square** can be used to predict the results of simple genetic crosses. Dominant alleles are often symbolized by a capital letter, recessive alleles by a small letter.

■ **INTERACTIVE QUESTION 14.1**

Fill in this diagram of a monohybrid cross of round- and wrinkled-seeded pea plants. The round allele (*R*) is dominant and the wrinkled allele (*r*) is recessive.

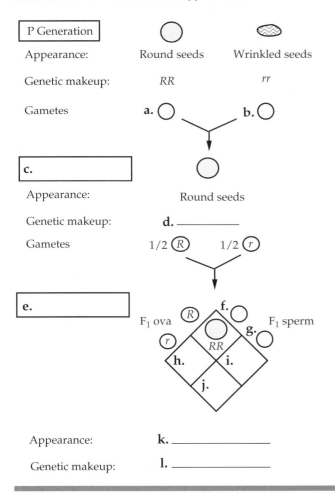

Some Useful Genetic Vocabulary An organism that has a pair of identical alleles is said to be **homozygous** for that gene. If the organism has two different alleles, it is said to be **heterozygous** for that gene. Homozygotes are true breeding; heterozygotes are not,

since they produce gametes with one or the other allele that can combine to produce homozygous dominant, heterozygous, and homozygous recessive offspring. **Phenotype** is an organism's expressed traits; **genotype** is its genetic makeup.

The Testcross In a testcross, an organism expressing the dominant phenotype is bred with a recessive homozygote to determine the genotype of this phenotypically dominant organism.

■ **INTERACTIVE QUESTION 14.2**

A tall pea plant is crossed with a recessive dwarf pea plant. What will the phenotypic and genotypic ratio of offspring be

a. if the tall plant was *TT*?

b. if the tall plant was *Tt*?

By the law of independent assortment, each pair of alleles segregates into gametes independently (252–254)

Mendel used **dihybrid crosses,** involving parents differing in two characters, to determine whether the two characters were transmitted independently of each other or together in the same combination in which they were inherited from the parent plants.

If the two pairs of alleles segregate independently, then gametes from an F_1 hybrid generation (*AaBb*) should contain four combinations of alleles in equal quantities (*AB, Ab, aB, ab*). The random fertilization of these four classes of gametes should result in 16 (4 × 4) gamete combinations that produce four phenotypic categories in a ratio of 9 : 3 : 3 : 1 (nine offspring showing both dominant traits, three showing one dominant trait and one recessive, three showing the opposite dominant and recessive traits, and one showing both recessive traits). Mendel obtained these ratios when he categorized the F_2 progeny of dihybrid crosses, providing evidence for his **law of independent assortment.** This principle states that pairs of alleles for different characters segregate independently of each other in the formation of gametes.

■ INTERACTIVE QUESTION   14.3

A true-breeding tall, purple-flowered pea plant (*TTPP*) is crossed with a true-breeding dwarf, white-flowered plant (*ttpp*).

a. What is the phenotype of the F_1 generation?

_____

b. What is the genotype of the F_1 generation?

_____

c. What four types of gametes are formed by F_1 plants?

_____ _____ _____ _____

d. Fill in the following Punnett square to show the offspring of the F_2 generation. Shade each phenotype a different color so you can see the ratio of offspring.

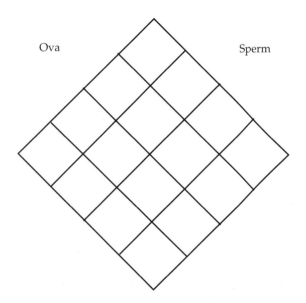

Ova   Sperm

e. List the phenotypes and ratios found in the F_2 generation.

_____   _____

_____   _____

f. What is the ratio of tall to dwarf plants? _____ Of purple- to white-flowered plants? _____ (Note that the alleles for each individual character segregate as in a monohybrid cross.)

Mendelian inheritance reflects rules of probability (254–255)

General rules of probability apply to the laws of segregation and independent assortment. The probability scale goes from 0 to 1; the probabilities of all possible outcomes must add up to 1. The probability of an event occurring is the number of times that event could occur over all the possible events; for example, the probability of drawing an ace from a deck of cards is 4/52. The outcome of independent events is not affected by previous or simultaneous trials.

The Rule of Multiplication   The rule of multiplication states that the probability that a certain combination of independent events will occur together is equal to the product of the separate probabilities of the independent events. The probability of a particular genotype being formed by fertilization is equal to the product of the probabilities of forming each type of gamete needed to produce that genotype.

■ INTERACTIVE QUESTION   14.4

Apply the rule of multiplication to a dihybrid cross. How would you determine the probability of getting an F_2 offspring that is homozygous recessive for both traits?

The Rule of Addition   If a genotype can be formed in more than one way, then the rule of addition states that its probability is equal to the sum of the separate probabilities of the different ways the event can occur. For example, a heterozygote offspring can occur if the ovum contains the dominant allele and the sperm the recessive ($\frac{1}{2} \times \frac{1}{2} = \frac{1}{4}$ probability) or vice versa ($\frac{1}{4}$). Therefore, the heterozygote offspring would be the predicted result from a monohybrid cross half of the time ($\frac{1}{4} + \frac{1}{4} = \frac{1}{2}$).

Using Rules of Probability to Solve Genetics Problems
Fairly complex genetics problems can be solved by applying the rules of multiplication and addition. The probability of a particular genotype arising from a cross can be determined by considering each gene involved as a separate monohybrid cross and then multiplying the probabilities of all the independent events involved in the final genotype.

Mendel discovered the particulate behavior of genes: *a review* (255)

Mendel's laws of segregation and independent assortment explain the inheritance of discrete hereditary units (alleles of genes) from generation to generation according to basic rules of probability. The larger the sample size, the more closely the results will conform to statistical predictions.

■ **INTERACTIVE QUESTION 14.5**

a. In the following cross, what is the probability of obtaining offspring that show all three dominant traits, *A_B_C* (_ indicates that the second allele can be either dominant or recessive without affecting the phenotype determined by the first dominant allele)?

$$AaBbcc \times AabbCC$$

probability of offspring that are *A_B_C_* = _____

b. What is the probability that the offspring of this *AaBbcc* × *AabbCC* cross will show at least two dominant traits? _____

Extending Mendelian Genetics

The relationship between genotype and phenotype is rarely simple (255–260)

Incomplete Dominance Some characters show **incomplete dominance,** in that the F_1 hybrids have a phenotype intermediate between that of the parents. The F_2 show a $1:2:1$ phenotypic and genotypic ratio.

What Is a Dominant Allele? In the case of an allele showing **complete dominance,** the phenotype of the heterozygote is indistinguishable from that of the homozygous dominant. At the other end of the continuum, alleles that exhibit **codominance** are both separately expressed in the phenotype, as in the case of the MN blood group. Intermediate phenotypes are characteristic of alleles showing incomplete dominance. Even when alleles exhibit dominance or incomplete dominance at the phenotypic level, both alleles may be expressed at the microscopic or molecular level.

Even though one allele may be considered "dominant," the two alleles in a cell do not interact at all. Also,

whether or not an allele is dominant or recessive has no relation to how common it is in a population.

Multiple Alleles Most genes exist in more than two allelic forms. The gene that determines human blood groups has three alleles. The alleles I^A and I^B are codominant with each other; each codes for a carbohydrate found on the surface of red blood cells. The allele *i* does not code for a blood protein and is recessive to I^A and I^B. Blood type is critical in transfusions because, if the carbohydrate coating on the donor's blood cells is foreign to the recipient, the recipient's antibodies will cause agglutination, or clumping, of the donated blood cells.

■ **INTERACTIVE QUESTION 14.6**

List the possible genotypes for the following blood groups.

a. A _____   b. B _____

c. AB _____   d. O _____

Pleiotropy **Pleiotropy** is the characteristic of a single gene having multiple phenotypic effects in an individual. Many hereditary diseases with complex sets of symptoms are caused by a single allele.

Epistasis In **epistasis,** a gene at one locus may affect the expression of another gene. F_2 ratios that differ from the typical $9:3:3:1$ often indicate epistasis.

■ **INTERACTIVE QUESTION 14.7**

A dominant allele *M* is necessary for the production of the black pigment melanin; *mm* individuals are white. A dominant allele *B* results in the deposition of a lot of pigment in an animal's hair, producing a black color. The genotype *bb* produces brown hair. Two black animals heterozygous for both genes are bred. Fill in the following table for the offspring of this *MmBb* × *MmBb* cross.

Phenotype	Genotype	Ratio
Black		
	M_bb	

Polygenic Inheritance   **Quantitative characters,** such as height or skin color, vary along a continuum in a population. Such variation is usually due to **polygenic inheritance,** in which two or more genes have an additive effect on one phenotypic character. Each dominant allele contributes one "unit" to the phenotype. A polygenic character may result in a normal distribution (forming a bell-shaped curve) of the character within a population.

■ INTERACTIVE QUESTION   14.8

The height of spike weed is a result of polygenic inheritance involving three genes, each of which can contribute an additional 5 cm to the base height of the plant. The base height of the weed is 10 cm, and the tallest plant can reach 40 cm.

a. If a tall plant (*AABBCC*) is crossed with a base-height plant (*aabbcc*), what is the height of the F₁ plants?

b. How many phenotypic classes will there be in the F₂?

.

Nature and Nurture: the Environmental Impact on Phenotype   The phenotype of an individual is the result of complex interactions between its genotype and the environment. Genotypes have a phenotypic range called a **norm of reaction** within which the environment influences phenotypic expression. Polygenic characters are often **multifactorial,** meaning that a combination of genetic and environmental factors influences phenotype.

Integrating a Mendelian View of Heredity and Variation   The phenotypic expression of most genes is influenced by other genes and by the environment. The theory of particulate inheritance and Mendel's principles of segregation and independent assortment, however, still form the basis for modern genetics.

Mendelian Inheritance in Humans

Pedigree analysis reveals Mendelian patterns in human inheritance (260–261)

A family **pedigree** is a family tree with the history of a particular trait shown across the generations. By convention, circles represent females, squares are used for males, and solid symbols indicate individuals that express the phenotype in question. Parents are joined by a horizontal line, and offspring are listed below parents from left to right in order of birth. The genotype of individuals in the pedigree can often be deduced by following the patterns of inheritance.

■ INTERACTIVE QUESTION   14.9

Consider this pedigree for the trait albinism (lack of skin pigmentation) in three generations of a family. (Solid symbols represent individuals who are albinos.) From your knowledge of Mendelian inheritance, answer the questions that follow.

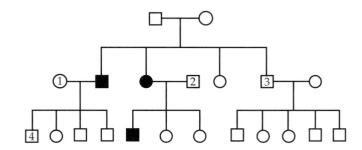

a. Is this trait caused by a dominant or recessive allele? How can you tell?

b. Determine the genotypes of the parents in the first generation. (Let *AA* and *Aa* represent normal pigmentation and *aa* be the albino genotype.) Genotype of father_____; of mother_____.

c. Determine the probable genotypes of the mates of the albino offspring in the second generation and the grandson 4 in the third generation. Genotypes: mate 1 _____   mate 2 _____   grandson 4 _____.

d. Can you determine the genotype of son 3 in the second generation? Why or why not?

Many human disorders follow Mendelian patterns of inheritance (261–264)

Recessively Inherited Disorders   Only homozygous recessive individuals express the phenotype for the thousands of genetic disorders that are inherited as simple recessive traits. **Carriers** of the disorder are heterozygotes who are phenotypically normal but may transmit the recessive allele to their offspring.

Cystic fibrosis is the most common lethal genetic disease in the United States; it is found more frequently in whites of European descent than in other groups. This recessive allele results in defective chloride channels in certain cell membranes. Accumulating extracellular chloride leads to the buildup of thickened mucus in various organs and a predisposition to bacterial infections.

Tay-Sachs disease is a lethal disorder in which the brain cells are unable to metabolize a type of lipid that then accumulates and damages the brain, resulting in an early death.

Sickle-cell disease is the most common inherited disease among African Americans. Due to a single amino acid substitution in the hemoglobin protein, red blood cells deform into a sickle shape when blood-oxygen concentration is low, triggering blood clotting and other pleiotropic effects. Heterozygous individuals are said to have sickle-cell trait but are usually healthy. The resistance to malaria that accompanies the sickle-cell trait may explain why this lethal recessive allele remains in relatively high frequency in areas where malaria is common.

The likelihood of two mating individuals carrying the same rare deleterious allele increases when the individuals have common ancestors. Consanguineous matings, between siblings or close relatives, are indicated on pedigrees by double lines.

■ INTERACTIVE QUESTION 14.10

a. What is the probability that a mating between two carriers will produce an offspring with a recessively inherited disorder? _____

b. What is the probability that a phenotypically normal child produced by a mating of two heterozygotes will be a carrier? _____

Dominantly Inherited Disorders A few human disorders are due to dominant genes. In achondroplasia, dwarfism is due to a single copy of a mutant allele. Homozygosity for this dominant allele is lethal.

Dominant lethal alleles are more rare than recessive lethals because the harmful allele cannot be masked in the heterozygote. A late-acting lethal dominant allele can be passed on if the symptoms do not develop until after reproductive age. Molecular geneticists have developed a method to detect the lethal gene for **Huntington's disease,** a degenerative disease of the nervous system that does not develop until later in life.

Multifactorial Disorders Many diseases have genetic (usually polygenic) and environmental components. These multifactorial disorders include heart disease, diabetes, cancer, and others.

Technology is providing new tools for genetic testing and counseling (264–266)

The risk of a genetic disorder being transmitted to offspring can sometimes be determined through genetic counseling and testing before a child is conceived or in the early stages of pregnancy. The probability of a child having a genetic defect may be determined by considering the family history of the disease.

■ INTERACTIVE QUESTION 14.11

If two prospective parents both have siblings who had a recessive genetic disorder, what is the chance that they would have a child who inherits the disorder?

Carrier Recognition Determining whether parents are carriers can determine the risk of passing on a genetic disorder. Biochemical tests that permit carrier recognition have been developed for a number of heritable disorders.

Fetal Testing **Amniocentesis** is a procedure that extracts a small amount of amniotic fluid from the sac surrounding the fetus. The fluid is analyzed biochemically, and fetal cells present in the fluid are cultured for several weeks and then tested for certain genetic disorders and karyotyped to check for chromosomal defects.

Chorionic villus sampling (CVS) is a technique in which a small amount of the fetal tissue is suctioned from the placenta. These rapidly growing cells can be karyotyped immediately, and this procedure can be performed at only 8 to 10 weeks of pregnancy, earlier than an amniocentesis can be performed. A new technique isolates fetal cells from maternal blood, which can then be cultured and tested.

Ultrasound is a simple, noninvasive procedure that can reveal major abnormalities. Fetoscopy, the insertion of a needle-thin viewing scope and light into the uterus, allows the fetus to be checked for anatomical problems.

Newborn Screening   Some genetic disorders can be detected at birth. Routine screening is now used to test for the recessively inherited disorder phenylketonuria (PKU) in newborns. If detected, dietary adjustments can lead to normal development.

WORD ROOTS

co- = together (*codominance*: phenotype in which both dominant alleles are expressed in the heterozygote)

-centesis = a puncture (*amniocentesis*: a technique for determining genetic abnormalities in a fetus by the presence of certain chemicals or defective fetal cells in the amniotic fluid, obtained by aspiration from a needle inserted into the uterus)

di- = two (*dihybrid cross*: a breeding experiment in which parental varieties differing in two traits are mated)

epi- = beside; **-stasis** = standing (*epistasis*: a phenomenon in which one gene alters the expression of another gene that is independently inherited)

geno- = offspring (*genotype*: the genetic makeup of an organism)

hetero- = different (*heterozygous*: having two different alleles for a trait)

homo- = alike (*homozygous*: having two identical alleles for a trait)

mono- = one (*monohybrid cross*: a breeding experiment that uses parental varieties differing in a single character)

pedi- = a child (*pedigree*: a family tree describing the occurrence of heritable characters in parents and offspring across as many generations as possible)

pheno- = appear (*phenotype*: the physical and physiological traits of an organism)

pleio- = more (*pleiotropy*: when a single gene impacts more than one characteristic)

poly- = many; **gene-** = produce (*polygenic*: an additive effect of two or more gene loci on a single phenotypic character)

STRUCTURE YOUR KNOWLEDGE

1. Relate Mendel's two laws of inheritance to the behavior of chromosomes in meiosis that you studied in Chapter 13.

2. Mendel worked with characters that exhibited two alternate forms: smooth or wrinkled, green or yellow, tall or short. In his F$_1$ generation, the dominant

allele was always expressed, with the recessive trait reappearing in the F$_2$ offspring. But not all allele pairs operate by complete dominance; some show incomplete dominance or codominance. And some that show dominance on the phenotypic level are actually incompletely dominant or codominant when observed at the microscopic or molecular level. Taking into account the mechanisms by which genotype becomes expressed as phenotype, explain how this range in dominance can occur.

GENETICS PROBLEMS

*One of the best ways to learn genetics is to work problems. You can't memorize genetic knowledge; you have to practice using it. Work through the problems presented below, methodically setting down the information you are given and what you are to determine. Write down the symbols used for the alleles and genotypes, and the phenotypes resulting from those genotypes. Avoid using Punnett squares; while useful for learning basic concepts, they are much too laborious and mistake-prone. Instead, break complex crosses into their monohybrid components and rely on the rules of multiplication and addition. **Always** look at the answers you get to see if they make logical sense. And remember that study groups are great for going over your problems and helping each other "see the light." Answers and explanations are provided at the end of the book.*

1. Summer squash are either white or yellow. To get white squash at least one of the parental plants must be white. Which color is dominant?

2. For the following crosses, determine the probability of obtaining the indicated genotype in an offspring.

Cross	Offspring	Probability
AAbb × *AaBb*	*AAbb*	**a.**
AaBB × *AaBb*	*aaBB*	**b.**
AABbcc × *aabbCC*	*AaBbCc*	**c.**
AaBbCc × *AaBbcc*	*aabbcc*	**d.**

3. True-breeding tall red-flowered plants are crossed with dwarf white-flowered plants. The resulting F$_1$ generation consists of all tall pink-flowered plants. Assuming that height and flower color are each determined by a single gene locus on different chromosomes, predict the results of an F$_1$ cross of the *TtRr* plants. List the phenotypes and predicted ratios for the F$_2$ generation.

4. Blood typing has often been used as evidence in paternity cases, when the blood type of the mother and child may indicate that a man alleged to be the father could not possibly have fathered the child. For the following mother and child combinations, indicate which blood groups of potential fathers would be exonerated.

Blood Group of Mother	Blood Group of Child	Man Exonerated if He Belongs to Blood Group(s)
AB	A	**a.**
O	B	**b.**
A	AB	**c.**
O	O	**d.**
B	A	**e.**

5. In rabbits, the homozygous *CC* is normal, *Cc* results in rabbits with deformed legs, and *cc* is lethal. For a gene for coat color, the genotype *BB* produces black, *Bb* brown, and *bb* a white coat. Give the phenotypic ratio of offspring from a cross of a deformed-leg, brown rabbit with a deformed-leg, white rabbit.

6. Polydactyly (extra fingers and toes) is due to a dominant gene. A father is polydactyl, the mother has the normal phenotype, and they have had one normal child. What is the genotype of the father? Of the mother? What is the probability that a second child will have the normal number of digits?

7. In dogs, black (*B*) is dominant to chestnut (*b*), and solid color (*S*) is dominant to spotted (*s*). What are the genotypes of the parents that would produce a cross with 3/8 black solid, 3/8 black spotted, 1/8 chestnut solid, and 1/8 chestnut spotted puppies? (*Hint*: first determine what genotypes the offspring must have before you deal with the fractions.)

8. When hairless hamsters are mated with normal-haired hamsters, about one-half the offspring are hairless and one-half are normal. When hairless hamsters are crossed with each other, the ratio of normal-haired to hairless is 1 : 2. How do you account for the results of the first cross? How would you explain the unusual ratio obtained in the second cross?

9. Two pigs whose tails are exactly 25 cm in length are bred over 10 years and they produce 96 piglets with the following tail lengths: 6 piglets at 15 cm, 25 at 20 cm, 37 at 25 cm, 23 at 30 cm, and 5 at 35 cm.
   a. How many pairs of genes are regulating the tail length character? *Hint:* count the number of

phenotypic classes, or determine the sum of the ratios of the classes. In a monohybrid cross, the F_2 ratios add up to 4 (3 : 1 or 1 : 2 : 1). In a dihybrid cross, the F_2 ratios add up to 16 (9 : 3 : 3 : 1 or some variation if the genes are epistatic or quantitative).
   b. What offspring phenotypes would you expect from a mating between a 15-cm and a 30-cm pig?

10. Fur color in rabbits is determined by a single gene locus for which there are four alleles. Four phenotypes are possible: black, Chinchilla (gray color caused by white hairs with black tips), Himalayan (white with black patches on extremities), and white. The black allele (*C*) is dominant over all other alleles, the Chinchilla allele (C^{ch}) is dominant over Himalayan (C^h), and the white allele (*c*) is recessive to all others.
    a. A black rabbit is crossed with a Himalayan, and the F_1 consists of a ratio of 2 black to 2 Chinchilla. Can you determine the genotypes of the parents?
    b. A second cross was done between a black rabbit and a Chinchilla. The F_1 contained a ratio of 2 black to 1 Chinchilla to 1 Himalayan. Can you determine the genotypes of the parents of this cross?

11. In Labrador retriever dogs, the dominant gene *B* determines black coat color and *bb* produces brown. A separate gene *E*, however, shows dominant epistasis over the *B* and *b* alleles, resulting in a "golden" coat color. The recessive *e* allows expression of *B* and *b*. A breeder wants to know the genotypes of her three dogs, so she breeds them and makes note of the offspring of several litters. Determine the genotypes of the three dogs.
    a. golden female (Dog 1) × golden male (Dog 2) offspring: 7 golden, 1 black, 1 brown
    b. black female (Dog 3) × golden male (Dog 2) offspring: 8 golden, 5 black, 2 brown

12. The ability to taste phenylthiocarbamide (PTC) is controlled in humans by a single dominant allele (*T*). A woman nontaster married a man taster and they had three children, two boy tasters and a girl nontaster. All the grandparents were tasters. Create a pedigree for this family for this trait. (Solid symbols should signify nontasters (*tt*).) Where possible, indicate whether tasters are *TT* or *Tt*.

13. Two true-breeding varieties of garden peas are crossed. One parent had red, axial flowers, and the other had white, terminal flowers. All F_1 individuals had red, terminal flowers. If 100 F_2 offspring were counted, how many of them would you expect to have red, axial flowers?

14. You cross true-breeding red-flowered plants with true-breeding white-flowered plants, and the F_1 are all red-flowered plants. The F_2, however, occur in a ratio of 9 red: 6 pale purple: 1 white. How many genes are involved in the inheritance of this color character? Explain why the F_1 are all red and how the 9:6:1 ratio of phenotypes in the F_2 occurred.

TEST YOUR KNOWLEDGE

MATCHING: *Match the definition with the correct term.*

_____ 1. codominance

_____ 2. homozygous

_____ 3. heterozygous

_____ 4. phenotype

_____ 5. polygenic (quantitative)

_____ 6. pleiotropy

_____ 7. epistasis

_____ 8. testcross

_____ 9. dihybrid cross

_____ 10. incomplete dominance

A. true-breeding variety

B. cross between two hybrids

C. cross that involves two different gene pairs

D. cross with homozygous recessive to determine genotype of unknown

E. the physical characteristics of an individual

F. genotype with two different alleles

G. genotype with multiple alleles for same locus

H. one gene influences the expression of another gene

I. both alleles are fully expressed in heterozygote

J. single gene with multiple phenotypic effects

K. heterozygote intermediate between homozygous phenotypes

L. two or more genes with additive effect on phenotype

MULTIPLE CHOICE: *Choose the one best answer.*

1. According to Mendel's law of segregation,
   a. there is a 50% probability that a gamete will get a dominant allele.
   b. gene pairs segregate independently of other genes in gamete formation.
   c. allele pairs separate in gamete formation.
   d. the laws of probability determine gamete formation.
   e. there is a 3 : 1 ratio in the F_2 generation.

2. After obtaining two heads from two tosses of a coin, the probability of tossing the coin and obtaining a head is
   a. ½.　　c. ⅙.　　e. ¹⁄₁₆.
   b. ¼.　　d. ⅛.

3. The probability of tossing three coins simultaneously and obtaining two heads and one tail is
   a. ½.　　c. ⅛.　　e. ⅜.
   b. ¾.　　d. ¹⁄₁₆.

4. A multifactorial disorder
   a. can usually be traced to consanguineous matings.
   b. is caused by recessively inherited lethal genes.
   c. has both genetic and environmental causes.
   d. has a collection of symptoms traceable to an epistatic gene.
   e. is usually associated with quantitative traits.

5. The F_2 generation
   a. has a phenotypic ratio of 3:1.
   b. is the result of the self-fertilization or crossing of F_1 individuals.
   c. can be used to determine the genotype of individuals with the dominant phenotype.
   d. has a phenotypic ratio that equals its genotypic ratio.
   e. has 16 different genotypic possibilities.

6. The base height of the dingdong plant is 10 cm. Four genes contribute to the height of the plant, and each dominant allele contributes 3 cm to height. If you cross a 10-cm plant (quadruply homozygous recessive) with a 34-cm plant, how many phenotypic classes will there be in the F_2?
   a. 4
   b. 5
   c. 8
   d. 9
   e. 64

7. A 1 : 1 phenotypic ratio in a testcross indicates that
   a. the alleles are dominant.
   b. one parent must have been homozygous recessive.
   c. the dominant phenotype parent was a heterozygote.
   d. the alleles segregated independently.
   e. the alleles are codominant.

8. Carriers of a genetic disorder
   a. are indicated by solid symbols on a family pedigree.
   b. are involved in consanguineous matings.
   c. will produce children with the disease.
   d. are heterozygotes for the gene that can cause the disorder.
   e. have a homozygous recessive genotype.

9. If both parents are carriers of a lethal recessive gene, the probability that their child will inherit and express the disorder is
   a. $\frac{1}{8}$.
   b. $\frac{1}{4}$.
   c. $\frac{1}{2}$.
   d. $\frac{1}{2} \times \frac{1}{2} \times \frac{1}{4}$, or $\frac{1}{16}$.
   e. $\frac{2}{3} \times \frac{2}{3} \times \frac{1}{4}$, or $\frac{1}{9}$.

10. Which phase of meiosis is most directly related to the law of independent assortment?
    a. prophase I
    b. prophase II
    c. metaphase I
    d. metaphase II
    e. anaphase II

11. You think that two alleles for coat color in mice show incomplete dominance. What is the best and simplest cross to perform in order to support your hypothesis?
    a. a testcross of a homozygous recessive mouse with a mouse of unknown genotype
    b. a cross of F_1 mice to look for a $1 : 2 : 1$ ratio in the offspring
    c. a reciprocal cross in which the sex of the mice of each coat color is reversed
    d. a cross of two true-breeding mice of different colors to look for an intermediate phenotype in the F_1
    e. a cross of F_1 mice to look for a $9 : 7$ ratio in the offspring

12. Which of the following human diseases is inherited as a simple recessive trait?
    a. Tay-Sachs disease
    b. cancer
    c. diabetes
    d. Alzheimer's disease
    e. cardiovascular disease

13. In a dihybrid cross of heterozygotes, what proportion of the offspring will be phenotypically dominant for both traits?
    a. $\frac{1}{16}$
    b. $\frac{3}{16}$
    c. $\frac{1}{4}$
    d. $\frac{9}{16}$
    e. $\frac{3}{4}$

14. A mother with type B blood has two children, one with type A blood and one with type O blood. Her husband has type O blood. Which of the following could you conclude from this information?
    a. The husband could not have fathered either child.
    b. The husband could have fathered both children.
    c. The husband must be the father of the child with type O blood and could be the father of the type A child.
    d. The husband could be the father of the child with type O blood, but not the type A child.
    e. Neither the mother nor the husband could be the biological parent of the type A child.

15. In guinea pigs, the brown coat color allele (*B*) is dominant over red (*b*), and the solid color allele (*S*) is dominant over spotted (*s*). The F_1 offspring of a cross between true-breeding brown, solid-colored guinea pigs and red, spotted pigs are crossed. What proportion of their offspring (F_2) would be expected to be red and solid colored?
    a. $\frac{1}{9}$
    b. $\frac{1}{16}$
    c. $\frac{3}{16}$
    d. $\frac{9}{16}$
    e. $\frac{3}{4}$

16. A dominant allele *P* causes the production of purple pigment; *pp* individuals are white. A dominant allele *C* is also required for color production; *cc* individuals are white. What proportion of offspring will be purple from a *ppCc* × *PpCc* cross?
    a. $\frac{1}{8}$
    b. $\frac{3}{8}$
    c. $\frac{1}{2}$
    d. $\frac{3}{4}$
    e. None

CHAPTER 15

THE CHROMOSOMAL BASIS
OF INHERITANCE

FRAMEWORK

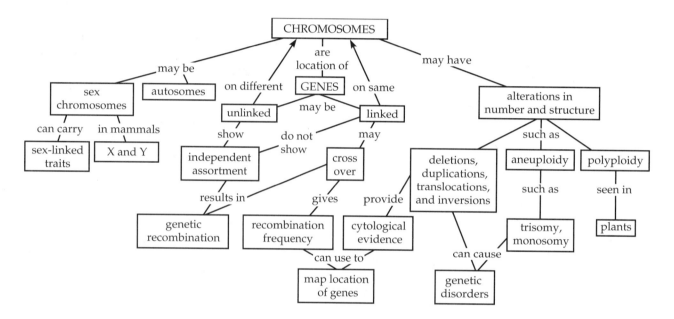

CHAPTER REVIEW

By 1900, three botanists had independently reproduced the plant-breeding results that Mendel had explained 35 years previously.

Relating Mendelism to Chromosomes

Mendelian inheritance has its physical basis in the behavior of chromosomes during sexual life cycles (269–270)

Mendel's principles, combined with cytological evidence of the processes of mitosis and meiosis gathered in the late 1800s, led to the **chromosome theory of inheritance** that Mendelian genes have loci on chromosomes, and that chromosomes undergo segregation and independent assortment in the process of gamete formation.

Morgan traced a gene to a specific chromosome: *the process of science* **(271–272)**

Morgan's Choice of Experimental Organism   Morgan, working with the fruit fly, *Drosophila melanogaster,* was the first to identify a specific gene with a specific chromosome. Fruit flies are prolific breeders and have only four pairs of chromosomes, which are easily distinguishable with a microscope. The sex chromosomes occur as XX in females and XY in male flies.

The normal phenotype found most commonly in nature for a character is called the **wild type,** whereas alternative traits, assumed to have arisen as mutations, are called **mutant phenotypes.**

Discovery of Sex Linkage   Morgan discovered a mutant white-eyed male fly that he mated with a wild-type red-eyed female. The F_1 were all red-eyed. In the F_2, however, all female flies were red-eyed, whereas half of the males were red-eyed and half were white-eyed. Morgan deduced that the gene for eye color was **sex-linked,** occurring only on the X sex chromosome. Males have only one X, so their phenotype is determined by the eye-color allele they inherit from their mother. This association of a specific gene with a chromosome provided evidence for the chromosome theory of inheritance.

■ INTERACTIVE QUESTION 15.1

Complete this summary of Morgan's crosses with the mutant white-eyed fly by filling in the Punnett square and showing the phenotypes and genotypes of the F₂ generation. (*w* stands for the mutant recessive white allele; *w⁺* for the wild-type red allele.)

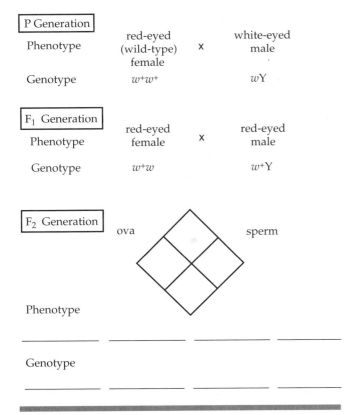

Linked genes tend to be inherited together because they are located on the same chromosome (272–273)

Genes that are located on the same chromosome are called **linked genes.** Morgan performed a testcross of heterozygote wild-type flies with flies that were homozygous recessive for black bodies and vestigial wings and found that the offspring were not in the

predicted $1:1:1:1$ phenotypic classes. Rather, most of the offspring were the same phenotypes as the P generation parents—either wild type (gray, normal wings) or double mutant (black, vestigial). Morgan deduced that these traits were inherited together because their genes were located on the same chromosome.

Independent assortment of chromosomes and crossing over produce genetic recombinants (273–275)

Genetic recombination results in offspring with new combinations of traits inherited from their parents.

The Recombination of Unlinked Genes: Independent Assortment of Chromosomes   In a dihybrid cross between a heterozygote and a homozygous recessive, one-half of the offspring will be **parental types** and have phenotypes like one or the other of the P generation parents, and one-half of the offspring, called **recombinants,** will have combinations of the two traits that are unlike the parents. This 50% frequency of recombination is observed when two genes are located on different chromosomes and it results from the random alignment of homologous chromosomes at metaphase I and the resulting independent assortment of alleles.

■ INTERACTIVE QUESTION 15.2

In a testcross between a heterozygote tall, purple-flowered pea plant and a dwarf, white-flowered plant,

a. what are the phenotypes of offspring that are parental types?

b. what are the phenotypes of offspring that are recombinants?

The Recombination of Linked Genes: Crossing Over   Linked genes do not assort independently, and one would not expect to see recombination of parental traits in the offspring. Recombination of linked genes does occur, however, due to crossing over, the reciprocal trade between nonsister chromatids of synapsed homologous chromosomes during prophase of meiosis I.

■ **INTERACTIVE QUESTION   15.3**

With unlinked genes, an equal number of parental and recombinant offspring are produced. With linked genes, (more/fewer) parentals than recombinants are produced. (Circle and then explain your answer.)

■ **INTERACTIVE QUESTION   15.4**

Recombination frequency is given below for several gene pairs. Create a linkage map for these genes, showing the map unit distance between loci.

j, k 12%   k, l 6%

j, m 9%   l, m 15%

Geneticists can use recombination data to map a chromosome's genetic loci (275–276)

Sturtevant suggested that recombination frequencies reflect the relative distance between genes; genes located farther apart have a greater probability that a crossover event will occur between them. Sturtevant used recombination data to order genes on a **genetic map,** defining one **map unit** (or centimorgan) as equal to a 1% recombination frequency.

The sequence of genes on a chromosome can be determined by finding the recombination frequency between different pairs of genes. Linkage cannot be determined if genes are 50 or more map units apart because they would have the 50% recombination frequency typical of unlinked genes. Distant genes on the same chromosome may be mapped by adding the recombination frequencies determined for intermediate genes.

Sturtevant and his co-workers found that the genes for the various known mutations of *Drosophila* clustered into four groups of linked genes, providing additional evidence that genes are located on chromosomes.

The frequency of crossing over may vary along the length of a chromosome, and a **linkage map** provides the sequence but not the exact location of genes on chromosomes. **Cytological mapping** locates gene loci in reference to visible chromosomal features.

Sex Chromosomes

The chromosomal basis of sex varies with the organism (276–277)

Sex is a phenotypic character usually determined by sex chromosomes. In humans and mammals, females, who are XX, produce ova that each contain an X chromosome. Males, who are XY, produce two kinds of sperm, each with either an X or a Y chromosome. Whether the gonads of an embryo develop into testes or ovaries depends on the presence or absence of the gene *Sry*, found on the Y chromosome, whose protein product regulates many other genes. A few other genes on the Y chromosome are necessary for normal functioning of the testes.

Other sex-determination systems include X-O (in grasshoppers and some other insects), Z-W (in birds and some fishes and insects), and haplo-diploid (in bees and ants).

Sex-linked genes have unique patterns of inheritance (277–279)

Sex chromosomes may carry genes for traits that are not related to sex. Males inherit their X-linked alleles (coding for what are usually called sex-linked traits) only from their mothers; daughters inherit sex-linked alleles from both parents. Recessive sex-linked traits are seen more often in males, since they are hemizygous for sex-linked genes.

Sex-Linked Disorders in Humans   **Duchenne muscular dystrophy** is a sex-linked disorder resulting from the lack of a key muscle protein. **Hemophilia** is a sex-linked recessive trait characterized by excessive bleeding due to the absence of one or more blood-clotting proteins.

X Inactivation in Female Mammals   Only one of the X chromosomes is fully active in most mammalian female somatic cells. The other X chromosome is contracted into a **Barr body** located inside the nuclear membrane. Lyon demonstrated that the selection of which X chromosome is inactivated is a random event occurring independently in embryonic cells. A female who is heterozygous for a sex-linked trait is a mosaic and will express one allele in approximately half her cells and the alternate in the other cells.

As a result of X-inactivation, both males and females have an equal dosage of X-linked genes. A gene called *XIST* is active on the X chromosome that forms the Barr body. Its RNA product may trigger DNA methylation and X-inactivation.

■ INTERACTIVE QUESTION 15.5

Two normal color-sighted individuals produce the following children and grandchildren. Fill in the probable genotype of the indicated individuals in this pedigree. *Squares are males, circles are females, and solid symbols represent color blindness. Choose an appropriate notation for the genotypes.*

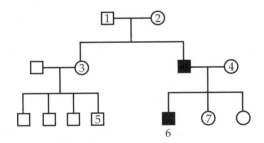

Genotypes:

1. _____   4. _____   7. _____

2. _____   5. _____

3. _____   6. _____

Errors and Exceptions in Chromosomal Inheritance

Alterations of chromosome number or structure cause some genetic disorders (279–282)

Alterations of Chromosome Number: Aneuploidy and Polyploidy   **Nondisjunction** occurs when a pair of homologous chromosomes does not separate properly in meiosis I or sister chromatids do not separate in meiosis II. As a result, a gamete receives either two or no copies of that chromosome. A zygote formed with one of these aberrant gametes has a chromosomal alteration known as **aneuploidy,** a nontypical number of chromosomes. The zygote will be either **trisomic**

for that chromosome (chromosome number is $2n + 1$) or **monosomic** ($2n - 1$). Aneuploid organisms usually have a set of symptoms caused by the abnormal dosage of genes. A mitotic nondisjunction early in embryonic development is likely to be harmful.

Polyploidy is a chromosomal alteration in which an organism has more than two complete chromosomal sets, as in triploid ($3n$) or tetraploid ($4n$) organisms. Polyploidy is common in the plant kingdom and has played an important role in the evolution of plants.

■ INTERACTIVE QUESTION 15.6

a. What is the difference between a trisomic and a triploid organism?

b. Which of these is likely to show the most deleterious effects of its chromosomal imbalance?

Alterations of Chromosome Structure   Chromosome breakage can result in chromosome fragments that are lost, called **deletion;** that join to a sister chromatid (during meiosis) or the homologous chromosome, called **duplication;** that join a nonhomologous chromosome, called **translocation;** or that rejoin the original chromosome in the reverse orientation, called **inversion.** Errors in crossing over can result in a deletion and duplication in nonsister chromatids, caused by nonequal exchange of chromatids.

A homozygous deletion is usually lethal. Duplications and translocations also are typically harmful. Even though all the genes are present in proper quantities in inversions and translocations, the phenotype may be altered due to the influence of neighboring genes on the expression of the relocated genes.

■ INTERACTIVE QUESTION 15.7

Two nonhomologous chromosomes have gene orders, respectively, of A-B-C-D-E-F-G-H-I-J and M-N-O-P-Q-R-S-T. What types of chromosome alterations would have occurred if daughter cells were found to have a gene sequence on the first chromosome of A-B-C-O-P-Q-G-J-I-H?

Human Disorders Due to Chromosomal Alterations
The frequency of aneuploid zygotes may be fairly high in humans, but development is usually so disrupted that the embryos spontaneously abort. Some genetic disorders, expressed as syndromes of characteristic traits, are the result of aneuploidy. Fetal testing can detect such disorders before birth.

Down syndrome, caused by trisomy of chromosome 21, results in characteristic facial features, short stature, heart defects, and mental retardation. The incidence of Down syndrome increases for older mothers.

XXY males exhibit Klinefelter syndrome, a condition in which the individual has abnormally small testes, is sterile, and is usually of normal intelligence.

Males with an extra Y chromosome do not exhibit any well-defined syndrome. Trisomy X results in females who are healthy and distinguishable only by karyotype. Monosomy X individuals (XO) exhibit Turner syndrome and are phenotypically female, sterile individuals with short stature and usually normal intelligence.

Structural alterations of chromosomes, such as deletions or translocations, may be associated with specific human disorders. The *cri du chat* syndrome is caused by a deletion in chromosome 5; chronic myelogenous leukemia is a cancer that is associated with a reciprocal chromosomal translocation.

■ INTERACTIVE QUESTION 15.8

What is an explanation for the observation that most sex chromosome aneuploidies have less deleterious effects than do autosomal aneuploidies?

The phenotypic effects of some mammalian genes depend on whether they were inherited from the mother or the father (imprinting) (282–283)

Some traits, including some genetic disorders, seem to depend on which parent supplied the alleles for the trait. A child with a deletion of a segment of chromosome 15 will exhibit Prader-Willi syndrome if the abnormal chromosome came from the father, or Angelman syndrome if the abnormal chromosome came from the mother. These phenomena provide evidence for **genomic imprinting,** a process in which certain genes are imprinted differently (either silenced or not), depending on the individual's sex. In the formation of gametes, chromosomes are reimprinted according to the sex of the individual. Most of the mammalian genes subject to imprinting identified so far are involved in embryonic development. The addition of methyl groups may inactivate the imprinted gene, assuring that the developing embryo has only one active copy.

Genomic imprinting may help to explain **fragile X syndrome,** a mental retardation more prevalent in males than females and seemingly linked to whether the fragile X chromosome was inherited from the mother.

Extranuclear genes exhibit a non-Mendelian pattern of inheritance (283)

Exceptions to Mendelian inheritance are found in the case of extranuclear genes located on small circles of DNA in mitochondria and plant plastids, which are transmitted to offspring in the cytoplasm of the ovum. Some rare human disorders are caused by mitochondrial mutations, and maternally inherited mitochondrial defects may contribute to diabetes, heart disease, and Alzheimer's disease.

WORD ROOTS

aneu- = without (*aneuploidy:* a chromosomal aberration in which certain chromosomes are present in extra copies or are deficient in number)

cyto- = cell (*cytological maps:* charts of chromosomes that locate genes with respect to chromosomal features)

hemo- = blood (*hemophilia:* a human genetic disease caused by a sex-linked recessive allele, characterized by excessive bleeding following injury)

mono- = one (*monosomic:* a chromosomal condition in which a particular cell has only one copy of a chromosome, instead of the normal two; the cell is said to be monosomic for that chromosome)

non- = not; **dis-** = separate (*nondisjunction:* an accident of meiosis or mitosis, in which both members of a pair of homologous chromosomes or both sister chromatids fail to move apart properly)

poly- = many (*polyploidy:* a chromosomal alteration in which the organism possesses more than two complete chromosome sets)

re- = again; **com-** = together; **bin-** = two at a time (*recombinant:* an offspring whose phenotype differs from that of the parent)

trans- = across (*translocation:* attachment of a chromosomal fragment to a nonhomologous chromosome)

tri- = three; **soma-** = body (*trisomic:* a chromosomal condition in which a particular cell has an extra copy of one chromosome, instead of the normal two; the cell is said to be trisomic for that chromosome)

STRUCTURE YOUR KNOWLEDGE

1. Mendel's law of independent assortment applies only to genes that are on different chromosomes. However, two of the genes Mendel studied were actually located on the same chromosome. Explain why genes located more than 50 map units apart behave as though they are not linked. How can one determine whether these genes are linked and what the relative distance is between them?

2. You have found a new mutant phenotype in fruit flies that you suspect is recessive and sex-linked. What is the single, best cross you could make to confirm your predictions?

3. Various human disorders or syndromes are related to chromosomal abnormalities. What explanation can you give for the adverse phenotypic effects associated with these chromosomal alterations?

GENETICS PROBLEMS

1. The following pedigree traces the inheritance of a genetic trait.

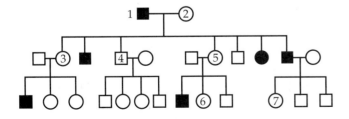

   a. What type of inheritance does this trait show?

b. Give the predicted genotype for the following individuals:

   1. _____   5. _____
   2. _____   6. _____
   3. _____   7. _____
   4. _____

   c. What is the probability that a child of individual #6 and a phenotypically normal male will have this trait?

2. The following recombination frequencies were found. Determine the order of these genes on the chromosome.

   a, c   10%   b, c   24%
   a, d   30%   b, d   16%

3. In guinea pigs, black (*B*) is dominant to brown (*b*), and solid color (*S*) is dominant to spotted (*s*). A heterozygous black, solid-colored pig is mated with a brown, spotted pig. The total offspring for several litters are black solid = 16, black spotted = 5, brown solid = 5, and brown spotted = 14. Are these genes linked or nonlinked? If they are linked, how many map units are they apart?

4. A woman is a carrier for a sex-linked lethal gene that causes spontaneous abortions. She has nine children. How many of these children do you expect to be boys?

5. A dominant sex-linked gene *B* produces white bars on black chickens, as seen in the Barred Plymouth Rock breed. A clutch of chicks has equal numbers of black and barred chicks. (Remember that sex is determined by the Z-W system in birds: ZZ are males, ZW are females.)

   a. If only the females are found to be black, what were the genotypes of the parents?

   b. If males and females are evenly represented in the black and barred chicks, what were the genotypes of the parents?

TEST YOUR KNOWLEDGE

MULTIPLE CHOICE: *Choose the one best answer.*

1. The chromosomal theory of inheritance states that
   a. genes are located on chromosomes.
   b. chromosomes and their associated genes undergo segregation during meiosis.
   c. chromosomes and their associated genes undergo independent assortment in gamete formation.
   d. Mendel's laws of inheritance relate to the behavior of chromosomes in meiosis.
   e. all of the above are correct.

2. A wild type is
   a. the phenotype found most commonly in nature.
   b. the dominant allele.
   c. designated by a small letter if it is recessive or a capital letter if it is dominant.
   d. a trait found on the X chromosome.
   e. your basic party animal.

3. Sex-linked traits
   a. are carried on an autosome but expressed only in males.
   b. are coded for by genes located on a sex chromosome.
   c. are found in only one or the other sex, depending on the sex-determination system of the species.
   d. are always inherited from the mother in mammals and fruit flies.
   e. depend on whether the gene was inherited from the mother or the father.

4. Linkage and cytological maps for the same chromosome
   a. are both based on mutant phenotypes and recombination data.
   b. may have different sequences of genes.
   c. have both the same sequence of genes and intergenic distances.
   d. have the same sequence of genes but different intergenic distances.
   e. are created using chromosomal abnormalities.

5. The genetic event that results in Turner syndrome (XO) is probably
   a. nondisjunction.
   b. deletion.
   c. parental imprinting.
   d. monoploidy.
   e. independent assortment.

6. A $1:1:1:1$ ratio of offspring from a dihybrid testcross indicates that
   a. the genes are linked.
   b. the dominant organism was homozygous.
   c. crossing over has occurred.
   d. the genes are 25 map units apart.
   e. the genes are not linked.

7. Genes *A* and *B* are linked 12 map units apart. A heterozygous individual, whose parents were *AAbb* and *aaBB*, would be expected to produce gametes in the following frequencies:
   a. 44% *AB*   6% *Ab*   6% *aB*   44% *ab*
   b. 6% *AB*   44% *Ab*   44% *aB*   6% *ab*
   c. 12% *AB*   38% *Ab*   38% *aB*   12% *ab*
   d. 6% *AB*   6% *Ab*   44% *aB*   44% *ab*
   e. 38% *AB*   12% *Ab*   12% *aB*   38% *ab*

8. A female tortoiseshell cat is heterozygous for the gene that determines black or orange coat color, which is located on the *X* chromosome. A male tortoiseshell cat
   a. is hemizygous at this locus.
   b. must have had a tortoiseshell mother.
   c. must have resulted from a nondisjunction and has a Barr body in his cells.
   d. must have three alleles for coat color, one from his father and two from his mother.
   e. would be hermaphroditic.

9. A color-blind son inherited this trait from his
   a. mother.
   b. father.
   c. mother only if she is color-blind.
   d. father only if he is color-blind.
   e. mother only if she is not color-blind.

10. Genomic imprinting
    a. explains cases in which the phenotypic effect of an allele depends on the gender of the parent from whom that allele is inherited.
    b. may involve the silencing of an allele by methylation so that offspring inherit only one active copy of a gene.
    c. is greatest in females because of the larger maternal contribution of cytoplasm.
    d. is more likely to occur in offspring of older mothers.
    e. involves both a and b.

11. A cross of a wild-type red-eyed female *Drosophila* with a violet-eyed male produces all red-eyed offspring. If the gene is sex-linked, what should the reciprocal cross (violet-eyed female × red-eyed male) produce? (Assume that the red allele is dominant to the violet allele.)
    a. all violet-eyed flies
    b. 3 red-eyed flies to 1 violet-eyed
    c. a $1:1$ ratio of red and violet eyes in both males and females
    d. red-eyed females and violet-eyed males
    e. all red-eyed flies

12. Which of the following chromosomal alterations does not alter genic balance but may alter phenotype because of differences in gene expression?
    a. deletion
    b. inversion
    c. duplication
    d. nondisjunction
    e. genomic imprinting

13. Two true-breeding *Drosophila* are crossed: a normal-winged, red-eyed female and a miniature-winged, vermilion-eyed male. The F$_1$ all have normal wings

and red eyes. F_1 offspring are crossed with miniature-winged, vermilion-eyed flies. The following offspring of that cross were counted:

233 normal wing, red eye

247 miniature wing, vermilion eye

7 normal wing, vermilion eye

13 miniature wing, red eye

From these results you could conclude that the alleles for miniature wings and vermilion eyes are

a. both X-linked and dominant.

b. located on autosomes and dominant.

c. recessive, and these genes are located 4 map units apart.

d. recessive, and these genes are located 20 map units apart.

e. recessive, and the deviation from the expected 9:3:3:1 ratio is due to epistasis.

14. Tests of social behavior given to Turner's syndrome volunteers (who are XO) found a correlation between scores on "behavioral inhibition" tasks and the source of the lone X chromosome. These test results appear to be an example of

a. parental imprinting of a gene on the X chromosome.

b. Y-linked inheritance.

c. meiotic nondisjunction.

d. chromosomal reciprocal translocation.

e. fragile X syndrome.

15. Suppose that a sex-linked character for wing shape in *Drosophila* shows incomplete dominance. The X^+ allele codes for pointed wings, the X^r for round wings, and $X^+ X^r$ individuals have oval wings. In a cross between an oval-winged female and a round-winged male, the following offspring were observed: oval-winged females, round-winged females, pointed-winged males, and round-winged males. A rare pointed-winged female was noted. What could account for this unusual offspring?

a. a crossover between the two X chromosomes

b. a crossover between the X and Y chromosomes

c. a nondisjunction in meiosis II between two X^+ chromatids

d. a nondisjunction in meiosis I between the X^+ and X^r chromosomes

e. an X^+O female formed when an X^+ ovum was fertilized by a sperm in which there was no sex chromosome due to a nondisjunction

16. What is the function of the gene *SRY*?

a. codes for an RNA molecule that coats the X chromosome and initiates X-inactivation

b. produces the fragile X syndrome when it is mutated

c. codes for a protein that regulates genes that control the development of ovaries

d. located on the Y chromosome and codes for a protein that regulates genes that control development of testes

e. located on the Y chromosome and required for production of normal sperm

17. A mutation in a mitochondrial gene has been linked to a rare muscle-wasting disease. This disease is

a. found more often in males than females.

b. found more often in females than males.

c. inherited in a simple Mendelian fashion.

d. caused by a translocation of a nuclear gene.

e. inherited from the mother.

18. In which of the following would you expect to find a Barr body?

a. an ovum

b. a sperm

c. a liver cell of a man

d. a liver cell of a woman

e. a mitochondrion

19. Fragile X syndrome, the most common form of mental retardation with a genetic basis, is more prevalent in males than in females. One of the factors involved in the inheritance of this syndrome is

a. sex-linked inheritance of a dominant allele on the X chromosome.

b. mitochondrial inheritance of a mutant allele from the mother.

c. genomic imprinting by the mother.

d. the formation of Barr bodies in the cells of female offspring.

e. the silencing of the mutant allele by *XIST* RNA in female offspring.

20. Which of the following statements is *not* true about genetic recombination:

a. Recombination of linked genes occurs by crossing over.

b. Recombination of unlinked genes occurs by independent assortment of chromosomes.

c. Genetic recombination results in offspring with combinations of traits that differ from the phenotypes of both parents.

d. Recombinant offspring outnumber parental-type offspring when two genes are 50 map units apart on a chromosome.

e. The number of recombinant offspring is proportional to the distance between two gene loci on a chromosome.

CHAPTER 16

THE MOLECULAR BASIS
OF INHERITANCE

FRAMEWORK

This chapter outlines the key evidence that was gathered to establish DNA as the molecular basis of inheritance. Watson and Crick's double helix, with its rungs of specifically paired nitrogenous bases and twisting side ropes of phosphate and sugar groups, provided the three-dimensional model that explained DNA's ability to encode a great variety of information and produce exact copies of itself through semiconservative replication. The replication of DNA is an extremely fast and accurate process involving many enzymes and proteins.

CHAPTER REVIEW

Deoxyribonucleic acid, or DNA, is the genetic material, the substance of genes, the basis of heredity. Nucleic acids' unique ability to direct their own replication allows for the precise copying and transmission of DNA to all the cells in the body and from one generation to the next. DNA encodes the blueprints that direct and control the biochemical, anatomical, physiological, and behavioral traits of organisms.

DNA as the Genetic Material

The search for the genetic material led to DNA: *the process of science* (287–290)

Chromosomes were shown to carry hereditary information and to consist of proteins and DNA. Until the 1940s, most scientists believed that proteins were the genetic material because of their specificity and heterogeneity. The role of DNA in heredity was first established through work with microorganisms—bacteria and viruses.

Evidence That DNA Can Transform Bacteria   In 1928, Griffith was working with two strains of *Streptococcus pneumoniae.* When he mixed the remains of heat-killed pathogenic bacteria with harmless bacteria, some bacteria were changed into disease-causing bacteria. These bacteria had incorporated external genetic material in a process called **transformation.**

Avery worked for over a decade to identify the transforming agent by purifying chemicals from heat-killed pathogenic cells. In 1944, Avery, McCarty, and MacLeod announced that DNA was the molecule that transformed the bacteria.

Evidence That Viral DNA Can Program Cells   Viruses consist of little more than DNA, or sometimes RNA, contained in a protein coat. They reproduce by infecting a cell and commandeering that cell's metabolic machinery. **Bacteriophages,** or **phages,** are viruses that infect bacteria. In 1952, Hershey and Chase showed that DNA was the genetic material of a phage known as T2 that infects the bacterium *Escherichia coli (E. coli).*

Additional Evidence That DNA Is the Genetic Material of Cells   Circumstantial evidence that DNA is the genetic material came from the observation that a eukaryotic cell doubles its DNA content prior to mitosis and that diploid cells have twice as much DNA as haploid gametes of the same organism.

Chargaff, in 1947, reported that the ratio of nitrogenous bases in the DNA from various organisms was species specific. Chargaff also determined that the number of adenines and thymines was approximately equal, and the number of guanines and cytosines was also equal in the DNA from all the organisms he studied. The A=T and G=C properties of DNA became known as Chargaff's Rules.

■ **INTERACTIVE QUESTION 16.1**

Hershey and Chase devised an experiment using radioactive isotopes to determine whether the phage's DNA or protein was transferred to the bacteria.

a. How did they label phage protein?

b. How did they label phage DNA?

Separate samples of *E. coli* were infected with the differently labeled T2 cells, then blended and centrifuged to isolate the bacterial cells from the lighter viral particles.

c. Where was the radioactivity found in the samples with labeled phage protein?

d. In the samples with labeled phage DNA?

e. What did Hershey and Chase conclude from these results?

Watson and Crick discovered the double helix by building models to conform to X-ray data: *the process of science* (290–292)

By the early 1950s, the arrangement of covalent bonds in a nucleic acid polymer was established, but the three-dimensional structure of DNA was yet to be determined.

An X-ray beam passed through a substance can expose photographic film to produce a pattern of spots that a crystallographer interprets as information about three-dimensional atomic structure. Watson saw an X-ray photo produced by Franklin that indicated the helical shape of DNA. He deduced that the helix had a width of 2 nm, suggesting that the helix consisted of two strands, thus the term **double helix.**

Watson and Crick constructed wire models to build a double helix that would conform to the X-ray measurements and the known chemistry of DNA. They finally arrived at a model that paired the nitrogenous bases on the inside of the helix with the sugar-phosphate chains on the outside. The helix makes one full turn every 3.4 nm; thus ten layers of nucleotide pairs, stacked 0.34 nm apart, are present in each turn of the helix.

To produce the molecule's uniform 2-nm width, a purine base must pair with a pyrimidine. The molecular arrangements of the side groups of the bases permit two hydrogen bonds to form between adenine and thymine and three hydrogen bonds between guanine and cytosine. This complementary pairing explains Chargaff's Rules. Van der Waals attractions between the closely stacked bases help to hold the molecule together.

In 1953, Watson and Crick published a paper in *Nature* reporting the double helix as the molecular model for DNA. (See Interactive Question 16.2, p. 118.)

DNA Replication and Repair

During DNA replication, base pairing enables existing DNA strands to serve as templates for new complementary strands (292–294)

Watson and Crick noted that the base-pairing rule of DNA sets up a mechanism for its replication. Each side of the double helix is an exact complement to the other. When the two sides of a DNA molecule separate in the replication process, each strand serves as a template for rebuilding a double-stranded molecule.

The **semiconservative model** of DNA replication predicts that the two daughter DNA molecules each have one parental strand and one newly formed strand. In contrast, a conservative model predicts that the parent double helix remains intact and the duplicated molecule is totally new, whereas a dispersive model predicts that all four strands of the two DNA molecules are a mixture of parental and new DNA.

Meselson and Stahl tested these models by growing *E. coli* in a medium with ^{15}N, a heavy isotope that the bacteria incorporated into their nitrogenous bases. Cells with labeled DNA were transferred to a medium with the lighter isotope, ^{14}N. Samples were removed after one and two generations of bacterial growth, and the DNA was extracted and centrifuged. The resulting locations of the density bands in the centrifuge tubes confirmed the semiconservative model of DNA replication.

■ INTERACTIVE QUESTION 16.2

Review the structure of DNA by labeling the following diagrams.

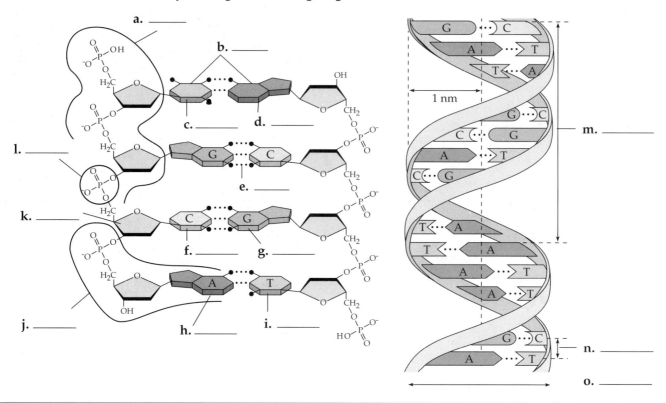

■ INTERACTIVE QUESTION 16.3

Using different colors for heavy (parental) and light (new) strands of DNA, sketch the results of two replication cycles when *E. coli* were moved from medium with ^{15}N to ^{14}N medium. Show the resulting density bands in the centrifuge tubes.

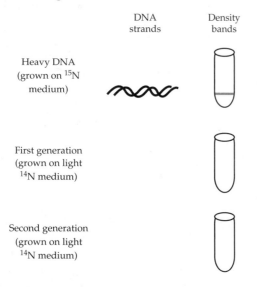

A large team of enzymes and other proteins carries out DNA replication (295–298)

Getting Started: Origins of Replication   Replication begins at special sites, called **origins of replication,** where proteins that initiate replication bind to a specific sequence of nucleotides and separate the two strands to form a replication "bubble." Replication proceeds in both directions in the Y-shaped **replication forks.**

Elongating a New DNA Strand   Enzymes called **DNA polymerases** connect nucleotides to the growing end of the new DNA strand. A nucleoside triphosphate lines up with its complementary base on the template strand; it loses two phosphate groups and the hydrolysis of this pyrophosphate to two inorganic phosphates ($\textcircled{P}_i$) provides the energy for polymerization.

The Antiparallel Arrangement of the DNA Strands   The two strands of a DNA molecule are antiparallel; their sugar-phosphate backbones run in opposite directions. The deoxyribose sugar of each nucleotide is connected to its own phosphate group at its 5′ carbon and connects to the phosphate group of the adjacent nucleotide by its 3′ carbon. Thus a strand of DNA has polarity, with a 5′ end where the first nucleotide's phosphate group is exposed, and a 3′ end where the nucleotide at the other end has a hydroxyl group attached to its 3′ carbon.

■ INTERACTIVE QUESTION 16.4

Look back to Interactive Question 16.2 and label the 5′ and 3′ ends of both strands of the DNA molecule.

DNA polymerases add nucleotides only to the 3′ end of a growing strand; DNA is replicated in a 5′ → 3′ direction. Because the DNA strands run in an antiparallel direction, the simultaneous synthesis of both strands presents a problem. The **leading strand** is the new 5′ → 3′ strand being formed along the template by a DNA polymerase in the progressing replication fork. The **lagging strand** is created as a series of short segments, called Okazaki fragments, that are formed in the 5′ → 3′ direction away from the replication fork. An enzyme called **DNA ligase** joins the sugar-phosphate backbones of the fragments.

Priming DNA Synthesis DNA polymerase cannot initiate synthesis of a DNA strand; it can only add nucleotides to an existing chain that is attached to a template strand. An enzyme called **primase** joins about 10 RNA nucleotides to form the **primer** needed to start the chain. Each fragment on the lagging strand requires a primer. A continuous strand of DNA is produced after another DNA polymerase replaces the RNA primer with DNA nucleotides and ligase joins the fragments.

Other Proteins Assisting DNA Replication Enzymes called **helicases** unwind the helix and separate the parent strand at replication forks, and **single-strand binding proteins** keep the separated strands apart.

■ INTERACTIVE QUESTION 16.5

In this diagram showing the replication of DNA, label the following items: leading and lagging strands, Okazaki fragment, DNA polymerase, DNA ligase, helicase, primase, single-strand binding proteins, RNA primer, replication fork, and 5′ and 3′ ends of parental DNA.

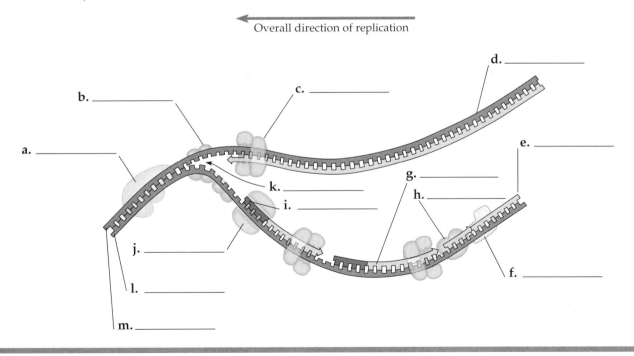

Overall direction of replication

b. _____ c. _____ d. _____

a. _____ e. _____

g. _____ h. _____

k. _____ i. _____

j. _____ f. _____

l. _____

m. _____

The DNA Replication Machine as a Stationary Complex The various proteins that function in DNA replication form a large complex. In eukaryotic cells, many such complexes may anchor to the nuclear matrix and the DNA polymerase molecules may pull the parental DNA through them.

Enzymes proofread DNA during its replication and repair damage in existing DNA (299)

Initial pairing errors in nucleotide placement may occur as often as 1 per 10,000 base pairs. The amazing accuracy of DNA replication is achieved by proofreading and correcting pairing errors. DNA polymerase checks each newly added nucleotide against its template and replaces incorrect nucleotides. Other enzymes also fix incorrectly paired nucleotides, called **mismatch repair.**

DNA molecules may be altered by reactive chemicals, radioactive emissions, X-rays, and UV light. These changes, or mutations, may be corrected by many different types of DNA repair enzymes. In **nucleotide excision repair,** the damaged strand is cut out by a **nuclease** and the gap is correctly filled through the action of DNA polymerase and ligase. In skin cells, nucleotide excision repair frequently corrects thymine dimers caused by ultraviolet rays of sunlight.

The ends of DNA molecules are replicated by a special mechanism (299–301)

Because DNA polymerase cannot attach nucleotides to the 5' end of a daughter DNA strand, repeated replications cause a progressive shortening of linear DNA molecules. Multiple repetitions of a short nucleotide sequence at the ends of chromosomes, called **telomeres,** protect an organism's genes from being shortened during successive DNA replications.

The shortening of telomeres may limit cell division. In germ line cells, however, the enzyme **telomerase,** which contains an RNA template for the telomere sequence, lengthens telomeres. Some somatic cancer cells and "immortal" strains of cultured cells produce telomerase and are thus capable of unregulated cell division.

WORD ROOTS

helic- = a spiral (*helicase:* an enzyme that untwists the double helix of DNA at the replication forks)

liga- = bound or tied (*DNA ligase:* a linking enzyme for DNA replication)

-phage = to eat (*bacteriophages:* viruses that infect bacteria)

semi- = half (*semiconservative model:* type of DNA replication in which the replicated double helix consists of one old strand, derived from the old molecule, and one newly made strand)

telos- = an end (*telomere:* the protective structure at each end of a eukaryotic chromosome)

trans- = across (*transformation:* a phenomenon in which external DNA is assimilated by a cell)

STRUCTURE YOUR KNOWLEDGE

1. Summarize the evidence and techniques Watson and Crick used to deduce the double-helix structure of DNA.

2. Review your understanding of DNA replication by describing the key enzymes and proteins (in the order of their functioning) that direct replication.

TEST YOUR KNOWLEDGE

MULTIPLE CHOICE: *Choose the one best answer.*

1. One of the reasons most scientists believed proteins were the carriers of genetic information was that
   a. proteins were more heat stable than nucleic acids.
   b. the protein content of duplicating cells always doubled prior to division.
   c. proteins were much more complex and heterogeneous molecules than nucleic acids.
   d. early experimental evidence pointed to proteins as the hereditary material.
   e. proteins were found in DNA.

2. Transformation involves
   a. the uptake of external genetic material, often from one bacterial strain to another.
   b. the creation of a strand of RNA from a DNA molecule.
   c. the infection of bacterial cells by phage.
   d. the type of semiconservative replication shown by DNA.
   e. the replication of DNA along the lagging strand.

3. The DNA of an organism has thymine as 20% of its bases. What percentage of its bases would be guanine?
   a. 20%   c. 40%   e. 80%
   b. 30%   d. 60%

4. In his work with pneumonia-causing bacteria, Griffith found that
   a. DNA was the transforming agent.
   b. the pathogenic and harmless strains mated.
   c. heat-killed harmless cells could cause pneumonia when mixed with heat-killed pathogenic cells.
   d. some heat-stable chemical was transferred to harmless cells to transform them into pathogenic cells.
   e. a T2 phage transformed harmless cells to pathogenic cells.

5. When T2 phages are grown with radioactive sulfur,
   a. their DNA is tagged.
   b. their proteins are tagged.
   c. their DNA is found to be of medium density in a centrifuge tube.
   d. they transfer their radioactivity to *E. coli* chromosomes when they infect the bacteria.
   e. their excision enzymes repair the damage caused by the radiation.

6. Meselson and Stahl
   a. provided evidence for the semiconservative model of DNA replication.
   b. were able to separate phage protein coats from *E. coli* by using a blender.
   c. found that DNA labeled with ^{15}N was of intermediate density.
   d. grew *E. coli* on labeled phosphorus and sulfur.
   e. found that DNA composition was species specific.

7. Watson and Crick concluded that each base could not pair with itself because
   a. there would not be room for the helix to make a full turn every 3.4 nm.
   b. the uniform width of 2 nm would not permit two purines or two pyrimidines to pair together.
   c. the bases could not be stacked 0.34 nm apart.
   d. identical bases could not hydrogen-bond together.
   e. they would be on antiparallel strands.

8. The joining of nucleotides in the polymerization of DNA requires energy from
   a. DNA polymerase.
   b. the hydrolysis of the terminal phosphate group of ATP.
   c. RNA nucleotides.
   d. the hydrolysis of GTP.
   e. the hydrolysis of the pyrophosphates removed from nucleoside triphosphates.

9. Continuous elongation of a new DNA molecule along one strand of DNA
   a. requires the action of DNA ligase as well as polymerase.
   b. occurs because DNA ligase can only elongate in the 5′ → 3′ direction.
   c. occurs on the leading strand.
   d. occurs on the lagging strand.
   e. a, b, and c are correct.

10. Which of the following statements about DNA polymerase is *incorrect?*
    a. It joins complementary base pairs to each other.
    b. It is able to proofread and correct for errors in base pairing.
    c. It is unable to initiate synthesis; it requires an RNA primer.
    d. It only works in the 5′ → 3′ direction.
    e. It is found in eukaryotes and prokaryotes.

11. Thymine dimers, covalent links between adjacent thymine bases in DNA, may be induced by UV light. When they occur, they are repaired by
    a. excision enzymes (nucleases).
    b. DNA polymerase.
    c. ligase.
    d. primase.
    e. a, b, and c are all needed.

12. How does DNA synthesis along the lagging strand differ from that on the leading strand?
    a. Nucleotides are added to the 5′ end instead of the 3′ end.
    b. Ligase is the enzyme that polymerizes DNA on the lagging strand.
    c. An RNA primer is needed on the lagging strand but not on the leading strand.
    d. Okazaki fragments, which each grow 5′ → 3′, must be joined along the lagging strand.
    e. Helicase synthesizes Okazaki fragments, which are then joined by ligase.

13. Which of the following enzymes or proteins is paired with an incorrect or inaccurate function?
    a. Helicase—unwind and separate parental double helix
    b Telomerase—add telomere repetitions to end of chromosomes
    c. Single-strand binding protein—hold strands of unwound DNA apart and straight
    d. Nuclease—cut out (excise) damaged DNA strand
    e. Primase—form DNA primer to start replication

Use the following diagram to answer Questions 14 through 17.

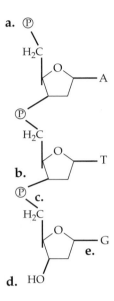

14. Which letter indicates the 5′ end of this single DNA strand?
    a.   b.   c.   d.   e.

15. At which letter would the next nucleotide be added?
    a.   b.   c.   d.   e.

16. Which letter indicates a phosphodiester bond formed by DNA polymerase?
    a.   b.   c.   d.   e.

17. The base sequence of the DNA strand made from this template would be (from top to bottom)
    a. A T C.
    b. C G A.
    c. T A C.
    d. U A C.
    e. A T G.

18. T2 phage is grown with radioactive phosphorus and then allowed to infect *E. coli*. The culture is blended to separate the viral coats from the bacterial cells and centrifuged. Which of the following best describes the expected results of such an experiment?
    a. Both viral and bacterial DNA are labeled; radioactivity is found in the supernatant.
    b. Both viral and bacterial proteins are labeled; radioactivity is present in both the supernatant and the pellet.

    c. Viral proteins are labeled; radioactivity is found in the supernatant but not in the pellet.
    d. Viral DNA is labeled; radioactivity is found in the pellet.
    e. The virus destroyed the bacteria; no pellet is formed.

19. What are telomeres and what do they do?
    a. ever-shortening tips of chromosomes that may signal cells to stop dividing at maturity
    b. highly repetitive sequences at tips of chromosomes that protect the lagging strand during replication
    c. repetitive sequences of nucleotides at the centromere region of a chromosome
    d. enzymes that are present in germ-line cells that allow these cells to undergo repeated divisions
    e. Both a and b are correct.

20. You are trying to support your hypothesis that DNA replication is *conservative;* i.e., parental strands separate; complementary strands are made, but these new strands join together to make a new DNA molecule and the parental strands rejoin. You take *E. coli* that had grown in a medium containing only heavy nitrogen (^{15}N) and transfer a sample to a medium containing light nitrogen (^{14}N). After allowing time for only one DNA replication, you centrifuge a sample and compare the density band(s) formed with control bands for bacteria grown on either normal ^{14}N or ^{15}N medium. Which band location would support your hypothesis of *conservative* DNA replication?

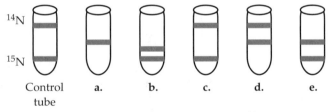

Bands after one replication on ^{14}N

21. Using the experiment explained in question 20, which centrifuge tube would represent the band distribution obtained after one replication showing that DNA replication is *semiconservative?*

CHAPTER 17

FROM GENE TO PROTEIN

FRAMEWORK

This chapter deals with the pathway from DNA to RNA to proteins. The triplet code instructions of DNA are transcribed to a sequence of codons in mRNA. In eukaryotes, mRNA is processed before it leaves the nucleus. Complexed with ribosomes, mRNA is translated into a sequence of amino acids in a polypeptide as tRNAs match their anticodons to the mRNA codons. Mutations, which alter the base pairs in DNA, usually alter the protein product.

CHAPTER REVIEW

The information of DNA is contained in specific sequences of nucleotides. These sequences are translated into proteins, which are responsible for the specific traits of an organism and thus form the link between genotype and phenotype.

The Connection Between Genes and Proteins

The study of metabolic defects provided evidence that genes specify proteins: *the process of science* (303–304)

In 1909, Garrod first suggested that genes determine phenotype through the action of enzymes. Garrod reasoned that inherited diseases were caused by an inability to make certain enzymes.

How Genes Control Metabolism: One Gene–One Enzyme   Beadle and Tatum, working with mutants of a bread mold, *Neurospora crassa,* demonstrated the relationship between genes and enzymes. They studied several nutritional mutants that could not grow on the minimal medium that sufficed for wild-type mold. By growing nutritional mutants on complete growth medium and then transferring samples to various combinations of minimal medium and one added nutrient, Beadle and Tatum were able to identify the specific metabolic defect for each mutant.

Three classes of *Neurospora* mutants that were unable to synthesize arginine were identified by supplementing different precursors of the pathway. Beadle and Tatum reasoned that the metabolic pathway of each class was blocked at a different enzymatic step. They formulated the *one gene–one enzyme hypothesis* that the function of a gene is to control production of a specific enzyme.

One Gene–One Polypeptide   Biologists revised Beadle and Tatum's idea to one gene–one protein, because not all proteins are enzymes. Because many proteins consist of more than one polypeptide chain, each of which is codified by its own gene, the axiom is now the **one gene–one polypeptide hypothesis.**

Transcription and translation are the two main processes linking gene to protein: *an overview* (304–306)

RNA is the link between a gene and the protein for which it codes. RNA differs from DNA in three ways: the sugar component of its nucleotides is ribose, rather than deoxyribose; uracil (U) replaces thymine as one of its nitrogenous bases; and RNA is usually single stranded.

Transcription is the transfer of information from DNA to **messenger RNA (mRNA)** or another type of RNA, using the "language" of nucleic acids. **Translation** transfers information from mRNA to a polypeptide, changing from the language of nucleotides to that of amino acids.

In prokaryotes, which lack a nucleus, transcription of DNA to mRNA and translation of mRNA by ribosomes into protein occur almost simultaneously. In eukaryotes, the mRNA must exit the nucleus before translation can occur. **RNA processing,** the modification of *pre-mRNA* or the **primary transcript** within the nucleus, occurs only in eukaryotes.

■ INTERACTIVE QUESTION   17.1

Fill in the sequence in the synthesis of proteins. Put the name of the process above each arrow.

$$\underline{\hspace{5cm}} \rightarrow \underline{\hspace{4cm}} \rightarrow \underline{\hspace{4cm}}$$

In the genetic code, nucleotide triplets specify amino acids (306–308)

A sequence of three nucleotides provides 4^3 or 64 possible unique sequences of nucleotides, more than enough to code for the 20 amino acids. The translation of nucleotides into amino acids uses a **triplet code** to specify each amino acid.

The base triplets along the **template strand** of a gene are transcribed into mRNA **codons.** The same strand of a long DNA molecule can be the template strand for one gene and the complementary strand for another. The mRNA is complementary to the DNA template since its bases follow the same base-pairing rules, with the exception that uracil substitutes for thymine in RNA. Thus, the base triplet ATG is transcribed as the mRNA codon UAC. Codons are read in the $5' \rightarrow 3'$ direction. Each mRNA codon specifies one of the 20 amino acids to be sequenced in the polypeptide chain.

Cracking the Genetic Code: the process of science In the early 1960s, Nirenberg synthesized artificial mRNA by linking uracil RNA nucleotides. Adding this "poly U" to a test tube containing all the biochemical ingredients necessary for protein synthesis, he obtained a polypeptide containing the single amino acid phenylalanine. Molecular biologists had deciphered all 64 codons by the mid-1960s. Three codons function as stop signals, or termination codons. The codon AUG both codes for methionine and functions as an initiation codon, a start signal for translation.

The code is often redundant, meaning that more than one codon may specify a single amino acid. The code is never ambiguous; no codon specifies two different amino acids.

The nucleotide sequence on mRNA is read in the correct **reading frame,** starting at a start codon and reading each triplet sequentially.

■ INTERACTIVE QUESTION   17.2

Practice using the dictionary of the genetic code in your textbook. Determine the amino acid sequence for a polypeptide coded for by the following mRNA transcript (written $5' \rightarrow 3'$):

A U G C C U G A C U U U A A G U A G

The genetic code must have evolved very early in the history of life (308–309)

The genetic code of codons and their corresponding amino acids is almost universal. A bacterial cell can translate the genetic messages of human cells. The near universality of a common genetic language provides compelling evidence of the antiquity of the code and the evolutionary connection of all living organisms.

The Synthesis and Processing of RNA

Transcription is the DNA-directed synthesis of RNA: *a closer look* (309–311)

Transcription begins when an **RNA polymerase** separates the two strands of DNA and links RNA nucleotides that base-pair along the template strand to the 3' end of the growing RNA polymer. The **promoter** is the DNA sequence where RNA polymerase attaches and initiates transcription; the **terminator** is the sequence that signals the end of transcription. A **transcription unit** is the sequence of DNA that is transcribed into one RNA molecule.

Bacteria have one type of RNA polymerase. Eukaryotes have three types; the one that synthesizes mRNA is called RNA polymerase II.

RNA Polymerase Binding and Initiation of Transcription The promoter determines which DNA strand is used as the template and includes recognition sequences, such as the **TATA box** common in eukaryotes, upstream from the start point. In eukaryotes, **transcription factors** must first recognize and bind to the promoter before RNA polymerase II can attach. Other transcription factors then join to form an assembly called the **transcription initiation complex.**

Elongation of the RNA Strand RNA polymerase untwists the double helix, exposing DNA nucleotides for base pairing with RNA nucleotides, and joins the nucleotides to the 3' end of the growing polymer. The new RNA peels away from the DNA template and the DNA double helix re-forms. Several molecules of RNA polymerase may be transcribing simultaneously from a single gene, enabling a cell to produce large quantities of a protein.

Termination of Transcription Transcription ends after RNA polymerase transcribes the terminator. In eukaryotes, polymerase continues past the termination signal AAUAAA before it cuts loose the pre-mRNA.

Review the key steps of transcription in eukaryotes:

a.

b.

c.

Eukaryotic cells modify RNA after transcription (311–313)

Alteration of mRNA Ends A modified guanine nucleotide is attached to the 5′ end of a pre-mRNA, and a string of adenine nucleotides, called a **poly(A) tail,** is added to the 3′ end. The **5′ cap** serves as part of a recognition site for ribosomes, and both the 5′ cap and poly(A) tail protect the ends of the mRNA from hydrolytic enzymes. The poly(A) tail also facilitates transport of the mRNA from the nucleus and probably aids ribosome attachment. The cap and tail are attached to nontranslated leader and trailer segments.

Split Genes and RNA Splicing Long segments of noncoding base sequences, known as **introns** or intervening sequences, occur within the boundaries of eukaryotic genes. The remaining coding regions are called **exons,** since they are expressed in protein synthesis. (The leader and trailer RNA sequences are not translated.) A primary transcript is made of the gene, but introns are removed and exons joined before the mRNA leaves the nucleus, in a process called **RNA splicing.**

Signals for RNA splicing are sets of a few nucleotides at either end of each intron. *Small nuclear ribonucleoproteins (snRNPs),* composed of proteins and *small nuclear RNA (snRNA),* are components of a molecular complex called a **spliceosome.** The spliceosome snips an intron out of the RNA transcript and connects the adjoining exons. In addition to splice-site recognition and spliceosome assembly, a function of snRNA may be catalytic in intron removal. RNA molecules that act as enzymes are called **ribozymes.**

Ribozymes Several other schemes of RNA splicing have been identified for primary transcripts other than pre-mRNA. In the protozoan *Tetrahymena,* splicing occurs with the intron RNA catalyzing its own removal.

How does the mRNA that leaves the nucleus differ from the primary transcript pre-mRNA?

The Functional and Evolutionary Importance of Introns Some introns are involved in regulating gene activity, and splicing may help regulate the export of mRNA from the nucleus. **Alternative RNA splicing** allows some genes to produce different polypeptides. Exons may code for polypeptide **domains,** functional segments of a protein, such as binding and active sites. Introns may facilitate recombination of exons between different alleles or even between different genes to create novel proteins.

The Synthesis of Protein

Translation is the RNA-directed synthesis of a polypeptide: *a closer look* (313–320)

Transfer RNA (tRNA) molecules are specific for the amino acid they carry to the ribosomes. They each have a specific base triplet, called an **anticodon,** that base-pairs with a complementary codon on mRNA, thus assuring that amino acids are arranged in the sequence prescribed by the transcription from DNA.

The Structure and Function of Transfer RNA As with other RNAs, transfer RNA is transcribed in the nucleus of a eukaryote and moves into the cytoplasm where it can be used repeatedly. These single-stranded, short RNA molecules are arranged into a clover-leaf shape by hydrogen bonding between complementary base sequences and then folded into a three-dimensional, roughly L-shaped structure. The anticodon is at one end of the L; the 3′ end is the attachment site for its amino acid.

Sixty-one codons for amino acids can be read from mRNA, but there are only about 45 different tRNA molecules. A phenomenon known as **wobble** enables the third nucleotide of some tRNA anticodons to pair with more than one kind of base in the codon. Thus, one tRNA can recognize more than one mRNA codon, all of which code for the same amino acid carried by that tRNA. The modified base inosine (I) is in the third position on several tRNAs and can pair with U, C, or A.

■ INTERACTIVE QUESTION 17.5

Using some of the codons and the amino acids you identified in Interactive Question 17.2, fill in the following table.

DNA Triplet $3' \to 5'$	mRNA Codon $5' \to 3'$	Anticodon $3' \to 5'$	Amino Acid
			methionine
		GGA	
TTC			
	UAG		

Aminoacyl-tRNA Synthetases   Each amino acid has a specific **aminoacyl-tRNA synthetase** that attaches it to its appropriate tRNA molecule to create an aminoacyl tRNA. The hydrolysis of ATP drives this process.

Ribosomes   Ribosomes facilitate the specific pairing of tRNA anticodons with mRNA codons during protein synthesis. They consist of a large and a small subunit, each composed of proteins and a form of RNA called **ribosomal RNA (rRNA)**. Subunits are constructed in the nucleolus in eukaryotes. Prokaryotic ribosomes are smaller and differ enough in molecular composition that some antibiotics can inhibit them without affecting eukaryotic ribosomes.

A large and small subunit join to form a ribosome when they attach to an mRNA molecule. Ribosomes have a binding site for mRNA, a **P site** (peptidyl-tRNA site) that holds the tRNA carrying the growing polypeptide chain, an **A site** (aminoacyl-tRNA site) that holds the tRNA carrying the next amino acid, and an **E site** (exit site) from which discharged tRNAs leave the ribosome. The attachment of an amino acid to the carboxyl end of the growing polypeptide chain is catalyzed by the ribosome.

According to the recently determined detailed structure of the small and large subunits of bacterial ribosomes, the protein components are mostly on the outside and the rRNA is in the interior at the interface between the subunits and at the A and P sites. This location supports the hypothesis that rRNA performs a ribosome's most important functions.

Building a Polypeptide   The three stages of protein synthesis—**initiation, elongation,** and **termination**—all require the aid of protein "factors." The first two stages also require energy, which is provided by the hydrolysis of GTP (guanosine triphosphate).

The initiation stage begins as the small subunit of the ribosome binds to the leader segment of an mRNA. An initiator tRNA, carrying methionine, attaches to the start codon AUG on the mRNA. With the aid of proteins called *initiation factors* and the expenditure of a GTP, the large subunit of the ribosome attaches to the small one, forming a translation initiation complex. The initiator tRNA fits into the P site.

The addition of amino acids in the elongation stage involves several proteins called *elongation factors* and occurs in a three-step cycle: codon recognition, peptide bond formation, and translocation.

In the codon recognition step, an elongation factor brings the correct tRNA into the A binding site of the ribosome, where the anticodon hydrogen-bonds with the mRNA codon. This step requires energy from the hydrolysis of 2 GTP.

In the second step, an RNA molecule of the large subunit catalyzes the formation of a peptide bond between the carboxyl end of the polypeptide held in the P site and the amino acid in the A site. The polypeptide is now held by the amino acid in the A site.

The tRNA from the P site is moved to the E site and then leaves the ribosome. The tRNA carrying the growing polypeptide is now translocated to the P site, a process requiring energy from the hydrolysis of a GTP molecule. The next mRNA codon moves into the A site as the mRNA ratchets through the ribosome.

Termination occurs when a stop codon—UAA, UAG, or UGA—reaches the A site of the ribosome. A *release factor* binds to the stop codon and hydrolyzes the bond between the polypeptide and the tRNA in the P site, freeing the completed polypeptide. The two ribosomal subunits and other components dissociate.

Polyribosomes   An mRNA may be translated simultaneously by several ribosomes in strings called **polyribosomes.**

From Polypeptide to Functional Protein   During and following translation, a polypeptide folds spontaneously into its secondary and tertiary structure. Chaperone proteins often facilitate the correct folding. The protein may need to undergo *posttranslational modifications:* Amino acids may be chemically modified; one or more amino acids at the beginning of the chain may be enzymatically removed; segments of the polypeptide may be excised; or several polypeptides may associate into a quaternary structure.

■ INTERACTIVE QUESTION 17.6

In the following diagrams of polypeptide synthesis, name the stages (**1–4**), identify the components (**a–l**), and then briefly describe what happens in each stage.

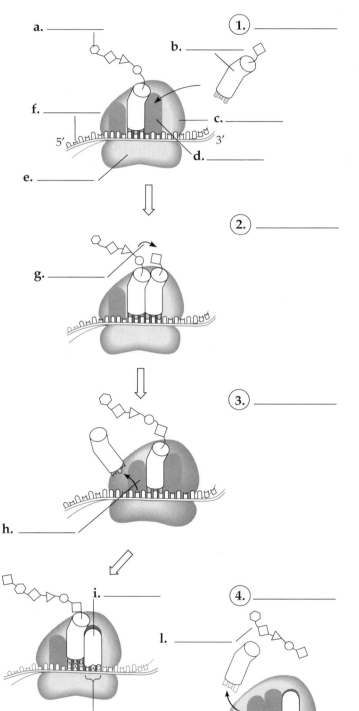

a. _____
b. _____
1. _____
f. _____
c. _____
5′ _____ 3′
d. _____
e. _____
2. _____
g. _____
3. _____
h. _____
i. _____
l. _____
j. _____
4. _____
k. _____

Signal peptides target some eukaryotic polypeptides to specific destinations in the cell (320)

All ribosomes are identical, whether they are free ribosomes that synthesize cytosolic proteins or ER-bound ribosomes that make membrane and secretory proteins. All protein synthesis begins in the cytoplasm. If a protein is destined for the endomembrane system or for secretion, its polypeptide chain will begin with a **signal peptide** that is recognized by a protein-RNA complex called a **signal-recognition particle (SRP),** which attaches the ribosome to a receptor protein that is part of a multiprotein complex on the ER membrane. As the growing polypeptide threads into the ER, the signal peptide is usually removed. Other signal peptides direct some proteins made in the cytosol to specific sites such as mitochondria, chloroplasts, or the interior of the nucleus.

■ INTERACTIVE QUESTION 17.7

What determines if a ribosome becomes bound to the ER?

RNA plays multiple roles in the cell: *a review* (321)

The versatility of RNA molecules stems from their ability to hydrogen-bond to other RNA or DNA molecules and to form specific three-dimensional shapes by forming intramolecular hydrogen bonds.

■ INTERACTIVE QUESTION 17.8

Fill in the functions for the following types of RNA molecules.

a. mRNA

b. tRNA

c. rRNA

d. snRNA

e. SRP RNA

Comparing protein synthesis in prokaryotes and eukaryotes: *a review* (321–322)

The basic processes of transcription and translation are the same in bacteria and eukaryotes, although the RNA polymerases and ribosomes differ. In bacteria, these processes occur almost simultaneously, whereas in eukaryotes, transcription is physically separated from translation by the nuclear envelope, allowing extensive RNA processing to occur before RNA leaves the nucleus. Eukaryotes also can target proteins to particular organelles.

Point mutations can affect protein structure and function (322–325)

Mutations, changes in the genetic information of a cell (or virus) may involve large portions of a chromosome or affect just one or a few nucleotides, as in a **point mutation.** If the mutation is in a gamete, it may be passed on to offspring.

Types of Point Mutations   A **base-pair substitution** replaces one nucleotide and its complementary partner with another pair of nucleotides. Due to the redundancy of the genetic code, some base-pair substitutions in the third nucleotide of a codon do not change the amino acid translation and are called *silent mutations.* A substitution may result in the insertion of a different amino acid without altering the character of the protein if the new amino acid has similar properties or is not located in a region crucial to that protein's function.

A base-pair substitution that results in a different amino acid in a critical portion of a protein, such as the active site of an enzyme, may significantly impair protein function.

A substitution that results in an incorrectly coded amino acid is called a **missense mutation. Nonsense mutations** occur when the point mutation changes an amino acid codon into a stop codon, prematurely halting the translation of the polypeptide chain and usually creating a nonfunctional protein.

Base-pair **insertions** or **deletions** that are not in multiples of three nucleotides alter the reading frame. All nucleotides downstream from the mutation will be improperly grouped into codons, creating extensive missense and usually prematurely ending in nonsense. These **frameshift mutations** almost always produce nonfunctional proteins.

Mutagens   Mutations can occur in a number of ways. *Spontaneous mutations* include base-pair substitutions, insertions, deletions, and longer mutations that occur during DNA replication, repair, or recombination. Physical agents, such as X-rays and UV light, and various chemical agents that cause mutations are called **mutagens.** Chemical mutagens include base analogs that substitute

for normal bases and then pair incorrectly in DNA synthesis, chemicals that insert into and distort the double helix, and other agents that chemically change DNA bases. Tests can measure the mutagenic effects of chemicals and thus their potential carcinogenic risk.

What is a gene? *Revisiting the question* (325)

Our definition of a gene has evolved from Mendel's heritable factors, to Morgan's loci along chromosomes, to the one gene–one polypeptide axiom.

Research continually refines our understanding of the structural and functional aspects of genes, which now include introns, promoters, and other regulatory regions. The best working definition of a gene is that it is a region of DNA whose final product is either a polypeptide or an RNA molecule.

■ INTERACTIVE QUESTION   17.9

Define the following, and explain what type of point mutation could cause each of these mutations.

a. silent mutation

b. missense mutation

c. nonsense mutation

d. frameshift mutation

WORD ROOTS

anti- = opposite (*anticodon:* a specialized base triplet on one end of a tRNA molecule that recognizes a particular complementary codon on an mRNA molecule)

exo- = out, outside, without (*exon:* a coding region of a eukaryotic gene that is expressed)

intro- = within (*intron:* a noncoding, intervening sequence within a eukaryotic gene)

muta- = change; **-gen** = producing (*mutagen:* a physical or chemical agent that causes mutations)

poly- = many (*poly(A) tail:* the modified end of the 3′ end of an mRNA molecule consisting of the addition of some 50 to 250 adenine nucleotides)

trans- = across; **-script** = write (*transcription:* the synthesis of RNA on a DNA template)

STRUCTURE YOUR KNOWLEDGE

1. Make sure you understand and can explain the processes of transcription and translation. You may find that filling in the table below in a study group helps you to review these processes.

2. What is the genetic code? Explain redundancy and the wobble phenomenon. What is the significance of the fact that the genetic code is nearly universal?

3. Prepare a concept map showing the types and consequences of point mutations.

TEST YOUR KNOWLEDGE

MULTIPLE CHOICE: *Choose the one best answer.*

1. In Beadle and Tatum's study of *Neurospora*, they were able to identify three classes of mutants that needed arginine added to minimal media in order to grow. The production of arginine includes the following steps: precursor → ornithine → citrulline → arginine. What nutrient(s) had to be supplied for the mutants with a defective enzyme for the precursor → ornithine step to grow?
   a. precursor only
   b. ornithine only
   c. citrulline only
   d. ornithine or citrulline
   e. precursor, ornithine, and citrulline

2. Transcription involves the transfer of information from
   a. DNA to RNA.
   b. RNA to DNA.
   c. mRNA to an amino acid sequence.
   d. DNA to an amino acid sequence.
   e. the nucleus to the cytoplasm.

3. If the 5′ → 3′ nucleotide sequence on the complementary (noncoding) DNA strand is CAT, what is the corresponding codon on mRNA?
   a. UAC
   b. CAU
   c. GUA
   d. GTA
   e. CAT

4. RNA polymerase
   a. is the protein responsible for the production of ribonucleotides.
   b. is the enzyme that creates hydrogen bonds between nucleotides on the DNA template strand and their complementary RNA nucleotides.
   c. is the enzyme that transcribes exons, but does not transcribe introns.
   d. is a ribozyme composed of snRNPs.
   e. begins transcription at a promoter sequence and moves along the template strand of DNA, elongating an RNA molecule in a 5′ → 3′ direction.

	Transcription	Translation
Template		
Location		
Molecules involved		
Enzymes involved		
Control—start and stop		
Product		
Product processing		
Energy source		

5. How is the template strand for a particular gene determined?
   a. It is the DNA strand that runs from the $5' \rightarrow 3'$ direction.
   b. It is the DNA strand that runs from the $3' \rightarrow 5'$ direction.
   c. It depends on the orientation of RNA polymerase, whose position is determined by particular sequences of nucleotides within the promoter.
   d. It doesn't matter which strand is the template because they are complementary and will produce the same mRNA.
   e. The template strand always contains the TATA box.

6. Which enzyme is responsible for the synthesis of tRNA?
   a. RNA replicase
   b. RNA polymerase
   c. aminoacyl-tRNA synthetase
   d. ribosomal enzymes
   e. ribozymes

7. Which of the following is true of RNA processing?
   a. Exons are excised before the mRNA is translated.
   b. The RNA transcript that leaves the nucleus may be much longer than the original transcript.
   c. Assemblies of protein and snRNPs, called spliceosomes, may catalyze splicing.
   d. Large quantities of rRNA are assembled into ribosomes.
   e. Signal peptides are added to the 5' end of the transcript.

8. Which of the following is *not* involved in the formation of a eukaryotic transcription initiation complex?
   a. TATA box
   b. transcription factors
   c. snRNA
   d. RNA polymerase II
   e. promoter

9. A prokaryotic gene 600 nucleotides long can code for a polypeptide chain of about how many amino acids?
   a. 100   c. 300   e. 1800
   b. 200   d. 600

10. All of the following are transcribed from DNA *except*
    a. exons.
    b. introns.
    c. tRNA.
    d. rRNA.
    e. promoter.

11. What might introns have to do with the evolution of new proteins?
    a. The excised introns may be transcribed and translated as new proteins by themselves.
    b. Introns are more likely to accumulate mutations than exons, and these mutations may result in the production of novel proteins.
    c. Introns that are self-excising may also function as hydrolytic enzymes for other nuclear processes.
    d. Introns provide more area where crossing over may occur (without interfering with the coding sequences) and thus increase the probability of exon shuffling between alleles.
    e. Introns often correspond to domains in proteins that fold independently and have specific functions. Changing domains between nonallelic genes could produce novel proteins.

12. A ribozyme is
    a. an exception to the one gene–one RNA molecule axiom.
    b. an enzyme that adds the 5' cap and poly(A) tail to mRNA.
    c. an example of rearrangement of protein domains caused by RNA splicing.
    d. an RNA molecule that functions as an enzyme.
    e. an enzyme that produces both small and large ribosomal subunits.

13. All of the following would be found in a prokaryotic cell *except*
    a. mRNA.
    b. rRNA.
    c. simultaneous transcription and translation.
    d. snRNA.
    e. RNA polymerase.

14. Which of the following is transcribed and then translated to form a protein product?
    a. gene for tRNA
    b. intron
    c. gene for a transcription factor
    d. leader and trailer
    e. gene for rRNA

15. Transfer RNA
    a. forms hydrogen bonds between its codon and the anticodon of an mRNA in the A site of a ribosome.
    b. binds to its specific amino acid in the active site of an aminoacyl-tRNA synthetase.
    c. uses GTP as the energy source to bind its amino acid.
    d. is translated from mRNA.
    e. is formed in the nucleolus.

16. Place the following events in the synthesis of a polypeptide in the proper order.
    1. A peptide bond forms.
    2. A tRNA matches its anticodon to the codon in the A site.
    3. A tRNA translocates from the A to the P site, and an unattached tRNA leaves the ribosome from the E site.
    4. The large subunit attaches to the small subunit and the initiator tRNA fits in the P site.
    5. A small subunit binds to an mRNA and an initiator tRNA.

    a. 4-5-3-2-1
    b. 4-5-2-1-3
    c. 5-4-3-2-1
    d. 5-4-1-2-3
    e. 5-4-2-1-3

17. Translocation involves
    a. the hydrolysis of a GTP molecule.
    b. the movement of the tRNA in the A site to the P site.
    c. the movement of the mRNA strand one triplet length in the A site.
    d. the release of the unattached tRNA from the E site.
    e. all of the above.

18. Which of the following catalyzes the formation of a peptide bond?
    a. RNA polymerase
    b. rRNA
    c. mRNA
    d. aminoacyl-tRNA synthetase
    e. protein ribosomal enzyme

19. Which of the following is *not* true of an anticodon?
    a. It consists of three nucleotides.
    b. It lines up in the $5' \rightarrow 3'$ direction along the $5' \rightarrow 3'$ mRNA strand.
    c. It extends from one end of a tRNA molecule.
    d. It may pair with more than one codon, especially if it has the base inosine in its third position.
    e. Its base uracil base-pairs with adenine.

20. Changes in a polypeptide following translation may involve
    a. the addition of sugars or lipids to certain amino acids.
    b. the action of enzymes to add amino acids at the beginning of the chain.

    c. the removal of poly(A) from the end of the chain.
    d. the addition of a 5′ cap of a modified guanosine residue.
    e. all of the above.

21. Several proteins may be produced at the same time from a single mRNA by
    a. the action of several ribosomes in a string called a polyribosome.
    b. several RNA polymerase molecules working sequentially.
    c. signal peptides that associate ribosomes with rough ER.
    d. containing several promoter regions.
    e. the involvement of multiple spliceosomes.

22. A signal peptide
    a. is most likely to be found on proteins produced by bacterial cells.
    b. directs an mRNA molecule into the cisternal space of the ER.
    c. is a sign to help bind the small ribosomal unit at the initiation codon.
    d. would be the first 20 or so amino acids of a protein destined for secretion from the cell.
    e. is part of the 5′ cap.

23. A base deletion early in the coding sequence of a gene may result in
    a. a nonsense mutation.
    b. a frameshift mutation.
    c. multiple missense mutations.
    d. a nonfunctional protein.
    e. all of the above.

24. Base-pair substitutions may have little effect on the resulting protein for all of the following reasons *except* which one?
    a. The redundancy of the code may result in a silent mutation.
    b. The substitution must involve three nucleotide pairs, or else the reading frame will be altered.
    c. The missense mutation may not occur in a critical part of the protein.
    d. The new amino acid may have similar properties to the replaced one.
    e. The wobble phenomenon could result in no change in translation.

CHAPTER 18

MICROBIAL MODELS: THE GENETICS OF VIRUSES AND BACTERIA

FRAMEWORK

A virus is an infectious particle consisting of a genome of single- or double-stranded DNA or RNA enclosed in a protein capsid. Viruses replicate using the metabolic machinery of their bacterial, animal, or plant host. Viral infections may destroy the host cell and cause diseases within the host organism. Viruses may have evolved from plasmids or transposons.

Bacteria have a circular chromosome and may have additional genes carried on plasmids. In conjunction with the genetic diversity generated by mutation, genetic recombination, and transposons, the short generation time of bacteria facilitates their adaptation to changing environments. Individual cells may alter their metabolic response to changing environmental conditions through feedback inhibition of enzyme activity and the regulation of gene expression.

CHAPTER REVIEW

The study of viruses and bacteria has provided information about the molecular biology of all organisms, an appreciation for the special genetic features of microbes, and an understanding of how viruses and bacteria cause disease, and it has led to the development of new, powerful techniques of manipulating genes that have had an immense impact on basic research and biotechnology.

The Genetics of Viruses

Researchers discovered viruses by studying a plant disease: *the process of science* **(328–329)**

The search for the cause of tobacco mosaic disease led to the discovery of viruses. No microbe could be identified, nor could the infectious agent be removed from infected sap by passing it through a filter designed to remove bacteria. Unlike bacteria, the infectious agent could not be cultivated on nutrient media nor was it

inactivated by alcohol. In 1935, Stanley crystallized the infectious particle, now known as tobacco mosaic virus (TMV). Since that time, many viruses have been seen with the electron microscope.

A virus is a genome enclosed in a protective coat (329–330)

Viral Genomes Viral genomes may be single- or double-stranded DNA or single- or double-stranded RNA. Viral genes are contained on a single linear or circular nucleic acid molecule.

Capsids and Envelopes The **capsid,** or protein shell, is built from a large number of often identical protein subunits (capsomeres) and may be rod-shaped (helical), polyhedral, or more complex in shape. **Viral envelopes,** derived from membranes of the host cell but also including viral proteins and glycoproteins, may cloak the capsids of viruses found in animals. Some viruses also contain a few viral enzymes.

Complex capsids are found among **bacteriophages,** or **phages,** the viruses that infect bacteria. Of the seven phages that infect the bacterium *E. coli*, T2, T4, and T6 have similar capsid structures consisting of a polyhedral head and a protein tail piece with tail fibers for attaching to a bacterium.

Viruses can reproduce only within a host cell: *an overview* **(330–331)**

Viruses are obligate intracellular parasites that lack metabolic enzymes and other equipment needed to express their genes and reproduce. Each virus type has a limited **host range** due to proteins on the outside of the virus that recognize only specific receptor molecules on the host cell surface.

Once the viral genome enters the host cell, the cell's enzymes, nucleotides, amino acids, ribosomes, ATP, and other resources are used to make copies of the viral genome and capsid proteins. DNA viruses use host DNA polymerases to copy their genome, whereas RNA viruses use virus-encoded polymerases for replicating their RNA genome.

After replication, viral nucleic acid and capsid proteins spontaneously assemble to form new virus particles within the host cell, a process called self-assembly. Hundreds or thousands of newly formed viruses are released, often destroying the host cell in the process.

Phages reproduce using lytic or lysogenic cycles (331–333)

The Lytic Cycle   A replication cycle of a virus that culminates in lysis of the host cell and release of newly produced phages is known as a **lytic cycle. Virulent phages** reproduce only by a lytic cycle.

The T4 phage uses its tail fibers to stick to a receptor site on the surface of an *E. coli* cell. The sheath of the tail contracts and thrusts its viral DNA into the cell, leaving the empty capsid behind. The *E. coli* cell begins to transcribe and translate phage genes, one of which codes for an enzyme that chops up host cell DNA. Nucleotides from the degraded bacterial DNA are used to create viral DNA. Capsid proteins are made and assembled into phage tails, tail fibers, and polyhedral heads. The viral components assemble into 100 to 200 phage particles that are released after a lysozyme is manufactured that digests the bacterial cell wall.

Mutations that change their receptor sites and restriction nucleases that chop up viral DNA once it enters the cell help to defend bacteria against viral infection.

The Lysogenic Cycle   In a **lysogenic cycle,** a virus reproduces its genome without killing its host. **Temperate phages** can reproduce by the lytic and lysogenic cycles.

When the phage lambda (λ) injects its DNA into an *E. coli* cell, it can begin a lytic cycle, or its DNA may be incorporated by genetic recombination into the host cell's chromosome and begin a lysogenic cycle as a **prophage.** Most of the genes of the inserted phage genome are repressed by a protein coded for by a prophage gene. Reproduction of the host cell replicates the phage DNA along with the bacterial DNA. The prophage can become virulent if it is excised from the bacterial chromosome, usually in response to environmental stimuli, and starts the lytic cycle.

■ INTERACTIVE QUESTION   18.1

In this diagram of a lytic and lysogenic cycle, describe steps **1–8** and label structures **a–e.**

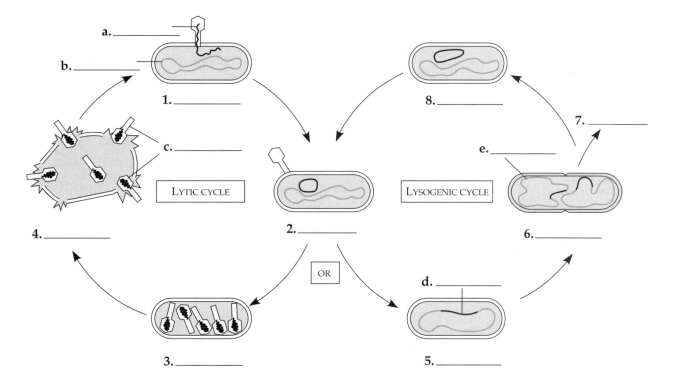

Expression of prophage genes may cause a change in the bacterial phenotype. Several disease-causing bacteria would be harmless except for the expression of prophage genes that code for toxins.

Animal viruses are diverse in their modes of infection and replication (333–338)

Reproductive Cycles of Animal Viruses   A viral envelope surrounds the capsid in many animal viruses. Glycoproteins extending from the viral membrane attach to receptor sites on a host cell plasma membrane, and the two membranes fuse, transporting the capsid into the cell. The viral genome replicates and directs the synthesis of viral proteins. Glycoproteins are deposited in patches in the plasma membrane, where new viruses bud off within an envelope derived from the host's plasma membrane.

The envelopes of herpesviruses come from the host nuclear membrane. The herpesvirus' double-stranded DNA can integrate into the cell's genome as a **provirus** and remain latent until it initiates herpes infections in times of stress.

RNA viruses infect animal hosts as well as plants and some bacteria. The single-stranded RNA of animal class IV viruses can serve directly as mRNA. The RNA genome of class V viruses must first be transcribed into a strand of complementary RNA (using a viral enzyme packaged inside the capsid) that then serves as mRNA and a template for making genome RNA.

In the complicated reproductive cycle of **retroviruses,** the viral RNA genome is transcribed into double-stranded DNA by a viral enzyme, **reverse transcriptase.** This viral DNA is then integrated into a chromosome, where it is transcribed by the host cell into viral RNA, which acts both as new viral genome and as mRNA for viral proteins. **HIV (human immunodeficiency virus)** is a retrovirus that causes **AIDS (acquired immunodeficiency syndrome).**

■ INTERACTIVE QUESTION   18.2

Summarize the flow of genetic information during replication of a retrovirus. Indicate the enzymes that catalyze this flow.

_____ → _____ → _____

Enzymes:

Causes and Prevention of Viral Diseases in Animals
The symptoms of a viral infection may be caused by toxins produced by infected cells, toxic components of the viruses themselves, cells killed or damaged by the virus, or the body's defense mechanisms fighting the infection.

Vaccines are variants or derivatives of pathogens that induce the immune system to react against the actual disease agent. In 1796, Jenner used the cowpox virus to vaccinate against smallpox. Vaccinations have greatly reduced the incidence of many viral diseases.

Unlike bacteria, viruses use the host's cellular machinery to replicate, and few drugs have been found to treat or cure viral infections. Some antiviral drugs interfere with viral nucleic acid synthesis.

Emerging Viruses   The emergence of "new" viral diseases may be linked to the mutation of an existing virus (as in influenza viruses), the spread from one host species to another (as in hantavirus), or the dissemination of an existing virus to a more widespread population (as in HIV).

Viruses and Cancer   Viruses that can cause cancer in animals are called *tumor viruses*. Cells become transformed into cancer cells when viral nucleic acid becomes integrated into host DNA. Viruses have been linked to certain types of human cancer.

Oncogenes are genes responsible for triggering cancerous transformation in cells. Many of these genes have been found not only in tumor viruses, but versions of them are also found within the genomes of normal cells. Some tumor viruses lack oncogenes but transform cells simply by turning on the cell's proto-oncogenes.

■ INTERACTIVE QUESTION   18.3

What do proto-oncogenes code for?

Plant viruses are serious agricultural pests (338)

Most plant viruses are RNA viruses. Plant viral diseases may spread through *vertical transmission* from a parent plant or through *horizontal transmission* from an external source. Plant injuries increase susceptibility to viral infections, and insects can act as carriers of viruses.

Viral particles spread easily through the plasmodesmata, the cytoplasmic connections between plant cells. Reducing the spread of disease and breeding resistant varieties are the best preventions of plant viral infections.

Viroids and prions are infectious agents even simpler than viruses (339)

Viroids are very small infectious molecules of naked circular RNA that can disrupt the metabolism of a plant cell and severely stunt plant growth.

Prions are protein infectious agents that may be linked to several degenerative brain diseases, such as "mad cow disease" and Creutzfeldt-Jacob disease in humans. Although proteins cannot reproduce, prions may spread disease by converting normal cellular proteins into the defective form of the prion.

Viruses may have evolved from other mobile genetic elements (339–340)

Viruses inhabit a gray area between life and nonlife, containing a genetic program that can be expressed only within a host cell.

The genomes of viruses are more similar to those of their host cells than to the genomes of viruses infecting other hosts. Viruses may have evolved from fragments of cellular nucleic acids that moved from one cell to another and eventually evolved special packaging. Sources of viral genomes may have been plasmids, self-replicating circles of DNA found in bacteria and yeast, and transposons, segments of DNA that can change locations within a cell's genome.

The Genetics of Bacteria

The short generation span of bacteria helps them adapt to changing environments (340–341)

The circular bacterial chromosome contains about 100 times more DNA than a typical virus, but only one-thousandth as much DNA as a typical eukaryotic genome. This double-stranded DNA molecule is tightly packed into a region of the cell called the **nucleoid.** Many bacteria also have plasmids, small rings of DNA with a few genes.

Replication of the bacterial chromosome proceeds bidirectionally from a single origin prior to binary fission. Most bacteria in a colony are genetically identical. Mutations, although statistically rare, produce a great deal of genetic diversity because bacteria can reproduce as often as once every 20 minutes. Genetically well-adapted bacteria clone themselves more rapidly than do less fit individuals, driving the evolution of bacterial populations.

Genetic recombination produces new bacterial strains (341–346)

Genetic recombination, the combining of genetic material from two individuals, also adds to the genetic diversity of bacterial populations. Evidence that genetic recombination occurs in bacteria is provided by growing two mutant strains of *E. coli,* each of which is unable to produce a particular amino acid, together in a liquid medium. When later transferred to minimal media, numerous colonies grow that are now

---

■ **INTERACTIVE QUESTION 18.4**

Complete the following concept map that summarizes these sections on viruses.

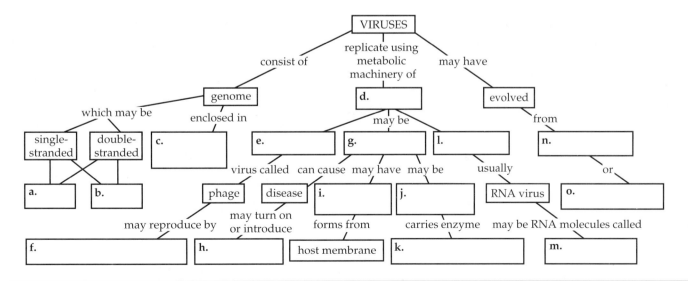

able to synthesize both amino acids. The mechanisms of genetic recombination in bacteria are different from the eukaryotic mechanisms of meiosis and fertilization.

Transformation Bacteria can take up segments of naked DNA in a process called **transformation.** The foreign DNA is integrated into the bacterial chromosome by an exchange involving crossing over. Many bacteria have surface proteins specialized for uptake of DNA from closely related species. *E. coli* can be artificially induced to take up foreign DNA, a procedure important to biotechnology.

Transduction Phages can transfer genes from one bacterium to another by a process called **transduction.** In **generalized transduction,** a random piece of host DNA is accidentally packaged within a phage capsid and introduced into a new bacterium; in **specialized transduction,** bacterial genes adjacent to a prophage insertion site are excised with the prophage from the bacterial chromosome. By either method, recombination occurs when the newly introduced bacterial genetic material replaces the homologous region of a bacterial chromosome.

■ INTERACTIVE QUESTION 18.5

What type of phage and reproductive cycle would cause specialized transduction?

Conjugation and Plasmids In **conjugation,** two cells temporarily join by appendages called sex pili and transfer DNA from one to the other. The ability to form pili and donate DNA usually results from the presence of an **F factor,** a special piece of DNA that is located on either the chromosome or a plasmid.

Plasmids are small, circular, self-replicating DNA molecules. Some plasmids, called **episomes,** can reversibly incorporate into the cell's chromosome. Temperate viruses are also considered to be episomes. Plasmid genes are not required for reproduction and survival under normal conditions, but may confer advantages to bacteria in stressful environments.

Bacterial cells containing the F factor on the **F plasmid** are called F$^+$. The F plasmid replicates in synchrony with the bacterial chromosome, and F$^+$ cells pass the trait to offspring cells. The plasmid also replicates before conjugation and is transferred to the recipient cell, changing it from an F$^-$ (female) to an F$^+$ (male) bacterium.

Cells in which the F factor is inserted into the bacterial chromosome are called Hfr cells, for "high frequency of recombination." When these cells undergo conjugation, the replicating F factor transfers an attached copy of the bacterial chromosome to the recipient cell. Movements disrupt the mating, usually resulting in a partial transfer of genes and F factor. Recombination between the new DNA and the recipient cell's chromosome produces new genetic combinations in this bacterium and its offspring.

R plasmids carry genes that code for antibiotic-destroying enzymes. R plasmids also have genes coding for sex pili and are transferred to nonresistant cells during conjugation, creating the medical problem of antibiotic-resistant pathogens.

Transposons Transposable genetic elements, or **transposons,** are mobile segments of DNA that may move within a chromosome and to and from plasmids. In cut-and-paste transposition, the transposon simply changes location; in replicative transposition, the transposon first replicates, thus remaining in its original position and also inserting in a new location. Unlike other forms of genetic recombination where alleles exchange between homologous regions, transposons can move genes to totally new areas.

Insertion sequences are the simplest transposons, consisting of only a transposase gene and inverted repeats, which serve as recognition sites for transposase. Transposase is the enzyme responsible for the cutting and ligating of DNA required for transposition. Other enzymes are also required, such as DNA polymerase, which forms direct repeats at both ends of a transposon when it is inserted between the single-stranded DNA that the staggered cut of transposase had produced at the target site. Insertion sequences cause mutations when they insert within a gene or in regulatory regions.

Composite transposons contain other genes between two insertion sequences. They may confer selective advantage by moving beneficial genes, such as those for antibiotic resistance, about in the bacterial genome.

Wandering DNA segments were first described by McClintock from her corn-breeding experiments in the 1940s and 50s. Transposons have since been found to be important components of eukaryotic genomes.

The control of gene expression enables individual bacteria to adjust their metabolism to environmental change (347)

Feedback inhibition allows regulation of enzyme activity in response to short-term environmental fluctuations, and gene expression can be regulated as metabolic needs change. The operon model for gene regulation was first described by Jacob and Monod in 1961.

Operons: The Basic Concept   Genes for the different enzymes of a single metabolic pathway may be grouped together into one transcription unit and served by a single promoter. An **operator** is a segment of DNA within the promoter region or between it and the enzyme-coding genes that controls the access of RNA polymerase to the genes. An **operon** is the DNA segment that includes the clustered genes, the promoter, and the operator.

Operators are normally "on." A **repressor** is a protein that binds to a specific operator, blocking attachment of RNA polymerase and thus turning the operon "off." **Regulatory genes** code for repressor proteins. These allosteric proteins, which may assume active or inactive shapes, are usually produced at a slow but continuous rate. The activity of the repressor protein may be determined by the presence or absence of a **corepressor.**

In the *trp* operon, tryptophan is the corepressor that binds to the repressor protein, changing its conformation into its active state, which has a high affinity for the operator and switches off the *trp* operon. Should the tryptophan concentration of the cell fall, repressor proteins are no longer bound with tryptophan, the operator is no longer repressed, RNA polymerase attaches to the promoter, and mRNA for the enzymes needed for tryptophan synthesis is produced.

Repressible versus Inducible Operons: Two Types of Negative Gene Regulation   The transcription of a *repressible operon,* such as the *trp* operon, is inhibited when a specific small molecule binds to a regulatory protein. The transcription of an *inducible operon* is stimulated when a specific small molecule interacts with a regulatory protein.

The *lac operon,* controlling lactose metabolism in *E. coli,* is an inducible operon that contains three genes. The *lac* repressor is innately active, binding to the *lac* operator and switching off the operon. Allolactose, an isomer of lactose, acts as an **inducer,** a small molecule that binds to and inactivates the repressor protein, so that the operon can be transcribed.

▩ INTERACTIVE QUESTION   18.6

Complete the following concept map that summarizes the genetic characteristics of bacteria.

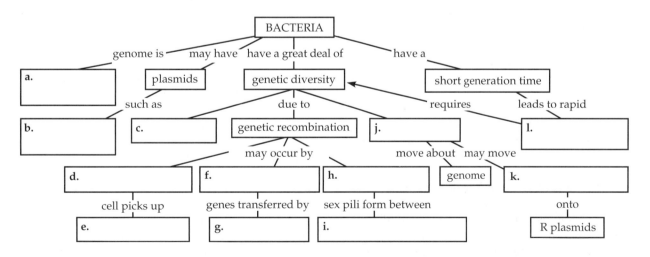

■ **INTERACTIVE QUESTION   18.7**

In the following diagram of the *lac* operon, an operon for inducible enzymes, identify components a through i.

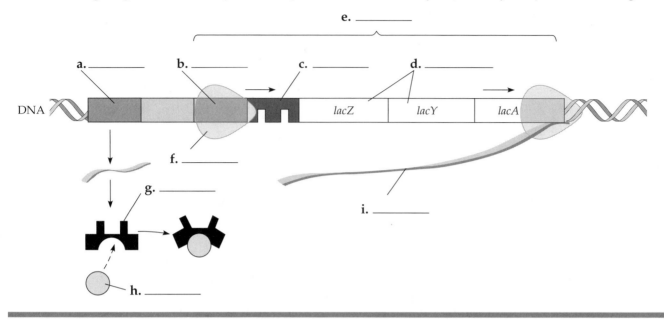

■ **INTERACTIVE QUESTION   18.8**

a. Repressible enzymes usually function in _____ pathways. The pathway's end product serves as a _____ to activate the repressor and turn off enzyme synthesis and prevent overproduction of the end product of the pathway. Genes for repressible enzymes are usually switched_____ and the repressor is synthesized in an_____ form.

b. Inducible enzymes usually function in _____ pathways. Nutrient molecules serve as _____ to stimulate production of the enzymes necessary for their breakdown. Genes for inducible enzymes are usually switched _____ and the repressor is synthesized in an _____ form.

An Example of Positive Gene Regulation   E. coli cells preferentially use glucose as their source of energy and carbon skeletons. Should glucose levels fall, transcription of operons for other catabolic pathways can be increased through the action of the allosteric regulatory protein **cAMP receptor protein (CRP). Cyclic AMP (cAMP)** accumulates in the cell when glucose is scarce and binds with CRP, changing it to its active shape. The active CRP attaches near the promoter region and stimulates transcription by facilitating the binding of RNA polymerase.

The regulation of the *lac* operon includes negative control by the repressor protein that is inactivated by the presence of lactose, and positive control by CRP when complexed with cAMP.

WORD ROOTS

capsa- = a box (*capsid:* the protein shell that encloses the viral genome)

conjug- = together (*conjugation:* in bacteria, the transfer of DNA between two cells that are temporarily joined)

lyto- = loosen (*lytic cycle:* a type of viral replication cycle resulting in the release of new phages by death or lysis of the host cell)

-oid = like, form (*nucleoid:* a dense region of DNA in a prokaryotic cell)

-phage = to eat (*bacteriophages:* viruses that infect bacteria)

pro- = before (*provirus:* viral DNA that inserts into a host genome)

retro- = backward (*retrovirus:* an RNA virus that reproduces by transcribing its RNA into DNA and then inserting the DNA into a cellular chromosome)

trans- = across (*transformation:* a phenomenon in which external DNA is assimilated by a cell)

virul- = poisonous (*virulent virus:* a virus that reproduces only by a lytic cycle)

STRUCTURE YOUR KNOWLEDGE

1. Create a concept map that describes the lytic and lysogenic cycles of a phage.

2. Create a concept map to develop your understanding of the mechanisms by which bacteria regulate their gene expression in response to varying metabolic needs. Distinguish repressible and inducible operons, which are both examples of negative control, and cAMP receptor protein, which illustrates positive control of gene expression.

TEST YOUR KNOWLEDGE

MULTIPLE CHOICE: *Choose the one best answer.*

1. The study of the genetics of viruses and bacteria has done all of the following *except*
   a. provide information on the molecular biology of all organisms.
   b. illuminate the sexual reproductive cycles of viruses.
   c. develop new techniques for manipulating genes.
   d. develop an understanding of the causes of cancer.
   e. show that genetic recombination occurs even in asexual bacteria.

2. Beijerinck concluded that the cause of tobacco mosaic disease was not a filterable toxin because
   a. the infectious agent could not be cultivated on nutrient media.
   b. a plant sprayed with filtered sap would develop the disease.
   c. the infectious agent could be crystallized.
   d. the infectious agent reproduced and could be passed on from a plant infected with filtered sap.
   e. the filtered sap was infectious even though microbes could not be found in it.

3. Viral genomes may be any of the following except
   a. single-stranded DNA.
   b. double-stranded RNA.
   c. misfolded infectious proteins.
   d. a linear single-stranded RNA molecule.
   e. a circular double-stranded DNA molecule.

4. Retroviruses have a gene for reverse transcriptase that
   a. uses viral RNA as a template for making complementary RNA strands.
   b. protects viral DNA from degradation by restriction enzymes.

c. destroys the host cell DNA.
   d. translates RNA into proteins.
   e. uses viral RNA as a template for DNA synthesis.

5. Virus particles are formed from capsid proteins and nucleic acid molecules
   a. by spontaneous self-assembly.
   b. at the direction of viral enzymes.
   c. by using host cell enzymes.
   d. using energy from ATP stored in the tail piece.
   e. by both b and d.

6. A virus has a base ratio of $(A + G)/(U + C) = 1$. What type of virus is this?
   a. a single-stranded DNA virus
   b. a single-stranded RNA virus
   c. a double-stranded DNA virus
   d. a double-stranded RNA virus
   e. a retrovirus

7. Vertical transmission of a plant viral disease may involve
   a. the movement of viral particles through the plasmodesmata.
   b. the inheritance of an infection from a parent plant.
   c. a bacteriophage transmitting viral particles.
   d. insects carrying viral particles between plants.
   e. the transfer of filtered sap.

8. Bacteria defend against viral infection through the action of
   a. antibiotics that they produce.
   b. restriction nucleases that chop up foreign DNA.
   c. their R plasmids.
   d. reverse transcriptase.
   e. episomes that incorporate viral DNA into the bacterial chromosome.

9. Drugs that are effective in treating viral infections
   a. induce the body to produce antibodies.
   b. inhibit the action of viral ribosomes.
   c. interfere with the synthesis of viral nucleic acid.
   d. change the cell-recognition sites on the host cell.
   e. produce vaccines that stimulate the immune system.

10. Which of the following is *not* true of tumor viruses?
    a. They can integrate viral nucleic acid into the host cell genome.
    b. They may transform cells and cause cancer.
    c. They may turn on the host cell's proto-oncogenes.
    d. They may carry oncogenes.
    e. They may be transferred between animals by phage.

11. The herpesvirus
    a. acts as a provirus when its DNA becomes incorporated into the host cell's genome.
    b. is a retrovirus that uses restriction enzymes to transcribe DNA from its RNA genome.
    c. has an envelope derived from the host cell's plasma membrane.
    d. is the retrovirus that has been linked to HIV, the virus that causes AIDS.
    e. can be used to vaccinate against hepatitis B.

12. The replication of the genome of an RNA virus uses
    a. DNA polymerase from the host.
    b. RNA replicating enzymes coded for by viral genes.
    c. reverse transcriptase to synthesize RNA.
    d. RNA polymerase from the host.
    e. restriction nucleases from the host.

13. The replication of the genome of a DNA virus uses
    a. DNA polymerase from the host.
    b. RNA replicating enzymes coded for by viral genes.
    c. reverse transcriptase to make an RNA copy from the DNA.
    d. RNA polymerase from the host.
    e. restriction nucleases from the host.

14. Which of the following would never be an episome?
    a. an F plasmid
    b. a prophage
    c. a provirus
    d. a retrovirus
    e. a proto-oncogene

15. Tiny molecules of naked RNA that may act as infectious agents are
    a. retroviruses.
    b. transposons.
    c. viroids.
    d. reoviruses.
    e. prions.

Choose from the following types of genetic variation in bacteria to answer questions 16–20.
    a. conjugation
    b. mutation
    c. transduction
    d. transformation
    e. transposon

16. When harmless *Streptococcus pneumoniae* are mixed with heat-killed, broken-open cells of pathogenic bacteria, live pneumonia-causing bacteria are found in the culture.

17. Transfer of genes by viruses is called _____.

18. Transfer of antibiotic-resistant genes to R plasmids may occur this way.

19. The source of most of the genetic variation found in bacterial populations is _____.

20. Two mutant *E. coli* strains, which cannot grow on minimal media, are grown together in complete media (all amino acids supplied). Samples are later transferred to minimal media and numerous colonies are able to grow.

21. Which of the following is *not* descriptive of conjugation between an Hfr and F⁻ bacterium?
    a. The Hfr cell has an F plasmid integrated into its chromosome.
    b. The Hfr cell forms sex pili and transfers a copy of its chromosome into an F⁻ cell.
    c. Random movements often break the conjugation bridge before the entire chromosome and F episome is transferred.
    d. Hfr conjugation is the way in which antibiotic resistance genes are transferred between bacteria.
    e. Crossing over between homologous genes creates a recombinant F⁻ cell.

22. A regulatory gene
    a. has its own promoter.
    b. is transcribed continuously.
    c. is not contained in the operon it controls.
    d. may code for active or inactive repressor proteins.
    e. is or does all of the above.

23. Inducible enzymes
    a. are usually involved in anabolic pathways.
    b. are produced when a small molecule inactivates the repressor protein.
    c. are produced when an activator molecule enhances the attachment of RNA polymerase with the operator.
    d. are regulated by inherently inactive repressor molecules.
    e. are regulated almost entirely by feedback inhibition.

24. In *E. coli*, tryptophan switches off the *trp* operon by
    a. inactivating the repressor protein.
    b. inactivating the gene for the first enzyme in the pathway by feedback inhibition.
    c. binding to the repressor and increasing the latter's affinity for the operator.
    d. binding to the operator.
    e. binding to the promoter.

25. A mutation that renders nonfunctional the product of a regulatory gene for an inducible operon would result in
    a. continuous transcription of the genes of the operon.
    b. complete blocking of the attachment of RNA polymerase to the promoter.
    c. irreversible binding of the repressor to the operator.
    d. no difference in transcription rate when an activator protein was present.
    e. negative control of transcription.

26. An insertion sequence
    a. carries a gene only for transposase between two inverted repeats.
    b. may transpose genes for antibiotic resistance to R plasmids.
    c. involves the exchange of homologous regions of DNA when it is a replicative transposition.
    d. is necessary for the F plasmid to incorporate into the bacterial chromosome to form an Hfr cell.
    e. consists of an inverted repeat sandwiched between two direct repeats.

MATCHING: *Match these components of the lac operon with their functions:*

_____ 1. β-galactosidase

_____ 2. active CRP

_____ 3. allolactose

_____ 4. operator

_____ 5. promoter

_____ 6. regulator gene

_____ 7. repressor

_____ 8. gene in operon

A. is inactivated when attached to allolactose

B. codes for repressor protein

C. hydrolyzes lactose

D. stimulates gene expression

E. repressor attaches here

F. RNA polymerase attaches here

G. acts as inducer that inactivates repressor

H. codes for an enzyme

CHAPTER 19

THE ORGANIZATION AND CONTROL OF EUKARYOTIC GENOMES

FRAMEWORK

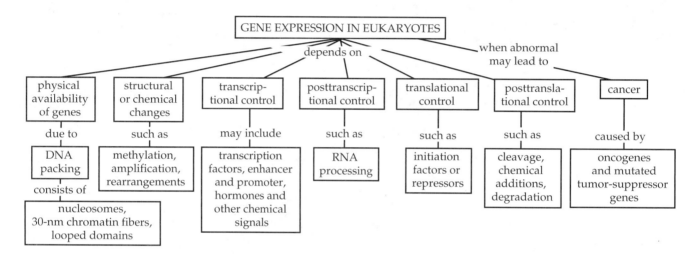

CHAPTER REVIEW

The control of gene expression is more complex in eu-karyotes due to the much greater size of the genome and the cell specialization found in multicellular eukaryotes.

Eukaryotic Chromatin Structure

Chromatin structure is based on successive levels of DNA packing (354–356)

In eukaryotes, each chromosome consists of a single, extremely long DNA double helix precisely com-plexed with a large amount of protein. During inter-phase, chromatin is extended within the nucleus, whereas discrete, condensed chromosomes appear during mitosis.

Nucleosomes, or "Beads on a String" **Histones** are small, positively charged proteins that bind tightly to the negatively charged DNA to make up chromatin.

Unfolded chromatin appears as a string of beads, each bead a **nucleosome** consisting of the DNA helix wound around a protein core of four pairs of differ-ent histone molecules. Nucleosomes are the basic unit of DNA packing. A fifth histone, H1, attaches to the DNA near the bead for the next level of packing.

Higher Levels of DNA Packing The *30-nm chromatin fiber* is a tightly coiled cylinder of nucleosomes orga-nized with the aid of histone H1. The *looped domain* is a loop of the 30-nm chromatin fiber attached to a non-histone protein scaffold. In the mitotic chromosome, looped domains coil and fold, further compacting the chromatin.

The looped domains of interphase chromosomes appear to be attached to discrete locations of the nu-clear lamina inside the nuclear envelope, perhaps helping to organize areas of transcription. Chromatin visible with the light microscope during interphase is called **heterochromatin,** a highly compacted DNA that is not actively transcribed. The more open form of interphase chromatin is called **euchromatin.**

■ **INTERACTIVE QUESTION 19.1**

List the multiple levels of packing in a metaphase chromosome in order of increasing complexity.

Genome Organization at the DNA Level

Repetitive DNA and other noncoding sequences account for much of a eukaryotic genome (357–358)

Unlike the prokaryote genome, in which most of the DNA contains uninterrupted codes for proteins (or RNA), the eukaryote genome contains mostly noncoding DNA, including regulatory sequences, introns that interrupt coding sequences, and sequences present in many copies, called **repetitive DNA.**

Tandemly Repetitive DNA   Highly repetitive, short sequences of nucleotides (up to 10 base pairs) make up 10–15% of the DNA in mammals. This so-called **satellite DNA** is classified into three types: regular satellite DNA, minisatellite, and microsatellite. The short microsatellite DNA is useful for DNA fingerprinting. Most of a genome's regular satellite DNA is located at the centromeres and telomeres.

■ **INTERACTIVE QUESTION 19.2**

Name the two human genetic diseases that have been linked to elongations of tandemly repeated nucleotide triplets.

Interspersed Repetitive DNA   As much as 40% of mammalian genomes may be interspersed repetitive DNA, similar but scattered sequences that are hundreds to thousands of base pairs long. In primates, at least 5% of the genome is made up of *Alu* **elements,** 300-nucleotide-long sequences, many of which are transcribed into RNA but are of unknown function. Most interspersed repetitive DNA sequences are transposons.

Gene families have evolved by duplication of ancestral genes (358–359)

Most genes are present in single copies as unique sequences. **Multigene families** are collections (often clustered but occasionally scattered throughout the genome) of similar or identical genes that probably evolved from a single ancestral gene. Multigene families usually code for RNA products. The identical genes coding for the three largest rRNA molecules are arranged in huge tandem arrays that enable cells to produce the millions of ribosomes needed for protein synthesis.

Examples of multigene families of nonidentical genes are the two families of genes that code for the α and β polypeptide chains of hemoglobin. Different versions of each globin subunit are clustered together on two different chromosomes and are expressed at the appropriate time during development. Both types appear to have evolved from a common ancestral globin.

Families of genes most likely arose by repeated gene duplication due to mistakes in DNA replication and recombination. **Pseudogenes,** sequences of DNA similar to real genes but lacking functional regulatory sequences needed for gene expression, may be present within gene families.

Gene amplification, loss, or rearrangement can alter a cell's genome during an organism's lifetime (359–362)

The DNA of somatic cells may be altered in ways that affect gene expression in certain cells and tissues.

Gene Amplification and Selective Gene Loss   In an amphibian ovum, a million or more extra copies of the rRNA genes are synthesized and exist as tiny DNA circles in nucleoli, enabling the developing egg to make the huge numbers of ribosomes needed for protein synthesis. Selective replication of a gene is called **gene amplification.** Cancer cells resistant to chemotherapeutic drugs contain amplified genes conferring drug resistance.

In certain insects, genes may be selectively lost, and whole or parts of chromosomes may be eliminated from some cells early in development.

Rearrangements in the Genome   **Transposons** may affect gene expression if they move into the middle of a coding sequence of a gene or insert into a sequence that regulates transcription. If a transposon inserts downstream from a promoter, a gene carried on the transposon may be activated.

Over 10% of the human genome is made up of transposons, most of which have been identified as **retrotransposons.** Retrotransposons contain the code for reverse transcriptase and an insertion enzyme, and move about the genome as an RNA transcript of the retrotransposon DNA.

B lymphocytes are white blood cells that produce highly specific antibodies, or **immunoglobulins,** that recognize and bind to specific foreign molecules. DNA segments of the genes for the light and heavy polypeptide chains of an antibody molecule are permanently rearranged during cellular differentiation. An **immunoglobulin gene** is pieced together from hundreds of different variable regions that are widely separated from several constant regions. Each differentiated B lymphocyte and its descendants possess a distinctive genome that will produce one specific antibody.

■ **INTERACTIVE QUESTION   19.3**

What accounts for the huge variety and specificity of immunoglobulins?

The Control of Gene Expression

Each cell of a multicellular eukaryote expresses only a small fraction of its genes (362)

Eukaryotic cells in multicellular organisms control the expression of their genes in response to their external and internal environments and also to direct the **cellular differentiation** necessary to create specialized cells. In both prokaryotes and eukaryotes, gene expression is most often regulated at the level of transcription by DNA-binding proteins. The more complex eukaryotic cell, however, provides more opportunities for control of gene activity.

The control of gene expression can occur at any step in the pathway from gene to functional protein: *an overview* (362)

In a eukaryotic cell, the stages in the expression of a protein-coding gene include DNA unpacking, transcription, RNA processing, transport to cytoplasm, translation, and protein alterations. Each of these stages presents a potential control point.

Chromatin modifications affect the availability of genes for transcription (362–364)

DNA packing and the location of genes relative to nucleosomes and attachment to the protein scaffold may help control which genes are available for transcription.

DNA Methylation   In a process called **DNA methylation,** methyl groups ($-CH_3$) are added to DNA bases (usually cytosine) after DNA synthesis. Genes are generally more heavily methylated in cells in which they are not expressed. DNA methylation may reinforce gene regulatory decisions of early development as methylation enzymes act during DNA replication to methylate daughter DNA strands and pass on methylation records. Such methylation patterns account for **genomic imprinting** in mammals in which either a maternal or paternal allele of certain genes is permanently turned off.

Histone Acetylation   The attachment of acetyl groups ($-COCH_3$) to amino acids in histone proteins, called **histone acetylation,** appears to change the shape of a nucleosome's histones such that transcription proteins may more easily access genes. Some enzymes that acetylate or deacetylate histones are associated with or part of transcription factors. Proteins that bind to methylated DNA have been found to interact with histone deacetylation enzymes.

■ **INTERACTIVE QUESTION   19.4**

a. Give an example of highly methylated and inactive DNA common in mammalian cells.

b. Would histone deacetylation increase or decrease the transcription of a gene located in that nucleosome?

Transcription initiation is controlled by proteins that interact with DNA and with each other (364–367)

Organization of a Typical Eukaryotic Gene   A typical gene consists of a promoter sequence, where a transcription initiation complex, including RNA polymerase, attaches, and a sequence of introns interspersed among the coding exons. After transcription, RNA processing removes the introns and usually adds the guanosine cap at the 5′ end and a poly(A) tail at the 3′ end. **Control elements** are noncoding sequences to which transcription factors bind that help regulate transcription.

The Role of Transcription Factors   For all protein-coding genes, one transcription factor recognizes the TATA box in the promoter and other transcription factors interact with each other and RNA polymerase, forming the initiation complex essential for transcription to begin. The binding of additional transcription factors to control elements, however, greatly increases transcription rates. *Proximal control elements* are located close to, and may be considered to include, the promoter. **Enhancers** are *distal control elements* located far upstream or downstream from the promoter. A bend in the DNA apparently brings transcription factors bound to enhancers into contact with the initiation complex, where they may function as **activators** to stimulate transcription.

There is some evidence of transcription factors that act as repressors when they bind to control elements called *silencers*, although most eukaryotic gene repression may occur at the level of chromatin modification.

Hundreds of transcription factors have been identified. Most have a **DNA-binding domain** and a protein-binding domain that recognizes another transcription factor. DNA-binding domains are based on only a few structural motifs, which contain α helices that fit into the major groove of the double helix and recognize specific nucleotide sequences within a control element.

Coordinately Controlled Genes   Unlike prokaryotic genes, which are often organized into operons that are controlled by the same promoter and control elements and transcribed together, eukaryotic genes coding for enzymes in the same metabolic pathway are often scattered throughout the chromosomes and separately transcribed. The integrated control of these scattered genes may involve a specific collection of control elements associated with each related gene for which the same transcription factor serves as activator.

For example, steroid hormones may activate multiple genes when they bind to specific receptor proteins within a cell that function as transcription activators for several genes. Other signal molecules bind to cell surface receptors and initiate signal transduction pathways that activate particular transcription factors. The same chemical signal will activate genes with the same control elements and thus coordinate gene expression.

■ **INTERACTIVE QUESTION   19.5**

Label the components of this diagram of how enhancers and transcription factors facilitate formation of a transcription initiation complex.

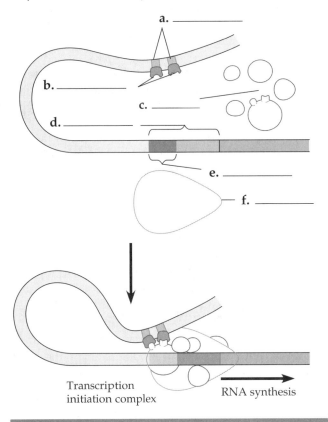

Transcription
initiation complex

RNA synthesis

Post-transcriptional mechanisms play supporting roles in the control of gene expression (367–368)

Gene expression, measured by the amount of functional protein that is made, can be blocked or stimulated at any posttranscriptional step. At the level of RNA processing, **alternative RNA splicing** may produce different mRNA molecules when regulatory proteins control which segments of the primary transcript are chosen as introns and which as exons.

Regulation of mRNA Degradation   In contrast to prokaryotic mRNA that is degraded after a few minutes, eukaryotic mRNA can last hours or even weeks. The trailer region of mRNA may contain nucleotide sequences that affect stability. The hydrolysis of mRNA by nucleases is usually preceded by the enzymatic shortening of the poly(A) tail and removal of the 5′ cap.

Control of Translation   The translation of certain eukaryotic mRNA can be delayed by the binding of regulatory proteins to the 5′ leader regions that prevent ribosome binding or by the inactivation of initiation factors. A great deal of mRNA is synthesized and stored in egg cells; translation is begun when translation initiation factors are activated following fertilization.

Protein Processing and Degradation   Following translation, polypeptides are often cleaved or chemical groups added to yield an active protein. Some proteins must be transported to target locations. Selective degradation of proteins may serve as a control mechanism in the cell. Molecules of the protein ubiquitin are added to mark proteins for destruction. **Proteasomes** recognize and degrade the marked proteins.

■ INTERACTIVE QUESTION   19.6

Complete the following five key points that summarize the control of gene expression in eukaryotes:

a. Different cell types of a multicellular organism

_____ .

b. Physical organization and modification of the chromatin makes certain genes _____

_____ .

c. _____ may occur at each step from gene to functional protein.

d. Control of transcription is especially important. The

_____

_____ activates transcription of specific genes.

e. Gene expression can also be regulated by _____

_____ .

The Molecular Biology of Cancer

Cancer results from genetic changes that affect the cell cycle (369)

A change in a cell cycle regulatory mechanism that leads to cancer is most often caused by carcinogens, physical mutagens, or viruses. **Oncogenes,** or cancer-causing genes, were first found in certain retroviruses. Similar genes were later recognized in the genomes of humans and other animals. Cellular **proto-oncogenes,** which code for proteins that stimulate cell growth and division, may become oncogenes by several mechanisms. Mutations may result in more copies of the gene being present than normal (amplification), transposition or chromosomal translocation (both of which may bring the gene under the control of a more active promoter or control element), or a change in nucleotide sequence that creates a more active or resilient protein.

Mutations in **tumor-suppressor genes** can contribute to the onset of cancer when they result in a decrease in the activity of proteins that prevent uncontrolled cell growth. Tumor-suppressor proteins have various functions, such as repair of damaged DNA, control of cell adhesion, and inhibition of the cell cycle.

Oncogene proteins and faulty tumor-suppressor proteins interfere with normal signaling pathways (369–371)

In about 30% of human cancers, the *ras* proto-oncogene is mutated. The ***ras* gene** codes for a G protein that connects a growth-factor receptor on the plasma membrane to a cascade of protein kinases that leads to the production of a cell cycle stimulating protein. A point mutation may create a *ras* oncogene that produces hyperactive versions of the Ras protein that relay a signal without binding of a growth factor.

The ***p53* gene** is mutated in about 50% of human cancers. It codes for a tumor-suppressor protein that is a transcription factor involved in the synthesis of several growth-inhibiting proteins. Such proteins may activate the *p21* gene, whose product binds to cyclin-dependent kinases, halting the cell cycle and allowing time for the cell to repair damaged DNA. The p53 transcription factor can also activate genes involved in DNA repair. Should DNA damage be irreparable, p53 activates "suicide genes" that initiate apoptosis.

Multiple mutations underlie the development of cancer (371–372)

More than one somatic mutation appears to be needed to produce a cancerous cell. Mutation of a single proto-oncogene can stimulate cell growth and division, but both tumor-suppressor alleles at a particular locus must be defective to allow uncontrolled cell growth. In many malignant tumors, the gene for telomerase becomes activated and cells become able to divide indefinitely.

Virus-associated cancers are thought to account for about 15% of human cancer cases. The virus might add an oncogene or its insertion may affect a proto-oncogene or tumor-suppressor gene.

A genetic predisposition to certain cancers may involve the inheritance of an oncogene or a recessive mutant allele for a tumor-suppressor gene. Approximately 15% of colorectal cancers involve inherited mutations, either in DNA repair genes or in the tumor-suppressor gene *APC*, which regulates cell migration and adhesion.

The 5–10% of breast cancer cases linked to family history have been linked to mutant alleles for either *BRCA1* or *BRCA2,* both of which appear to be tumor-suppressor genes.

WORD ROOTS

eu- = true (*euchromatin:* the more open, unraveled form of eukaryotic chromatin)

hetero- = different (*heterochromatin:* nontranscribed eukaryotic chromatin that is so highly compacted that it is visible with a light microscope during interphase)

immuno- = safe, free (*immunoglobulin:* one of the class of proteins comprising the antibodies)

nucleo- = the nucleus; **-soma** = body (*nucleosome:* the basic beadlike unit of DNA packaging in eukaryotes)

proto- = first, original; **onco-** = tumor (*proto-oncogene:* a normal cellular gene corresponding to an oncogene)

pseudo- = false (*pseudogenes:* DNA segments very similar to real genes but which do not yield functional products)

retro- = backward (*retrotransposons:* transposable elements that move within a genome by means of an RNA intermediate, a transcript of the retrotransposon DNA)

STRUCTURE YOUR KNOWLEDGE

1. Fill in the table below to help you organize the major mechanisms that can regulate the expression of eukaryotic genes.

2. **a.** What are proto-oncogenes? How do they become oncogenes?

   b. What is the role of tumor-suppressor genes in the development of cancer?

TEST YOUR KNOWLEDGE

MULTIPLE CHOICE: *Choose the one best answer.*

1. The control of gene expression is more complex in eukaryotic cells because
   a. DNA is associated with protein.
   b. gene expression may differentiate specialized cells.
   c. the chromosomes are linear and more numerous.
   d. operons are controlled by more than one promoter region.
   e. inhibitory or activating molecules may help regulate transcription.

2. Histones are
   a. small, positively charged proteins that bind tightly to DNA.
   b. small bodies in the nucleus involved in rRNA synthesis.
   c. basic units of DNA packing consisting of DNA wound around a protein core.
   d. repeating arrays of six nucleosomes organized around an H1 molecule.
   e. proteins responsible for producing repeating sequences at telomeres.

3. Heterochromatin
   a. has a higher degree of packing than does euchromatin.
   b. is visible with a light microscope during interphase.
   c. is not actively involved in transcription.
   d. makes up metaphase chromosomes.
   e. is all of the above.

4. DNA methylation of cytosine residues
   a. can be induced by drugs that reactivate genes.
   b. may contribute to long-term gene inactivation.
   c. produces the promoter regions that specifically bind RNA polymerase.
   d. makes satellite DNA a different density so it can be separated by ultracentrifugation.
   e. may be related to the transformation of proto-oncogenes to oncogenes.

Level of Control	Examples
Chromatin modification	**a.**
Transcriptional control	**b.**
Post-transcriptional control	**c.**
Translational control	**d.**
Post-translational control	**e.**

5. Which of the following best describes what pseudogenes and introns have in common?
   a. They are RNA molecules that are not translated into proteins.
   b. They are DNA segments that lack promoter or other control regions.
   c. They are not expressed—they do not produce a functional product.
   d. They code for RNA products, not proteins.
   e. They appear to have arisen from retrotransposons.

6. Which of the following is *not* true of enhancers?
   a. They may be located thousands of nucleotides upstream from the genes they affect.
   b. When bound with activators, they interact with the promoter region and other transcription factors to increase the activity of a gene.
   c. They may complex with steroid-activated receptor proteins and thus selectively activate specific genes.
   d. They may coordinate the transcription of enzymes involved in the same metabolic pathway.
   e. They are located within the promoter, and, when complexed with a steroid or other small molecule, they release an inhibitory protein and thus make DNA more accessible to RNA polymerase.

7. Which of the following is *not* an example of the control of gene expression that occurs after transcription?
   a. mRNA stored in the cytoplasm needing a control signal to initiate translation
   b. the length of time mRNA lasts before it is degraded
   c. rRNA genes amplified in tandem arrays
   d. alternative RNA splicing before mRNA exits from the nucleus
   e. splicing or modification of a polypeptide

8. Pseudogenes are
   a. tandem arrays of rRNA genes that enable actively synthesizing cells to create enough ribosomes.
   b. genes that can become oncogenes when mutated by carcinogens.
   c. genes of multigene families that are expressed at different times during development.
   d. sequences of DNA that are similar to real genes but lack regulatory sequences necessary for gene expression.
   e. both c and d.

9. Which of the following might a proto-oncogene code for?
   a. DNA polymerase
   b. reverse transcriptase
   c. receptor proteins for growth factors

d. immunoglobulins
e. transcription factors that inhibit cell division genes

10. A gene can develop into an oncogene when it
    a. is present in more copies than normal.
    b. undergoes a translocation that removes it from its normal control region.
    c. develops a mutation that creates a more active or resistant protein.
    d. is transposed to a new location where its expression is enhanced.
    e. does any of the above.

11. Gene amplification involves
    a. the production of stereotypic loud genes.
    b. the enlargement of chromosomal areas undergoing transcription.
    c. the presence of multiple copies of genes.
    d. the transformation of genes into tumor-suppressor genes.
    e. an enhancer or promoter that is activated and increases transcription rate.

12. A tumor-suppressor gene could cause the onset of cancer if
    a. both alleles have mutations that decrease the activity of the gene product.
    b. only one allele has a mutation that alters the gene product.
    c. it is inherited in mutated form from a parent.
    d. a proto-oncogene has also become an oncogene.
    e. both a and d have happened.

13. What is apoptosis?
    a. a cell suicide program that may be initiated by p53 protein in response to DNA damage
    b. metastasis, or the spread of cancer cells to a new location in the body
    c. the transformation of a normal cell to a cancer cell
    d. the vascularization of a tumor by the growth of blood vessels
    e. the transformation of a proto-oncogene to an oncogene by a point mutation

14. Which of the following would you expect to find as part of a receptor protein that binds with a steroid hormone?
    a. a TATA box located within the promoter region
    b. a zinc finger, leucine zipper, or helix-turn-helix domain that binds to DNA
    c. an activated operator region that allows attachment of RNA polymerase
    d. an enhancer sequence located at some distance upstream or downstream from the promoter
    e. a G protein that activates a cascade of protein kinases

15. A eukaryotic gene typically has all of the following features *except*
    a. a promoter.
    b. an operator.
    c. enhancers.
    d. introns and exons.
    e. control elements.

16. What are proteasomes?
    a. complexes of snRNA and proteins that excise introns
    b. small, extrachromosomal circles of rRNA genes
    c. small, positively charged proteins that form the core of nucleosomes
    d. enormous protein complexes that degrade unneeded proteins in the cell
    e. complexes of transcription factors whose protein–protein interactions are required for enhancing gene transcription

17. How is the incredible diversity of antibodies that can be produced by different B cells made possible?
    a. There are over 20 genes for the light and heavy chains, and they each have 20 alleles, producing 20^{20} antibody possibilities.
    b. Alternative RNA splicing can produce millions of different combinations of exons.
    c. During cellular differentiation, different variable regions are permanently rearranged and connected to constant regions to produce the genes for the light and heavy chains.
    d. Mutations are more common in the genes for immunoglobulins.
    e. Gene duplication and transposons have created multigene families of antibody genes on several different chromosomes.

18. Which of the following would most likely account for a family history of colorectal cancer?
    a. inheritance of a proto-oncogene
    b. a family diet that is low in fats and high in fiber
    c. a family history of breast cancer

    d. inheritance of the *ras* oncogene that locks the G protein in an active configuration
    e. inheritance of one mutated *APC* allele that regulates cell adhesion and migration

19. Interspersed repetitive DNA
    a. is found at the centromeres and telomeres of eukaryotic chromosomes.
    b. is composed of long similar nucleotide sequences that, as transposons, became scattered throughout the genome.
    c. includes tandemly repeated nucleotide triplets that are linked with several genetic disorders.
    d. is needed to create artificial chromosomes in the laboratory that are capable of replication.
    e. is microsatellite DNA that is useful for DNA fingerprinting.

20. Which of the following appears to be attached to the nuclear lamina in a precise and organized fashion?
    a. nucleosomes
    b. heterochromatin
    c. 30-nm fiber
    d. looped domains of interphase chromosomes
    e. enhancer regions of actively transcribed genes

21. How are *Alu* elements unusual as retrotransposons?
    a. They do not include the coding sequence for reverse transcriptase, but move about the genome using the enzymes coded for by other retrotransposons.
    b. They are not transcribed back into DNA from RNA before they are inserted into a new area of the genome.
    c. They "jump" from one location to another, without leaving a copy of themselves in the first location.
    d. Most retrotransposons are interspersed repetitive DNA; *Alu* elements are not.
    e. They did not arise from infections by retroviruses; most retrotransposons did evolve from viruses.

CHAPTER 20

DNA TECHNOLOGY AND GENOMICS

FRAMEWORK

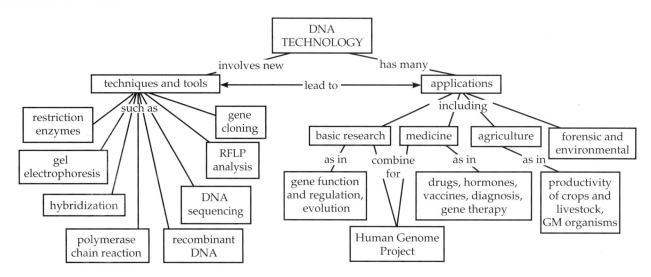

CHAPTER REVIEW

Genetic engineering, the direct manipulation of genetic material for practical purposes, has begun an industrial revolution in **biotechnology.** This use of living organisms to manufacture desirable products dates back centuries, but advances in DNA technology have resulted in hundreds of new products. Genes from different organisms and species are now routinely combined *in vitro* to form **recombinant DNA** and inserted into living cells in which these genes are replicated and may be expressed and their products studied or harvested. DNA technology has led to major advances in all fields of biology and in our knowledge of the humane genome.

DNA Cloning

Techniques for **gene cloning** are used to prepare multiple identical copies of pieces of DNA.

DNA technology makes it possible to clone genes for basic research and commercial applications: *an overview* (376–377)

One of the approaches used to clone pieces of DNA makes use of the plasmids of bacterial cells. Recombinant DNA may be made by inserting foreign DNA into plasmids. These plasmids are put back into bacterial cells where they will replicate as the bacteria reproduce. Such cloned DNA provides multiple copies of the gene and may also be used to produce protein coded for by the foreign DNA.

Restriction enzymes are used to make recombinant DNA (377–378)

Restriction enzymes protect bacteria from the DNA of phages or other bacteria by cutting up foreign DNA in a process called *restriction*. Most restriction enzymes recognize short nucleotide sequences and cut at specific points within them. The cell protects its own DNA from restriction by methylating nucleotide bases within its own recognition sequences.

A recognition sequence, or **restriction site,** is usually symmetrical, the same sequence of four to eight nucleotides running in opposite directions on the two strands. The restriction enzyme usually cuts phosphodiester bonds in a staggered way, between the same adjacent nucleotides on both strands, leaving **sticky ends** of short single-stranded sequences on both sides of the resulting **restriction fragment.**

DNA from different sources can be combined in the laboratory when the DNA is cut by the same restriction enzyme and the complementary bases on the sticky ends of the restriction fragments pair by hydrogen bonding. **DNA ligase** is used to seal the strands together.

■ **INTERACTIVE QUESTION   20.1**

Which of these DNA sequences would function best as a restriction site for a restriction enzyme? Why?

··CAGCAG··   ··GTGCTG··   ··GAATTC··
··GTCGTC··   ··CACGAC··   ··CTTAAG··

Genes can be cloned in recombinant DNA vectors: *a closer look* (378–381)

Cloning vectors are DNA molecules that can move foreign DNA into a cell and replicate there. Recombinant plasmids returned to bacterial cells will replicate cloned DNA as the bacteria reproduce. The ease with which DNA can be isolated from and returned to bacterial cells make bacteria the most common host for gene cloning.

Procedure for Cloning a Eukaryotic Gene in a Bacterial Plasmid   The plasmid method of gene cloning involves treating antibiotic-resistant plasmids with a restriction enzyme that cuts the DNA ring at a single restriction site and disrupts a gene whose activity is easily determined, such as *lacZ,* the gene for β-galactosidase. The clipped plasmids are mixed with DNA containing the gene of interest, which has also been treated with the same restriction enzyme and has complementary sticky ends. The sticky ends form hydrogen bonds with each other, and DNA ligase seals the recombinant molecules.

The plasmids are introduced by transformation into bacterial cells that are *lacZ⁻* and thus unable to produce β-galactosidase. Bacteria are plated onto a medium containing ampicillin and X-gal, a compound

that is cleaved by β-galactosidase and yields a blue product. Colonies that are able to grow on the medium (because they contain a plasmid with the *amp^R* gene) and are not blue (because of the foreign DNA inserted in the middle of the β-galactosidase gene) are carrying a recombinant plasmid.

The colonies are tested to find the ones that contain the gene of interest. If the gene is expressed, the presence of the protein product can be determined by its activity or structure (using antibodies).

The gene itself can be detected by **nucleic acid hybridization,** using a **nucleic acid probe,** which has complementary sequences to segments of the gene and can hybridize with the DNA of the gene after **denaturation** of the cell's DNA by heat or chemicals produces single-stranded DNA. The probe is located by its radioactively labeled molecules or fluorescent tag. Once the desired clone is identified, it can be grown in liquid culture and the gene of interest isolated in large quantities.

Cloning and Expressing Eukaryotic Genes: Problems and Solutions   Differences between prokaryotic and eukaryotic mechanisms for gene expression can be overcome by using an **expression vector,** a vector that has a prokaryotic promoter just upstream from the eukaryotic gene insertion site.

Creating artificial genes eliminates the problem caused by large introns that can make bacterial cells, which lack RNA-processing enzymes, unable to express eukaryotic genes. Using reverse transcriptase from retroviruses, mRNA—which has no introns—is used as a template to produce **complementary DNA (cDNA).**

■ **INTERACTIVE QUESTION   20.2**

What steps can be taken to facilitate the expression of a eukaryotic gene in bacteria?

Yeast cells are eukaryotic hosts used to clone eukaryotic genes; they are easy to grow and have plasmids that serve as vectors. **Yeast artificial chromosomes (YACs)** have also been constructed that carry foreign DNA and contain an origin for DNA replication, a centromere, and two telomeres. These vectors behave normally in mitosis and are able to clone large pieces of DNA. Plant and animal cells in culture can serve as hosts and may be necessary when a protein must be modified following translation.

■ INTERACTIVE QUESTION   20.3

This schematic diagram shows the steps in plasmid cloning of a gene. Identify components **a–j**. Briefly describe the five steps of the process. How are bacterial clones that have picked up the recombinant plasmid identified?

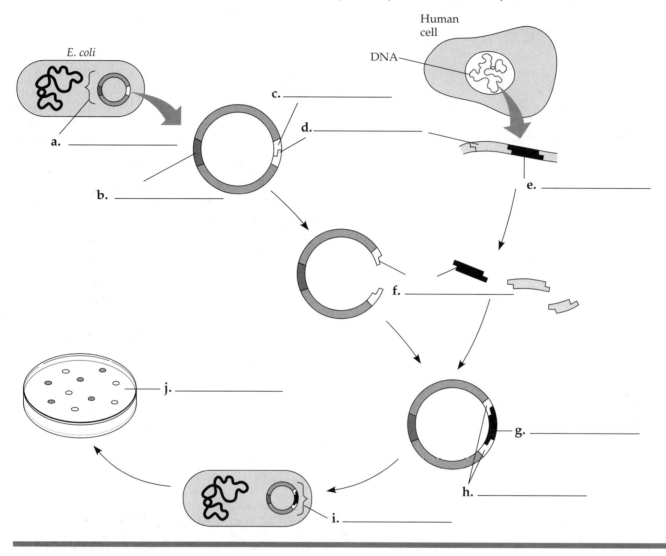

Yeasts and other eukaryotic cells can take up foreign DNA, which may become incorporated into a chromosome through recombination. More efficient means for introducing DNA into eukaryotic cells include **electroporation,** in which an electric pulse briefly opens holes in the plasma membrane through which DNA can enter, injection into cells using microscopically thin needles, or firing into plant cells on microscopic metal particles using a gene gun.

Cloned genes are stored in DNA libraries (381)

The genes used for cloning can be isolated from an organism by cutting its DNA into thousands of pieces with restriction enzymes and inserting them into plasmids. The collection of the thousands of clones of bacteria containing recombinant plasmids derived from this "shotgun" approach is called a **genomic library.**

Bacteriophages are also used as vectors for creating genomic libraries. DNA fragments are spliced into phage DNA, which is packaged into capsids and used to infect bacteria. The production of new phage clones the foreign DNA.

A partial genomic library can be produced using the mRNA molecules isolated from a cell. Such a **cDNA library** contains only the genes that are expressed (transcribed) within the cell.

The polymerase chain reaction (PCR) clones DNA entirely *in vitro* (382–383)

A technique developed in 1985, known as **polymerase chain reaction (PCR),** has contributed to molecular biology by being able to produce billions of copies of a section of DNA *in vitro* in only a few hours. DNA containing the region of interest is incubated with the four nucleotides, a special heat-resistant type of DNA polymerase, and specially synthesized primers that bind upstream from the target sequence. The solution is heated to separate the DNA strands, then cooled so the primers can hydrogen-bond. DNA polymerase adds nucleotides to the 3′ ends of the primers. The solution is heated again and the process repeated. The desired DNA segment does not need to be purified from the starting material, and very small samples can be used.

■ **INTERACTIVE QUESTION 20.4**

a. Why is PCR often used prior to cloning a gene in cells?

b. Why even bother cloning genes in cells, since PCR produces so many copies so fast?

DNA Analysis and Genomics

Producing cloned segments of DNA allows scientists to make comparisons between cells, individuals, and different species. Sequencing and studying whole sets of genes and their interactions is an approach known as **genomics.**

Many of the methods for analyzing and comparing DNA make use of the technique called **gel electrophoresis** that separates nucleic acids and proteins on the basis of their size and electrical charge. Due to the negative charge of their phosphate groups, DNA molecules migrate through the electric field produced in a thin slab of gel toward the positive electrode. Molecules move at a rate inversely proportional to their size, producing band patterns in the gel of fragments of decreasing size.

Restriction fragment analysis detects DNA differences that affect restriction sites (383–386)

Cutting a long DNA molecule with a particular restriction enzyme and separating the resulting restriction fragments by gel electrophoresis produces a characteristic pattern of bands. Pure samples of such bands can be recovered from the gel and retain their biological activity.

Two DNA samples, such as cloned alleles of a gene, will produce different patterns of bands when differences in their DNA sequences add or delete restriction sites. The addition of nucleic acid hybridization with a probe allows two or more unpurified samples of DNA (such as the entire genome) to be compared for the presence and band location of a particular DNA sequence. In a technique known as Southern hybridization or **Southern blotting,** DNA is treated with a restriction enzyme; the fragments are separated on a gel and then transferred to nitrocellulose or nylon paper by blotting; labeled probes of single-stranded DNA are added and hybridize with complementary DNA sequences; and the restriction fragment bands of interest are identified by autoradiography.

Samples of noncoding DNA treated with restriction enzymes also produce different band patterns due to differences in nucleotide sequences in restriction sites. Such **restriction fragment length polymorphisms (RFLPs)** can serve as genetic markers on a chromosome. A particular RFLP marker often occurs in several variants in a population. RFLPs are detected by Southern blotting, and the entire genome can be used as the DNA starting material.

■ **INTERACTIVE QUESTION 20.5**

A bloody crime has occurred. Police have collected blood samples from the victim, two suspects, and blood found at the scene. Briefly list the steps the lab went through to produce the following autoradiograph.

a.

b.

c.

d.

e.

Which suspect would you charge with the crime?

Entire genomes can be mapped at the DNA level (386–389)

The international effort to map the human genome is called the **Human Genome Project** and included three stages: genetic (linkage) mapping, physical mapping, and DNA sequencing. The project also included the mapping of important research organisms, such as *E. coli*, yeast, the nematode *C. elegans*, *Drosophila*, and mouse.

Genetic (Linkage) Mapping   The first stage in mapping a large genome is to develop a linkage map of several thousand genetic markers, either genes or other identifiable sequences such as RFLPs or short repetitive sequences (microsatellites). Testing for genetic linkage to these known markers enables researchers to place other genes or markers.

Physical Mapping: Ordering DNA Fragments   In physical mapping, the actual distances between markers are determined. The DNA of each chromosome is cut into restriction fragments, which are then ordered along the chromosome. First large fragments are cut, cloned, and ordered. Then those fragments are cut and their fragments ordered. Finally, fragments are cut, cloned, and ordered that are small enough to be sequenced.

Cloning vectors include yeast artificial chromosomes, which can carry long fragments, and **bacterial artificial chromosomes (BACs)**, artificial bacterial chromosomes that carry shorter inserts.

The ordering of fragments is facilitated by the technique of **chromosome walking.** Two sets of overlapping fragments (cut by two different restriction enzymes) are prepared. A probe made from the 3' end of a known sequence is used to identify the fragment in the other set that overlaps that sequence. A 3' probe from that fragment is used to search the first library for the next overlapping fragment, continuing to "walk" down the chromosome.

DNA Sequencing   Clones of short DNA fragments are ultimately used to determine the nucleotide sequence of an entire genome. In the Sanger method of DNA sequencing, samples of a single-stranded restriction fragment are incubated with modified nucleotides (dideoxyribonucleotides) that randomly block further synthesis when they are incorporated into a growing DNA strand. The sets of radioactively or fluorescently tagged strands of varying lengths are separated by gel electrophoresis, and the nucleotide sequence is read from the sequence of bands.

Faster sequencing machines and better computer software were developed during the Human Genome Project.

■ **INTERACTIVE QUESTION   20.6**

The following gel was produced from four samples of a single-stranded DNA fragment that were incubated with radioactively labeled primer, DNA polymerase, A, T, C, and G nucleotides, and a different one of the four dideoxy nucleotides. Starting from the bottom of the gel, what is the sequence of nucleotides in the original single-stranded DNA fragment?

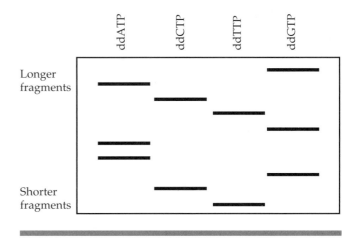

Alternative Approaches to Whole-Genome Sequencing   J. C. Venter, founder of the company Celera Genomics, developed a whole-genome shotgun approach that relies on powerful computer programs to order the large number of random short sequences cut from the whole genome. In February 2001, Celera (as well as the public consortium) announced that it had sequenced over 90% of the human genome. As of mid-2001 the genomes of about 50 species had been sequenced.

■ **INTERACTIVE QUESTION   20.7**

Compare the strategies of the public consortium of the Human Genome Project and of Celera Genomics for whole-genome sequencing.

Genome sequences provide clues to important biological questions (389–393)

Genomics bases the study of genomes on their DNA sequences, allowing researchers to address questions about gene expression, growth and development, and evolution.

Analyzing DNA Sequences Computer data banks of DNA sequences are available to researchers on the Internet. Computer software can scan these sequences for signs of genes, such as start and stop codons, RNA-splicing sites, and expressed sequence tags, or ESTs (sequences similar to those of known genes).

The number of putative genes revealed by the analysis of the human genome is surprisingly small, perhaps only 30–40 thousand. Large amounts of non-coding DNA (in particular, repetitive DNA) and unusually long introns make up the majority of the human genome. Alternative splicing of exons, which creates different polypeptides from the same gene, and greater polypeptide complexity due to more combinations of protein domains, may explain how this relatively small number of genes can produce the complexity of humans (and vertebrates in general).

The sequences of unknown genes are compared with sequences of known genes of that species and other species to look for similarities that might indicate the function of the new gene. Many of the putative genes identified so far have never been encountered before, but many others have been found in even distantly related organisms.

Studying Gene Expression In order to study patterns of gene expression, researchers isolate the mRNA made in different cells, create a cDNA library using reverse transcriptase, and then use the cDNA as probes to explore DNA collections. Using this cDNA in **DNA microarray assays,** scientists can test all the genes expressed in a tissue for hybridization with thousands of short, single-stranded DNA fragments from different genes arrayed on a microscope slide (called a DNA chip). The intensity with which the hybridized spots fluoresce indicates the relative amount of mRNA that was in the tissue. Gene expression in different tissues and at different stages of development can be determined. DNA microarray assays are used to compare gene expression in cancerous and noncancerous tissues.

Determining Gene Function The function of unknown genes can be studied using *in vitro* **mutagenesis,** in which changes are made to a cloned gene, the gene is returned to the cell, and changes in physiology or developmental patterns that result from the altered gene product are monitored. A new way to stop the expression of selected genes in nonmammalian organisms is called **RNA interference (RNAi).** Synthetic double-stranded RNA molecules that match a gene sequence somehow trigger the breakdown of that gene's mRNA.

Future Directions of Genomics **Proteomics** is the challenging identification and study of entire protein sets coded for by a genome. **Bioinformatics** applies computer science and mathematics to these and other new sources of genetic and biological information.

The human species is comparatively young, and its genetic variation is small. Human DNA sequences are about 99.9% identical. Most variation seems to be one-base-pair changes called **single nucleotide polymorphisms (SNPs),** occurring at about 3 million sites in the human genome. Identifying these sites will provide genetic markers for studying human evolution and differences between human populations, as well as for locating health-related genes.

Practical Applications of DNA Technology

DNA technology is reshaping medicine and the pharmaceutical industry (393–395)

DNA technology and the Human Genome Project are identifying genes responsible for genetic diseases, hopefully leading to new ways to diagnose, treat, and even prevent those disorders.

Diagnosis of Diseases PCR and labeled DNA probes are being used to identify pathogens and to diagnose infectious diseases and genetic disorders. Even if a gene has not yet been cloned, a disease allele may be diagnosed when closely linked with an RFLP marker. Comparisons within a family must be used to determine which variant of the RFLP marker is linked to the abnormal allele.

Human Gene Therapy **Gene therapy** may provide the means for correcting genetic disorders in individuals by replacing or supplementing defective genes. New genes would be introduced into somatic cells of the affected tissue. For the correction to be permanent, the cells must be types that actively reproduce within the body, such as bone marrow cells.

To date, little effective gene therapy has been reported. Current research efforts are aimed at using gene therapy to fight heart disease and cancer.

■ **INTERACTIVE QUESTION 20.8**

a. Why is it easier to perform a test for Huntington's disease now that the gene has been cloned?

b. What are some of the practical and ethical considerations in human gene therapy?

Pharmaceutical Products   Many pharmaceutical proteins are produced using DNA technology. Large quantities of proteins can be made by inserting a gene into vector DNA with a highly active promoter. Engineering host cells to secrete a protein as it is made simplifies its purification. Insulin and human growth hormone were among the first hormones made by recombinant DNA techniques.

Recombinant DNA techniques have been used to make large amounts of protein molecules from the surfaces of pathogens, which can be used as **vaccines** if the protein subunit triggers an immune response. Genetic engineering methods can also directly modify the genome of a pathogen so as to attenuate it (make it nonpathogenic) so it can be used as a vaccine.

DNA technology offers forensic, environmental, and agricultural applications (395–399)

Forensic Uses of DNA Technology   RFLP analysis can be used in criminal cases to compare the **DNA fingerprint,** or specific pattern of RFLP bands, of a victim, suspect, and crime sample. Variations in the number of tandem repeated base sequences (**simple tandem repeats** or **STRs**) found in satellite DNA are now commonly used in forensic DNA fingerprint analysis. Forensic tests use only about five small regions of the genome that are known to be highly variable from one person to the next in order to provide a high statistical probability that matching DNA fingerprints come from the same individual.

Environmental Uses of DNA Technology   Genetically engineered microorganisms that are able to extract heavy metals, such as copper, lead, and nickel, may become important in mining and cleaning up mining waste. Microbes are used in sewage treatment plants to degrade many organic compounds into nontoxic form. Engineering organisms to degrade chlorinated hydrocarbons and other toxic compounds is an area of active research. Environmental disasters such as oil spills and waste dumps are other areas for which detoxifying microbes are being developed.

Agricultural Uses of DNA Technology   Vaccines and growth hormones for use in farm animals are being produced with DNA technology.

Transgenic organisms containing genes from other species are being developed for potential agricultural use. One use is for increased productivity. Another use, however, is as a **"pharm" animal** engineered to produce large quantities of a pharmaceutical product,

often by secretion in the animal's milk. Transgenic animals are produced by fertilizing egg cells *in vitro* and injecting their nuclei with a foreign gene. The eggs are then transplanted into surrogate mothers.

The ability to regenerate plants from single cells growing in tissue culture has made plant cells easier to genetically manipulate than animal cells. The bacterium *Agrobacterium tumefaciens* produces crown gall tumors in the plants it infects when its **Ti plasmid** (tumor inducing) integrates a segment of its DNA into the plant chromosomes. Using DNA technology, foreign genes are inserted into Ti plasmids whose disease-causing properties have been eliminated, and the recombinant plasmids are introduced into plant cells growing in culture. When these cells regenerate whole plants, the foreign gene is included in the plant genome.

Only dicots are susceptible to infection by *Agrobacterium.* Techniques such as electroporation and DNA guns are being used to transfer foreign genes into important monocots such as corn and wheat.

Half of the soybean and corn crops in the United States have been genetically modified, many with genes for herbicide resistance. Crop plants are being engineered to be resistant to infectious pathogens and insects. Research efforts are also focusing on producing more nutritional transgenic crops, as in the "golden" rice enriched with beta-carotene.

The production of bacteria with increased nitrogen-fixing potential and the engineering of plants that can fix nitrogen themselves are examples of a valuable potential use of DNA technology for improving plant productivity and reducing the use of expensive and polluting fertilizers.

DNA technology raises important safety and ethical questions (399)

Safety regulations have focused on potential hazards of engineered microbes. Most public concern, however, now centers on agricultural **genetically modified (GM) organisms,** which contain artificially acquired genes from the same or different species. Some fear that GM foodstuffs may be hazardous to human health. One environmental risk of transgenic plants is that their herbicide- or insect-resistant genes might pass to wild plants, creating "superweeds." An international Biosafety Protocol requires exporters to label GM organisms in bulk food shipments so that importing countries can decide on potential environmental or health risks. In the United States, several federal agencies evaluate and regulate new products and procedures.

Ethical questions about human genetic information include who should have access to information about a person's genome and how that information should be used. Potential ethical, environmental, and health issues must be considered in the development of these powerful genetic techniques and remarkable products of biotechnology.

WORD ROOTS

liga- = bound, tied (*DNA ligase:* a linking enzyme essential for DNA replication)

electro- = electricity (*electroporation:* a technique to introduce recombinant DNA into cells by applying a brief electrical pulse to a solution containing cells)

muta- = change; **-genesis** = origin, birth (*in vitro mutagenesis:* a technique to discover the function of a gene by introducing specific changes into the sequence of a cloned gene, reinserting the mutated gene into a cell, and studying the phenotype of the mutant)

poly- = many; **morph-** = form (*Single nucleotide polymorphisms:* one-base-pair variations in the genome sequence)

STRUCTURE YOUR KNOWLEDGE

1. Fill in the table below on the basic tools of gene manipulation used in DNA technology.
2. Describe several examples of the many possible applications for DNA technology in agriculture and in medicine.

TEST YOUR KNOWLEDGE

MULTIPLE CHOICE: *Choose the one best answer.*

1. The role of restriction enzymes in DNA technology is to
   a. provide a vector for the transfer of recombinant DNA.
   b. produce cDNA from mRNA.
   c. produce a cut (usually staggered) at specific recognition sequences on DNA.
   d. reseal "sticky ends" after base pairing of complementary bases.
   e. denature DNA into single strands that can hybridize with complementary sequences.

Technique or Tool	Brief Description	Some Uses in DNA Technology
Restriction enzymes	a.	
Gel electrophoresis	b.	
cDNA	c.	
Labeled probe	d.	
Southern blotting	e.	
DNA sequencing	f.	
PCR	g.	
RFLP analysis	h.	

2. This segment of DNA has restriction sites I and II, which create restriction fragments a, b, and c. Which of the following gels produced by electrophoresis would represent the separation and identity of these fragments?

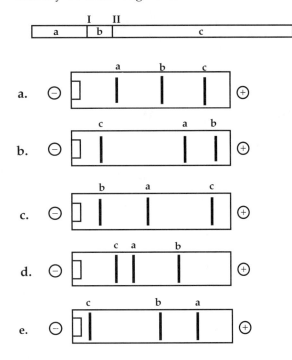

3. Yeast has become important in genetic engineering because it
   a. has RNA splicing machinery.
   b. has plasmids that can be genetically engineered.
   c. allows the study of eukaryotic gene regulation and expression.
   d. grows readily and rapidly in the laboratory.
   e. does all of the above.

4. Which of the following DNA sequences would most likely be a restriction site?
   a. AACCGG
      TTGGCC
   b. GGTTGG
      CCAACC
   c. AAGG
      TTCC
   d. AATTCCGG
      TTAAGGCC
   e. CTGCAG
      GACGTC

5. What is genomics?
   a. the public consortium effort to sequence the human genome
   b. the Celera shotgun approach to sequencing the human genome
   c. the sequencing and systematic study of whole genomes

   d. the use of gene therapy in the treatment of human diseases
   e. the use of nucleotide sequences to determine the function of all proteins encoded by a genome

6. Petroleum-lysing bacteria are being engineered for the treatment of oil spills. What is the most realistic danger of these bacteria to the environment?
   a. mutations leading to the production of a strain pathogenic to humans
   b. extinction of natural microbes due to the competitive advantage of the "petro-bacterium"
   c. destruction of natural oil deposits
   d. poisoning of the food chain
   e. contamination of the water

7. You are attempting to introduce a gene that imparts larval moth resistance to bean plants. Which of the following vectors are you most likely to use?
   a. phage DNA
   b. *E. coli* plasmid
   c. Ti plasmid
   d. yeast plasmid
   e. bacterial artificial chromosome

8. An attenuated virus
   a. is a virus that is nonpathogenic.
   b. in an elongated viral particle.
   c. can transfer recombinant DNA to other viruses.
   d. will not produce an immune response.
   e. is made with cDNA.

9. Which of the following is a difficulty in getting prokaryotic cells to express eukaryotic genes?
   a. The signals that control gene expression are different and prokaryotic promoter regions must be added to the vector.
   b. The genetic code differs between the two because prokaryotes substitute the base uracil for thymine.
   c. Prokaryotic cells cannot transcribe introns because their genes do not have them.
   d. The ribosomes of prokaryotes are not large enough to handle long eukaryotic genes.
   e. The RNA splicing enzymes of bacteria are different from those of eukaryotes.

10. Complementary DNA does not create as complete a gene library as the shotgun approach because
   a. it has eliminated introns from the genes.
   b. a cell produces mRNA for only a small portion of its genes.
   c. the shotgun approach produces more restriction fragments.

d. cDNA is not as easily integrated into plasmids or phage genomes.

e. restriction enzymes are not used to create cDNA.

11. Which of the following is *not* true of recognition sequences?
    a. Modification by methylation of bases within them prevents restriction of bacterial DNA.
    b. They are usually symmetrical sequences of four to eight nucleotides.
    c. They signal the attachment of RNA polymerase.
    d. Each recognition sequence is cut by a specific restriction enzyme.
    e. Cutting a recognition sequence in the middle of a functional and identifiable gene is used to screen clones that have taken up foreign DNA.

Use the following choices to answer questions 12–14.
    a. restriction enzyme
    b. reverse transcriptase
    c. ligase
    d. DNA polymerase
    e. RNA polymerase

12. Which is the first enzyme used in the production of cDNA?

13. Which enzyme is used in the polymerase chain reaction?

14. Which is the first enzyme used in the production of DNA fragments for DNA fingerprinting?

15. A plasmid has two antibiotic resistance genes, one for ampicillin and one for tetracycline. It is treated with a restriction enzyme that cuts in the middle of the ampicillin gene. DNA fragments containing a human globin gene were cut with the same enzyme. The plasmids and fragments are mixed, treated with ligase, and used to transform bacterial cells. Clones that have taken up the recombinant DNA are the ones that
    a. are blue and can grow on plates with both antibiotics.
    b. can grow on plates with ampicillin but not with tetracycline.
    c. can grow on plates with tetracycline but not with ampicillin.
    d. cannot grow with any antibiotics.
    e. can grow on plates with tetracycline and are not blue.

16. STRs are a valuable tool for
    a. forming overlapping sections in chromosome walking.
    b. infecting eudichotomous plant cells with recombinant DNA.

c. acting as probes in Southern blots.
d. DNA fingerprinting.
e. PCR to produce multiple copies of a DNA segment.

17. You have affixed the chromosomes from a cell onto a microscope slide. Which of the following would *not* make a good radioactively labeled probe to help map a particular gene to one of those chromosomes? (Assume DNA of chromosomes and probes is single stranded.)
    a. cDNA made from the mRNA transcribed from the gene
    b. a portion of the amino acid sequence of that protein
    c. mRNA transcribed from the gene
    d. a piece of the restriction fragment on which the gene is located
    e. a sequence of nucleotide bases determined from the genetic code needed to produce a known sequence of amino acids found in the protein product of the gene

18. If the first three nucleotides in a six-nucleotide recognition sequence are CTG, what would the next three nucleotides most likely be?
    a. AGG
    b. GTC
    c. CTG
    d. CAG
    e. GAC

19. Computer software is used to identify putative genes in the nucleotide sequences in data banks. Which of the following is the software probably *not* scanning for?
    a. promoters
    b. RNA-splicing sites
    c. STRs (simple tandem repeats)
    d. ESTs (expressed sequence tags)
    e. stop signals for transcription

20. The human genome appears to have only two times the number of genes as the simple nematode, *C. elegans*. How can humans be so much more complex than nematodes?
    a. The unusually long introns in human genes are involved in regulation of gene expression.
    b. More than one polypeptide can be produced from a gene by alternative splicing.
    c. Human genes code for many more types of domains.
    d. The human genome has a high proportion of noncoding DNA.
    e. The large number of SNPs (single nucleotide polymorphisms) in the human genome provides for a great deal of genetic variability.

21. What is a "pharm" animal?
    a. a transgenic animal that produces large quantities of a pharmaceutical product
    b. an animal used by the pharmaceutical industry to test new medical treatments
    c. a cloned animal that was produced from an adult cell nucleus inserted into an ovum
    d. a genetically engineered animal that produces more meat or milk
    e. a genetically modified organism whose production is permitted in the United States but not in the European Union

22. Which of the following processes or procedures does *not* involve any nucleic acid hybridization?
    a. separation of fragments by gel electrophoresis
    b. Southern blotting
    c. polymerase chain reaction
    d. DNA fingerprinting
    e. DNA microarray assay

23. Which of the following genomes has been completely (or almost completely) sequenced?
    a. nematode *(C. elegans)*
    b. human
    c. *E. coli* and yeast *(Saccharomyces cerevisiae)*
    d. fruit fly *(Drosophila melanogaster)*
    e. all of the above

24. Which of the following techniques has been used for ordering restriction fragments along a chromosome in the construction of a physical map?
    a. DNA microarray assay
    b. DNA sequencing by the Sanger method
    c. Southern blotting
    d. genetic linkage using recombination frequencies
    e. chromosome walking

25. Which of the following techniques can be used to determine the function of a newly identified gene?
    a. comparisons with genes of known functions that have similar sequences and whose products have similar domains
    b. *in vitro* mutagenesis or RNA interference
    c. DNA microarray assay
    d. both a and b
    e. a, b, and c

26. This restriction fragment contains a gene whose recessive allele is lethal. The normal allele has restriction sites for the restriction enzyme *PST*I at sites I and II. The recessive allele lacks restriction site I. An individual who had a sister with the lethal trait is being tested to determine if he is a carrier of that allele. Indicate which of these band patterns would be produced on a gel if he is a carrier (*heterozygous* for the gene)?

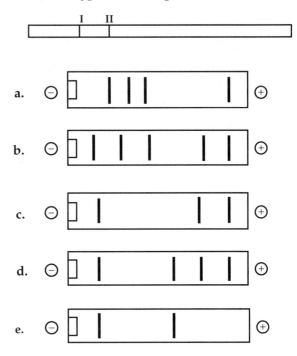

CHAPTER 21

THE GENETIC BASIS OF DEVELOPMENT

FRAMEWORK

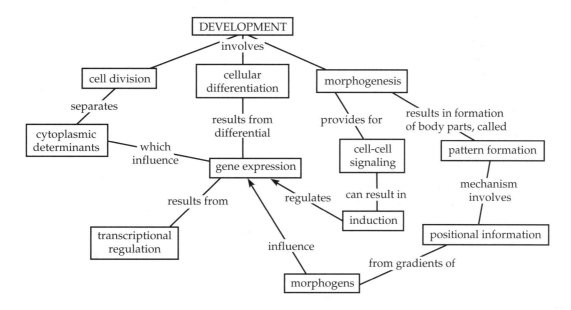

CHAPTER REVIEW

The application of molecular genetics and DNA technology, along with the study of developmental mutants, has revolutionized the field of development.

From Single Cell to Multicellular Organism

The steps from a zygote to a multicellular organism include differentiation of cell types as well as organization of cells into tissues, organs, and organ systems.

Embryonic development involves cell division, cell differentiation, and morphogenesis (403)

The three key processes of embryonic development are cell division, the production of large numbers of cells; **differentiation,** the formation of cells specialized in structure and function; and **morphogenesis,** the physical processes that produce body shape and form. In animal development, morphogenetic movements of cells and tissues produce body form. Growth

and morphogenesis continue throughout the life of a plant, relying on the perpetually embryonic regions in root and shoot tips known as **apical meristems** to produce new roots, shoots, and leaves.

Researchers study development in model organisms to identify general principles (403–406)

In order to study broad biological principles, researchers often choose a **model organism** that is representative of a larger group and well suited for answering particular types of research questions.

The fruit fly *Drosophila melanogaster* has a long history as a model organism for genetic studies. Although its early embryology differs from that of other animals, research on *Drosophila* development has provided key information about animal development.

The transparent nematode *Caenorhabditis elegans* has several advantages as a model organism: its ease of culture and short generation time, its sequenced genome, its hermaphroditic reproduction that allows for easy detection of recessive mutations, and its small number of cells that has allowed researchers to follow every cell division and determine a **cell lineage** by reconstructing the ancestry of every adult cell.

Researchers have already developed a great deal of genetic knowledge of the mouse *Mus musculus* and are now able to manipulate genes to make transgenic mice and mice in which certain genes are "knocked out" by mutation. Their large genomes and in utero development are disadvantages.

The zebrafish *Danio rerio* reproduces in large numbers and has transparent embryos that develop outside the mother.

Arabidopsis thaliana, a small plant in the mustard family, is easily grown in test tubes, is self-fertilizing, produces large numbers of offspring, and has a very small, already sequenced genome. Cells are easily grown in culture where they can be induced to take up foreign DNA.

■ INTERACTIVE QUESTION   21.1

What are some of the important criteria for model organisms chosen for the study of developmental genetics?

Differential Gene Expression

Cellular differentiation arises not from differences in the cells' genomes but from differences in gene expression as regulatory mechanisms turn specific genes on and off during development.

Different types of cells in an organism have the same DNA (406–410)

Nearly all cells of an organism have the same genes; they have *genomic equivalence.*

Totipotency in Plants   **Cloning** involves producing genetically identical individuals (**clones**) from a single somatic cell of a multicellular organism. Genomic equivalence was first demonstrated in plants when Steward was able to clone new carrot plants from root cells. In plants, differentiation does not result in irreversible changes in DNA, and most cells remain **totipotent,** retaining the ability to form a complete new organism.

Nuclear Transplantation in Animals   Early evidence of genomic equivalence in animals was provided by the work of Briggs, King, and Gurdon, in which

nuclei from embryonic and tadpole cells were transplanted into enucleated frog egg cells. The ability of the transplanted nucleus to direct normal development was inversely related to its developmental age.

In 1997, Wilmut and his colleagues reported their success at cloning an adult sheep by transplanting a nucleus from a fully differentiated udder cell into an unfertilized egg cell, and then implanting the resulting early embryo into a surrogate mother. The mammary cell was induced to dedifferentiate by culturing in a nutrient-poor medium that forced the cells into the G_0 phase of the cell cycle.

■ INTERACTIVE QUESTION   21.2

Although numerous mammals have now been cloned successfully, only a very small percentage of cloned embryos develop normally. What is a likely cause of this developmental failure?

The Stem Cells of Animals   **Stem cells** are relatively unspecialized cells that continue to reproduce and can, under proper conditions, differentiate into one or more types of cells. Adult stem cells are present in bone marrow and have been isolated from various tissues and grown in culture. Cells capable of producing multiple types of cells are called *pluripotent. Embryonic stem cells* taken from early embryos can be cultured indefinitely. Both types of stem cells can be induced to differentiate into specialized cells. Stem cell research has the potential to provide cells to repair organs that are damaged or diseased.

Different cell types make different proteins, usually as a result of transcriptional regulation (410–411)

A cell's developmental history leads to its eventual differentiation as a cell with a specific structure and function. **Determination** is a term used to describe the condition when a cell is irreversibly committed to its fate. When a cell becomes differentiated, it expresses genes for *tissue-specific proteins,* and that expression is usually controlled at the level of transcription.

The embryonic precursor cells from which muscle cells arise have the potential to develop into a number of different cell types. Researchers have isolated mRNA from *myoblasts,* prepared a library of cDNA genes, positioned the cloned genes next to an active promoter, and inserted them into separate embryonic precursor cells in order to identify muscle-determination genes. One of these master regulatory genes is *myoD*, which codes for a transcription factor named MyoD that binds to control elements and initiates transcription of other muscle-specific transcription factors. These secondary transcription factors then activate muscle-protein genes. MyoD also turns on genes that block the cell cycle and stop cell division.

■ INTERACTIVE QUESTION 21.3

The MyoD protein has been shown to be able to transform some, but not all, differentiated cells into muscle cells. Why doesn't it work on all kinds of cells?

Transcriptional regulation is directed by maternal molecules in the cytoplasm and signals from other cells (411–412)

The cytoplasm of an unfertilized egg cell contains maternal mRNA, proteins, and other substances that are unevenly distributed, and the first few mitotic divisions separate these components and expose cell nuclei to different environments. These maternal components of the egg cell that influence early development by regulating gene expression are called **cytoplasmic determinants.**

The other important source of developmental control comes from signals received from other embryonic cells. Change in the gene expression of target cells resulting from chemical or physical signals from other cells is called **induction.**

Genetic and Cellular Mechanisms of Pattern Formation

Pattern formation is the ordering of cells and tissues into their characteristic structures and locations. In animals, pattern formation takes place in embryos and juveniles; in plants, it occurs continually in the apical meristems.

The animal's basic body plan is laid out early in development. Cells and their progeny develop in response to molecular cues called **positional information** that tell a cell where it is located relative to the body axes and neighboring cells.

Genetic analysis of *Drosophila* reveals how genes control development: *an overview* (412–413)

The Life Cycle of Drosophila   A fruit fly's body consists of a series of segments grouped into the head, thorax, and abdomen. The anterior-posterior and dorsal-ventral axes are determined by positional information provided by cytoplasmic determinants present in the unfertilized egg. After fertilization, positional information establishes the proper number of segments and then triggers the formation of each segment's characteristic structures.

Each egg cell in the mother's ovary is surrounded by nurse cells and follicle cells that supply nutrients and other molecules needed for development. Eggs are laid following fertilization, and the first ten mitotic divisions occur without cytokinesis, yielding a multinucleated embryo. Nuclei migrate to the periphery, and plasma membranes divide the nuclei into separate cells. Segments become visible, organs form, and a wormlike larva hatches out of the egg shell. Following three larval stages and molts, the larva becomes a pupa. Metamorphosis produces an adult fly, with distinct segments bearing characteristic appendages.

Genetic Analysis of Early Development in Drosophila   In the 1940s, Lewis studied developmental mutants and was able to map certain mutations that control development in the late embryo to specific genes. In the late 1970s, Nüsslein-Volhard and Wieschaus undertook a search for the genes that control segment formation. They studied mutations that were **embryonic lethals,** which prevented the development of viable larvae. They performed a *saturation screen* in which they exposed flies to a chemical mutagen and then performed many thousands of crosses to detect recessive mutations causing death of embryos or larvae with abnormal segmentation. They identified 1,200 genes essential for development, 120 of which were involved in pattern formation leading to normal segmentation.

Gradients of maternal molecules in the early embryo control axis formation (414–415)

Maternal effect genes code for proteins or mRNA that are deposited in the unfertilized egg. These genes are also called **egg-polarity genes** because they determine the anterior-posterior and dorsal-ventral axes of the egg and consequently the embryo.

One egg-polarity gene is *bicoid*, and offspring of a mother defective for this gene have two tail regions and lack the front half of the body. The product of the *bicoid* gene is a **morphogen** concentrated at one end of the larva and responsible for determining its anterior end.

Researchers used the cloned *bicoid* gene as a probe and found *bicoid* mRNA concentrated in the most anterior end of egg cells. Following fertilization, the mRNA is translated into bicoid protein, which diffuses posteriorly, forming a gradient in the early embryo. Proteins whose gradients determine the posterior end and establish the dorsal-ventral axis have also been identified.

■ INTERACTIVE QUESTION 21.4

In saturation screening for mutations affecting development, the phenotypic effect of a maternal effect gene does not show up in the F_1 or F_2 generations, but is seen by studying the F_3 generation, in which some of the offspring die as embryos. Explain these results.

A cascade of gene activations sets up the segmentation pattern in *Drosophila*: a closer look (416)

The products of the egg-polarity genes are transcription factors. Gradients of these morphogens affect the expression of the embryo's **segmentation genes** that direct the formation of segments. In a cascade of gene activations, products of the **gap genes** first determine the basic subdivisions along the embryo; **pair-rule genes** establish the modular pattern in terms of segment pairs; and then the **segment polarity genes** determine the anterior-posterior axis of each segment. The products of many of the segmentation genes are also transcription factors that directly activate the next genes in the hierarchy of gene expression responsible for the segmentation pattern of the embryo. Some segmentation genes code for the cell-signaling molecules or their receptors, which are necessary for the cell-cell communication needed once plasma membranes have separated the cells of the embryo.

Homeotic genes direct the identity of body parts (417)

Master regulatory genes called **homeotic genes** determine each segment's anatomical fate by encoding transcription factors that control the expression of genes that build a segment's characteristic structures.

■ INTERACTIVE QUESTION 21.5

a. What could cause a mutant fly embryo to develop two tail ends but no head?

b. What could cause a mutant embryo to have half the normal number of segments?

c. What could cause a fruit fly to have legs growing out of its head in place of antennae?

Homeobox genes have been highly conserved in evolution (417)

A sequence of 180 nucleotides called a **homeobox** has been found in each *Drosophila* homeotic gene. The same or similar homeobox nucleotide sequences have been identified in genes often called *Hox* genes of many animals. Related sequences are found in regulatory genes of yeast and even prokaryotes. These similarities indicate that the homeobox must have arisen early and been conserved through evolution as part of genes involved in regulation of gene expression and development.

The homeobox sequence is translated into a 60-amino-acid sequence known as a *homeodomain*. This portion of the resulting transcription factor binds to DNA, while other, more variable domains interact with other transcription factors to recognize specific enhancers or promoters. Proteins with homeodomains probably coordinate the transcription of groups of developmental genes and thus control pattern formation.

■ INTERACTIVE QUESTION 21.6

In *Drosophila*, homeobox sequences have been found not only in the homeotic genes, but also in the egg-polarity gene *bicoid*, in several segmentation genes, and in the master regulatory gene for eye development. Is this just a coincidence? Explain.

Neighboring cells instruct other cells to form particular structures: cell signaling and induction in the nematode (418–421)

In induction, signaling from one group of cells to nearby cells influences the gene expression that leads to cellular differentiation.

Induction in Vulval Development Researchers have applied genetics, biochemistry, and embryology to study the development of the vulva in the nematode *C. elegans.* Six cells on the ventral surface of the second-stage larva are destined to give rise to the vulva. From studying mutants with abnormal vulval development, researchers have identified multiple genes and their products involved in vulval formation.

An *anchor cell* of the embryonic gonad secretes a signal protein that reaches the closest precursor cell in high levels. The signal is a growth-factor–like protein that activates a tyrosine-kinase receptor, a Ras protein, and a protein-kinase cascade in a pathway that leads to transcriptional regulation. This inducer stimulates cell division and differentiation to form the inner part of the vulva and also the production of a cell surface signaling protein that acts as an inducer to the two adjacent precursor cells. These cells are then stimulated to divide and differentiate into the outer vulva. The remaining three cells are too distant to receive either signal; they develop into epidermal cells.

■ INTERACTIVE QUESTION 21.7

The study of vulval development in *C. elegans* supports the following generalizations regarding the role of induction in development:

a. The pathway to organ formation often follows a series of steps that involve _____

_____.

b. The effect of an inducer on its possible target cells often depends on the _____ of the inducer.

c. Inducers often initiate _____

_____ in the target cell that lead to _____ _____ of important developmental genes.

Programmed Cell Death (Apoptosis) Lineage analysis of *C. elegans* has established that cell suicide or **apoptosis** occurs numerous times during normal development. Genetic screening has identified three key genes involved in cell death: *ced-9* that produces a protein that inhibits the activity of the protein products of *ced-3* and *ced-4.* In the case of apoptosis, protein activity is regulated, not transcription or translation. When a cell receives a "death" signal on its membrane receptor, Ced-9 protein becomes inactivated and the other suicide proteins initiate an activation cascade leading to the enzymatic hydrolysis of the cell's proteins and DNA. Ced-3 is the main caspase (protease) of apoptosis in the nematode. Signals from the dying cell stimulate neighboring cells to engulf and digest the remains. The more complicated apoptosis pathways in mammals involve proteins that are released from mitochondria.

■ INTERACTIVE QUESTION 21.8

a. What can one conclude from the fact that the apoptosis genes of nematodes and mammals are similar?

b. Give some examples of programmed cell death in humans.

Plant development depends on cell signaling and transcriptional regulation (421–423)

The processes of development evolved independently in plants and animals due to their ancient divergence. The rigid cell walls of plants restrict movement, and morphogenesis relies more on the orientation of cell divisions and cell enlargement. Embryonic development occurring in a seed is difficult to study, but research on apical meristems has provided insight into the genetics and control of plant development.

Cell Signaling in Flower Development Environmental cues trigger signal-transduction pathways through which shoot meristems are transformed to floral meristems. A floral meristem consists of three cell layers, from which the organs of the flower—the carpels, stamens, petals, and sepals—arise. Grafting stems of mutant tomato plants onto wild-type plants produces shoots from the graft site that are **chimeras,** having a mixture of cells from different genetic backgrounds. The number of organs produced in flowers of the chimeras depends on the parental source of the innermost cell layer, which must somehow induce the two outer layers to produce its genetically determined number of organs.

■ INTERACTIVE QUESTION 21.9

This diagram shows the active genes in each whorl and the resulting anatomy of a wild-type flower.

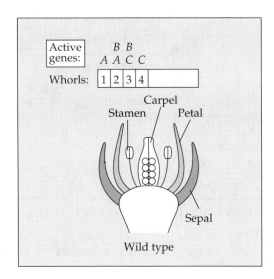

a. Fill in the table below to show which organs are produced in the whorls in a normal flower. In a mutant that lacks a functional gene *A*, what gene expression pattern and resulting flower organ arrangement would be produced? (Remember that a lack of *A* activity removes the inhibition of gene *C*.)

Whorl	Genes Active	Organs in Normal	Genes Active in Mutant *A*	Organs in Mutant *A*
1	*A*			
2	*AB*			
3	*BC*			
4	*C*			

b. If you had a double-mutant plant that had no gene activity for *B* or *C*, what would the resulting flower look like?

Organ Identity Genes in Plants Analogous to homeotic genes in animals are plant **organ identity genes** that determine what type of structure develops from a whorl of floral meristem. These genes have been most extensively studied with floral mutants in *Arabidopsis*. Three classes of genes, *A, B,* and *C*, were first identified. A model predicts that each gene affects two adjacent whorls. Using cloned genes as *in situ* probes, researchers found that mRNA from the predicted transcribed genes was located in the expected whorls. These results explained the phenotypes of mutants for organ identity genes, with the additional observation that activity of the *A* or *C* gene inhibited the other gene's activity. A mutant lacking a functional *A* gene would have *C* expression in whorls where *A* should be expressed. The organ identity genes contain a sequence that encodes a DNA-binding domain and appear to act as master regulatory genes that control the transcription of many other genes responsible for organ development. A fourth class of organ identity genes, class *E*, was discovered in 2001.

WORD ROOTS

apic- = tip (*apical meristem:* embryonic plant tissue in the tips of roots and in the buds of shoots that supplies cells for the plant to grow in length)

morph- = form; **-gen** = produce (*morphogen:* a substance that provides positional information in the form of a concentration gradient along an embryonic axis)

toti- = all; **-potent** = powerful (*totipotent:* the ability of a cell to form all parts of the mature organism)

STRUCTURE YOUR KNOWLEDGE

1. Describe the cascade of gene activations that leads to the development of a fruit fly.

2. How might the mechanism for transcriptional regulation differ for cytoplasmic determinants and the cell-cell signaling involved in induction?

TEST YOUR KNOWLEDGE

MULTIPLE CHOICE: *Choose the one best answer.*

1. Which of the following is descriptive of a cell that is differentiated?
   a. The cell's development fate has been decided, although it may not look any different from another precursor cell.
   b. The cell has a definite anterior-posterior and dorsal-ventral axis.
   c. The cell has developed cell surface receptors that allow it to receive signals from other cells.
   d. The cell has been induced to transcribe all its genes.
   e. The cell is producing tissue-specific proteins and has its characteristic structure.

2. Morphogenesis in plants results from
   a. migration of cells from the apical meristems to produce three tissue layers.
   b. differences in the plane of cell division and the direction of cell expansion.
   c. cytoplasmic determinants that were deposited in the egg cell.
   d. the differentiation of cells within the apical meristem.
   e. the production of cell walls and plasmodesmata that communicate between cells.

3. In which of these model organisms has it been possible to create a cell lineage?
   a. *Drosophila melanogaster* (fruit fly)
   b. *Danio rerio* (zebrafish)
   c. *Caenorhabditis elegans* (nematode)
   d. *Mus musculus* (mouse)
   e. *Arabidopsis thaliana* (common wall cress)

4. Cytoplasmic determinants are
   a. unevenly distributed cytoplasmic components of an unfertilized egg.
   b. often involved in transcriptional regulation.
   c. often separated in the first few mitotic divisions following fertilization.
   d. maternal contributions that help to direct the initial stages of development.
   e. all of the above.

5. For which of the following does the cloning of Dolly the sheep provide evidence?
   a. totipotency of most adult animal cells
   b. determination of most adult animal cells
   c. embryonic nature of mammary cells
   d. genomic equivalence of most animal cells
   e. immune tolerance of embryos in mammals

6. The fact that transplanted nuclei from most tadpole cells were unable to direct normal development in an enucleated frog cell gives evidence for
   a. the differentiation of adult cells resulting in the deletion of certain key genes.
   b. the totipotency of adult cells even when they are differentiated.
   c. changes in chromatin during development that may make genes no longer available for transcription.
   d. the dedifferentiation of embryonic cells.
   e. the need for cytoplasmic determinants to direct initial developmental stages.

7. Pattern formation in animals is based on
   a. the first few mitotic divisions.
   b. the induction of cells by the initial fertilized egg.

c. the locations and activity of apical meristems.
   d. differentiation of cells, which then migrate together to form tissues and organs.
   e. positional information a cell receives from gradients of morphogens.

8. Which of the following developmental processes involves apoptosis?
   a. the development of the vulva in *C. elegans*
   b. the sequential activation of segmentation genes in *Drosophila*
   c. the production of genetically engineered mice with "knockout" genes
   d. the development of separate fingers and toes during mammalian development
   e. the induction of tissue layers in the production of the organs from a floral meristem

9. A fruit fly that has two sets of wings growing from its thorax (instead of a single pair of wings and a pair of small balancing organs) would probably have mutations in its
   a. gap genes.
   b. segment polarity genes.
   c. homeotic genes.
   d. egg-polarity genes.
   e. pair-rule genes.

10. What would be the fate of a *Drosophila* larva that inherits two copies of a mutant *bicoid* gene (one mutant allele from each heterozygous parent)?
    a. It develops two heads, one at each end of the larva.
    b. It develops two tails, one at each end of the larva.
    c. It develops normally but produces mutant larvae that have two tail regions.
    d. It develops into an adult with legs growing out of its head.
    e. It receives no *bicoid* mRNA from the nurse cells of its mother.

11. A highly conserved nucleotide sequence that has been found in master regulatory genes in many diverse organisms is called a
    a. homeobox.
    b. homeodomain.
    c. transcription factor.
    d. homeotic gene.
    e. morphogen.

12. The gene *ced-9* codes for a protein that inactivates the proteins of suicide genes found in the genome of *C. elegans*. For development to proceed normally, the *ced-9* gene
    a. should be activated in all cells, but its protein product must remain inactive.
    b. should be activated in all cells, but its product will be inactivated when cells programmed to die receive the proper signal.
    c. should be inactivated in all cells except when a cell receives a signal to die.
    d. should be inactivated only in those cells that must die for proper development to occur.
    e. should code for a transcription factor that attaches to the enhancers of other suicide genes.

13. Once the developmental fate of a cell is set, the cell is said to be
    a. determined.
    b. differentiated.
    c. totipotent.
    d. genomically equivalent.
    e. regulated.

14. Which of the following is *not* true of adult stem cells?
    a. They have been found not only in bone marrow, but also in other tissues, including the adult brain.
    b. Although more difficult to grow than embryonic stem cells, they have been successfully grown in culture and made to differentiate into specialized cells.
    c. They are differentiated cells that can be induced to dedifferentiate and become totipotent.
    d. If pluripotent, they are capable of developing into different types of cells under appropriate conditions.
    e. These relatively unspecialized cells continually reproduce themselves.

15. In this hypothetical embryo, a high concentration of a morphogen called morpho is needed to activate gene *P*; gene *Q* is active at medium concentrations of morpho or above; and gene *R* is expressed as long as there is any quantity of morpho present. A different morphogen called phogen has the following effects: activates gene *S* and inactivates gene *Q* when at medium to high concentrations. If morpho and phogen are diffusing from where they are produced at the opposite ends of the embryo, which genes will be expressed in region 2 of this embryo? (Assume diffusion through the three regions from high at source to medium to low concentration.)

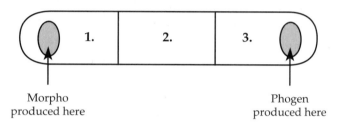

Morpho produced here   Phogen produced here

    a. genes *P, Q, R,* and *S*
    b. genes *P, Q,* and *R*
    c. genes *Q* and *R*
    d. genes *R* and *S*
    e. gene *R*

16. What do most master regulatory genes do?
    a. produce mRNAs that function as cytoplasmic determinants
    b. code for tissue-specific proteins that differentiate cells
    c. produce the proteins that function as inducers to neighboring cells
    d. produce transcription factors that coordinate the expression of other transcription factors and groups of developmental genes
    e. either c or d

MECHANISMS OF EVOLUTION

CHAPTER 22

DESCENT WITH MODIFICATION: A DARWINIAN VIEW OF LIFE

FRAMEWORK

This chapter describes Darwin's formulation of evolution—descent from a common ancestor modified by the mechanism of natural selection, resulting in the evolution of species adapted to their environments. The scientific and philosophical climate of Darwin's day was quite inhospitable to the implications of evolution, but most biologists accepted the theory of evolution quite rapidly. Only later was natural selection recognized as a mechanism of evolution. Evidence for evolution is drawn from homologies (structural, developmental, and molecular), biogeography, and the fossil record.

CHAPTER REVIEW

Charles Darwin presented the first convincing case for evolution in his book *On the Origin of Species by Means of Natural Selection,* published in 1859. Darwin made two major claims: The species present on Earth today descended from ancestral species, and **natural selection** is the mechanism for evolution. Natural selection leads to **evolutionary adaptation,** as individuals with beneficial heritable traits leave more offspring, and the frequency of such traits in a population increases over generations. **Evolution** may be defined as the changes in a population's genetic composition over time.

The Historical Context for Evolutionary Theory

Western culture resisted evolutionary views of life (429–430)

Darwin's book challenged both the prevailing scientific views and the world view that had been held for centuries in Western culture.

The Scale of Nature and Natural Theology The Greek philosopher Plato believed in two worlds, an ideal and eternal real world and the illusory world perceived by the senses. Aristotle believed that all living forms could be arranged on a "scale of nature" of increasing complexity in which each group of organisms was permanent and perfect.

The Judeo-Christian account of creation embedded the idea of the fixity of species in Western thought. Biology in the 1700s was dominated by **natural theology,** the study of nature to reveal the Creator's plan. One of the goals of natural theology was to classify the species that God had created. Linnaeus developed both a binomial system for naming organisms according to their genus and species, and a hierarchy of classification groupings. **Taxonomy,** the branch of biology that names and classifies organisms, originated in the work of Linnaeus.

Cuvier, Fossils, and Catastrophism **Fossils** are remnants or impressions of organisms laid down in rock, usually **sedimentary rocks** formed through the compression of layers of sand and mud into superimposed layers called strata. Fossils from strata of different ages reveal that a succession of organisms has existed on Earth.

Cuvier may be considered the father of **paleontology,** the study of fossils. Advocating **catastrophism,** he maintained that the differences he observed in the fossils found in different strata were the result of local catastrophic events such as floods or drought and were not indicative of evolution.

Theories of geologic gradualism helped clear the path for evolutionary biologists (430–431)

Gradualism, the idea that immense change is the cumulative result of slow but continuous processes, was proposed by Hutton in 1795 to explain the geologic state of the Earth. Lyell, a contemporary of Darwin, extended gradualism to a theory of **uniformitarianism,** proposing that the rates and effects of geologic processes have remained the same through Earth's history.

Darwin took two ideas from the observations of Hutton and Lyell: The Earth must be very old if geologic change is slow and gradual, and very slow processes can produce substantial change.

Lamarck placed fossils in an evolutionary context (431)

Lamarck published a theory of evolution in 1809. He explained the mechanism of evolution with two principles: The use or disuse of body parts leads to their development or deterioration, and acquired characteristics can be inherited. With present genetic knowledge, Lamarck's ideas are sometimes ridiculed. His theory, however, presented several key evolutionary ideas: that evolution is the best explanation for the fossil record and the diversity of life, that Earth is very old, and that adaptation to the environment is the main result of evolution.

The Darwinian Revolution

Field research helped Darwin frame his view of life: *the process of science* (432–433)

The Voyage of the Beagle   Darwin was 22 years old when he sailed from Great Britain on the H.M.S. *Beagle.* He spent the voyage collecting thousands of specimens of the fauna and flora of South America, observing the various adaptations of organisms living in very diverse habitats, and making special note of the geographic distribution of the taxonomically related species of South America. He was particularly struck by the uniqueness of the fauna of the Galápagos Islands. Darwin also read and was influenced by Lyell's *Principles of Geology.*

■ **INTERACTIVE QUESTION   22.1**

a. Match the theory or philosophy and its proponent(s) with the following descriptions.

A. catastrophism	**a.** Aristotle
B. inheritance of acquired characteristics	**b.** Cuvier
C. gradualism	**c.** Darwin
D. natural selection	**d.** Hutton
E. natural theology	**e.** Lamarck
F. scale of nature	**f.** Linnaeus
G. uniformitarianism	**g.** Lyell

Theory   Proponent

1. ____ ____ Discovery of the Creator's plan through the classification of species

2. ____ ____ History of Earth marked by floods or droughts that resulted in extinctions

3. ____ ____ Early explanation of mechanism of evolution

4. ____ ____ Profound change is the cumulative product of slow but continuous processes

5. ____ ____ Fixed species on a continuum from simple to complex

6. ____ ____ Differential reproductive success leads to adaptation to environment and evolution

7. ____ ____ Geologic processes have constant rates throughout time

b. Now place 1 through 7 in chronological order.

____ ____ ____ ____ ____ ____ ____

Darwin's Focus on Adaptation   Darwin began to link the origin of new species to the process of adaptation to different environments. In 1844, he wrote an essay on the origin of species and natural selection but did not publish it. In 1858, Darwin received Wallace's manuscript describing an identical theory of natural selection. Wallace's paper and extracts of Darwin's unpublished essay were jointly presented to the Linnaean Society, and Darwin published *On the Origin of Species* the next year. Within a decade, Darwin's book and its defenders had convinced the majority of biologists that evolution was the best explanation for the diversity of life.

The Origin of Species developed two main points: the occurrence of evolution and natural selection as its mechanism (434–437)

Descent with Modification   Darwin's concept of **descent with modification** included the notion that all organisms were related through descent from some unknown ancestor and had developed increasing modifications as they adapted to various habitats. The history of life is analogous to a tree with a common ancestor at the fork of each new branch. The taxonomy developed by Linnaeus provided a hierarchical organization of groups that suggested to Darwin this branching tree of life. Species along one branch are grouped into the same family or order or class (depending on where on the branching sequence one looks) and are more closely related to each other than to species from a different branch.

Natural Selection and Adaptation   Evolutionary biologist Ernst Mayr described Darwin's theory of natural selection as follows:

Observation 1:   Species have the potential for their population size to increase exponentially.

Observation 2:   Most population sizes are stable.

Observation 3:   Environmental resources are limited.

Inference 1:   Since only a fraction of offspring survive, there is a struggle for limited resources.

Observation 4:   Individuals vary within a population.

Observation 5:   Much of this variation is inherited.

Inference 2:   Individuals whose inherited characteristics fit them best to the environment are likely to leave more offspring.

Inference 3:   Unequal reproduction leads to the gradual accumulation of favorable characteristics in a population over generations.

■ **INTERACTIVE QUESTION   22.2**

Summarize in your own words Darwin's theory of natural selection as the mechanism of evolution.

Darwin found support for the struggle for existence and the capacity of organisms to overproduce in the essay on human population growth published by Malthus in 1798.

Artificial selection used in the breeding of domesticated plants and animals provided Darwin with evidence that selection among the variations present in a population can lead to substantial changes. He reasoned that natural selection, working over thousands of generations, could gradually create the modifications essential for the present diversity of life. Gradualism is basic to the Darwinian view of evolution.

Natural selection results in the evolution of populations, groups of interbreeding individuals of the same species in a common geographic area. Evolution is measured only as change in the relative proportions of variations in a population over time. Natural selection affects only those traits that are heritable—acquired characteristics cannot evolve. And natural selection is a local and temporal phenomenon, depending on the specific environmental factors present in a region at a given time.

Examples of natural selection provide evidence of evolution (437–438)

Natural Section in Action: The Evolution of Insecticide-Resistant Insects   The evolution of insecticide resistance in insects illustrates two facets of natural selection: It is an editing, not a creative, mechanism that selects for variations already present in a population. And it is regional and temporal, selecting for traits that fit the local environment at that current time. Spraying with insecticides may initially kill 99% of the insects, but those few that survive because of a genetic trait that makes them resistant will reproduce and pass that trait on to their offspring, creating a population of insecticide-resistant organisms.

■ **INTERACTIVE QUESTION   22.3**

The Evolution of Drug-Resistant HIV. Within a few weeks of treatment with the drug 3TC, a patient's HIV population consists entirely of 3TC-resistant HIV. Explain how this rapid evolution of drug resistance is an example of natural selection.

Other evidence of evolution pervades biology (438–441)

Homology   Homology is the term for a similarity resulting from common ancestry. The forelimbs of all mammals are **homologous structures,** containing the same skeletal elements regardless of function or external shape. Comparative anatomy illustrates that evolution is a remodeling process in which ancestral structures become modified for new functions. Some homologous structures that differ greatly in adult form and function are more evident during embryonic development.

Vestigial organs are rudimentary structures, of little or no value to the organism, that are historical remnants of ancestral structures.

Homologies can be seen on a molecular level. DNA, RNA, and an essentially universal genetic code, which have been passed along through all branches of evolution, are important evidence that all forms of life descended from the earliest organisms and are thus related. Homology is evident on different hierarchical levels, reflecting evolutionary history and the degree of relationship among organisms. Closely related species have a larger proportion of DNA and proteins in common than do more distantly related species.

Biogeography   The geographic distribution of species, or **biogeography,** provides evidence for evolution. Islands often have **endemic** species that are related to species on the nearest island or mainland. Widely separated areas having similar environments are more likely to have species that are taxonomically related to those of their region, regardless of environment, than to each other. Biogeographic distribution patterns are explained by evolution; modern species are found where they are because they evolved from ancestors that inhabited those regions.

The Fossil Record   The major branches of evolutionary descent established with evidence from homologies and molecular biology are also supported by the sequence of fossil forms found in the fossil record. Paleontologists continue to discover transitional fossils linking modern species to their ancestral forms.

What is theoretical about the Darwinian view of life? (441–442)

The evolution of modern species from ancestral forms is supported by historical facts such as fossils, biogeography, and molecular biology. The second of Darwin's claims, that natural selection is the main

---

■ INTERACTIVE QUESTION   22.4

Complete the following concept map that summarizes the main sources of evidence for evolution.

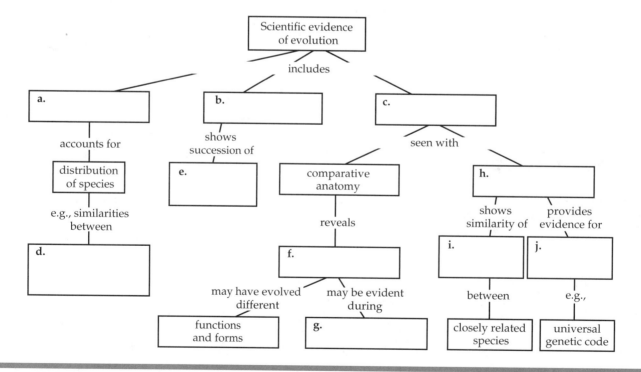

mechanism of evolution, is a theory that explains the historical facts of evolution. A scientific "theory" is a unifying concept with broad explanatory power and predictions that have been and continue to be tested by experiments and observations.

WORD ROOTS

bio- = life; **geo-** the Earth (*biogeography:* the study of the past and present distribution of species)

end- = within (*endemic:* a type of species that is found only in one region and nowhere else in the world)

homo- = like, resembling (*homology:* similarity in characteristics resulting from a shared ancestry)

paleo- = ancient (*paleontology:* the scientific study of fossils)

taxo- = arrange (*taxonomy:* the branch of biology concerned with naming and classifying the diverse forms of life)

vestigi- = trace (*vestigial organs:* structures of marginal, if any, importance to an organism, historical remnants of structures that had important functions in ancestors)

STRUCTURE YOUR KNOWLEDGE

1. Briefly state the main components of Darwin's theory of evolution.

TEST YOUR KNOWLEDGE

MULTIPLE CHOICE: *Choose the one best answer.*

1. The classification of organisms into hierarchical groups is called
   a. the scale of nature.
   b. taxonomy.
   c. natural theology.
   d. biogeography.
   e. natural selection.

2. The study of fossils is called
   a. phylogeny.
   b. gradualism.
   c. paleontology.
   d. anthropology.
   e. biogeography.

3. To Cuvier, the differences in fossils from different strata were evidence for
   a. changes occurring as a result of cumulative but gradual processes.
   b. divine creation.
   c. evolution by natural selection.
   d. continental drift.
   e. local catastrophic events such as droughts or floods.

4. Darwin proposed that new species evolve from ancestral forms by
   a. the gradual accumulation of adaptations to changing or different environments.
   b. the inheritance of acquired adaptations to the environment.
   c. the struggle for limited resources.
   d. the accumulation of mutations.
   e. the exponential growth of populations.

5. The best description of natural selection is
   a. the survival of the fittest.
   b. the struggle for existence.
   c. the reproductive success of the members of a population best adapted to the environment.
   d. the overproduction of offspring in environments with limited natural resources.
   e. a change in the proportion of inheritable variations within a population.

6. The remnants of pelvic and leg bones in a snake
   a. are vestigial structures.
   b. show that lizards evolved from snakes.
   c. are homologous structures.
   d. provide evidence for inheritance of acquired characteristics.
   e. resulted from artificial selection.

7. The hypothesis that whales evolved from land-dwelling ancestors is supported by
   a. evidence from the biogeographic distribution of whales.
   b. molecular comparisons of whales, fish, and reptiles.
   c. historical accounts of walking whales.
   d. the ability of captive whales to be trained to walk.
   e. fossils of extinct whales found in Egypt and Pakistan that had small hind limbs.

8. Darwin's claim that all of life descended from a common ancestor is best supported with evidence from
   a. the fossil record.
   b. comparative embryology.
   c. taxonomy.
   d. molecular biology.
   e. comparative anatomy.

9. The smallest unit that can evolve is
   a. a genome.
   b. an individual.
   c. a species.
   d. a population.
   e. a community.

10. Which of the following would *not* be considered part of the process of natural selection?
    a. Many of the variations among individuals in a population are heritable.
    b. More offspring are produced than are able to survive and reproduce.
    c. Individuals with traits best adapted to the environment are likely to leave more offspring.
    d. Many adaptive traits may be acquired during an individual's lifetime, helping that individual to evolve.
    e. Differential reproductive success leads to gradual change in a population.

11. When cytochrome *c* molecules are compared, yeasts and molds are found to differ by approximately 46 amino acids per 100 residues (amino acids making up a protein), whereas insects and vertebrates are found to differ by approximately 29 amino acids per 100 residues. What can one conclude from these data?
    a. Very little can be concluded unless the DNA sequence for the cytochrome *c* genes are compared.
    b. Yeasts evolved from molds, but vertebrates did not evolve from insects.
    c. Insects and vertebrates diverged from a common ancestor more recently than did yeasts and molds.
    d. Yeasts and molds diverged from a common ancestor more recently than did insects and vertebrates.
    e. The evolution of cytochrome *c* occurred more rapidly in yeasts and molds than in insects and vertebrates.

12. All of the following influenced Darwin as he synthesized the theory of evolution by natural selection *except*
    a. the biogeographic distribution of species such as the finches on the Galápagos Islands.
    b. Lyell's book, *Principles of Geology*, on the gradualness of geologic changes.
    c. Linnaeus' hierarchical classification of species, which could be interpreted as evidence of evolutionary relationships.

d. examples of artificial selection that produce rapid changes in domesticated species.
e. Mendel's paper in which he described his "laws of inheritance."

13. What might you conclude from the observation that the bones in your arm and hand are similar to the bones that make up a bat's wing?
    a. The bones in the bat's wing are vestigial structures, no longer useful as "arm" bones.
    b. The bones in a bat's wing are homologous to your arm and hand bones.
    c. Bats and humans evolved in the same geographic area.
    d. Bats lost their opposable digits during the course of evolution.
    e. Our ancestors could fly.

14. The best description of endemic species are species that are
    a. found only on islands.
    b. found in the same geographic area.
    c. found only on mainlands.
    d. found only in that location and nowhere else on Earth.
    e. disease causing and pesticide resistant.

15. Which of the following is an example of convergent evolution?
    a. remnants of pelvic girdles and limb bones found in whales
    b. two very different plants that are found in different habitats, but evolved from a fairly recent common ancestor
    c. similarities between the marsupial sugar glider and the eutherian flying squirrel
    d. the remodeling of a vertebrate forelimb in the evolution of a bird wing
    e. the many different bill sizes and shapes of finches on the Galápagos Islands

16. Chimpanzees and humans share many of the same genes, indicating that
    a. the two groups belong to the same species.
    b. the two groups belong to the same phylum.
    c. the two groups share a relatively recent common ancestor.
    d. humans evolved from chimpanzees.
    e. chimpanzees evolved from humans.

CHAPTER 23

THE EVOLUTION OF POPULATIONS

FRAMEWORK

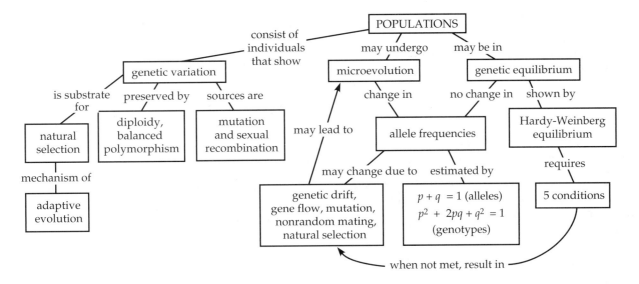

CHAPTER REVIEW

Although it is individuals that are selected for or against by natural selection, it is populations that actually evolve. Differential reproductive success of individuals with favorable characteristics leads to microevolution, changes in the allele frequencies of a population.

Population Genetics

The modern evolutionary synthesis integrated Darwinian selection and Mendelian inheritance: *the process of science* **(446)**

Most biologists rapidly accepted evolution but not Darwin's proposal of natural selection as its mechanism. Without a theory of genetics, the occurrence of chance variations and their transmission from parents to offspring could not be explained. When Mendel's work was rediscovered in the early 1900s, many geneticists believed that Darwin's focus on the inheritance of quantitative traits that vary on a continuum

could not be explained by the inheritance of discrete Mendelian traits.

The emergence of **population genetics** in the 1930s, with its emphasis on quantitative inheritance and genetic variation within populations, reconciled Mendelism with Darwinism.

In the early 1940s, scientists developed a comprehensive theory of evolution, known as the **modern synthesis,** that emphasized the importance of populations as the units of evolution, the essential role of natural selection, and the gradualness of evolution.

A population's gene pool is defined by its allele frequencies (446–447)

A **population** is a localized group of individuals of the same species. A **species** is a group of populations whose members have the ability to interbreed in nature. Within the geographic range of a species, populations may be totally isolated or contiguous with members concentrated in population centers.

The **gene pool** is the term for all the genes present in a population at any given time. For individuals of a diploid species, the pool includes two alleles for each gene locus. If all individuals are homozygous for the same allele, the allele is said to be *fixed*. More often, two or more alleles are present in the gene pool in some relative proportion or frequency.

■ INTERACTIVE QUESTION   23.1

In a population of 200 mice, 98 are homozygous dominant for brown coat color (*BB*), 84 are heterozygous (*Bb*), and 18 are homozygous recessive (*bb*).

a. The allele frequencies of this population are _____ *B* allele _____ *b* allele.

b. The genotype frequencies of this population are _____ *BB* _____ *Bb* _____ *bb*.

The Hardy-Weinberg theorem describes a nonevolving population (447–449)

In the absence of selection pressure and other agents of change, the gene pool of a population will remain constant from one generation to the next, in spite of the shuffling of alleles by meiosis and random fertilization. This stasis is formulated as the **Hardy-Weinberg theorem,** named for its originators.

Hardy-Weinberg Equilibrium   The allele frequency within a population determines the proportion of gametes that will contain that allele. The random combination of gametes will yield offspring with genotypes that reflect and reconstitute the allele frequencies of the previous generation. The frequencies of both alleles and genotypes will remain stable in a population that is in **Hardy-Weinberg equilibrium.**

The Hardy-Weinberg Equation   With the **Hardy-Weinberg equation,** the frequencies of alleles within a population can be calculated from the genotype frequencies and vice versa. In a simple case of having only two alleles at a particular gene locus, the letters p and q represent the proportions of the two alleles within the population, and their combined frequencies must equal $1 : p + q = 1$. The frequencies of the genotypes in the offspring reflect the frequencies of the alleles and the probability of each combination. According to the rule of multiplication, the probability that two gametes containing the same allele will come together in a zygote is equal to ($p \times p$) or p^2, or ($q \times q$) or q^2. A p and q allele can combine in two different ways, depending on which parent contributes which allele; therefore, the frequency of a heterozygous offspring is equal to $2pq$. The sum of the frequencies of all possible genotypes in the population adds up to $1 : p^2 + 2pq + q^2 = 1$.

■ INTERACTIVE QUESTION   23.2

Use the allele frequencies you determined in Interactive Question 23.1 to predict the genotype frequencies of the next generation.

Frequencies of

B (p) = _____ b (q) = _____

$BB = p^2 =$ _____ $Bb = 2pq =$ _____ $bb = q^2 =$ _____

Population Genetics and Health Science   If the frequency of homozygous recessive individuals is known (q^2), then the frequency of q may be estimated as the square root of q^2 (assuming the population is in Hardy-Weinberg equilibrium for that gene). From our example above, $q = \sqrt{0.09} = 0.3$. If the frequency of individuals with a recessively inherited disease is known (q^2), then the frequency of carriers of that recessive allele in the population can be calculated.

■ INTERACTIVE QUESTION   23.3

Practice using the Hardy-Weinberg equation so that you can easily determine genotype frequencies from allele frequencies and vice versa.

a. The allele frequencies in a population are $A = 0.6$ and $a = 0.4$. Predict the genotype frequencies for the next generation.

   *AA*_____   *Aa* _____   *aa* _____

b. What would the allele frequencies be for the generation you predicted above in part a.?

   *A*_____   *a* _____

c. Suppose that one gene locus determines stripe pattern in skunks. *SS* skunks have two broad stripes; *Ss* skunks have two narrow stripes; *ss* skunks have white speckles down their backs. A sampling of a population of skunks found 65 broad-striped skunks, 14 narrow-striped, and 1 speckled. Determine the allele frequencies.

   *S*_____   *s*_____

The Hardy-Weinberg Theorem and Genetic Variation   The Hardy-Weinberg theorem explains how Mendelian inheritance maintains genetic variation within a population. Hereditary factors are not blended in offspring; the alleles remain independent and are passed from generation to generation, preserving the genetic variation necessary for natural selection to act.

The Assumptions of the Hardy-Weinberg Theorem
Hardy-Weinberg equilibrium is maintained only if all of the following five conditions are met: a very large population, no migration between populations, no net changes in the gene pool due to mutation, random mating, and no natural selection.

Causes of Microevolution

Microevolution is a generation-to-generation change in a population's allele frequencies (450)

Hardy-Weinberg equilibrium serves as a baseline against which allele frequencies of a population can be compared to determine whether evolution is occurring. Evolution is the generation-to-generation change in allele frequencies. These changes in a population's gene pool are referred to as **microevolution.**

The two main causes of microevolution are genetic drift and natural selection (450–452)

The four causes of microevolution arise from departures from one of the conditions of Hardy-Weinberg equilibrium. Natural selection is the only factor that tends to increase the fitness of a population to its environment by selecting for favorable traits. Nonrandom mating can affect genotype frequencies, but does not affect allele frequencies.

Genetic Drift Chance deviations from expected results are more likely to occur in a small sample due to sampling error. Microevolution may result from **genetic drift,** a change in a small population's allele frequencies due to chance.

The **bottleneck effect** occurs when some disaster or other factor reduces the population size dramatically, and the few surviving individuals are unlikely to represent the genetic makeup of the original population. Genetic drift will remain a factor until the population grows large enough for chance events to be less significant. A bottleneck usually reduces variability because some alleles are lost from the gene pool.

Genetic drift that occurs when only a few individuals colonize a new area is known as the **founder effect.** Allele frequencies in the small sample are unlikely to be representative of the parent population, and genetic drift will affect the gene pool of the new population until it is larger.

Natural Selection For the Hardy-Weinberg equilibrium to be maintained, there must be no **natural selection,** no differential success in reproduction—a condition that is probably never met. Individuals that are more successful in producing viable, fertile offspring pass their alleles to the next generation in disproportionate number. Natural selection is likely to be adaptive; favorable genotypes are increased and maintained in a population.

■ INTERACTIVE QUESTION 23.4

Fill in the following concept map that summarizes the main causes of microevolution. Better still, create your own concept map to help you review microevolution.

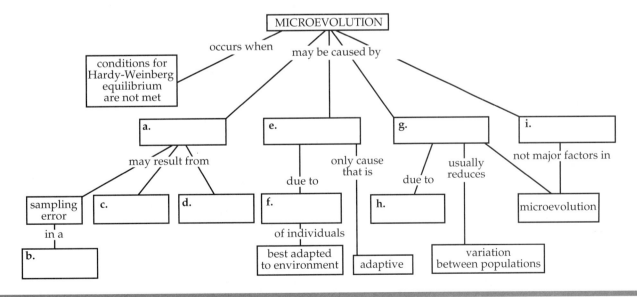

Gene Flow   **Gene flow,** the migration of individuals or the transfer of gametes between populations, may change allele frequencies. Differences in allele frequencies between populations, which may have developed by natural selection or genetic drift, tend to be reduced by gene flow.

Mutation   Changes in the nucleotide sequence of DNA are called **mutations.** The altering of allele frequency due to mutation is probably of little importance in microevolution due to very low mutation rates for most gene loci. Mutation is central to evolution, however, because it is the original source of genetic variation.

Genetic Variation, the Substrate for Natural Selection

Genetic variation occurs within and between populations (453–454)

Variation Within Populations   Individual variation, the slight differences between individuals as a result of their unique genomes, is the raw material for natural selection. *Quantitative characters,* those that are additively affected by two or more gene loci, provide most of the heritable variation within a population. *Discrete characters* vary categorically as distinct phenotypes and usually are determined by a single gene locus. **Polymorphism** occurs when two or more forms of a discrete character—called *morphs*—are evident in a population.

Quantitative measures of genetic variation include **gene diversity,** the average percent of loci that are heterozygous in individuals of a population, and **nucleotide diversity,** the average percent of differences in nucleotide sites between individuals of a population. Human gene diversity is about 14%, whereas nucleotide diversity in only about 0.1%.

Variation Between Populations   **Geographic variations** are regional genetic differences between populations. These variations may be due to differing environmental selection factors. A **cline,** or graded variation within a species along a geographic axis, may parallel an environmental gradient.

Mutation and sexual recombination generate genetic variation (454–456)

Mutation   New alleles originate by mutations, most of which occur in somatic cells and cannot be passed on to the next generation. Most point mutations occur in silent DNA or do not change protein function and produce little effect. Those mutations that do alter a protein's function are most often harmful. Rarely, however,

a new mutant allele may increase an individual's fitness, or a mutation already present in the population may be selected for when the environment changes.

Chromosomal mutations are most often deleterious. Occasionally, a translocation of a chromosomal piece may bring alleles together that are beneficial in combination. Duplication of chromosomal segments is usually harmful, but those that do not upset the genetic balance within cells may provide an expanded genome with extra loci that could eventually take on new functions by mutation. The shuffling of exons may also result in new genes.

Mutation produces genetic variation very rapidly in HIV because of its short generation span (high replication rate) and higher rate of mutation in its RNA genome.

Sexual Recombination   The genetic differences in a sexually reproducing population arise from the reshuffling of alleles by crossing over and independent assortment of chromosomes in the formation of gametes and by the random union of gametes from two parents.

■ INTERACTIVE QUESTION   23.5

a. What is a major source of genetic variation for bacteria and microorganisms?

b. What is the major source of genetic variation for plants and animals?

c. Explain why your answers to a. and b. are different.

Diploidy and balanced polymorphism preserve variation (456–457)

Diploidy   The diploidy of most eukaryotes maintains genetic variation by hiding recessive alleles in heterozygotes, enabling them to persist in the population and be selected for, should the environment change.

Balanced Polymorphism   Natural selection may maintain variation at some gene loci, resulting in **balanced polymorphism.** When individuals heterozygous at a certain gene locus are reproductively more fit, this **heterozygote advantage** tends to maintain two or more alleles at this locus.

In **frequency-dependent selection,** a morph's reproductive success declines if it becomes too common in the population. Host-parasite interactions often result in frequency-dependent selection for varying morphs in both populations.

Neutral Variation Some of the diversity seen in populations may be **neutral variations** that do not confer a selective advantage or disadvantage. Such neutral alleles will not be affected by natural selection and will change randomly by genetic drift. There is no consensus on how much variation is truly neutral.

■ **INTERACTIVE QUESTION 23.6**

a. Why is the highly deleterious sickle-cell allele still present in the gene pool of the U.S. population?

b. Why is this allele at such a relatively high frequency in the gene pool of some African populations?

A Closer Look at Natural Selection as the Mechanism of Adaptive Evolution

Adaptive evolution is a combination of chance and sorting: the chance occurrence of new genetic variation by mutation and sexual recombination, and the selection of those variations most fit for the environment.

Evolutionary fitness is the relative contribution an individual makes to the gene pool of the next generation (457–458)

Darwinian fitness is a measure of an individual's relative contribution to the gene pool of the next generation. Population geneticists speak in terms of the **relative fitness** of a genotype as its contribution to the next generation in comparison with the contribution of other genotypes for this locus. The most fecund variants are said to have a relative fitness of 1, whereas the fitness of another genotype is the proportion of offspring it produces compared to the most successful variant.

Factors that contribute to both survival and fertility determine an individual's evolutionary fitness.

Selection acts on phenotype—the physical, biochemical, and behavioral traits of an organism—and thus indirectly adapts a population to its environment

as beneficial genotypes are increased or maintained in the gene pool. An organism is an integrated collection of many phenotypic features. The relative fitness of an allele depends on the entire genetic context of the individual.

■ **INTERACTIVE QUESTION 23.7**

A gene locus has two alleles, *B* and *b*. The genotype *BB* has a relative fitness of 0.5 and *bb* has a relative fitness of 0.25.

a. What is the relative fitness of the genotype *Bb*?

b. If a new generation has 100 *BB* individuals, how many *Bb* and *bb* would you expect to find?

   Bb _____   *bb* _____

c. What might account for the different relative fitness values for these genotypes?

The effect of selection on a varying characteristic can be directional, diversifying, or stabilizing (458–459)

The frequency of a trait, especially a quantitative trait, may be affected by three modes of selection. **Directional selection** occurs most frequently during periods of environmental change, when individuals deviating in one direction from the average for some phenotypic character may be favored. **Diversifying selection** occurs when the environment favors individuals on both extremes of a phenotypic range. Balanced polymorphism may result from diversifying selection. **Stabilizing selection** acts against extreme phenotypes and favors more intermediate forms, tending to reduce phenotypic variation.

Natural selection maintains sexual reproduction (459–460)

From a reproductive output standpoint, asexual reproduction is far superior, and sexual reproduction should, in theory, be selected against. One explanation for how sexual reproduction is maintained by natural selection is the selective advantage of genetic variation in disease resistance. For example, maintaining high variation in the cell surface molecules would help a population's offspring resist pathogens such as viruses, bacteria, and parasites that key in on such receptor molecules.

Sexual selection may lead to pronounced secondary differences between the sexes (460–461)

Sexual dimorphism is the distinction between males and females on the basis of secondary sexual characteristics. In vertebrates, the male is usually the larger and showier sex. Sexual selection is the selection for traits that may not be adaptive to the environment but do enhance reproductive success by increasing an individual's success in competing for (**intrasexual selection**) or attracting (**intersexual selection**) a mate. Also called mate choice, intersexual selection may be based on traits that reflect the general health of the male and thus the fitness of his alleles.

Natural selection cannot fashion perfect organisms (461)

There are at least four reasons why evolution does not produce perfect organisms. First, each species has evolved from a long line of ancestral forms, many of whose structures have been co-opted for new situations. Second, adaptations are often compromises between the need to do several different things, such as swim and walk, or be agile and strong. Third, the evolution that occurs as the result of chance events, such as genetic drift, is not adaptive. Fourth, natural selection can act only on variations that are available; new alleles do not arise when needed.

WORD ROOTS

inter- = between (*intersexual selection:* individuals of one sex are choosy in selecting their mates from individuals of the other sex, also called mate choice)

intra- = within (*intrasexual selection:* a direct competition among individuals of one sex for mates of the opposite sex)

micro- = small (*microevolution:* a change in the gene pool of a population over a succession of generations)

muta- = change (*mutation:* a change in the DNA of genes that ultimately creates genetic diversity)

poly- = many; **morph-** = form (*polymorphism:* the coexistence of two or more distinct forms of individuals in the same population)

STRUCTURE YOUR KNOWLEDGE

1. **a.** What is the Hardy-Weinberg theorem?
   b. Define the variables of the Hardy-Weinberg equation. Make sure you can use this equation to determine allele frequencies and predict genotype frequencies.

2. It seems that natural selection would work toward genetic unity; the genotypes that are most fit produce the most offspring, increasing the frequency of adaptive alleles and eliminating less beneficial alleles from the population. Yet there remains a great deal of variability within populations of a species. Describe some of the factors that contribute to this genetic variability.

3. You collect 100 samples from a large butterfly population. Fifty specimens are dark brown, 20 are speckled, and 30 are white. Coloration in this species of butterfly is controlled by one gene locus: *BB* individuals are brown, *Bb* are speckled, and *bb* are white. What are the allele frequencies for the coloration gene in this population? Is this population in Hardy-Weinberg equilibrium? Explain your answer.

TEST YOUR KNOWLEDGE

MULTIPLE CHOICE: *Choose the one best answer.*

1. Darwinian fitness is a measure of
   a. survival.
   b. number of matings.
   c. adaptation to the environment.
   d. successful competition for resources.
   e. number of viable offspring.

2. According to the Hardy-Weinberg theorem,
   a. the allele frequencies of a population should remain constant from one generation to the next if the population is large and there is no natural selection.
   b. only natural selection, resulting in unequal reproductive success, will cause evolution.
   c. the square root of the frequency of individuals showing the recessive trait will always equal the frequency of q.
   d. genetic drift, gene flow, and mutations are always maladaptive.
   e. all of the above are correct.

3. If a population has the following genotype frequencies, $AA = 0.42$, $Aa = 0.46$, and $aa = 0.12$, what are the allele frequencies?
   a. $A = 0.42$ $a = 0.12$
   b. $A = 0.6$ $a = 0.4$
   c. $A = 0.65$ $a = 0.35$
   d. $A = 0.76$ $a = 0.24$
   e. $A = 0.88$ $a = 0.12$

4. In a population with two alleles, B and b, the allele frequency of b is 0.4. What would be the frequency of heterozygotes if the population is in Hardy-Weinberg equilibrium?
   a. 0.16
   b. 0.24
   c. 0.48
   d. 0.6
   e. You cannot tell from this information.

5. In a population that is in Hardy-Weinberg equilibrium for two alleles, C and c, 16% of the population show a recessive trait. Assuming C is dominant to c, what percent show the dominant trait?
   a. 36%
   b. 48%
   c. 60%
   d. 84%
   e. 96%

6. Genetic drift is likely to be seen in a population
   a. that has a high migration rate.
   b. that has a low mutation rate.
   c. in which natural selection is occurring.
   d. that is very small.
   e. for which environmental conditions are changing.

7. Gene flow often results in
   a. populations that move to better environments.
   b. an increase in sampling error in the formation of the next generation.
   c. adaptive microevolution.
   d. a decrease in allele frequencies.
   e. a reduction of the allele frequency differences between populations.

8. The existence of two distinct phenotypic forms in a species is known as
   a. geographic variation.
   b. stabilizing selection.
   c. heterozygote advantage.
   d. polymorphism.
   e. directional selection.

9. Human gene diversity is estimated to be about 14%, which means that
   a. 86% of our genes are identical.
   b. on average, 14% of an individual's gene loci are heterozygous.
   c. only 14% of nucleotide sites differ between individuals.
   d. nucleotide diversity must be very great between individuals.
   e. the human population never experienced a bottleneck effect.

10. Mutations are rarely the cause of microevolution because
    a. they are most often harmful and do not get passed on.
    b. they do not directly produce most of the genetic variation present in a diploid population.
    c. they occur very rarely.
    d. they are only passed on when they occur in gametes.
    e. of all of the above.

11. In a study of a population of field mice, you find that 48% of the mice have a coat color that indicates that they are heterozygous for a particular gene. What would be the frequency of the dominant allele in this population?
    a. 0.24
    b. 0.48
    c. 0.50
    d. 0.60
    e. You cannot estimate allele frequency from this information.

12. In a random sample of a population of shorthorn cattle, 73 animals were red ($C^R C^R$), 63 were roan, a mixture of red and white ($C^R C^r$), and 13 were white ($C^r C^r$). Estimate the allele frequencies of C^R and C^r, and determine whether the population is in Hardy-Weinberg equilibrium.
    a. $C^R = 0.64$, $C^r = 0.36$; because the population is large and a random sample was chosen, the population is in equilibrium.
    b. $C^R = 0.7$, $C^r = 0.3$; the genotype ratio is not what would be predicted from these frequencies and the population is not in equilibrium.
    c. $C^R = 0.7$, $C^r = 0.3$; the genotype ratio is what would be predicted from these frequencies and the population is in equilibrium.
    d. $C^R = 1.04$, $C^r = 0.44$; the allele frequencies add up to greater than 1 and the population is not in equilibrium.
    e. You cannot estimate allele frequency from this information.

13. A scientist observes that the height of a certain species of asters decreases as the altitude on a mountainside increases. She gathers seeds from samples at various altitudes, plants them in a uniform environment, and measures the height of the new plants. All of her experimental asters grow to approximately the same height. From this she concludes that
    a. height is not a quantitative trait.
    b. the cline she observed was due to genetic variations.
    c. the differences in the parent plants' heights were due to directional selection.
    d. the height variation she initially observed was an example of nongenetic environmental variation.
    e. stabilizing selection was responsible for height differences in the parent plants.

14. Sexual selection will
    a. select for traits that enhance an individual's chance of mating.
    b. increase the size of individuals.
    c. result in individuals better adapted to the environment.
    d. result in stabilizing selection.
    e. result in a relative fitness of more than 1.

15. The greatest source of genetic variation in plant and animal populations is from
    a. mutations.
    b. sexual recombination.
    c. selection.
    d. polymorphism.
    e. recessive masking in heterozygotes.

16. A plant population is found in an area that is becoming more arid. The average surface area of leaves has been decreasing over the generations. This trend is an example of
    a. a cline.
    b. directional selection.
    c. disruptive selection.
    d. gene flow.
    e. genetic drift.

17. Mice that are homozygous for a lethal recessive allele die shortly after birth. In a large breeding colony of mice, you find that a surprising 5% of all newborns die from this trait. In checking lab records, you discover that the same proportion of offspring have been dying from this trait in this colony for the past three years. (Mice breed several times a year and have large litters.) How might you explain the persistence of this lethal allele at such a high frequency?
    a. Homozygous recessive mice have a reproductive advantage.
    b. A large mutation rate keeps producing this lethal allele.
    c. There is some sort of heterozygote advantage and perhaps selection against the homozygous dominant trait.
    d. Genetic drift has kept the recessive allele at this high frequency in the population.
    e. Since this is a diploid species, the recessive allele cannot be selected against when it is in the heterozygote.

18. In breeding experiments with *Drosophila,* you count the offspring produced by three different genotypes and determine that flies with the genotype *AA* have a relative fitness of 1. What does that mean?
    a. *AA* flies have a lower Darwinian fitness than do flies that are *Aa* or *aa.*
    b. *AA* flies produce more viable offspring than do *Aa* or *aa* flies.
    c. This fly population must be in Hardy-Weinberg equilibrium.
    d. *AA* flies live longer than do *Aa* or *aa* flies.
    e. The *A* and *a* alleles must both have a frequency of 0.5.

19. All of the following would tend to maintain balanced polymorphism in a population *except*
    a. hybrid vigor.
    b. directional selection.
    c. diversifying selection.
    d. heterozygote advantage.
    e. frequency-dependent selection.

20. Genetic analysis of a large population of mink inhabiting an island in Michigan revealed an unusual number of loci where one allele was fixed. Which of the following is the most probable explanation for this genetic homogeneity?
    a. The population exhibited nonrandom mating, producing homozygous genotypes.
    b. The gene pool of this population never experienced mutation or gene flow.
    c. A very small number of mink may have colonized this island, and this founder effect and subsequent genetic drift could have fixed many alleles.
    d. Natural selection has selected for and fixed the best adapted alleles at these loci.
    e. The colonizing population may have had much more genetic diversity, but genetic drift in the last year or two may have fixed these alleles by chance.

21. Directional selection would be most likely to occur when
    a. a population's environment has changed.
    b. a population's environment has two very different habitats.
    c. frequency-dependent selection is acting on a population.
    d. a population's environment is very harsh.
    e. a population is small and its environment is stable.

22. If an allele is recessive and lethal in homozygotes,
    a. the allele is present in the population at a frequency of 0.001.
    b. the allele will be removed from the population by natural selection in approximately 1,000 years.
    c. the relative fitness of the homozygous recessive genotype is 0.
    d. the allele will most likely remain in the population at a low frequency because it cannot be selected against when in a heterozygote.
    e. Both c and d are correct.

23. Sexual reproduction may be maintained by natural selection because
    a. it produces the greatest number of offspring.
    b. intrasexual selection produces the strongest males.

    c. intersexual selection allows females to choose their mates.
    d. maintaining a high variability in a population for traits such as cell surface markers protects against pathogens such as viruses, bacteria, and parasites.
    e. it maintains high genetic variability in a population's gene pool so that the population can adapt should environmental conditions change.

24. Which of the following is an example of frequency-dependent selection?
    a. the genetic drift that occurs within a founder population
    b. a ritualized contest in which one male wins reproductive access to a group of females
    c. host-parasite interactions in which parasites in a region are genetically adapted to infect the most common morph of host
    d. the very rapid evolution of HIV due to its short generation time and high rate of mutation
    e. the high frequency of intermediate phenotypes that results from stabilizing selection

CHAPTER 24

THE ORIGIN OF SPECIES

FRAMEWORK

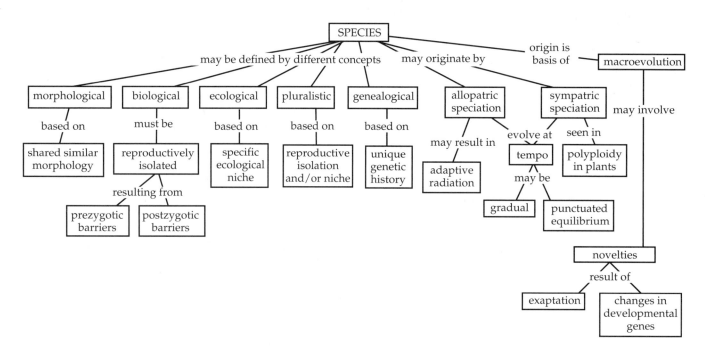

CHAPTER REVIEW

Microevolution explains evolutionary changes within a population. **Macroevolution** considers the origin of new taxonomic groups (from species to kingdom), and the process begins with **speciation**, the origin of a new species. Anagenesis, or phyletic evolution, involves the transformation of one species into a new species. In cladogenesis, or branching evolution, new species arise from parent species that continue to exist. Cladogenesis is the process that increases biological diversity.

What Is a Species?

Species are most often characterized by their physical form or morphology, although taxonomists now consider physiology, biochemistry, behavior, and genetics in order to distinguish species.

The biological species concept emphasizes reproductive isolation (465)

According to the **biological species concept,** developed by Ernst Mayr in 1942, a species is a population or group of populations of individuals that have the potential to interbreed in nature and produce viable, fertile offspring, but which do not successfully interbreed with other species. Biological species are genetically isolated from other species.

Prezygotic and postzygotic barriers isolate the gene pools of biological species (465–467)

Any intrinsic mechanism that prevents two species from producing viable, fertile hybrids is a reproductive barrier serving to preserve the genetic integrity of a biological species.

Prezygotic Barriers   **Prezygotic barriers** function before the formation of a zygote by preventing mating between species or successful fertilization should gametes meet. Prezygotic barriers include *habitat isolation*, in which two species that live in the same area occupy different habitats; *temporal isolation*, in which two species breed at different times; *behavioral isolation*, in which courtship rituals and behavioral signals are species specific; *mechanical isolation*, in which anatomical incompatibilities prevent mating with members of other species; and *gametic isolation*, in which the gametes of different species fail to fuse, often due to the lack of specific recognition molecules on the surfaces of gametes.

Postzygotic Barriers   Should a hybrid zygote form, **postzygotic barriers** prevent it from developing into a viable, fertile adult. Postzygotic barriers include *reduced hybrid viability*, in which a hybrid zygote fails to survive embryonic development; *reduced hybrid fertility*, in which a viable hybrid individual is sterile, often due to the inability to produce normal gametes in meiosis; or *hybrid breakdown*, in which the hybrids are viable and fertile, but their offspring are feeble or sterile.

■ **INTERACTIVE QUESTION 24.1**

Name the type of reproductive barrier and whether it is pre- or postzygotic for the following examples.

Type of Barrier	Pre- or Post-	Example
a.	b.	Two species of frogs are mated in the lab and produce viable, but sterile, offspring.
c.	d.	Two species of sea urchins release gametes at the same time, but no cross fertilization occurs.
e.	f.	Two orchid species with different length nectar tubes are pollinated by different moths.
g.	h.	Two species of mayflies emerge during different weeks in spring.
i.	j.	Two species of salamanders will mate and produce offspring, but their offspring are sterile.
k.	l.	Two similar species of birds have different mating rituals.
m.	n.	Embryos of two species of mice bred in the lab usually abort.
o.	p.	Peepers breed in woodland ponds; leopard frogs breed in swamps.

The biological species concept has some major limitations (468)

The biological species concept does not work for species that are completely asexual, such as bacteria. Extinct species also cannot be grouped based on the criterion of interbreeding. Even most living species are identified by comparative morphology and not the more difficult-to-obtain data on interbreeding.

Evolutionary biologists have proposed several alternative concepts of species (468)

The **ecological species concept** defines species on the basis of their ecological niche, the role they play and resources they use in the specific environments in which they are found. According to the **pluralistic species concept,** the mechanisms that maintain a species as a cohesive unit may include reproductive barriers, its specific niche, or a combination of both. Most species have been identified on the basis of physical characteristics, an approach called the **morphological species concept.** The emphasis of the **genealogical species concept** is on evolutionary lineages through which a unique genetic history gives each species a set of unique genetic markers.

■ **INTERACTIVE QUESTION 24.2**

Fill in the following table to review the five approaches that evolutionary biologists have proposed for conceptualizing a species.

Concept	Emphasis
biological	a.
b.	anatomical differences, most commonly used
c.	unique roles in specific environments
d.	both reproductive isolation and ecological niche
genealogical	e.

Modes of Speciation

Speciation requires the interruption of gene flow between populations. **Allopatric speciation** occurs when two populations are geographically separated. In **sympatric speciation**, biological barriers prevent gene flow between overlapping populations.

Allopatric speciation: geographic barriers can lead to the origin of species (469–473)

Conditions for Allopatric Speciation Geologic change can isolate populations. The extent of the geographic barrier necessary to maintain genetic separation depends on the ability of the organisms to disperse. Colonization of a new area may also geographically isolate populations. If the isolated population is small, its gene pool is more easily changed by genetic drift and natural selection.

Ring Species: Allopatric Speciation in Progress? When a species spreads around some geographic barrier, the populations may gradually diverge. In a ring species, populations that eventually meet at the other end of the barrier may have evolved enough that they are no longer able to interbreed and can thus be considered separate species.

Adaptive Radiation on Island Chains Allopatric speciation may occur on island chains when small founding populations evolve in isolation and under somewhat differing environmental conditions. These new species may then recolonize the original island and coexist with the parent species. **Adaptive radiation** is the evolution of numerous, variously adapted species from a common ancestor.

■ **INTERACTIVE QUESTION 24.3**

What factors have contributed to the adaptive radiation of the thousands of endemic species of the Hawaiian Archipelago?

How Do Reproductive Barriers Evolve? Geographic separation alone is not a reproductive barrier in the biological species sense. Such intrinsic reproductive barriers, however, may arise coincidentally as allopatric populations undergo genetic drift and natural selection.

Laboratory studies have shown the evolution of a prezygotic barrier to reproduction in fruit fly populations reared for several generations on different food sources. Perhaps the allele that affects nutrient digestion also influences the odor of the fly cuticle, one of the factors involved in mate choice.

Hybridization studies between various populations of monkey flowers have shown that hybrid offspring from close populations were usually fertile, whereas hybrids from distant populations were often sterile.

This postzygotic reproductive barrier may have evolved because there was no selection pressure to eliminate alleles that may have harmed the reproductive success of hybrids from distant populations, which would never naturally interbreed.

Summary of Allopatric Speciation The genetic drift and natural selection that affect an isolated population's gene pool may result in the evolution of reproductive barriers as a side effect of such genetic change.

Sympatric speciation: a new species can originate in the geographic midst of the parent species (473–475)

Polyploid Speciation in Plants Mistakes during cell division may lead to **polyploidy** in plants. An **autopolyploid** has more than two sets of chromosomes that have all come from the same species. Meiotic failure can produce tetraploids (4*n*), which can fertilize themselves or other tetraploids but cannot successfully breed with diploids from the parent population, resulting in reproductive isolation in just one generation.

Polyploid species arise more commonly through **allopolyploidy**. Interspecific hybrids may propagate asexually, but are usually sterile due to difficulties in the meiotic production of gametes. Future mitotic or meiotic nondisjunctions can result in the production of a fertile polyploid.

Speciation of polyploids has been frequent and important in plant evolution. Many of our agricultural plants are polyploids, and plant geneticists now hybridize plants and induce meiotic and mitotic errors to create new species.

■ **INTERACTIVE QUESTION 24.4**

a. A new plant species B forms by autopolyploidy from species A, which had a chromosome number of 2*n* = 10. How many chromosomes would species B have?

b. If species A were to hybridize by allopolyploidy with species C (2*n* = 14) and produce a new, fertile species, D, how many chromosomes would species D have?

Sympatric Speciation in Animals Sympatric speciation in animals may involve isolation within the geographic range of the parent population based on different resource usage. Nonrandom mating in a polymorphic population could also lead to sympatric speciation. In a laboratory study of two sympatric species of cichlids, mate choice based on coloration was shown to be the reproductive barrier that normally separates the two species.

Summary of Sympatric Speciation   In sympatric speciation, some reproductive barrier isolates the gene pool of a subgroup of a population within the same geographic range as the parent population. Hybridization between closely related species followed by mitotic or meiotic errors that produce fertile polyploids is the common mechanism in plants. A change in resource use or mating preference may reproductively isolate a subset of an animal population.

The punctuated equilibrium model has stimulated research on the tempo of speciation (475–476)

The traditional evolutionary concept of the origin of species involves the gradual divergence of populations by microevolution, with each newly formed species continuing to evolve over long periods of time. However, in the fossil record, new forms often appear rather suddenly, persist unchanged for a long time, and then disappear. Part of this nongradual appearance of fossils may be explained by the allopatric model of speciation. New species diverge in a location separate from the parent species and may then return to the parental location (and show up in the fossil record) in an already differentiated form.

According to the model of evolution known as **punctuated equilibrium,** long periods of stasis are punctuated by episodes of relatively rapid speciation and change. In geologic time, a few thousand years for a species to evolve is small compared with the few millions of years a successful species may exist. Long periods of stasis may be the result of stabilizing selection in an unchanging environment. Some gradualists maintain, however, that fossils show stasis only in external anatomy and that changes in internal anatomy, physiology, and behavior go unrecorded.

■ **INTERACTIVE QUESTION   24.5**

Compare the gradual and punctuated equilibrium models of evolution.

From Speciation to Macroevolution

The generation-to-generation change in allele frequencies in a population is considered microevolution. Macroevolution includes the genetic changes that result in reproductive isolation of new species as well as the evolution of the defining characteristics of higher taxonomic groups.

Most evolutionary novelties are modified versions of older structures (477)

Often, very complex organs, such as the eyes of vertebrates and mollusks, have evolved gradually from simpler structures that served similar needs in ancestral species.

Evolutionary novelties that define higher taxa may evolve by the gradual modification of existing structures for new functions. **Exaptation** is the term for structures that evolved and functioned in one setting and were then co-opted for a new function.

■ **INTERACTIVE QUESTION   24.6**

Give examples of reptilian structures that were exaptations for flight in birds.

"Evo-devo": genes that control development play a major role in evolution (477–479)

The combination of the fields of evolutionary and developmental biology, called "evo-devo," explores how slight changes in developmental genes can result in major morphological differences between species.

Allometric growth, the differing rates of growth of various parts of the body, leads to the final shape of the organism. A minor genetic alteration that affects allometric growth can produce a very differently proportioned adult form.

Heterochrony is an evolutionary change in the rate or timing of development. **Paedomorphosis** is the retention in the adult of juvenile traits of ancestral organisms and can occur when genetic changes speed up the development of reproductive organs relative to the development of body form.

■ INTERACTIVE QUESTION   24.7

a. Fetal skulls of humans and chimpanzees have similar shapes. The quite distinctive differences in adult skull shape results from different patterns of _____.

b. A salamander species that retains its gills (a larval trait) when it is full grown and sexually mature is an example of _____.

c. The shorter feet of tree-dwelling salamanders may result from an evolutionary change in a regulatory gene that switches off growth of the foot sooner than in ground-dwelling salamanders. These three cases (a., b., and c.) are all examples of _____.

Changes in genes that alter the spatial arrangement of body parts have also been important in macroevolution. Mutations in *Hox* genes, which provide positional information in embryos, can drastically alter body form. Duplications of the *Hox* complex of invertebrates may have been central to the evolution of vertebrates.

An evolutionary trend does not mean that evolution is goal oriented (480–481)

Equus, the modern horse, descended from its much smaller, browsing, four-toed ancestor, *Hyracotherium,* through a series of speciation episodes that produced many different species and diverging trends, as documented in the fossil record.

According to S. Stanley's model of **species selection,** an evolutionary trend is analogous to a trend in a population produced by natural selection. The successful species that last the longest before extinction and bud off the most new species will determine the direction of the trend. Evolutionary trends are ultimately dictated by environmental conditions; if conditions change, an evolutionary trend may end or change.

WORD ROOTS

allo- = other; **-metron** = measure (*allometric growth:* the variation in the relative rates of growth of various parts of the body, which helps shape the organism)

ana- = up; **-genesis** = origin, birth (*anagenesis:* a pattern of evolutionary change involving the transformation of an entire population, sometimes to a state different enough from the ancestral population to justify renaming it as a separate species)

auto- = self; **poly-** = many (*autopolyploid:* a type of polyploid species resulting from one species doubling its chromosome number to become tetraploid)

clado- = branch (*cladogenesis:* a pattern of evolutionary change that produces biological diversity by budding one or more new species from a parent species that continues to exist)

hetero- = different (*heterochrony:* evolutionary changes in the timing or rate of development)

macro- = large (*macroevolution:* evolutionary change on a grand scale, encompassing the origin of novel designs, evolutionary trends, adaptive radiation, and mass extinction)

paedo- = child (*paedomorphosis:* the retention in the adult organism of the juvenile features of its evolutionary ancestors)

post- = after (*postzygotic barrier:* any of several species-isolating mechanisms that prevent hybrids produced by two different species from developing into viable, fertile adults)

sym- = together; **-patri** = father (*sympatric speciation:* a mode of speciation occurring as a result of a radical change in the genome that produces a reproductively isolated subpopulation in the midst of its parent population)

STRUCTURE YOUR KNOWLEDGE

1. How are speciation and microevolution different?
2. Describe two major mechanisms through which evolutionary novelties may originate.

TEST YOUR KNOWLEDGE

MULTIPLE CHOICE: *Choose the one best answer.*

1. The type of evolution that results in the greatest increase in biological diversity is
   a. anagenesis.
   b. cladogenesis.
   c. phyletic evolution.
   d. microevolution.
   e. Both a and c are correct.

2. Which of the following is *not* a type of intrinsic reproductive isolation?
   a. mechanical isolation
   b. behavioral isolation
   c. geographic isolation
   d. gametic isolation
   e. temporal isolation

3. Two species of frogs occasionally mate, but the offspring do not complete development. What type of barrier isolates these gene pools?
   a. gametic isolation
   b. prezygotic barrier
   c. hybrid breakdown
   d. reduced hybrid viability
   e. reduced hybrid fertility

4. For which of the following is the biological species concept least appropriate?
   a. plants
   b. animals
   c. bacteria
   d. fossils
   e. both c and d

5. A horse ($2n = 64$) and a donkey ($2n = 62$) can mate and produce a mule. How many chromosomes would there be in a mule's cells?
   a. 31   **c.** 63   **e.** 126
   b. 62   **d.** 64

6. What prevents horses and donkeys from hybridizing to form a new species?
   a. reduced hybrid fertility
   b. reduced hybrid viability
   c. mechanical isolation
   d. gametic isolation
   e. behavioral isolation

7. Which of the following species concepts identifies species based on their similarities resulting from their unique niche or role in the environment?
   a. biological
   b. ecological
   c. genealogical
   d. morphological
   e. pluralistic

8. Allopatric speciation is more likely to occur when an isolated population
   a. is large and thus has more genetic variation.
   b. is reintroduced to its original homeland.
   c. is exposed to different selection pressures in its new habitat.
   d. inhabits an island close to its parent species' mainland.
   e. All of the above contribute to allopatric speciation.

9. A tetraploid plant species (with four identical sets of chromosomes) is probably the result of
   a. allopolyploidy.
   b. autopolyploidy.
   c. hybridization and nondisjunction.
   d. allopatric speciation.
   e. a and c.

10. All of the following are examples of the evolution of reproductive barriers *except*
    a. pleiotropic effects of genes that may allow specialization on different resources but also change cuticle odor in fruit flies reared on different nutrient sources.
    b. loss of selection for genes that maintain hybrid viability between widely separated populations of monkey flowers.
    c. lack of gene flow and different selection pressures for populations at the ends of opposite arms of the distribution of a ring species.
    d. the ability of two widely separated subspecies of mice to produce viable, fertile hybrids when bred in a laboratory.
    e. hybridization and cell division mistakes that create a fertile allopolyploid that cannot reproduce with either parent species.

11. There are 28 morphologically diverse species of a group of sunflowers called silverswords found on the Hawaiian Archipelago. These species are an example of
    a. a geographical cline.
    b. adaptive radiation.
    c. allopolyploidy.
    d. the bottleneck effect.
    e. sympatric speciation.

12. Which of the following is descriptive of the punctuated equilibrium model?
    a. Long periods of stasis are punctuated by episodes of relatively rapid speciation and change.
    b. Microevolution is the driving force of speciation.
    c. Most rapid speciation events involve polyploidy in plants.
    d. Evolution occurs gradually as the environment gradually changes.
    e. In the framework of geologic time periods, speciation events occur very slowly and the equilibrium of species is punctuated by frequent extinctions.

13. A new plant species C formed from hybridization of species A ($2n = 18$) with species B ($2n = 12$) would probably produce gametes with a chromosome number of
    a. 12.   **c.** 18   **e.** 60.
    b. 15.   **d.** 30.

14. Which of the following would *not* contribute to allopatric speciation?
    a. geographic separation
    b. genetic drift
    c. gene flow
    d. different selection pressures
    e. founder effect

15. What is meant by the concept of species selection?
    a. Reproductive isolating mechanisms (both prezygotic and postzygotic) are responsible for maintaining the integrity of individual species.
    b. Characteristics that increase the probability of selection as a mate or success in competition for mates will be selected for and increase in frequency in a population.
    c. The species that last the longest and speciate the most often will determine the direction of evolutionary trends.
    d. A new species accumulates most of its unique features as it originates and then maintains long periods of stasis in body form.
    e. The colonization of new and diverse habitats can lead to the adaptive radiation of numerous species.

16. Allometric growth
    a. results in paedomorphosis.
    b. results in an evolutionary trend of increasing body size.
    c. results from differences in the locations of expression of *Hox* genes and thus the placement of body parts.
    d. is usually associated with polyploidy in plants.
    e. is the relative differences in growth rates of different body parts.

17. The evolution of lungs from the swim bladder of ancestral fish is an example of
    a. heterochrony.
    b. paedomorphosis.
    c. exaptation.
    d. changes in homeotic gene expression.
    e. punctuated equilibrium.

18. Which of the following is thought to be a critical event in the evolution of vertebrates from an invertebrate ancestor?
    a. exaptation of body segments to vertebrae
    b. duplication of the *Hox* complex (homeotic genes)
    c. allometric growth of tail segments
    d. allopatric speciation
    e. sympatric speciation

19. A botanist identifies a new species of plant that has 32 chromosomes. It grows in the same habitat with three similar species: species A ($2n = 14$), species B ($2n = 16$), and species C ($2n = 18$). Suggest a possible speciation mechanism for the new species.
    a. allopatric divergence by development of a reproductive isolating mechanism
    b. change in a key developmental gene that causes the plants to flower at different times
    c. autopolyploidy, perhaps due to a nondisjunction in the formation of gametes of species B
    d. allopolyploidy, a hybrid formed from species A and C
    e. Either answer c or d could account for the formation of this new plant species.

20. Which concept of species would be most useful to a field biologist identifying new species in a tropical rain forest?
    a. biological
    b. ecological
    c. genealogical
    d. morphological
    e. pluralistic

CHAPTER 25

PHYLOGENY AND SYSTEMATICS

FRAMEWORK

The evolutionary history of life is chronicled in the fossil record. The four eras of the geologic time scale are marked by major biological transitions, often linked with continental drift and mass extinctions.

A goal of systematics is to reconstruct the phylogenetic history of species. Cladistic analysis is used to identify monophyletic taxa based on shared derived characters. The best phylogenetic hypotheses are parsimonious trees based on molecular homologies as well as on morphological and fossil evidence.

CHAPTER REVIEW

Phylogeny is the evolutionary history of a species. The study of the diversity of life and its phylogenetic history is called systematics.

The Fossil Record and Geologic Time

Fossils are preserved impressions or remnants of past organisms. Paleontologists reconstruct evolutionary history by collecting and interpreting the succession of organisms found in the **fossil record.**

Sedimentary rocks are the richest source of fossils (484–486)

The strata of sedimentary rocks form from the compression of layers of sand and silt deposits. Hard parts of animals, such as bones, teeth, or shells, may be preserved as fossils. Replacement of organic tissue with dissolved minerals may turn the fossil to stone. Sometimes enough organic material is retained in the fossil that biochemical analysis and electron microscopy of cells can be done.

Molds of organisms, left when their bodies decay and minerals dissolved in water crystallize in the cast, are a common type of fossil. Trace fossils, such as footprints, burrows, or other impressions of animal activities, give clues to how animals moved or lived. Some fossils are the entire body of an organism preserved in a substance, such as ice, peat, or amber, that retards decomposition.

Paleontologists use a variety of methods to date fossils (486–488)

Relative Dating The order in which fossils appear in the strata of sedimentary rocks indicates their relative age. *Index fossils,* such as shells of widespread animals, are used to correlate strata from different locations. Gaps may appear in the sequence, indicating that an area was above sea level or underwent erosion during some period of geologic time.

Geologists have developed a **geologic time scale** with four eras—the Precambrian, Paleozoic, Mesozoic, and Cenozoic—that are delineated by major transitions, such as mass extinction and extensive radiation. The eras are subdivided into periods and epochs, which are associated with particular evolutionary developments shown in the fossil record.

Absolute Dating **Radiometric dating** is used to determine the ages of rocks and fossils. During an organism's lifetime, it accumulates radioactive isotopes in proportions equal to the relative abundance of the isotopes in the environment. After the organism dies, the isotopes decay at a fixed rate, known as the **half-life,** or the number of years it takes for one-half of the radioactive isotopes in the specimen to decay. Carbon-14, with a half-life of 5,730 years, is used for determining the age of relatively young fossils; uranium-238, with a half-life of 4.5 billion years, can be used to date volcanic rock and thus infer the age of fossils associated with those rocks.

During an organism's life, only L-amino acids are synthesized, but following death, L-amino acids are slowly converted to D-amino acids. The proportion of L- and D-amino acids and the rate of this racemization can be used to date some fossils.

■ INTERACTIVE QUESTION 25.1

a. A fossil has one-eighth of the atmospheric ratio of C-14 to C-12. Estimate the age of this fossil.

b. Potassium-40 has a half-life of 1.3 billion years. If an organism had 1 mg of potassium-40 when it died and its fossil now has 0.25 mg, how old is this fossil?

The fossil record is a substantial, but incomplete, chronicle of evolutionary history (488)

The formation of a fossil is an unlikely occurrence. A large number of species that lived probably left no fossils; geologil processes destroy many fossils; and only a fraction of existing fossils have been found.

■ INTERACTIVE QUESTION 25.2

What types of organisms are most likely to appear in the fossil record?

Phylogeny has a biogeographic basis in continental drift (488–490)

Biogeography, the geographic distribution of species, correlates with the geologic history of the Earth. The continents rest on great plates of crust that float on the molten mantle. Earthquakes, volcanoes, and mountains are formed in regions where these shifting plates abut.

Large-scale continental drift brought all the land masses together into a supercontinent named **Pangaea** about 250 million years ago (mya), near the end of the Paleozoic era. This tremendous change undoubtedly had a great environmental impact as shorelines were eliminated, shallow coastal areas were drained, and the majority of the land became part of the continental interior. Species that had evolved in isolation came together and competed, many species became extinct, and new opportunities for remaining species became available. About 180 mya, during the Mesozoic era, Pangaea broke up and the continents drifted apart, creating a huge geographic isolation event.

■ INTERACTIVE QUESTION 25.3

Marsupials evolved in what is now North America, yet their greatest diversity is found in Australia. How can you account for this biogeographic distribution?

The history of life is punctuated by mass extinctions (490–492)

A species may become extinct due to a change in its physical or biological environment. Mass extinctions have occurred during periods of major environmental change.

The Permian mass extinction, occurring at the boundary between the Paleozoic and Mesozoic eras about 250 mya, claimed over 90% of marine and terrestrial species. These extinctions coincided with the formation of Pangaea and may be related to environmental changes associated with that event. Massive volcanic eruptions in Siberia may have contributed to the Permian extinctions.

The Cretaceous mass extinction, which marks the boundary between the Mesozoic and Cenozoic eras about 65 mya, claimed over one-half of the marine species, many families of terrestrial plants and animals, and the dinosaurs. The climate was cooling during that time, and shallow seas were receding from continental lowlands. According to the impact hypothesis, an asteroid or comet collided with Earth and caused the Cretaceous extinctions. The thin layer of clay rich in iridium that separates Mesozoic from Cenozoic sediments may have been the fallout from a huge cloud of dust created by the collision. This cloud would have blocked sunlight and severely affected weather. The huge Chicxulub crater on the Yucatán coast of Mexico may have been the site of the impact, from which a devastating firestorm instantly spread to the North American continent. Some researchers maintain that environmental changes were primarily responsible for the Cretaceous extinctions, not the collision with an asteroid.

■ INTERACTIVE QUESTION 25.4

Why do extensive adaptive radiations often follow mass extinctions?

Systematics: Connecting Classification to Phylogeny

Systematics, which uses an evolutionary context to study diversity, traces phylogeny using the fossil record, molecular data, and other approaches. Taxonomy is part of systematics, and this naming and classification of species and groups of species dates back to Linnaeus and the 18th century.

Taxonomy employs a hierarchical system of classification (493–494)

The Binomial Each species is assigned a two-part latinized name—a **binomial**—consisting of the **genus** and the **specific epithet,** or **species.**

Hierarchical Classification Systematics also organizes species into broader categories. Related genera are grouped into **families,** which are then grouped into **orders, classes, phyla,** and finally into **kingdoms.**

A taxonomic unit at any level is called a **taxon.** Both words of the binomial are italicized, and the names for all taxa at the genus level and higher are capitalized. A **phylogenetic tree** represents this taxonomic hierarchy.

Modern phylogenetic systematics is based on cladistic analysis (494–497)

Most systematists use cladistic analysis to help them reconstruct evolutionary history and then construct phylogenetic trees.

Clades: Monophyletic Groups A **cladogram** is a phylogenetic tree with a series of two-way branches, or dichotomies, each of which represents the divergence of two species from a common ancestor. Each branch in a cladogram, called a **clade,** consists of an ancestral species and all of its descendant species and is said to be **monophyletic.** A polyphyletic taxon includes groups having different ancestry, and a paraphyletic taxon excludes some species that share a common ancestor with other species in the taxon.

Constructing a Cladogram Systematists determine the branching sequence in a cladogram based on similarities that reflect relatedness.

Homology is likeness based on shared ancestry; **analogy** is similarity due to evolutionary convergence. In **convergent evolution,** unrelated species develop similar features because they have similar ecological roles and natural selection has led to similar adaptations. In general, the more homology between two species, the more closely they are related. When two similar structures are very complex, they are more likely to be homologous and have a shared origin.

■ **INTERACTIVE QUESTION 25.5**

What two complications may make it difficult to determine phylogenetic relationships based on similarities between species?

After separating similar characters that are homologous from those that are analogous, systematists identify **shared primitive characters,** which are common to more inclusive taxa and evolved at an earlier branch point, and **shared derived characters,** which are evolutionary novelties that are unique to a clade.

To determine the branching sequence of a group of related species, the group is compared to an **outgroup,** a species or group of species that is closely related to the group being studied. Any homologies that are common to both the outgroup and the **ingroup,** the taxa to be grouped, are shared primitive characters that were present in an ancestor common to both groups. A comparison of the numbers of characters that are present in each taxon of the ingroup indicates the sequence in which shared derived characters evolved and determines the branch points used to produce a cladogram.

■ **INTERACTIVE QUESTION 25.6**

Place the taxa (outgroup, A, B, C, and D) on the cladogram based on the presence or absence of the characters 1–4 as shown in this table. Indicate before each branch point which shared derived character evolved in the ancestor of that clade.

	Outgroup O	A	B	C	D
1	0	1	1	1	1
2	0		1		1
3	0	1	1		1
4	0		1		

Taxa, Characters

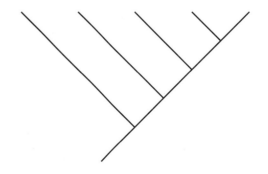

Summary of Cladistic Analysis By identifying homologies and shared derived characters within a set of organisms, systematists infer a chronological sequence of evolutionary branching. A cladogram does not indicate the age of branch points, only the order in which members of each clade last shared a common ancestor. Also, an individual species within a clade that branched off earlier may have evolved more recently than one in a newer clade.

Many evolutionary trees combine fossil evidence with cladistic analysis to represent the timing of phylogenetic branching. Cladistic analysis indicates a taxonomic hierarchy of evolutionary groups. Some systematists, however, propose a nonhierarchical classification system, called **phylocode,** that simply names clades.

Systematics can infer phylogeny from molecular data (497–499)

The amino acid sequences for many proteins and the nucleotide sequences for an ever-increasing number of genomes are available in Internet databases for researchers studying evolutionary relationships. Molecular comparisons of sequences of DNA nucleotides and proteins provide fundamental evidence for inferring evolutionary relatedness and divergence. Molecular systematics makes it possible both to discriminate between two very similar species and to trace phylogeny among very distantly related species.

Phylogenetic Data from DNA Sequences Comparing nucleotide sequences in homologous regions of DNA provides quantitative data for constructing cladograms. Changes in base sequences (mutations) are used as the shared derived characters that mark each clade. The genes for ribosomal RNA change very slowly, and differences in these sequences can be used to infer evolutionary relationships between groups that diverged millions of years ago. Mutations occur more frequently in mitochondrial DNA, and these changes can be used to investigate phylogeny among species, or even populations, that diverged more recently.

Aligning the DNA Sequences Insertion or deletion mutations change the lengths of homologous regions of DNA. Computer programs are used to identify similar sequences and then insert gaps in order to align homologous DNA segments properly for nucleotide comparisons.

■ **INTERACTIVE QUESTION   25.7**

Base deletions have changed the lengths and alignments of these two homologous regions of DNA. Determine the best possible fit between these two DNA sequences. How many deletions and base changes have occurred in these DNA segments?

A C G T G C A C G

A G T G A G G

The principle of parsimony helps systematists reconstruct phylogeny (499)

Each base change between homologous DNA segments counts as an evolutionary event. Systematists use such data to choose among many possible phylogenetic trees using the principle of **parsimony**—the smallest number of evolutionary changes is the simplest explanation and thus the best estimate of relationships among species.

Phylogenetic trees are hypotheses (502–503)

Heredity is based on the exact transmission of genetic information; molecular and morphological change is less common than stasis. The most parsimonious cladogram based on the fewest number of evolutionary changes represents the best hypothesis of the relationships among a set of species. The more data that can be compared (a large database of DNA sequence comparisons and several derived characters used to define each clade), the more reliable a parsimonious hypothesis of phylogenetic relationships becomes.

■ **INTERACTIVE QUESTION   25.8**

According to the principle of parsimony, the evolution of the four-chambered heart should place birds and mammals in the same clade. Why does the most accepted evolutionary tree show them as separate branches from the reptilian line?

Molecular clocks may keep track of evolutionary time (503)

Some regions of DNA appear to evolve at constant rates, and comparisons of the number of nucleotide or amino acid substitutions among taxa can serve as **molecular clocks** to estimate the relative times of their divergence.

Using Molecular Clocks to Measure Absolute Time
Some genes appear to have a reliable average rate of evolution. Graphs that plot nucleotide or amino acid differences against times for known evolutionary branch points can be used to estimate phylogenetic branchings that are not evident from the fossil record.

Evolutionists disagree about the extent of neutral mutations (which underlie genes that show consistent rates of evolution) and thus about whether molecular clocks are better used to determine the sequence of branches in phylogeny but not to determine the actual dates for the origin of taxa.

Using a Molecular Clock to Date the Origin of HIV
By comparing DNA sequences of an HIV gene from samples taken from patients across the past 20 years, researchers have observed a relatively consistent rate of evolution for that gene. They used a graph of that data to estimate that the HIV-IM strain first infected humans in the 1930s.

Modern systematics is flourishing with lively debate (503–505)

Cladistic analysis and molecular systematics, coupled with paleontology and comparative biology, are facilitating the reassessment and refinement of phylogenetic hypotheses. Many sources of evidence now indicate that crocodiles are more closely related to birds than to lizards and snakes (thus the traditional class Reptilia is paraphyletic).

Fossil evidence indicates that most modern mammalian orders evolved about 60 million years ago, whereas molecular clocks estimate an origin about 100 million years ago. The **phylogenetic fuse** hypothesis reconciles these differences by suggesting that mammalian orders did not become abundant and widespread enough to show up in the fossil record until after the extinction of the dinosaurs.

WORD ROOTS

analog- = proportion (*analogy:* similarity due to convergence)
bi- = two; **nom-** = name (*binomial:* a two-part latinized name of a species)

clado- = branch (*cladogram:* a dichotomous phylogenetic tree that branches repeatedly)
homo- = like, resembling (*homology:* similarity in characteristics resulting from a shared ancestry)
mono- = one (*monophyletic:* pertaining to a taxon derived from a single ancestral species that gave rise to no species in any other taxa)
parsi- = few (*principle of parsimony:* the premise that a theory about nature should be the simplest explanation that is consistent with the facts)
phylo- = tribe; **-geny** = origin (*phylogeny:* the evolutionary history of a taxon)

STRUCTURE YOUR KNOWLEDGE

1. The geologic time scale divides the history of life on Earth into four eras, delineated by major changes recorded in the fossil record. List these four eras and briefly describe some factors that may have contributed to the transitions between eras. (*Hint:* Consider continental drift, mass extinctions, and adaptive radiations.)

2. Develop a concept map to organize your understanding of systematics, including taxonomy and cladistic analysis. Compare your map with those of your classmates or the suggested one in the Answer Section, but remember that the value of this exercise is in the process of creation.

TEST YOUR KNOWLEDGE

MULTIPLE CHOICE: *Choose the one best answer.*

1. The richest source of fossils is found
   a. in coal and peat moss.
   b. along gorges.
   c. within sedimentary rock strata.
   d. encased in volcanic rocks.
   e. in amber.

2. Which of the following is *least* likely to leave a fossil?
   a. a soft-bodied land organism such as a slug
   b. a marine organism with a shell such as a mussel
   c. a vascular plant embedded in layers of mud
   d. a freshwater snake
   e. a human

3. Index fossils are fossils of
   a. unique organisms that are used to determine the relative rates of evolution in different areas.
   b. widespread organisms that allow geologists to correlate strata of rocks from different locations.
   c. transitional forms that link major evolutionary groups.
   d. extinct organisms that indicate the separation of different eras.
   e. organisms with new derived characters.

4. The half-life of carbon-14 is 5,730 years. A fossil that is 22,920 years old would have what amount of the normal proportion of C-14 to C-12?
   a. ½   c. ⅙   e. ¹⁄₁₆
   b. ¼   d. ⅛

5. The Permian mass extinction between the Paleozoic and Mesozoic eras
   a. affected a large percentage of marine and terrestrial species.
   b. coincided with the breaking apart of Pangaea.
   c. made way for the adaptive radiation of mammals, birds, and pollinating insects.
   d. appears to have been caused by a large asteroid striking Earth.
   e. did all of the above.

6. Related families are grouped into the next highest taxon called a
   a. class.
   b. order.
   c. phylum.
   d. genus.
   e. kingdom.

7. Which of the following would provide the best data for determining the phylogeny of three very similar species?
   a. the fossil record
   b. a comparison of embryological development
   c. a quantitative analysis of their morphological similarities and differences
   d. comparisons of amino acid sequences in proteins or DNA sequences
   e. comparison of their taxonomic assignments

8. Convergent evolution may result
   a. when older structures are co-opted for new functions.
   b. when homologous structures are adapted for different functions.
   c. from adaptive radiation.
   d. when species are widely separated geographically.
   e. when species have similar ecological roles.

9. *Lystrasaurus*, a fossil dinosaur, is found on most continents. This distribution is best explained by
   a. the relative ease of finding dinosaur bones.
   b. convergent evolution.
   c. divergent evolution.
   d. dispersal capabilities of ancient reptiles.
   e. biogeography.

10. The temperature at which hybrid DNA melts is indicative of the degree of homology between the DNA sequences. The more extensive the pairing, the higher the temperature required to separate the strands. You are trying to determine the phylogenetic relationships among species A, B, and C. You mix single-stranded DNA from all three species (in test groups of two) and measure the temperatures at which the hybrid DNA melts (separates). You find that hybrid BC has the highest melting temperature, AC the next highest, and AB the lowest. From these data you conclude that
    a. species A and B are most closely related, whereas B and C are least closely related.
    b. B and C must be the same species, and A is more closely related to C than to B.
    c. species B and C must have diverged most recently, and A is more closely related to C than to B.
    d. A hybridizes most easily with B, and they must have a more recent common ancestor than do A and C.
    e. these tests are inconclusive and you had better go back and check the fossil record.

11. Shared derived characters are
    a. used to characterize all the species on a branch of a dichotomous phylogenetic tree.
    b. determined through a computer comparison of all anatomical characteristics of a group of species.
    c. homologous structures that develop during adaptive radiation.
    d. characters found in the outgroup but not in the species to be classified.
    e. used to identify species but not higher taxa.

12. A comparative study of which of the following would provide the best data on the early evolution of fungi and plants?
    a. mitochondrial DNA
    b. DNA from a chloroplast
    c. the amino acid sequence of chlorophyll
    d. DNA for ribosomal RNA
    e. the morphology of present-day specimens

13. A comparative study of which of the following would provide the best data on the ancestry of people from Germany, Italy, and Spain?
    a. mitochondrial DNA
    b. nuclear DNA
    c. ribosomal RNA
    d. the sequence of amino acids in cytochrome *c*
    e. the fossil record

14. A taxon such as the class Reptilia, which does not include its relatives, the birds, is
    a. really an order.
    b. a clade.
    c. monophyletic.
    d. polyphyletic.
    e. paraphyletic.

15.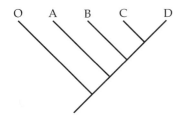

    According to this dichotomous phylogenetic tree created using cladistic analysis, C and D are most closely related because they
    a. do not share a common ancestor with O, A, or B.
    b. are monophyletic.
    c. evolved from a common ancestor a long time ago.
    d. have the most shared derived characters in common.
    e. have the greatest number of anatomical similarities as shown by statistical analysis.

16. A biologist is studying the evolution of four similar species of birds. Which approach would allow her to choose the best phylogenetic tree from all possible phylogenies?
    a. Draw the simplest cladogram and choose that one.
    b. From a comparison of nucleotide sequences, determine the number of evolutionary events required for each tree and choose the most parsimonious tree.

    c. Compare the entire genomes of each species; the two most similar genomes are the two species that are most closely related.
    d. Determine which species can interbreed; those that can evolved from a common ancestor most recently.
    e. Choose the tree that has the most evolutionary changes required as the most probable explanation for why these similar birds have evolved into four distinct species.

TRUE OR FALSE: *Indicate T or F, and then correct the false statements.*

_____ 1. A monophyletic taxon includes only species that share a common ancestor.

_____ 2. The more the sequences of amino acids in homologous proteins vary, the more recently the two species have diverged.

_____ 3. Phylogenetic trees determined on the basis of similar structures may be inaccurate when adaptive radiations have created large differences or when convergent evolution has created misleading analogies.

_____ 4. A cladogram represents both the chronology and the age of evolutionary branches, as well as the degree of divergence between taxa.

_____ 5. Phylogenetic trees designed with the concept of parsimony should show the most likely evolutionary relationships.

_____ 6. Carbon-14 dating is not a good technique for sequencing fossils from the Carboniferous period (363–290 mya) because carbon is too abundant in that period to allow accurate dating.

_____ 7. The layer of clay rich in iridium between the sediments of the Mesozoic and Cenozoic eras is indicative of volcanic activity.

_____ 8. The strongest support for a phylogenetic hypothesis comes from the concurrence of molecular and morphological evidence.

THE EVOLUTIONARY HISTORY OF BIOLOGICAL DIVERSITY

CHAPTER 26

EARLY EARTH AND THE ORIGIN OF LIFE

FRAMEWORK

This chapter presents a possible scenario for the chemical evolution of life on Earth between 4.0 and 3.5 billion years ago. Conditions on the primitive Earth are thought to have favored the spontaneous formation of organic monomers, the linking of these monomers into polymers, the development of self-replicating molecules, and the grouping of aggregates of organic molecules into droplets (called protobionts) that became capable of metabolism and reproduction.

From this proposed beginning, an incredible biological diversity has evolved. This heterogeneous group of organisms has been classified into five kingdoms: Monera, Protista, Plantae, Fungi, and Animalia. Phylogenetic relationships point to a three-domain system—Archaea, Bacteria, and Eukarya—and alternative numbers of kingdoms.

CHAPTER REVIEW

Life on Earth has been evolving for over 3.5 billion years. Geologic events have altered the course of biological evolution, and life has changed the Earth. The fossil record documents much of the episodic development of the diversity of life.

Introduction to the History of Life

Life on Earth originated between 3.5 and 4.0 billion years ago (512)

Earth formed about 4.5 billion years ago and was bombarded by huge rocks until about 3.9 billion years ago. Fossils resembling filamentous prokaryotes have been found in 3.5 billion-year-old rocks. It is possible that the earliest life forms may have emerged as early as 3.9 billion years ago.

Prokaryotes dominated evolutionary history from 3.5 to 2.0 billion years ago (512–513)

Many of the oldest prokaryotic fossils are found in **stromatolites,** which are fossil mats that resemble the banded mats of sediment that form around the jelly-like coats of certain modern-day bacterial colonies. Fossils of 3.2 billion-year-old prokaryotes have been found in sediments that formed around hydrothermal vents. The two main branches of prokaryotes, the bacteria and archaea, diverged early in evolutionary history.

Oxygen began accumulating in the atmosphere about 2.7 billion years ago (513–514)

Prokaryotes capable of non-oxygen-producing photosynthesis probably appeared early in prokaryotic history. Cyanobacteria may have evolved as early as 3.5 billion years ago. Banded iron formations in marine sediments and the rusting of iron in terrestrial rocks from about 2.7 billion years ago provide evidence of oxygen accumulation from the photosynthesis of cyanobacteria. The relatively rapid accumulation of atmospheric oxygen around 2.2 billion years ago caused the extinction of many prokaryotic groups and relegated others to anaerobic habitats.

■ INTERACTIVE QUESTION 26.1

What adaptation evolved in some prokaryotes that enabled them to thrive in this new oxygen-rich atmosphere?

Eukaryotic life began by 2.1 billion years ago (514)

Chemical traces of eukaryotes may date back to 2.7 billion years ago. Fossils that are generally accepted as eukaryotic date from 2.1 billion years ago. The larger and more complex eukaryotic cells may have evolved from symbiotic prokaryotes living within larger prokaryotic cells. Eukaryotic chloroplasts and mitochondria both relate to the oxygen revolution taking place during this time period.

Multicellular eukaryotes evolved by 1.2 billion years ago (514–515)

A variety of unicellular eukaryotes evolved, as well as the ancestors of multicellular algae, plants, fungi, and animals. Although molecular clocks estimate that the common multicellular ancestor arose 1.5 billion years ago, the oldest known fossils are small algae from 1.2 billion years ago. Larger animals appear in the fossil record in the late Precambrian era about 600 million years ago.

   According to the **snowball Earth** hypothesis, a severe ice age that occurred from 750 to 570 million years ago accounted for the limited diversity and distribution of multicellular eukaryotes, with the first major diversification occurring after the Earth thawed.

Animal diversity exploded during the early Cambrian period (515)

The Cambrian period begins the Paleozoic era 540 million years ago. Most modern animal phyla first appear in the fossil record during the first 30 million years of the Cambrian period.

Plants, fungi, and animals colonized the land about 500 million years ago (515)

Plants, fungi, and animals moved onto land during the early Paleozoic era. Fossil evidence indicates that fungi and plants, in symbiotic association, colonized land together. Arthropods became widespread and diverse. The fossil record documents the evolution of terrestrial vertebrates: amphibians from fish, reptiles from amphibians, and both birds and mammals from reptiles.

■ INTERACTIVE QUESTIONS 26.2

Review some of the key episodes in the history of life by placing them on this time line: origin of life (between 3,900 and 3,500 mya), oldest prokaryotic fossils (3,500), accumulation of atmospheric oxygen (2,700), oldest eukaryotic fossils (2,100), origin of multicellular eukaryotes (1,200), oldest animal fossils (600), plants, fungi, and then animals colonize land (500).

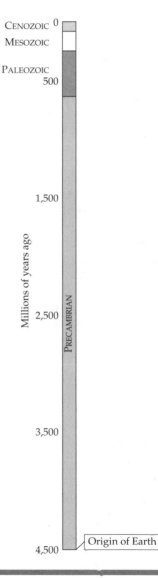

The Origin of Life

Between 4.0 billion years ago, when Earth's crust began to solidify, and 3.5 billion years ago, when records of stromatolites were deposited, life began.

The first cells may have originated by chemical evolution on a young Earth: *an overview* (516)

Most biologists believe that life evolved on Earth from nonliving materials that became ordered into collections of molecules capable of self-replication and metabolism.

Resolving the Biogenesis Paradox   The idea of **spontaneous generation** was finally rejected in 1862 with Pasteur's experiments that showed that microorganisms would not emerge spontaneously in sterilized broth placed in sealed or swan-necked flasks. **Biogenesis** is the principle that all life arises from preexisting life. The spontaneous origin of life on a young Earth is attributed to the extreme environmental conditions present at that time.

A Four-Stage Hypothesis for the Origin of Life   Life may have resulted from chemical evolution in four stages: the abiotic synthesis of small organic molecules; the joining of these monomers into polymers; the origin of self-replicating molecules that provided for inheritance; and the packaging of these molecules into membrane-enclosed droplets.

Abiotic synthesis of organic monomers is a testable hypothesis (516–518)

In the 1920s, Oparin and Haldane independently hypothesized that conditions on the primitive Earth, in particular the reducing atmosphere, lightning, and intense ultraviolet radiation, favored the synthesis of organic compounds from inorganic precursors available in the atmosphere and seas.

In 1953, Miller and Urey tested the Oparin-Haldane hypothesis with an apparatus that simulated the hypothetical conditions of early Earth. After a week of applying sparks (lightning) to a warmed flask of water (the primeval sea) in an atmosphere of H_2O, H_2, CH_4, and NH_3, Miller and Urey found a variety of amino acids and other organic compounds in the flask.

Numerous laboratory replications of the early Earth, using various combinations of atmospheric gases, have been able to produce organic compounds. One line of research focuses on volcanoes and deep-sea vents as the possible locations of these early reactions and sources of chemical precursors. It is also possible that

organic compounds could form in clouds of gas between stars and reach Earth on meteorites.

Laboratory simulations of early-Earth conditions have produced organic polymers (518–519)

The abiotic synthesis of polymers may have occurred at deep sea vents or on land when dissolved organic monomers splashed onto hot lava or rocks. Researchers have created polypeptides by dripping dilute solutions of organic monomers onto hot sand or rock in the laboratory.

RNA may have been the first genetic material (519–520)

RNA may have functioned as the first hereditary material and the first enzymes.

Molecular Replication in an RNA World   Short strands of RNA have been shown in the laboratory to be capable of self-replication. Cech's discovery of RNA catalysts, called **ribozymes,** indicates that RNA molecules may have been capable of ribozyme-catalyzed replication in the prebiotic world.

Natural Selection in an RNA World   In populations of RNA molecules in the laboratory, those molecules with the most stable three-dimensional conformations and greatest autocatalytic activity within a particular environment successfully compete for monomers and generate families of similar sequences, some of which may be even better adapted for self-replication. Such natural selection may have operated among early populations of RNA molecules.

RNA-directed protein synthesis may have begun with the weak binding of specific amino acids to bases along RNA molecules and their linkage to form a short polypeptide. This polypeptide may then have behaved as an enzyme to help the RNA molecule replicate.

Protobionts can form by self-assembly (520)

Protobionts, aggregates of abiotically produced organic molecules, probably preceded living cells. Laboratory experiments indicate that protobionts could have formed spontaneously. When a mixture of organic ingredients includes certain lipids, droplets, called liposomes, become surrounded by a lipid bilayer, which acts as a selectively permeable membrane. An energy-storing membrane potential may develop across this membrane. If enzymes are included in the mix, the protobionts are capable of absorbing substrates, catalyzing reactions, and releasing the products.

Natural selection could refine protobionts containing hereditary information (520–521)

When early RNA and polypeptide molecules became packaged into protobionts, molecular cooperation could become more efficient due to the concentration of components and the potential for the protobiont to evolve as a unit.

As protobionts grew, split, and distributed copies of their genes to offspring, errors in the copying of RNA would have led to variations, some of which could have had metabolic or genetic advantages. Differential reproductive success of these offspring would have accumulated metabolic and hereditary improvements. DNA, a more stable genetic molecule, eventually replaced RNA as the carrier of genetic information.

INTERACTIVE QUESTION 26.3

Describe the four steps of chemical evolution that are proposed as a possible route to the first cells.

a.

b.

c.

d.

Debate about the origin of life abounds (521–522)

Laboratory simulations cannot establish that this sequence actually occurred in the evolution of life, but they have shown that some of the key steps are possible. Alternative explanations include the idea that some organic compounds may have reached the early Earth on meteorites and comets. Because the surface of early Earth was such an inhospitable environment, some scientists speculate that life began on the sea floor, where deep-sea vents may have provided energy, chemical precursors, and organic compounds such as acetyl CoA.

Scientists are also exploring whether life exists or has existed on other planets. Via whatever route, at some point membrane-enclosed compartments capable of metabolism and genetic replication moved past the gray border separating them from true cells so that, by 3.5 billion years ago, prokaryotes had evolved.

The Major Lineages of Life

The five-kingdom system reflected increased knowledge of life's diversity (522–523)

The kingdom has heretofore been the most inclusive taxonomic category. Intuitively and historically, we have divided the diversity of life into two kingdoms—plants and animals. Whittaker proposed a five-kingdom system in 1969. The prokaryotes are set apart from the eukaryotes and placed in the kingdom Monera. The kingdoms Plantae, Fungi, and Animalia consist of multicellular eukaryotes, defined partly by characteristics of nutrition. Plants are autotrophic organisms; fungi are absorptive heterotrophic organisms; and most animals ingest and digest their food. The kingdom Protista includes unicellular eukaryotes or simple multicellular organisms that are believed to have descended from them.

Arranging the diversity of life into the highest taxa is a work in progress (523)

As systematists increase their understanding of phylogenetic origins, alternative classification systems are proposed. The current **three-domain system** places the three major evolutionary lineages into a super-kingdom taxon. The domains Bacteria and Archaea represent the early evolutionary divergence within the prokaryotes, and the domain Eukarya includes all the kingdoms of eukaryotes. Using molecular systematics and cladistics, systematists have split the kingdom Protista into five or more kingdoms and assigned some former members to the Plantae, Fungi, and Animalia. New data will continue to be used in the challenge of constructing a classification system that reflects the evolutionary history of life.

WORD ROOTS

bio- = life; **-genesis** = origin, birth (*biogenesis:* the principle that "life comes from life")

proto- = first (*protobionts:* aggregates of abiotically produced molecules)

stromato- = something spread out; **-lite** = a stone (*stromatolite:* rocks made of banded domes of sediment in which are found the most ancient forms of life)

STRUCTURE YOUR KNOWLEDGE

1. What do scientists think the primitive Earth was like? How could life possibly evolve in such an inhospitable environment?

2. Develop a simple concept map of Whittaker's five-kingdom system, indicating the key characteristics that are used to differentiate the kingdoms.

3. Now draw a map that shows the current three-domain system. Why do systematists keep changing the classification system?

TEST YOUR KNOWLEDGE

MULTIPLE CHOICE: *Choose the one best answer.*

1. The primitive atmosphere of Earth may have favored the synthesis of organic molecules because it
   a. was highly oxidative.
   b. was reducing and had energy sources in the form of lightning and UV radiation.
   c. had a great deal of methane and organic fuels.
   d. had plenty of water vapor, carbon, and nitrogen, providing the C, H, O, and N needed for organic molecules.
   e. had no oxygen.

2. Life on Earth is thought to have begun
   a. 540 million years ago at the beginning of the Paleozoic era.
   b. 700 million years ago during the Precambrian era.
   c. 2.7 billion years ago when oxygen began to accumulate in the atmosphere.
   d. between 3.5 and 4.0 billion years ago.
   e. 4.5 billion years ago when the Earth formed.

3. Stromatolites are
   a. early prokaryotic fossils found in sediments around hydrothermal vents.
   b. a type of protobiont that forms when lipids assemble into a bilayer surrounding organic molecules.
   c. members of the kingdom Protista that are both motile and photosynthetic.
   d. fossils appearing to be eukaryotes that are about twice the size of bacteria.
   e. fossilized mats that contain the oldest known fossils of prokaryotes.

4. Liposomes are
   a. spontaneously forming droplets of organic molecules surrounded by a lipid membrane.
   b. prokaryotes that lived as endosymbionts in larger prokaryotic cells.
   c. polypeptides formed in the laboratory by dripping organic monomers onto hot rocks.
   d. abiotically produced lipids.
   e. RNA self-replicating molecules.

5. Which of the following is a proposed hypothesis for the origin of genetic information?
   a. Early DNA molecules coded for RNA, which then catalyzed the production of proteins.
   b. Early polypeptides became associated with RNA bases and catalyzed their linkage into RNA molecules.
   c. Short RNA strands were capable of self-replication and evolved by the natural selection of molecules that were most stable and autocatalytic.
   d. As protobionts grew and split, they distributed copies of their molecules to their offspring.
   e. Early RNA molecules coded for DNA, which then dictated the order of amino acids in a polypeptide.

6. Evidence that protobionts may have formed spontaneously comes from
   a. the discovery of ribozymes, showing that prebiotic RNA molecules may have been autocatalytic.
   b. the laboratory synthesis of liposomes.
   c. the fossil record.
   d. the abiotic synthesis of polymers from monomers dripped onto hot rocks.
   e. the production of organic compounds in the Miller-Urey experimental apparatus.

7. Banded iron formations in marine sediments provide evidence of
   a. the first prokaryotes around 3.5 billion years ago.
   b. oxidized iron layers in terrestrial rocks.
   c. the accumulation of oxygen in the seas from the photosynthesis of cyanobacteria.
   d. the evolution of photosynthetic archaea near deep-sea vents.
   e. the crashing of meteorites onto Earth, possibly transporting abiotically produced organic molecules from space.

8. Simple animals such as jellies and worms first appear in the fossil record around 600 million years ago. This time period may correspond to
   a. the thawing of snowball Earth.
   b. the first appearance of multicellular eukaryotes in the fossil record.
   c. the beginning of the Cambrian explosion of animal forms.
   d. the colonization of land by plants and fungi, providing the food source for animals to follow.
   e. the rapid increase in atmospheric oxygen that allowed for the active lifestyle of animals.

9. The first fossil evidence of eukaryotic cells
   a. dates from 1.2 billion years ago.
   b. appears in sediments around deep-sea vents.
   c. is found in stromatolites.
   d. dates from 2.7 billion years ago when oxygen accumulated in the atmosphere.
   e. appears after prokaryotes had been evolving on Earth for 1.5 billion years.

10. In the five-kingdom system of classification,
    a. prokaryotes are placed in the kingdom Protista.
    b. the fungi include absorptive heterotrophs.
    c. single-celled or simple multicellular eukaryotes are placed in the kingdom Monera.
    d. the kingdoms Animalia, Plantae, and Fungi contain the only multicellular organisms.
    e. autotrophic multicellular prokaryotes are placed in the kingdom Plantae.

11. In order to place a newly identified species into one of the traditional five kingdoms, the first piece of information you would need to know is
    a. the way in which it acquires food.
    b. its life cycle.

c. the absence or presence of chloroplasts.
d. the absence or presence of a nuclear membrane.
e. whether the organism is multicellular.

12. Why is the diversity of life now organized into three domains?
    a. The domains Bacteria and Archaea reflect the early evolutionary divergence within the prokaryotes.
    b. Molecular evidence indicates that the protists really include three separate lineages.
    c. The eukaryotes are more alike than are the prokaryotes and thus belong in one big group.
    d. The division into plants, animals, and bacteria is more intuitive and accessible to most people.
    e. The origin of life involved three distinct stages—protobiont, prokaryote, and eukaryote—and each domain represents one of those stages.

MATCHING: *Match the scientist with the theory or experiment.*

_____ 1. produced organic compounds in a lab appartus

_____ 2. discovered RNA catalysts called ribozymes

_____ 3. developed the five-kingdom system

_____ 4. claimed that conditions of primitive Earth favored synthesis of organic compounds

A. Cech

B. Crick

C. Miller and Urey

D. Oparin and Haldane

E. Whittaker

CHAPTER 27

PROKARYOTES AND THE ORIGINS OF METABOLIC DIVERSITY

FRAMEWORK

Most systematists now place the prokaryotes of the kingdom Monera into separate domains, Bacteria and Archaea, due to their early evolutionary divergence. This chapter presents the morphology, phylogeny, and ecology of these generally single-celled organisms. Every mode of nutrition and most metabolic pathways evolved within the prokaryotes. Various lines of evidence point to an early evolution of photosynthesis.

CHAPTER REVIEW

The history of prokaryotes spans more than 3.5 billion years, and they evolved alone for the first 1.5 billion years.

The World of Prokaryotes

They're (almost) everywhere! *An overview of prokaryotic life (526–527)*

Prokaryotes greatly outnumber and outweigh all eukaryotes combined and flourish in all habitats, including ones that are too harsh for any other forms of life.

The collective impact of these microscopic organisms is huge. The chemical cycles necessary for life depend on prokaryotes. Many prokaryotes live in symbiotic relationships. About 5,000 species have been identified, with estimates of prokaryotic diversity as high as 4 million.

Bacteria and archaea are the two main branches of prokaryotic evolution (527–528)

Many archaea are found in extreme habitats reminiscent of early Earth. Archaea differ from bacteria in structure, biochemistry, and physiology, and the two branches diverged early in evolutionary history. Woese and other researchers favor using **domains,** a taxonomic unit above kingdoms, to represent the three major evolutionary lineages.

■ INTERACTIVE QUESTION 27.1

Fill in the domains on this phylogenetic tree that represents the hypothesis that Archaea is more closely related to Eukarya than to Bacteria.

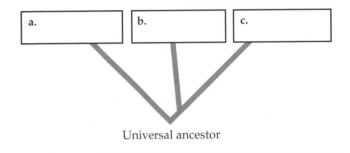

Universal ancestor

The Structure, Function, and Reproduction of Prokaryotes

Prokaryotes exist primarily as single cells, although some species may form transient groupings or permanent colonies, and some form associations of specialized cells. Prokaryotic cells are usually 1–5 μm in diameter, one-tenth the size of eukaryotic cells. The three most common cell shapes are spheres (cocci), rods (bacilli), and helices (spirilla and spirochetes).

Nearly all prokaryotes have a cell wall external to the plasma membrane (528–529)

Prokaryotic cell walls maintain cell shape and protect the cell from bursting in a hypotonic surrounding. The cell walls of bacteria (but not archaea) contain **peptidoglycan,** a matrix composed of polymers of sugars cross-linked by short polypeptides.

The **Gram stain** is an important tool for identifying bacteria as **gram-positive** (bacteria with walls containing a thick layer of peptidoglycan) or **gram-negative** (bacteria with more complex walls including an outer lipopolysaccharide membrane). Pathogenic gram-negative bacteria are often more harmful because their lipopolysaccharides may be toxic and this outer membrane also protects them. Many antibiotics, such as penicillin, inhibit the synthesis of cross-links in peptidoglycan and prevent wall formation.

Many prokaryotes secrete a sticky **capsule** outside the cell wall that serves as protection and a glue for adhering to a substratum or other prokaryotes. Some bacteria may also attach by means of surface appendages called **pili**, which may be specialized to hold bacteria together during conjugation.

Many prokaryotes are motile (529–530)

Many bacteria are equipped with flagella, either scattered over the cell surface or concentrated at one or both ends of the cell. These flagella differ from eukaryotic flagella in size, lack of a plasma membrane cover, structure, and function. Spirochetes have filaments that spiral around the cell under the cell wall and cause the cell to move in a corkscrew fashion. Other motile bacteria glide on slimy threads that they secrete.

Many motile bacteria exhibit **taxis**, an oriented movement in response to chemical, light, magnetic, or other stimuli.

The cellular and genomic organization of prokaryotes is fundamentally different from that of eukaryotes (530–531)

Extensive internal membrane systems are not found in prokaryotic cells, although some may have infoldings of the plasma membrane that function in respiration, photosynthesis, or other metabolic processes.

The circular double-stranded DNA chromosome is concentrated in a **nucleoid region.** It contains one one-thousandth as much DNA as a eukaryotic genome and has relatively little protein associated with it. Smaller rings of DNA, called **plasmids**, may carry genes for antibiotic resistance, metabolism of unusual nutrients, or other functions. They replicate independently and may be transferred between bacteria during conjugation.

The ribosomes of prokaryotes are smaller than eukaryotic ribosomes and differ in their protein and RNA content. Some antibiotics work by binding to prokaryotic ribosomes and blocking protein synthesis.

Populations of prokaryotes grow and adapt rapidly (531–532)

Prokaryotes reproduce asexually by **binary fission**, producing a colony of progeny. Despite the lack of meiosis and sexual cycles, there are three mechanisms of genetic recombination: **transformation**, the uptake of genes from the environment; **conjugation**, the di-

rect transfer of genetic material between prokaryotes; and **transduction**, the transfer of genes by a virus. Mutations are the major source of genetic variation in prokaryotes. Due to short generation times, favorable mutations and beneficial genomes resulting from recombination are rapidly propagated, facilitating adaptive evolution to environmental changes.

Prokaryotic growth rate in an optimal environment is geometric, with generation times from 20 minutes to 3 hours. The growth of colonies usually stops due to the exhaustion of nutrients or the toxic accumulation of wastes. Some prokaryotes, as well as certain protists and fungi, release **antibiotics**, which inhibit the growth of competitors.

Some bacteria form **endospores**, which are tough-walled cells that can resist even boiling water; thus microbiologists must use autoclaves to sterilize laboratory equipment and media.

■ INTERACTIVE QUESTION 27.2

Fill in the following table with a brief description of the characteristics of prokaryotic cells.

Property	Description
Cell shape	**a.**
Cell size	**b.**
Cell surface	**c.**
Motility	**d.**
Internal membranes	**e.**
Genome	**f.**
Reproduction and growth	**g.**

Nutritional and Metabolic Diversity

Prokaryotes can be grouped into four categories according to how they obtain energy and carbon (532–534)

Nutrition refers to how an organism obtains energy (photo- or chemotroph) and the carbon it uses for synthesizing organic compounds (auto- or heterotroph).

Thus, there are four major nutritional categories: (1) **photoautotrophs,** which use light energy and CO_2 to synthesize organic compounds; (2) **chemoautotrophs,** which obtain energy by oxidizing inorganic substances (such as H_2S, NH_3, or Fe^{2+}) and need only CO_2 as a carbon source; (3) **photoheterotrophs,** which use light energy to generate ATP but must obtain carbon in organic form; and (4) **chemoheterotrophs,** which use organic molecules as both an energy and a carbon source.

Nutritional Diversity Among Chemoheterotrophs Most known prokaryotes are either **saprobes,** which decompose and absorb nutrients from dead organic matter, or **parasites,** which absorb nutrients from living hosts.

Some prokaryotes require all 20 amino acids, vitamins, and other organic compounds in order to grow, whereas others may flourish with glucose as their only organic nutrient. Almost any organic compound can serve as food for some prokaryote;

only a few classes of synthetic organic compounds are nonbiodegradable.

Nitrogen Metabolism Nitrogen, an essential component of proteins and nucleic acids, is cycled through the ecosystem with the aid of prokaryotes that are able to metabolize various nitrogenous compounds. Through **nitrogen fixation,** some cyanobacteria and a few other species of prokaryotes convert atmospheric N_2 to ammonium, making nitrogen available to plants.

Metabolic Relationships to Oxygen **Obligate aerobes** need oxygen for cellular respiration; **facultative anaerobes** can use oxygen but also can grow in anaerobic conditions using fermentation; **obligate anaerobes** cannot use—and are poisoned by—oxygen. Some obligate anaerobes use **anaerobic respiration** to break down nutrients, with inorganic molecules other than oxygen serving as the final electron acceptor.

Photosynthesis evolved early in prokaryotic life (534–535)

All major metabolic pathways evolved in prokaryotes before there were eukaryotes, probably within the first billion years of life on Earth. Evidence for the early evolution of photosynthesis comes from molecular systematics, metabolic comparisons of extant prokaryotes, and geology.

■ **INTERACTIVE QUESTION 27.3**

Complete the following concept map that summarizes the four categories of nutrition.

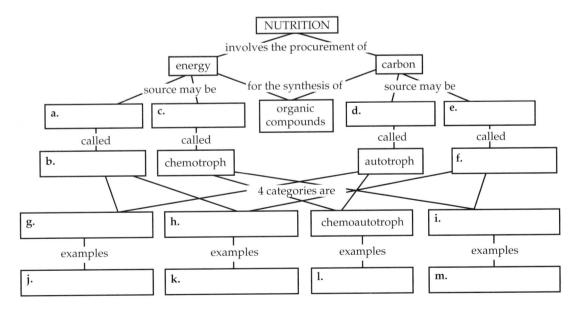

The very first prokaryotes were probably heterotrophs that absorbed organic compounds from their surroundings. Nearly all extant organisms use glycolysis as a metabolic pathway to generate ATP, implying that it was one of the first metabolic pathways to evolve.

Diverse groups of prokaryotes are photosynthetic. It is likely that these complex metabolic processes evolved only once in an early ancestor. Molecular data suggest that autotrophic species represent older lineages than closely related heterotrophic species, implying that many of today's heterotrophs descended from photosynthetic ancestors.

Cyanobacteria, the only autotrophic prokaryotes that perform oxygenic photosynthesis, were probably well established by the time atmospheric O_2 began to accumulate 2.7 billion years ago. Fossils in stromatolites that resemble modern cyanobacteria date back 3.5 billion years. The common components of the nonoxygenic, single photosystem of some modern prokaryotes and those of cyanobacteria suggest that cyanobacteria evolved from an ancestor with simpler, nonoxygenic photosynthesis.

Cellular respiration was an important evolutionary adaptation to the oxidizing atmosphere created by the oxygen revolution. Both photosynthesis and cellular respiration use electron transport chains and chemiosmosis to generate ATP, suggesting that cellular respiration evolved by the modification of photosynthetic machinery.

■ INTERACTIVE QUESTION   27.4

Photosynthesis is scattered throughout the phylogenetic tree of prokaryotes. Why is it more parsimonious to assume that photosynthesis evolved very early in prokaryotic life?

A Survey of Prokaryotic Diversity

Molecular systematics is leading to a phylogenetic classification of prokaryotes (535)

While shape, staining characteristics, nutritional mode, and motility have been helpful characters for clinically identifying bacteria, they have not provided a phylogenetic classification. Molecular comparisons of base sequences of the small-subunit ribosomal RNA (SSU-rRNA) revealed unique **signature sequences** that have been used to define taxonomic groups that diverged very early in the history of life.

Researchers are identifying a great diversity of archaea in extreme environments and in the oceans (535–537)

Archaea have some characteristics in common with bacteria, some with eukaryotes, and some unique characteristics. Systematists have identified two major taxa, the Euryarchaeota and Crenarchaeota. Many archaea are **extremeophiles** and have been grouped based on their ecology.

Methanogens have a unique energy metabolism in which CO_2 is used to oxidize H_2, producing the waste product methane (CH_4). These strict anaerobes often live in swamps and marshes, are important decomposers in sewage treatment, and are gut inhabitants that contribute to the nutrition of cattle and other herbivores.

The **extreme halophiles** (salt-lovers) live in extremely saline waters. Their photosynthetic pigment **bacteriorhodopsin** gives them a purple-red color.

Extreme thermophiles may be found oxidizing sulfur for energy in hot sulfur springs and near deep-sea hydrothermal vents. Extreme thermophiles are thought by some to have been the first prokaryotes to evolve.

Not all archaea are extremeophiles; numerous archaea have now been identified in more moderate marine habitats. The methanogens and halophiles are placed into the **Euryarchaeota.** Most thermophiles fit into the **Crenarchaeota.** The newly identified marine archaea are spread between the two taxa.

Most known prokaryotes are bacteria (537–539)

Bacteria show great diversity in their modes of nutrition and metabolism. Five taxa are described in the text. Most prokaryotic systematists assign the major groups of domains Archaea and Bacteria to the status of kingdoms.

Proteobacteria is a nutritionally diverse group of aerobic and anaerobic gram-negative bacteria. Its five subgroups are **Alpha Proteobacteria,** many of which are mutual symbionts or parasites of eukaryotes such as *Rhizobium, Agrobacterium,* and rickettsias; **Beta Proteobacteria,** which include the important soil bacteria *Nitrosomonas*; **Gamma Proteobacteria,** which include photosynthetic sulfur bacteria and some serious pathogens and enterics (intestinal inhabitants such as *E. coli*); **Delta Proteobacteria,** including the colony-forming myxobacteria; and **Epsilon Proteobacteria,** the group that includes the stomach ulcer–causing *Helicobacter pylori.*

Chlamydias are obligate intracellular animal parasites. One of these gram-negative species is a common cause of blindness and nongonococcal urethritis, a common sexually transmitted disease.

Spirochetes are helical heterotrophs that move in a corkscrew fashion. Spirochetes cause syphilis and Lyme disease.

The **gram-positive bacteria** is a diverse clade, even including some gram-negative species. The subgroup actinomycetes, once mistaken for fungi, includes the species that cause tuberculosis and leprosy. Most actinomycetes, however, are branching soil bacteria, some of which are cultured to produce antibiotics. There are diverse solitary species of gram-positive bacteria, including the spore-forming species responsible for anthrax and botulism, and the species of *Staphylococcus* and *Streptococcus*. Mycoplasmas, the smallest of all cells, are the only bacteria that lack cell walls.

The **cyanobacteria** have plantlike photosynthesis and are important producers in freshwater and marine ecosystems. Some filamentous cyanobacteria have cells specialized for nitrogen fixation.

■ INTERACTIVE QUESTION 27.5

a. Identify the large structure inside this *Bacillus* cell. What is its significance?

Bacillus

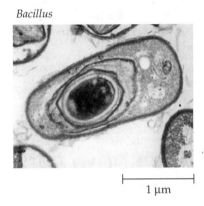

├─────── 1 μm ───────┤

b. Identify the major bacterial group (kingdom) to which this filamentous, photosynthetic species belongs. What is the function of the spherical cell indicated by the arrow?

Anabaena

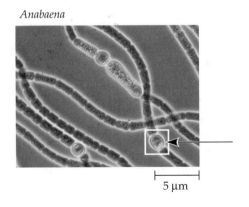

├─── 5 μm ───┤

The Ecological Impact of Prokaryotes

Prokaryotes are indispensable links in the recycling of chemical elements in ecosystems (540)

Prokaryotes are indispensable in recycling chemical elements between the biological and physical worlds. As **decomposers,** they return carbon, nitrogen, and other elements to the environment for assimilation into new living forms. Autotrophic bacteria bring carbon from CO_2 into the food chain and release O_2 to the atmosphere, and some cyanobacteria fix atmospheric nitrogen, supplying plants with the nitrogen they need to make proteins.

Many prokaryotes are symbiotic (540)

Symbiosis is an ecological relationship involving direct contact between organisms of two different species. The organisms are called symbionts and, if one is much larger than the other, it is called the **host.** In **mutualism,** both symbionts benefit. Nodules on the roots of legumes house symbiotic bacteria that fix nitrogen for the host and, in return, receive nutrients. In **commensalism,** one symbiont benefits while the other is neither harmed nor helped. Many of the bacteria found in and on the human body are commensal, although some may be mutualistic. In **parasitism,** the symbiont, now called a **parasite,** benefits at the expense of the host. Parasitic bacteria that cause disease are called pathogens.

Pathogenic prokaryotes cause many human diseases (540–542)

About one-half of all human diseases are caused by pathogenic prokaryotes that manage to invade the body, avoid internal defenses, and reproduce enough to harm the host. **Opportunistic** pathogens are normal inhabitants of the human body that cause illness when the body's defenses are weakened.

Koch's postulates include four criteria for establishing a pathogen as the cause of a disease: (1) find the same pathogen in each diseased individual; (2) isolate and grow the pathogen in a pure culture; (3) induce the disease in healthy experimental animals using the cultured pathogen; and (4) isolate the same pathogen from the experimental animal after it develops the disease.

Some prokaryotes cause disease when they invade host tissues. Pathogens more commonly cause disease by producing toxins. **Exotoxins,** among the most potent poisons known, are proteins secreted by prokaryotes that cause such diseases as botulism and cholera. **Endotoxins,** which are components of the outer membrane of certain gram-negative bacteria, cause such diseases as typhoid fever and *Salmonella* food poisoning.

In the past century, improved hygiene and sanitation and the development of antibiotics have decreased the incidence of bacterial disease, reduced infant mortality, and extended life expectancy in developed countries. More than half of our antibiotics come from the genus *Streptomyces*, an actinomycete. The evolution of antibiotic-resistant strains of pathogenic bacteria poses a serious health threat.

Humans use prokaryotes in research and technology (542)

Research using *E. coli* and other bacteria has expanded our understanding of molecular biology, and DNA technology using bacteria creates both basic knowledge and practical applications. The diverse metabolic capabilities of prokaryotes have been used to digest organic wastes, produce chemical products, make vitamins and antibiotics, and produce food products such as yogurt and cheese. **Bioremediation** is the use of organisms to remove environmental pollutants. Prokaryotes are used to treat sewage, clean up oil spills, and decompose pesticides.

WORD ROOTS

-gen = produce (*methanogen:* microorganisms which obtain energy by using carbon dioxide to oxidize hydrogen, producing methane as a waste product)

-oid = like, form (*nucleoid:* a dense region of DNA in a prokaryotic cell)

an- = without, not; **aero-** = the air (*anaerobic:* lacking oxygen; referring to an organism, environment, or cellular process that lacks oxygen and may be poisoned by it)

anti- = against; **-biot** = life (*antibiotic:* a chemical that kills bacteria or inhibits their growth)

bi- = two (*binary fission:* the type of cell division by which prokaryotes reproduce; each dividing daughter cell receives a copy of the single parental chromosome)

chemo- = chemical; **hetero-** = different (*chemoheterotroph:* an organism that must consume organic molecules for both energy and carbon)

endo- = inner, within (*endotoxin:* a component of the outer membranes of certain gram-negative bacteria responsible for generalized symptoms of fever and ache)

exo- = outside (*exotoxin:* a toxic protein secreted by a bacterial cell that produces specific symptoms even in the absence of the bacterium)

halo- = salt; **-philos** = loving (*halophile:* microorganisms which live in unusually highly saline environments such as the Great Salt Lake or the Dead Sea)

mutu- = reciprocal (*mutualism:* a symbiotic relationship in which both the host and the symbiont benefit)

photo- = light; **auto-** = self; **-troph** = food, nourish (*photoautotroph:* an organism that harnesses light energy to drive the synthesis of organic compounds from carbon dioxide)

sapro- = rotten (*saprobe:* an organism that acts as a decomposer by absorbing nutrients from dead organic matter)

sym- = with, together; **-bios** = life (*symbiosis:* an ecological relationship between organisms of two different species that live together in direct contact)

thermo- = temperature (*thermophiles:* microorganisms which thrive in hot environments, often 60–80°C)

trans- = across (*transformation:* a phenomenon in which external DNA is assimilated by a cell)

STRUCTURE YOUR KNOWLEDGE

1. What is the rationale for separating the archaea, the bacteria, and all the eukaryotes into three domains?

2. Describe four positive ways in which prokaryotes have an impact on our lives and on the world around us.

TEST YOUR KNOWLEDGE

MULTIPLE CHOICE: *Choose the one best answer.*

1. Which of the following is *not* true of plasmids?
   a. They replicate independently of the main chromosome.
   b. They may carry genes for antibiotic resistance.
   c. They are essential for the existence of bacterial cells.
   d. They may be transferred between bacteria during conjugation.
   e. They may carry genes for special metabolic pathways.

2. Gram-positive bacteria
   a. have peptidoglycan in their cell walls, whereas gram-negative bacteria do not.
   b. have an outer membrane around their cell walls.
   c. have lipopolysaccharides in their cell walls and thus may be more pathogenic than gram-negative bacteria.
   d. are members of domain Archaea.
   e. have simpler, thick peptidoglycan cell walls.

3. Many prokaryotes secrete a sticky capsule outside the cell wall that
   a. allows them to glide along a slime thread.
   b. serves as protection from host defenses and glue for adherence.
   c. reacts with the Gram stain.
   d. is used for attaching cells during conjugation.
   e. is composed of peptidoglycan, polymers of modified sugars cross-linked by short polypeptides.

4. Chemoautotrophs
   a. are photosynthetic.
   b. use organic molecules for an energy and carbon source.
   c. oxidize inorganic substances for energy and use CO_2 as a carbon source.
   d. use light to generate ATP but need organic molecules for a carbon source.
   e. use light energy to extract electrons from H_2S.

5. The major source of genetic variation in prokaryotes is
   a. transduction.
   b. mutation.
   c. conjugation.
   d. plasmid exchange.
   e. transformation.

6. A symbiotic relationship in which one symbiont benefits and the host is neither harmed nor helped is called
   a. saprocism.
   b. opportunism.
   c. mutualism.
   d. commensalism.
   e. parasitism.

7. Facultative anaerobes
   a. can survive with or without oxygen.
   b. are poisoned by oxygen.
   c. are anaerobic chemoautotrophs.
   d. are able to fix atmospheric nitrogen to make NH_4, which is then available for plant growth.
   e. include the methanogens that reduce CO_2 to CH_4.

8. Some prokaryotes have specialized membranes that
   a. contain ribosomes and function in protein synthesis.
   b. arose from endosymbiosis of smaller prokaryotes, creating mitochondria and chloroplasts.
   c. form from infoldings of the plasma membrane and may function in cellular respiration or photosynthesis.
   d. are produced by the endoplasmic reticulum, but differ in composition from eukaryotic membranes.
   e. enclose the nucleoid region and separate plasmids from the single prokaryotic chromosome.

9. Glycolysis
   a. is the only metabolic pathway common to nearly all modern organisms and may have been one of the first metabolic pathways to evolve.
   b. breaks down organic molecules and generates ATP by chemiosmosis.
   c. uses the electron transport chain to create a proton gradient to produce ATP.
   d. is an aerobic process that did not evolve until after photosynthesis.
   e. is found in all anaerobic species, but is replaced by cellular respiration in aerobic species.

10. Some cyanobacteria are capable of nitrogen fixation. This process
    a. oxidizes nitrogen-containing compounds to produce ATP.
    b. allows these bacteria to live in anaerobic environments.
    c. removes soil nitrite or nitrate and returns N_2 gas to the atmosphere.
    d. converts N_2 to ammonium, making nitrogen available to plants for incorporation into proteins and nucleic acids.
    e. is an essential part of the nitrogen cycle that rescues nitrogen trapped in mineral deposits and makes it available to organisms for their nitrogen metabolism.

11. Archaea
    a. are believed to be more closely related to eukaryotes than to bacteria.
    b. have cell walls that lack peptidoglycan.
    c. are often found in harsh habitats, reminiscent of the environment of early Earth.
    d. include the methanogens, extreme halophiles, and extreme thermophiles.
    e. are all of the above.

12. Organizing prokaryotes into major taxa within the domains Archaea and Bacteria has been based on
    a. fossil records of prokaryotes that have been recently discovered.
    b. molecular comparisons, particularly of signal sequences identified from small-subunit rRNA (SSU-rRNA).
    c. the Gram stain and colony characteristics when grown on solid media.
    d. shape and motility.
    e. nutritional modes and ecology.

13. Photosynthesis may have evolved very early in the history of life on Earth. Evidence for this hypothesis includes
    a. the occurrence of photosynthetic species in several diverse branches of prokaryotes.
    b. fossils that resemble modern cyanobacteria in stromatolites dating from 3.5 billion years ago.
    c. molecular systematics indicating that autotrophic species are from older lineages than closely related heterotrophic species.
    d. similarities between the oxygenic photosynthesis of cyanobacteria (which produced the oxygen revolution 2.7 billion years ago) and the nonoxygenic, single photosystem of some modern prokaryotes.
    e. All of the above provide evidence of an early evolution of photosynthesis.

14. The newly discovered archaea found in more moderate marine environments
    a. have been placed in the domain Bacteria because they are not extremeophiles.
    b. are grouped with the methanogens even though they are not anaerobic.
    c. are spread between the Euryarchaeota and Crenarchaeota based on molecular comparisons.
    d. are grouped with the extreme halophiles because of their photosynthetic pigment bacteriorhodopsin.
    e. are grouped with the extreme thermophiles and believed to be descendants of the first prokaryotes.

15. Which of the following groupings is mismatched?
    a. Proteobacteria—diverse gram-negative bacteria including pathogens such as rickettsias, *Salmonella, Helicobacter pylori*
    b. Chlamydias—intracellular parasites, including a species that causes blindness and nongonococcal urethritis
    c. Spirochetes—helical heterotrophs, including pathogens causing syphilis and Lyme disease

    d. Gram-positive bacteria—diverse group that includes actinomycetes, mycoplasmas, and pathogens that cause anthrax, botulism, and tuberculosis
    e. Cyanobacteria—photosynthetic filamentous colonies; some solitary species use the pigment bacteriorhodopsin and inhabit saline waters

FILL IN THE BLANKS

_____ 1. the name for spherical prokaryotes

_____ 2. region in which the prokaryotic chromosome is found

_____ 3. common laboratory technique for identifying bacteria

_____ 4. surface appendages of prokaryotes used for adherence to substrate

_____ 5. an oriented movement in response to light or chemical stimuli

_____ 6. resistant cell that can survive harsh conditions

_____ 7. organisms that decompose and absorb nutrients from dead organic matter

_____ 8. proteins that are secreted by pathogens and are potent poisons

_____ 9. use of organisms to remove environmental pollutants

_____ 10. responsible for oxygen revolution

CHAPTER 28

THE ORIGINS OF EUKARYOTIC DIVERSITY

FRAMEWORK

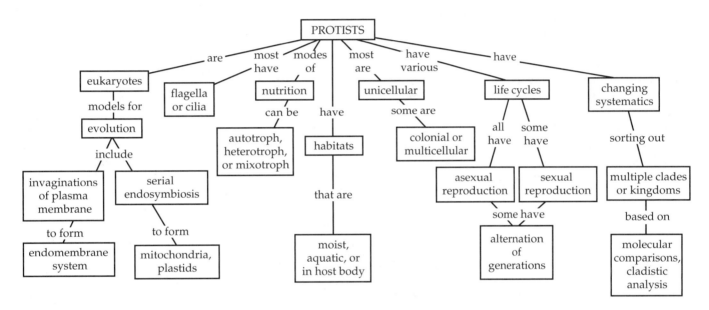

CHAPTER REVIEW

Introduction to the Protists

The oldest fossils of eukaryotes date from 2.1 billion years ago; chemical signs of eukaryotes date back 2.7 billion years. These early eukaryotes diverged into a variety of protists for almost 2 billion years before the evolution of plants, fungi, and animals.

Systematists have split protists into many kingdoms (546)

The traditional kingdom Protista has been shown by molecular systematics and cladistics to be an unworkable classification: It included diverse lineages that were not closely related and excluded others that shared a common ancestor. As many as 20 kingdoms have been proposed in place of the single kingdom Protista. Most biologists still use the informal term "protist" for this wide-ranging group.

Protists are the most diverse of all eukaryotes (546–548)

Protists are an exceedingly diverse group of eukaryotic primarily unicellular organisms, with a few colonial and multicellular forms. These unicellular organisms are exceedingly complex at the cellular level.

Nutrition   Nearly all protists are aerobic and use mitochondria for their respiration. Nutritionally, protists can be photoautotrophs, heterotrophs, or **mixotrophs,** which are both photosynthetic and heterotrophic. These modes of nutrition are spread throughout the various protist lineages. Based on their nutrition and not phylogeny, protists are often grouped for convenience into three categories: photosynthetic, plantlike protists (**algae**); ingestive, animal-like protists (**protozoa**); and absorptive, funguslike protists.

Motility   Most protists have flagella or cilia at some point during their lifetime. These structures are extensions of the plasma membrane that encase a 9 + 2 arrangement of microtubules.

Life Cycles   Mitosis occurs in most phyla of protists, but details of the process vary. Protists reproduce asexually; some also have sexual reproduction, and others may use meiosis and **syngamy** (the union of gametes) to exchange genes between two otherwise asexually reproducing individuals. All three basic types of sexual life cycles, and variations on those types, are found among protists. Often, however, the only diploid cell is the zygote. Many life histories include the formation of resistant **cysts.** Ruptured cysts are some of the oldest eukaryotic fossils.

Protistan Habitats   Protists abound almost anywhere there is water—they occupy freshwater, marine, and moist terrestrial habitats or live symbiotically within the bodies of hosts. They form an important part of **plankton,** the drifting community of mostly microscopic organisms found in bodies of water. Most aquatic food chains are based on **phytoplankton,** planktonic algae and cyanobacteria.

■ INTERACTIVE QUESTION   28.1

Fill in this summary table of protist characteristics.

Characteristic	Brief Description
Fossil evidence	**a.**
Habitats	**b.**
Nutrition	**c.**
Locomotion	**d.**
Reproduction, life cycles	**e.**

The Origin and Early Diversification of Eukaryotes

How the complex eukaryotic cell evolved from simpler prokaryotic cells is a fundamental question in biology.

Endomembranes contributed to larger, more complex cells (548–549)

The increase in complexity and organization of prokaryotes followed three separate trends: the move toward multicellular prokaryotes, such as filamentous cyanobacteria with specialized cells; the evolution of prokaryotic communities in which species benefited from the metabolic specialties of other species; and the trend to compartmentalize different functions within a cell. Invaginations of the plasma membrane may have created the nuclear membrane, endoplasmic reticulum, and other membranes of the endomembrane system, and endosymbiosis may have led to mitochondria and **plastids,** the general name for photosynthetic organelles.

Mitochondria and plastids evolved from endosymbiotic bacteria (549–550)

According to the theory of **serial endosymbiosis,** mitochondria and **plastids** evolved from small prokaryotes that had entered larger cells as prey or parasites. Mitochondria developed from aerobic heterotrophic prokaryotes and chloroplasts are descendants of photosynthetic prokaryotes. The endomembrane system and cytoskeleton of the ancestral eukaryote may have evolved first, allowing the cell to engulf and package the endosymbiont.

Several lines of evidence, including modern endosymbiotic relationships, support the endosymbiotic origin of mitochondria and plastids. Similarities exist between modern bacteria and the chloroplasts and mitochondria of eukaryotes, including size, membrane enzymes and transport systems, circular DNA molecules not associated with proteins, and the process of division. Chloroplasts and mitochondria contain ribosomes and other equipment needed to transcribe and translate their DNA into proteins. These ribosomes more closely resemble prokaryotic ribosomes than they do cytoplasmic ribosomes.

Researchers still need to develop hypotheses for the evolution of the cytoskeleton, including the 9 + 2 microtubule arrangement in flagella and cilia, and the processes of mitosis and meiosis, which also rely on microtubules.

The eukaryotic cell is a chimera of prokaryotic ancestors (550–551)

The nuclear genome of the eukaryotic ancestor came from the host cell; mitochondria arose from one type of bacteria, whereas plastids derived from another.

■ INTERACTIVE QUESTION 28.2

a. How is the endomembrane system of eukaryotic cells thought to have evolved?

b. Review the evidence for the serial endosymbiotic model of eukaryotic cell evolution.

c. Explain the "serial" part of the hypothesis.

The Ancestors of Mitochondria and Chloroplasts All organisms contain the gene for the small ribosomal subunit RNA (SSU-rRNA). Comparisons of nucleotide sequences between prokaryotes and mitochondria indicate that prokaryotes of the alpha proteobacteria group are most closely related to mitochondria. Sequence comparisons of the plastids of various photosynthetic eukaryotes show that the cyanobacteria are most closely related to these plastids.

Gene Transfer to the Nucleus Plastids and mitochondria are not genetically self-sufficient; some of their proteins are coded for by nuclear DNA. Some other proteins, including mitochondrial ATP synthase, are chimeras of organelle-made polypeptides and polypeptides imported from the cytoplasm (and coded by nuclear DNA). During the transition from a symbiotic relationship of prokaryotes to an integrated eukaryotic cell, some endosymbiont DNA was transferred to the host cell's genome.

Secondary endosymbiosis increased the diversity of algae (551–552)

Eukaryotes with plastids are scattered throughout the phylogenetic tree, and these plastids may differ in structure. Plants and the protistan green algae both have plastids surrounded by two membranes, whereas some algal groups have plastids with three or four membranes. Some algal groups may have acquired their plastids by **secondary endosymbiosis,** when a protist engulfed a eukaryotic alga. Most often, the endosymbiont lost most of its parts, except for the plastid, which became a plastid for the host cell. An exception are the cryptomonad algae, whose plastids contain a nucleomorph and some ribosomes that remain from the eukaryotic endosymbiont.

■ INTERACTIVE QUESTION 28.3

a. Explain the occurrence of multiple membranes found around the plastids of some protist groups.

b. What was the ultimate origin of all plastids?

Research on the relationships between the three domains is changing ideas about the deepest branching in the tree of life (552–553)

According to the conventional model, bacteria and archaea diverged from an ancestral prokaryote (the last universal common ancestor or LUCA), and, since archaea and eukaryotes share many similarities, the host cell that engulfed the bacterial ancestor of mitochondria (and later cyanobacteria) was an archaean. Thus, the only bacterial DNA that should be found in the eukaryotic genome should have been transferred from the mitochondrial or plastid endosymbiont. But molecular comparisons have revealed many bacterial DNA sequences within the eukaryotic nucleus, as well as in modern archaea. A new model for the deepest branching in the tree of life replaces a single ancestor with a community of primitive cells that exchanged DNA freely and eventually differentiated into the three domains.

The origin of eukaryotes catalyzed a second great wave of diversification (553–554)

The increased complexity of the eukaryotic cell made possible more structural diversity. Thus, the metabolic diversification of prokaryotes was followed by a eukaryotic adaptive radiation, which then led to a third wave of diversification with the origin of multicellular bodies in some lineages.

Eukaryotic groups are clustered into clades based on comparisons of cell structure, life cycles, and nucleic acid and proteins, in particular SSU-rRNA sequences and amino acid sequences of cytoskeleton proteins. Although fungi, plants, and animals are given kingdom status, debate continues around protistan phylogeny and which of the protistan major clades are designated as kingdoms.

A Sample of Protistan Diversity

Diplomonadida and Parabasala: diplomonads and parabasalids lack mitochondria (555)

According to the "archaezoa hypothesis," a few lineages of protists diverged before the evolution of mitochondria. The presence of mitochondrial genes in the genomes of diplomonads and parabasalids, however, supports the hypothesis that these groups lost their mitochondria after they diverged earliest in eukaryotic history.

Diplomonads, such as the intestinal parasite *Giardia lamblia,* have two nuclei, multiple flagella, a simple cytoskeleton, and no mitochondria or plastids. The **parabasalids** include protists called trichomonads, of which *Trichomonas vaginalis* is the best known.

Euglenozoa: the euglenozoa includes both photosynthetic and heterotrophic flagellates (555)

The clade Euglenozoa includes the **euglenoids,** which are characterized by one or two flagella that emerge from an anterior chamber and by the storage polymer paramylon. A single large mitochondrion associated with a kinetoplast that contains extranuclear DNA is characteristic of the group named kinetoplastids.

Alveolata: the alveolates are unicellular protists with subcellular cavities (alveoli) (556–559)

The alveoli that are characteristic of the groups in the **Alveolata** may help to stabilize the cell surface or to osmoregulate.

Dinoflagellates (Dinoflagellata) **Dinoflagellates** make up a large proportion of the phytoplankton at the base of most marine and many freshwater food chains. Most species are unicellular, each with a characteristic shape. The beating of two flagella in perpendicular grooves between the internal cellulose plates of the cell produces a characteristic spinning movement.

Blooms of dinoflagellates are responsible for the often harmful red tides, the brownish-red color coming from their xanthophyll pigments. Some dinoflagellates are carnivorous, such as *Pfiesteria piscicida,* the cause of recent fish kills in eastern U.S. coastal waters.

Some dinoflagellates are bioluminescent. Photosynthesis by symbiotic dinoflagellates, living in small coral cnidarians, provides the main food source for coral reef communities.

Apicomplexans (Apicomplexa) All **apicomplexans** are animal parasites, and most have complex life cycles that often include sexual and asexual stages and several host species. The parasites spread by tiny infectious cells called **sporozoites.** The control of malaria, caused by the apicomplexan *Plasmodium,* is complicated by the development of insecticide resistance in *Anopheles* mosquitoes, which spread the disease, and drug resistance in *Plasmodium,* as well as by the sequestering of the parasite within human liver and blood cells and the ability of the parasite to change the surface proteins

presented to the host's immune system. New antimalarial drugs may be developed to target the now nonphotosynthetic plastid found in *Plasmodium.*

Ciliates (Ciliaphora) These protists are characterized by the cilia they use to move and feed. Their numerous cilia, coordinated by a submembrane system of microtubules, may be widespread or clumped. Most **ciliates** are complex unicellular organisms found in fresh water.

Ciliates have two types of nuclei—a large macronucleus, which controls everyday functions of the cell, has 50 or more copies of the genome, and splits during binary fission (asexual reproduction); and many small micronuclei, which are exchanged in a process called **conjugation** and provide for genetic recombination.

■ INTERACTIVE QUESTION　28.4

Label the indicated structures in this diagram of a *Paramecium.* To which group and protistan clade does this organism belong?

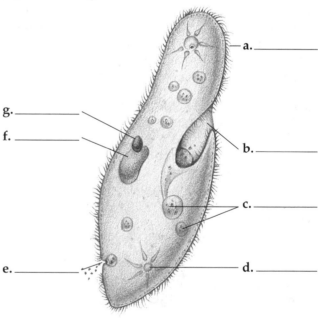

Stramenopila: the stramenopile clade includes the water molds and the heterokont algae (560–562)

The clade **Stramenopila** includes several heterotrophs and various groups of photosynthetic protists (algae), all characterized by hairlike projections on the flagella. In most cases, only reproductive cells are flagellated, having one smooth and one "hairy" flagellum.

Water Molds and Their Relatives (Oomycota) The **water molds, white rusts,** and **downy mildews** are heterotrophic stramenopiles. Many **oomycotes** resemble fungi in appearance, but they have cell walls made of cellulose, a predominant diploid stage in their life cycles, and biflagellated cells in their life cycles. Molecular systematics has confirmed that the oomycotes and fungi are not closely related. Convergent evolution has led to similar filamentous body plans, providing extensive surface area for absorption of nutrients.

The large egg cell of a water mold is fertilized by a smaller sperm nucleus. Zygotes, following dormancy within resistant walls, germinate to form hyphae, which are tipped by zoosporangia that asexually produce flagellated zoospores. Water molds are important decomposers in aquatic ecosystems. White rusts and downy mildews can be destructive parasites of land plants.

Overview of Heterokont Algae This collection of mostly photosynthetic stramenopiles includes the diatoms, golden algae, and brown algae. Their four-membraned plastids evolved by secondary endosymbiosis, probably from a red alga, whose own plastid came from a cyanobacterium by primary endosymbiosis.

Diatoms (Bacillariophyta) **Diatoms** are common unicellular marine and freshwater plankton with unique boxlike silica walls. Diatoms usually reproduce asexually, although eggs and sperm may be produced. Food reserves are in the form of laminarin, a glucose polymer, although some diatoms also produce an oil. Massive quantities of fossilized shells make up diatomaceous earth.

Golden Algae (Chrysophyta) The color of **golden algae** results from yellow and brown carotene and xanthophyll accessory pigments. Most golden algae are unicellular, biflagellated, and found among marine and freshwater plankton. Some species are mixotrophic. Many species form long-lasting resistant cysts when population densities become too high.

■ **INTERACTIVE QUESTION   28.5**

a. In which protistan clade are the heterokont algae placed?

b. Where does the name heterokont come from?

c. What three groups of algae are heterokonts?

d. Describe the plastids of the heterokont algae.

Brown Algae (Phaeophyta) The mostly marine **brown algae** are the largest and most complex algae. Their color is due to accessory pigments in their plastids.

Structural and biochemical adaptations help seaweeds survive and reproduce at the ocean's margins (562–563)

Large marine brown, red, and green algae are called seaweeds. Some seaweeds have tissues and organs that are analogous to those found in plants. Seaweeds may have a plantlike **thallus,** or body, consisting of a rootlike **holdfast** and a stemlike **stipe** that supports leaflike **blades.** Some brown algae have floats, which keep the blades near the surface. The giant, fast-growing brown algal kelps live in deeper waters. Biochemical adaptations to intertidal habitats include cell walls containing gel-forming polysaccharides that protect seaweeds from the abrasive action of waves.

Humans use some seaweeds for food. Commercial uses of the gel-forming substances in their cell walls include thickeners for processed foods, lubricants, and the culture medium, agar.

Some algae have life cycles with alternating multicellular haploid and diploid generations (563–564)

Some algae have an **alternation of generations.** The multicellular, haploid **gametophyte** produces gametes. After syngamy, the diploid zygote grows into the multicellular **sporophyte,** which then produces spores by meiosis. The gametophyte and sporophyte may be similar in appearance (**isomorphic**) or distinct (**heteromorphic**).

Rhodophyta: red algae lack flagella (565)

The **red algae** have no flagellated stages in their life cycle. Their accessory pigment phycoerythrin belongs to the family of pigments called phycobilins, also found in cyanobacteria, from which their plastids evolved by primary endosymbiosis. Phycobilins are able to absorb the wavelengths of light that penetrate into deep water. Most red algae are marine and multicellular, often with filamentous, delicately branched thalli. Alternation of generations is common, although life cycles are diverse.

Chlorophyta: green algae and plants evolved from a common photoautotrophic ancestor (565–567)

The chloroplasts of **green algae** (Chlorophyta and charophytes) resemble those of plants, and molecular systematics indicate that green algae and plants share a common ancestor whose chloroplasts derived from endosymbiotic cyanobacteria. Some systematists favor including green algae, at least the charophytes, in the kingdom Plantae.

Most species live in fresh water, although many are marine. Unicellular forms may be planktonic, inhabitants of damp soil, symbionts in other eukaryotes, or mutualistic partners with fungi—forming associations known as **lichens.** There are colonial species and multicellular filamentous ones. Some large marine forms are considered seaweeds.

Size and complexity have increased in the green algae in three ways: the formation of colonies; repeated nuclear divisions to produce multinucleated filaments; and cell division and differentiation to produce true multicellular forms.

The life cycles of most green algae include sexual (with biflagellated gametes) and asexual stages. Some multicellular green algae have an alternation of generations, as in the seaweed *Ulva.*

A diversity of protists use pseudopodia for movement and feeding (567–570)

The text presents three groups of uncertain phylogeny that use **pseudopodia** to move and often to feed on bacteria, other protists, and dead organic matter called detritus. Some symbiotic species are human parasites.

Rhizopods (Rhizopoda Amoebas)   The **amoebas** are naked or shelled unicellular protists that move and feed with pseudopodia. Microtubules and microfilaments of the cytoskeleton function in the formation of these cellular extensions. Most amoebas are found free-living in freshwater, marine, or soil habitats. *Entamoeba histolytica* is the parasite that causes amoebic dysentery.

Actinopods (Actinopoda Heliozoans and Radiolarians)   The slender projections called axopodia help these planktonic organisms to remain buoyant and to phagocytize microscopic food organisms. **Heliozoans** are mostly freshwater actinopods. **Radiolarians** are primarily marine and have delicate silica shells.

Foraminiferans (Foraminifera Forams)   These marine organisms are known for their porous, calcareous shells. Cytoplasmic strands extending through the pores function in swimming, shell formation, and feeding. Many **forams** derive nourishment from symbiotic algae. Foram fossils are useful as index fossils for dating sedimentary rocks.

■ INTERACTIVE QUESTION   28.6

Why are the amoebas, actinopods, and forams not included in the phylogenetic tree of eukaryotes presented in your text?

Mycetozoa: slime molds have structural adaptations and life cycles that enhance their ecological role as decomposers (570–571)

The resemblance of the slime molds to fungi is the result of convergent evolution to similar lifestyles. Molecular comparisons place them branching from the lineage leading to the fungi and animals, and systematists mostly agree that the slime molds belong in a distinct eukaryotic kingdom, **Mycetozoa.**

Plasmodial Slime Molds (Myxogastrida)   **Plasmodial slime molds** are heterotrophic, engulfing food particles by phagocytosis as they grow through leaf litter or rotting logs. A multinucleate amoeboid mass called a **plasmodium** is the feeding stage. Under harsh conditions, sporangia on erect stalks produce resistant spores by meiosis. Spores germinate, the amoeboid or flagellated haploid cells fuse, and the diploid nucleus repeatedly divides to form a new plasmodium.

Cellular Slime Molds (Dictyostelida)   During the feeding stage of the life cycle, **cellular slime molds** consist of haploid solitary amoeboid cells. In the absence of food, the individual cells congregate into a sluglike mass. Asexual fruiting bodies produce resistant spores. In sexual reproduction, two amoebas fuse to form a zygote that develops into a giant cell enclosed in a protective wall. Following meiosis and mitosis, haploid amoebas are released.

■ INTERACTIVE QUESTION   28.7

Compare the following aspects of the life cycles of plasmodial slime molds and cellular slime molds.

Phylum	Haploid or Diploid	Common Body Form	Gametes or Sexual Cells
Plasmodial slime molds	a.	b.	c.
Cellular slime molds	d.	e.	f.

Multicellularity originated independently many times (572)

Eukaryotic organization allowed for the development of more complex structures. The step from unicellular to multicellular forms opened new opportunities for specialization and adaptation. Multicellularity evolved in early eukaryotes several times, creating the diverse multicellular algae (seaweeds) and the ancestors of plants, fungi, and animals.

WORD ROOTS

-phyte = plant (*gametophyte:* the multicellular haploid form in organisms undergoing alternation of generations)

con- = with, together (*conjugation:* in bacteria, the transfer of DNA between two cells that are temporarily joined)

helio- = sun; **-zoan** = animal (*heliozoan:* sun animals that live in fresh water; they have skeletons made of siliceous or chitinous unfused plates)

hetero- = different; **-morph** = form (*heteromorphic:* a condition in the life cycle of all modern plants in which the sporophyte and gametophyte generations differ in morphology)

iso- = same (*isomorphic:* alternating generations in which the sporophytes and gametophytes look alike, although they differ in chromosome number)

pseudo- = false; **-podium** = foot (*pseudopodium:* a cellular extension of amoeboid cells used in moving and feeding)

thallos- = sprout (*thallus:* a seaweed body that is plantlike but lacks true roots, stems, and leaves)

STRUCTURE YOUR KNOWLEDGE

1. List the general characteristics of protists.
2. Describe some of the current issues concerning the phylogenetic relationships of protistan groups and their classification.

TEST YOUR KNOWLEDGE

MULTIPLE CHOICE: *Choose the one best answer.*

1. Mixotrophs are
   a. plasmodial slime molds.
   b. cellular slime molds.
   c. gametophyte and sporophyte generations that differ in morphology.
   d. organisms that can be both heterotrophic and autotrophic.
   e. organisms that can be both parasitic and free-living.

2. Phytoplankton
   a. form the basis of most marine food chains.
   b. include the multicellular green, red, and brown algae.
   c. are mutualistic symbionts that provide food for coral reef communities.
   d. are unicellular heterotrophs that float near the ocean surface.
   e. Both a and d are correct.

3. According to the theory of serial endosymbiosis,
   a. multicellularity evolved when primitive cells incorporated prokaryotic cells that then took on specialized functions.
   b. the symbiotic associations found in lichens resulted from the incorporation of algal protists into the ancestors of fungi.
   c. the mitochondria and plastids of eukaryotic cells were originally prokaryotic endosymbionts that became permanent parts of the host cell.
   d. the infoldings and specializations of the plasma membrane led to the evolution of the endomembrane system.
   e. the nuclear membrane evolved first, then mitochondria, and then plastids.

4. Comparisons of nucleotide sequences of the gene for the small ribosomal subunit RNA (SSU-rRNA) indicate that the prokaryotes of the alpha proteobacteria group are
   a. the ancestors of the trichomonads and diplomonads.
   b. closely related to amoebas and may help to determine the phylogenetic position of that protist group.
   c. the closest relatives to mitochondria.
   d. the closest relatives to chloroplasts.
   e. the last common universal ancestor of bacteria and archaea.

5. Mitochondria and plastids contain DNA and ribosomes and make some, but not all, of their proteins. Some proteins are coded for by nuclear DNA and produced in the cytoplasm. What may explain this division of labor?
   a. Over the course of evolution, some of the original endosymbiont's genes were transferred to the host cell's nucleus.
   b. The host cell's genome always included genes for making mitochondrial and plastid proteins.
   c. These organelles do not have sufficient resources to make all their proteins and rely on the help of the cell.
   d. Some mitochondria and plastid genes were contributed by early bacterial prokaryotes that shared genes with other primitive cells.
   e. The smaller prokaryotic ribosomes in these organelles cannot produce the eukaryotic proteins required for their functions.

6. The diplomonads and parabasalids are unique among eukaryotes in that
   a. they lack nuclear membranes, even though diplomonads have two nuclei.
   b. they lack ribosomes.
   c. they lack plastids, and all other protist lineages have at least some members that are autotrophic.
   d. they lack mitochondria, although their genomes do have some mitochondrial genes.
   e. they were an early branch from the bacteria rather than the archaea.

7. Genetic variation is generated in the ciliate *Paramecium* when
   a. a micronucleus replicates its genome many times and becomes a macronucleus.
   b. zoospores, which are gametes, fuse.
   c. micronuclei are exchanged in conjugation.
   d. mutations occur in the many copies of the genome contained in the macronucleus and are passed on to offspring.
   e. plasmids are exchanged in conjugation.

8. The Chlorophyta, or green algae, are believed to share a common ancestor with plants because
   a. they are the only multicellular algal protists.
   b. they do not have flagellated gametes.
   c. they are the only protists whose plastids evolved from cyanobacteria.
   d. their chloroplasts are similar in ultrastructure and pigment composition to those of plants.
   e. they are the only algae that exhibit alternation of generations.

9. Which of the following is *not* true of seaweeds?
   a. They are found in the red, green, and brown algal groups.
   b. They have true roots that anchor them tightly to withstand the turbulence of waves.
   c. They are a source of food and commercial products.
   d. The gel-forming polysaccharides in their cell walls protect them from the abrasive action of waves.
   e. They are multicellular photoautotrophs.

10. Examples of mutualistic symbiotic relationships include
   a. dinoflagellates producing red tides.
   b. dinoflagellates living within cnidarians in coral reefs.
   c. sporozoans in their multiple hosts.
   d. lichens.
   e. both b and d.

11. The protistan clade Alveolata would include
   a. the diatoms, golden algae, and brown algae.
   b. the dinoflagellates, apicomplexans, and ciliates.
   c. the slime molds and water molds that are the ancestors of fungi.
   d. the diplomonads and trichomonads.
   e. the amoebas and other protists that move using pseudopods.

12. Evidence that the three domains of life may have evolved from a community of primitive cells that readily exchanged DNA comes from
   a. the unexpected presence of bacterial genes unrelated to the functions of mitochondria or plastids in eukaryotic genomes.
   b. comparisons of SSU-rRNA sequences among the three domains of life.
   c. evidence of the transfer of genes from endosymbiotic mitochondria or plastids to the host cell's nucleus.
   d. evidence of "chemical signatures" of eukaryotes that date from 2.7 billion years ago.
   e. the universal presence of ribosomes in all cells.

MATCHING: *Match the protistan clades with their descriptions.*

____ **1.** Giardia; lacks mitochondria

____ **2.** slime molds; resemble fungi

____ **3.** amoebas; pseudopodia

____ **4.** red algae; pigment phycoerythrin

____ **5.** water molds, heterokont algae; "hairy" flagella

____ **6.** green algae; common ancestor with plants

____ **7.** dinoflagellates; subsurface cavities

____ **8.** euglenoids, kinetoplastids; flagellates

   A. Alveolata
   B. Chlorophyta
   C. Diplomonadida
   D. Euglenozoa
   E. Mycetozoa
   F. Rhizopoda
   G. Rhodophyta
   H. Stramenopila

CHAPTER 29

PLANT DIVERSITY I: HOW PLANTS COLONIZED LAND

FRAMEWORK

The evolution of plants involved adaptations to terrestrial habitats. The kingdom Plantae includes multicellular, photoautotrophic eukaryotes that develop from embryos nourished by the parent plant. Plants exhibit an alternation of generations in which the diploid sporophyte is the more conspicuous stage in all groups except the bryophytes. This chapter outlines the major periods of plant evolution in which bryophytes and seedless vascular plants appeared and radiated.

CHAPTER REVIEW

For the first 3 billion years, life was confined to marine and aquatic environments. Plants began to move onto land about 500 million years ago, changing the biosphere as their various groups evolved and adapted to changing terrestrial habitats.

An Overview of Land Plant Evolution

Evolutionary adaptations to terrestrial living characterize the four main groups of land plants (576)

Bryophytes, which include the mosses, differ from algae by the terrestrial adaptation of a multicellular embryo attached to and nourished by the parent plant. The clade known as the **vascular plants** differs from bryophytes by derived characters that include **vascular tissue,** tubes of cells that transport water and nutrients. The **pteridophytes,** which include the ferns, lack a seed stage and are sometimes called "seedless plants."

A **seed** is an embryo contained with its food supply within a protective coat. The first seed plants evolved in the late Devonian period around 360 million years ago. **Gymnosperms,** including the conifers, are considered "naked" seed plants. **Angiosperms,** the flowering plants, contain their seeds within protective ovaries. The diversification of angiosperms began in the early Cretaceous period, around 130 mya.

■ INTERACTIVE QUESTION 29.1

Consult the hypothetical phylogenetic tree in the text figure 29.1. What four major evolutionary episodes define the history of land plants?

a.

b.

c.

d.

Charophyceans are the green algae most closely related to land plants (576–577)

Land plants are multicellular, photoautotrophic eukaryotes, with chloroplasts containing chlorophylls a and b and cell walls of cellulose. These characteristics, however, are shared with several algal groups and cannot be used to distinguish plants from algae. Two ultrastructural features link plants to their closest green algal relatives, the **charophyceans: rosette cellulose-synthesizing complexes,** which produce the cellulose microfibrils of their cell walls, and **peroxisomes,** enzyme-containing organelles associated with chloroplasts that help minimize photorespiration losses. Other derived homologies in charophyceans and land plants include similarities in sperm cells (in those land plants that have flagellated sperm) and the formation of a **phragmoplast** in the synthesis of a cell plate in mitosis.

Several terrestrial adaptations distinguish land plants from charophycean algae (578–582)

Apical Meristems, Producers of a Plant's Tissues Regions of cell division at the tips of roots and shoots, called **apical meristems,** produce linear growth and cells that differentiate into a protective epidermis and internal plant tissues.

Multicellular, Dependent Embryos   Plant embryos, which are retained in the female plant, have **placental transfer cells** that facilitate the transfer of nutrients from parental tissues. This derived character is the basis for referring to land plants as **embryophytes.**

Alternation of Generations   Whereas this type of reproductive cycle did evolve in various groups of algae, **alternation of generations** does not occur in the charophyceans and appears to be a derived character of land plants.

The multicellular haploid **gametophyte** alternates with the multicellular diploid **sporophyte.** The sporophyte produces **spores,** which are reproductive cells that directly develop into new organisms. In all land plants, these generations differ in morphology. In all but the mosses, the sporophyte is the more conspicuous or dominant stage.

■ **INTERACTIVE QUESTION   29.2**

The gametophyte produces **a.** _____ by **b.** _____. Following fertilization, the **c.** _____ divides by **d.** _____ to develop into the **e.** _____. The sporophyte produces **f.** _____ by **g.** _____. The spores germinate and develop into the **h.** _____.

Walled Spores Produced in Sporangia   The tough polymer **sporopollenin** protects spores as they disperse on land. Haploid spores are produced within multicellular **sporangia** on the sporophyte plant when **spore mother cells** undergo meiosis.

Multicellular Gametangia   Gametes are produced within multicellular **gametangia** in the bryophytes, pteridophytes, and gymnosperms. Sperm are produced in gametangia called **antheridia.** A single egg cell is produced and fertilized within the **archegonium,** where the zygote develops into an embryo.

Other Terrestrial Adaptations Common to Many Land Plants   Most land plants have a waxy **cuticle** coating their leaves and stems, which protects from water loss and microbial attack. **Stomata** are pores that allow for the exchange of gases between the air and the leaf interior.

Except in the bryophytes, the vascular tissues **xylem** and **phloem** provide the conducting vessels to connect the true roots, stems, and leaves of land plants.

■ **INTERACTIVE QUESTION   29.3**

a. What is the structure and function of xylem?

b. What is the structure and function of phloem?

Plants produce many unique **secondary compounds** through side branches of their main metabolic pathways. These compounds may discourage herbivory or microbial attack, protect from UV radiation, or serve as signals. Lignin is a secondary compound that hardens cell walls.

The Origin of Land Plants

Land plants evolved from charophycean algae over 500 million years ago (582–583)

Evidence that the green algae, specifically the charophyceans, are the closest relatives to the plant kingdom includes the following: homologous chloroplasts with accessory pigments chlorophyll *b* and beta-carotene, grana, and closely matched chloroplast DNA; homologous cellulose cell walls with cellulose-manufacturing rosettes; peroxisomes; phragmoplasts in cell plate formation; homologous sperm; and molecular systematics in comparisons of nuclear genes for ribosomal RNA and proteins of the cytoskeleton.

Fossils that appear to be plant spores have been dated at 550 million years old. Widespread fossilized plant spores appear by 460 mya, and vascular plants left an excellent fossil record by 425 mya.

Alternation of generations in plants may have originated by delayed meiosis (583)

In charophyceans, zygotes remain attached to the haploid parental plant and undergo meiosis to produce haploid spores. There is no diploid sporophyte generation.

■ **INTERACTIVE QUESTION   29.4**

How might alternation of generations have originated in the ancestor of plants?

Adaptations to shallow water preadapted plants for living on land (583–584)

Ancient charophycean algae living along the edges of bodies of water may have given rise to the first plants to colonize land. The development of sporopollenin that protected charophycean zygotes during fluctuations in water levels helped to open the adaptive zone of land.

Plant taxonomists are reevaluating the boundaries of the plant kingdom (584)

Molecular systematics, cladistic analysis, and data from morphology, life cycles, and cell ultrastructure are all contributing to the reevaluation of plant classification. The international movement called **"deep green"** is working to identify the major plant clades. Botanists are debating three different ways in which to establish the boundaries of the plant kingdom: the traditional embryophyte clade including just the land plants; the **kingdom Streptophyta,** which includes the charophyceans; or the **kingdom Viridiplantae,** which includes the chlorophytes as well as the charophyceans with the embryophytes. This textbook uses the traditional embryophyte taxon in its discussion of **kingdom Plantae.**

The plant kingdom is monophyletic (584)

The variations in life cycles found in the various plant phyla evolved from the ancestral cycle of alternation of generations as plants diversified and adapted to life on land.

Bryophytes

The three phyla of bryophytes are mosses, liverworts, and hornworts (585)

The three separate phyla of nonvascular plants, commonly referred to as bryophytes, are **phylum Hepatophyta (liverworts), phylum Anthocerophyta (hornworts),** and **phylum Bryophyta (mosses).** These three phyla appear to have diverged independently before the origin of vascular plants.

The gametophyte is the dominant generation in the life cycles of bryophytes (585–587)

When moss spores germinate in a moist habitat, they grow into a mass of green filaments called a **protonema.** Meristem-containing buds produce the mature body form known as a **gametophore.** Bryophyte gametophytes are generally only a few cells thick and lack lignin-coated vascular tissue. They may be anchored by **rhizoids.**

Hornwort and some liverwort gametophytes are low growing and flattened. Although moss and some liverwort gametophytes have leaf-like appendages, they do not qualify as true stems and leaves due to their lack of lignified vascular cells. Some mosses, such as *Polytrichum,* have leaf ridges coated with cuticle and conducting tissues in their "stems." Plant biologists have yet to determine whether these tissues are homologous or analogous with vascular plant xylem and phloem.

Gametangia are enclosed in jackets of protective tissue. Sperm are produced in antheridia, from which they swim to eggs retained in archegonia. Zygotes develop into embryos, which rely on nutrients transported by placental nutritive cells to develop into sporophytes.

Bryophyte sporophytes disperse enormous numbers of spores (587–588)

Although they are usually photosynthetic when young, bryophyte sporophytes remain attached to and dependent on their parental gametophytes. Liverworts have tiny, simple sporophytes. Hornwort sporophytes have a cuticle, and sporophytes of hornworts and mosses have epidermal stomata.

A moss sporophyte has a **foot** that obtains nutrients from the gametophyte, a stalk **(seta)** that elongates for spore dispersal, and a **sporangium** or **capsule** in which millions of spores are produced by meiosis. The protective **calyptra** covers the immature capsule. The specialized upper **peristome** gradually releases spores to be dispersed by wind currents.

▧ INTERACTIVE QUESTION 29.5

Review the life cycle of a typical moss plant by filling in the following blanks.

The dominant generation is the **a.** _____.

Female gametophytes produce eggs in **b.** _____.

Male gametophytes produce sperm in **c.** _____.

Sperm **d.** _____ through the damp environment to fertilize the egg. The zygote remains in the archegonium and grows into the **e.** _____ still attached to the female gametophyte. Spores are formed by the process of **f.** _____ in the **g.** _____ When shed, spores develop into the **h.** _____.

Bryophytes provide many ecological and economic benefits (588–589)

Bryophytes are widely distributed and common in moist habitats. Some are adapted to very cold or dry habitats due to their ability to dehydrate and rehydrate without dying. *Sphagnum,* or peat moss, is an abundant wetland moss that forms undecayed deposits known as **peat.** Peatlands are extensive boreal regions occupied by *Sphagnum* that serve as huge stores of organic carbon and help to stabilize atmospheric carbon dioxide concentrations. The phenolic compounds in moss cell walls protect from UV radiation and resist decay. *Sphagnum* is harvested for use as a soil conditioner and plant packing material.

The Origin of Vascular Plants

Vascular plants have **branched sporophytes** that become independent of the gametophyte parent. Pteridophytes (ferns and related plants) are **seedless vascular plants.**

Additional terrestrial adaptations evolved as vascular plants descended from mosslike ancestors (589)

Meristems, gametangia, sporophytes, stomata, cuticle, and sporopollenin-walled spores probably came from the mosslike ancestors of vascular plants. The evolution of branched, independent sporophytes is chronicled in Silurian fossils called **protracheophyte polysporangiophytes.** The branching made possible more complex bodies and multiple sporangia.

A diversity of vascular plants evolved over 400 million years ago (589)

Cooksonia, an extinct small, dichotomously branching plant with sporangia at the tips of some branches, had true stems with lignified transport cells and is the oldest known vascular plant. By the Devonian period (408–362 mya), diverse vascular plants had evolved.

Pteridophytes: Seedless Vascular Plants

The two phyla of extant seedless vascular plants, phylum Lycophyta (lycophytes) and phylum Pterophyta (ferns, whisk ferns, and horsetails), probably evolved separately from early vascular plants.

Pteridophytes provide clues to the evolution of roots and leaves (590–591)

The resemblance between the true roots of pteridophytes and the fossilized stems of early vascular plants suggests that pteridophyte roots evolved from underground stems. The roots of seed plants may be homologous or may have evolved independently.

The single-veined leaves of lycophytes, known as **microphylls,** probably originated as small tissue flaps growing out from stems. The branching vascular systems of **megaphylls,** typical of other modern vascular plants, provide for more efficient water supply and sugar export and thus support larger leaves with greater photosynthetic capacity. Fossil evidence suggests that megaphylls evolved by the joining of closely lying branches.

A sporophyte-dominant life cycle evolved in seedless vascular plants (591–592)

In all vascular plants, the diploid sporophyte is the larger and more complex plant. **Homosporous** sporophytes produce only one kind of spore, which develops into bisexual gametophytes with both archegonia and antheridia. Plants that are **heterosporous** produce two kinds of spores: **megaspores** that develop into female gametophytes and **microspores** that develop into male gametophytes.

The sperm of the seedless vascular plants must swim through a film of water to reach eggs. Their fragile gametophytes also tend to restrict these plants to relatively damp habitats.

Lycophyta and Pterophyta are the two phyla of modern seedless vascular plants (592–594)

Phylum Lycophyta (Lycophytes)   Lycophytes were a major part of the landscape during the Carboniferous period (360 to 290 mya). One evolutionary line, the giant lycophytes, became extinct when the climate became cooler and drier. The other line of small lycophytes is represented today by the club mosses or ground pines. Many tropical species grow on trees as epiphytes—plants that anchor to other organisms but are not parasites.

Sporangia develop on **sporophylls**—specialized leaves that bear sporangia. Spores germinate and grow into inconspicuous, subterranean gametophytes, which may depend on symbiotic fungi for nutrition.

Phylum Pterophyta (Ferns and Their Relatives) The dichotomous branching and lack of true leaves and roots of *Psilotum,* the whisk fern, initially caused plant biologists to consider them descendants of an early vascular plant lineage. Molecular analysis and ultrastructure comparisons now indicate that **psilophytes** are closely related to ferns, and the lack of true roots and leaves evolved secondarily.

Sphenophytes are also now considered to be closely related to ferns. Sphenophytes grew as tall plants during the Carboniferous period. Today only the genus *Equisetum* is found. Commonly called **horsetails,** these small, upright plants grow in damp locations. Clusters of sporophylls form cones at the tips of some of the green, jointed stems.

Ferns were also found in the great forests of the Carboniferous period and are the most numerous pteridophytes in the modern flora. The large fern leaves, or fronds, often grow from horizontal rhizomes.

Sporangia are arranged into clusters called **sori** on the undersides of sporophylls (reproductive leaves). The wind disperses sporopollenin-protected spores. Most ferns are homosporous, although the archegonia and antheridia on the small, photosynthetic gametophyte mature at different times. Sperm swim to fertilize the egg in neighboring archegonia, and the young sporophyte grows out from the archegonium.

Seedless vascular plants formed vast "coal forests" during the Carboniferous period (594)

The seedless vascular plants of the Carboniferous forests left behind extensive beds of coal. Dead plants did not completely decay in the stagnant swamp waters, and great accumulations of peat developed. When the sea later covered the swamps and marine sediments piled on top, heat and pressure converted the peat to coal.

■ INTERACTIVE QUESTION 29.6

In the following diagram of the life cycle of a fern, label the processes (in the boxes) and structures (on the lines). Indicate which portion of the life cycle is haploid and which is diploid.

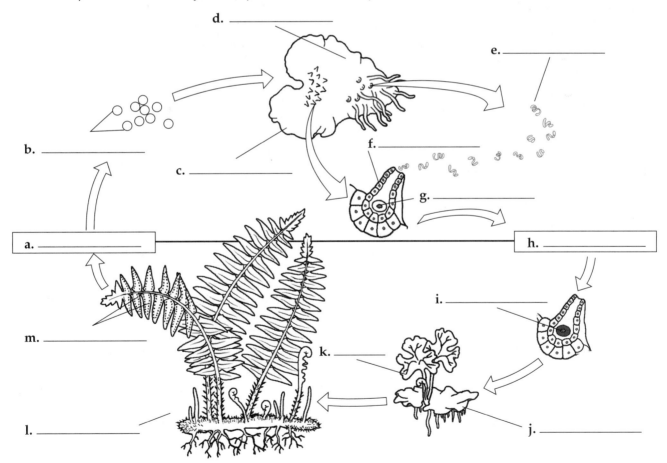

WORD ROOTS

-angio = vessel (*gametangia:* the reproductive organ of bryophytes, consisting of the male antheridium and female archegonium; a multichambered jacket of sterile cells in which gametes are formed)

-phore = bearer (*gametophore:* the mature gamete-producing structure of a gametophyte body of a moss)

bryo- = moss; **-phyte** = plant (*bryophytes:* the mosses, liverworts, and hornworts; a group of nonvascular plants that inhabit the land but lack many of the terrestrial adaptations of vascular plants)

gymno- = naked; **-sperm** = seed (*gymnosperm:* a vascular plant that bears naked seeds not enclosed in any specialized chambers)

hetero- = different; **-sporo** = a seed (*heterosporous:* referring to plants in which the sporophyte produces two kinds of spores that develop into unisexual gametophytes, either female or male)

homo- = like (*homosporous:* referring to plants in which a single type of spore develops into a bisexual gametophyte having both male and female sex organs)

mega- = large (*megaspores:* a spore from a heterosporous plant that develops into a female gametophyte bearing archegonia)

micro- = small; **-phyll** = leaf (*microphylls:* the small leaves of lycophytes that have only a single, unbranched vein)

peri- = around; **-stoma** = mouth (*peristome:* the upper part of the moss capsule often specialized for gradual spore discharge)

phragmo- = a partition; **-plast** = formed, molded (*phragmoplast:* an alignment of cytoskeletal elements and Golgi-derived vesicles across the midline of a dividing plant cell)

pro- = before; **poly-** = many (*protracheophyte polysporangiophytes:* a group of Silurian mosslike ancestors that were like bryophytes in lacking lignified vascular tissue but were different in having independent, branched sporophytes that were not dependent on gametophytes for their growth)

proto- = first; **-nema** = thread (*protonema:* a mass of green, branched, one-cell-thick filaments produced by germinating moss spores)

pterido- = fern (*pteridophytes:* seedless plants with true roots with lignified vascular tissue; the group includes ferns, whisk ferns, and horsetails)

rhizo- = root; **-oid** = like, form (*rhizoids:* long tubular single cells or filaments of cells that anchor bryophytes to the ground)

STRUCTURE YOUR KNOWLEDGE

1. The evolution of land plants involved adaptations to the terrestrial habitat. List some of these adaptations that evolved in the bryophytes and the seedless vascular plants.

2. On page 227, fill in blanks **a–d** with the evolutionary events and blanks **e–j** with the modern plant groups (include some common name examples) on this hypothetical phylogenetic tree that shows the major lines of plant evolution on a geological time frame.

TEST YOUR KNOWLEDGE

MULTIPLE CHOICE: *Choose the one best answer.*

1. Adaptations for terrestrial life seen in all plants are
   a. chlorophylls *a* and *b*.
   b. cell walls of cellulose and lignin.
   c. sporopollenin, protection and nourishment of embryo by gametophyte.
   d. vascular tissue and stomata.
   e. alternation of generations.

2. Plants are thought to be most closely related to charophyceans based on
   a. peroxisomes.
   b. rosette cellulose-synthesizing complexes in their plasma membranes.
   c. phragmoplast formation of cell plates in mitosis
   d. similarities in sperm cells (in those land plants with flagellated sperm).
   e. all of the above.

3. Megaphylls
   a. are large leaves.
   b. are gametophyte plants that develop from megaspores.
   c. are leaves specialized for reproduction.
   d. are leaves with branching vascular systems.
   e. were a dominant group of the great coal forests.

4. Bryophytes differ from the other land plant groups because
   a. their gametophyte generation is dominant.
   b. they are lacking gametangia.
   c. they have flagellated sperm.
   d. they are not embryophytes.
   e. of all of the above.

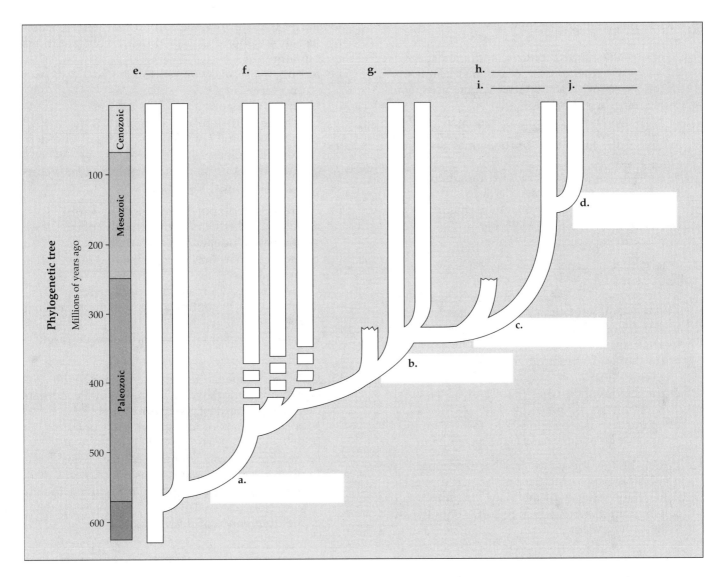

Phylogenetic tree

5. Which of the following plant groups is *incorrectly* paired with its gametophyte generation?
   a. charophycean—none, no alternation of generations
   b. lycophyte—either green or nonphotosynthetic underground plants
   c. moss—green matlike plant
   d. fern—frond growing from rhizome
   e. liverwort—flattened, low-growing green plant

6. Which of the following is most likely the closest relative of the vascular plants?
   a. charophyceans
   b. red algae
   c. bryophytes
   d. pteridophytes
   e. lycophytes

7. The evolution of sporopollenin was important to the movement of plants onto land because it
   a. provided a mechanism for the dispersal of spores and pollen.
   b. provided the structural support necessary to withstand gravity.
   c. enclosed developing gametes and embryos in maternal tissue and prevented desiccation.
   d. provided a tough coating for spores so they could disperse on land.
   e. initiated the alternation of generations that is characteristic of all plants.

8. If a plant's life cycle includes both a male and female gametophyte, the sporophyte plant must be
   a. heterosporous.
   b. homosporous.
   c. homologous.
   d. analogous.
   e. megasporous.

9. Xylem and phloem are found in
   a. all plants.
   b. bryophytes, ferns, conifers, and angiosperms.
   c. only the gametophytes of vascular plants.
   d. the vascular plants, which include ferns, conifers, and angiosperms.
   e. only the vascular plants with seeds.

10. Which of the following functions may secondary compounds serve?
    a. protect plant from ultraviolet radiation
    b. support plant cell walls (lignin)
    c. discourage herbivory or inhibit microbes
    d. function as signaling molecules
    e. all of the above

11. Alternation of generations may have evolved in land plants when
    a. the zygote was retained and protected within the gametophyte plant.
    b. the diploid generation became the more dominant stage.
    c. the haploid gametophyte generation became multicellular.
    d. the zygote underwent mitotic divisions before spores were produced by meiosis.
    e. spores divided by mitosis to form a multicellular stage before meiosis produced gametes.

12. Which of the following has been proposed by plant biologists as the deepest branch that establishes the boundary of the plant kingdom?
    a. kingdom Plantae, which includes only the embryophyte clade
    b. kingdom Streptophyta, which also includes the charophyceans
    c. kingdom Viridiplantae, which also includes all green algae (chlorophytes)
    d. kingdom Vasculata, which includes only the vascular plants
    e. a, b, and c have all been proposed as the clade that defines the plant kingdom

13. Large stores of organic carbon that help to stabilize atmospheric carbon dioxide concentrations are found in
    a. huge tropical swamps dominated by mosses and ferns.
    b. boreal peatlands.
    c. coal deposits left by Carboniferous forests.
    d. the abundant epiphytic lycophytes found in tropical regions.
    e. large tracts of small seedless vascular plants found in boreal regions.

14. Protracheophyte polysporangiophytes are
    a. Silurian fossils of nonvascular plants with branched, independent sporophytes.
    b. fossils such as *Cooksonia* of the first known vascular plants.
    c. the first known fossil pteridophytes.
    d. fossils of bryophytes that had branching sporophytes.
    e. fossils of lycophytes that did not have vascular tissues.

15. If you could take a time machine back to the Carboniferous period, which of the following scenarios would you most likely encounter?
    a. creeping mats of low-growing bryophytes
    b. fields of tall grasses swaying in the wind
    c. swampy forests dominated by large lycophytes, horsetails, and ferns
    d. huge forests of naked-seed trees filling the air with pollen
    e. the dominance of flowering plants

CHAPTER 30

PLANT DIVERSITY II:
THE EVOLUTION OF SEED PLANTS

FRAMEWORK

The reduction and protection of the gametophyte within the parent sporophyte, the use of seeds as a means of dispersal, and the development of pollen for the transfer of sperm are three reproductive adaptations of seed plants that enhanced their success on land.

Gymnosperms bear their seeds "naked" on modified sporophylls. Angiosperms are the most diverse and widespread of plants, owing much of their success to their efficient reproductive apparatus—the flower—and their dispersal mechanism—the fruit. Agriculture is based almost entirely on angiosperms.

CHAPTER REVIEW

Overview of Seed Plant Evolution

Gymnosperms and angiosperms, known as **seed plants,** are vascular plants that produce seeds.

Reduction of the gametophyte continued with the evolution of seed plants (598)

The extremely reduced female gametophyte of seed plants is retained within the sporangium, thus protected and nourished by the sporophyte plant. The retention of this reduced gametophyte generation within the life cycle may allow for the screening out of harmful mutations in this haploid stage. The gametophyte also helps to nourish the sporophyte embryo in its early development.

Seeds became an important means of dispersing offspring (599)

Spores, which are resistant cells that can develop into new organisms, are the mechanism for withstanding harsh environmental conditions and for dispersal in bryophytes and seedless vascular plants. In seed plants, a multicellular **seed,** consisting of a sporophyte embryo surrounded by a food supply and enclosed within a protective coat, provides the resistant stage of the life cycle and the means of dispersal. The gametophyte generations are retained within the walls of the microspore or megaspore that produced them.

The megasporangium of a seed plant is surrounded by **integuments** derived from sporophyte tissue. The **ovule,** which later develops into a seed, consists of the integuments, megasporangium, and megaspore. The female gametophyte develops within the megaspore and produces one or more egg cells. A fertilized egg develops into a sporophyte embryo contained within the seed. The protective coat of the seed, derived from the integuments, enables a seed to withstand harsh conditions.

■ INTERACTIVE QUESTION   30.1

Label the parts in this generalized diagram of a gymnosperm ovule and seed. Indicate whether structures are diploid or haploid tissues.

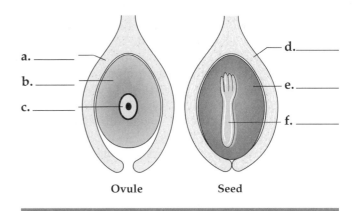

Ovule   Seed

Pollen eliminated the liquid-water requirement for fertilization (600)

Within the microsporangium, microspores develop into pollen grains surrounded by sporopollenin-protected tough coats. Pollen grains mature into the male gametophyte and may be dispersed by wind or animals to an ovule, where they release one or more sperm to the female gametophyte. This transfer of pollen to ovules is called **pollination.**

■ INTERACTIVE QUESTION   30.2

Why is the evolution of pollen an important terrestrial adaptation?

The two clades of seed plants are gymnosperms and angiosperms (600)

New data now suggest that the phylogeny of seed plants includes two monophyletic clades—the gymnosperms and the angiosperms—that probably evolved from different ancestors that were members of an extinct group known as **progymnosperms.**

Gymnosperms

Gymnosperms are named for their naked seeds that develop on the surface of sporophylls.

The Mesozoic era was the age of gymnosperms (600)

Gymnosperms probably originated from the progymnosperms, some of whose species had evolved seeds by the end of the Devonian period. The gymnosperms radiated during the Carboniferous and Permian. Warmer and drier conditions gave gymnosperms a selective advantage by the end of the Permian at the close of the Paleozoic era. Conifers and the great palmlike cycads supported the giant dinosaurs of the Mesozoic. With the environmental changes at the end of the Mesozoic, the dinosaurs became extinct, but many gymnosperms persisted.

The four phyla of extant gymnosperms are ginkgo, cycads, gnetophytes, and conifers (600–603)

Three gymnosperm phyla are relatively small: **phylum Ginkgophyta,** with the deciduous, ornamental ginkgo tree; **phylum Cycadophyta,** which includes the palm-like cycads; and **phylum Gnetophyta,** which includes three genera: the bizarre *Welwitschia,* the tropical *Gnetum,* and the American desert shrub, *Ephedra.*

The largest gymnosperm phylum is **phylum Coniferophyta.** The name **conifer** refers to the reproductive structure, the cone. Most conifers are evergreens, with needle-shaped leaves covered with a thick cuticle. Coniferous trees are some of the largest and oldest living organisms.

The life cycle of a pine demonstrates the key reproductive adaptations of seed plants (603–606)

The pine tree is a heterosporous sporophyte. Pollen cones consist of many scalelike sporophylls that bear sporangia. Meiosis gives rise to microspores that develop into pollen grains enclosing the male gametophytes. Scales of the ovulate cone hold ovules, each of which contains a megasporangium. A megaspore mother cell undergoes meiosis, and one of the resulting megaspores undergoes repeated divisions to produce a female gametophyte in which a few archegonia develop.

Pollination occurs when a pollen grain is drawn through the micropyle, an opening in the integuments. A pollen tube grows and digests its way through the megasporangium. Fertilization occurs when a sperm nucleus joins with the egg nucleus. The zygote develops into a sporophyte embryo, which is nourished by the remaining female gametophyte tissue and is enclosed in a seed coat derived from integuments of the parent sporophyte. Seed production takes three years.

Angiosperms (Flowering Plants)

Angiosperms, the vascular seed plants that produce flowers and fruits, are the most diverse and widespread of modern plants.

Systematists are identifying the angiosperm clades (606–608)

The **phylum Anthophyta** was divided into two main classes until quite recent DNA comparisons and cladistic analysis revealed that, although the **monocots** are a monophyletic group, the traditional **dicots** are not. The clade **eudicots** includes most previously classified dicots, but earlier branches from the angiosperm line gave rise to the water lilies, to other early angiosperm lines, and to the oldest branch represented today by one species, *Amborella trichopoda.*

Both gymnosperms and angiosperms transport water through tracheids, tapered cells that also function

■ INTERACTIVE QUESTION 30.3

In the diagram of the life cycle of a pine, label the indicated structures and show where meiosis, pollination, and fertilization take place. Which structures represent the gametophyte generation?

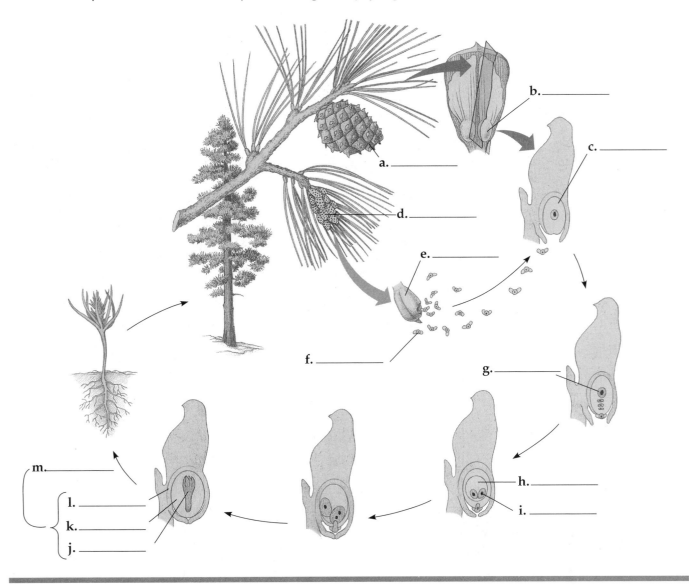

in mechanical support. In the xylem of angiosperms, fiber cells are specialized for support and shorter, wider vessel elements are arranged end to end to form more efficient water transport tubes called xylem vessels.

The flower is the defining reproductive adaptation of angiosperms (608)

Many angiosperms rely on insects or other animals for transferring pollen from flower to flower, an advance over the more random wind pollination of most gymnosperms. Some angiosperms, however, are also wind-pollinated.

The **flower,** the reproductive structure of an angiosperm, is a compressed shoot with four circles of modified leaves. The outer **sepals** are usually green, whereas **petals** are brightly colored in most flowers that are pollinated by insects and birds. **Stamens** are male sporophylls that produce microspores, which develop into male gametophytes. Megaspores develop into female gametophytes and are produced by **carpels,** the female sporophylls.

A stamen has a stalk, called a **filament,** and a terminal **anther,** in which pollen is produced. The carpel has a sticky **stigma,** which receives pollen, and a **style,** which leads to the **ovary.** The ovary contains ovules, which develop into seeds.

The carpel may have evolved from a seed-bearing leaf that rolled into a tube. The term "pistil" is used by some botanists to refer to both single and more complex, fused carpels.

Fruits help disperse the seeds of angiosperms (608–609)

A **fruit** is a mature ovary that functions in the protection and dispersal of seeds. Fruits may be modified in various ways to disperse seeds, enlisting the aid of wind or animals.

Hormonal changes following pollination cause the ovary to enlarge, its wall becoming the **pericarp,** or thickened wall of the fruit. Fruit usually does not set if a flower has not been pollinated. A **simple fruit** develops from a single ovary. An **aggregate fruit** results from a single flower that has several separate carpels. A **multiple fruit,** such as pineapple, develops from a group of tightly clustered flowers called an inflorescence.

Humans have selectively bred edible fruits. Cereal grains—the wind-dispersed fruits of grasses—are essential human foods.

■ INTERACTIVE QUESTION   30.4

a. Name the four types of modified leaves that make up a flower.

b. What is a fruit?

The life cycle of an angiosperm is a highly refined version of the alternation of generations common to all plants (610)

Pollen grains, which develop from microspores within the anthers, contain the immature male gametophytes, consisting of two haploid cells. **Ovules** contain the female gametophyte, called an **embryo sac,** which consists of a few cells.

Most flowers have some mechanism to ensure **cross-pollination.** A pollen grain germinates on the stigma and extends a pollen tube down the style to the ovule, where it releases two sperm cells into the embryo sac. In a process called **double fertilization,** one sperm unites with the egg to form the zygote and the other sperm nucleus fuses with the large cell containing the polar nuclei in the center of the female gametophyte.

The zygote divides and forms the sporophyte embryo, consisting of a rudimentary root and one (in monocots) or two (in eudicots) seed leaves, called **cotyledons.** The **endosperm,** which develops from the triploid ($3n$) nucleus, serves as a food reserve for the embryo.

■ INTERACTIVE QUESTION   30.5

a. What constitutes the gametophyte generation of an angiosperm?

b. What does a seed consist of?

c. What is a possible function of double fertilization?

The radiation of angiosperms marks the transition from the Mesozoic era to the Cenozoic era (610–611)

Angiosperms appear in the fossil record about 130 million years ago. The Cretaceous was a period of climatic change and extinctions, marking the boundary between the Mesozoic and Cenozoic eras. By the end of the Cretaceous period, 65 million years ago, angiosperms had radiated and become the dominant plants.

Angiosperms and animals have shaped one another's evolution (611–612)

Animals influenced the evolution of plants and vice versa. Animal predation on plants may have provided selective pressure for plants to keep spores and vulnerable gametophytes on the plant. As flowers and fruits evolved, some predators became beneficial as pollinators and seed dispersers. The **coevolution,** or mutual evolutionary influence, of angiosperms and their pollinators is seen in the diversity of flowers, whose color, fragrance, and shape are often matched to the sense of sight and smell or to the particular morphology of a group of pollinators.

Plants and Human Welfare

Agriculture is based almost entirely on angiosperms (612)

All of our fruit and vegetable crops are angiosperms. Grains are grass fruits; their endosperm is the main food source for most people and their domesticated animals. Agriculture is a unique evolutionary relationship between plants and the humans who breed and cultivate them.

Plant diversity is a nonrenewable resource (612–613)

The growing human population with its demand for space, food, and natural resources is leading to the extinction of hundreds of species each year. Tropical rain forests, where plant diversity is greatest, could be eliminated within 25 years if the pace of destruction continues. Humans are destroying their potential supply of new food crops and medicines. Preserving plant diversity is an urgent problem.

WORD ROOTS

co- = with, together (*coevolution:* the mutual influence on the evolution of two different species interacting with each other and reciprocally influencing each other's adaptations)

endo- = inner (*endosperm:* a nutrient-rich tissue formed by the union of a sperm cell with two polar nuclei during double fertilization, which provides nourishment to the developing embryo in angiosperm seeds)

peri- = around; **-carp** = fruit (*pericarp:* the thickened wall of a fruit)

pro- = before; **gymno-** = naked; **-sperm** = seed (*progymnosperm:* an extinct group of plants that is probably ancestral to gymnosperms and angiosperms)

STRUCTURE YOUR KNOWLEDGE

1. **a.** From what extinct group did the gymnosperms and angiosperms probably independently evolve?

    b. List the four phyla that are considered gymnosperms.

    c. The angiosperms are grouped into a single phylum, Anthophyta, which traditionally contained the classes, monocots and dicots. List four of the clades that have recently been distinguished in this phylum.

2. List the characteristics of seed plants that were evolutionary adaptations to their terrestrial habitat.

3. What adaptations helped angiosperms to become the most successful and widespread land plants?

TEST YOUR KNOWLEDGE

MULTIPLE CHOICE: *Choose the one best answer.*

1. In which of the following groups do sperm no longer have to swim to reach the female gametophyte?
    a. bryophytes
    b. ferns
    c. gymnosperms
    d. angiosperms
    e. Both c and d are correct.

2. What provides food for a developing sporophyte embryo in a gymnosperm seed?
    a. endosperm
    b. female gametophyte tissue
    c. female sporophyte tissue
    d. male gametophyte tissue
    e. pine cone

3. Which of the following is the correct path that a pollen tube takes to reach the female gametophyte in an angiosperm?
    a. stigma, style, ovary, ovule, embryo sac
    b. anther, stigma, filament, ovule, ovum
    c. stigma, filament, carpel, ovary, ovule
    d. carpel, pistil, ovary, ovule, embryo sac
    e. stigma, style, pistil, ovule, ovary

4. How many generations are represented in a pine seed?
    a. one: the new sporophyte generation
    b. two: seed coat and food supply from female gametophyte and sporophyte embryo
    c. two: seed coat from integuments of parent sporophyte and new sporophyte embryo
    d. three: seed coat from parent sporophyte, food supply from gametophyte, and sporophyte embryo
    e. three: seed coat from female gametophyte, food supply from parent sporophyte, and sporophyte embryo

5. Gymnosperms rose to dominance during which of the following periods?
    a. Carboniferous, when they formed important components of the great "coal forests"
    b. Devonian, when they successfully competed with the short-statured bryophytes
    c. Permian, when they replaced the seedless vascular plants as the climate became drier with the formation of the supercontinent Pangaea
    d. Cretaceous, when cooler climates favored their growth over the ferns and other seedless vascular plants
    e. Cretaceous, when they replaced angiosperms as the largest and most widespread group of plants

6. If the angiosperm gametophyte generation has been reduced to so few cells, why hasn't it been eliminated from the life cycle?
   a. The gametophyte generation produces the resistant spores that allow flowering plants to withstand harsh conditions and to disperse on land.
   b. In the progymnosperm ancestors of angiosperms, the gametophyte was the dominant generation and thus natural selection has retained it.
   c. The gametophyte generation produces both the protective seed coat and nourishment for the developing embryo.
   d. The gametophyte may allow for elimination of harmful mutations that are expressed in these haploid cells, and it helps to provide nourishment during early development of the embryo.
   e. Natural selection is not effective at eliminating such microscopic structures.

7. Which of the following is a key difference between seedless vascular plants and plants with seeds?
   a. The gametophyte generation is dominant in the seedless plants, whereas the sporophyte is dominant in the seed plants.
   b. The spore is the agent of dispersal in the first, whereas the seed functions in dispersal in the second.
   c. The gametophyte is photoautotrophic in all seedless plants but dependent on the sporophyte generation in seed plants.
   d. The embryo is unprotected in the seedless plants but retained within the female reproductive structure in the seed plants.
   e. Sporopollenin is not found in the seedless plants.

8. An example of coevolution is
   a. wind pollination in conifers.
   b. a flower with nectar guides that direct bees to its nectaries.
   c. the synchronization of nutrient development and fertilization resulting from double fertilization.
   d. the retention of the gametophyte generation to weed out harmful mutations.
   e. the development of alternation of generations independently in land plants and some algal groups.

9. Where would you find a microsporangium in the life cycle of a pine?
   a. within the embryo sac in an ovule
   b. in the pollen sacs in an anther
   c. at the base of a sporophyll in a pollen cone
   d. on a scalelike sporophyll found in an ovulate cone
   e. forming a seed coat surrounding a pine seed

10. Which of these plants is believed to be the only survivor of the oldest branch of the angiosperm lineage?
    a. the ornamental ginkgo tree
    b. the water lilies
    c. the eudicot poppies
    d. the star anise
    e. the small shrub *Amborella*

TRUE OR FALSE: *Indicate T or F, and then correct the false statements.*

_____ 1. All photoautotrophic, multicellular eukaryotes are plants.

_____ 2. Heterosporous plants produce male and female gametophytes.

_____ 3. The gametophyte generation is most reduced in the gymnosperms.

_____ 4. The *Ginkgo*, cycads, and conifers are naked-seed plants.

_____ 5. A sporangium produces spores, no matter what group it is found in.

_____ 6. A fruit consists of an embryo, nutritive material, and a protective coat.

_____ 7. Tracheids are wide, specialized cells arranged end to end for water transport and are found in angiosperms.

_____ 8. A stamen consists of a filament and anther in which pollen is produced.

_____ 9. The female gametophyte in angiosperms consists of haploid cells in which a few archegonia develop.

_____10. The male gametophyte in angiosperms is contained within a pollen grain.

CHAPTER 31

FUNGI

FRAMEWORK

This chapter describes the morphology, life cycles, and ecological and economic importance of the kingdom Fungi. Lichens are symbiotic complexes of fungi and algae. Fungi play an important ecological role, both as decomposers and by their mycorrhizal association with plant roots. A flagellated protistan may have been the common ancestor to fungi and animals.

CHAPTER REVIEW

Introduction to the Fungi

Fungi were once classified with plants, but these eukaryotic, mostly multicellular organisms are now placed in their own kingdom.

Absorptive nutrition enables fungi to live as decomposers and symbionts (616–617)

Fungi are heterotrophs that obtain their nutrients by **absorption;** they secrete digestive enzymes, called **exoenzymes,** into the surrounding food media and absorb the resulting small organic molecules. Fungi may be saprobes that decompose nonliving organic material, parasites that absorb nutrients from living cells, or mutualistics that feed on, but also benefit, their hosts.

Extensive surface area and rapid growth adapt fungi for absorptive nutrition (617–618)

Most fungi are composed of **hyphae** that form a network called a **mycelium** that provides an extensive surface area for absorption. The hyphae of most fungi are divided into cells by cross-walls called **septa,** which usually have pores through which nutrients and cell organelles can pass. Cell walls are composed of **chitin.** Some hyphae are aseptate, and these **coenocytic** fungi consist of a continuous cytoplasmic mass with many nuclei. Parasitic fungi may penetrate host cells with modified hyphae called **haustoria.**

Although fungi are nonmotile, they rapidly enter into new food territory by the rapid growth of their hyphae.

Fungi disperse and reproduce by releasing spores that are produced sexually or asexually (618)

Fungi release huge quantities of spores, produced either sexually or asexually, that aid in dispersal.

Many fungi have a heterokaryotic stage (618–619)

Nuclei of hyphae and spores of most species are haploid. Genetic heterogeneity may exist when hyphae with different nuclei join, forming a **heterokaryon.** The haploid nuclei may stay in separate regions of the mycelia, or they may mingle and engage in a process similar to crossing over.

The sexual cycle of many fungi contains two separate events. **Plasmogamy** is cytoplasmic fusion of parental mycelia. The two nuclei may pair up and divide in tandem, forming **dikaryotic** cells. After a period of time, the nuclei fuse **(karyogamy),** and the zygote undergoes immediate meiosis.

■ INTERACTIVE QUESTION 31.1

Briefly define the following terms that relate to the structure and reproduction of fungi:

a. mycelium

b. septa

c. coenocytic

d. heterokaryon

e. plasmogamy

f. karyogamy

Diversity of Fungi

Phylum Chytridiomycota: chytrids may provide clues about fungal origins (619–620)

The **chytrids** were previously classified as protists because they form uniflagellated zoospores. Molecular comparisons, however, support their inclusion with the fungi and their probable earliest divergence from the fungal branch of the eukaryotic tree. Chytrids are primarily aquatic saprobes or parasites. Most form coenocytic hyphae.

Phylum Zygomycota: zygote fungi form resistant structures during sexual reproduction (620–622)

Zygomycetes or **zygote fungi,** are mostly terrestrial fungi living in soil or on decaying matter. One group forms important mutualistic associations called **mycorrhizae** with plant roots. Hyphae are coenocytic. Resistant **zygosporangia** are formed in sexual reproduction.

Rhizopus stolonifer, the black bread mold, is a common zygomycete that spreads horizontal hyphae across its food substrate and erects hyphae with bulbous black sporangia containing hundreds of haploid spores. Sexual reproduction between mycelia of opposite mating types produces a heterokaryotic zygosporangium with a tough protective coat that remains dormant until conditions are favorable. Karyogamy occurs between paired nuclei, followed by meiosis to produce haploid spores.

Phylum Ascomycota: sac fungi produce sexual spores in saclike asci (622–623)

The ascomycetes, or **sac fungi,** range in complexity from unicellular yeasts to elaborate cup fungi and morels. They are found in a wide variety of habitats. Many species are in mutualistic relationships with algae to form lichens or with plants to form mycorrhizae.

Plasmogamy between two parental hyphae produces a heterokaryotic ascogonium, which produces a macroscopic fruiting body called an **ascocarp.** Karyogamy occurs in terminal cells, which become the saclike **asci,** and meiosis yields haploid spores called ascospores. Asexual reproduction involves asexual spores, called **conidia,** that are produced in chains or clusters at the ends of hyphae.

Phylum Basidiomycota: club fungi have long-lived dikaryotic mycelia (624–626)

The basidiomycetes, which include the mushrooms, shelf fungi, puffballs, and rusts, produce a club-shaped **basidium** during a short diploid stage in the life cycle. The **club fungi** include important saprobes, mycorrhizae-forming mutualists, and plant parasites.

In response to environmental stimuli, an elaborate fruiting body called a **basidiocarp** is formed from the long-lived dikaryotic mycelium. Karyogamy and meiosis occur in terminal cells called basidia, producing haploid spores. Asexual reproduction is less common in basidiomycetes than in ascomycetes.

■ INTERACTIVE QUESTION 31.2

Indicate whether the following diagrams (e, h, l) are from a zygomycete, ascomycete, or basidiomycete life cycle. Identify the labeled structures.

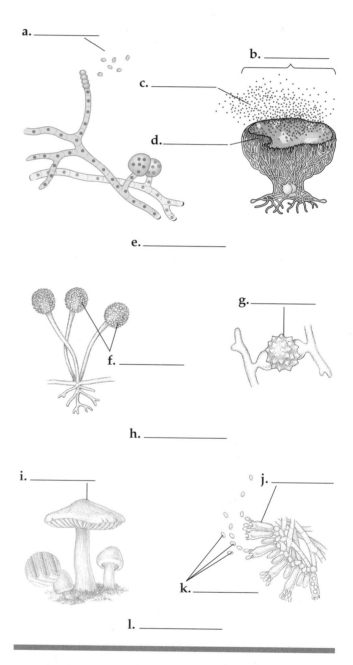

a. _____

b. _____

c. _____

d. _____

e. _____

f. _____

g. _____

h. _____

i. _____

j. _____

k. _____

l. _____

Molds, yeasts, lichens, and mycorrhizae are specialized lifestyles that evolved independently in diverse fungal phyla (626–629)

Molds **Molds** are rapidly growing, asexually reproducing fungi that are saprobes or parasites on a variety of substrates. In later life stages, sexual reproduction may occur in zygosporangia, ascocarps, or basidiocarps. If no sexual stage is known, the molds are called deuteromycetes, or **imperfect fungi,** in reference to their lack of a sexual stage. Some molds are sources of antibiotics, such as penicillin.

Yeasts **Yeasts** are unicellular fungi that grow in liquid or moist habitats. Reproduction is commonly asexual by cell division or budding, but some yeasts produce asci or basidia and are classified accordingly. Yeasts placed in the imperfect fungi have no known sexual stages. *Saccharomyces cerevisiae* is used in baking, brewing, and molecular genetic research.

Lichens **Lichens** are symbiotic associations of millions of photosynthetic organisms in a lattice of fungal hyphae. The fungus is usually an ascomycete, and the partner is usually unicellular or filamentous green algae or cyanobacteria. The alga provides the fungus with food and, in the case of cyanobacteria, nitrogen. The fungus creates most of the mass of the lichen and provides protection for the alga and absorbs water and minerals.

Lichens reproduce asexually, either as fragments or as tiny clusters called **soredia.** In addition, it is common for the fungal component to reproduce sexually and for the algal component independently to reproduce asexually.

Lichens are important colonizers of bare rock and soil and can withstand desiccation and great cold. Lichens cannot, however, tolerate air pollution.

Mycorrhizae **Mycorrhizae** are very common and important mutualistic associations of plant roots and fungi in which the fungal hyphae provide the plant with minerals absorbed from the soil and the plant provides organic nutrients to the fungus. The fungi periodically form fruiting bodies for sexual reproduction.

■ **INTERACTIVE QUESTION 31.3**

a. List some of the beneficial roles played by lichens.

b. Why are mycorrhizae so important?

Ecological Impacts of Fungi

Ecosystems depend on fungi as decomposers and symbionts (629)

Wood-rotting fungi became a dominant group during the Permian extinctions about 250 mya, when they decomposed the vast forests that were destroyed, releasing the nutrients needed for recolonization.

Fungi and bacteria are the principal decomposers of organic matter, making possible the essential recycling of chemical elements between living organisms and their abiotic surroundings. Invasive hyphae, exoenzymes that work on cellulose and lignin, and prolific production of colonizing spores make fungi excellent decomposers of plant material.

As well as decomposing the organic litter in our ecosystem, fungi work on our food, clothing, and the wood used in buildings and boats. Fungi destroy a large proportion of the world's fruit harvest each year.

Some fungi are pathogens (629–630)

Fungal diseases of plants are common, killing elm and chestnut trees and spoiling grain crops. An ascomycete forms ergots on rye that can cause serious symptoms when accidentally milled into flour. Lysergic acid, the raw material of LSD, is one of the toxins in the ergots.

Pathogenic fungi cause athlete's foot, vaginal yeast infections, and lung infections. Ringworm is a common skin **mycosis,** or fungal infection. Systemic mycoses are very serious.

Fungi are commercially important (630)

Commercially cultivated mushrooms are eaten, and fungi are used to ripen some cheeses.

Evolution of Fungi

Fungi colonized land with plants (630–631)

The oldest undisputed fossils of fungi date back 460 million years. The first vascular plant fossils have petrified mycorrhizae, indicating that plants and fungi moved onto land together.

The flagellated stage found in the most ancient lineage, the chytrids, is indicative of a protist ancestor. The fungal phyla Zygomycota, Ascomycota, and Basidiomycota apparently lost their flagellated stages as they developed reproductive and dispersal adaptations for life on land.

Fungi and animals evolved from a common protistan ancestor (631)

Comparisons of several proteins and ribosomal RNA indicate that animals and fungi diverged from a common protistan ancestor.

WORD ROOTS

-osis = a condition of (*mycosis:* the general term for a fungal infection)

coeno- = common; **-cyto** = cell (*coenocytic:* referring to a multinucleated condition resulting from the repeated division of nuclei without cytoplasmic division)

di- = two; **-karyo** = nucleus (*dikaryotic:* a mycelium with two haploid nuclei per cell, one from each parent)

exo- = out, outside (*exoenzymes:* powerful hydrolytic enzymes secreted by a fungus outside its body to digest food)

hetero- = different (*heterokaryon:* a mycelium formed by the fusion of two hyphae that have genetically different nuclei)

myco- = fungus; **rhizo-** = root (*mycorrhizae:* mutualistic associations of plant roots and fungi)

plasmo- plasm; **-gamy** = marriage (*plasmogamy:* the fusion of the cytoplasm of cells from two individuals; occurs as one stage of syngamy)

STRUCTURE YOUR KNOWLEDGE

1. Fill in the table below that summarizes the characteristics of the four fungal phyla.

2. The kingdom Fungi contains members with saprobic, parasitic, and mutualistic modes of nutrition. How do these types of nutrition relate to the ecological and economic importance of this group?

3. What is the basis for saying that fungi and animals evolved from a common protistan ancestor?

TEST YOUR KNOWLEDGE

FILL IN THE BLANKS

_____ 1. division between cells in fungal hyphae

_____ 2. cells with nuclei from two parents

_____ 3. component of cell walls in most fungi

_____ 4. produced by fungi for absorptive nutrition

_____ 5. club-shaped reproductive structure found in mushrooms

_____ 6. sacs that contain sexual spores in sac fungi

_____ 7. mutualistic associations between plant roots and fungi

_____ 8. resistant structures produced by zygomycetes

_____ 9. hyphae with many nuclei

_____ 10. most primitive fungal group

MULTIPLE CHOICE: *Choose the one best answer.*

1. The major difference between fungi and plants is that fungi
   a. have an absorptive form of nutrition.
   b. do not have a cell wall.
   c. are not eukaryotic.
   d. are multinucleate but not multicellular.
   e. reproduce by spores.

Phylum	Examples	Morphology	Asexual Reproduction	Sexual Reproduction
Chytridiomycota	a.	b.	c.	
Zygomycota	d.	e.	f.	g.
Ascomycota	h.	i.	j.	k.
Basidiomycota	l.	m.	n.	o.

2. Chytrids have been previously classified with protists because they
   a. do not have chitin in their cell walls.
   b. do not have absorptive nutrition.
   c. form flagellated spores.
   d. have metabolic pathways that resemble those of protists.
   e. are aquatic, and fungi are terrestrial.

3. A fungus that is both a parasite and a saprobe is one that
   a. digests only the nonliving portions of its host's body.
   b. lives off the sap within its host's body.
   c. first lives as a parasite but then consumes the host after it dies.
   d. lives as a mutualistic symbiont on its host.
   e. causes athlete's foot and vaginal infections.

4. The fact that karyogamy does not immediately follow plasmogamy
   a. is necessary to create coenocytic hyphae.
   b. allows for the development of more genetic variation.
   c. allows fungi to reproduce asexually most of the time.
   d. creates heterokaryotic cells that may benefit from the presence of two different genomes.
   e. is characteristic of yeasts.

5. Imperfect fungi
   a. represent the most ancient lineage of fungi.
   b. include the fungal components of lichens.
   c. have abnormal forms of sexual reproduction.
   d. are fungi that are predatory.
   e. include molds and other types of fungi whose sexual stage is lacking or unknown.

6. In the Ascomycota,
   a. sexual reproduction occurs by conjugation.
   b. spores often line up in a sac in the order they were formed by meiosis.
   c. asexual spores form in sporangia on erect hyphae.
   d. most hyphae are dikaryotic.
   e. sexual spores are produced in conidia.

7. Lichens are symbiotic associations that
   a. usually involve an ascomycete and a green alga or cyanobacterium.
   b. can reproduce sexually by forming soredia.
   c. require moist environments to grow.
   d. fix nitrogen for absorption by plant roots.
   e. are unusually resistant to air pollution.

8. The name given to three of the fungal phyla is based on
   a. the structure in which karyogamy occurs during sexual reproduction.
   b. the location of plasmogamy during sexual reproduction.
   c. the location of the heterokaryotic stage in the life cycle.
   d. the structure that produces asexual spores.
   e. their ancestral origin.

9. Fungi and animals appear to have evolved from a common ancestor
   a. because neither of them are photosynthetic.
   b. based on similarities in cell structure.
   c. based on molecular analysis of proteins and ribosomal RNA.
   d. about the time that fungi and plants moved onto land.
   e. based on homologous ultrastructure of their flagella.

CHAPTER 32

INTRODUCTION
TO ANIMAL EVOLUTION

FRAMEWORK

Animals are multicellular eukaryotic heterotrophs that ingest their food. The traditional phylogenetic tree of animals is based on body plan grades and includes four key branch points: the parazoa–eumetazoa split based on absence or presence of true tissues; the radiata–bilateria split based on radial versus bilateral body symmetry; the acoelomate–coelomate split based on absence or presence of a body cavity; and the protostome–deuterostome split based on differences in embryological development.

Molecular systematics supports the early branch points, but splits the protostomes into clades Ecdysozoa and Lophotrochozoa, and distributes the acoelomate, pseudocoelomate, and lophophorates within these clades.

Fossil evidence for the origin and rapid divergence of animals is limited. A colonial choanoflagellate is the probable ancestor, and most animal phyla diversified during the 20 million years of the Cambrian explosion.

CHAPTER REVIEW

What Is an Animal?

Structure, nutrition, and life history define animals (633–634)

Animals are multicellular heterotrophic eukaryotes, most of which use **ingestion** as their mode of nutrition. Animal cells lack walls and may have desmosomes, gap junctions, and tight junctions connecting adjacent cells. Structural proteins, such as collagen, bind cells together. Muscle and nervous tissues are unique to animals.

The diploid stage is usually dominant and reproduction is primarily sexual, with a flagellated sperm fertilizing a larger, nonmotile egg. The zygote undergoes a series of mitotic divisions, called **cleavage,** usually passing through a **blastula** stage during embryonic development. The process of **gastrulation** produces layers of embryonic tissues, resulting in the

gastrula stage. The life cycles of many animals includes a **larva**—a free-living, sexually immature form. **Metamorphosis** transforms a larva into an adult.

Animals appear to be the only organisms that have *Hox* genes—regulatory genes that contain DNA sequences called homeoboxes and that function in pattern formation.

The animal kingdom probably evolved from a colonial, flagellated protist (634)

The monophyletic animal kingdom probably originated over 700 million years ago in the Precambrian era from a common protist ancestor that may have been related to the choanoflagellates. One hypothesis for the origin of animals is that an ancestral colonial flagellate first became a hollow sphere of cells, then developed cell specialization, and later underwent invagination to form a gastrula-like "protoanimal."

Two Views of Animal Diversity

The traditional animal phylogenetic tree was based on anatomical features and embryonic development. Molecular systematics is reshaping this tree.

The remodeling of phylogenetic trees illustrates the process of scientific inquiry (635)

Science is a process of inquiry that refines old hypotheses and develops new ones based on the best currently available evidence. Molecular biology and cladistics are producing new data that support a remodeling of phylogenetic trees.

The traditional phylogenetic tree of animals is based mainly on grades in body "plans" (635–639)

Each major branch point of the traditional phylogenetic tree is called a **grade** and is defined by certain body plan features shared by all members of that branch.

The Parazoa–Eumetazoa Dichotomy The sponges (phylum Porifera) are called the **parazoans** and are separated from other animals on the basis of anatomical simplicity and unique development. Nearly all other animal phyla have tissues and are called the **eumetazoans.**

The Radiata–Bilateria Dichotomy The eumetazoans are divided into two branches: The **radiata** have **radial symmetry** and include phylum Cnidaria (hydras, jellies, sea anemones) and phylum Ctenophora (comb jellies). The **bilateria** include animals with **bilateral symmetry,** having distinct **anterior** (head) and **posterior** (tail) ends and left and right sides. Bilateral animals also have **dorsal** (top) and **ventral** (bottom) sides.

Bilateral symmetry is associated with **cephalization,** the concentration of sensory organs in the head end, which is an adaptation for unidirectional movement. A central nervous system is usually concentrated in the head, with a nerve cord extending toward the tail end.

During gastrulation, a eumetazoan embryo develops concentric layers of cells called **germ layers: Ectoderm** develops into the outer body covering and, in some phyla, into the central nervous system; **endoderm** lines the developing digestive tube, or **archenteron,** and gives rise to the lining of the digestive tract and associated organs. The radiata (cnidarians and ctenophores) are **diploblastic,** forming only these two germ layers. The bilateria are **triploblastic,** producing a middle layer, the **mesoderm,** from which arise muscles and most other organs.

The Acoelomate, Pseudocoelomate, and Coelomate Grades Triploblastic animals that have solid bodies are called **acoelomates.** This group includes the flatworms (phylum Platyhelminthes). A tube-within-a-tube body plan with a fluid-filled **body cavity** separating the digestive tract from the outer body wall is found in the other triploblastic animals. A complete digestive tract and some sort of circulatory system are found in most animals with body cavities. Most acoelomates, as well as cnidarians and ctenophores, have a gastrovascular cavity (an incomplete digestive tract with one opening), which functions in both digestion and circulation.

In **pseudocoelomates,** including the rotifers (phylum Rotifera) and roundworms (phylum Nematoda), the cavity is not completely lined by tissue derived from mesoderm and is called a **pseudocoelom. Coelomates** have a true **coelom,** a body cavity completely lined by mesodermally derived tissue. A fluid-filled body cavity cushions internal organs, allows organs to grow and move independently of the outer body wall, and also functions as a hydrostatic skeleton in soft-bodied animals.

The Protostome–Deuterostome Dichotomy Among Coelomates The coelomates divide into two grades—the **protostomes** (mollusks, annelids, and arthropods) and the **deuterostomes** (echinoderms and chordates).

Many protostomes have **spiral cleavage,** in which the planes of cell division are diagonal and newly formed cells fit in the grooves between cells of adjacent tiers. The **determinate cleavage** of some protostomes sets the developmental fate of each embryonic cell very early. Many deuterostomes exhibit **radial cleavage,** in which parallel and perpendicular cleavage planes result in aligned tiers of cells. **Indeterminate cleavage** in most deuterostomes means that cells from early cleavage divisions retain the capacity to develop into complete embryos.

In protostomes, the coelom forms from splits within solid masses of mesoderm, called **schizocoelous** development. In deuterostomes, the mesoderm begins as buds from the archenteron, called **enterocoelous** development.

The **blastopore** is the opening of the developing archenteron. In typical protostomes ("first mouth"), the blastopore develops into the mouth, and a second opening forms at the end of the archenteron to produce an anus. In deuterostomes, the blastopore becomes the anus, and the second opening develops into the mouth.

Complete Interactive Question 32.1 on page 242 to review traditional animal phylogeny based on body plan grades.

Molecular systematists are moving some branches around on the phylogenetic tree of animals (639–642)

According to cladistic methods, phylogenetic trees are a hierarchy of clades, defined by shared derived characters unique to the monophyletic taxa making up each clade. The traditional phylogenetic tree assumes that grades in body plans are indicators of clades, as long as the anatomical and embryological characteristics are common and unique to all phyla on that branch.

Molecular systematics uses unique monomer sequences in certain genes and their products to provide a new set of shared derived characters. Nucleotide sequences in the small subunit ribosomal RNA (SSU-rRNA) have been used to establish a molecular-based tree of animal phylogeny. Sequences from some *Hox* genes in various animals also support this tree.

■ **INTERACTIVE QUESTION   32.1**

This diagram shows the traditional view of animal diversity based on branch points determined by body plan grades.
Fill in the names of these groups, and then list below the major phyla that each terminal branch includes.

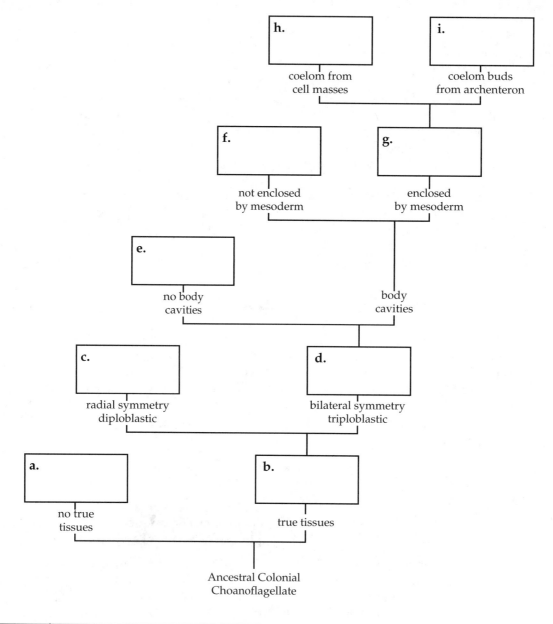

Group	Major Phyla
a.	
c.	
e.	
f.	
h.	
i.	

How Are the Two Views of Animal Phylogeny Alike? Molecular systematics agrees with the deepest branches of animal phylogeny—the parazoa–eumetazoa and the radiata–bilateria dichotomies, and with the deuterostomes as a monophyletic sub-branch of the coelomate–bilateria.

How Are the Two Views of Animal Phylogeny Different? The molecular data distinguish two protostome clades: **Lophotrochozoa,** which includes the annelids and mollusks, and **Ecdysozoa,** which includes the arthropods. Based on embryology or anatomy, some zoologists used the similar larval stage called the **trochophore larva** to link annelids closer to mollusks, whereas others pointed to body segmentation as evidence of an annelid-arthropod lineage.

Molecular evidence places the acoelomate phylum Platyhelminthes (flatworms) within the lophotrochozoan clade of the protostomes, implying that flatworms lost the coelom later in evolution. The protostome clade also includes the pseudocoelomate phyla Rotifera and Nematoda, with rotifers grouped with the lophotrochozoans and nematodes with the ecdysozoans.

■ INTERACTIVE QUESTION 32.2

Which phyla are placed in the clade Ecdysozoa?

Where does the name *Ecdysozoa* come from?

The traditional lophophorate phyla (Bryozoa, Phoronida, and Brachiopoda) are named for their common feeding apparatus called a **lophophore** and share characteristics with both protostomes and deuterostomes. Molecular data place them as protostomes among the lophotrochozoans.

■ INTERACTIVE QUESTION 32.3

Which phyla are placed in the clade Lophotrochozoa?

Where does the name *Lophotrochozoa* come from?

Summary of the Two Views of Animal Diversity Molecular systematics establishes two clades within the protostomes—Ecdysozoa and Lophotrochozoa. The acoelomates, pseudocoelomates, and lophophorate phyla are distributed between these two clades. Additional molecular evidence will help to determine whether this new tree is a reasonable hypothesis about the evolutionary history of animal life.

The Origins of Animal Diversity

Most animal phyla originated in a relatively brief span of geologic time (642–643)

Systematists group animals into about 35 phyla. The diversification of these phyla occurred within about 40 million years (about 565 to 525 mya during the late Precambrian and early Cambrian eras).

Fossils from the **Ediacaran period** of the Precambrian (565 to 545 mya) appear to be cnidarians and a few soft-bodied mollusks, although fossilized burrows indicate that worms may have evolved in this period. Some evidence indicates a much earlier origin of animal life—fossilized animal embryos in strata 570 million years old and what may be fossilized burrows of animals 1.1 billion years old. Molecular systematics also points to an animal origin about a billion years ago.

During the **Cambrian explosion** (545 to 525 mya), nearly all major body plans of animals evolved. The bizarre-looking fossils of the famous Burgess Shale may represent extinct "experiments" or, more likely, be ancient forms of modern phyla.

"Evo-devo" may clarify our understanding of the Cambrian diversification (643–644)

The three main hypotheses to explain the Cambrian explosion involve ecological, geologic, or genetic causes. The emergence of predator-prey relationships may have changed community dynamics and triggered various evolutionary adaptations. The accumulation of sufficient atmospheric oxygen to support the active metabolisms of mobile animals is a geologic explanation. The evolution of the *Hox* complex of regulatory genes, and then changes in their spatial and temporal expression during embryonic development, may have produced the body plan differences that appear during the Cambrian explosion.

Molecular data indicate that the three main branches of bilateral animals may have diverged very early, possibly associated with the evolution of the *Hox* complex. The differences in SSU-rRNA sequences within the Lophotrochozoa, Ecdysozoa, and Deuterostomia are much less, indicative of a rapid diversification of phyla within each of these three major clades. Ecologic and geologic changes may have driven this rapid radiation during the early Cambrian. During the past half-billion years, animal evolution has mainly involved variations on the designs that originated during the Cambrian explosion.

WORD ROOTS

a- = without; **-koilos** = a hollow (*acoelomate:* the condition of lacking a coelom)

arch- = ancient, beginning (*archenteron:* the endoderm-lined cavity, formed during the gastrulation process, that develops into the digestive tract of an animal)

bi- = two (*Bilateria:* the branch of eumetazoans possessing bilateral symmetry)

blast- = bud, sprout; **-pore** = a passage (*blastopore:* the opening of the archenteron in the gastrula that develops into the mouth in protostomes and the anus in deuterostomes)

cephal- = head (*cephalization:* an evolutionary trend toward the concentration of sensory equipment on the anterior end of the body)

deutero- = second (*deuterostome:* one of two distinct evolutionary lines of coelomates characterized by radial, indeterminate cleavage, enterocoelous formation of the coelom, and development of the anus from the blastopore)

di- = two (*diploblastic:* having two germ layers)

ecdys- = an escape (*Ecdysozoa:* one of two distinct clades within the protostomes; it includes the arthropods)

ecto- = outside; **-derm** = skin (*ectoderm:* the outermost of the three primary germ layers in animal embryos)

endo- = within (*endoderm:* the innermost of the three primary germ layers in animal embryos)

entero- = the intestine, gut (*enterocoelous:* the type of development found in deuterostomes; the coelomic cavities form when mesoderm buds from the wall of the archenteron and hollows out)

gastro- = stomach, belly (*gastrulation:* the formation of a gastrula from a blastula)

in- = into; **-gest** = carried (*ingestion:* a heterotrophic mode of nutrition in which other organisms or detritus are eaten whole or in pieces)

lopho- = a crest, tuft; **-trocho** = a wheel; (*Lophotrochozoa:* one of two distinct clades within the protostomes that includes annelids and mollusks)

meso- = middle (*mesoderm:* the middle primary germ layer of an early embryo)

meta- = boundary, turning point; **-morph** = form (*metamorphosis:* the resurgence of development in an animal larva that transforms it into a sexually mature adult)

para- = beside; **-zoan** = animal (*parazoan:* members of the subkingdom of animals consisting of the sponges)

proto- = first; **-stoma** = mouth (*protostomes:* a member of one of two distinct evolutionary lines of coelomates characterized by spiral, determinate cleavage, schizocoelous formation of the coelom, and development of the mouth from the blastopore)

pseudo- = false (*pseudocoelom:* a body cavity that is not completely lined by mesoderm)

radia- = a spoke, ray (*Radiata:* the radially symmetrical animal phyla, including cnidarians)

schizo- = split (*schizocoelous:* the type of development found in protostomes; initially, solid masses of mesoderm split to form coelomic cavities)

tri- = three (*triploblastic:* having three germ layers)

STRUCTURE YOUR KNOWLEDGE

1. Fill in this table to review some of the differences in the early embryological development of protostomes and deuterostomes.

	Protostomes	Deuterostomes
Cleavage	a.	b.
Coelom formation	c.	d.
Blastopore fate	e.	f.

2. Which three major phylogenies do the grade-based and molecular-based phylogenetic trees agree on?

3. How does the molecular-based tree differ from the grade-based tree?

TEST YOUR KNOWLEDGE

MULTIPLE CHOICE: *Choose the one best answer.*

1. Sponges differ from the rest of the animals because
   a. they are completely sessile.
   b. they have radial symmetry and are suspension feeders.
   c. their simple body structure has no true tissues and they have a unique embryology.
   d. they are not multicellular.
   e. they have no flagellated cells.

2. An insect larva
   a. is a miniature version of the adult.
   b. is transformed into an adult by molting.
   c. ensures more genetic variation in the insect life cycle.
   d. is a sexually immature organism specialized for eating and growth.
   e. is all of the above.

3. Cephalization
   a. is the development of bilateral symmetry.
   b. is the formation of a coelom by budding from the archenteron.
   c. is a diagnostic characteristic of deuterostomes.
   d. is common in radially symmetrical animals.
   e. is associated with motile animals that concentrate sensory organs in a head region.

4. A true coelom
   a. is found in deuterostomes.
   b. is found in most protostomes.
   c. is a fluid-filled cavity completely lined by mesoderm.
   d. may be used as a hydrostatic skeleton by soft-bodied coelomates.
   e. is all of the above.

5. Which of the following is descriptive of the embryonic development of most protostomes?
   a. radial and determinate cleavage, blastopore becomes mouth
   b. spiral and indeterminate cleavage, coelom forms as split in solid mass of mesoderm
   c. spiral and determinate cleavage, blastopore becomes mouth, schizocoelous development
   d. spiral and indeterminate cleavage, blastopore becomes mouth, enterocoelous development
   e. radial and determinate cleavage, enterocoelous development, blastopore becomes anus

6. A gastrovascular cavity
   a. functions in both digestion and circulation and has a single opening.
   b. develops as outpocketings of the archenteron.
   c. is found in the phyla Cnidaria, Platyhelminthes, and Rotifera.
   d. develops from the hollow blastula stage.
   e. forms from a split in mesoderm.

7. Which of the following is *not* descriptive of a pseudocoelomate?
   a. a body cavity incompletely lined by mesoderm
   b. bilateral symmetry
   c. triploblastic
   d. true tissues
   e. schizocoelous formation of body cavity

8. Which of the following characteristics is found only in animals?
   a. homeobox-containing genes
   b. flagellated sperm
   c. heterotrophic nutrition
   d. *Hox* genes
   e. All of the above are exclusive animal traits.

9. The early branching of the bilateria into two protostome clades and the deuterostomes was most likely related to
   a. the accumulation of sufficient atmospheric oxygen to support active animals.
   b. the evolution of bilateral symmetry and cephalization.
   c. the evolution of the *Hox* complex, which led to variations in pattern formation during development.
   d. the origin of predator-prey relationships that led to such evolutionary adaptations as protective shells and locomotion.
   e. a basic change in embryonic development that stemmed from differences in spiral and radial cleavage patterns.

10. Some of the oldest known animal fossils are
    a. colonies of flagellated protists.
    b. worms that are 1 billion years old.
    c. sponges, because they were the first animals to evolve.
    d. soft-bodied cnidarians from the Ediacaran period.
    e. bizarre-looking animals from the Burgess Shale.

11. According to the grade-based animal phylogeny, the acoelomate condition
    a. is primitive and the acoelomates branched off before the origin of the coelom.
    b. is the first branch point, separating the acoelomates from the eumetazoa.
    c. arose secondarily as the coelom was lost during the evolution of the flatworms.
    d. led to the origin of the lophophore as a feeding adaptation.
    e. was associated with a sessile or planktonic lifestyle and thus linked with radial symmetry.

12. According to the traditional grade-based tree, the three lophophorate phyla were grouped with
    a. radiata.
    b. pseudocoelomates.
    c. protostomes.
    d. deuterostomes.
    e. They had an uncertain phylogeny because they have both protostome and deuterostome characteristics.

13. Which of the following groupings of phyla within a clade is supported by molecular data?
    a. annelids and arthropods
    b. annelids and mollusks
    c. lophophorate phyla and deuterostomes
    d. pseudocoelomates (rotifers and nematodes) and arthropods
    e. acoelomates (flatworms) and cnidarians (jellies and hydras)

14. The grade-based and molecular-based phylogenetic trees agree in which of the following ways?
    a. the acoelomate condition as a branch point before the divergence of the two protostome clades
    b. the placement of the three lophophorate phyla into the protostomes

c. parazoa as the probable ancestor of the animal kingdom
d. the deepest branch points of the parazoa–eumetazoa and radiata–bilateria dichotomies, and the deuterostomes as a monophyletic clade
e. the origin of the *Hox* complex as the probable cause of the Cambrian explosion

15. The rapid diversification of animal phyla during the Cambrian explosion is most likely linked with
    a. the movement of animals onto land.
    b. the origin of the triploblastic body plan that allowed for the development of organs.
    c. the origin of the first homeobox-containing genes.
    d. the evolution of bilateral symmetry and cephalization that permitted more efficient movement and processing of sensory data.
    e. ecological changes such as the development of predator-prey relationships and geologic causes such as the accumulation of atmospheric oxygen.

CHAPTER 33

INVERTEBRATES

FRAMEWORK

This chapter surveys the characteristics and representatives of the major animal phyla. Refer to this orientation diagram of the molecular-based phylogenetic tree to help you organize this wonderful diversity of animals.

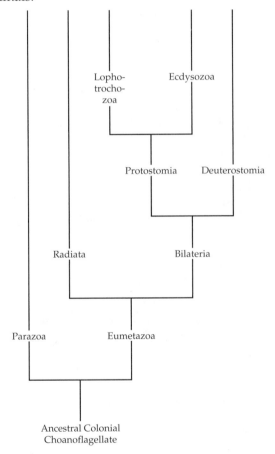

CHAPTER SUMMARY

Most of the 35 phyla of animals consist mainly of aquatic species. Terrestrial habitats have been extensively exploited by only a few animal phyla, notably the vertebrates and arthropods. Over 95% of all animal species—all except the one subphylum of vertebrates—are **invertebrates,** animals that lack backbones.

Parazoa

The sponges of clade Parazoa have relatively unspecialized cells and lack true tissues.

Phylum Porifera: sponges are sessile with porous bodies and choanocytes (647–648)

Sponges are sessile, primarily marine animals. Although they lack nerves and muscles, their individual cells can sense and respond to environmental changes. Water is drawn through pores in the body wall of this saclike animal into a central cavity, the **spongocoel,** and flows out through the **osculum.** Sponges are suspension feeders, collecting food particles by the action of collared, flagellated **choanocytes** lining the spongocoel or internal water chambers. Choanocytes resemble the cells of choanoflagellates, and molecular evidence indicates that animals originated from a choanoflagellate ancestor.

In the **mesohyl,** or gelatinous matrix between the two body-wall layers, are **amoebocytes.** These cells take up food from the water and from choanocytes, digest it, and carry nutrients to other cells. Amoebocytes also form skeletal fibers, which may be sharp spicules or flexible fibers.

Most sponges are **hermaphrodites,** producing both eggs and sperm. Sperm, carried out through the osculum, cross-fertilize eggs retained in the mesohyl of neighboring sponges. Flagellated larvae disperse to a suitable substratum and develop into sessile adults. Sponges are capable of extensive regeneration, replacing damaged body parts and reproducing asexually from fragments.

■ INTERACTIVE QUESTION 33.1

Give the locations and functions of the following:

a. choanocytes

b. amoebocytes

Radiata

The oldest clade of the eumetazoans, animals with true tissues, is Radiata, characterized by radial symmetry and diploblastic bodies. Phyla Cnidaria and Ctenophora may have originated from different parazoan ancestors.

Phylum Cnidaria: cnidarians have radial symmetry, a gastrovascular cavity, and cnidocytes (648–650)

The cnidarians include hydras, jellies, sea anemones, and coral animals. Their simple anatomy consists of a sac with a central **gastrovascular cavity** and a single opening serving as both mouth and anus. This body plan has two forms: **polyps,** which are sessile, cylindrical forms with mouth and tentacles extending upward, and **medusae,** which are flattened, mouth-down polyps that move by passive drifting and weak body contractions. Both these body forms occur in the life histories of some cnidarians.

Cnidarians use their ring of tentacles, armed with **cnidocytes,** to capture prey. Cnidocytes contain cnidae, which are capsules that can evert; stinging capsules are called **nematocysts.** A gelatinous mesoglea is sandwiched between the epidermis and gastrodermis. Cells have bundles of microfilaments arranged into contractile fibers acting as simple muscles. A nerve net is associated with simple sensory receptors and coordinates the contraction of cells against the hydrostatic skeleton of the gastrovascular cavity, producing movement.

Class Hydrozoa   Most hydrozoans alternate between an asexually reproducing polyp and a sexually reproducing medusa form. The common, freshwater *Hydra* exists only in polyp form.

Class Scyphozoa   The medusa stage is more prevalent in the scyphozoans. The sessile polyp stage often does not occur in the jellies of the open ocean.

Class Anthozoa   Sea anemones and coral animals occur only as polyps. Coral polyps secrete calcified external skeletons, and the accumulation of such skeletons produces coral.

Phylum Ctenophora: comb jellies possess rows of ciliary plates and adhesive colloblasts (650–651)

Comb jellies, or ctenophores, are the largest animals that use cilia for locomotion. Nerves running from a sensory organ to the combs of fused cilia coordinate movement of these small marine animals. A pair of retractable tentacles bear **colloblasts,** which release a sticky thread to capture planktonic prey.

■ **INTERACTIVE QUESTION   33.2**

Name these two cnidarian body plans and identify the indicated structures.

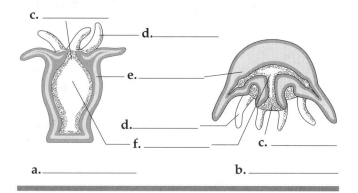

Protostomia: Lophotrochozoa

Molecular data support the hypothesis that the bilateral animals (clade Bilateria) originated from a common ancestor, and that **urbilateria,** the earliest bilateral animals, had true body cavities and were relatively complex. The acoelomates and pseudocoelomates, according to this view, evolved secondarily from coelomates. Molecular data indicate a Precambrian origin for bilateria and support the traditional division into protostomes and deuterostomes. Molecular-based phylogeny also splits the protostomes into the clades Lophotrochozoa and Ecdysozoa. All bilateria are triploblastic.

Phylum Platyhelminthes: flatworms are acoelomates with gastrovascular cavities (652–654)

Free-living flatworms live in marine, freshwater, and damp terrestrial habitats, and many are parasitic. Their mesoderm gives rise to organs and true muscles. Like cnidarians, however, typical flatworms have a gastrovascular cavity with only one opening, and they lack a body cavity.

Class Turbellaria   Turbellarians include freshwater **planarians** and other, mostly marine, free-living flatworms. Their branching gastrovascular cavity functions in both digestion and distribution of food. Gas exchange and diffusion of nitrogenous wastes occur across the body wall. Ciliated flame cells function in osmoregulation. Planarians move using cilia to glide on secreted mucus or may use body undulations to swim.

Eyespots on the head detect light, and lateral head flaps function for smell. Their centralized nervous system enables planarians to modify their behavior. Planarians reproduce asexually by regeneration or sexually by copulation between hermaphroditic worms.

Classes Monogenea and Trematoda   Adapted to live as parasites in or on other animals, flukes (trematodes) have a tough outer covering, suckers, and extensive reproductive organs. The life cycles of trematodes are complex, usually including asexual and sexual stages and intermediate hosts in which larvae develop. Most monogeneans are external parasites of fishes and have a simple life cycle with ciliated larvae dispersing to new hosts.

Class Cestoidea   As parasites, mostly of vertebrates, tapeworms consist of a scolex, with suckers and hooks for attaching to the host's intestinal lining, and a ribbon of proglottids packed with reproductive organs. Predigested food is absorbed from the host. The life cycles of tapeworms may include intermediate hosts.

■ INTERACTIVE QUESTION   33.3

a. Describe the digestive system of a planarian.

b. Why do tapeworms, which are also flatworms, lack a digestive system?

Phylum Rotifera: rotifers are pseudocoelomates with jaws, crowns of cilia, and complete digestive tracts (654)

Rotifers are smaller than many protists but have a **complete digestive tract** with separate mouth and anus, and other organ systems. A pseudocoelom, a body cavity not completely lined by mesoderm, functions as a hydrostatic skeleton and a circulatory system. A crown of cilia draws microscopic food into the mouth. The pharynx bears jaws that grind the ingested microorganisms. Some species reproduce by **parthenogenesis,** in which female offspring develop from unfertilized eggs. In other species, two types of egg develop parthenogenically: one type forming females and the other developing into degenerate males that produce sperm. The resulting resistant zygotes survive harsh conditions in a dormant state.

The lophophorate phyla: bryozoans, phoronids, and brachiopods are coelomates with ciliated tentacles around their mouths (654–655)

The three phyla of **lophophorate animals** (Bryozoa, Phoronida, and Brachiopoda) all have a **lophophore,** a horseshoe-shaped or circular fold bearing ciliated tentacles, which surrounds the mouth and functions in suspension feeding. This complex structure suggests that all three phyla are related. Molecular systematics places these phyla in the protostome branch.

Bryozoans are tiny, mostly marine animals living in colonies that are often encased in a hard exoskeleton, with pores through which their lophophores extend. **Phoronids** are marine, tube-dwelling worms that often live buried in sand, extending and withdrawing their lophophore from the tube opening. **Brachiopods,** or lamp shells, attach to the substrate by a stalk and open their hinged shell to allow water to flow through the lophophore.

Phylum Nemertea: proboscis worms are named for their prey-capturing apparatus (655)

Most ribbon or proboscis worms are marine. Although their body is acoelomate, a fluid-filled sac that may be a reduced true coelom allows these worms to hydraulically operate an extensible proboscis to capture prey. Their excretory, sensory, and nervous systems are similar to those of flatworms, but they differ in having a complete digestive tract and a **closed circulatory system.**

Phylum Mollusca: mollusks have a muscular foot, a visceral mass, and a mantle (656–659)

Mollusks are soft-bodied, mostly marine animals, most of which are protected by a shell. The molluscan body plan has three main parts: a muscular **foot** used for movement, a **visceral mass** containing the internal organs, and a **mantle** that covers the visceral mass and may secrete a shell. The **mantle cavity,** the water-filled chamber formed by the extension of the mantle, encloses the gills, anus, and excretory pores. A rasping **radula** is used for feeding by many mollusks.

Most mollusks have separate sexes. Many marine mollusks have a life cycle that includes a ciliated larva called the **trochophore,** also found in marine annelids. Four of the eight molluscan classes are discussed in the text.

Class Polyplacophora   Chitons are oval marine animals with shells that are divided into eight dorsal plates. Chitons creep slowly over rocks in the intertidal zone, where they feed on algae.

Class Gastropoda   Most members of this largest molluscan class are marine, although there are many freshwater species, and some snails and slugs are terrestrial. A distinctive feature of this class is **torsion,** the embryonic rotation of the visceral mass that results in the anus and mantle cavity being above the head. Most gastropods have single coiled shells, although slugs and nudibranchs have no shells. Many gastropods have distinct heads with eyes at the tips of tentacles. Moving by the rippling of the elongated foot, most gastropods graze on plant material. Land snails lack gills; their vascularized mantle cavity functions as a lung.

Class Bivalvia Clams, oysters, mussels, and scallops have the two halves of their shell hinged at the mid-dorsal line. Most bivalves are suspension feeders; water flows into and out of the mantle cavity through siphons, and food particles are trapped in the mucus that coats the gills and then are swept to the mouth by cilia.

Class Cephalopoda Squids and octopuses are rapid-moving carnivores. The mouth has beaklike jaws to bite prey and is surrounded by tentacles. The shell is reduced and internal in squids, absent in octopuses, and external only in the chambered nautilus. In squid, the foot has been modified to form parts of the tentacles and head and the muscular siphon used to jet-propel the squid when water from the mantle cavity is expelled.

Cephalopods are the only mollusks with a closed circulatory system. They have a well-developed nervous system, sense organs, and a complex brain—important features for active predators. The ancestors of octopuses and squids were probably shelled, predaceous mollusks. Shelled **ammonites** were the dominant invertebrate predators until their extinction at the end of the Cretaceous period.

■ INTERACTIVE QUESTION 33.4

a. Describe the three main parts of the molluscan body plan.

1.

2.

3.

b. Compare the feeding behavior and activity level of snails, clams, and squid.

1. Snails:

2. Clams:

3. Squid:

Phylum Annelida: annelids are segmented worms (659–661)

Annelids are segmented worms found in marine, freshwater, and damp soil habitats. The earthworm has a complete digestive system with specialized regions and a closed circulatory system. The dorsal vessel and five pairs of vessels encircling the esophagus pump the hemoglobin-containing blood. Respiration occurs across the moist, highly vascularized skin. Septa partition the coelom into segments, in each of which is found a pair of excretory **metanephridia,** which filter metabolic wastes from the blood and coelomic fluid.

The nervous system consists of a pair of cerebral ganglia, a subpharyngeal ganglion, and fused segmental ganglia along the ventral nerve cords. Earthworms are hermaphrodites; sperm are exchanged between worms during mating. A mucous cocoon, secreted by the clitellum, slides off the worm after picking up its eggs and stored sperm.

Noncompressible coelomic fluid enclosed by septa within the body segments serves as a hydrostatic skeleton; circular and longitudinal muscles contract alternately to extend the body and pull it forward.

Class Oligochaeta This class includes the earthworms and several aquatic species. Earthworm castings improve soil texture. Label the strucutres shown in Interactive Question 33.5.

Class Polychaeta These mostly marine worms have parapodia on each segment that function in gas exchange and locomotion. Polychaetes may be planktonic, bottom burrowers, or tube-dwellers.

Class Hirudinea Most leeches inhabit fresh water. Many feed on small invertebrates, whereas others are parasites that temporarily attach to animals, slit or digest a hole through the skin, and suck the blood of their host.

Annelids exhibit two evolutionary adaptations. A coelom provides a hydrostatic skeleton, space for complex organ development, a cushion for internal organs, and a separation of the actions of body wall and digestive tract muscles. Segmentation allows for regional specialization.

Protostomia: Ecdysozoa

The ecdysozoan phyla are grouped into a clade based on molecular data but named for their characteristic of ecdysis, the shedding of outgrown exoskeletons.

■ INTERACTIVE QUESTION 33.5

Identify the structures shown in this body segment of an earthworm.

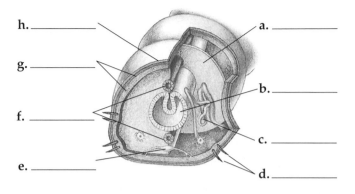

h. _____   a. _____

g. _____

                                                  b. _____

f. _____

                                                  c. _____

e. _____   d. _____

Phylum Nematoda: roundworms are nonsegmented pseudocoelomates covered by tough cuticles (661–662)

Roundworms are among the most widespread of all animals, found inhabiting water, soil, and the bodies of plants and animals. These cylindrical worms are covered by a tough exoskeleton called a cuticle, which is periodically shed as they grow. They have a complete digestive tract. Their thrashing movement is produced by contraction of longitudinal muscles. Fluid in the pseudocoelom circulates nutrients. Reproduction is usually sexual, fertilization is internal, and most zygotes form resistant cells.

Numerous nematode species are ecologically important decomposers. Other nematodes are serious agricultural pests and animal parasites.

Arthropods are segmented coelomates with exoskeletons and jointed appendages (662–672)

In terms of species diversity, distribution, and vast numbers, arthropods are the most successful group of animals.

General Characteristics of Arthropods   Characteristics of **arthropods** include segmentation, which allows for regional specialization, a hard exoskeleton, and jointed appendages. Appendages are modified for walking, feeding, sensing, mating, and defense. A **cuticle** of chitin and protein completely covers the body as an **exoskeleton,** providing protection and points of attachment for the muscles that move the appendages. To grow, an arthropod must **molt,** shedding its old exoskeleton and secreting a larger one. Arthropods have extensive cephalization and well-developed sensory organs.

A heart pumps hemolymph through an **open circulatory system** consisting of short arteries and a network of sinuses known as the hemocoel. The embryonic coelom becomes reduced during development, and the hemocoel becomes the main body cavity. Respiratory gas exchange in most aquatic species occurs through gills, whereas terrestrial arthropods have tracheal systems of branching internal ducts.

Arthropod Phylogeny and Classification   Molecular systematics generally supports the anatomical and fossil evidence indicating that arthropods diverged into four evolutionary lines: **trilobites, chelicerates, uniramians,** and **crustaceans.** The crabs, lobsters, shrimps, and others of the primarily aquatic crustaceans are believed to have evolved in the ocean. Uniramians (centipedes, millipedes, and insects) and most extant chelicerates diversified on land.

Chelicerates (horseshoe crabs, scorpions, spiders, and ticks) have clawlike **chelicerae** used for feeding, whereas uniramians and crustaceans have jawlike **mandibles.** Uniramians have one pair of sensory **antennae** and unbranched (uniramous) appendages; crustaceans have two pairs of antennae and branched appendages. Both groups usually have a pair of **compound eyes.**

The exoskeleton of early marine arthropods helped arthropods move onto land by providing support and protection from desiccation. During the late Silurian and early Devonian periods, chelicerates, insects, millipedes, and centipedes spread onto land. The oldest evidence of terrestrial animals is tracks of extinct chelicerates about 450 million years old.

All arthropods were traditionally grouped into phylum **Arthropoda,** but many zoologists now favor separate phyla for the four major lineages: **phylum Trilobita, phylum Chelicerata, phylum Uniramia,** and **phylum Crustacea.** Some systematists interpret the molecular and anatomical data as indicating that Uniramia is not a monophyletic clade, and that insects may be more closely related to crustaceans than to centipedes and millipedes. Phylogenetic hypotheses are being reconstructed based on molecular data and cladistic analysis.

Trilobites   **Trilobites,** with pronounced segmentation and uniform appendages, were common early arthropods throughout the Paleozoic era but disappeared in the Permian extinctions.

Spiders and Other Chelicerates   Existing during and beyond this same period were large predatory **eurypterids,** or water scorpions. The horseshoe crab is one of the few marine chelicerates that survive today. The chelicerate body is divided into a cephalothorax and abdomen, with the most anterior appendages, the chelicerae, modified as pincers or fangs.

Most modern chelicerate species are in the terrestrial **class Arachnida.** Most ticks are blood-sucking parasites on reptiles, birds, or mammals. Many mites are also parasites. Scorpions were among the first terrestrial carnivores. In most spiders, **book lungs,** consisting of stacked internal plates, function in gas exchange. Many spiders spin characteristic webs of silk from special abdominal glands.

In arachnids, the cephalothorax has six pairs of appendages: four pairs of walking legs, the chelicerae, and sensing or feeding appendages called pedipalps.

■ INTERACTIVE QUESTION 33.6

a. What are chelicerae?

b. How do spiders trap, kill, and eat their prey?

Millipedes and Centipedes The millipedes of **class Diplopoda** are wormlike, segmented vegetarians with two pairs of walking legs per segment. Millipedes may have been one of the first land animals. The centipedes of **class Chilopoda** are terrestrial carnivores with appendages modified as jawlike mandibles and poison claws. Each segment of the trunk has one pair of legs.

Insects **Class Insecta** is divided into about 26 orders and has more species than all other forms of life combined. The study of insects, called **entomology,** is a large field with many subspecialties.

The oldest insect fossils are from the Devonian period, but a major diversification occurred in the Carboniferous and Permian periods with the evolution of flight and modification of mouthparts for specialized feeding on plants. The major diversification of insects appears to have preceded and probably influenced the radiation of flowering plants.

Flight is a major key to the success of insects. Many insects have one or two pairs of wings that are extensions of the cuticle of the dorsal thorax. Dragonflies were among the first flying insects.

The complete digestive tract of an insect has several specialized regions. The circulatory system is open. **Malpighian tubules** are outpocketings of the digestive tract that function as excretory organs, removing metabolic wastes from the hemolymph. Gas exchange is accomplished by a **tracheal system** of chitin-lined tubes that ramify throughout the body and open to the outside through spiracles.

The nervous system consists of a pair of ventral nerve cords with segmental ganglia and a cerebral ganglion or brain.

In **incomplete metamorphosis,** the young are smaller versions of the adult and pass through several molts. In **complete metamorphosis,** the larvae look entirely different from the adult. The larvae eat and grow; adults reproduce and disperse. Metamorphosis occurs in a pupal stage. Reproduction is usually sexual; fertilization is usually internal.

Insects affect humans as pollinators of crops, vectors of disease, and competitors for food.

■ INTERACTIVE QUESTION 33.7

Describe the three regions of a generalized insect body.

a.

b.

c.

Crustaceans The mostly aquatic crustaceans have two pairs of antennae, three or more pairs of mouthpart appendages, including mandibles, walking legs on the thorax, and appendages on the abdomen. Larger crustaceans have gills. A heart pumps hemolymph through arteries into sinuses that bathe the organs. Nitrogenous wastes pass by diffusion through thin areas of the cuticle, and a pair of glands regulates salt balance. Sexes usually are separate. One or more swimming larval stages occur in most aquatic crustaceans.

Lobsters, crayfish, crabs, and shrimp are relatively large crustaceans called **decapods.** Their cuticle is hardened by calcium carbonate, and a carapace covers the dorsal side of their cephalothorax. **Isopods** are mostly small marine crustaceans but include terrestrial sow bugs and pill bugs. The small, very numerous **copepods** are important members of marine and freshwater plankton communities. Barnacles are sessile crustaceans that strain food from the water with their appendages.

■ INTERACTIVE QUESTION 33.8

Compare and contrast the following features for insects and crustaceans.

Feature	Insects	Crustaceans
Habitat	a.	b.
Locomotion	c.	d.
Respiration	e.	f.
Excretion	g.	h.
Antennae #	i.	j.
Appendages	k.	l.

How Many Times Did Segmentation Evolve in the Animal Kingdom? Most biologists supported the hypothesis that arthropods evolved from an annelid ancestor, or that arthropods and annelids shared a common segmented ancestor. This close annelid-arthropod relationship based on segmentation is now challenged by the molecular-based separation of lophotrochozoan and ecdysozoan clades.

In the embryos of bilateral animals, differential expression of key regulatory genes blocks out regions along an anterior to posterior axis in which various body parts will form. In segmented animals, genes first establish body segmentation and then genes of the *Hox* complex determine what organs will develop in each segment. Even sponges and cnidarians, however, have *Hox* genes, so the origin of these development-regulating genes predates the origin of the bilaterians.

Body segmentation evolved in all three clades of bilaterians: the arthropods in Ecdysozoa, the annelids in Lophotrochozoa, and the chordates in Deuterostomia. Each clade also includes non-segmented phyla. This scattered distribution may be explained by a separate evolutionary origin of segmentation in each clade; two separate origins, one for the protostomes and one for the deuterostomes; or a single origin in an ancestor common to all three bilaterian clades. Several "evo-devo" groups, who combine research in evolutionary and developmental biology, are comparing the functions of *Hox* genes in the development of segmented bodies in these phyla to test whether segmentation evolved once, twice, or three times in the animal kingdom.

Deuterostomia

The echinoderms and chordates are grouped together based on their common embryological traits of radial cleavage, coelom formation from the archenteron, and origin of the mouth opposite the blastopore. Molecular systematics also supports Deuterostomia as a clade.

Phylum Echinodermata: echinoderms have a water vascular system and secondary radial anatomy (672–674)

Most **echinoderms** are sessile or slow-moving marine animals. They have a thin skin covering an endoskeleton of calcareous plates. A **water vascular system** with a network of hydraulic canals controls extensions called **tube feet** that function in locomotion, feeding, and gas exchange. Sexual reproduction usually involves separate sexes and external fertilization. Bilateral larvae metamorphose into radial adults.

Class Asteroidea Sea stars have five arms radiating from a central disk and use tube feet lining the undersurfaces of these arms to creep slowly and to grasp and open prey. They evert their stomach through their mouth and slip it into a slightly opened bivalve shell. Sea stars can regenerate lost arms.

Class Ophiuroidea Brittle stars have distinct central disks and move by lashing their long, flexible arms. They may be predators, scavengers, or suspension feeders.

Class Echinoidea Sea urchins and sand dollars have no arms but are able to move slowly using their five rows of tube feet. Long spines also aid a sea urchin's movement. A sea urchin's mouth is ringed by complex jawlike structures used to eat seaweeds and other foods.

Class Crinoidea Sea lilies may live attached to the substrate by stalks. Their long, flexible arms extend upward from around the mouth and are used in suspension feeding. Their form has changed little in 500 million years of evolution.

Class Holothuroidea Sea cucumbers are elongated animals that bear little resemblance to other echinoderms other than having five rows of tube feet.

■ **INTERACTIVE QUESTION 33.9**

List the key characteristics that distinguish the phylum Echinodermata.

a.

b.

c.

Phylum Chordata: the chordates include two invertebrate subphyla and all vertebrates (674–675)

Echinoderms and chordates share deuterostome developmental characteristics but have existed as separate phyla for at least 500 million years.

WORD ROOTS

arthro- = jointed; **-pod** = foot (*Arthropoda:* segmented coelomates with exoskeletons and jointed appendages)

arachn- = a spider class (*Arachnida:* the animal group that includes scorpions, spiders, ticks, and mites)

brachio- = the arm (*brachiopod:* also called lamp shells, these animals superficially resemble clams and other bivalve mollusks, but the two halves of the brachiopod shell are dorsal and ventral to the animal rather than lateral, as in clams)

bryo- = moss; **-zoa** = animal (*bryozoan:* colonial animals that superficially resemble mosses)

cheli- = a claw (*chelicerae:* clawlike feeding appendages characteristic of the chelicerate group)

choano- = a funnel; **-cyte** = cell (*choanocyte:* flagellated collar cells of a sponge)

cnido- = a nettle (*cnidocytes:* unique cells that function in defense and prey capture in cnidarians)

-coel = hollow (*spongocoel:* the central cavity of a sponge)

cope- = an oar (*copepods:* a group of small crustaceans that are important members of marine and freshwater plankton communities)

cuti- = the skin (*cuticle:* the exoskeleton of an arthropod)

deca- = ten (*decapod:* a relatively large group of crustaceans that includes lobsters, crayfish, crabs, and shrimp)

diplo- = double (*Diplopoda:* the millipede group of animals)

echino- = spiny; **-derm** = skin (*echinoderm:* sessile or slow-moving animals with a thin skin that covers an exoskeleton; the group includes sea stars, sea urchins, brittle stars, crinoids, and basket stars)

entom- = an insect; **-ology** = the study of (*entomology:* the study of insects)

eury- = broad, wide; **-pter** = a wing, a feather, a fin (*eurypterid:* mainly marine and freshwater, extinct, chelicerates; these predators, also called water scorpions, ranged up to 3 meters long)

exo- = outside (*exoskeleton:* a hard encasement on the surface of an animal)

gastro- stomach; **-vascula** = a little vessel (*gastrovascular cavity:* the central digestive compartment, usually with a single opening that functions as both mouth and anus)

hermaphrod- = with both male and female organs (*hermaphrodite:* an individual that functions as both male and female in sexual reproduction by producing both sperm and eggs)

in- = without (*invertebrates:* an animal without a backbone)

iso- = equal (*isopods:* one of the largest groups of crustaceans, primarily marine, but including pill bugs common under logs and moist vegetation next to the ground)

lopho- = a crest, tuft; **-phora** = to carry (*lophophore:* a horseshoe-shaped or circular fold of the body wall bearing ciliated tentacles that surround the mouth)

meso- = the middle; **-hyl** = matter (*mesohyl:* a gelatinous region between the two layers of cells of a sponge.)

meta- = change; **-morph** = shape (*metamorphosis:* the resurgence of development in an animal larva that transforms it into a sexually mature adult)

nemato- = a thread; **-cyst** = a bag (*nematocysts:* the stinging capsules in cnidocytes, unique cells that function in defense and capture of prey)

nephri- = the kidney (*metanephridium:* in annelid worms, a type of excretory tubule with internal openings called nephrostomes that collect body fluids and external openings called nephridiopores)

oscul- = a little mouth (*osculum:* a large opening in a sponge that connects the spongocoel to the environment)

partheno- = without fertilization; **-genesis** = producing (*parthenogenesis:* a type of reproduction in which females produce offspring from unfertilized eggs)

plan- = flat or wandering (*planarians:* carnivores that prey on smaller animals or feed on dead animals)

tri- = three; **-lobi** = a lobe (*trilobite:* an extinct group of arthropods with pronounced segmentation)

trocho- = a wheel (*trochophore:* a ciliated larva common to the life cycle of many mollusks, it is also characteristic of marine annelids and some other lophotrochozoans)

uni- = one; **-rami** = a branch (*uniramia:* the animal group that includes centipedes, millipedes, and insects)

ur- = earliest (*urbilateria:* the original group of bilateral animals that were relatively complex with true coeloms)

STRUCTURE YOUR KNOWLEDGE

1. The following phylogenetic tree is based on molecular systematics. Fill in the major clades that determine each branch. The small sketches at the top represent the 15 animal phyla surveyed in this chapter. Identify and give common name examples for each of these phyla. The following list of phyla should help:

Annelida
Arthropoda
Brachiopoda
Bryozoa
Chordata
Cnidaria
Ctenophora
Echinodermata

Mollusca
Nematoda
Nemertea
Phoronida
Platyhelminthes
Porifera
Rotifera

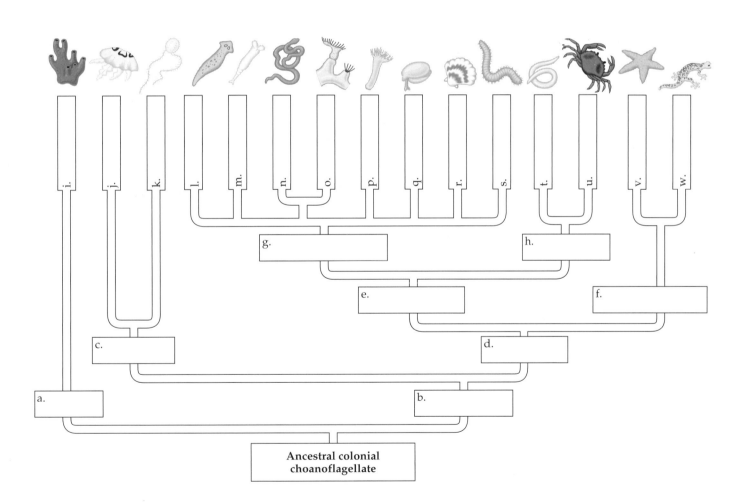

TEST YOUR KNOWLEDGE

MATCHING: *Match the following organisms with their phylum and class. Answers may be used more than once or not at all.*

Organism	Phylum	Class
1. jelly (jellyfish)	_____	_____
2. crayfish	_____	_____
3. snail	_____	_____
4. leech	_____	_____
5. tapeworm	_____	_____
6. cricket	_____	_____
7. scallop	_____	_____
8. tick	_____	_____
9. sea urchin	_____	_____
10. hydra	_____	_____
11. planaria	_____	_____
12. chambered nautilus	_____	_____

Phyla	**Classes**
A. Annelida	a. Arachnida
B. Arthropoda	b. Bivalvia
C. Cnidaria	c. Cephalopoda
D. Echinodermata	d. Cestoidea
E. Mollusca	e. Crustacea
F. Nematoda	f. Echinoidea
G. Nemertea	g. Gastropoda
H. Platyhelminthes	h. Hirudinea
I. Porifera	i. Hydrozoa
J. Rotifera	j. Insecta
	k. Oligochaeta
	l. Scyphozoa
	m. Turbellaria

MULTIPLE CHOICE: *Choose the one best answer.*

1. Invertebrates include
   a. all animals except for the phylum Vertebrata.
   b. all animals without backbones.
   c. only animals that use hydrostatic skeletons.
   d. members of the parazoa, radiata, and protostomes, but not of the deuterostomes.
   e. all the animals that evolved in the Cambrian explosion.

2. Which of the following is the best description of the phylum Porifera?
   a. radial symmetry, diploblastic, cnidocytes for capturing prey
   b. radial symmetry, triploblastic, nematocysts for protection
   c. radial symmetry, without true tissues, choanocytes for trapping food particles
   d. bilateral symmetry, pseudocoel, flame cells for excretion
   e. bilateral symmetry, osculum and spongocoel for filtering water

3. Which of the following does *not* have a gastrovascular cavity for digestion?
   a. flatworm
   b. hydra
   c. polychaete worm
   d. sea anemone
   e. fluke

4. The oldest clade of Eumetazoa is
   a. Bilateria.
   b. Ecdysozoa.
   c. Lophotrochozoa.
   d. Parazoa.
   e. Radiata.

5. Hermaphrodites
   a. contain male and female sex organs but usually cross-fertilize.
   b. include sponges, earthworms, and most insects.
   c. are characteristically found in parthenogenic rotifers.
   d. are both a and b.
   e. are a, b, and c.

6. Which of the following is *not* true of cnidarians?
   a. An alternation of medusa and polyp stage is common in class Hydrozoa.
   b. They use a ring of tentacles armed with stinging cells to capture prey.
   c. They include hydras, jellies, sponges, and sea anemones.
   d. They have a nerve net that coordinates contraction of microfilaments for movement.
   e. They have a gastrovascular cavity.

7. Which of the following combinations of phylum and characteristics is *incorrect*?
   a. Nemertea—proboscis worm, complete digestive tract
   b. Rotifera—parthenogenesis, crown of cilia, microscopic animals
   c. Nematoda—gastrovascular cavity, tough cuticle, ubiquitous
   d. Annelida—segmentation, closed circulatory system, hydrostatic skeleton
   e. Echinodermata—radial anatomy, endoskeleton, water vascular system

8. Which of the following is an excretory or osmoregulatory structure that is *incorrectly* matched with its class?
   a. metanephridia—Oligochaeta
   b. Malpighian tubules—Echinoidea
   c. flame cells—Turbellaria
   d. thin region of cuticle—Crustacea
   e. diffusion across cell membranes—Hydrozoa

9. Torsion
   a. is embryonic rotation of the visceral mass that results in a U-shaped digestive tract in gastropods.
   b. is characteristic of mollusks.
   c. is responsible for the spiral growth of bivalve shells.
   d. describes the thrashing movement of nematodes.
   e. is responsible for the metamorphosis of insects.

10. Bivalves differ from other mollusks in that they
    a. are predaceous.
    b. have no heads and are suspension feeders.
    c. have shells.
    d. have open circulatory systems.
    e. use a radula to feed as they burrow through sand.

11. The exoskeleton of arthropods
    a. functions in protection and anchorage for muscles.
    b. is composed of chitin and cellulose.
    c. is absent in millipedes and centipedes.
    d. expands at the joints when the arthropod grows.
    e. functions in respiration and movement.

12. Which of the following is true of the uniramians?
    a. The horseshoe crab is the one surviving marine member of this group.
    b. It was the first lineage of arthropods and thus shows the most primitive segmentation.
    c. It contains insects, centipedes, and millipedes, characterized by their unbranched appendages.
    d. It is characterized by jawlike mandibles, antennae, compound eyes, and includes all arthropods except the chelicerates.
    e. It includes the extinct trilobites and eurypterids.

13. What do nematodes and arthropods have in common?
    a. They are both segmented.
    b. They are both pseudocoelomates.
    c. They are both well adapted to terrestrial habitats.
    d. They both have exoskeletons and undergo ecdysis.
    e. Both a and d are correct.

14. Which of the following structures is *not* associated with prey capture?
    a. scolex of tapeworm
    b. colloblasts of comb jelly
    c. cnidocytes of hydra
    d. proboscis of ribbon or proboscis worm
    e. tube feet of sea star

15. Molecular data support the hypothesis that urbilateria, the original bilateral animals, were
    a. colonial choanoflagellates.
    b. acoelomate, diploblastic animals.
    c. pseudocoelomate burrowing worms.
    d. relatively complex coelomates.
    e. triploblastic cnidarians.

16. Many animals are parasitic. Which of the following is an *incorrect* description of one of these parasites?
    a. Ticks are blood-sucking parasites belonging to class Arachnida.
    b. Many roundworms (Nematoda) are internal parasites of humans.
    c. Lice are wingless ectoparasites in phylum Insecta.
    d. Flukes are flatworms and may have complex life cycles.
    e. Tapeworms are annelids that reproduce by shedding proglottids.

17. Which of the following does *not* function in suspension feeding?
    a. lophophore of brachiopod
    b. radula of snails
    c. choanocytes of sponges
    d. mucus-coated gills of clam
    e. crown of cilia of rotifer

18. Many systematists now favor dividing phylum Anthropoda into four separate phyla representing its four major lineages. Which of the following does *not* describe one of those lineages?
    a. Trilobites–extinct early arthropods with uniform appendages
    b. Chelicerates–horseshoe crabs, scorpions, spiders, and ticks with anterior appendages modified as pincers or fangs and no antennae
    c. Insects–26 orders of primarily terrestrial arthropods with compound eyes, mandibles and other appendages modified for feeding, three pairs of walking legs, and usually two pairs of wings
    d. Crustaceans–crabs, lobsters, shrimps, and many others with two pairs of antennae, branched appendages, and jawlike mandibles
    e. Uniramians–centipedes, millipedes, and insects with one pair of sensory antennae, unbranched appendages, and jawlike mandibles

CHAPTER 34

VERTEBRATE EVOLUTION
AND DIVERSITY

FRAMEWORK

This chapter focuses on phylum Chordata with emphasis on the origin and characteristics of the subphylum Vertebrata and its classes. One version of chordate phylogeny is illustrated on p. 260. Fossil evidence and hypotheses about human ancestry are described.

CHAPTER REVIEW

Invertebrate Chordates and the Origin of Vertebrates

The deuterostome branch of the animal kingdom includes the echinoderms and **chordates.** The chordates include two invertebrate groups, the urochordates and the cephalochordates, and the **vertebrates,** which have a backbone.

Four anatomical features characterize phylum Chordata (679–680)

Chordates have four distinguishing anatomical features, which may appear only during the embryonic stage: (1) a flexible rod called a **notochord** that provides skeletal support the length of the animal; (2) a dorsal, hollow nerve cord that develops into the brain and spinal cord; (3) pharyngeal slits that open from the pharynx to the outside and function in suspension feeding or are modified for gas exchange and other functions; and (4) a postanal tail containing skeletal elements and muscles.

Invertebrate chordates provide clues to the origin of vertebrates (680–682)

Subphylum Urochordata **Tunicates,** or sea squirts, are the common names for these mostly sessile marine **urochordates.** Water moves through the saclike, tunic-covered animal from the incurrent siphon, through pharyngeal slits where food is filtered by a mucous net, to the excurrent siphon. The larval traits of notochord, nerve cord, and tail are lost in the adult animal.

Subphylum Cephalochordata **Cephalochordates,** known as **lancelets,** are tiny marine animals that retain all four chordate characteristics into the adult stage. These suspension feeders burrow backward into the sand and filter food particles through a mucous net secreted across the pharyngeal slits. The lancelet swims by coordinated contractions of serial muscles that flex the notochord from side to side. Blocks of mesoderm, called **somites,** develop into these muscle segments.

■ **INTERACTIVE QUESTION   34.1**

What is the common name for the animal shown in this diagram, and to which subphylum does it belong? Identify the labeled structures and indicate which are the four chordate characteristics.

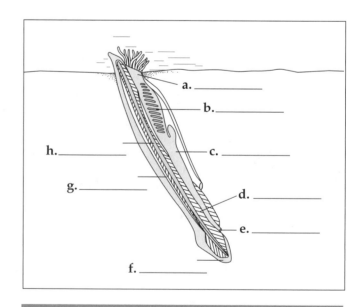

The Relationship between Invertebrate Chordates and Vertebrates   Molecular systematics indicates that vertebrates are most closely related to cephalochordates.

Two stages in vertebrate evolution are hypothesized: the evolution of a cephalochordate from an ancestral form resembling a urochordate larva, and the evolution of a vertebrate from a cephalochordate. The first stage may have occurred by **paedogenesis,** the development of sexual maturity in a larva. Fossil evidence appears to support the second stage—the evolution of vertebrates from a cephalochordate. Newly discovered 530-million-year-old fossils appear to be transitional stages: animals that resemble lancelets, but with vertebrate characters such as eyes, mineralized toothlike structures, and an enlarged brain. Pushing the origin of vertebrates back to the Cambrian explosion, other fossils from the same Chinese site had all the vertebrate characteristics, including a cranium (skull)

Introduction to the Vertebrates

Neural crest, pronounced cephalization, a vertebral column, and a closed circulatory system characterize subphylum Vertebrata (683)

A group of embryonic cells called the **neural crest** is found along the margin of the embryonic folds that meet to form the dorsal, hollow nerve cord. These cells migrate in the embryo and contribute to various unique vertebrate structures, including bones and cartilage of the cranium (brain case).

Vertebrates show a high degree of cephalization with sensory structures and an enlarged brain enclosed within a cranium. Hagfishes have these vertebrate characteristics, but lack a backbone. They are included with the vertebrates in a larger clade, **Craniata,** and represent an ancient branch of that clade.

The axial skeleton includes the cranium and the vertebral column, which protects the nerve cord and provides for support and movement. Many vertebrates also have ribs and an appendicular skeleton supporting two pairs of limbs. The vertebrate skeleton is made of bone and/or cartilage, which is secreted and maintained by living cells and can grow with the animal.

Increased activity levels require a high rate of cellular respiration, supported by adaptations of the respiratory and circulatory systems. Vertebrates have a closed circulatory system with a ventral, chambered heart pumping blood to capillaries that ramify through body tissues. The blood is oxygenated in capillaries of gills or lungs. An active lifestyle is also supported by adaptations for feeding, digestion, and nutrient absorption.

■ **INTERACTIVE QUESTION   34.2**

List the three shared derived features of vertebrates that facilitate increased size and activity.

a.

b.

c.

An overview of vertebrate diversity (683)

The hagfishes and the lampreys lack hinged jaws. All other vertebrates are grouped in the clade **gnathostomes,** also distinguished by two sets of paired appendages. These appendages are fins in the two classes of fishes: Chondrichthyes, sharks and rays with cartilaginous skeletons, and Osteichthyes, fishes with bony skeletons. In **tetrapods,** the paired appendages are legs that can support the animals on land. Amphibians are tetrapods, as are members of the clade identified as **amniotes,** named for the water-retaining egg adapted for reproduction on land. Mammals are one clade within the amniotes; the other monophyletic clade includes reptiles and birds.

Jawless Vertebrates

The hagfishes and the lampreys are the two extant classes of **agnathans,** ancient jawless vertebrate lineages that do not have paired fins, teeth, or mineralized bones.

Class Myxini: hagfishes are the most primitive living "vertebrates" (685)

The marine hagfishes are mostly bottom-dwelling scavengers. Their cartilaginous skeleton includes a cranium and a notochord; they lack vertebrae. These most primitive vertebrates (or craniates) are believed to have diverged from the vertebrate lineage about 530 million years ago.

Class Cephalaspidomorphi: lampreys provide clues to the evolution of the vertebral column (685–686)

Lampreys are suspension feeders as larvae and blood-sucking parasites as adults. The notochord is the main axial skeleton. The notochord is surrounded by a cartilaginous pipe with pairs of projections that partially enclose the nerve cord, perhaps a vestige of an early evolutionary stage of the vertebral column.

■ INTERACTIVE QUESTION   34.3

This cladogram represents one set of hypotheses of vertebrate phylogeny. As you work through this chapter, fill in the boxes with the derived characters that define these clades, identify the three large clades across the top, and fill in the names and/or representative animals of the vertebrate classes.

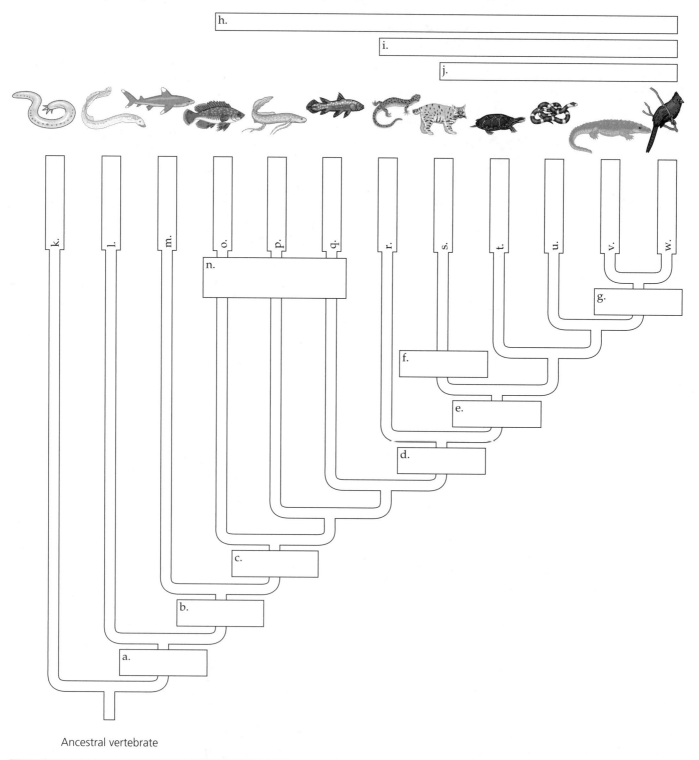

Ancestral vertebrate

The brain and cranium appear to have evolved first in the vertebrate lineage, followed by the vertebral column, and then the jaws, ossified skeletons, and paired appendages of the gnathostomes.

Some extinct jawless vertebrates had ossified teeth and bony armor (686)

An informal group of taxa called **ostracoderms** are common in the fossil record from about 450 to 375 million years ago. Most species were small bottom dwellers that lacked paired fins and jaws, feeding by filtering sediments through their gill slits. Their armors of bony plates provide evidence of the early evolution of ossification, the mineralization of connective tissue. Fossils of the ancient **conodonts**, dating back 510 million years, had ossified toothlike structures. The ostracoderms and most other agnathans disappeared during the Devonian period. Jawed vertebrates probably evolved from an ostracoderm lineage.

Fishes and Amphibians

Vertebrates with jaws (gnathostomes) replaced most agnathans during the late Silurian and early Devonian periods. The extant groups of fishes and the armored **placoderms** appeared. These vertebrates had two pairs of paired appendages and hinged jaws. Both these traits facilitated a predatory lifestyle that opened up new food resources.

Vertebrate jaws evolved from skeletal supports of the pharyngeal slits (687)

Jaws evolved from the skeletal rods of the anterior pharyngeal slits. The remaining gill slits functioned in gas exchange. The Devonian period is known as the "age of fishes." Placoderms and **acanthodians**, another class of jawed fishes, radiated and then disappeared almost completely by the Carboniferous period, although their ancestors may have given rise to the sharks.

Class Chondrichthyes: sharks and rays have cartilaginous skeletons (687–688)

The vertebrates of class **Chondrichthyes** have well-developed jaws, paired fins, and endoskeletons made of cartilage, parts of which are strengthened by mineralized granules. The cartilaginous skeleton evolved secondarily, perhaps by a modification of normal development in which a cartilaginous skeleton becomes ossified.

Sharks swim using the powerful muscles in their caudal fin. The paired pectoral and pelvic fins provide lift. Swimming maintains buoyancy and moves water past their gills. Although most sharks are carnivores, the largest sharks and rays are suspension feeders on plankton. Shark teeth probably evolved from the jagged scales that cover the skin.

Sharks have keen senses: sharp vision, nostrils, and skin regions in the head that can detect electric fields generated by muscle contractions of nearby animals. The **lateral line system** is a row of microscopic organs along the sides of the body that are sensitive to water pressure changes. The shark's body transmits sound waves to the inner ear.

Following internal fertilization, **oviparous** species of sharks lay eggs that hatch outside the mother; **ovoviviparous** species retain the fertilized eggs until they hatch; and, in the few **viviparous** species, developing young are nourished by a placenta until born.

Rays, which propel themselves with their enlarged pectoral fins, are primarily bottom dwellers that feed on mollusks and crustaceans.

■ INTERACTIVE QUESTION 34.4

a. What is the function of the **spiral valve** in a shark's intestine?

b. What is a **cloaca**?

Osteichthyes: the extant classes of bony fishes are the ray-finned fishes, the lobe-finned fishes, and the lungfishes (688–690)

Bony fishes had previously been combined in a single class, Osteichthyes, but cladistic analysis supports separation into three extant classes. Abundant in the oceans and nearly all fresh waters, bony fishes have endoskeletons reinforced with a hard matrix of calcium phosphate. The flattened, bony scales that cover the skin are coated with mucus to reduce drag. They have a lateral line system similar to that of sharks. The movement of muscles surrounding the gill chambers and the **operculum,** or flap over the gill chamber, forces water across the gills so that a bony fish can breathe while not moving. Most bony fishes also have a **swim bladder,** an air sac that controls buoyancy.

The flexible fins of bony fish are better for steering and propulsion than are the stiffer fins of sharks. Most species of fish are oviparous and fertilization is external.

Bony fishes probably originated in fresh water. The swim bladder was modified from lungs used to augment gills in stagnant water. The three extant classes had diverged by the end of the Devonian. The **ray-finned fishes (class Actinopterygii)** include most of the familiar modern fish, many of which spread to the seas during their evolution and some of which returned to fresh water. The **lobe-finned fishes (class Actinistia),** known mainly from the fossil record, have paired muscular fins supported by extensions of the skeleton that may have enabled these bottom-dwelling fish to "walk." Two populations of coelacanths are the only present-day lobe-finned fishes. Three genera of **lungfishes (class Dipnoi)** are found in the Southern Hemisphere, inhabiting stagnant ponds and swamps where they may augment their gas exchange by gulping air into their lungs. One of the Devonian lungfishes was probably the ancestor of amphibians and the other tetrapods.

■ **INTERACTIVE QUESTION 34.5**

a. How and why are the body shapes of sharks, bony fishes, and aquatic mammals similar?

b. Go back to the cladogram in Interactive Question 34.3 and fill in the information for the vertebrate classes up to letter **q.**

Tetrapods evolved from specialized fishes that inhabited shallow water (690–691)

If tetrapods are defined as vertebrates with skeleton-supported legs, then the ancestral fishes that lived in shallow water must be included in the tetrapod clade. During the Devonian, various fishes resembling modern lobe-fins and lungfishes had evolved, probably using their leglike appendages to crawl through shallow waters filled with dense vegetation and supplementing their gas exchange by forcing air into lungs by buccal pumping. Many newly discovered fossils from 400 to 350 million years ago document the transition from sturdy-finned fishes to tetrapod fish whose walking limbs had the same skeletal elements as terrestrial tetrapods, to amphibians. Amphibians, as the first tetrapods to exploit the terrestrial habitat, underwent an adaptive radiation during the early Carboniferous period. Their numbers and diversity declined during the late Carboniferous; by the Mesozoic era about 245 million years ago, the survivors resembled modern amphibians.

Class Amphibia: salamanders, frogs, and caecilians are the three extant amphibian orders (691–693)

Present-day amphibians (**class Amphibia**) include salamanders (**order Urodela**), frogs (**order Anura**), and caecilians (**order Apoda**).

Salamanders, which may be aquatic or terrestrial, swim or walk with a lateral bending of the body. Frogs are more specialized for moving on land, using their powerful hind legs for hopping. Protective adaptations of frogs include camouflage coloring and distasteful or poisonous skin mucus that may be advertised by bright coloration. Tropical caecilians are burrowing, legless, nearly blind, wormlike animals.

Many frogs undergo a metamorphosis from a tadpole—the aquatic, herbivorous larval form with gills and a tail—to the carnivorous, terrestrial adult form with legs and lungs. Many amphibians do not have an aquatic tadpole stage, and some species are exclusively aquatic or terrestrial. Amphibians are most abundant in damp habitats, since some of their gas exchange occurs across their moist skin or mouth.

Fertilization is generally external. Oviparous species lay their eggs in aquatic or moist environments. Some species show varying amounts of parental care. During breeding season, some frogs communicate with vocalizations. Some species migrate to specific breeding sites, using various forms of communication and navigation.

■ **INTERACTIVE QUESTION 34.6**

a. From what group are amphibians thought to have evolved?

b. When was the "age of amphibians"?

AMNIOTES

Evolution of the amniotic egg expanded the success of vertebrates on land (693)

The amniotic egg, enclosed in a waterproof shell, permitted vertebrates to complete their life cycles on land. **Extraembryonic membranes,** which develop from embryonic tissue layers, are involved in gas exchange, waste storage, and transfer of nutrients to the embryo. The amnion encases the embryo in amniotic fluid. Birds, most reptiles, and some mammals have shelled, amniotic eggs, and placental mammals retain the extraembryonic membranes during embryonic development within the uterus. Other terrestrial adaptations of the amniotes include a waterproof skin and increased use of the rib cage to ventilate the lungs.

■ INTERACTIVE QUESTION 34.7

Identify the four extraembryonic membranes in this sketch of an amniotic egg. What are the functions of these membranes?

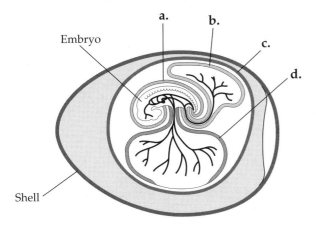

a.

b.

c.

d.

Vertebrate systematists are reevaluating the classification of amniotes (694–695)

The amniotes are a monophyletic clade; all modern reptiles, birds, and mammals share a common ancestor. Three main groups radiated during the early Mesozoic era: **synapsids, anapsids,** and **diapsids** (named for differences in their skulls). Mammals evolved from reptiles called therapsids, which were part of the synapsid lineage. The anapsids are probably all extinct. Turtles were previously placed in this lineage, but molecular data indicate that they should be grouped with the diapsids. The diapsids include the modern reptiles and diverse extinct reptiles. The

diapsid lineage split into two branches during the early Mesozoic: the **lepidosaurs** (including lizards and snakes) and the **archosaurs** (including crocodiles and alligators, dinosaurs, and birds).

Classical taxonomy divides the amniotes into the reptiles, birds, and mammals, a classification that emphasizes the evolutionary adaptations within the avian lineage. Cladistic-based alternative taxonomies range from a "lumpers" version of only two classes, Mammalia and Reptilia, to a "splitters" scheme that creates six classes, giving class status to mammals, birds, and all four reptile groups.

A reptilian heritage is evident in all amniotes (695–698)

Reptilian Characteristics   The skin of reptiles is covered with keratinized, waterproof scales. Lungs are the vehicle for gas exchange. Most species lay shelled amniotic eggs, and fertilization is internal. Some species are viviparous, in which case the extraembryonic membranes form a placenta.

Reptiles are **ectotherms.** They absorb external heat rather than generating their own, although they may regulate their body temperature through behavioral adaptations.

The Origin and Evolutionary Radiation of Reptiles   Reptiles were the dominant terrestrial vertebrates during the Mesozoic era. The oldest reptilian fossils date from the late Carboniferous period, 300 million years ago. During the Permian period, the first reptilian radiation gave rise to the three evolutionary branches: Synapsida, Anapsida, and Diapsida.

The second great radiation occurred a little over 200 million years ago during the late Triassic period, when the terrestrial **dinosaurs** and flying **pterosaurs** originated and diversified. The highly diverse dinosaurs included the largest animals ever to live on land. The pterosaurs had wings formed from a membrane of skin stretched between an elongated finger and the body. Evidence indicates that many dinosaurs were active, agile, and, in some species, social. Some scientists postulate that at least some dinosaurs were **endothermic**—able to use metabolic heat to warm the body.

During the Cretaceous period at the end of the Mesozoic, dinosaurs became extinct.

Modern Reptiles   The four groups of extant reptiles are **Testudines,** the turtles; **Sphenodontia,** the tuataras; **Squamata,** the lizards and snakes; and **Crocodilia,** the alligators and crocodiles. Turtles evolved during the Mesozoic era and have changed little since then. They are protected by their hard shell.

Lizards, the most numerous and diverse extant reptiles, are relatively small. Snakes, apparently descended from lizards that adopted a burrowing lifestyle, are limbless, although primitive species have vestigial pelvic and limb bones. Snakes are carnivorous, with adaptations for locating (heat-detecting and olfactory organs), killing (sharp teeth, sometimes that inject toxins), and swallowing (loosely articulated jaws) their prey.

Crocodiles and alligators, among the largest living reptiles, spend most of their time in water. Crocodilians are the reptiles (as traditionally defined) most closely related to the dinosaurs.

■ **INTERACTIVE QUESTION   34.8**

a. List some characteristics of reptiles.

b. What are the four groups of extant reptiles?

c. Why is the traditional class Reptilia not a monophyletic taxon?

Birds began as feathered reptiles (698–701)

Birds evolved during the reptilian radiation of the Mesozoic era. The amniote egg and scales on the legs are two of the reptilian traits that birds display.

Characteristics of Birds   Bird anatomy has evolved to enhance flight. Weight is reduced by honeycombed bones, air sacs off the lungs, and lack of teeth (food is ground in the gizzard). The keratinized beak has assumed a great variety of shapes associated with different diets. Birds are endothermic; feathers and a fat layer in some species help retain metabolic heat. The four-chambered heart and efficient respiratory system facilitate a high metabolic rate.

Birds have excellent vision. Their relatively large brains permit detailed visual processing, motor coordination, and complex behavior, particularly in courtship rituals and parenting. Eggs are brooded during development by the female and/or male bird.

Aerodynamic wings are flapped by large pectoral muscles attached to a keel on the sternum. Feathers made of keratin, a diagnostic characteristic of birds, are extremely light and strong, and shape the wing into an airfoil. Feathers also provide insulation, perhaps their first function during the evolution of endothermy.

The evolution of strong flying ability enhanced hunting, foraging, escape from predators, and migration to favorable habitats.

The Origin of Birds   The closest relatives of birds were a group of small, bipedal, carnivorous dinosaurs called the **theropods.** The ancestor of birds was probably a feathered theropod.

Fossils of *Archaeopteryx,* an ancient feathered creature with clawed forelimbs, teeth, and long tail, have been found from 150 million years ago. This weak flyer was probably a tree-dwelling glider, perhaps the earliest mode of flight.

Feathered, but flightless dinosaur fossils have recently been found in Chinese sediments. These fossils, which are 20 million years younger than *Archaeopteryx,* may be descendants of an ancient line of feathered, flightless dinosaurs, or may have evolved from early flying birds, losing flight secondarily.

Modern Birds   Flightless birds collectively called **ratites,** such as the ostrich, kiwi, and emu, lack a keeled breastbone. **Carinates** have a sternal keel to which attach large breast muscles. **Passeriforms,** or perching birds, are an order of carinate birds that includes 60 percent of living bird species.

■ **INTERACTIVE QUESTION   34.9**

List several adaptations for flight found in birds.

Mammals diversified extensively in the wake of the Cretaceous extinctions (701–706)

Mammalian Characteristics   Vertebrates of **class Mammalia** are characterized by mammary glands which produce milk to nourish the young. Mammals have hair made of keratin, which helps to insulate these endothermic animals. Active metabolism is provided for by an efficient respiratory system that uses a diaphragm to help ventilate the lungs and a circulatory system with a four-chambered heart.

Fertilization is internal, and the embryo develops in the uterus. In eutherian (placental) mammals and marsupials, a **placenta,** formed from the uterine lining and extraembryonic membranes of the embryo, nourishes the developing embryo.

Mammals generally have large brains and many are capable of learning. Extended parental care of the young provides time to learn survival skills.

Mammalian teeth come in a diverse assortment of shapes and sizes that are specialized for eating a variety of foods. Part of the evolutionary remodeling of the mammalian jaw produced a stronger, single-bone lower jaw and the incorporation of two jaw bones into the middle ear.

The Evolution of Mammals The oldest fossils of mammals date back to the Triassic period, 220 million years ago. Reptilian **therapsids,** part of the synapsid branch of reptiles, were mammalian ancestors. Small Mesozoic mammals, which were probably nocturnal and insectivorous, coexisted with dinosaurs. Mammals underwent an extensive radiation in the wake of the Cretaceous extinctions. There are three major groups of mammals today.

Monotremes The platypus and echidnas (spiny anteaters) are the only extant **monotremes,** egg-laying mammals. The egg is reptilian in structure, but monotremes have hair and produce milk for their young. Monotremes seem to have descended from a very early mammalian branch.

Marsupials **Marsupials,** including opossums, kangaroos, bandicoots, and koalas, complete their embryonic development in a maternal pouch called a marsupium. Born very early in development, the neonate crawls into the marsupium, fixes its mouth to a nipple, and completes its development while nursing.

Australian marsupials have radiated and filled the niches occupied by eutherian mammals in other parts of the world. Marsupials apparently originated in what is now North America, spreading southward. After the breakup of Pangaea, South America and Australia became island continents where marsupials diversified in isolation from placental mammals. Placental mammals reached South America in the Cenozoic era, whereas Australia has remained isolated.

Eutherian (Placental) Mammals **Eutherian mammals** complete development attached to a placenta within the maternal uterus. According to fossil evidence, eutherians and marsupials may have diverged from a common ancestor 80 to 100 million years ago.

The major orders of eutherian mammals radiated during the late Cretaceous and early Tertiary periods.

There appear to be at least four main evolutionary clades. The first branch, referred to as the **Afrotheria,** includes the elephants, manatees, and several lesser known African mammals. The second branch radiated in South America and includes the sloths, anteaters, and armadillos. The third clade includes many groups: the bats, certain "core insectivores" such as shrews and moles, carnivores, two ungulate orders (the hoofed artiodactyls and perissodactyls), and the cetaceans (dolphins and whales). Recent evidence indicates that the cetaceans are a branch within the artiodactyls. The fourth clade includes the lagomorphs (rabbits), rodents, and primates.

■ INTERACTIVE QUESTION 34.10

a. List the main characteristics of mammals.

b. Now go back to the evolutionary tree in Interactive Question 34.3 and fill in the information for the rest of the classes and derived characters.

Primates and the Evolution of Homo sapiens

Primate evolution provides a context for understanding human origins (707–709)

Some General Primate Characteristics Order Primates includes humans and their closest relatives. Most primates have grasping hands and feet, large brains, short jaws, forward-looking eyes, and flat nails on their digits, and exhibit complex social behavior and extensive parental care. Many of these traits were shaped by natural selection in the tree-dwelling early primates. All primates, except *Homo,* have a wide separation between the big toe and the other toes, allowing them to grasp branches with their feet. The anthropoid primates have an **opposable thumb,** which functions in a grasping "power grip" in monkeys and apes, but is adapted for precise manipulation in humans.

Modern Primates There are two suborders of Primates. **Prosimians,** such as lemurs, lorises, and tarsiers, probably resemble early arboreal primates. The

anthropoids include monkeys, apes, and humans. The oldest known anthropoid fossils are 45 million years old and point to tarsiers as the closest prosimian relative.

Fossils indicate that monkeys were established in both the Old World and the New World by 40 million years ago, somehow reaching the New World after first evolving in the Old World. The strictly arboreal New World monkeys have been evolving separately from Old World monkeys for millions of years. Most monkeys in both groups are diurnal and live in social bands.

The anthropoid suborder also includes four genera of apes: gibbons, orangutans, gorillas, and chimpanzees and bonobos, all of which live in the Old World tropics. Apes evolved from Old World monkeys about 25–30 million years ago. Modern apes generally are larger than monkeys, with relatively long arms, short legs, and no tails. Ape behavior is more flexible than that of monkeys, and gorillas and chimpanzees are highly social.

Humanity is one very young twig on the vertebrate tree (709–715)

Paleoanthropology is the study of human origins and evolution. Paleoanthropologists use the term **hominoid** to refer to great apes and humans, whereas the term **hominid** refers to branches of the evolutionary tree that are more closely related to humans than to chimpanzees, gorillas, or orangutans. The two main groups of hominids are the extinct australopithecines and the genus *Homo,* all the species of which are extinct except for *Homo sapiens.*

Some Common Misconceptions   Humans are not descended from chimpanzees; the two represent divergent branches from a common, anthropoid ancestor. Human evolution has not occurred within an unbranched hominid line; there have been times when several different human species coexisted. Different human features have evolved at different rates, known as **mosaic evolution.** Thus, for example, bipedalism or erect posture evolved before an enlarged brain developed.

Early Anthropoids: A Change of Environment   About 20 million years ago, the Himalayan range rose, the climate became drier, and the forests of Africa and Asia contracted, presenting an increased savannah habitat that influenced anthropoid evolution.

Fossil evidence and DNA comparisons indicate that humans and chimpanzees diverged from a common hominoid ancestor only about 5 to 7 million years ago.

Some Major Features of Human Evolution   The following five characteristics have changed during the course of human evolution. The brain size of hominoids of about 6 million years ago was equivalent to that of modern chimpanzees; modern humans have a brain volume three times as large. Chimpanzees have retained the **prognathic jaw** of the hominoid ancestor; human jaws are shorter, producing a flatter face, more pronounced chin, and changed dentition. Hominoid ancestors walked on all four limbs, as do all modern apes; bipedal posture required skeletal modifications evident in early hominid fossils. Compared to the great apes, sexual dimorphism in size difference between the sexes is greatly reduced in humans, with males only about 1.2 times the weight of females. Comparing extant hominoid and human social behavior, researchers find that humans differ in monogamy and long-term pair bonding, exceptional dependence of newborn infants on their mothers, and greater duration of parental care.

Australopithecines: Early Hominids and the Origin of Bipedal Posture   Most of the oldest hominid fossils are teeth and skeletal fragments, from which paleontologists attempt to reconstruct human phylogeny. Hypotheses change as new fossils are found and data from molecular systematics are gathered.

In 1924, an anthropologist found a skull of an early human that he called *Australopithecus africanus.* It appears that *A. africanus* was an upright hominid, with humanlike hands and teeth but a small brain. Various species of *Australopithecus* existed for about 3 million years, beginning about 4 million years ago.

In 1974, a petite *Australopithecus* skeleton was discovered in the Afar region of Ethiopia. Lucy, as this 3.24 million-year-old fossil was named, and similar fossils have been designated *Australopithecus afarensis,* a species which appears to have existed for about 1 million years. Although bipedal, *A. afarensis* had a brain size similar to that of a chimpanzee, a prognathic skull, more apelike sexual dimorphism, and long arms that suggest a capacity for arboreal locomotion.

Even older fossils have been found and assigned to *Australopithecus anamensis,* which lived over 4 million years ago, and some fossils of putative hominids go back 6 million years.

An apparent adaptive radiation began about 3 million years ago that produced several species, including *A. africanus.* A later lineage included the heavier boned "robust" australopithecines, apparently an evolutionary dead end. The other australopithecines had lighter jaws and teeth, and these "gracile" forms probably included the ancestors of *Homo.*

■ **INTERACTIVE QUESTION 34.11**

a. How long ago did humans and chimpanzees diverge from a common ancestor?

b. List five characteristics that changed over the course of the evolution of humans from a hominoid ancestor.

Homo: *The Evolution of Larger Brains and the Global Dispersion of Humans* Larger brained fossils dating from 2.5 to 1.6 million years old are the first to be placed in the genus *Homo.* Sometimes found with sharp stone tools, *Homo habilis,* or "handy man," had some modern hominid characters. A fairly complete fossil known as "Turkana Boy" dates from 1.6 million years ago and had a brain size between that of *H. habilis* and *Homo erectus.*

Homo erectus was the first hominid to migrate into Asia and Europe. *H. erectus* lived from about 1.8 million years ago to 500,000 years ago. *H. erectus* had greater height and brain capacity than did *H. habilis.*

The Neanderthals were among the descendants of *H. erectus.* They lived in Europe from about 200,000 to 40,000 years ago. Some paleoanthropologists group post-*H. erectus* fossils dating from 500,000 to 100,000 years ago in Africa, Europe and Asia as the earliest forms of our species and call them "archaic *Homo sapiens.*" Other researchers give these regional descendants of *H. erectus* separate species names, such as *Homo neanderthalensis.*

The Origin of Anatomically Modern Humans Two different hypotheses attempt to explain the origin of **anatomically modern humans.** According to the **multiregional hypothesis,** regional groups of "archaic *Homo sapiens*" evolved into modern humans in parallel in different parts of the world, with occasional interbreeding maintaining the genetic similarity among the populations. The **"Out of Africa" hypothesis,** also called the **replacement hypothesis,** maintains that a second migration out of Africa occurred about 100,000 years ago, and these anatomically modern humans replaced all the regional descendants of *H. erectus.*

The two hypotheses differ on when the last common ancestor to the world's modern populations lived in Africa: over 1.5 million years ago, according to the multiregional hypothesis, or 100,000 years ago, according to the replacement hypothesis. A third hypothesis is that *H. sapiens* dispersing from Africa 100,000 years ago interbred with the regional descendants of the first migration.

Genetic data appear to support the replacement hypothesis. Comparisons of mitochondrial DNA from various regional populations date the time of genetic divergence from about 100,000 years ago. Studies of genetic markers in the nucleus also support this date. A recent comparison of mtDNA extracted from Neanderthal bones with that of modern Europeans suggests that Neanderthals did not contribute to the ancestry of anatomically modern humans in Europe. And recent genetic data from comparisons of Y chromosomes of males from different geographic regions indicates a divergence from a common African ancestor less than 100,000 years ago.

Some researchers interpret some Asian fossils as intermediate between older fossils of *H. erectus* and modern Asians, as predicted by the multiregional hypothesis or the interbreeding hypothesis. The fossil evidence in western Europe, however, is consistent with the replacement hypothesis. Neanderthals were totally replaced about 40,000 years ago by the anatomically modern humans in the area, known as Cro-Magnons, and there is no skeletal evidence for interbreeding between the groups.

WORD ROOTS

a- = without (*agnathan:* a member of a jawless class of vertebrates represented today by the lampreys and hagfishes)

an- = without; **-apsi** = a juncture (*anapsids:* one of three groups of amniotes based on key differences between their skulls)

arch = ancient (*archosaurs:* the reptilian group which includes crocodiles, alligators, dinosaurs, and birds)

cephalo- = head (*cephalochordates:* a chordate without a backbone, represented by lancelets, tiny marine animals)

aktin- = a ray; **-pterygi** = a fin (*Actinopterygii:* the class of ray-finned fishes)

crani- = the skull (*craniata:* the chordate subgroup that possess a cranium)

crocodil- = a crocodile (*Crocodilia:* the reptile group that includes crocodiles and alligators)

di- = two (*diapsids:* one of three groups of amniotes based on key differences between their skulls)

dino- terrible; **-saur** = lizard (*dinosaurs:* an extremely diverse group of ancient reptiles varying in body shape, size, and habitat)

endo- = inner; **therm-** = heat (*endotherm:* an animal that uses metabolic energy to maintain a constant body temperature, such as a bird or mammal)

eu- = good (*eutherian mammals:* placental mammals; those whose young complete their embryonic development within the uterus, joined to the mother by the placenta)

extra- = outside, more (*extaembryonic membranes:* four membranes that support the developing embryo in reptiles, birds, and mammals)

gnantho- = the jaw; **-stoma** = the mouth (*gnathostomes:* the vertebrate subgroup that possesses jaws)

homin- = man (*hominid:* a term that refers to mammals that are more closely related to humans than to any other living species)

lepido- = a scale (*lepidosaurs:* the reptilian group which includes lizards, snakes, and two species of New Zealand animals called tuataras)

marsupi- = a bag, pouch (*marsupial:* a mammal, such as a koala, kangaroo, or opossum, whose young complete their embryonic development inside a maternal pouch called the marsupium)

mono- = one (*monotremes:* an egg-laying mammal, represented by the platypus and echidna)

neuro- = nerve (*neural crest:* a band of cells along the border where the neural tube pinches off from the ectoderm)

noto- = the back; **-chord** = a string (*notochord:* a longitudinal, flexible rod formed from dorsal mesoderm and located between the gut and the nerve cord in all chordate embryos)

opercul- = a covering, lid (*operculum:* a protective flap that covers the gills of fishes)

osteo- = bone; **ichthy-** = fish (*Osteichthyes:* the vertebrate class of bony fishes)

ostraco- = a shell; **-derm** = skin (*ostracoderm:* an extinct agnathan; a fishlike creature encased in an armor of bony plates)

ovi- = an egg; **-parous** = bearing (*oviparous:* referring to a type of development in which young hatch from eggs laid outside the mother's body)

paedo- = a child; **-genic** = producing (*paedogenesis:* the precocious development of sexual maturity in a larva)

paleo- = ancient; **anthrop-** = man; **-ology** = the science of (*paleoanthropology:* the study of human origins and evolution)

passeri- = a sparrow; **form-** = shape (*passeriformes:* the order of perching birds)

placo- = a plate (*placoderm:* a member of an extinct class of fishlike vertebrates that had jaws and were enclosed in a tough, outer armor)

pro- = before; **-simi** = an ape (*prosimians:* a suborder of primates, the premonkeys, that probably resemble early arboreal primates)

ptero- = a wing (*pterosaurs:* winged reptiles that lived during the time of dinosaurs)

ratit- = flat-bottomed (*ratites:* the group of flightless birds)

soma- = body (*somites:* blocks of mesoderm that give rise to muscle segments in chordates)

syn- = together (*synapsids:* one of three groups of amniotes based on key differences between their skulls)

tetra- = four; **-podi** = foot (*tetrapod:* a vertebrate possessing two pairs of limbs, such as amphibians, reptiles, birds, and mammals)

tunic- = a covering (*tunicates:* members of the subphylum Urochordata)

uro- = the tail (*urochordate:* a chordate without a backbone, commonly called a tunicate, a sessile marine animal)

uro- = tail; **-del** = visible (*Urodela:* the order of salamanders that includes tetrapod amphibians with tails)

vivi- = alive (*ovoviviparous:* referring to a type of development in which young hatch from eggs that are retained in the mother's uterus)

STRUCTURE YOUR KNOWLEDGE

1. Review the cladogram on page 260. Make note of the derived characters that define each branch point.

2. Describe several examples from vertebrate evolution that illustrate the common evolutionary theme that new adaptations usually evolve from preexisting structures.

3. How did various vertebrate groups meet the challenges of a terrestrial habitat?

4. Contrast the multiregional and replacement hypotheses of the evolution of anatomically modern humans. What evidence is used to support each of these models?

TEST YOUR KNOWLEDGE

FILL IN THE BLANKS

_____ 1. chordate subphylum of sessile, suspension-feeding marine animals

_____ 2. chordate subphylum that includes lancelets

_____ 3. blocks of mesoderm along notochord that develop into muscles

_____ **4.** clade of jawed vertebrates

_____ **5.** flap over the gills of bony fishes

_____ **6.** adaptation that allows reptiles to reproduce on land

_____ **7.** reptilian ancestor of lizards, snakes, crocodilians, dinosaurs, and birds

_____ **8.** group from which birds probably descended

_____ **9.** egg-laying mammals

_____ **10.** structure that helps mammals ventilate their lungs

_____ **11.** suborder that includes monkeys, apes, and humans

_____ **12.** genus in which the fossil Lucy is placed

MULTIPLE CHOICE: *Choose the one best answer.*

1. Pharyngeal slits appear to have functioned first as
   a. suspension-feeding devices.
   b. gill slits for respiration.
   c. components of the jaw.
   d. portions of the inner ear.
   e. mouth openings.

2. Which of the following characteristics differentiates the vertebrates from the invertebrate chordates?
   a. notochord
   b. extensive cephalization
   c. postanal tail
   d. dorsal hollow nerve cord
   e. motility

3. Modern agnathans are represented by
   a. caecilians.
   b. sharks and rays.
   c. the lobe-finned coelocanth.
   d. lungfishes.
   e. hagfishes and lampreys.

4. Which of the following is more accurately classified in the clade Craniata than in the Vertebrata?
   a. lampreys
   b. sharks and rays
   c. placoderms
   d. hagfishes
   e. lobe-finned fishes

5. Which of the following evolutionary stages may have occurred by paedogenesis?
   a. the origin of cephalochordates from a urochordate larva
   b. the evolution of amphibians from a tetrapod fish
   c. the secondary loss of flight in ratites
   d. the origin of vertebrates from cephalochordates
   e. the evolution of jaws from skeletal rods of anterior pharyngeal slits

6. Which of the following is *incorrectly* paired with its gas exchange mechanism?
   a. amphibians—skin and lungs
   b. lungfishes—gills and lungs
   c. reptiles—lungs
   d. bony fishes—swim bladder
   e. mammals—lungs with diaphragm

7. Reptiles have lower caloric needs than do mammals of comparable size because they
   a. are ectotherms.
   b. have waterproof scales.
   c. have an amniotic egg.
   d. move by bending their vertebral column back and forth.
   e. have a more efficient respiratory system.

8. Which of the following best describes the earliest mammals?
   a. large, herbivorous
   b. large, carnivorous
   c. small, insectivorous
   d. small, herbivorous
   e. small, carnivorous

9. Oviparity is a reproductive strategy that
   a. allows mammals to bear well-developed young.
   b. is used by both birds and reptiles.
   c. is necessary for vertebrates to reproduce on land.
   d. is a necessity for all flying vertebrates.
   e. protects the embryo inside the mother and uses the food resources of the egg.

10. In Australia, marsupials fill the niches that eutherian (placental) mammals fill in other parts of the world because
    a. they are better adapted and have outcompeted eutherians.
    b. their offspring complete their development attached to a nipple in a marsupium.
    c. they originated in Australia.
    d. they evolved from monotremes that migrated to Australia about 12 million years ago.
    e. after Pangaea broke up, they diversified in isolation from eutherians.

11. Which of the following adaptations permitted vertebrates to complete their life cycle on land?
    a. amniotic egg
    b. internal fertilization
    c. placenta
    d. milk
    e. parental care

12. In accordance with the model of mosaic evolution,
    a. biogeography reflects the adaptive radiation of many groups.
    b. erect posture preceded the enlargement of the brain in human evolution.
    c. modern humans evolved in parallel in different parts of the world, laying the groundwork for geographic differences more than a million years ago.
    d. modern humans first evolved in Africa and later dispersed to other regions.
    e. the rapid divergence of hominids about 2.5 million years ago was associated with climatic changes.

13. Which hominid species is believed to be the direct ancestor of *Homo sapiens*?
    a. *Australopithecus afarensis*
    b. *A. africanus*
    c. *Homo habilis*
    d. *H. erectus*
    e. *H. neanderthalensis*

14. According to the replacement hypothesis, anatomically modern humans
    a. have been diverging from *H. erectus* for over 1 million years.
    b. evolved from many regional descendants of *H. erectus*.
    c. interbred with regional groups of "archaic" *Homo sapiens* and then replaced them.
    d. evolved from African descendants of *H. erectus* and dispersed to other regions about 100,000 years ago.
    e. descended from the following sequence: *Australopitheus afarensis, A. anamensis, Homo habilis, H. erectus*.

15. Which of the following is *not* a characteristic of *all* vertebrates?
    a. cranium
    b. neural crest in embryonic development
    c. two pairs of paired appendages
    d. closed circulatory system
    e. pronounced cephalization

PLANT FORM AND FUNCTION

CHAPTER 35

PLANT STRUCTURE AND GROWTH

FRAMEWORK

A rather large new vocabulary is needed to name the specialized cells and structures in a study of plant structure and growth. Focus your attention on how the roots, stems, and leaves of a plant are specialized to function in absorption, support, transport, protection, and photosynthesis. Plants exhibit indeterminate growth. Apical meristems at the tips of roots and shoots create primary growth; the primary meristems produce dermal, ground, and vascular tissues. The lateral meristems, vascular cambium and cork cambium, create secondary growth that adds girth to stems and roots. New techniques and model systems such as *Arabidopsis* are allowing researchers to explore the molecular bases for plant growth, morphogenesis, and cellular differentiation.

CHAPTER REVIEW

The Plant Body

Both genes and environment affect plant structure (720–721)

Plants show adaptation to their environments on two time scales: the evolutionary interactions that have shaped plants as terrestrial organisms and the structural and physiological responses of individual plants to their environments. As always, the correlation of structure and function will be obvious in our study of plants.

Plants have three basic organs: roots, stems, and leaves (721–724)

To be able to acquire dispersed resources in a terrestrial environment, plants have evolved an underground **root system** for obtaining water and minerals from the soil and an aerial **shoot system** of stems and leaves for absorbing light and carbon dioxide for photosynthesis.

The Root System The functions of roots include anchorage, absorption, and storage of food. A **taproot** system is found commonly in dicots, whereas monocots generally have **fibrous root** systems. Most absorption of water and minerals occurs through the tiny **root hairs** that are clustered near the root tips. Mycorrhizae, symbiotic associations of roots and fungi, contribute greatly to water and mineral absorption. Some plants have root nodules that house nitrogen-fixing symbiotic bacteria. **Adventitious** roots arise from stems or leaves and may serve as props to help support stems.

The Shoot System: Stems and Leaves The shoot system includes vegetative shoots, which bear leaves, and reproductive shoots, which bear flowers.

A stem consists of alternating **nodes,** the points at which leaves are attached, and segments between nodes, called **internodes.** An axillary bud is found in the angle (axil) between a leaf and the stem. A **terminal bud,** consisting of developing leaves and compacted nodes and internodes, is found at the tip or apex of a shoot. The terminal bud may exhibit **apical dominance,** inhibiting the growth of axillary buds and producing a taller plant. Axillary buds may develop into vegetative branches with their own terminal buds, leaves, and axillary buds.

Modifications of shoots include stolons, horizontal stems that grow along the surface of the ground; rhizomes, horizontal stems growing underground that may end in enlarged tubers; and bulbs, vertical underground shoots functioning in food storage.

Leaves, the main photosynthetic organs of most plants, usually consist of a flattened **blade** and a **petiole,** or stalk. Many monocots lack petioles. Monocot leaves usually have parallel major veins, whereas dicot leaves have networks of branched veins.

■ INTERACTIVE QUESTION   35.1

a. What is a node?

b. Where are axillary buds found?

c. List some of the differences between monocots and dicots. Add to this list as you go through the chapter.

Monocots	Dicots

Plant organs are composed of three tissue systems: dermal, vascular, and ground (724–726)

The **dermal tissue,** or **epidermis,** is a single layer of cells that covers and protects the young parts of the

plant. Root hairs are extensions of epidermal cells near root tips that increase the surface area for absorption. The epidermis of leaves and most stems is covered with a **cuticle,** a waxy coating that prevents excess water loss. **Vascular tissue** consists of **xylem** and **phloem** and functions in transport and support.

The water-conducting cells of xylem die at functional maturity, leaving behind their secondary walls, which are interrupted only by scattered **pits. Tracheids** are long, thin, tapered cells with lignin-strengthened walls. Water passes through pits from cell to cell. **Vessel elements** are wider, shorter, and thinner walled, with perforations in their end walls. Vessel cells align to form long **xylem vessels.**

In the phloem, sucrose, other organic compounds, and some mineral ions flow through chains of cells called **sieve-tube members,** which remain alive at functional maturity but lack nuclei, ribosomes, and vacuoles. In angiosperms, fluid flows through pores in the **sieve plates** in the end walls between cells. The nucleus and ribosomes of an adjacent **companion cell,** which is connected to a sieve-tube member by numerous plasmodesmata, may serve both cells.

Ground tissue functions in photosynthesis, support, and storage. In dicots, ground tissue internal to vascular tissue is called **pith;** outside the vascular tissue it is called **cortex.**

Plant tissues are composed of three basic cell types: parenchyma, collenchyma, and sclerenchyma (726–728)

Plant cells have adaptations of the cell wall and/or of the **protoplast,** the cell contents excluding the cell wall, that specialize them for different functions.

Parenchyma Cells   These relatively unspecialized cells usually lack secondary walls and have large central vacuoles. **Parenchyma cells** carry on many metabolic functions, such as photosynthesis and food storage. All developing plant cells begin as unspecialized parenchyma cells. Mature parenchyma cells usually do not divide but retain the ability to divide and differentiate into other types of plant cells.

Collenchyma Cells   These cells lack secondary walls but have thickened primary walls. Strands or cylinders of **collenchyma cells** function in support for young parts of the plant and elongate along with the plant.

Sclerenchyma Cells   **Sclerenchyma cells** have thick secondary walls strengthened with lignin. These specialized supporting cells often lose their protoplasts at maturity. **Fibers** are long, tapered cells that usually occur in groups. **Sclereids** are shorter and irregular in shape. Vessel elements and tracheids are sclerenchyma cells that also function in transport.

■ **INTERACTIVE QUESTION   35.2**

a. Which types of plant cells are dead at functional maturity?

b. Which types of plant cells lack nuclei at functional maturity?

The Process of Plant Growth and Development

Growth and development continue throughout the life of a plant. **Growth** is the increase in size resulting from cell division and expansion. **Development** includes all the changes that progressively produce a specific body organization.

Meristems generate cells for new organs throughout the lifetime of a plant: *an overview of plant growth* (729–730)

Most plants exhibit indeterminate growth, continuing to grow as long as they live. Plant organs such as leaves and flowers, as well as animals, have determinate growth and stop growing after reaching a certain size. Plants may be **annuals,** which complete their life cycle in one growing season or year; **biennials,** which have a life cycle spanning two years; or **perennials,** which live many years.

Plants have tissues called **meristems** that perpetually divide to form new cells. Cells that remain to divide in the meristem region are called initials, whereas cells that will specialize into developing tissue are called derivatives.

Apical meristems, located at the tips of roots and in the buds of shoots, produce **primary growth,** resulting in the elongation of roots and shoots. **Secondary growth** is an increase in diameter as new cells are produced by **lateral meristems,** creating a thicker secondary dermal tissue and new layers of vascular tissue.

■ **INTERACTIVE QUESTION   35.3**

What types of growth occur in woody plants?

Primary growth: Apical meristems extend roots and shoots by giving rise to the primary plant body (730–734)

The **primary plant body** is produced by primary growth in herbaceous plants and the youngest parts of a woody plant.

Primary Growth of Roots   The meristem of the root tip is protected by a **root cap** that secretes a polysaccharide slime to lubricate the growth route. The **zone of cell division** includes the apical meristem and the primary meristems. At the center of the apical meristem is the **quiescent center,** a slowly dividing, damage-resistant reserve of meristematic tissue. Derived from the apical meristem are three concentric cylinders of tissue: the **protoderm, procambium,** and **ground meristem,** the primary meristems that continue to divide and give rise to the cells of the three tissue systems of the root.

In the **zone of elongation,** cells lengthen to many times their original size, which pushes the root tip through the soil. Cells begin to specialize in structure and function as the zone of elongation grades into the **zone of maturation.**

The primary meristems produce the tissue systems of the root. The protoderm gives rise to the single cell layer of the epidermis. The procambium forms a central vascular cylinder or **stele.** In most dicots, xylem cells radiate from the center of the stele in spokes, with phloem in between. The stele of a monocot may have pith, a central core of undifferentiated parenchyma cells, inside the xylem and phloem.

The ground meristem produces the ground tissue system, filling the cortex, or region of the root between the stele and epidermis. These parenchyma cells store food and participate in the uptake of minerals from the soil solution entering the root. The innermost layer of cortex is the one-cell-thick **endodermis,** which regulates the passage of materials into the stele.

Lateral roots may develop from the **pericycle,** a layer of cells just inside the endodermis with meristematic potential. A lateral root pushes through the cortex and retains its vascular connection with the central stele.

■ **INTERACTIVE QUESTION   35.4**

Label the tissues in the cross section of a monocot root and list their functions. Identify the three primary tissues.

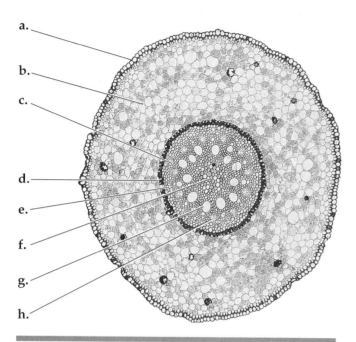

a. _____
b. _____
c. _____
d. _____
e. _____
f. _____
g. _____
h. _____

Primary Growth of Shoots   The dome-shaped mass of apical meristem in the terminal bud gives rise to the primary meristems: protoderm, procambium, and ground meristem. Leaf primordia form on the sides of the apical dome, and axillary buds develop from clumps of meristematic cells left at the bases of the leaf primordia.

Elongation of the shoot occurs by cell division and cell elongation within young internodes. In the grasses and some other plants, internodes have intercalary meristems that enable shoots to continue elongation. Axillary buds may connect to the vascular tissue located near the stem's surface and form branches later in development.

A shoot has a modular construction of serial segments of stem with leaves and axillary buds, resulting from the development of successive nodes and internodes from the apical meristem.

Unlike the single central stele of a root, vascular tissue runs through the stem in several **vascular bundles.** In dicots, the vascular bundles may be arranged in a ring, with the pith inside the ring connected by thin rays of ground tissue to the cortex outside the ring. Xylem is located internal to the phloem in the vascular bundles. In most monocot stems, the vascular bundles are scattered throughout the ground

tissue. A layer of collenchyma just beneath the epidermis and fiber cells of sclerenchyma within vascular bundles may strengthen the stem. An epidermis derived from the protoderm covers the stem and leaves.

The leaf is covered by the wax-coated, tightly interlocking cells of the epidermis. **Stomata,** tiny pores flanked by **guard cells,** permit both gas exchange and transpiration, the evaporation of water from the leaf.

Mesophyll consists of parenchyma ground tissue cells containing chloroplasts. In many dicot leaves, columnar palisade parenchyma is located above spongy parenchyma, which has loosely packed, irregularly shaped cells surrounding many air spaces.

A leaf trace, which is a branch of a vascular bundle in the stem, continues into the petiole and divides repeatedly within the blade of the leaf, providing support and vascular tissue to the photosynthetic mesophyll.

■ **INTERACTIVE QUESTION   35.5**

Name the indicated structures in this diagram of a leaf.

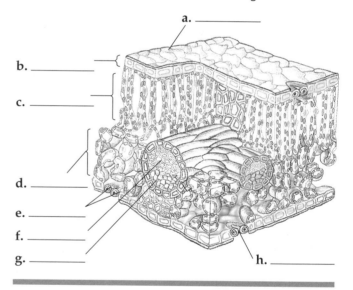

a. _____
b. _____
c. _____
d. _____
e. _____
f. _____
g. _____
h. _____

Secondary growth: lateral meristems add girth by producing secondary vascular tissue and periderm (734–738)

The **secondary plant body** consists of the tissues produced by secondary growth. **Vascular cambium** is the lateral meristem that produces secondary xylem and phloem, and **cork cambium** produces a tough covering for stems and roots. Secondary growth occurs in all gymnosperms and most dicots.

■ **INTERACTIVE QUESTION 35.6**

Return to the table in Interactive Question 35.1 on page 272 to add anatomical differences between monocot and dicot roots and stems.

Secondary Growth of Stems   Vascular cambium forms a continuous cylinder from a band of parenchyma cells that become meristematic, located between the primary xylem and phloem of each vascular bundle, and in the rays of ground tissue between bundles. **Ray initials** produce radial lines of parenchyma cells called xylem rays and phloem rays, which function in storage and lateral transport of water and nutrients. **Fusiform initials** produce secondary xylem to the inside and secondary phloem to the outside.

Wood is the accumulation of secondary xylem cells with thick, lignified walls. Annual growth rings result from the seasonal cycle of cambium dormancy, early wood production (usually in the spring), and late wood production (in the summer) in temperate regions.

The epidermis splits off during secondary growth and is replaced by new protective tissues produced by the cork cambium, a meristematic cylinder that first forms in the outer cortex and later in the secondary phloem. The cork cambium produces cork cells with suberin-impregnated walls. The protective coat formed by the layers of cork and cork cambium is called the **periderm. Lenticels** are splits in the periderm through which gas exchange occurs. **Bark** refers to phloem and periderm. As secondary growth continually splits the periderm, new cork cambia develop, eventually forming from parenchyma cells in the secondary phloem.

In older trees, a central supportive column of heartwood consists of older xylem with resin-filled cell cavities; the sapwood consists of secondary xylem that still functions in transport.

■ **INTERACTIVE QUESTION 35.7**

Starting from the outside, place the letters of the tissues in the order in which they are located in a woody tree trunk.

___ ___ ___ ___ ___ ___ ___ ___

A. primary phloem   E. pith

B. secondary phloem   F. cork cambium

C. primary xylem   G. vascular cambium

D. secondary xylem   H. cork cells

Secondary Growth of Roots   The two lateral meristems also function in the secondary growth of roots. Vascular cambium produces xylem internal and phloem external to itself, and a cork cambium forms from the pericycle and produces the periderm. Older roots resemble old stems, with annual rings and a thick, tough bark.

Mechanisms of Plant Growth and Development

Molecular biology is revolutionizing the study of plants (738–739)

Modern plant biology is using new methods and new research organisms, such as *Arabidopsis thaliana*, to study the genetic control of plant development. The entire genome of *Arabidopsis* has been sequenced, and researchers are working to determine the functions of all of its 25,000 (15,000 different) genes.

Growth, morphogenesis, and differentiation produce the plant body (739)

Growth, or the increase in mass, results from cell division and expansion. The organization of cells into organs and tissues is called **morphogenesis,** the development of body form. **Differentiation** creates cells with specific structural and functional features. The developmental phase of a plant, such as vegetative or reproductive, affects the growth, morphogenesis, and differentiation of its parts.

Growth involves both cell division and cell expansion (739–742)

The Plane and Symmetry of Cell Division   The plane and symmetry of plant cell division is important in determining form. In **asymmetrical cell division,** one daughter cell receives more cytoplasm than the other, often leading to a key developmental event.

The plane of cell division is determined by the **preprophase band,** a ring of cytoskeletal microtubules that forms during late interphase. The microtubules disperse, but leave behind an ordered array of actin microfilaments that first orient the nucleus and later direct the movement of the vesicles that form the cell plate.

The Orientation of Cell Expansion   About 90% of plant cell growth is due to the uptake of water. When enzymes break the cross-links in the cell wall, the turgor pressure is reduced and water enters the cell by osmosis. This economical means of cell elongation produces the rapid growth of shoots and roots.

Cells expand in a direction perpendicular to the orientation of the cellulose microfibrils in the inner layers of the cell wall. The orientation of cellulose microfibrils parallels the orientation of microtubules on the other side of the plasma membrane. These microtubules may control the movement of cellulose-producing enzymes and thus determine the alignment of microfibrils in the wall.

The Importance of Cortical Microtubules in Plant Growth In the squat-bodied *fass* mutants of *Arabidopsis*, preprophase bands are not formed in mitosis and cortical microtubules are randomly distributed, resulting in cells that divide in a random arrangement and expand equally in all directions.

■ **INTERACTIVE QUESTION 35.8**

Review the role of microtubules in the orientation of plant cell division and expansion.

Morphogenesis depends on pattern formation (742)

Pattern formation is the characteristic development of structures in specific locations. Pattern formation appears to depend on **positional information,** signals that indicate a cell's location within an embryonic structure or developing organ and thus direct its growth and differentiation. An embryonic cell may detect its location by gradients of molecules, probably proteins, that diffuse from specific locations in a developing structure.

Plants typically have an axial **polarity** with a root end and a shoot end. The asymmetric first division of the zygote establishes the shoot–root polarity.

Plants have homeotic genes that regulate key developmental events. In many plants, the *KNOTTED-1* homeotic gene influences leaf morphology. Its overexpression in tomato plants results in "super-compound" leaves.

Cellular differentiation depends on the control of gene expression (743)

Differential gene expression within genetically identical cells leads to their differentiation into the diverse cell types in a plant. In the root epidermis of *Arabidopsis*, the homeotic gene *GLABRA2* is expressed in epidermal cells that are in contact with only one underlying cortical cell, and these cells do not develop root hairs. The gene is not expressed in those cells in contact with two cortical cells, and they differentiate into root hair cells.

Clonal analysis of the shoot apex emphasizes the importance of a cell's location in its developmental fate (743–744)

Using clonal analysis, researchers are able to identify cells in the apical meristem and follow their lineages to pinpoint when the developmental fate of a cell becomes determined. Apparently the cells of the meristem are not committed to the formation of specific organs and tissues. A cell's final position in a developing organ determines what type of cell it will become.

Phase changes mark major shifts in development (744)

In a process known as a **phase change,** the apical meristem can switch from one developmental phase to another, such as from producing juvenile vegetative nodes and internodes to laying down mature modules. A module's developmental phase is fixed when it is produced, and thus both juvenile and mature regions can coexist along the shoots of a plant.

Genes controlling transcription play key roles in a meristem's change from a vegetative to a floral phase (744–745)

In the transition from a vegetative shoot tip to a floral meristem, **meristem identity genes** are expressed that code for transcription factors that activate the genes necessary for floral meristem development.

Some of the **organ identity genes** that determine floral pattern have been identified, and their expression seems to be influenced by positional information. The transcription factors they code for then influence the transcription of those genes that control development of a specific organ in its proper location. Mutations of organ-identity genes result in the placement of one type of floral organ where another type would normally develop.

WORD ROOTS

apic- = the tip; **meristo-** = divided (*apical meristems:* embryonic plant tissue on the tips of roots and in the buds of shoots that supplies cells for the plant to grow)

a- = not, without; **-symmetr** = symmetrical (*asymmetric cell division:* cell division in which one daughter cell receives more cytoplasm than the other during mitosis)

bienn- = every 2 years (*biennial:* a plant that requires two years to complete its life cycle)

coll- = glue; **-enchyma** = an infusion (*collenchyma cell:* a flexible plant cell type that occurs in strands or cylinders that support young parts of the plant without restraining growth)

endo- = inner; **derm-** = skin (*endodermis:* the innermost layer of the cortex in plants roots)

epi- = over (*epidermis:* the dermal tissue system in plants; the outer covering of animals)

fusi- = a spindle (*fusiform initials:* the cambium cells within the vascular bundles; the name refers to the tapered ends of these elongated cells)

inter- = between (*internode:* the segment of a plant stem between the points where leaves are attached)

meso- = middle; **-phyll** = a leaf (*mesophyll:* the ground tissue of a leaf, sandwiched between the upper and lower epidermis and specialized for photosynthesis)

morpho- = form; **-genesis** = origin (*morphogenesis:* the development of body shape and organization during ontogeny)

perenni- = through the year (*perennial:* a plant that lives for many years)

peri- = around; **-cycle** = a circle (*pericycle:* a layer of cells just inside the endodermis of a root that may become meristematic and begin dividing again)

phloe- = the bark of a tree (*phloem:* the portion of the vascular system in plants consisting of living cells arranged into elongated tubes that transport sugar and other organic nutrients throughout the plant)

pro- = before (*procambium:* a primary meristem of roots and shoots that forms the vascular tissue)

proto- = first; **-plast** = formed, molded (*protoplast:* the contents of a plant cell exclusive of the cell wall)

quiesc- = quiet, resting (*quiescent center:* a region located within the zone of cell division in plant roots, containing meristematic cells that divide very slowly)

sclero- = hard (*sclereid:* a short, irregular sclerenchyma cell in nutshells and seed coats and scattered through the parenchyma of some plants)

trachei- = the windpipe (*tracheids:* a water-conducting and supportive element of xylem composed of long, thin cells with tapered ends and walls hardened with lignin)

trans- = across (*transpiration:* the evaporative loss of water from a plant)

vascula- = a little vessel (*vascular tissue:* plant tissue consisting of cells joined into tubes that transport water and nutrients throughout the plant body)

xyl- = wood (*xylem:* the tube-shaped, nonliving portion of the vascular system in plants that carries water and minerals from the roots to the rest of the plant)

STRUCTURE YOUR KNOWLEDGE

1. Fill in the following diagram of the meristems and the primary and secondary tissues they produce in a woody stem.

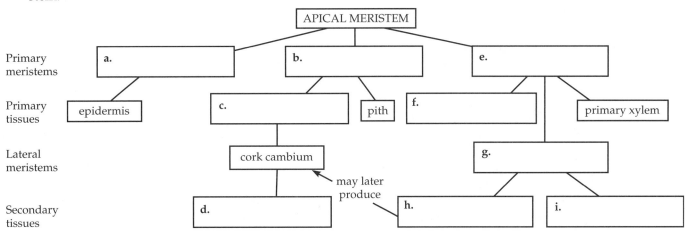

2. In this cross section of a young stem, label the indicated structures. Is this a stem of a monocot or a dicot? How can you tell?

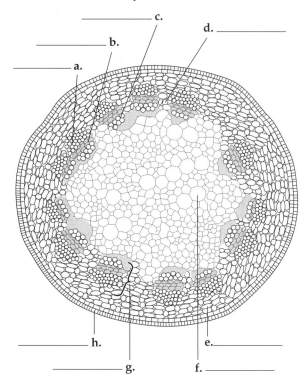

TEST YOUR KNOWLEDGE

MATCHING: *Match the plant tissue with its description.*

_____	**1.** sclerenchyma	**A.**	layer from which lateral roots originate
_____	**2.** collenchyma	**B.**	tapered xylem cells with lignin in cell walls
_____	**3.** tracheids	**C.**	parenchyma cells with chloroplasts in leaves
_____	**4.** fibers	**D.**	protective coat made of cork and cork cambium
_____	**5.** fusiform initials	**E.**	bundles of long sclerenchyma cells
_____	**6.** pericycle	**F.**	supporting cells with thickened primary walls
_____	**7.** mesophyll	**G.**	parenchyma cells inside vascular ring in dicot stem
_____	**8.** periderm	**H.**	supporting cells with thick secondary walls
_____	**9.** endodermis	**I.**	cambium cells that produce secondary xylem and phloem
_____	**10.** pith	**J.**	cell layer in root regulating movement into stele

MULTIPLE CHOICE: *Choose the one best answer.*

1. Which of the following is *incorrect?* Monocots typically have
   a. a taproot rather than a fibrous root system.
   b. leaves with parallel veins rather than branching venation.
   c. no secondary growth.
   d. scattered vascular bundles in the stem rather than bundles in a ring.
   e. pith in the center of the vascular stele in the root.

2. Which of the following is *incorrectly* paired with its function?
   a. ray initials—form radial xylem and phloem rays
   b. lenticels—gas exchange in woody stem
   c. root hairs—absorption of water and dissolved minerals
   d. root cap—protects root as it pushes through soil
   e. procambium—meristematic tissue that forms protective layer of cork

3. Axillary buds
   a. may exhibit apical dominance over the terminal bud.
   b. form at nodes in the angle where leaves join the stem.
   c. grow out from the pericycle layer.
   d. are formed from intercalary meristems.
   e. only develop into vegetative shoots.

4. A leaf trace is
   a. a petiole.
   b. the outline of the vascular bundles in a leaf.
   c. a branch from a vascular bundle that extends into a leaf.
   d. a tiny bulge on the flank of the apical dome that grows into a leaf.
   e. a system of plant identification based on leaf morphology.

5. Ground meristem
   a. produces the root system.
   b. produces the ground tissue system.
   c. produces secondary growth.
   d. is meristematic tissue found at the nodes in monocots.
   e. develops from protoderm.

6. The zone of cell elongation
   a. is responsible for pushing a root through the soil.
   b. has a quiescent center that can undergo mitosis should the apical meristem be destroyed.
   c. produces the protoderm, procambium, and ground meristem tissues.

d. is further from the root tip than the zone of maturation.

e. is or does all of the above.

7. Which of the following is *incorrectly* paired with its type of life cycle?

a. pine tree—perennial

b. rose bush—perennial

c. carrot—biennial

d. marigold—annual

e. wheat—biennial

8. Secondary xylem and phloem are produced in a root by the

a. pericycle.

b. endodermis.

c. vascular cambium.

d. apical meristem.

e. ray initials.

9. Bark consists of

a. secondary phloem.

b. periderm.

c. cork cells.

d. cork cambium.

e. all of the above.

10. Sieve-tube members

a. are responsible for lateral transport through a woody stem.

b. control the activities of phloem cells that have no nuclei or ribosomes.

c. have spiral thickenings that allow the cell to elongate along with a young shoot.

d. are transport cells with sieve plates in the end walls between cells.

e. are tapered water transport cells with pits.

11. Which of the following cells are dead at functional maturity?

a. tracheids

b. cork cells

c. vessel elements

d. sclerenchyma cells

e. all of the above

12. Clonal analysis of cells of the shoot apex indicates that

a. organ-identity gene mutations substitute one body part for another.

b. a cell's developmental fate is more influenced by position effects than by its meristematic lineage.

c. cellular differentiation results from regulation of gene expression resulting in production of different proteins in different cells.

d. all cells were derived from the same parent cell.

e. the apical meristem of the root tip produces the protoderm, ground meristem, and procambium.

13. In what direction does a plant cell enlarge?

a. toward the basal end as a result of positional information

b. parallel to the orientation of the preprophase band of microtubules

c. perpendicular to the orientation of cellulose microfibrils in the cell wall

d. in the direction from which water flows into the cell

e. perpendicular to the internodes

14. What do organ identity genes code for?

a. the three primary meristems

b. transcription factors for fruit-ripening enzymes

c. transcription factors that control development of floral organs

d. signals that change a vegetative shoot into a floral meristem

e. the root and shoot meristems

15. What is a usual sign of the change of the apical meristem from the juvenile to the mature phase?

a. the production of a flower

b. the initiation of secondary growth

c. the production of longer internodes

d. the activation of axillary buds

e. a change in the morphology of leaves produced at nodes

16. Which of the following is essential to establishing the axial polarity of a plant?

a. expression of different homeotic genes in the shoot and root meristems

b. orientation of cortical microtubules perpendicular to the ground

c. the expression of the *GLABRA2* gene in the shoot but not the root

d. the asymmetric first division of the zygote

e. the proper orientation of microtubules that control the movement of cellulose-producing enzymes

CHAPTER 36

TRANSPORT IN PLANTS

FRAMEWORK

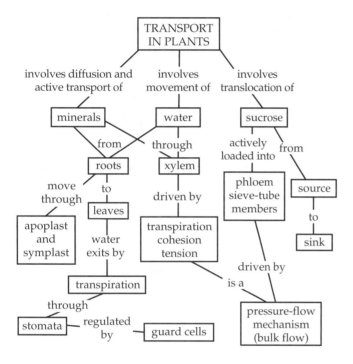

CHAPTER REVIEW

The move onto land involved the evolutionary specialization of roots for the uptake of water and minerals, shoots for the uptake of CO_2 and exposure to light, and vascular tissue for the transport of materials between roots and shoots.

An Overview of Transport Mechanisms in Plants

Transport in plants involves the absorption and loss of water and solutes by individual cells, the short-distance movement of substances from cell to cell within tissues, and the long-distance transport of sap in xylem and phloem.

Transport at the cellular level depends on the selective permeability of membranes (749)

Solutes may move across the selectively permeable plasma membrane by passive transport when they diffuse down their concentration gradients. **Transport proteins** speed passive transport and may be either specific carrier proteins, which selectively bind and transport a solute, or **selective channels,** which allow specific solutes to cross the membrane. Some channels are gated: They open and close in response to certain stimuli.

Active transport requires the cell to expend energy to move a solute against its electrochemical gradient.

Proton pumps play a central role in transport across plant membranes (749–750)

A plant **proton pump** uses ATP to pump hydrogen ions out of the cell, generating an energy-storing proton gradient and a membrane potential due to the separation of charges. The proton gradient and membrane potential are used by plant cells to drive the transport of many solutes. The membrane potential helps drive positively charged ions down their electrochemical gradient and into the negatively charged cell. In **cotransport,** a solute can be pumped against its concentration gradient when a transport protein links its passage to the "downhill" movement of H^+. **Chemiosmosis** links energy-releasing and energy-consuming processes using a transmembrane proton gradient like that created by the proton pump.

Differences in water potential drive water transport in plant cells (750–752)

The passive transport of water across a membrane is called **osmosis.** In animal cells, water moves by osmosis in the hypotonic → hypertonic direction. **Water potential,** designated by Ψ *(psi),* is a useful measurement for predicting the direction that water will move when a plant cell is surrounded by a particular solution. It takes into account both solute concentration and the physical pressure exerted by the plant cell wall. Water will flow from a region of higher water potential to one of lower water potential. Water potential is measured in **megapascals (MPa);** 1 MPa is equal to about 10 atmospheres of pressure.

How Solutes and Pressure Affect Water Potential
The water potential of pure water in an open container is 0 MPa. Adding solutes lowers water potential; therefore, a solution at atmospheric pressure has a negative water potential. The addition of pressure increases water potential because it increases the potential for water to move. A **tension,** or negative pressure, will lower Ψ.

Quantitative Analysis of Water Potential The combined effect of pressure and solute concentration is shown by the equation for water potential: $\Psi = \Psi_P + \Psi_S$ (Ψ_P = pressure potential, Ψ_S = solute potential or osmotic potential). Ψ_P can be a positive or negative value, whereas Ψ_S is always a negative number.

A flaccid plant cell bathed in a solution more concentrated than the cell will lose water by osmosis because the solution has a lower Ψ. When bathed in pure water, the cell has the lower Ψ. Water will enter the cell until enough **turgor pressure** builds up so that Ψ_P and Ψ_S are equal and opposite in magnitude, and $\Psi = 0$ both inside and outside the cell. Net movement of water will then stop.

A **flaccid** cell has no pressure potential. Plant cells are usually **turgid;** they have a greater solute concentration than their extracellular environment and turgor pressure keeps them firm.

■ **INTERACTIVE QUESTION 36.1**

a. A flaccid plant cell has a water potential of -0.6 MPa. Fill in the water potential equation for this cell.

$$\Psi_P =$$
$$+\Psi_S =$$
$$\overline{\Psi \ =}$$

b. The cell is then placed in a beaker of distilled water ($\Psi = 0$). What happens to the water potential of this cell? Fill in the equation for the cell after it is bathed in pure water.

$$\Psi_P =$$
$$+\Psi_S =$$
$$\overline{\Psi \ =}$$

c. What would happen to the same cell if it is placed in a solution that has a water potential of -0.8 MPa? Fill in the equation for the cell after it is moved to this more concentrated solution and reaches equilibrium.

$$\Psi_P =$$
$$+\Psi_S =$$
$$\overline{\Psi \ =}$$

Aquaporins affect the rate of water transport across membranes (752–753)

Although small water molecules can move relatively freely across membranes, the rate at which they move is too specific and rapid to be attributed entirely to diffusion. Water-specific transport proteins called **aquaporins,** now identified in both plant and animal membranes, increase the rate of water diffusion.

Vacuolated plant cells have three major compartments (753–754)

The three compartments of most mature plant cells are the cell wall, cytosol, and vacuole. The plasma membrane regulates traffic between the cell wall and cytosol.

The **tonoplast,** the membrane of the central vacuole, regulates solute movement between the cytosol and the cell sap of the vacuole. Its proton pumps move H^+ from the cytoplasm into the vacuole, helping to maintain a low cytosolic H^+ concentration.

The cytosolic compartments of plant cells are connected by plasmodesmata, forming a cytoplasmic continuum called the **symplast.** The continuum of cell walls within a plant tissue is called the **apoplast.**

Both the symplast and the apoplast function in transport within tissues and organs (754)

Short-distance or lateral transport of water and solutes within plant tissues can occur by three routes: transmembrane, by crossing plasma membranes and cell walls; symplastic, moving through plasmodesmata; and apoplastic, the extracellular pathway along cell walls and extracellular spaces. Solutes and water may change routes during transit.

Bulk flow functions in long-distance transport (754)

Long-distance transport throughout plants occurs by **bulk flow,** the movement of fluid driven by pressure. Water and minerals move through xylem vessels as the result of tension created by transpiration. Sap is forced through phloem sieve tubes by hydrostatic pressure.

Sieve-tube members lack most cellular organelles, and vessel elements and tracheids are dead (and empty) at maturity, both facilitating more efficient bulk flow.

Absorption of Water and Minerals by Roots

Root hairs, mycorrhizae, and a large surface area of cortical cells enhance water and mineral absorption (754–756)

Much of the absorption of water and minerals occurs along young root tips where root hairs are located.

The soil solution soaks into the hydrophilic walls of epidermal cells and moves along the apoplast into the root cortex, exposing a large surface area of plasma membrane for the uptake of water and minerals. Selective absorption of minerals, using transport proteins in the plasma membrane and tonoplast, allows cells to accumulate essential minerals. **Mycorrhizae,** symbiotic associations of plant roots and fungal hyphae, greatly increase the surface area for absorption of water and selected minerals.

The endodermis functions as a selective sentry between the root cortex and vascular tissue (765)

The **endodermis,** the innermost layer of cortex cells surrounding the stele, selectively screens all minerals entering the vascular tissue. A ring of suberin around each endodermal cell, called the **Casparian strip,** prevents water from the apoplast from entering the stele without passing through the selectively permeable plasma membrane of an endodermal cell. Water and minerals that had already entered the symplast through a cortex or epidermal cell pass through plasmodesmata of endodermal cells into the stele.

By a combination of diffusion and active transport, minerals move from the symplast to the apoplast and enter the nonliving xylem tracheids and vessel elements along with water.

■ INTERACTIVE QUESTION   36.2

Label the diagram of the cell layers and routes of transport of water and minerals from the soil through the root. (**a** and **b** refer to routes of transport; letters **c–i** refer to cell layers or structures)

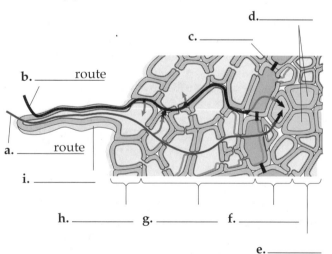

Transport of Xylem Sap

The branching of xylem veins provides water to the cells of each leaf. Through **transpiration,** the loss of water vapor from leaves, plants lose a tremendous amount of water that must be replaced by water transported up from the roots.

The ascent of xylem sap depends mainly on transpiration and the physical properties of water (756–758)

Pushing Xylem Sap: Root Pressure   As minerals are actively pumped into the stele and prevented from leaking out by the endodermis, water potential in the stele is lowered. Water flows in from the cortex by osmosis, resulting in **root pressure,** which forces fluid up the xylem. Root pressure may cause **guttation,** the exudation of water droplets from leaves when more water is forced up the xylem than is transpired by the plant.

Pulling Xylem Sap: The Transpiration-Cohesion-Tension Mechanism   By transpiration, water vapor from saturated air spaces within a leaf exits to drier air outside the leaf by way of stomata. The thin layer of water that coats the mesophyll cells lining the air spaces begins to evaporate. The adhesion of the remaining water to the hydrophilic walls and the cohesion between water molecules causes menisci to form that increase the tension of this water layer. This negative pressure pulls water from the xylem and through the apoplast and symplast of the mesophyll to the cells and surface film lining the air spaces. Water moves along a gradient of decreasing water potential from xylem to neighboring cells to air spaces to the drier air outside the leaf.

The transpirational pull on xylem sap is transmitted from the leaves to the root tips by the cohesiveness of water that results from hydrogen bonding between molecules. Adhesion of water molecules to the hydrophilic walls of the narrow xylem elements and tracheids also contributes to overcoming the downward pull of gravity.

The upward transpirational pull on the cohesive sap creates tension within the xylem, further decreasing water potential so that water flows passively from the soil, across the cortex, and into the stele.

A break in the chain of water molecules by the formation of a water vapor pocket in a xylem vessel, called cavitation, breaks the transpirational pull, and the vessel cannot function in transport.

Xylem sap ascends by solar-powered bulk flow: *a review* (758)

The transpiration-cohesion-tension mechanism results in the bulk flow of water from roots to leaves. Water potential differences caused by solute concentration and pressure contribute to the movement of water from cell to cell, but solar-powered tension caused by transpiration is responsible for long-distance transport of water and minerals.

■ INTERACTIVE QUESTION 36.3

Explain the contribution of each of the following to the long-distance transport of water.

a. Transpiration:

b. Cohesion:

c. Adhesion:

d. Tension:

The Control of Transpiration

Guard cells mediate the photosynthesis-transpiration compromise (759–761)

The Photosynthesis-Transpiration Compromise A plant's tremendous requirement for water is partly a consequence of making food by photosynthesis. To obtain sufficient CO_2 for photosynthesis, leaves must exchange gases through the stomata and provide a large internal surface area for CO_2 uptake, but also from which water may evaporate.

A common **transpiration-to-photosynthesis ratio** is 600 grams of water transpired for each gram of CO_2 incorporated into carbohydrate (600:1). Plants such as corn that use the more efficient C_4 pathway for photosynthesis have ratios of 300:1 or less.

Transpiration assists in transferring minerals and other substances to the leaves and results in evaporative cooling. When transpiration exceeds the water available, leaves wilt as cells lose turgor pressure.

How Stomata Open and Close Guard cells regulate the size of stomatal openings and control the rate of transpiration. When the kidney-shaped guard cells of dicots become turgid and swell, their radially oriented microfibrils cause them to buckle outward and increase the size of the gap between them. When the guard cells become flaccid, they sag and close the space. The dumbbell-shaped monocot guard cells operate by the same basic mechanism.

Guard cells can lower their water potential by actively accumulating potassium ions (K^+), which leads to osmosis and the resulting increase in turgor pressure. The exodus of K^+ leads to a loss of turgor. The regulation of aquaporins may also vary the membrane's permeability to water.

The movement of K^+ across the guard cell membrane is probably coupled with the generation of membrane potentials by proton pumps that transport H^+ out of the cell. A technique called patch clamping allows plant physiologists to study proton pumps and K^+ channels in guard cells; a small patch of membrane is held across the opening of a micropipette while an electrode records ion fluxes across the membrane.

The opening of stomata at dawn is related to at least three factors: Light stimulates guard cells to accumulate K^+, perhaps triggered by the illumination of blue-light receptors which activate the proton pumps of the plasma membrane. Light also drives photosynthesis in the guard cell chloroplasts, making ATP available for proton pumping. Second, stomata are stimulated to open when CO_2 within air spaces of the leaf is depleted as photosynthesis begins in the mesophyll. The third factor is a daily rhythm of opening and closing that is endogenous to guard cells. Cycles that have intervals of approximately 24 hours are called **circadian rhythms.**

Environmental stress can cause stomata to close during the day. Guard cells lose turgor when water is in short supply. A hormone called abscisic acid, produced by mesophyll cells in response to a lack of water, signals guard cells to close stomata. High temperatures induce closing, probably because increasing cellular respiration raises CO_2 concentrations in the air spaces in the leaf.

■ INTERACTIVE QUESTION 36.4

How are proton pumps involved in stomatal opening?

Xerophytes have evolutionary adaptations that reduce transpiration (762)

Many **xerophytes,** plants adapted to arid climates, have leaves that are small and thick, limiting water loss by reducing their surface-to-volume ratio. Leaf cuticles may be thick, and the stomata may be sheltered in depressions. Some desert plants lose their leaves in the driest months.

Succulent plants of the family Crassulaceae and some other plant families assimilate CO_2 into organic acids during the night by a pathway known as CAM (crassulacean acid metabolism) and then release it for photosynthesis during the day. Thus, the stomata are open at night and are closed during the day when water loss would be greatest.

Translocation of Phloem Sap

Translocation, the transport of photosynthetic products throughout the plant, occurs in the end-to-end sieve-tube members as sap flows through porous sieve plates. Phloem sap may have a sucrose concentration as high as 30% and may also contain minerals, amino acids, and hormones.

Phloem translocates its sap from sugar sources to sugar sinks (762–763)

Phloem sap flows from a **sugar source,** where it is produced by photosynthesis or the breakdown of starch, to a **sugar sink,** an organ that consumes or stores sugar. The direction of transport in any one sieve tube depends on the location of the source and sink connected by that tube, and the direction may change with the season or needs of the plant.

■ INTERACTIVE QUESTION 36.5

Explain how a root or tuber can serve as both a sugar source and a sugar sink.

Phloem Loading and Unloading   In some species, sugar in the leaf moves through the symplast of the mesophyll cells to sieve-tube members. In other species, sugar first moves through the symplast and then into the apoplast in the vicinity of sieve-tube members and companion cells, which actively accumulate sugar. In some plants, companion cells are specialized as **transfer cells** with ingrowths of their wall that increase surface area for movement of solutes from apoplast to symplast.

Phloem loading requires active transport in plants that accumulate high sugar concentration in the sieve tubes. Proton pumps and the cotransport of sucrose through membrane proteins along with the returning protons is the mechanism used for active transport. Sugar is moved by various mechanisms out of sieve tubes at the sink end. The concentration gradient favors this movement because sugar is either being used or converted into starch within sink cells.

■ INTERACTIVE QUESTION 36.6

Label the components of this diagram of the chemiosmotic mechanism used to actively transport sucrose into companion cells or sieve-tube members.

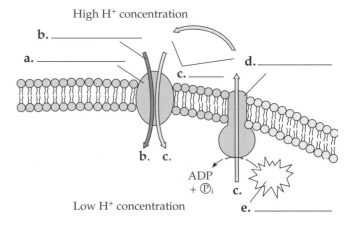

High H^+ concentration

b. _____

a. _____

c. _____

d. _____

b.   c.

ADP
+ $\circled{P}_i$

c. _____

Low H^+ concentration

e. _____

Pressure flow is the mechanism of translocation in angiosperms (763–764)

The rapid movement of phloem sap from source to sink is due to a pressure flow mechanism. High solute concentration at the source lowers water potential, and the resulting movement of water into the sieve tube produces hydrostatic pressure. At the sink end, the osmotic loss of water following the exodus of sucrose into the surrounding tissue results in a lower hydrostatic pressure. The difference in these pressures causes sap to move by bulk flow from source to sink; xylem recycles water from sink to source. Innovative tests of this model support it as the explanation for the flow of sap in the phloem of angiosperms.

WORD ROOTS

apo- = off, away; **-plast** = formed, molded (*apoplast:* in plants, the nonliving continuum formed by the extracellular pathway provided by the continuous matrix of cell walls)

aqua- = water; **-pori** = a pore, small opening (*aquaporin:* a transport protein in the plasma membranes of a plant or animal cell that specifically facilitates the diffusion of water across the membrane)

chemo- = chemical (*chemiosmosis:* the production of ATP using the energy of hydrogen-ion gradients across membranes to phosphorylate ADP)

circa- = a circle (*circadian rhythm:* a physiological cycle of about 24 hours, present in all eukaryotic organisms, that persists even in the absence of external cues)

co- = together; **trans-** = across; **-port** = a gate, door (*cotransport:* the coupling of the "downhill" diffusion of one substance to the "uphill" transport of another against its own concentration gradient)

endo- = within, inner; **-derm** = skin (*endodermis:* the innermost layer of the cortex in plant roots)

gutt- = a drop (*guttation:* the exudation of water droplets caused by root pressure in certain plants)

mega- = large, great (*megapascal:* a unit of pressure equivalent to 10 atmospheres of pressure)

myco- = a fungus; **-rhizo** = a root (*mycorrhizae:* mutualistic associations of plant roots and fungi)

osmo- = pushing (*osmosis:* the diffusion of water across a selectively permeable membrane)

sym- = with, together (*symplast:* in plants, the continuum of cytoplasm connected by plasmodesmata between cells)

turg- = swollen (*turgor pressure:* the force directed against a cell wall after the influx of water and the swelling of a walled cell due to osmosis)

xero- = dry; **-phyto** = a plant (*xerophytes:* plants adapted to arid climates)

STRUCTURE YOUR KNOWLEDGE

1. Describe the ways in which solutes may move across the plasma membrane in plants.

2. Both xylem sap and phloem sap move by bulk flow in an angiosperm. Compare and contrast the mechanisms for their movement.

TEST YOUR KNOWLEDGE

MULTIPLE CHOICE: *Choose the one best answer.*

1. Which of the following is *not* a component of the symplast?
   a. sieve-tube members
   b. xylem tracheids
   c. endodermal cells
   d. cortex cells
   e. companion cells

2. Proton pumps in the plasma membranes of plant cells may
   a. generate a membrane potential that helps drive cations into the cell through their specific carriers.
   b. be coupled to the movement of K^+ into guard cells.
   c. drive the accumulation of sucrose in sieve-tube members.
   d. contribute to the movement of anions through a cotransport mechanism.
   e. be involved in all of the above.

3. The Casparian strip prevents water and minerals from entering the stele through the
   a. plasmodesmata.
   b. endodermal cells.
   c. symplast.
   d. apoplast.
   e. xylem vessels.

4. The water potential of a plant cell
   a. is equal to 0 when the cell is in pure water and is turgid.
   b. is equal to that of air.
   c. is equal to −0.23 MPa.
   d. becomes greater when K^+ ions are actively moved into the cell.
   e. becomes 0 due to loss of turgor pressure in a hypertonic solution.

5. Guttation results from
   a. the pressure flow of sap through phloem.
   b. a water vapor break in the column of xylem sap.
   c. root pressure causing water to flow up through xylem faster than it can be lost by transpiration.
   d. a higher water potential of the leaves than of the roots.
   e. specialized structures in transport cells that accumulate sucrose.

6. Which of these is *not* a major factor in the movement of xylem sap up a tall tree?
   a. transpiration
   b. plasmodesmata
   c. adhesion
   d. cohesion
   e. tension

7. Adhesion is a result of
   a. hydrogen bonding between water molecules.
   b. the pull on the water column as water evaporates from the surfaces of mesophyll cells.
   c. tension within the xylem caused by a lowered water potential.
   d. attraction of water molecules to hydrophilic walls of narrow xylem tubes.
   e. the high surface tension of water.

8. A plant with a low transpiration-to-photosynthesis ratio
   a. would lose more water through transpiration for each gram of CO_2 fixed than would a plant with a high ratio.
   b. could be a C_4 plant.
   c. could be a CAM plant.
   d. b and c are correct.
   e. a, b, and c are correct.

9. Which of these does *not* stimulate the opening of stomata?
   a. release of abscisic acid by mesophyll cells
   b. depletion of CO_2 in the air spaces of the leaf
   c. stimulation of proton pumps that results in the movement of K^+ into the guard cells
   d. the circadian rhythm of the guard cell opening
   e. an increase in the turgor of guard cells

10. Your favorite spider plant is wilting. What is the most likely cause and remedy for its declining condition?
    a. Water potential is too low; apply sugar water.
    b. The stomata won't open; no remedy available.
    c. Plasmolysis of its cells; water the plant.
    d. Cavitation; perform a xylem vessel bypass.
    e. Circadian rhythm has stomata closed; place it in bright light.

11. A turgid plant cell placed in a solution in an open beaker becomes flaccid.
    a. The water potential of the cell was higher than that of the solution.
    b. The water potential of the cell was equal to that of the solution.
    c. The pressure (Ψ_P) of the cell was initially lower than that of the solution.
    d. The cell was hypertonic to the solution.
    e. Turgor pressure disappears since the cell no longer needs support.

12. What facilitates the movement of K^+ into epidermal cells of the root?
    a. cotransport through a membrane protein
    b. bulk flow of water into the root
    c. passage through selective channels, aided by the membrane potential created by proton pumps
    d. active transport through a potassium pump
    e. simple diffusion across the cell membrane down its concentration gradient

13. What are aquaporins?
    a. cytoplasmic connections between cortical cells
    b. pores through the ends of sieve-tube members through which phloem sap flows
    c. openings in the lower epidermis of leaves through which water vapor escapes
    d. openings into root hairs through which water enters
    e. water-specific channels in membranes that may regulate the rate of osmosis

14. The formation of a meniscus along the cell walls surrounding the air space of a leaf contributes to water transport by
    a. creating a more positive water potential than in the surrounding mesophyll cells.
    b. creating tension, thus lowering pressure and the water potential of the leaf.
    c. raising the water potential of the surrounding saturated air.
    d. increasing the adhesion of water molecules to the cell walls.
    e. increasing the rate of transpiration from the leaf.

15. The technique of patch clamping can be used to study
    a. the speed of phloem sap flow.
    b. the opening and closing of stomata when a plant is kept in a constant dark environment.
    c. ion flow across membranes.
    d. the water potential of leaves as measured by the amount of pressure needed to push water back to the cut surface of a branch.
    e. the rate of transpiration from a leaf.

16. All of the following increase the surface area available for absorption of water and minerals by a root *except*
    a. mycorrhizae.
    b. numerous branch roots.
    c. root hairs.
    d. cytoplasmic extensions of the endodermis.
    e. the large surface area of cortical cells.

17. By what method do most mineral anions (negatively charged ions) enter root cells?
    a. apoplastic route
    b. symplastic route
    c. diffusion
    d. cotransport using a proton gradient
    e. bulk flow

18. What mechanism explains the movement of sucrose from source to sink?
    a. evaporation of water and active transport of sucrose from the sink
    b. osmotic movement of water into the sucrose-loaded sieve-tube members creating a higher hydrostatic pressure in the source than in the sink
    c. tension created by the differences in hydrostatic pressure in the source and sink
    d. active transport of sucrose through the sieve-tube cells driven by proton pumps
    e. the hydrolysis of starch to sucrose in the mesophyll cells that raises their water potential and drives the bulk flow of sap to the sink

19. Considering an animal cell (first group) and a plant cell (second group) placed in test solutions, which of the following choices gives the *correct* direction for water flow by osmosis?
    a. hypertonic $\rightarrow$ hypotonic; higher Ψ $\rightarrow$ lower Ψ
    b. hypertonic $\rightarrow$ hypotonic; lower Ψ $\rightarrow$ higher Ψ
    c. hypotonic $\rightarrow$ hypertonic; lower Ψ $\rightarrow$ higher Ψ
    d. hypotonic $\rightarrow$ hypertonic; higher Ψ $\rightarrow$ lower Ψ
    e. One cannot tell unless told the Ψ_p for the plant cell.

20. What is a function of the tonoplast?
    a. regulate movement of solutes between cells joined by plasmodesmata
    b. help move water and minerals past the Casparian strip into the stele
    c. help maintain low cytosolic H^+ concentration by pumping H^+ into the vacuole
    d. increase the surface area for pumping H^+ out of transfer cells
    e. control the expansion of turgid guard cells so that the space between them opens up

CHAPTER 37

PLANT NUTRITION

FRAMEWORK

The nutritional requirements of plants include essential macronutrients and micronutrients. Carbon dioxide enters the plant through the leaves, but water and minerals must be absorbed through the roots. Soil fertility is influenced by its texture and composition. Nitrogen assimilation by plants is made possible by the decomposition of humus by microbes and nitrogen fixation by bacteria. Mycorrhizae are important associations between fungi and plant roots that increase mineral and water absorption.

CHAPTER REVIEW

Plants, as photoautotrophs, make their own organic compounds and are critical to the energy flow and chemical cycling of an ecosystem. Roots, shoots, and leaves are structurally adapted to obtain water, minerals, and carbon dioxide from the soil and air.

Nutritional Requirements of Plants

The chemical composition of plants provides clues to nutritional requirements (767–768)

Early scientists speculated on whether soil, water, or the air provides the substance for plant growth. **Mineral nutrients,** essential inorganic ions absorbed from the soil, make only a small contribution to the mass of a plant. Water makes up over three fourths of the weight of a nonwoody plant and supplies most of the hydrogen and some of the oxygen incorporated into organic compounds. More than 90% of the water absorbed is lost by transpiration, however, and most of the water retained is used as a solvent, for growth by cell elongation, or for support through turgor pressure. By weight, CO_2 from the air is the source of most of the organic material of a plant.

Organic substances, most of which are carbohydrates (especially cellulose), make up 95% of the dry weight of plants. Thus, carbon, oxygen, and hydrogen are the most abundant elements. Nitrogen, sulfur, and phosphorus—also ingredients of organic compounds—are relatively abundant.

Plants require nine macronutrients and at least eight micronutrients (768–769)

Essential nutrients are those required for a plant to complete its life cycle from a seed to an adult that produces more seeds. Hydroponic culture has been used to determine which of the mineral elements found in plants are essential nutrients. Seventeen elements have been identified as essential in all plants.

Nine **macronutrients** are required by plants in relatively large amounts and include the six major elements of organic compounds as well as calcium, potassium, and magnesium.

Eight **micronutrients** have been identified as needed by plants in very small amounts, functioning mainly as cofactors of enzymatic reactions.

■ INTERACTIVE QUESTION 37.1

a. What is a function of the macronutrient magnesium in plants?

b. What is a function of the macronutrient phosphorus?

c. What is a function of the micronutrient iron?

The symptoms of a mineral deficiency depend on the function and mobility of the element (769–770)

A mobile nutrient will move to young, growing tissues, so that a deficiency will show up first in older parts of the plant. Symptoms of a mineral deficiency may be distinctive enough for the cause to be diagnosed by a plant physiologist or farmer. Soil and plant analysis can confirm a specific deficiency. Nitrogen, potassium, and phosphorus deficiencies are most common.

■ INTERACTIVE QUESTION 37.2

Where would you expect a deficiency of a relatively immobile element to be seen first?

■ INTERACTIVE QUESTION 37.3

Describe the characteristics of a fertile soil.

The Role of Soil in Plant Nutrition

Soil characteristics are key environmental factors in terrestrial ecosystems (770–772)

Texture and Composition of Soils The formation of soil begins with the weathering of rock and accelerates with the secretion of acids by lichens, fungi, bacteria, and plant roots. **Topsoil** is a mixture of broken down rock, living organisms, and humus (decomposing organic matter). Several other distinct soil layers, or **horizons,** are found under the topsoil layer.

The size of soil particles varies from coarse sand to fine clay. **Loams,** made up of a mixture of sand, silt, and clay, are often the most fertile soils, having enough fine particles to provide a large surface area for retaining water and minerals but enough coarse particles to provide air spaces with oxygen for respiring roots.

The activities of the numerous soil inhabitants, such as bacteria, fungi, algae, other protists, insects, worms, nematodes, and plant roots, affect the physical and chemical properties of soil.

Humus builds a crumbly soil that retains water, provides good aeration of roots, and supplies mineral nutrients.

The Availability of Soil Water and Minerals Water containing dissolved minerals binds to hydrophilic soil particles and is held there in small spaces, available for uptake by plant roots. Positively charged minerals, such as K^+, Ca^{2+}, and Mg^{2+}, adhere to the negatively charged surfaces of finely divided clay particles. Negatively charged minerals, such as nitrate (NO_3^-), phosphate ($H_2PO_4^-$), and sulfate (SO_4^{2-}), tend to leach away more quickly. The release of H^+ and CO_2 by roots facilitates **cation exchange,** in which hydrogen ions displace positively charged mineral ions from the clay particles, making the ions available for absorption.

Soil conservation is one step toward sustainable agriculture (772–774)

Without good soil conservation, agriculture can quickly destroy the fertility of a soil that has built up over centuries. Agriculture diverts essential elements from the chemical cycles when crops are harvested, and many crops use more water than the natural vegetation.

Fertilizers Historically, farmers used manure to fertilize their crops. Today in developed nations, commercially produced fertilizers, usually containing nitrogen, phosphorus, and potassium, are used. Manure, fishmeal, and compost are called organic fertilizers because they contain organic material that is in the process of decomposing. These fertilizers decompose into inorganic nutrients, so that they are taken up by the plant in the same form supplied by commercial fertilizers. Commercial fertilizers may be rapidly leached from the soil, polluting streams and lakes.

The acidity of the soil affects cation exchange and can alter the chemical form of minerals and thus their ability to be absorbed by the plant. Managing the pH of soil is an important aspect of maintaining fertility.

Irrigation Irrigation can make farming possible in arid regions, but it places a huge drain on water resources and raises soil salinity. New methods of irrigation and new varieties of plants that can tolerate less water may reduce some of these problems.

Erosion In the U.S., topsoil from thousands of acres of farmland is lost to water and wind erosion each year. Agricultural use of cover crops, windbreaks, and terracing can minimize erosion.

■ INTERACTIVE QUESTION 37.4

What is **sustainable agriculture?**

Phytoremediation The use of plants to extract heavy metals and other pollutants from contaminated soils is an emerging technology known as **phytoremediation.**

The Special Case of Nitrogen as a Plant Nutrient

Nitrogen is the mineral that usually limits plant growth and crop yields.

The metabolism of soil bacteria makes nitrogen available to plants (774–775)

To be absorbed by plants, nitrogen must be converted to nitrate (NO_3^-) or ammonium (NH_4^+) by the action of microbes, such as ammonifying bacteria, that decompose humus. Some nitrate is lost to the atmosphere by the action of denitrifying bacteria. **Nitrogen-fixing bacteria** convert atmospheric nitrogen into ammonia through the process of **nitrogen fixation.**

Bacteria capable of nitrogen fixation contain **nitrogenase,** an enzyme complex that reduces N_2 by adding H^+ and electrons to form ammonia (NH_3). This process is energetically expensive, and nitrogen-fixing bacteria rely on organic material in soils or symbiotic relationships with plant roots to supply fuel for their cellular respiration. In the soil solution, ammonia forms ammonium, which plants can absorb. Nitrifying bacteria oxidize ammonium, producing nitrate, the form of nitrogen most readily absorbed by roots. Most plants incorporate nitrogen into amino acids or other organic compounds in the roots and then transport it through the xylem to shoots.

Improving the protein yield of crops is a major goal of agricultural research (775)

Protein deficiency is the most common form of human malnutrition. Agricultural research attempts to improve the quality and quantity of proteins in crops. Varieties of corn, wheat, and rice have been developed that are enriched in protein, but they require the addition of large quantities of expensive nitrogen fertilizer. Other strategies for increasing protein yields include improving the productivity of symbiotic nitrogen fixation.

Nutritional Adaptations: Symbiosis of Plants and Soil Microbes

Symbiotic nitrogen fixation results from intricate interactions between roots and bacteria (776–778)

Plants of the legume family have root swellings, called **nodules,** composed of plant cells with nitrogen-fixing bacteria in a form called **bacteroids** contained in vesicles. The plant provides the bacteria with carbohydrates and other organic molecules. The cells of the root nodules use most of the symbiotically fixed nitrogen to make amino acids, which are then transported throughout the plant.

Symbiotic Nitrogen Fixation and Agriculture Nonlegume crops are often rotated with a legume crop, which may be plowed under as "green manure." Legume seeds may be soaked or dusted with their specific *Rhizobium* before planting. Rice farmers culture a water fern that has symbiotic cyanobacteria that fix nitrogen, improving the fertility of rice paddies.

■ **INTERACTIVE QUESTION 37.5**

Fill in the types of bacteria (a–d) that participate in the nitrogen nutrition of plants. Indicate the form (e) in which nitrogen is transported in xylem to the shoot system.

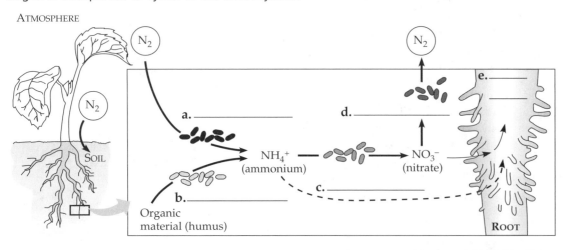

The Molecular Biology of Root Nodule Formation in Legumes A chemical dialogue between the roots of a particular legume species and its particular *Rhizobium* species leads to the development of a nodule. Roots secrete flavonoids that are detected and absorbed only by a certain *Rhizobium* species. The plant signal activates a gene-regulating protein that turns on the bacterial genes called *nod* (for "nodulation") genes. The enzymes produced from these genes catalyze the production of Nod factors that are secreted by the bacterial cells. Probably through signal-transduction pathways, the Nod factors activate plant early nodulin genes that cause the root to form an infection thread through which the bacteria enter the root cortex and to begin forming the root nodule from dividing cortex and pericycle cells. Vascular tissue later connects the nodule to the stele.

Molecular analysis shows that Nod factors are similar to chitins. Plants also produce chitinlike substances that may function as growth regulators. Thus, Nod factors mimic plant growth regulators and stimulate roots to grow nodules. Molecular manipulation of the root–*Rhizobium* relationship may allow researchers to increase nitrogen fixation and protein production by crops.

Mycorrhizae are symbiotic associations of roots and fungi that enhance plant nutrition (778–779)

Most plants have modified roots called **mycorrhizae.** The fungus receives food from the plant, while providing a large surface area for the absorption of water and minerals, which it provides to the plant. The fungus secretes growth hormones that stimulate root growth and branching and may help protect the plant from some soil pathogens.

The fossil record shows that some of the earliest land plants had mycorrhizae, and these symbiotic associations may have been an important adaptation helping plants to colonize land.

The Two Main Types of Mycorrhizae Especially common in woody plants, **ectomycorrhizae** form a dense sheath or mantle of mycelium over the root surface. Hyphae extending from the mantle provide a huge surface area for the absorption of water and minerals, especially phosphate. Hyphae also grow into extracellular spaces in the root cortex and facilitate exchange between plant and fungus.

Endomycorrhizae do not form a sheath but extend fine fungal hyphae into the soil. Hyphae also penetrate into root cell walls and form tubes that invaginate the root cell's membrane. Some of these tubes form dense branched structures called arbuscles that facilitate nutrient exchange. Ninety percent of plant species have endomycorrhizae.

Agricultural Importance of Mycorrhizae Most plants form mycorrhizae when they grow in their natural habitat. When seeds are planted in foreign soil where their particular species of fungus may not be found, or when soil fungi are poisoned, plants show signs of malnutrition. Foresters now inoculate seeds with spores of their mycorrhizal fungi to improve the growth of seedlings.

Mycorrhizae and root nodules may have an evolutionary relationship (779)

The early nodulin genes that are activated during early stages of root nodule formation are also activated during development of endomycorrhizae, and the signal-transduction pathways involved in the activation may have common components. The signals themselves may be similar: The Nod factors secreted by *Rhizobium* bacteria are related to chitins, and chitins make up the cell walls of fungi. These common molecular mechanisms indicate that root nodule development may have been adapted from a signaling pathway already in use in mycorrhizae.

Nutritional Adaptations: Parasitism and Predation by Plants

Parasitic plants extract nutrients from other plants (779–780)

Parasitic plants, such as the mistletoe or dodder, produce haustoria that may invade a host plant and siphon xylem or phloem sap from its vascular tissue. Epiphytes are plants that grow on the surface of another plant but do not take nourishment from it.

Carnivorous plants supplement their mineral nutrition by digesting animals (780)

Living in acid bogs or other nutrient-poor soils, carnivorous plants obtain nitrogen and minerals by killing and digesting insects that are caught in traps formed from modified leaves.

WORD ROOTS

ecto- = outside; **-myco-** = a fungus; **-rhizo** = a root (*ectomycorrhizae:* a type of mycorrhizae in which the mycelium forms a dense sheath, or mantle, over the surface of the root; hyphae extend from the mantle into the soil, greatly increasing the surface area for water and mineral absorption)

endo- = inside (*endomycorrhizae:* a type of mycorrhizae that unlike ectomycorrhizae, do not have a dense mantle ensheathing the root; instead, microscopic fungal hyphae extend from the root into the soil)

macro- = large (*macronutrient:* elements required by plants and animals in relatively large amounts)

micro- = small (*micronutrient:* elements required by plants and animals in very small amounts)

-phyto = a plant (*phytoremediation:* an emerging, non-destructive technology that seeks to cheaply reclaim contaminated areas by taking advantage of the remarkable ability of some plant species to extract heavy metals and other pollutants from the soil and to concentrate them in easily harvested portions of the plant)

STRUCTURE YOUR KNOWLEDGE

1. Develop a concept map that organizes your understanding of the basic nutritional requirements of plants.

2. What are the differences between root nodules and mycorrhizae? How are each beneficial to plants?

3. What are the similarities between root nodules and mycorrhizae?

TEST YOUR KNOWLEDGE

MULTIPLE CHOICE: *Choose the one best answer.*

1. The inorganic compound that contributes most of the mass to a plant's organic matter is
   a. H_2O.
   b. CO_2.
   c. NO_3^-.
   d. O_2.
   e. $C_6H_{12}O_6$.

2. The effects of mineral deficiencies involving fairly mobile nutrients will first be observed in
   a. older portions of the plant.
   b. new leaves and shoots.
   c. the root system.
   d. the color of the leaves.
   e. the flowers.

3. Most macronutrients are
   a. cofactors in enzymes.
   b. readily available from air and water.
   c. components of organic compounds.
   d. identified by hydroponic culture.
   e. components of cytochromes.

4. The most fertile type of soil is usually
   a. sand because its large particles allow room for air spaces.
   b. loam, which has a mixture of fine and coarse particles.
   c. clay, because the fine particles provide much surface area to which minerals and water adhere.
   d. humus, which is decomposing organic material.
   e. wet and alkaline.

5. Chlorosis is
   a. a symptom of a mineral deficiency indicated by yellowing leaves due to decreased chlorophyll production.
   b. the uptake of the micronutrient chlorine by a plant, which is facilitated by symbiotic bacteria.
   c. the production of chlorophyll within the thylakoid membranes of a plant.
   d. a contamination of glassware in hydroponic culture.
   e. a mold of roots caused by wet soil conditions.

6. Negatively charged minerals
   a. are released from clay particles by cation exchange.
   b. are reduced by cation exchange before they can be absorbed.
   c. are converted into amino acids before they are transported through the plant.
   d. are bound when roots release acids into the soil.
   e. are leached away by the action of rainwater more easily than positively charged minerals.

7. Nitrogenase
   a. is an enzyme complex that reduces atmospheric nitrogen to ammonia.
   b. is found in *Rhizobium* and other nitrogen-fixing bacteria.
   c. catalyzes the energy-expensive fixation of nitrogen.
   d. provides a model for chemical engineers to design catalysts to make nitrogen fertilizers.
   e. is or does all of the above.

8. Epiphytes
   a. have haustoria for anchoring to their host plants and obtaining xylem or phloem sap.
   b. are symbiotic relationships between roots and fungi.
   c. live in poor soil and digest insects to obtain nitrogen.
   d. grow on other plants but do not obtain nutrients from their hosts.
   e. are able to fix their own nitrogen.

9. The nitrogen content of some agricultural soils may be improved by
   a. the synthesis of leghemoglobin by ammonifying bacteria.
   b. mycorrhizae on legumes.
   c. water ferns with symbiotic cyanobacteria, or other plants with nitrogen-fixing bacteria.
   d. cation exchange.
   e. the action of denitrifying bacteria.

10. An advantage of organic fertilizers over chemical fertilizers is that they
    a. are more natural.
    b. release their nutrients over a longer period of time and are less likely to be lost to runoff.
    c. provide nutrients in the forms most readily absorbed by plants.
    d. are easier to mass produce and transport.
    e. are all of the above.

11. The early nodulin genes
    a. code for cytokinin plant hormones.
    b. produce Nod factors that are signals released from bacteria to roots.
    c. are turned on by Nod factor signals from *Rhizobium* and initiate nodule development in roots.
    d. code for receptors for the chitinlike signals released by *Rhizobium* and fungi.
    e. are fungal genes that control the development of arbuscles within root cells.

12. Which of the following takes the form of a mycelial sheath over plant roots with hyphae extending out that increase the surface area for absorption?
    a. root nodules
    b. haustoria
    c. endomycorrhizae
    d. ectomycorrhizae
    e. infection threads

13. Which of the following is an example of phytoremediation?
    a. dusting legume seeds with spores of *Rhizobium* to increase nodule formation
    b. inoculating seeds with fungal spores to ensure mycorrhizal formation
    c. using plants to remove toxic heavy metals from contaminated soils
    d. genetic engineering of plants to increase protein content
    e. seeding the ocean with iron to create algal blooms that reduce atmospheric CO_2 levels

14. What created the Dust Bowl in the Great Plains in the l930s?
    a. several years of drought
    b. building dams that removed the normal supply of water to the region
    c. removal of prairie grasses for wheat and cattle farming
    d. lack of crop rotation so that soil fertility was destroyed
    e. Both a and c were important factors.

15. Place these steps in the sequence of root nodule formation in proper order:
    1. production of infection thread through which bacteria enter root
    2. secretion of flavonoids by root
    3. continuation of nodule growth and connection to vascular stele of root
    4. activation of Nod D, which turns on bacterial *nod* genes
    5. activation of plant early nodulin genes by bacterial Nod factors
    6. bacteria wrapped in vesicles in cortical cells, becoming bacteroids

    a. 3, 5, 6, 1, 4, 2
    b. 2, 4, 5, 1, 6, 3
    c. 2, 1, 4, 5, 6, 3
    d. 5, 4, 2, 1, 3, 6
    e. 1, 2, 4, 5, 6, 3

CHAPTER 38

PLANT REPRODUCTION AND BIOTECHNOLOGY

FRAMEWORK

This chapter describes the sexual and asexual reproduction of flowering plants. The flower, with leaves modified for reproduction, produces the haploid gametophyte stages of the life cycle: Microspores in the anther develop into pollen grains, and a megaspore in the ovule produces an embryo sac. Pollination and the double fertilization of egg and polar nuclei are followed by the development of a seed with a quiescent embryo and endosperm, protected in a seed coat and housed within a fruit. Seed dormancy is broken following proper environmental cues and the imbibition of water.

Vegetative propagation allows successful plants to clone themselves. Agriculture makes extensive use of this type of plant reproduction by using cuttings, grafts, and test-tube cloning.

Plant biotechnologists are creating genetically modified (GM or transgenic) plants that have such traits as insect and disease resistance, herbicide tolerance, and improved nutritional value. Opposition to the development of GM organisms focuses on human health concerns and the unknown dangers of introducing transgenic plants into the environment.

CHAPTER REVIEW

Sexual Reproduction

Sporophyte and gametophyte generations alternate in the life cycles of plants: *a review* (783–784)

Plants exhibit an **alternation of generations** between haploid (*n*) and diploid (2*n*) generations. The diploid plant, the **sporophyte,** produces haploid spores by meiosis. Spores develop into multicellular haploid male or female **gametophytes** that produce gametes by mitosis. Fertilization yields diploid zygotes that grow into new sporophyte plants. In angiosperms, the male and female gametophytes have become reduced

to only a few cells that develop within the anthers and ovaries of the flower.

Flowers are specialized shoots bearing the reproductive organs of the angiosperm sporophyte (784–785)

Flowers are determinant shoots. These reproductive shoots usually contain four whorls of modified leaves call floral organs: **sepals, petals, stamens,** and **carpels,** which attach to the shoot at the **receptacle.** Sepals enclose and protect the unopened floral bud. Petals are generally more brightly colored and may attract pollinators. Stamens consist of a filament and an **anther,** which contains pollen sacs. A carpel consists of a sticky stigma at the top of a slender style, which leads to an **ovary.** The ovary encloses one or more **ovules.** A flower may have a single carpel, or multiple or fused carpels.

The pollen sacs and ovules contain the sporangia in which first spores and then gametophytes develop. **Pollen grains** are the sperm-producing male gametophytes, and **embryo sacs** are the egg-producing female gametophytes.

Pollination is the arrival of pollen onto a stigma. The pollen grain grows a tube down the style, releasing its sperm within the embryo sac.

Following fertilization, the zygote develops into an embryo as the surrounding ovule develops into a seed. The entire ovary forms a fruit, which aids in seed dispersal.

A **complete flower** has sepals, petals, stamens, and carpels. **Incomplete flowers** have eliminated one or more of these floral organs. A **bisexual flower** ("perfect") has both stamens and carpels; a **unisexual flower** ("imperfect") is missing one of these two. Imperfect flowers may be either staminate or carpellate. In **monoecious** plant species, both staminate and carpellate flowers are on the same plant; in **dioecious** species, these flowers are on separate plants. Floral variations include inflorescences and composite flowers, as well as diverse shapes, colors, and odors adapted to attract different pollinators.

■ INTERACTIVE QUESTION 38.1

Identify the flower parts in the following diagram. Indicate where pollen is produced and where pollination and fertilization occur.

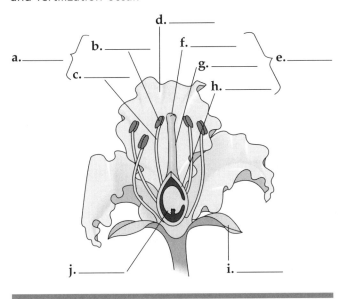

Pollination Pollination is the arrival of pollen onto the stigma, carried there by wind or animals.

■ INTERACTIVE QUESTION 38.2

a. Describe the male gametophyte.

b. Describe the female gametophyte.

Male and female gametophytes develop within the anthers and ovaries, respectively: pollination brings them together (786–788)

Development of the Male Gametophyte (Pollen Grain) Diploid cells called microsporocytes undergo meiosis to form four haploid **microspores.** A microspore divides once by mitosis to produce a generative cell and a tube cell. The wall surrounding the two cells thickens into the durable sculptured coat of the pollen grain. A pollen grain is an immature male gametophyte, which becomes mature when the generative cell divides to form two sperm cells, usually after the tube cell begins to form the pollen tube.

Development of the Female Gametophyte (Embryo Sac) Ovules form within the ovary. The megasporocyte in the sporangium of each ovule undergoes meiosis to form four haploid **megaspores,** only one of which usually survives. This megaspore grows and divides by mitosis three times, forming the female gametophyte, called the **embryo sac,** which typically consists of eight nuclei contained in seven cells. At one end of the embryo sac, an egg cell is lodged between two cells called synergids; three antipodal cells are at the other end; and two nuclei, called polar nuclei, are in a large central cell. The ovule consists of the embryo sac and its surrounding protective sporophyte layers called integuments.

Plants have various mechanisms that prevent self-fertilization (788–789)

Self-fertilization may be prevented by temporal or structural mechanisms. In flowers that are **self-incompatible,** a biochemical block prevents the development of pollen that does land on a stigma of the same plant.

The ability of flowers to reject their own pollen, or that of closely related individuals, depends on what are called S-genes. A plant population may have as many as 50 alleles at the S-locus, and if the pollen's allele matches either of the two alleles of the stigma, a pollen tube does not develop.

Self-incompatibility genes have evolved independently in various plant families, and thus the molecular mechanisms of the blockage differ. In gametophytic self-incompatibility, a matching allele in the developing pollen tube allows RNases from the carpel to enter and hydrolyze the pollen's RNA. In sporophytic incompatibility, self-recognition activates a signal-transduction pathway in stigma cells that blocks pollen germination.

Further research on the molecular basis of self-incompatibility may allow plant breeders to manipulate crop species to assure hybridization.

Double fertilization gives rise to the zygote and endosperm (789–790)

A pollen grain that lands on a receptive sigma absorbs moisture and germinates. The pollen tube grows through the style, and the generative cell divides to form two sperm, the male gametes. The pollen tube probes through the micropyle, an opening through the integuments of the ovule, and releases its two sperm within the embryo sac. By **double fertilization,** one sperm fertilizes the egg to form the zygote, and the other combines with the polar nuclei to form a triploid nucleus, which will develop into a food-storing tissue called the **endosperm.**

Researchers have recently been able to observe fertilization of egg cells *in vitro.* Gamete fusion is immediately followed by an increase in cytoplasmic Ca^{2+} levels in the egg and the establishment of a block to polyspermy. The opening of ion channels may produce a fast block, and the deposition of a cell wall may lead to the slow block to polyspermy.

■ INTERACTIVE QUESTION   38.3

What function may double fertilization serve?

The ovule develops into a seed containing an embryo and a supply of nutrients (790–792)

Endosperm Development   The triploid nucleus divides to form the endosperm, a multicellular mass rich in nutrients (proteins, oils, and starch), which are provided to the developing embryo and may be stored for later use by the seedling. In many dicots, endosperm is transferred to the cotyledons before the seed matures.

Embryo Development   In the zygote, the transverse first mitotic division creates a basal cell and a terminal cell. The basal cell divides to produce a thread of cells, called the suspensor, that anchors the embryo and transfers nutrients to it. The terminal cell divides to form a spherical proembryo, on which the cotyledons begin to form as bumps. The embryo elongates and apical meristems develop at the apexes of the embryonic shoot and root. Early development also establishes the radial arrangement of the primary meristems—protoderm, ground meristem, and procambium.

Structure of the Mature Seed   As it matures, the seed dehydrates and the embryo becomes dormant. The embryo and its food supply, the endosperm and/or enlarged cotyledons, are enclosed in a **seed coat** formed from the ovule integuments.

In a dicot seed, such as a bean, the embryo is an elongated embryonic axis attached to fleshy cotyledons. The axis below the cotyledonary attachment is called the **hypocotyl;** it terminates in the **radicle,** or embryonic root. The upper axis is the **epicotyl;** it terminates as a plumule, a shoot tip with a pair of leaves.

The monocot seed found in members of the grass family has a single thin cotyledon, called a **scutellum,** which absorbs nutrients from the endosperm during germination. A sheath called a **coleorhiza** covers the root, and a **coleoptile** encloses the young shoot.

■ INTERACTIVE QUESTION   38.4

Label the parts in these diagrams of a bean and a corn seed.

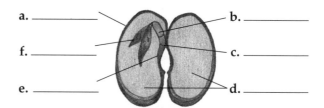

a. _____
f. _____
e. _____
b. _____
c. _____
d. _____

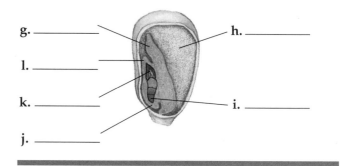

g. _____
l. _____
k. _____
j. _____
h. _____
i. _____

The ovary develops into a fruit adapted for seed dispersal (792–793)

The ovary of the flower develops into a **fruit,** which both protects and helps to disperse the seeds. Other floral parts (such as the receptacle) may contribute to what we commonly call a fruit. Hormonal changes following pollination cause the ovary to enlarge, its wall becoming the **pericarp,** or thickened wall of the fruit. Fruit usually does not set if a flower has not been pollinated.

Fruits usually ripen as the seeds are completing their development. Fruits are adapted to disperse seeds, enlisting the aid of wind or animals.

Humans have selectively bred edible fruits. Cereal grains—the wind-dispersed fruits of grasses—are staple foods for humans.

■ INTERACTIVE QUESTION   38.5

What changes usually occur when a fleshy fruit ripens?

Evolutionary adaptations of seed germination contribute to seedling survival (793–794)

At germination, the plant resumes the growth and development that was suspended when the seed matured and entered **dormancy.**

Seed Dormancy Dormancy increases the chances that the seed will germinate when and where the embryo has a good chance of surviving. The specific cues for breaking dormancy vary with the environment and may include heavy rain, intense heat from fires, cold, light, or chemical breakdown of the seed coat. The viability of a dormant seed may vary from a few days to decades or longer.

From Seed to Seedling Imbibition, the absorption of water by the dry seed, causes the seed to expand, rupture its coat, and begin a series of metabolic changes. Stored compounds are digested by enzymes, and nutrients are sent to growing regions.

The radicle emerges from the seed first, followed by the shoot tip. In many dicots, a hook that forms in the hypocotyl is pushed up through the ground, pulling the delicate shoot and cotyledons behind it. Light stimulates the straightening of the hook, and the first foliage leaves begin photosynthesis.

■ **INTERACTIVE QUESTION 38.6**

a. What happens to a bean seedling grown in the dark?

b. How does the shoot tip break through the soil in a germinating pea? In maize or other grasses?

Asexual Reproduction

Many plants clone themselves by asexual reproduction (794–795)

Many plant species clone themselves through asexual or **vegetative reproduction**—an extension of the indeterminate growth of plants in which meristematic tissues can grow indefinitely and parenchyma cells can divide and differentiate into specialized cells. A common type of vegetative reproduction is **fragmentation,** the separation of a plant into parts that then form whole plants. In some dicot species, the root system gives rise to many adventitious shoots that develop into a clone with separate shoot systems. Some plants, such as dandelions, can produce seeds asexually, a process called **apomixis.**

■ **INTERACTIVE QUESTION 38.7**

What would be an advantage of apomixis?

Sexual and asexual reproduction are complementary in the life histories of many plants (795)

Sexual reproduction in plants generates variation in a population, an advantage when the environment changes. Sexual reproduction also produces seeds, a means of dispersal to new locations and dormancy during harsh conditions. By asexual reproduction, plants well suited to a certain environment can clone exact copies. The progeny of vegetative propagation are usually not as frail as seedlings. Both modes of reproduction have been useful in the evolutionary adaptation of plants to their environments.

Vegetative propagation of plants is common in agriculture (795–797)

Clones from Cuttings New plants develop from stem cuttings when a **callus,** or mass of dividing cells, forms at the cut end of the shoot and adventitious roots develop from the callus. Adventitious roots can also form from a node in the shoot fragment. Some plants can be propagated from leaves or pieces of storage stems.

Twigs or buds of one plant can be grafted onto a plant of a different variety or closely related species. The plant that provides the root system is called the **stock,** and the twig is called the **scion.** Grafting can combine the best qualities of different plants.

Test-Tube Cloning and Related Techniques In test-tube cloning, whole plants can develop from pieces of tissue, called explants, or even from single parenchyma cells. A single plant can be cloned into thousands of plants by subdividing the undifferentiated calluses as they grow in tissue culture. Stimulated by proper hormone balances, calluses sprout shoots and roots and develop into plantlets, which can be transferred to soil to develop. With the use of a gene gun or other techniques, foreign DNA is inserted into individual plant cells, which then grow into plants by test-tube culture.

A technique called **protoplast fusion** is being coupled with tissue culture to create new plant varieties. Protoplasts are cells whose cell walls have been enzymatically removed. Protoplasts from different species can be fused and cultured to form hybrid plantlets.

Plant Biotechnology

Plant biotechnology refers both to the age-old use of plants to make products for human use and to the use of genetically modified (GM) organisms in agriculture.

Neolithic humans created new plant varieties by artificial selection (797–798)

Almost all of our crop species were first domesticated by Neolithic (late Stone Age) humans about 10,000 years ago. Natural hybridization between different species of plants is common, and humans have exploited such genetic variations using selective breeding and artificial selection to develop and improve crops.

Whereas it took plant breeders using traditional breeding methods 20 years to develop a desirable *opaque-2* maize (richer in two amino acids and with a hard endosperm), modern plant breeders could have directly inserted such desirable genes into hard-endosperm varieties. Genetic engineering techniques can also transfer genes between nonrelated species to produce "transgenic" plants.

Biotechnology is transforming agriculture (798)

The serious malnutrition of millions of people may be due to insufficient food production or inequities in distribution of food resources, and may signal that the world is already overpopulated. Estimates of population growth indicate that grain production must increase 40% per hectare to feed the human population in 2020.

The planting of transgenic crops has increased dramatically over the past several years. Cotton, maize, and potatoes have been engineered to contain genes from *Bacillus thuringiensis* that code for *Bt* toxin, reducing the need for spraying chemical insecticides. Other transgenic crops have been developed that are resistant to a number of herbicides, allowing farmers to "weed" crops without heavy tillage. Some transgenic plants are more resistant to disease or have improved nutritional quality.

Plant biotechnology has incited much public debate (799)

GM organisms may present an unknown risk to human health or the environment. Some of the major concerns include the transfer of allergens to a food source, the effect of GM crops on nontarget organisms, and the escape of herbicide or disease resistance genes through crop-to-weed hybridization that may create "superweeds." Various techniques are being developed to help reduce the ability of transgenic crops to hybridize, such as planting nontransgenic plant borders around fields, developing male sterility so that transgenic pollen is not produced, or introducing genes into chloroplast DNA, which is not present in pollen.

WORD ROOTS

a- = without; **-pomo** = fruit (*apomixis:* the asexual production of seeds)

anth- = a flower (*anther:* the terminal pollen sac of a stamen, inside which pollen grains with male gametes form in the flower of an angiosperm)

bi- = two (*bisexual flower:* a flower equipped with both stamens and carpels)

carp- = a fruit (*carpel:* The female reproductive organ of a flower, consisting of the stigma, style, and ovary)

coleo- = a sheath; **-rhiza** = a root (*coleorhiza:* the covering of the young root of the embryo of a grass seed)

di- = two (*dioecious:* referring to a plant species that has staminate and carpellate flowers on separate plants)

dorm- = sleep (*dormancy:* a condition typified by extremely low metabolic rate and a suspension of growth and development)

endo- = within (*endosperm:* a nutrient-rich tissue formed by the union of a sperm cell with two polar nuclei during double fertilization, which provides nourishment to the developing embryo in angiosperm seeds)

epi- = on, over (*epicotyl:* the embryonic axis above the point at which the cotyledons are attached)

gamet- = a wife or husband (*gametophyte:* the multicellular haploid form in organisms undergoing alternation of generations, which mitotically produces haploid gametes that unite and grow into the sporophyte generation)

hypo- = under (*hypocotyl:* the embryonic axis below the point at which the cotyledons are attached)

mega- = large (*megaspore:* a large, haploid spore that can continue to grow to eventually produce a female gametophyte)

micro- = small (*microspore:* a small, haploid spore that can give rise to a haploid male gametophyte)

mono- = one; **- ecious** = house (*monoecious:* referring to a plant species that has both staminate and carpellate flowers on the same individual)

peri- = around; **-carp** = a fruit (*pericarp:* the thickened wall of fruit)

proto- = first; **-plast** = formed, molded (*protoplast:* the contents of a plant cell exclusive of the cell wall)

scutell- = a little shield (*scutellum:* a specialized type of cotyledon found in the grass family)

sporo- = a seed; **= -phyto** = a plant (*sporophyte:* the multicellular diploid form in organisms undergoing alternation of generations that results from a union of gametes and that meiotically produces haploid spores that grow into the gametophyte generation)

stam- = standing upright (*stamen:* the pollen-producing male reproductive organ of a flower, consisting of an anther and filament)

uni- = one (*unisexual flower:* a flower missing either stamens or carpels)

STRUCTURE YOUR KNOWLEDGE

1. Draw yourself a diagram of the major events in the life cycle of an angiosperm.

2. List the advantages and disadvantages of sexual and asexual reproduction in plants.

3. List some of the potential benefits and dangers of plant biotechnology.

TEST YOUR KNOWLEDGE

FILL IN THE BLANKS

_____ 1. structure from which fruit typically develops

_____ 2. generation that produces spores by meiosis

_____ 3. species with male and female flowers on the same plant

_____ 4. female gametophyte of angiosperms

_____ 5. embryonic root

_____ 6. embryonic axis above attachment of cotyledon

_____ 7. protects dicot shoot as it breaks through the soil

_____ 8. twig or stem portion of a graft

_____ 9. plant cell from which cell wall is removed

_____ 10. mass of dividing cells at cut end of a shoot

MULTIPLE CHOICE *Choose the one best answer.*

1. A flower on a dioecious plant would be
   a. complete.
   b. bisexual.
   c. unisexual.
   d. asexual.
   e. carpellate.

2. Which of the following structures is haploid?
   a. embryo sac
   b. anther
   c. endosperm
   d. microsporocyte
   e. both a and b

3. The terminal cell of an early plant embryo
   a. develops into the shoot apex of the embryo.
   b. forms the suspensor that anchors the embryo and transfers nutrients.
   c. develops into the endosperm when fertilized by a sperm nucleus.
   d. divides to form the proembryo.
   e. develops into the cotyledons.

4. In angiosperms, sperm are formed by
   a. meiosis in the anther.
   b. meiosis in the pollen grain.
   c. mitosis in the anther.
   d. mitosis in the pollen tube.
   e. double fertilization in the embryo sac.

5. The endosperm
   a. may be absorbed by the cotyledons in the seeds of dicots.
   b. is usually a triploid tissue.
   c. is digested by enzymes in monocot seeds following hydration.
   d. develops in concert with the embryo as a result of double fertilization.
   e. is or does all of the above.

6. A seed consists of
   a. an embryo, a seed coat, and a nutrient supply.
   b. an embryo sac.
   c. a gametophyte and a nutrient supply.
   d. an enlarged ovary.
   e. an ovule.

7. Which structure protects a corn shoot as it breaks through the soil?
   a. hypocotyl hook
   b. epicotyl hook
   c. coleoptile
   d. coleorhiza
   e. aleurone

8. Which of the following is a form of asexual or vegetative reproduction?
   a. apomixis
   b. grafting
   c. test-tube cloning
   d. fragmentation
   e. all of the above

9. Protoplast fusion
   a. is used to study the fertilization of plant egg and sperm.
   b. is the method used to produce test-tube plantlets.
   c. can be used to form new plant species.
   d. occurs within a callus.
   e. is done with a gene gun.

10. In the plant embryo, the suspensor
    a. is produced by vertical mitotic divisions within the proembryo.
    b. connects the early root and shoot apexes.
    c. develops into the endosperm.
    d. is analogous to the umbilical cord in mammals.
    e. is the point of attachment of the cotyledons.

11. Flower organs have evolved from modified
    a. leaves.
    b. branches.
    c. sporangia.
    d. sporophytes.
    e. apical meristems.

12. Self-incompatibility
    a. prevents cross-pollination.
    b. involves the S-locus that contains as many as 50 genes.
    c. occurs when the pollen secretes RNases that hydrolyze RNA in the carpel.
    d. occurs when pollen and stigma have different S-locus alleles.
    e. occurs by different mechanisms because it evolved independently in several plant families.

13. Into what does a microspore develop in an angiosperm?
    a. the male gametophyte
    b. a pollen grain
    c. the male sporophyte
    d. the embryo sac
    e. Both a and b are correct.

14. Many plants form clones that develop when shoots emerge from the same root system. What is an advantage of forming such clones?
    a. provide a strong start for new plants
    b. disperse offspring to new habitats
    c. provide a period of dormancy until specific environmental cues signal regrowth
    d. provide a strong root system for a genetically altered shoot system
    e. allow for the production of seeds without fertilization

15. What does self-incompatibility provide for a plant?
    a. means of transferring pollen to another plant
    b. a means of coordinating the fertilization of an egg with the development of stored nutrients
    c. a means of destroying foreign pollen before it fertilizes the egg cell
    d. a biochemical block to self-fertilization so that cross-fertilization is assured
    e. a means of producing seeds without the need for fertilization

16. An immature male gametophyte differs from a mature male gametophyte in that it
    a. has not yet left the pollen sac.
    b. still consists of a microsporocyte.
    c. is a microspore that has not yet divided by mitosis to form a generative and a tube cell.
    d. has not yet produced double fertilization of the egg cell and polar nuclei.
    e. has not yet germinated and its generative cell has not divided to form two sperm.

17. Why did it take nearly 20 years for plant breeders to convert the *opaque-2* mutant maize (with higher levels of two essential amino acids) into a variety that had a more durable endosperm?
    a. Such genetic recombination between species was restricted by government regulations.
    b. Its development was delayed because of concern that the new variety, intended for swine feed, would get mixed with maize intended for human consumption.
    c. Traditional plant breeding using hybridization and artificial selection is a time-intensive process.
    d. Plant breeders were trying to combine two varieties that were not closely enough related.
    e. Few people saw the benefit of improving the protein content of maize, and funding was severely lacking for such research.

18. Which of the following is a technique being developed to reduce the threat of introduced genes for herbicide or insect resistance escaping to closely related weed species?
    a. planting a nontransgenic plant border around crop fields to reduce crop-to-weed gene transfer
    b. breeding male sterility into transgenic plants so that they have no pollen to be transferred to nearby weeds
    c. engineering the gene of interest into chloroplast DNA, which is inherited from the maternal plant and is not transferred by pollen
    d. engineering crops, such as soybeans, that have no weedy relatives nearby, or choosing to introduce genes for beneficial crop traits that would actually reduce the fitness of hybrid weeds
    e. All of the above would reduce the risk of crop-to-weed transgene escape.

CHAPTER 39

PLANT RESPONSES TO INTERNAL AND EXTERNAL SIGNALS

FRAMEWORK

Environmental stimuli and internal signals are linked by signal-transduction pathways to cellular responses such as changes in gene expression and activation of enzymes. Plant hormones—auxin, cytokinins, gibberellins, abscisic acid, ethylene, and brassinosteroids—control growth, development, flowering, and senescence, as plants respond and adapt to their environments. Plant movements in response to environmental stimuli include phototropism, gravitropism, and thigmotropism. The biological clock of plants controls circadian rhythms, such as stomatal opening and sleep movements. Phytochromes function as photoreceptors and are involved in the photoperiodic control of flowering. Plants have various physiological responses to environmental stresses and pathogens.

CHAPTER REVIEW

Various mechanisms have evolved in plants that enable them to sense and adaptively respond to their environments, generally by altering their patterns of growth and development. The intricate control systems of plants rely on internal signals.

Signal Transduction and Plant Responses

Signal-transduction pathways link cellular responses to plant hormonal signals and environmental stimuli (803–806)

The growth pattern of a sprouting seed or potato shoot assures that the shoot breaks ground before its stored food is exhausted. When the shoot reaches sunlight, stem elongation slows, leaves expand, the root system elongates, and chlorophyll production begins—all part of a process known as **greening.**

Reception Signals are detected by receptors, proteins that change shape in response to a specific stimulus. A phytochrome, a protein with an attached light-absorbing pigment, is the receptor involved in greening, and it is located in the cytosol. Researchers have studied the role of phytochrome in greening using the tomato mutant *aurea,* which has lower than normal levels of phytochrome.

Transduction **Second messengers** are small molecules the plant produces that amplify the signal from the receptor and carry it to proteins that produce the specific response. Each activated phytochrome may lead to the production of hundreds of second messenger molecules, each of which may activate hundreds of specific enzymes.

Light-activated phytochrome activates a G protein (guanine-nucleotide-binding protein), which then activates other enzymes, such as guanylyl cyclase, the enzyme that produces the second messenger cyclic GMP. Second messengers such as cGMP and cAMP may bind to membrane ion channels or may activate protein kinases. Phytochrome signal transduction also involves increases in cytoplasmic Ca^{2+}, which binds to proteins called calmodulins. Calcium-calmodulin complexes activate certain protein kinases.

Response The response to a signal-transduction pathway usually involves the activation of specific enzymes, either by stimulating gene expression for those enzymes or by activating existing enzymes. Several transcription factors are activated during phytochrome-induced greening, some by cGMP and others by Ca^{2+}-calmodulin. Changes in gene expression may involve the activation of positive transcriptional factors or the deactivation of negative transcriptional factors.

Post-translational modification of existing proteins usually involves phosphorylations, catalyzed by protein kinases. Cascades of protein kinase activations may lead to the phosphorylation of transcription factors, and thus, ultimately, to changes in gene expression. Protein phosphatases are enzymes that dephosphorylate specific proteins, allowing signal pathways to be turned off when a signal is no longer present.

The greening process involves changes in levels of growth-regulating hormones and the activation or new production of enzymes involved in producing chlorophyll or in photosynthesis.

■ **INTERACTIVE QUESTION   39.1**

In the process known as greening,

a. what is the signal and the receptor?

b. Briefly describe some of the steps in the transduction of this signal.

c. What is the plant's response?

Plant Responses to Hormones

Hormones, chemical signals that coordinate the parts of an organism, are translocated through the body, where minute concentrations are able to trigger responses in target cells and tissues.

Research on how plants grow toward light led to the discovery of plant hormones (806–807)

A **tropism** is a growth response of plant organs toward or away from stimuli. The growth of a shoot toward light is called positive **phototropism.** A coleoptile, enclosing the shoot of a grass seedling, bends toward the light when illuminated from one side because of the elongation of cells on the darker side.

Darwin and his son observed that a grass seedling would not bend toward light if its tip were removed or covered by an opaque cap. They postulated that a signal must be transmitted from the tip to the elongating region of the coleoptile. Boysen-Jensen demonstrated that the signal was a mobile substance, capable of being transmitted through a block of gelatin separating the tip from the rest of the coleoptile.

In 1926, Went placed coleoptile tips on blocks of agar to extract the chemical messenger. From his studies he concluded that the chemical produced in the tip, which he called auxin, promoted growth and that it was in higher concentration on the side away from the light.

Researchers have not found a light-induced asymmetrical distribution of auxin in sunflowers and other dicots, but certain substances that may act as growth inhibitors have been shown to be more concentrated on the lighted sides of such stems.

Plant hormones help coordinate growth, development, and responses to environmental stimuli (808–817)

Several major classes of plant hormones have been identified. Depending on the site of action, the developmental stage of the plant, and relative hormone concentrations, the effect of a hormone will vary. Very low concentrations of hormones, acting through signal-transduction pathways, may affect the expression of genes, the activity of enzymes, or the properties of membranes.

Auxin   **Auxins** include any substance that stimulates elongation of coleoptiles. The natural auxin extracted from plants is indoleacetic acid (IAA). It appears to be produced in shoot tips.

Auxin is transported through parenchyma tissue from the shoot tip down the shoot. This polar transport involves a chemiosmotic mechanism of ATP-driven proton pumps that generate a membrane potential favoring the exit of auxin anions through specific carriers located only at the basal ends of cells. In the more acidic environment outside the cell, auxin picks up a hydrogen ion and the now-neutral auxin molecule can move across the plasma membrane into the next parenchyma cell.

According to the acid growth hypothesis, auxin initiates cell growth in the region of elongation by binding to a plasma membrane receptor and stimulating proton pumps. The proton pumps lower pH in the cell wall, activating enzymes called **expansins** that break cross-links between cellulose microfibrils. The proton pumps also increase the membrane potential, enhancing ion uptake and the resulting osmotic uptake of water, which is facilitated by the increased plasticity of the cell wall. For continued growth after this relatively fast elongation, the cell must produce more cytoplasm and wall material, processes that rely on changes in gene expression that are also stimulated by auxin.

Auxin is involved in root branching and is used commercially to enhance formation of adventitious roots at the cut base of stems. Synthetic auxins, such as 2,4-D, are used as herbicides, killing dicot (broadleaf) weeds with a hormonal overdose. Auxin stimulates cell division in the vascular cambium and differentiation of secondary xylem. Auxin produced by developing seeds promotes fruit growth; synthetic auxins can induce seedless fruit development.

Cytokinins In tissue culture, coconut milk and degraded DNA were found to induce plant cell growth; later, **cytokinins** were identified as the active ingredients. Cytokinins are modified forms of adenine, named because they stimulate cytokinesis. Cytokinin-producing enzymes have not yet been identified, and one hypothesis is that cytokinins are produced by symbiotic prokaryotes that live in plant tissues. There may be two classes of cytokinin receptors, a cytoplasmic one that stimulates gene transcription and a cell surface one that may open Ca^{2+} channels.

Cytokinins are produced in actively growing roots, embryos, and fruits. Acting with auxin, they stimulate cell division and affect differentiation.

According to the direct inhibition hypothesis, the control of apical dominance involves the interaction between auxin, transported down from the terminal bud, which restrains axillary bud development, and cytokinins, transported up from the roots, which stimulate bud growth. Several lines of evidence support this hypothesis. Biochemical analyses, however, have shown that removal of the apical bud leads to an increase in auxin levels in the axillary buds, exactly opposite the prediction of the direct inhibition hypothesis.

Cytokinins can retard aging of some plant organs, because they stimulate RNA and protein synthesis, mobilize nutrients, and inhibit protein breakdown.

Gibberellins In the 1930s, Japanese scientists determined that the fungus *Gibberella* secreted a chemical that caused the hyperelongation of rice stems or "foolish seedling disease." Over 100 different naturally occurring **gibberellins** have now been identified.

Gibberellins, which are produced by roots and young leaves, stimulate growth in both leaves and stem, affecting cell division and elongation in stems. Gibberellins may promote cell elongation by stimulating cell wall-loosening enzymes, thus facilitating the penetration of expansins into the cell wall. (Remember that auxin acidifies the cell wall, which activates expansins.)

Gibberellin applied to dwarf plants may cause them to grow to normal height. Bolting, the growth of an elongated floral stalk, is caused by a surge of gibberellins. In many plants, both auxin and gibberellins contribute to fruit set. The release of gibberellins from the embryo signals seeds of many plants to break dormancy.

Abscisic Acid The hormone **abscisic acid (ABA)** generally slows growth. The high concentration of ABA in maturing seeds inhibits germination and stimulates production of proteins that protect the seeds during dehydration. For dormancy to be broken in some seeds, ABA must be removed or inactivated, or the ratio of gibberellins to ABA must increase.

ABA also reduces drought stress. In a wilting plant, ABA causes stomata in the leaves to close. ABA may be produced in the roots in response to water shortage and transported to the leaves.

Ethylene Plants produce the gas **ethylene** in response to stress and during fruit ripening and programmed cell death. Ethylene production is induced by a high concentration of auxin.

The mechanical stress of a seedling pushing against an obstacle as it grows upward through the soil induces the production of ethylene. Ethylene then initiates a growth pattern called the **triple response,** consisting of a slowing of stem elongation, a thickening of the stem, and initiation of horizontal growth. When the growing tip no longer detects a solid object, ethylene production decreases and normal upward growth resumes. Researchers have identified *Arabidopsis* mutants that are ethylene insensitive *(ein)*, ethylene overproducing *(eto)*, and that undergo the triple response in the absence of ethylene. In these latter constitutive triple response *(ctr)* mutants, the ethylene signal-transduction pathway is permanently turned on. Their mutant gene codes for a protein kinase, suggesting that the normal kinase product is a negative regulator of ethylene signal transduction. Binding of ethylene to the ethylene receptor may normally lead to the inactivation of the negative kinase, which allows the synthesis of the proteins involved in the triple response.

Apoptosis, or programmed cell death, requires the synthesis of new enzymes that break down many cellular components, which the plant may salvage. Ethylene is almost always associated with this programmed death of cells or organs, or of the entire plant.

Deciduous leaf loss protects against winter dehydration. Before leaves abscise in the autumn, many of their compounds are stored in the stem awaiting recycling to new leaves. A change in the balance of auxin and ethylene initiates changes in the abscission layer located near the base of the petiole, including the production of enzymes that hydrolyze polysaccharides in cell walls. A layer of cork forms a protective covering on the twig side of the abscission layer.

Ethylene initiates breakdown of cell walls and conversion of starches to sugars associated with fruit ripening. In a rare example of positive feedback, ethylene triggers ripening, and ripening triggers even more ethylene production. Many commercial fruits are ripened in huge containers perfused with ethylene gas. A gene for antisense RNA has been developed that blocks the expression of one of the genes involved in ethylene synthesis.

Brassinosteroids　Similar to cholesterol and animal sex hormones, **brassinosteroids** were first isolated in 1979. Their effects are very similar to those of auxin: They promote cell elongation and division, retard leaf abscission, and promote xylem differentiation. Identification of a brassinosteroid-deficient mutant of *Arabidopsis* helped to establish these compounds as nonauxin plant hormones.

■ INTERACTIVE QUESTION　39.2

Fill in the name of the hormone that is responsible for each of the following functions:

a. _____　promotes fruit ripening; initiates triple response; involved in apoptosis

b. _____　stimulate cell division, growth, and germination; anti-aging

c. _____　inhibits growth; maintains dormancy; closes stomata during water stress

d. _____　stimulates stem elongation, root branching, fruit development; apical dominance

e. _____　promote cell elongation and division, xylem differentiation; retard leaf abscission

f. _____　promote stem elongation, seed germination; contributes to fruit set

---

Plant Responses to Light

The effect of light on plant growth and development is called **photomorphogenesis**. An **action spectrum** graphs a physiological response across different wavelengths of light. An absorption spectrum shows the wavelengths of light a pigment absorbs. Close correlation between an action spectrum for a plant response and the absorption spectrum of a pigment may indicate that the pigment is the photoreceptor involved in the response. Red and blue light have the most influence on a plant's photomorphogenesis.

Blue-light photoreceptors are a heterogeneous group of pigments (817–818)

Molecular biologists have determined that plants use at least three different types of pigments to detect blue light: **cryptochrome** (for inhibition of hypocotyl elongation when a seedling breaks ground), **phototropin** (for phototropism), and **zeaxanthin** (for stomatal opening).

Phytochromes function as photoreceptors in many plant responses to light (818–819)

The Phytochrome Switch and Seed Germination
Studies of lettuce seed germination in the 1930s determined that red light (660 nm wavelength) increased germination the most and far-red light (730 nm) inhibited germination. The effects of red and far-red light are reversible, with the seed's response determined by the last flash of light it receives.

　Five different phytochromes have been identified in *Arabidopsis*. A phytochrome is a photoreceptor that consists of a protein component, which has protein kinase activity, bonded to a nonprotein light-absorbing chromophore. The chromophore alternates between two isomers, one of which absorbs red light and the other, far-red light. These two variations of phytochrome are photoreversible; the P_r to P_{fr} interconversion acts as a switch, controlling various events in the life of the plant.

　Phytochrome tells the plant that light is present by the conversion of P_r, which is the form the plant synthesizes, to P_{fr} in the presence of sunlight. P_{fr} triggers many plant responses to light, such as breaking seed dormancy.

The Phytochrome Switch and Shade Avoidance　The amount of available light, specifically the relative amounts of red and far-red light, is communicated to a plant by the ratio of the two forms of phytochrome. Canopy trees absorb red light, and a shaded tree will have a higher ratio of P_r to P_{fr}, inducing the tree to use its resources to grow taller.

■ INTERACTIVE QUESTION　39.3

One domain of the protein component of a phytochrome has protein kinase activity. What does this suggest about how this photoreceptor functions?

Biological clocks control circadian rhythms in plants and other eukaryotes (819–820)

A **circadian rhythm** is a physiological cycle with about a 24-hour frequency that is not directly paced by an environmental factor. These rhythms persist, even when the organism is sheltered from environmental cues. Research indicates that the oscillator for circadian rhythms is endogenous, although the clock is set (entrained) to a 24-hour period by daily environmental signals.

■ INTERACTIVE QUESTION 39.4

What are free-running periods? How are they determined?

The molecular mechanism of the biological clock, which may be common to all eukaryotes, may be the cyclical changes in the concentration of a protein that is a transcription factor that inhibits the expression of its own gene.

Researchers have identified clock mutants of *Arabidopsis* by splicing the gene for luciferase to the promoter for genes for photosynthesis-related proteins that follow a circadian rhythm in their production. When the biological clock turned on the promoter for these genes, the plants glowed, and plants that glowed for a nonnormal amount of time were isolated as clock mutants. Some of these mutants had defects in proteins that normally bind photoreceptors, pointing to a light-dependent mechanism that sets the biological clock.

Light entrains the biological clock (820–821)

Both blue-light photoreceptors and phytochrome can entrain the biological clock in plants. In darkness, the phytochrome ratio shifts toward P_r, in part because P_{fr} is converted to P_r in some plants, and also because P_{fr} is degraded and new pigment is synthesized as P_r. When the sun rises, P_r is rapidly converted to P_{fr}, resetting the biological clock each day at dawn.

Photoperiodism synchronizes many plant responses to changes of season (821–823)

Seasonal events in the life cycle of plants usually are cued by photoperiod, the relative length of night and day. A physiological response to photoperiod is called **photoperiodism.**

Photoperiodism and the Control of Flowering Garner and Allard discovered that a variety of tobacco plant flowered only when the day length was 14 hours or shorter. They termed it a **short-day plant. Long-day plants** flower when days are longer than a certain number of hours, and the flowering of **day-neutral plants** is unaffected by photoperiod.

Researchers have found that it is night length, not day length, that controls flowering and other photoperiod responses. If the dark period is interrupted by even a few minutes of light, a short-day (long-night) plant such as the cocklebur will not flower. Photoperiodic responses thus depend on a critical night length. Short-day plants require a minimum number of hours of uninterrupted darkness, and long-day plants will flower only if they receive less than a critical length of darkness.

Red light was found to be the most effective in interrupting a plant's perception of night length. A brief exposure to red light breaks a dark period of sufficient length and prevents short-day plants from flowering, whereas a flash of red light during a dark period longer than the critical length will induce flowering in a long-day plant. Due to the photoreversibility of phytochrome, a subsequent flash of far-red light negates the effect of the red light.

Some plants bloom after a single exposure to the required photoperiod. Others respond to photoperiod only after exposure to another environmental stimulus. The need for pretreatment with cold before flowering is called vernalization.

Is There a Flowering Hormone? Leaves detect the photoperiod. The signal for flowering travels from leaves to buds and is believed to be a hormone or change in the concentration of two or more hormones, but it has yet to be identified.

Meristem Transition from Vegetative Growth to Flowering In order for flowering to occur, the bud meristem must transition from a vegetative to a flowering state, a change that requires the activation of meristem-identity genes and organ-identity genes. Researchers are looking for the signal-transduction pathways that link photoperiod and hormonal cues to such changes in gene expression.

■ INTERACTIVE QUESTION 39.5

Indicate whether a short-day plant **(a–e)** and a long-day plant **(f–j)** would flower or not flower under the indicated light conditions.

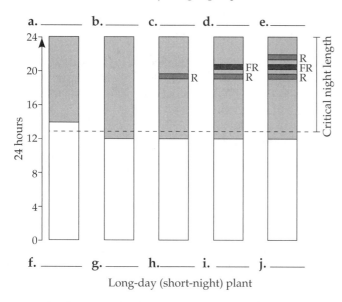

Short-day (long-night) plant

Long-day (short-night) plant

How do these results demonstrate the red/far-red photoreversibility of phytochrome?

Plant Responses to Environmental Stimuli Other than Light

Plants respond to environmental stimuli through a combination of developmental and physiological mechanisms (823–827)

Responses to Gravity Roots exhibit positive **gravitropism,** whereas shoots show negative gravitropism. According to one hypothesis, the settling of **statoliths,** plastids containing dense starch grains, in cells of the root cap triggers movement of calcium, which causes the lateral transport of auxin. Both of these accumulate on the lower side of the growing root, where the high concentration of auxin inhibits cell elongation, causing the root to curve downward. The settling of the protoplast and large organelles may distort the cytoskeleton and also signal gravitational direction.

Responses to Mechanical Stimuli The stunting of growth in height and increase in girth of plants that are exposed to wind or mechanical stimulation is called **thigmomorphogenesis.** Mechanical stimuli initiate a signal-transduction pathway that results in specific gene activations that alter growth patterns.

Most climbing plants have tendrils that coil rapidly around supports, exhibiting **thigmotropism** or directional growth in response to touch.

The sensitive plant *Mimosa* folds its leaves after being touched due to the rapid loss of turgor by cells in specialized motor organs called pulvini, located at the joints of the leaf. These cells lose potassium when stimulated, resulting in osmotic water loss. The message travels through the plant from the point of stimulation, perhaps as the result of electrical impulses, called **action potentials.** These electrical messages may be used in plants as a form of internal communication.

■ INTERACTIVE QUESTION 39.6

a. Name the types of tropisms that a stem exhibits.

b. Name and describe the growth mechanism that produces the coiling of a tendril.

Responses to Stress An environmental stress may be severe enough to threaten a plant's growth, reproduction, and survival.

Mechanisms that reduce transpiration help a plant respond to water deficit. Guard cells lose turgor and stomata close. Abscisic acid acts on guard cell membranes to keep stomata closed. Growth of young leaves is inhibited by the lack of cell-expanding water, and wilted leaves may roll up, further reducing transpiration. All of these responses, however, reduce photosynthesis. During a drought, root growth in shallow, dry soil decreases while deeper roots in moist soil continue to grow.

Plants adapted to wet habitats may have aerial roots that provide oxygen to their submerged roots. When roots of other plants are in waterlogged soils, oxygen deprivation may stimulate ethylene production, causing some root cortical cells to undergo apoptosis, opening up air tubes within the roots.

Excess salts in the soil may lower the water potential of the soil solution below that of roots, causing roots to lose water. The plasma membranes of root cells can reduce the uptake of sodium and some other ions that are toxic to plants in high concentrations. Plants may respond to moderate soil salinity by producing compatible solutes that lower the water potential of root cells. Special adaptations for dealing with high soil salinity have evolved in halophytes.

Transpiration creates evaporative cooling for a plant, but this effect may be lost on hot, dry days when stomata close to reduce water loss. Above critical temperatures, plant cells produce **heat-shock proteins** that may provide temporary support to reduce protein denaturation.

Plants respond to cold stress by increasing the proportion of unsaturated fatty acids in membrane lipids in order to maintain the fluidity of cell membranes. Subfreezing temperatures cause ice to form in cell walls, lowering the extracellular water potential and causing cells to dehydrate. Plants adapted to cold winters have adaptations to deal with freezing stress, such as changing the solute composition of the cytosol.

■ INTERACTIVE QUESTION   39.7

a. A plant is exposed to a spell of very hot and dry weather. How might it survive this stress?

b. How might a plant adapt to an unusually cold and very wet fall?

Plant Defense: Responses to Herbivores and Pathogens

Plants deter herbivores with both physical and chemical defenses (827–828)

Physical defenses, such as thorns, and chemical defenses, such as distasteful or toxic compounds, may help plants cope with herbivory. Some plants produce **canavanine,** which resembles arginine and may be incorporated into an insect's proteins, altering protein conformation and causing death.

Some plants recruit predators of their herbivores. A combination of a compound in a caterpillar's saliva and the damage caused by its eating stimulates leaves to release volatile compounds that attract parasitoid wasps.

The wasps lay their eggs inside the caterpillar, and the wasp larvae eat their host. These volatile compounds may also signal neighboring plants to activate defense genes. These gene activations are similar in pattern to those produced by exposure to the important plant defense molecule, **jasmonic acid.**

Plants use multiple lines of defense against pathogens (828–829)

The physical barrier of the plant's epidermis and/or periderm is the first line of defense against pathogenic viruses, bacteria, and fungi. Should these pathogens enter the plant due to injuries or through openings such as stomata, the plant mounts a chemical defense to either destroy or contain the pathogen.

Gene-for-Gene Recognition   Virulent pathogens are those to which the plant has no specific defense. Plants are generally resistant to most pathogens. These **avirulent** pathogens do not extensively harm or kill the plant.

Specific resistance is based on **gene-for-gene recognition** between a plant's dominant *R* allele, which probably codes for a specific receptor protein, and an avirulent pathogen's *Avr* allele, which codes for some pathogen molecule that also serves as a ligand to which the receptor can bind. Such binding triggers a signal-transduction pathway that produces a plant defense response. Plants have many different *R* genes, corresponding to the many different potential pathogens.

Hypersensitive Response   In response to chemical signals released from plant cells damaged by a pathogen, the plant mounts a localized attack. **Elicitors,** often cellulose fragments called **oligosaccharins** released from damaged cell walls, stimulate the production of antimicrobial **phytoalexins.** Infection also activates genes for **PR proteins** (some of which are antimicrobial, while others serve as alarm signals to neighboring cells) and stimulates strengthening of the cell walls to slow the spread of the pathogen. A signal-transduction pathway triggers a **hypersensitive response (HR)** when the pathogen is avirulent and there is an *R-Avr* match. This more vigorous defense includes increased production of phytoalexins and PR proteins and a more effective sealing of the area.

Systemic Acquired Resistance   Alarm hormones are released from infected cells before they destroy themselves. These signal molecules trigger phytoalexin and PR protein production throughout the plant, creating a **systemic acquired resistance (SAR).** This generalized defense response helps protect uninfected tissue. One of the hormones involved in activating SAR is probably **salicylic acid,** which humans have modified as the active ingredient in aspirin.

WORD ROOTS

aux- = grow, enlarge (*auxins:* a class of plant hormones, including indoleacetic acid, having a variety of effects, such as phototropic response through the stimulation of cell elongation, stimulation of secondary growth, and the development of leaf traces and fruit)

circ- = a circle (*circadian rhythm:* a physiological cycle of about 24 hours, present in all eukaryotic organisms, that persists even in the absence of external cues)

crypto- = hidden; **-chromo** = color (*cryptochrome:* the name given to the unidentified blue-light photoreceptor)

cyto- = cell; **-kine** = moving (*cytokinins:* a class of related plant hormones that retard aging and act in concert with auxins to stimulate cell division, influence the pathway of differentiation, and control apical dominance)

gibb- = humped (*gibberellins:* a class of related plant hormones that stimulate growth in the stem and leaves, trigger the germination of seeds and breaking of bud dormancy, and stimulate fruit development with auxin)

hyper- = excessive (*hypersensitive response:* a vigorous, localized defense response to a pathogen that is avirulent based on an *R-Avr* match)

photo- = light; **-trop** = turn, change (*phototropism:* growth of a plant shoot toward or away from light)

phyto- = a plant; **-alexi** to ward off (*phytoalexin:* an antibiotic, produced by plants, that destroys microorganisms or inhibits their growth)

stato- = standing, placed; **-lith** = a stone (*statolith:* specialized plastids that help a plant tell up from down)

thigmo- = a touch; **morpho-** = form; **-genesis** = origin (*thigmomorphogenesis:* a response in plants to chronic mechanical stimulation, resulting from increased ethylene production; an example is thickening stems in response to strong winds)

zea- = a grain; **-xantho** = yellow (*zeaxanthin:* a blue-light photoreceptor involved in stomatal opening)

STRUCTURE YOUR KNOWLEDGE

1. List some agricultural uses of plant hormones.

2. Develop a concept map to illustrate your understanding of the photoperiodic control of flowering. Do not forget to include the role of the biological clock.

3. Briefly describe steps 1–7 in this diagram of the hypersensitive response and systemic acquired resistance that result from a plant's encounter with an avirulent pathogen.

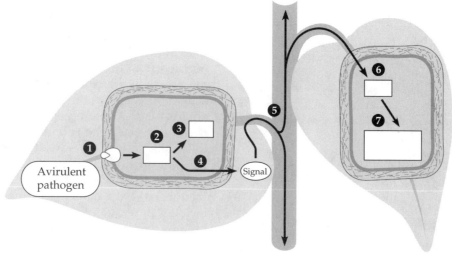

① _____

② _____

③ _____

④ _____

⑤ _____

⑥ _____

⑦ _____

TEST YOUR KNOWLEDGE

TRUE OR FALSE: *Indicate T or F and then correct the false statements.*

_____ 1. Abscisic acid is necessary for a seed to break dormancy.

_____ 2. Roots exhibit negative phototropism and positive gravitropism.

_____ 3. Gibberellins, synthesized in the root, counteract apical dominance.

_____ 4. Thigmomorphogenesis is a growth response of a seedling that encounters an obstacle while pushing upward through the soil.

_____ 5. The application of gibberellins or auxin may induce the development of seedless fruit.

_____ 6. Action potentials are electrical impulses that may be used as internal communication in plants.

_____ 7. A physiological response to day or night length is called a circadian rhythm.

_____ 8. Vernalization is the need for pretreatment with cold before flowering.

_____ 9. A virulent pathogen is one to which a plant has a specific resistance based on gene-for-gene recognition.

_____ 10. A cryptochrome is the light-absorbing portion of a phytochrome that reverts between two isomeric forms.

MULTIPLE CHOICE: *Choose the one best answer.*

1. The body form within a species of plants may vary more than that within a species of animals because
   a. growth in animals is predominantly indeterminate.
   b. plants respond adaptively to their environments by altering their patterns of growth and development.
   c. plant growth and development are governed by many hormones.
   d. plants can respond to environmental stress.
   e. all of the above are true.

2. Polar transport of auxin involves
   a. the accumulation of higher concentrations on the side of a shoot away from light.
   b. the reverse movement of auxin from roots to shoots.
   c. movement of auxin ions through carrier proteins located at the basal end of cells, facilitated by the membrane potential.
   d. the unidirectional active transport of auxin into and out of parenchyma cells.
   e. the settling of statoliths.

3. According to the acid growth hypothesis,
   a. auxin stimulates membrane proton pumps.
   b. a lowered pH outside the cell activates expansins that break cross-links between cellulose microfibrils.
   c. the membrane potential created by the proton pumps enhances ion uptake, increasing the osmotic movement of water into cells.
   d. cells elongate when they take up water by osmosis.
   e. all of the above are involved in cell elongation.

4. Which of the following situations would most likely stimulate the development of axillary buds?
   a. a large quantity of auxin traveling down from the shoot and a small amount of cytokinin produced by the roots
   b. a small amount of auxin traveling down from the shoot and a large amount of cytokinin traveling up from the roots
   c. an equal ratio of gibberellins to auxin
   d. the absence of cytokinins caused by the removal of the terminal bud
   e. an increase in the concentration of brassinosteroids produced by leaves

5. The growth inhibitor in seeds is usually
   a. abscisic acid.
   b. ethylene.
   c. gibberellin.
   d. a small amount of ABA combined with a larger concentration of gibberellins.
   e. a high cytokinin-to-auxin ratio.

6. A circadian rhythm
   a. is controlled by an external oscillator.
   b. is a physiological cycle of approximately a 24-hour frequency.
   c. involves an internal biological clock that is not set by daily environmental signals.
   d. provides the signal for seasonal flowering.
   e. involves all of the above.

7. Which of the following is *not* a component of the signal-transduction pathway involved in the greening process when a shoot breaks ground?
   a. reception of light by a phytochrome located in the cytosol, which activates a G protein
   b. inhibition of the triple response and initiation of phototropism as the shoot bends toward light
   c. production of second messengers such as cGMP and calcium-calmodulin complexes
   d. cascades of protein kinase activations that phosphorylate transcription factors
   e. activation of genes that code for photosynthesis-related enzymes

8. Injecting phytochrome into cells of the tomato mutant *aurea*
   a. causes the plant to undergo the triple response above ground when exposed to ethylene.
   b. causes the plant to undergo the triple response above ground without exposure to ethylene.
   c. helps researchers identify clock mutants.
   d. initiates flowering even when nights are longer than a critical period.
   e. causes the plant to develop a normal greening response to light.

9. A flash of far-red light during a critical-length dark period
   a. will induce flowering in a long-day plant.
   b. will induce flowering in a short-day plant.
   c. will not influence flowering.
   d. will increase the P_{fr} level suddenly.
   e. will be negated by a flash of red light.

10. The conversion of P_r to P_{fr}
   a. occurs slowly at night.
   b. may be the way in which plants sense daybreak and serves to entrain the biological clock.
   c. is the molecular mechanism responsible for the biological clock.
   d. occurs when P_r absorbs far-red light.
   e. occurs within flower buds and induces flowering.

11. A plant may withstand salt stress by
   a. releasing abscisic acid that closes stomata to salt accumulation.
   b. the production of compatible solutes that lower the water potential of root cells.
   c. wilting, which reduces water and salt uptake by reducing transpiration.
   d. producing canavanine that reduces the toxic effect of sodium ions.
   e. releasing ethylene that leads to apoptosis of damaged cells.

12. Which of the following hormones would be sprayed on barley seeds to speed germination in the production of malt for making beer?
   a. abscisic acid
   b. auxin

c. cytokinin
d. ethylene
e. gibberellin

13. Many plants will flower in response to a specific
   a. flowering hormone that is produced in the apical bud.
   b. elicitor, which is a cellulose fragment.
   c. minimal temperature that appears to signal a seasonal change.
   d. photoperiod, which seems to be measured by the length of darkness to which the leaves of the plant are exposed.
   e. combination of a high level of auxin and a low level of cytokinin.

14. Which of the following is *not* a plant defense against herbivory?
   a. production of distasteful compounds.
   b. production of toxic compounds such as canavanine.
   c. physical defenses such as thorns.
   d. initiation of a hypersensitive response with production of phytoalexins and PR proteins.
   e. release of volatile compounds that recruit parasitoid wasps.

15. Which of the following is *not* true of the hypersensitive response?
   a. It relies on an *R-Avr* recognition between a specific plant cell receptor protein and a pathogen molecule.
   b. It increases the production of PR proteins that may have antimicrobial or signaling functions.
   c. It enhances the production of ethylene that serves as a signal molecule transported throughout the plant to activate SAR (systemic acquired resistance).
   d. It contains an infection by stimulating crosslinking of cell wall molecules and production of lignin.
   e. The plant cells involved in the defense destroy themselves, leaving lesions that indicate the site of the contained infection.

UNIT SEVEN

ANIMAL FORM AND FUNCTION

CHAPTER 40

AN INTRODUCTION TO ANIMAL STRUCTURE AND FUNCTION

FRAMEWORK

The body structures of an animal include organs that are composed of specialized cells grouped into the four basic tissues: epithelial, connective, muscle, and nervous. Organs function together in organ systems. Structure correlates with function in these hierarchical levels of organization. The functions of an animal are powered by chemical energy derived from food. Metabolic rate, the amount of energy used in a unit of time, is higher for endothermic animals and inversely related to body size.

All cells must be bathed in an aqueous solution. Compact animal bodies have highly folded exchange surfaces and a circulatory system that distributes materials throughout the body. The internal environment is carefully regulated by the process of homeostasis.

CHAPTER REVIEW

Functional Anatomy: An Overview

Animal form and function reflect biology's major themes (834–835)

The comparative study of animals illustrates several key biological themes: evolution, regulation, bioenergetics, and the correlation of structure and function.

Anatomy is the study of an organism's structure; **physiology** is the study of function.

Function correlates with structure in the tissues of animals (835–839)

Hierarchical levels of organization characterize life. Multicellular organisms have specialized cells grouped into tissues, which may be combined into organs. Various organs may function together in organ systems.

Tissues are collections of cells with a common structure and function, held together by a sticky extracellular matrix or fibers. Tissues are classified into four categories.

Epithelial Tissue   **Epithelial tissue** lines the outer and inner surfaces of the body in protective sheets of tightly packed cells. Cells at the base of an epithelium are attached to a **basement membrane,** a dense layer of extracellular matrix. A **simple epithelium** has one layer, whereas a **stratified epithelium** has multiple layers of cells. The shape of cells at the free surface may be **squamous** (flat), **cuboidal** (boxlike), or **columnar** (pillarlike).

Glandular epithelia are specialized for absorption or secretion. **Mucous membranes** lining the digestive and respiratory tracts secrete mucus. Cilia on the epithelium lining the air passages sweep particles trapped in mucus away from the lungs.

■ INTERACTIVE QUESTION 40.1

Name the two types of epithelium illustrated below. One of these epithelial types forms the outer skin and the other lines the digestive tract. Explain why each would be found in its location.

a. _____   b. _____

Connective Tissue   **Connective tissue** connects and supports other tissues and is characterized by having relatively few cells suspended in an extracellular matrix of fibers, which may be embedded in a liquid, jellylike, or solid substance.

Connective tissue fibers are of three types: **Collagenous fibers** are made of collagen and have tensile strength that resists stretching. **Elastic fibers,** made of the protein elastin, can stretch and provide resilience. Branched and thin **reticular fibers** are composed of collagen and form a tightly woven connection with adjacent tissues.

Loose connective tissue, made of loosely woven fibers, attaches epithelia to underlying tissues and holds organs in place. The most common types of cells enmeshed in loose connective tissue are **fibroblasts,** which secrete the protein of the extracellular fibers, and **macrophages,** amoeboid cells that engulf bacteria and cellular debris.

Adipose tissue is a special form of loose connective tissue that pads and insulates the body and stores fat. Adipose cells each contain a large fat droplet.

Fibrous connective tissue, with its dense arrangement of parallel collagenous fibers, is found in **tendons,** which attach muscles to bones, and in **ligaments,** which join bones together at joints.

Cartilage is composed of collagenous fibers embedded in a rubbery substance called chondroitin sulfate, both secreted by **chondrocytes.** Cartilage is a strong but somewhat flexible support material, making up the skeleton of sharks and vertebrate embryos.

Bone is a mineralized connective tissue formed by **osteoblasts** that deposit a matrix of collagen and calcium, magnesium, and phosphate ions, which hardens into hydroxyapatite. **Osteons** (Haversian systems) consist of concentric layers of matrix deposited

around a central canal containing blood vessels and nerves. Osteoblasts are called osteocytes, once they are trapped in the matrix.

Blood is a connective tissue that has a liquid extracellular matrix called plasma. Erythrocytes (red blood cells) carry oxygen; leukocytes (white blood cells) function in defense; and cell fragments called platelets are involved in the clotting of blood.

■ INTERACTIVE QUESTION 40.2

Identify the types of connective tissue and their components in the following three diagrams.

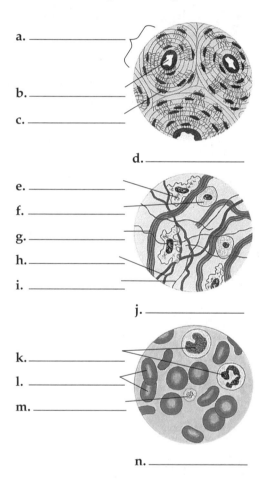

a. _____

b. _____

c. _____

d. _____

e. _____

f. _____

g. _____

h. _____

i. _____

j. _____

k. _____

l. _____

m. _____

n. _____

Nervous Tissue   **Nervous tissue** senses stimuli and transmits electrical signals. The **neuron,** or nerve cell, consists of a cell body and two or more processes that conduct impulses toward (dendrites) and away from (axons) the cell body.

Muscle Tissue   **Muscle tissue** consists of long, contractile cells called muscle fibers that are packed with myofibrils of actin and myosin. **Skeletal muscle**—also called **striated muscle**—is responsible for voluntary body movements. **Cardiac muscle,** forming the wall of the heart, is also striated, but its cells are branched, joined at their ends by intercalated discs. **Smooth muscle** is composed of spindle-shaped cells lacking striations. It is found in the walls of the digestive tract, arteries, and other internal organs, and is under involuntary control.

■ INTERACTIVE QUESTION 40.3

Identify these types of vertebrate muscle. What are the dark bands between fibers in figure **a,** and what is their function?

a. _____

b. _____

The organ systems of an animal are interdependent (839)

In all animals but sponges and some cnidarians, tissues are organized into **organs.** Organs often consist of a layered arrangement of tissues. Many vertebrate organs are suspended by **mesenteries** in moist or fluid-filled body cavities. Mammals have a **thoracic cavity** separated by a muscular diaphragm from an **abdominal cavity.**

Groups of organs are integrated into **organ systems,** which perform the major functions required for life. The organ systems are coordinated to create the functional integration needed by an organism.

■ INTERACTIVE QUESTION 40.4

There are 11 organ systems in mammals. How many of them can you name?

Body Plans and the External Environment

Physical laws constrain animal form (840)

Physical laws constrain the evolution of an animal's body plan or design. The laws of hydrodynamics drove the natural selection of a fusiform body shape for both fast-swimming fish and mammals, an example of convergent evolution.

Body size and shape affect interactions with the environment (840–842)

Every cell must be in an aqueous medium to maintain the integrity of the plasma membrane and allow for exchange across it. Cell size is limited by the need for a sufficient surface area-to-volume ratio. Single-celled organisms or animals with two-layered saclike bodies or thin flat bodies can maintain sufficient cellular contact with the aqueous environment.

Animals with compact bodies, however, must provide extensively branched or folded internal membranes for exchanging materials with the environment. The circulatory system connects these exchange surfaces with the aqueous environment bathing the body's cells.

■ INTERACTIVE QUESTION 40.5

List two advantages of a compact, complex body form in which most of the body cells are not in contact with the external environment.

Regulating the Internal Environment

Mechanisms of homeostasis moderate changes in the internal environment (842)

The internal environment of vertebrates is the **interstitial fluid** surrounding the cells, through which oxygen, nutrients, and wastes are exchanged with blood in capillaries. The "steady state" or internal balance of this environment is called **homeostasis.**

Homeostasis depends on feedback circuits (843–844)

The mechanisms by which animals maintain home-ostasis involve a *receptor* that detects a change in the internal environment and a *control center* that processes information and directs an *effector* to re-spond. Most homeostatic control mechanisms operate by **negative feedback.** When some variable moves above or below a set point, a control mechanism is turned on or off to return the condition to normal.

Positive feedback in a physiological function is a mechanism in which a change in a variable serves to amplify rather than reverse the activity. The stimula-tion of uterine contractions during childbirth is an example.

Homeostatic mechanisms allow for regulated change in the body's internal environment when necessary.

■ INTERACTIVE QUESTION 40.6

One of the negative feedback mechanisms controlling body temperature in humans involves the **a.**_____ _____, which are signaled to increase their activity when the **b.**_____ senses a rise in body temperature. When body temperature falls below the **c.**____ _____, the thermostat in the brain stops sending "sweat" signals.

Introduction to the Bioenergetics of Animals

Animals are heterotrophs that harvest chemical energy from the food they eat (844)

Animals, as heterotrophs, obtain their chemical en-ergy from organic molecules synthesized by other or-ganisms. The fuel molecules obtained from the diges-tion of food are absorbed into body cells and used to generate ATP for cellular work and as carbon skele-tons for biosynthesis.

Metabolic rate provides clues to an animal's bioenergetic "strategy" (844–845)

The total energy an animal uses in a unit of time is its **metabolic rate.** Energy is measured in calories (cal) or kilocalories (kcal). Metabolic rate can be measured by placing an animal in a calorimeter and measuring heat loss. The rate of oxygen consumption, also a measure of metabolic rate, can be determined with a respirometer.

Birds and mammals are **endothermic,** warming their bodies with metabolic heat. This high-energy

bioenergetic strategy allows for high activity levels over a range of environmental temperatures. **Ectotherms,** such as most fishes, amphibians, reptiles, and inverte-brates, require less energy and do not produce enough metabolic heat to generate and maintain a higher body temperature than that of the environment.

Metabolic rate per gram is inversely related to body size among similar animals (845)

The energy required to maintain each gram of body weight is inversely related to body size. Smaller ani-mals have higher metabolic rates, breathing rates, rel-ative volume of blood, and heart rates. With a greater surface-to-volume ratio, small endotherms may have a higher energy cost to maintain a stable body tem-perature. But the factors that contribute to this inverse relationship among ectotherms as well as endotherms are not fully understood.

Animals adjust their metabolic rates as conditions change (845–846)

The minimal metabolic rate for a nongrowing en-dotherm at rest, fasting, and nonstressed is called the **basal metabolic rate (BMR).** A human adult male's rate is about 1,600 to 1,800 kcal/day; a female is about 1,300–1,500 kcal/day. The **standard metabolic rate (SMR)** is the metabolic rate of a resting, fasting, non-stressed ectotherm determined at a specific temperature.

Maximal metabolic rates occur during intense ac-tivity and are inversely related to the duration of the activity. Metabolic rates are influenced by age, sex, size, activity level, time of day, and other variables. Most terrestrial animals have an average daily rate of energy consumption that is 2–4 times BMR or SMR.

■ INTERACTIVE QUESTION 40.7

Why can an endotherm sustain intense activity for a much longer time period than an ectotherm can?

Energy budgets reveal how animals use energy and materials (846–847)

Different species vary in the allocation of their food energy. For most animals, however, the majority of food is used to produce ATP, with little going to growth and reproduction. Endotherms require much more energy per kilogram than do ectotherms; and a small animal requires more energy per kilogram than does a large animal of the same taxonomic class.

WORD ROOTS

chondro- = cartilage; **-cyte** = cell (*chondrocytes:* cartilage cells)

ecto- = outside; **-therm** = heat (*ectothermic:* organisms that do not produce enough metabolic heat to have much effect on body temperature)

endo- = inside (*endothermic:* organisms with bodies that are warmed by heat generated by metabolism. This heat is usually used to maintain a relatively stable body temperature higher than that of the external environment)

fibro- = a fiber (*fibroblast:* a type of cell in loose connective tissue that secretes the protein ingredients of the extracellular fibers)

homeo- = same; **-stasis** = standing, posture (*homeostasis:* the steady-state physiological condition of the body)

inter- = between (*interstitial fluid:* the internal environment of vertebrates, consisting of the fluid filling the space between cells)

macro- = large (*macrophage:* an amoeboid cell that moves through tissue fibers, engulfing bacteria and dead cells by phagocytosis)

osteo- = bone; **-blast** = a bud, sprout (*osteoblasts:* bone-forming cells that deposit a matrix of collagen)

STRUCTURE YOUR KNOWLEDGE

1. Fill in the table below on the structure and function of the four types of animal tissues.

2. Organs are composed of layers of several different tissues. Which of the four major animal tissues do you think would be included in all organs? In what types of organs would you predict the remaining tissues would be included? Give an example of an organ composed of several layers of tissues.

TEST YOUR KNOWLEDGE

MULTIPLE CHOICE: *Choose the one best answer.*

1. Which of the following is *not* an organ system?
   a. skeletal
   b. connective
   c. digestive
   d. excretory
   e. immune and lymphatic

2. A stratified squamous epithelium would be composed of
   a. several layers of flat cells attached to a basement membrane.
   b. a layer of ciliated, mucus-secreting, flattened cells.
   c. a hierarchical arrangement of boxlike cells.
   d. an irregularly arranged layer of pillarlike cells.
   e. several layers of flat cells underneath columnar cells.

3. Which of the following is *not* true of connective tissue?
   a. It consists of few cells surrounded by fibers in a matrix.
   b. It includes such diverse tissues as bone, cartilage, tendons, adipose, and loose connective tissue.
   c. It connects and supports other tissues.
   d. It forms the internal and external lining of many organs.
   e. It can have a matrix that is a liquid, gel, or solid.

Tissue	Structural Characteristics	General Functions	Specific Examples

4. Which of the following are *incorrectly* paired?
   a. blood—erythrocytes, leukocytes, and platelets in plasma
   b. bone—osteocytes embedded in hydroxyapatite in osteon
   c. loose connective tissue—collagenous, elastic, reticular fibers
   d. adipose tissue—loose connective tissue with fat-storing cells
   e. fibrous connective tissue—chondrocytes embedded in chondroitin sulfate

5. Which of the following is the best description of smooth muscle?
   a. striated, branching cells; involuntary control
   b. spindle-shaped cells; involuntary control
   c. spindle-shaped cells connected by intercalated disks
   d. striated cells containing overlapping filaments; involuntary control
   e. spindle-shaped striated cells; voluntary control

6. The diaphragm
   a. is a mesentery.
   b. increases the surface area of the lungs.
   c. is part of the mammalian reproductive system.
   d. separates the thoracic and abdominal cavities in mammals.
   e. separates the lungs from the heart cavity.

7. The interstitial fluid of vertebrates
   a. is the internal environment within cells.
   b. bathes cells and provides for the exchange of nutrients and wastes.
   c. makes up the plasma of blood.
   d. surrounds unicellular and flat, thin animals.
   e. is less abundant in ectotherms than in endotherms.

8. Negative feedback circuits are
   a. mechanisms that most commonly maintain homeostasis.
   b. activated only when a physiological variable rises above a set point.
   c. analogous to a radiator that heats a room.
   d. involved in maintaining contractions during childbirth.
   e. found in endotherms but not in ectotherms.

9. The basal metabolic rate
   a. is constant for each species.
   b. may vary depending on the sex or size of an organism.
   c. is highest when an animal is actively exercising.
   d. is lower than the standard metabolic rate for ectotherms.
   e. may be measured from the quantity of food an animal eats.

10. Which of the following would most likely have the highest metabolic rate?
    a. whale
    b. dog
    c. snake
    d. tuna
    e. bat

11. Dolphins, penguins, sharks, and seals all have a fusiform body shape. What is the best explanation for this similarity?
    a. This shape is an example of divergent evolution.
    b. They are all vertebrates and evolved from a common ancestor.
    c. The physical laws of hydrodynamics drove the natural selection for this shape.
    d. All aquatic animals have this shape, an example of convergent evolution.
    e. This shape is determined by the thermodynamics of maintaining body temperature in cold water.

12. Of the following animals, whose percentage of the energy budget available for growth would be the largest?
    a. deer mouse in temperate forest
    b. penguin from Antarctica
    c. human from temperate climate
    d. python from tropical India
    e. They would all be the same, because most energy is used to maintain their basic metabolic rate.

CHAPTER 41

ANIMAL NUTRITION

FRAMEWORK

A nutritionally adequate diet provides sufficient calories, essential amino acids, vitamins, and minerals.

Animals eat other organisms to obtain fuel for respiration, organic raw materials for biosynthesis, and essential nutrients. Digestion is the enzymatic hydrolysis of macromolecules into monomers that can be absorbed across cell membranes.

Gastrovascular cavities are digestive sacs in which some extracellular digestion takes place before food particles are phagocytosed by cells lining the cavity. Alimentary canals are one-way tracts with specialized regions for mechanical breakdown of food, storage, digestion, absorption of nutrients, and elimination of wastes.

This chapter details the structures, functions, enzymes, and hormones of the human digestive tract.

CHAPTER REVIEW

Nutritional Requirements

Animals are heterotrophs that require food for fuel, carbon skeletons, and essential nutrients: *an overview* **(850)**

A nutritionally adequate diet provides chemical energy for cellular work, raw materials for biosynthesis, and essential nutrients in prefabricated form.

Homeostatic mechanisms manage an animal's fuel (850–852)

The monomers of carbohydrates, fats, and proteins can be used as fuel to produce ATP by cellular respiration, although the first two are used preferentially.

Glucose Regulation as an Example of Homeostasis in Nutrition When an animal consumes more calories than are needed to meet its energy requirements, the excess can be used for biosynthesis, or stored in the liver and muscles as glycogen or in adipose tissue as fat when the glycogen stores are full. The level of glucose in the blood is carefully regulated by pancreatic hormones.

Caloric Imbalance An **undernourished** person or other animal has a diet insufficient in calories. With severe deficiency, the body breaks down its own proteins for energy, eventually causing irreversible damage.

Overnourishment, or obesity, is becoming increasingly common in affluent nations. Excess fat calories in the diet are readily converted into fat stores.

Obesity The human body seems to have complex feedback mechanisms that stabilize weight. The concentration of leptin, a hormone produced by adipose cells, increases with an increase in adipose tissue, signaling the brain to decrease appetite and increase energy use. A decrease in adipose tissue, and thus leptin levels, results in increased appetite and weight gain. Some of the chemical signals involved in fat homeostasis may be developed as obesity drugs.

Sometimes obesity is adaptive, as in the obese chicks of some petrel species.

An animal's diet must supply essential nutrients and carbon skeletons for biosynthesis (852–856)

Animals can fabricate most of the organic molecules they need from the carbon skeletons and organic nitrogen acquired from food.

Molecules that an animal requires but cannot make are called **essential nutrients.** These requirements vary from species to species. When the diet is lacking one or more essential nutrients, the animal is said to be **malnourished.**

Essential Amino Acids Eight of the 20 amino acids required to make proteins are **essential amino acids** in the adult human diet. Protein deficiency develops from a diet that lacks one or more essential amino acids and produces retarded physical and perhaps mental development in children.

Meat, eggs, and cheese contain complete proteins with all essential amino acids in proportions that meet human requirements. Most plant proteins are incomplete, and diets built on a single staple, such as corn, beans, or rice, can result in protein deficiency.

■ INTERACTIVE QUESTION   41.1

How can vegetarians avoid protein deficiencies? Explain why such dietary consideration is necessary.

Essential Fatty Acids   Animals are able to make most of the fatty acids they need. Linoleic acid is required in the human diet. Deficiencies of **essential fatty acids** are rare.

Vitamins   **Vitamins** are essential organic molecules required in small amounts in the diet. Thirteen vitamins essential to humans have been identified. Water-soluble vitamins include the B-complex, most of which function as coenzymes, and vitamin C, required for the production of connective tissue. The fat-soluble vitamins are A, incorporated into visual pigments; D, aiding in calcium absorption and bone formation; E, seeming to protect phospholipids in membranes from oxidation; and K, required for blood clotting.

Minerals   **Minerals** are inorganic nutrients, usually needed in very small amounts. Vertebrates require relatively large quantities of calcium and phosphorus for bone construction. Calcium is also needed for normal nerve and muscle function, and phosphorus is needed for ATP and nucleic acids. Iron is a component of the cytochromes and hemoglobin. Other minerals function as cofactors of enzymes. Iodine is needed by vertebrates to make the metabolism-regulating thyroid hormones. Sodium, potassium, and chlorine are important in nerve function and osmotic balance.

Food Types and Feeding Mechanisms

Most animals are opportunistic feeders (856)

Herbivores eat autotrophs; **carnivores** eat animals; and **omnivores** consume both autotrophs and animals. Most animals are opportunistic and eat foods from different categories when they are available.

Diverse feeding adaptations have evolved among animals (856–857)

Many aquatic animals are **suspension-feeders,** sifting small food particles from the water. **Substrate-feeders** live in or on their food, eating their way through it. **Deposit-feeders** are substrate-feeders that consume decaying organic matter. **Fluid-feeders** suck fluids from a living plant or animal host. Most animals are **bulk-feeders,** eating relatively large pieces of food.

Overview of Food Processing

The four main stages of food processing are ingestion, digestion, absorption, and elimination (857–858)

Ingestion is the act of eating. **Digestion** splits macromolecules into monomers by **enzymatic hydrolysis,** the addition of a water molecule when the bond between monomers is broken. Mechanical fragmentation often precedes chemical digestion.

In the last two stages of food processing, monomers and small molecules are **absorbed** into the cells of the animal, and the undigested remainder of the food is **eliminated.**

■ INTERACTIVE QUESTION   41.2

Why must macromolecules be digested into monomers?

Digestion occurs in specialized compartments (858–859)

Intracellular Digestion   **Intracellular digestion** of food molecules in food vacuoles is typical of heterotrophic protists and sponges. Food vacuoles fuse with enzyme-containing lysosomes, and digestion occurs safely within a membrane-enclosed compartment.

Extracellular Digestion   Most animals break down their food, at least initially, by **extracellular digestion** within a separate compartment of the body that connects to the external environment.

Single-opening **gastrovascular cavities** function in both digestion and transport of nutrients throughout the body. In the small cnidarian hydra, digestive enzymes secreted into the gastrovascular cavity initiate food breakdown. Gastrodermal cells take in food particles by phagocytosis, and hydrolysis of macromolecules occurs within food vacuoles. Undigested materials are expelled through the mouth/anus.

Most animals have **complete digestive tracts** or **alimentary canals,** with two openings, a mouth and an anus, and specialized regions allowing for the sequential digestion and absorption of nutrients. Food, ingested through the mouth and pharynx, passes through an esophagus that leads to either a crop, stomach, or gizzard—organs specialized for storing or grinding food. In the intestine, digestive enzymes hydrolyze macromolecules and nutrients are absorbed across the tube lining. Undigested material exits through the anus.

■ **INTERACTIVE QUESTION 41.3**

Label and list the functions for the five organs of an earthworm's digestive tract. What is the function of the typhlosole?

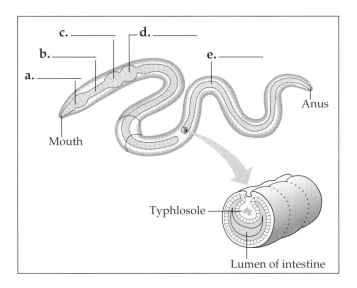

The Mammalian Digestive System

The mammalian digestive system consists of the alimentary canal and accessory glands. Rhythmic waves of muscular contraction called **peristalsis** push food through the tract. Ringlike valves called **sphincters** regulate the passage of material between some segments. Accessory glands, the **salivary glands, pancreas,** and **liver** with its **gallbladder,** secrete digestive juices into the alimentary canal through ducts.

The following sections describe the human digestive system.

The oral cavity, pharynx, and esophagus initiate food processing (860–861)

The Oral Cavity Physical and chemical digestion begins in the mouth, where teeth grind food to expose a greater surface area to enzyme action. The presence of food in the **oral cavity** triggers the release of saliva.

Saliva has several components: mucin, a glycoprotein that protects the mouth lining from abrasion and lubricates food for swallowing; buffers to neutralize acidity; antibacterial agents; and **salivary amylase,** which begins the hydrolysis of starch and glycogen into smaller polysaccharides and maltose. The tongue tastes and manipulates food, and pushes the food ball, or **bolus,** into the pharynx for swallowing.

The Pharynx The **pharynx** is the intersection leading to both the esophagus and the trachea. During swallowing, the top of the windpipe moves up so that its opening is blocked by the cartilaginous **epiglottis.**

The Esophagus Food moves down through the **esophagus** to the stomach, squeezed along by a wave of peristalsis.

The stomach stores food and performs preliminary digestion (861–863)

The expandable **stomach** stores food so we do not have to eat constantly.

The epithelium lining the stomach secretes **gastric juice,** a digestive fluid containing hydrochloric acid that breaks down food tissues, kills bacteria, and denatures proteins, and **pepsin,** an enzyme that hydrolyzes specific peptide bonds in proteins. Pepsin is synthesized and secreted in an inactive form called **pepsinogen.** It is activated by hydrochloric acid and by pepsin itself—an example of positive feedback.

The mucous coating secreted by the epithelium protects the stomach lining from digestion. Gastric ulcers, mainly caused by bacteria, may worsen when the lining is eroded faster than it can be regenerated.

Smooth muscles mix the contents of the stomach. **Acid chyme** is the nutrient broth produced by the action of the stomach and its secretions on ingested food. The stomach is usually closed off by two sphincters: one at the cardiac orifice prevents backflow into the esophagus, and the **pyloric sphincter** regulates passage of acid chyme into the intestine.

■ INTERACTIVE QUESTION 41.4

List the two types of macromolecules that have been partially digested by the time acid chyme moves into the intestine. Where did this digestion take place and what enzymes were involved?

a.

b.

The small intestine is the major organ of digestion and absorption (863–866)

Most enzymatic hydrolysis of macromolecules and nutrient absorption into the blood takes place in the **small intestine.**

Digestive juices (from the pancreas, liver, gallbladder, and gland cells of the intestinal wall) are mixed with the chyme in the **duodenum,** the first section of the small intestine. The pancreas produces digestive enzymes and a bicarbonate-rich alkaline solution that offsets the acidity of the chyme. The liver produces **bile,** which is stored in the gallbladder until needed. Bile aids in the digestion of fats and contains pigments that are by-products of the breakdown of red blood cells in the liver.

Enzymatic Action in the Small Intestine The digestion of starch and glycogen into disaccharides is continued by pancreatic amylases. Disaccharidases, enzymes specific for hydrolysis of different disaccharides, are built into the membranes and extracellular matrix of epithelial cells, facilitating monomer absorption through the intestinal wall.

Protein digestion is completed in the small intestine by **trypsin** and **chymotrypsin,** enzymes specific for peptide bonds adjacent to certain amino acids; **carboxypeptidase,** which splits amino acids off the free carboxyl end; and **aminopeptidase,** which works from the amino end. **Dipeptidases** are attached to the intestinal epithelium and split small peptides. The pancreatic protein-digesting enzymes are secreted in inactive form and activated by **enteropeptidase.**

Nucleases are a group of enzymes that hydrolyze DNA and RNA into their nucleotide monomers. Other enzymes dismantle nucleotides.

The digestion of fats is aided by bile salts, which coat or **emulsify** tiny fat droplets so they do not coalesce, leaving a greater surface area for **lipase** to hydrolyze the fat molecules.

Most digestion is completed while the chyme is still in the duodenum. The **jejunum** and **ileum** are regions of the small intestine that function in nutrient and water absorption.

■ INTERACTIVE QUESTION 41.5

List the enzymes that hydrolyze the following macromolecules in the small intestine. Indicate which enzymes are attached to the epithelial cells of the intestine.

a. polysaccharides

b. polypeptides

c. DNA, RNA

d. fats

Absorption of Nutrients Circular folds of the small intestine lining are covered with fingerlike projections called **villi,** on which the epithelial cells have microscopic extensions called **microvilli,** creating a huge surface area adapted for absorption.

The core of each villus has a net of capillaries and a lymph vessel called a **lacteal.** Nutrients are absorbed across the epithelium of the villus and then across the single-celled epithelium of the capillaries or lacteal. Transport may be passive by diffusion or active by pumping against a gradient.

Glycerol and fatty acids are absorbed by epithelial cells where they recombine to form fats and are mixed with cholesterol and coated with proteins to make tiny globules called **chylomicrons.** These packages are transported by exocytosis out of the epithelial cells and into a lacteal, then transported by the lymphatic system to veins near the heart.

Absorbed amino acids and sugars enter capillaries. The nutrient-laden blood from the small intestine is carried directly to the liver by the **hepatic portal vessel.** The liver interconverts molecules and regulates the nutrient content of the blood.

Digestive Efficiency and Cost From a diet typical of developed countries, 80–90% of the organic matter consumed is absorbed. Cellulose is the main undigestible material. Depending on the animal and the diet, the energy cost of digestion and absorption can range from 3% to 30% of the energy of the meal.

Hormones help regulate digestion (866)

The sight, smell, or taste of food sends a nervous message from the brain to the stomach that initiates secretion of gastric juice. Food then stimulates the stomach wall to release the hormone **gastrin** into the circulatory system, stimulating continued secretion of gastric juice. If the pH of the stomach contents becomes too low, the release of gastrin is inhibited.

Enterogastrones are hormones produced by the duodenum that coordinate the release of digestive secretions: **Secretin** is released in response to the acidic pH of the chyme and stimulates the pancreas to release bicarbonate. **Cholecystokinin (CCK),** produced in response to amino acids or fatty acids, stimulates gallbladder contraction and the release of pancreatic enzymes. A fat-rich chyme causes the duodenum to release other enterogastrones that inhibit peristalsis in the stomach, slowing the release of chyme.

Reclaiming water is a major function of the large intestine (866)

The small intestine leads into the **large intestine,** or **colon,** at a T-shaped junction with a sphincter. A pouch called the **cecum,** with a fingerlike extension, the **appendix,** attaches at this juncture. The colon finishes the reabsorption of the large quantity of water secreted with digestive enzymes into the digestive tract.

Escherichia coli and other mostly harmless bacteria live on organic material in the **feces.** Some of these bacteria produce vitamins such as several B vitamins and vitamin K, which are absorbed by the host. The feces contain cellulose, other undigested ingredients of food, salts excreted by the colon, and a large proportion of intestinal bacteria. Feces are stored in the **rectum.**

Evolutionary Adaptations of Vertebrate Digestive Systems

Structural adaptations of digestive systems are often associated with diet (867)

Dentition, the type and arrangement of teeth, correlates with diet.

Herbivores have longer alimentary canals because plant material is more difficult to digest than meat. Specialized structures, such as a shark's spiral valve, functionally increase intestinal length.

Symbiotic microorganisms help nourish many vertebrates (868)

Many herbivorous mammals have special fermentation chambers filled with symbiotic bacteria and protists. These microorganisms, often housed in the cecum, digest cellulose into simple sugars and produce a variety of essential nutrients for the animal.

Ruminants have an elaborate system involving several stomachs, regurgitation and rechewing of the cud, and digestion of their symbiotic bacteria to maximize the nutrient yield of their grass or hay diet.

■ INTERACTIVE QUESTION 41.6

Compare and contrast the dentition and alimentary canals of carnivores and herbivores.

WORD ROOTS

chylo- = juice; **micro-** = small (*chylomicron:* small globules composed of fats that are mixed with cholesterol and coated with special proteins)

chymo- = juice; **-trypsi** = wearing out (*chymotrypsin:* an enzyme found in the duodenum; it is specific for peptide bonds adjacent to certain amino acids)

di- = two (*dipeptidase:* an enzyme found attached to the intestinal lining; it splits small peptides)

entero- = the intestines (*enterogastrones:* a category of hormones secreted by the wall of the duodenum)

epi- = over; **-glotti** = the tongue (*epiglottis:* a cartilaginous flap that blocks the top of the windpipe, the glottis, during swallowing)

extra- = outside (*extracellular digestion:* the breakdown of food outside cells)

gastro- = stomach; **-vascula** = a little vessel (*gastrovascular cavities:* an extensive pouch that serves as the site of extracellular digestion and a passageway to disperse materials throughout most of an animal's body)

herb- = grass; **-vora** = eat (*herbivore:* a heterotrophic animal that eats plants)

hydro- = water; **-lysis** = to loosen (*hydrolysis:* a chemical process that lyses or splits molecules by the addition of water)

intra- = inside (*intracellular digestion:* the joining of food vacuoles and lysosomes to allow chemical digestion to occur within the cytoplasm of a cell)

micro- = small; **-villi** = shaggy hair (*microvilli:* many fine, fingerlike projections of the epithelial cells in the lumen of the small intestine that increase its surface area)

omni- = all (*omnivore:* a heterotrophic animal that consumes both meat and plant material)

peri- = around; **-stalsis** = a constriction (*peristalsis:* rhythmic waves of contraction of smooth muscle that push food along the digestive tract)

STRUCTURE YOUR KNOWLEDGE

1. Food provides fuel, organic raw materials, and essential nutrients. Complete the following concept map that summarizes the nutritional needs of animals.

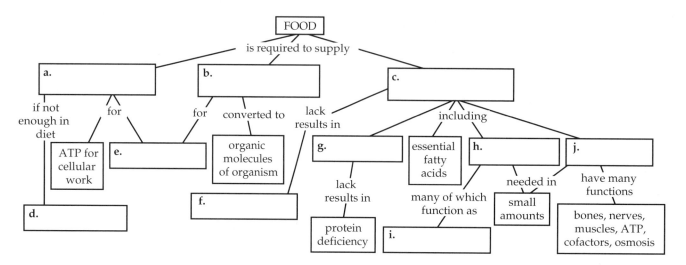

2. Label the indicated structures in this diagram of the human digestive system. Review the functions of these structures.

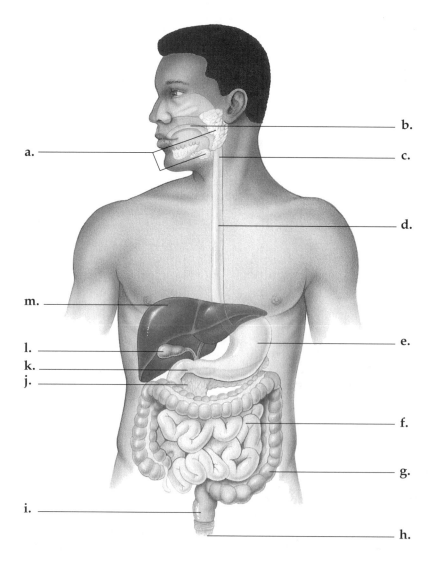

TEST YOUR KNOWLEDGE

MATCHING: *Match the description with the correct enzyme or hormone.*

_____ 1. enzyme that hydrolyzes peptide bonds, works in the stomach

_____ 2. hormone that stimulates secretion of gastric juice

_____ 3. hormone that stimulates release of bile and pancreatic enzymes

_____ 4. enzyme that begins digestion of starch in the mouth

_____ 5. enzymes specific for hydrolyzing various disaccharides

_____ 6. hormones that inhibit peristalsis in the stomach

_____ 7. enzyme that hydrolyzes fats

_____ 8. intestinal enzyme that activates protein-digesting enzymes

_____ 9. hormone that stimulates secretion of bicarbonate ions from the pancreas

_____ 10. intestinal enzyme specific for peptide bonds adjacent to specific amino acids

A. aminopeptidase

B. amylase

C. bile salts

D. cholecystokinin

E. chymotrypsin

F. disaccharidases

G. enterogastrones

H. enteropeptidase

I. gastrin

J. lipase

K. maltase

L. nuclease

M. pepsin

N. secretin

MULTIPLE CHOICE: *Choose the one best answer.*

1. Deposit-feeders such as earthworms
   a. feed mostly on mineral substrates.
   b. filter small organisms from water.
   c. eat autotrophs.
   d. feed on detritus.
   e. are bulk feeders.

2. The energy content of fats
   a. is released by bile salts.
   b. may be lost unless an herbivore eats some of its feces.
   c. is approximately two times that of carbohydrates or proteins.
   d. can reverse the effects of malnutrition.
   e. Both c and d are correct.

3. Nucleosidases are
   a. hormones that stimulate release of pancreatic enzymes.
   b. enzymes attached to the intestinal epithelium that hydrolyze nucleosides.
   c. hydrolytic enzymes manufactured in inactive forms to protect the cells that produce them.
   d. protein-digesting enzymes that are activated by hydrochloric acid.
   e. enzymes that hydrolyze DNA.

4. Which of the following statements is *false?*
   a. The average human has enough stored fat to supply calories for several weeks.
   b. An increase in leptin levels leads to an increase in appetite and weight gain.
   c. Conversion of glucose and glycogen takes place in the liver.
   d. After glycogen stores are filled, excessive calories are stored as fat, regardless of their original food source.
   e. Carbohydrates and fats are preferentially used as fuel before proteins are used.

5. The purpose of antidiarrhea medicine would most likely be to
   a. speed up peristalsis in the small intestine.
   b. speed up peristalsis in the large intestine.
   c. kill *E. coli* in the intestine.
   d. increase water reabsorption in the large intestine.
   e. increase salt secretion into the feces.

6. Incomplete proteins are
   a. lacking in essential vitamins.
   b. a cause of undernourishment.
   c. found in meat, eggs, and cheese.
   d. lacking in one or more essential amino acids.
   e. a result of overcooking vegetables.

7. Which of the following is a *true* statement about vitamins?
   a. They may be produced by intestinal microorganisms.
   b. They are the same from one species to the next.
   c. They are stored in large quantities in the liver.
   d. They are inorganic nutrients, needed in small amounts, that usually function as cofactors.
   e. They are all water soluble and must be replaced every day.

8. A distinct advantage of extracellular digestion over intracellular digestion is that
   a. polymers are hydrolyzed to monomers by digestive enzymes.
   b. there is a greater surface area for absorption of digested nutrients.
   c. larger pieces of food can be ingested and then digested.
   d. all four types of macromolecules can be digested instead of just glucose.
   e. the products of extracellular digestion can be absorbed into all body cells, without the need for a transport system.

9. Ruminants
   a. have teeth adapted for an omnivorous diet.
   b. use microorganisms to digest cellulose.
   c. eat their feces to obtain nutrients digested from cellulose by microorganisms.
   d. house symbiotic bacteria and protists in a cecum.
   e. get all of their nutrition from digested plant material.

10. How do molting penguins obtain enough protein to produce new feathers?
    a. They become obese right before molting in order to store protein.
    b. They combine different protein sources to make up for the incomplete proteins in their food.
    c. They become malnourished during the period of new feather production.
    d. They feed on fish, which are an excellent protein source.
    e. They stockpile extra muscle proteins prior to molting, which then supply amino acids for protein synthesis.

11. The acid pH of the stomach
    a. hydrolyzes proteins.
    b. is regulated by the release of gastrin.
    c. is neutralized by gastric juice.
    d. is produced by pepsin.
    e. triggers the release of enterogastrones.

12. After a meal of greasy french fries, which enzymes would you expect to be most active?
    a. salivary and pancreatic amylase, disaccharidases, lipase
    b. lipase, lactase, maltase
    c. pepsin, trypsin, chymotrypsin, dipeptidases
    d. gastric juice, bile, bicarbonate
    e. sucrase, lipase, bile

13. Chylomicrons are
    a. lipoproteins transported by the circulatory system.
    b. small branches of the lymphatic system.
    c. protein-coated fat globules excreted out of epithelial cells into a lacteal.
    d. small peptides acted upon by chymotrypsin.
    e. fats emulsified by bile salts.

14. Which of the following is *not* a common component of feces?
    a. intestinal bacteria
    b. cellulose
    c. saturated fats
    d. bile pigments
    e. salts

15. The hepatic portal vessel
    a. supplies the capillaries of the intestines.
    b. carries absorbed nutrients to the liver for processing.
    c. carries blood from the liver to the heart.
    d. drains the lacteals of the villi.
    e. supplies oxygenated blood to the liver.

16. An organism that exclusively uses extracellular digestion would
    a. be a sponge or a heterotrophic protist.
    b. have a gastrovascular cavity.
    c. not need a circulatory system.
    d. have to be very thin and elongated.
    e. have a specialized body compartment with digestive enzymes.

17. One would expect to find a gastrovascular cavity in a (an)
    a. hydra.
    b. fish.
    c. insect.
    d. earthworm.
    e. bird.

18. What moves a bolus of food down the esophagus to the stomach?
    a. the closing of the epiglottis over the glottis
    b. a lower pressure in the abdominal cavity compared to the oral cavity
    c. the contraction of the diaphragm and the opening of the pyloric sphincter
    d. peristalsis caused by waves of smooth muscle contraction
    e. the action of cilia coated by mucus

19. What are villi and what do they do?
    a. folds in the stomach that allow the stomach to expand
    b. extensions of the lymphatic system that pick up digested fats for transport to the circulatory system and then to the liver
    c. fingerlike projections of the small intestine lining that increase the surface area for absorption
    d. microscopic extensions of epithelial cells lining the small intestine that provide more surface area for digestion
    e. projections of the intestinal capillaries that join to form the hepatic portal vein

20. Which of the following is mismatched with its function?
    a. most B vitamins—coenzymes
    b. vitamin E—antioxidant
    c. vitamin K—blood clotting
    d. iron—component of thyroid hormones
    e. phosphorus—bone formation, nucleotide synthesis

21. Why does salivary amylase not hydrolyze starch in the duodenum?
    a. Starch is completely hydrolyzed into maltose in the oral cavity.
    b. The acid pH of the stomach denatures salivary amylase and pepsin begins hydrolyzing it.
    c. Salivary amylase is produced by salivary glands and never leaves the oral cavity.
    d. Pancreatic amylase is a more effective enzyme in the pH of the duodenum.
    e. Salivary amylase can hydrolyze glycogen but not starch.

22. Why do many vegetarians combine different protein sources in the same meal or eat some animal products such as eggs or milk products?
    a. to make sure they obtain sufficient calories
    b. to provide sufficient vitamins
    c. to make sure they ingest all essential fatty acids
    d. to make their diet more interesting
    e. to provide all essential amino acids at the same time

CHAPTER 42

CIRCULATION AND GAS EXCHANGE

FRAMEWORK

This chapter surveys the basic approaches to circulation and gas exchange found in the animal kingdom, with special attention to the human systems and the problems of cardiovascular disease. The following concept map organizes some of the chapter's key ideas.

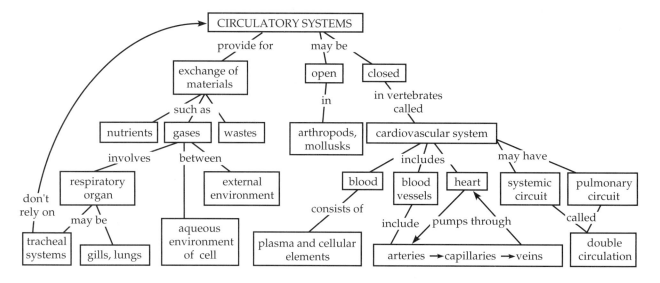

CHAPTER REVIEW

Every living cell must reside in an aqueous environment, which provides oxygen and nutrients and permits disposal of carbon dioxide and metabolic wastes. Most animals have organ systems that exchange materials with the environment, and many have an internal transport system to service the body's cells.

Circulation in Animals

Transport systems functionally connect the organs of exchange with the body cells: *an overview* (871–872)

The circulatory system provides an internal transport system to move substances across long distances in animals. Blood exchanges materials with the external environment across the thin and extensive epithelia of organs specialized for gas exchange, nutrient absorption, and waste removal and then carries materials throughout the body for delivery to the interstitial fluid that bathes the body cells.

Most invertebrates have a gastrovascular cavity or a circulatory system for internal transport (872–873)

Gastrovascular Cavities The central gastrovascular cavity inside the two-cell-thick body wall of cnidarians serves for both digestion and transport of materials. The internal fluid exchanges directly with the aqueous environment through the single opening. Flatworms also have gastrovascular cavities that branch throughout the thin and flat body.

Open and Closed Circulatory Systems The circulatory systems of more complex animals consist of **blood** pumped by a **heart** through **blood vessels. Blood pressure** is the hydrostatic force that moves blood. In the **open circulatory system** found in insects, other arthropods, and in most mollusks, **hemolymph in sinuses,** or spaces between organs, bathes the internal tissues, providing for chemical exchange.

Annelids, some mollusks, and vertebrates have **closed circulatory systems,** in which the blood remains in vessels and exchanges materials with the interstitial fluid bathing the cells.

■ INTERACTIVE QUESTION 42.1

a. How is the hemolymph of an open circulatory system moved throughout the body?

b. How is the blood of a closed circulatory system circulated through the body?

Vertebrate phylogeny is reflected in adaptations of the cardiovascular system (873–875)

In the **cardiovascular system** of vertebrates, the heart has one or more **atria,** which receive blood, and one or more **ventricles,** which pump blood out of the heart.

Arteries, carrying blood away from the heart, branch into tiny **arterioles** within organs, which then divide into the microscopic **capillaries.** Exchange of substances between blood and interstitial fluid occurs within **capillary beds.** Capillaries converge to form **venules,** which meet to form the **veins** that return blood to the heart.

The ventricle of a fish's two-chambered heart pumps blood first to the capillary beds of the gills (the **gill circulation**), from which the oxygen-rich blood flows through a vessel to the capillary beds in the other organs (the **systemic circulation**). Veins return the oxygen-poor blood to the atrium. Passage through two capillary beds slows the flow of blood, but body movements help to maintain circulation.

In the three-chambered heart of amphibians, the single ventricle pumps blood through a forked artery into the **pulmocutaneous circulation,** which leads to lungs and skin capillaries and then back to the left atrium, and the **systemic circulation,** which carries blood to the rest of the body and back to the right atrium. This **double circulation** repumps blood after it returns from the capillary beds of the lungs or skin, ensuring a strong flow of oxygen-rich blood to the brain, muscles, and body organs. A ridge in the ventricle helps to direct oxygen-rich blood from the left atrium into the systemic circuit and oxygen-poor blood into the pulmocutaneous circuit.

The three-chambered reptilian heart has a partially divided ventricle that helps to separate blood flow through a **pulmonary circuit** to the gas exchange tissues in the lungs and through a systematic circuit to the body tissues.

Delivery of oxygen for cellular respiration is most efficient in birds and mammals, which, as endotherms, have high oxygen demands. The left side of the large and powerful four-chambered heart handles only oxygen-rich blood, whereas the right side receives and pumps oxygen-poor blood.

Double circulation in mammals depends on the anatomy and pumping cycle of the heart (875–877)

The Mammalian Heart: a closer look The human heart, located just beneath the sternum, is composed mostly of cardiac muscle. The atria have relatively thin muscular walls, whereas the ventricles have thicker walls. The **cardiac cycle** consists of the **systole,** during which the cardiac muscle contracts and the chambers pump blood, and the **diastole,** when the heart chambers are relaxed and filling with blood.

The **cardiac output,** or volume of blood pumped per minute into the systemic circuit, depends on **heart rate** and **stroke volume,** or quantity of blood pumped by each contraction of the left ventricle.

Atrioventricular (AV) valves between each atrium and ventricle are snapped shut when blood is forced back against them as the ventricles contract. The **semilunar valves** at the exit of the aorta and pulmonary artery are forced open by ventricular contraction and close when the ventricles relax and blood starts to flow back toward the ventricles.

Heart rate can be measured by taking the **pulse,** which is caused by the rhythmic stretching of arteries as the ventricles contract and pump blood through them.

The heartbeat sounds are caused by the closing of the heart valves. A **heart murmur** is the detectable hissing sound of blood leaking back through a defective valve.

■ **INTERACTIVE QUESTION   42.2**

The diagram below will help you review the flow of blood through a mammalian circulatory system. Label the indicated parts, color the vessels that carry oxygen-rich blood red, and then trace the flow of blood by numbering the circles from 1–11. Start with number 1 in the right ventricle.

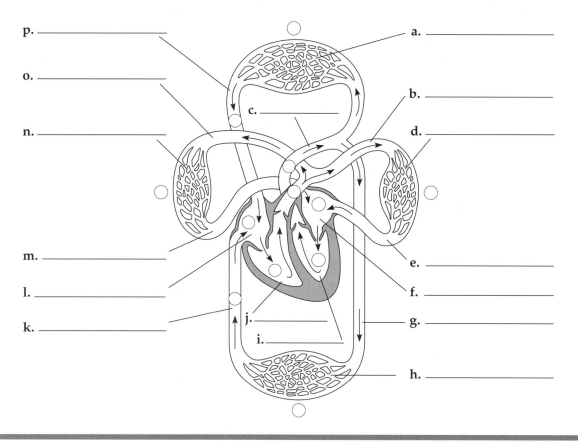

p. _____

o. _____

c. _____

n. _____

m. _____

l. _____

k. _____

j. _____

i. _____

a. _____

b. _____

d. _____

e. _____

f. _____

g. _____

h. _____

Maintaining the Heart's Rhythmic Beat   Certain cells of cardiac muscle are self-excitable; they have an intrinsic ability to contract. The rhythm of contractions is coordinated by the **sinoatrial (SA) node,** or **pacemaker,** a region of specialized muscle tissue located in the wall of the right atrium near the superior (anterior) vena cava. The SA node initiates an electrical impulse that spreads via the intercalated disks of the cardiac muscle cells, and the two atria contract. The **atrioventricular (AV) node,** located between the right atrium and ventricle, relays the impulse after a 0.1-second delay through fibers to all parts of the ventricular walls, causing contraction.

The electrical currents produced during the heart cycle can be detected by electrodes placed on the skin and recorded in an **electrocardiogram (EKG or ECG).**

The SA node is controlled by two sets of nerves with antagonistic signals and is influenced by hormones, temperature, and exercise.

■ **INTERACTIVE QUESTION   42.3**

a. The valve between an atrium and ventricle is

_____.

b. The valve between a ventricle and the aorta or pulmonary artery is _____.

c. The node that directly controls contraction of the atria is _____.

d. The node that directly controls contraction of the ventricles is _____.

Structural differences of arteries, veins, and capillaries correlate with their different functions (877–878)

The wall of an artery or vein consists of three layers: an outer connective tissue zone with elastic fibers that allow the vessel to stretch and recoil; a middle layer of smooth muscle and more elastic fibers; and an inner lining of **endothelium,** a single layer of flattened cells. The outer two layers are thicker in arteries, which must be stronger and more elastic. One-way valves in the larger veins assure that blood flows back toward the heart. Capillaries have only the endothelial layer.

Physical laws governing the movement of fluids through pipes affect blood flow and blood pressure (878–880)

Blood Flow Velocity Following the *law of continuity,* the flow of blood decelerates with the increasing cross-sectional area of the branching vessel system. The enormous number of capillaries creates a large total diameter, and blood flow is very slow in capillaries, improving opportunity for exchange with interstitial fluid. Blood flow speeds up within venules and veins, due to the decrease in total cross-sectional area.

Blood Pressure Blood pressure, the hydrostatic force exerted against the wall of a blood vessel, is much greater in arteries than in veins and is greatest during systole **(systolic pressure).** This hydrostatic pressure drives blood from the heart to the capillary beds. **Peripheral resistance,** caused by the narrow openings of the arterioles impeding the exit of blood from arteries, causes the swelling of the arteries during systole. The recoiling of the stretched elastic arteries during diastole creates **diastolic pressure** and maintains a continuous blood flow into arterioles and capillaries.

Arterial blood pressure may be measured with a sphygmomanometer. The first number is the pressure during systole (when blood first spurts through the artery that was closed off by the cuff), and the second is the pressure during diastole.

Together, cardiac output and peripheral resistance determine blood pressure. Contraction of smooth muscles in arteriole walls increases resistance and thus increases blood pressure, whereas dilation of arterioles as smooth muscles relax increases blood flow into the arterioles and thus lowers blood pressure. Nervous and hormonal signals control these muscles.

Blood pressure drops to almost zero beyond the capillary beds. The one-way valves in veins and the contraction of skeletal muscles between which veins are embedded force blood to flow back to the heart. Pressure changes during breathing also draw blood into the large veins in the thoracic cavity.

■ **INTERACTIVE QUESTION   42.4**

Complete the following concept map to help you organize your understanding of blood pressure.

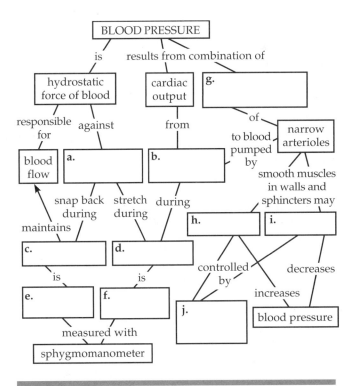

Transfer of substances between the blood and the interstitial fluid occurs across the thin walls of capillaries (880–881)

Blood Flow Through Capillary Beds Only about 5% to 10% of the body's capillaries have blood flowing through them at any one time. Capillaries branch off thoroughfare channels that are direct connections between arterioles and venules. The distribution of blood to capillary beds varies with need and is regulated by nerve signals and hormones that control contraction of smooth muscle in arteriole walls and precapillary sphincters at the entrance to capillary beds.

Capillary Exchange The exchange of substances between blood and interstitial fluid may involve transport by endocytosis and exocytosis by the endothelial cells, passive diffusion of small molecules, and bulk flow through the clefts between cells due to hydrostatic pressure. Water, sugars, salts, oxygen, and urea move through capillary clefts, whereas blood cells and proteins are too large to fit through. Hydrostatic pressure (blood pressure) at the upstream end of a capillary forces fluid out, whereas osmotic pressure at the downstream end tends to draw about 85% of that fluid back into the capillary.

The lymphatic system returns fluid to the blood and aids in body defense (881–882)

The fluid that does not return to the capillary and any proteins that may have leaked out are returned to the blood through the **lymphatic system.** Fluid diffuses into lymph capillaries intermingled in the blood capillary net. The fluid, called **lymph,** moves through lymph vessels with one-way valves, mainly as a result of the movement of skeletal muscles, and is returned to the circulatory system near the junction of the venae cavae with the right atrium. In **lymph nodes,** the lymph is filtered, and white blood cells attack viruses and bacteria. Recall that the lymphatic system transports fats absorbed from the intestine to the circulatory system.

Blood is a connective tissue with cells suspended in plasma (882–884)

Cells and cell fragments make up about 45% of the volume of blood; the rest is a liquid matrix called **plasma.**

Plasma The plasma consists of a large variety of solutes dissolved in water. The collective and individual concentrations of the dissolved ions of inorganic salts are important to osmotic balance, pH levels, and the functioning of muscles and nerves. The kidney is responsible for the homeostatic regulation of these electrolytes.

Plasma proteins function in the osmotic balance of the blood and as buffers, antibodies, escorts for lipids, and clotting factors called fibrinogens. Blood plasma from which clotting factors have been removed is called serum. Nutrients, metabolic wastes, gases, and hormones are transported in the plasma.

Cellular Elements **Red blood cells (erythrocytes)** transport oxygen. They lack mitochondria and generate their ATP by anaerobic metabolism. The red blood cells of mammals lack nuclei. Their small size and biconcave shape create a large surface area of plasma membrane across which oxygen can diffuse. Erythrocytes are packed with **hemoglobin,** an iron-containing protein that binds oxygen. Hemoglobin also binds nitric oxide (NO). The release of NO relaxes the walls of capillaries, probably helping to deliver O_2 from the expanded capillaries.

There are five major types of **leukocytes,** or **white blood cells,** all of which fight infections. Some white cells are phagocytes that engulf bacteria and cellular debris; the lymphocytes give rise to cells that produce the immune response. The majority of leukocytes are located in interstitial fluid and lymph nodes.

Platelets, pinched-off fragments of large cells in the bone marrow, are involved in the blood-clotting mechanism.

Stem Cells and the Replacement of Cellular Elements Erythrocytes circulate for about 3 to 4 months before they are phagocytosed by cells in the liver and spleen and their components recycled. Red and white blood cells and platelets are produced from **pluripotent stem cells** in red bone marrow. Isolated and purified

---

■ **INTERACTIVE QUESTION 42.5**

Complete this concept map of the components of vertebrate blood and their functions.

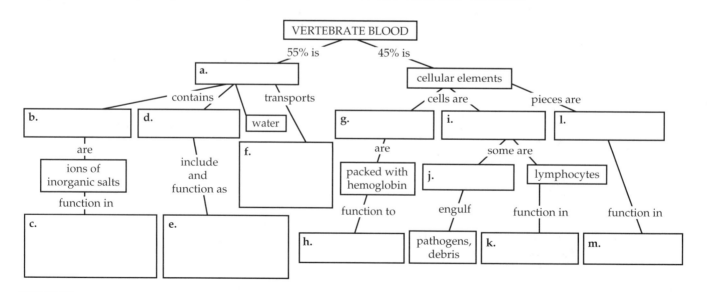

pluripotent stem cells grown in tissue culture may soon be used to treat human diseases such as leukemia.

The production of red blood cells is controlled by a negative feedback mechanism involving the hormone **erythropoietin,** which is produced by the kidney in response to low oxygen supply in tissues.

Blood Clotting The clotting process usually begins when platelets, clumped together along a damaged endothelium, release clotting factors. By a series of steps, the plasma protein **fibrinogen** is converted to its active form, **fibrin.** The threads of fibrin form a patch. An inherited defect in any step of the complex clotting process causes **hemophilia.** Anticlotting factors normally prevent clotting of blood in the absence of injury. A **thrombus** is a clot that occurs within a blood vessel and blocks the flow of blood. Such clots may form in individuals with cardiovascular disease.

Cardiovascular diseases are the leading cause of death in the United States and most other developed nations (884–885)

Diseases of the heart and blood vessels are called **cardiovascular diseases.** Blockage of coronary arteries leads to the death of cardiac muscle in a **heart attack;** blockage of arteries in the head leads to a **stroke,** the death of nervous tissue in the brain. A heart attack or stroke may result from blockage by a thrombus or an embolus, which is a clot formed elsewhere that moves through the circulatory system.

Most heart-attack and stroke victims had been suffering from a chronic cardiovascular disease known as **atherosclerosis,** in which growths called plaques develop within arteries and narrow the vessels. Plaques that become hardened by calcium deposits result in **arteriosclerosis,** or hardening of the arteries. An embolus is more likely to be trapped in narrowed vessels, and plaques are common sites of thrombus formation. Occasional chest pains, known as angina pectoris, may warn that a coronary artery is partially blocked.

Hypertension, or high blood pressure, is thought to damage the endothelium and initiate plaque formation, promoting atherosclerosis and increasing the risk of heart attack and stroke. This condition can be easily diagnosed and controlled by drugs, diet, and exercise. Hypertension and atherosclerosis may have a genetic component.

Cholesterol is carried in the blood as **low-density lipoproteins (LDLs)** and this type is associated with cholesterol deposits in plaques. **High-density lipoproteins (HDLs)** are cholesterol particles that appear to reduce cholesterol deposition in plaques.

■ **INTERACTIVE QUESTION 42.6**

What lifestyle choices have been correlated with increased risk of cardiovascular disease?

Gas Exchange in Animals

Gas exchange supplies oxygen for cellular respiration and disposes of carbon dioxide: *an overview* (886–887)

Gas exchange, or respiration, is the uptake of O_2 and the discharge of CO_2 and usually involves both the respiratory system and the circulatory system. The **respiratory medium** is air for terrestrial animals and water for aquatic ones. The **respiratory surface,** where gas exchange with the respiratory medium occurs, must be moist, thin, and large enough to supply the whole body.

In protists, sponges, cnidarians, and flatworms, gas exchange takes place across the plasma membranes of cells exposed to the aqueous environment. In animals with denser bodies, a localized region of the body surface is usually specialized as a respiratory surface with a thin, moist epithelium separating a rich blood supply and the respiratory medium. Most animals have an extensively branched or folded respiratory organ.

Gills are respiratory adaptations of most aquatic animals (887–888)

Gills are outfoldings of the body surface, ranging from simple bumps on echinoderms to the more complex gills of mollusks, crustaceans, and fishes. Due to the low oxygen concentration in a water environment, gills usually require **ventilation,** or movement of the respiratory medium across the respiratory surface—often an energy-intensive process due to the density of water.

In the gills of a fish, blood flows through capillaries in a direction opposite to the flow of water. This arrangement sets up a **countercurrent exchange,** in which the diffusion gradient favors the movement of oxygen into the blood throughout the length of the capillary.

Tracheal systems and lungs are respiratory adaptations of terrestrial animals (889–892)

The advantages of air as a respiratory medium include its higher concentration of oxygen, faster diffusion rate of O_2 and CO_2, and lower density. To prevent the disadvantageous water loss from large, moist surfaces, the respiratory surfaces of most terrestrial organisms are invaginated.

Tracheal Systems   The **tracheal systems** of insects are tiny air tubes that branch throughout the body to come into contact with nearly every cell. Thus, the open circulatory system does not function in O_2 and CO_2 transport. Rhythmic body movements of large insects and the contraction and relaxation of flight muscles serve to ventilate tracheal systems.

Lungs   **Lungs** are invaginated respiratory surfaces restricted to one location from which oxygen is transported to the rest of the body by the circulatory system. The lungs of amphibians are small, and gas exchange is supplemented by diffusion across the skin. Birds, mammals, and most reptiles rely totally on lungs for gas exchange.

Mammalian Respiratory Systems: a closer look   The lungs of mammals are located in the thoracic cavity and have a spongy texture with a honeycombed moist epithelium that provides a large surface area for gas exchange.

Air, entering through the nostrils, is filtered, warmed, humidified, and smelled in the nasal cavity. Air passes through the pharynx and enters the respiratory tract through the glottis. The **larynx** moves up and tips the epiglottis over the glottis when food is swallowed. The larynx functions as a voicebox in most mammals; exhaled air vibrates a pair of **vocal cords**. The **trachea,** or windpipe, branches into two **bronchi,** which then branch repeatedly into **bronchioles** within the lungs. Ciliated, mucus-coated epithelium lines much of the respiratory tree. Multilobed air sacs encased in a web of capillaries are at the tips of the tiniest bronchioles. Gas exchange takes place across the thin, moist epithelium of these **alveoli.**

Ventilating the Lungs   Vertebrate lungs are ventilated by **breathing,** the alternate inhalation and exhalation of air. A frog uses **positive pressure breathing.** It lowers its jaw, expanding its mouth cavity and drawing air into the mouth; raising the jaw then pushes air into the lungs.

Mammals ventilate their lungs by **negative pressure breathing.** The lungs are enclosed in a double-walled sac. Surface tension holds the two layers together, and they adhere to the lungs and the chest-cavity wall. Contraction of the rib muscles and the **diaphragm** expands the chest cavity and increases the volume of the lungs. Air pressure is reduced within this increased volume, and air is drawn through the nostrils down to the lungs. Relaxation of the rib muscles and diaphragm compresses the lungs, increasing the pressure and forcing air out.

Tidal volume is the volume of air inhaled and exhaled by an animal during normal breathing. The maximum volume that can be inhaled and exhaled by forced breathing is called **vital capacity.** The **residual volume** is the air that remains in the alveoli and lungs after forceful exhaling.

Birds have air sacs penetrating their abdomen, neck, and wings that act as bellows to maintain air flow through the lungs and also lower the density of the bird. The air sacs and lungs are ventilated through a circuit that includes a one-way passage through tiny, gas-exchange channels, called **parabronchi,** in the lungs.

Control centers in the brain regulate the rate and depth of breathing (892–893)

Breathing is under automatic regulation by the **breathing control centers** in the medulla oblongata and the pons. Nerve impulses from the medulla instruct the rib muscles and diaphragm to contract. In a negative feedback loop, stretch sensors in the lungs

---

■ **INTERACTIVE QUESTION   42.7**

For the following animals, indicate the respiratory medium, respiratory surface, and means of ventilation.

Animal	Respiratory Medium	Respiratory Surface	Ventilation
Fish	a.	b.	c.
Grasshopper	d.	e.	f.
Frog	g.	h.	i.
Human	j.	k.	l.

respond to the expansion of the lungs and send nervous impulses that inhibit the breathing control center in the medulla. A drop in the pH of the blood and cerebrospinal fluid when CO_2 concentration increases is sensed by the medulla's control center, which increases the depth and rate of breathing. Although breathing centers respond primarily to CO_2 levels, they are alerted by oxygen sensors in key arteries that react to severe deficiencies of O_2.

Gases diffuse down pressure gradients in the lungs and other organs (893–894)

The concentration of a gas in air or dissolved in water is measured as **partial pressure.** At sea level, the partial pressure of oxygen, which makes up 21% of the atmosphere, is 160 mm Hg (0.21×760 mm—atmospheric pressure at sea level). The partial pressure of CO_2 is 0.23 mm Hg. A gas will diffuse from a region of higher partial pressure to lower partial pressure. Blood entering the capillaries of the lungs has a lower P_{O_2} and a higher P_{CO_2} than does the air in the alveoli, so oxygen diffuses into the capillaries and carbon dioxide diffuses out. In the tissue capillaries, pressure differences favor the diffusion of O_2 out of the blood into the interstitial fluid and CO_2 into the blood.

Respiratory pigments transport gases and help buffer the blood (894–896)

Oxygen Transport   In most animals, O_2 is carried by **respiratory pigments** in the blood. Copper is the oxygen-binding component in the blue respiratory protein **hemocyanin,** common in arthropods and many mollusks.

Hemoglobin, contained in red blood cells, is the respiratory pigment of almost all vertebrates. Hemoglobin is composed of four subunits, each of which has a cofactor called a heme group with iron at its center. The binding of O_2 to the iron atom of one subunit induces a change in shape of the other subunits, and their affinity for oxygen increases. Likewise, the unloading of the first O_2 lowers the other subunits' affinity for oxygen.

The **dissociation curve** for hemoglobin shows the relative amounts of oxygen bound to hemoglobin under varying oxygen partial pressures. In the steep part of this S-shaped curve, a slight change in partial pressure will cause hemoglobin to load or unload a substantial amount of oxygen.

Carbon Dioxide Transport   About 70% of CO_2 is transported in blood as bicarbonate ions. Carbon dioxide enters into the red blood cells where it first combines with H_2O to form carbonic acid (catalyzed by carbonic anhydrase) and then dissociates into H^+ and a bicarbonate ion. Bicarbonate moves into the plasma for transport. The hydrogen ions are bound to hemoglobin and other proteins, and the pH value of the blood is not lowered during the transport of carbon dioxide. In the lungs, the diffusion of CO_2 out of the blood shifts the equilibrium in favor of the conversion of bicarbonate back to CO_2, which is unloaded from the blood.

■ INTERACTIVE QUESTION   42.8

The following graph shows dissociation curves for hemoglobin at two different pH values. Explain the significance of these two curves.

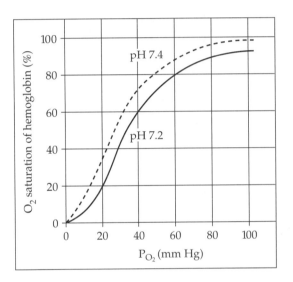

Deep-diving air-breathers stockpile oxygen and deplete it slowly (896–897)

Special physiological adaptations have enabled some air-breathing animals to make sustained underwater dives. Weddell seals store twice the amount of O_2/kg body weight as do humans, mostly by having a larger volume of blood, a huge spleen that stores blood, and a higher concentration of **myoglobin,** an oxygen-storing muscle protein. Oxygen-conserving adaptations during a dive include decreasing the heart rate, rerouting blood to the brain and essential organs, and restricting blood supply to the muscles, which can use fermentation to produce ATP during long dives.

WORD ROOTS

alveol- = a cavity (*alveoli:* one of the dead-end, multilobed air sacs that constitute the gas exchange surface of the lungs)

arterio- = an artery; **-sclero** = hard (*arteriosclerosis:* a cardiovascular disease caused by the formation of hard plaques within the arteries)

atrio- = a vestibule; **-ventriculo** = ventricle (*atrioventricular node:* a region of specialized muscle tissue between the right atrium and right ventricle; it generates electrical impulses that primarily cause the ventricles to contract)

cardi- = heart; **-vascula** = a little vessel (*cardiovascular system:* the closed circulatory system characteristic of vertebrates)

counter- = opposite (*countercurrent exchange:* the opposite flow of adjacent fluids that maximizes transfer rates)

endo- = inner (*endothelium:* the innermost, simple squamous layer of cells lining the blood vessels; the only constituent structure of capillaries)

erythro- = red; **-poiet** = produce (*erythropoietin:* a hormone produced in the kidney when tissues of the body do not receive enough oxygen. This hormone stimulates the production of erythrocytes)

fibrino- = a fiber; **-gen** = produce (*fibrinogen:* the inactive form of the plasma protein that is converted to the active form fibrin, which aggregates into threads that form the framework of a blood clot)

hemo- = blood; **-philia** = loving (*hemophilia:* a human genetic disease caused by a sex-linked recessive allele, characterized by excessive bleeding following injury)

leuko- = white; **-cyte** = cell (*leukocyte:* a white blood cell)

myo- = muscle (*myoglobin:* an oxygen-storing, pigmented protein in muscle cells)

para- = beside, near (*parabronchi:* the sites of gas exchange in bird lungs; they allow air to flow past the respiratory surface in just one direction)

pluri- = more, several; **-potent** = powerful (*pluripotent* stem cell: a cell within bone marrow that is a progenitor for any kind of blood cell)

pulmo- = a lung; **-cutane** = skin (*pulmocutaneous:* the route of circulation that directs blood to the skin and lungs)

semi- = half; **-luna** = moon (*semilunar valve:* a valve located at the two exits of the heart, where the aorta leaves the left ventricle and the pulmonary artery leaves the right ventricle)

thrombo- = a clot (*thrombus:* a clump of platelets and fibrin that block the flow of blood through a blood vessel)

STRUCTURE YOUR KNOWLEDGE

1. Identify the labeled structures in this diagram of a human heart. Draw arrows to trace the flow of blood.

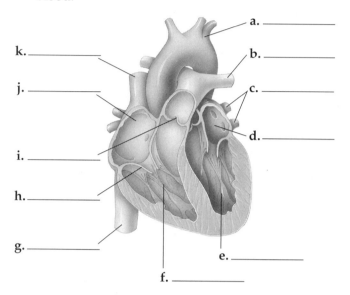

a. _____
b. _____
c. _____
d. _____
e. _____
f. _____
g. _____
h. _____
i. _____
j. _____
k. _____

2. Trace the path of a molecule of oxygen being inhaled through the nasal cavity to delivery to the kidney.

TEST YOUR KNOWLEDGE

MULTIPLE CHOICE: *Choose the one best answer.*

1. A gastrovascular cavity
   a. is found in cnidarians and annelids.
   b. functions to pump fluid in short vessels from which it spreads throughout the body.
   c. functions in both digestion and distribution of nutrients.
   d. has a single opening for ingestion and elimination and a separate opening for gas exchange in larger animals.
   e. involves all of the above.

2. Which of the following is *not* a similarity between open and closed circulatory systems?
   a. Some sort of pumping device helps to move blood through the body.
   b. Some of the circulation of blood is a result of movements of the body.
   c. The blood and interstitial fluid are distinguishable from each other.
   d. All tissues come into close contact with the circulating body fluid so that the exchange of nutrients and wastes can take place.
   e. All of these apply to both open and closed circulatory systems.

3. In a system with double circulation,
   a. blood is pumped at two separate locations as it circulates through the body.
   b. there is a countercurrent exchange within the gills.
   c. hemolymph circulates both through a pumping blood vessel and through body sinuses.
   d. blood is pumped to the gas-exchange organ and returning blood is then pumped through systemic vessels.
   e. there are always two ventricles and often two atria.

4. During diastole,
   a. the atria fill with blood.
   b. blood flows passively into the ventricles.
   c. the elastic recoil of the arteries maintains hydrostatic pressure on the blood.
   d. semilunar valves are closed but atrioventricular valves are open.
   e. all of the above are occurring.

5. An atrioventricular valve prevents the backflow or leakage of blood from
   a. the right ventricle into the right atrium.
   b. the left atrium into the left ventricle.
   c. the aorta into the left ventricle.
   d. the pulmonary vein into the right atrium.
   e. the right atrium into the vena cava.

6. Breathing rate will increase when ____ CO_2 in your blood causes a ____ in pH.
   a. increased/rise
   b. increased/drop
   c. decreased/rise
   d. decreased/drop
   e. It is the level of O_2 in the blood that normally signals changes in breathing rate.

7. A heartbeat is initiated by the
   a. SA node.
   b. AV node.
   c. right atrium.
   d. anterior and posterior venae cavae.
   e. left ventricle.

8. Blood flows more slowly in the arterioles than in the arteries because the arterioles
   a. have thoroughfare channels to venules that are often closed off.
   b. collectively have a larger cross-sectional area than do the arteries.
   c. must provide opportunity for exchange with the interstitial fluid.
   d. have sphincters that restrict flow to capillary beds.
   e. are narrower than arteries.

9. If all the body's capillaries were open at the same time,
   a. blood pressure would fall dramatically.
   b. peripheral resistance would increase.
   c. blood would move too rapidly through the capillary beds.
   d. the amount of blood returning to the heart would increase.
   e. the increased gas exchange would allow for strenuous exercise.

10. Which of the following is *not* a factor in the exchange of substances in capillary beds?
    a. endocytosis and exocytosis
    b. passive diffusion
    c. hydrostatic pressure
    d. bulk flow through clefts between endothelial cells
    e. active transport by white blood cells

11. Edema involves the swelling of body tissues. It can result from
    a. swollen lymph glands.
    b. a loss of interstitial fluid.
    c. too much lymph being returned to the circulatory system.
    d. too low a concentration of blood proteins.
    e. swollen feet.

12. The functions of the kidney include
    a. production of the hormone erythropoietin to increase production of red blood cells.
    b. maintenance of electrolyte balance.
    c. destruction and recycling of red blood cells.
    d. both a and b.
    e. a, b, and c.

13. Fibrinogen is
    a. a blood protein that escorts lipids through the circulatory system.
    b. a cell fragment involved in the blood-clotting mechanism.
    c. a blood protein that is converted to fibrin to form a blood clot.
    d. a leukocyte involved in trapping viruses and bacteria.
    e. a lymph protein that regulates osmotic balance in the tissues.

14. A thrombus
    a. may form at a region of plaque in an artery.
    b. is a traveling embolism that may cause a heart attack or stroke.
    c. may cause hypertension.
    d. may calcify and result in arteriosclerosis.
    e. is often associated with high levels of HDLs in the blood.

15. In countercurrent exchange,
   a. the flow of fluids in opposite directions maintains a favorable diffusion gradient along the length of an exchange surface.
   b. oxygen is exchanged for carbon dioxide.
   c. double circulation keeps oxygenated and de-oxygenated blood separate.
   d. oxygen moves from a region of high partial pressure to one of low partial pressure, but carbon dioxide moves in the opposite direction.
   e. the capillaries of the lung pick up more oxygen than do tissue capillaries.

16. The tracheae of insects
   a. are stiffened with chitinous rings and lead into the lungs.
   b. are filled by positive pressure breathing.
   c. ramify along capillaries of the circulatory system for gas exchange.
   d. are highly branched, coming into contact with almost every cell for gas exchange.
   e. provide an extensive, fluid-filled system of tubes for gas exchange.

17. Which of the following is *not* involved in speeding up breathing?
   a. signals from the aorta to the medulla in response to low blood pH.
   b. stretch receptors in the lungs
   c. impulses from the breathing centers in the medulla
   d. severe deficiencies of oxygen
   e. a drop in the pH of cerebrospinal fluid

18. The binding of an O_2 atom to the first iron atom of a hemoglobin molecule
   a. occurs at a very low partial pressure of oxygen.
   b. produces a conformational change that lowers the other subunits' affinity for oxygen.
   c. is an example of cooperativity as O_2 readily binds to the other subunits.
   d. occurs more readily at a lower pH value.
   e. is the signal for hemoglobin to release H^+ so that bicarbonate converts to CO_2.

19. Fluid moves back into the capillaries at the venous end of a capillary bed as a result of
   a. active transport.
   b. a hydrostatic pressure that is higher than the osmotic pressure.
   c. an osmotic pressure that is higher than the hydrostatic pressure.
   d. a partial pressure that is lower than that at the arteriole end.
   e. the squeezing of surrounding muscles on the interstitial fluid.

20. As a general rule, blood leaving the right ventricle of a mammal's heart will pass through how many capillary beds before it returns to the right ventricle?
   a. one
   b. two
   c. three
   d. one or two, depending on the circuit it takes
   e. at least two, but most often three

21. Gas exchange between maternal and fetal blood takes place in the placenta. Fetal and adult hemoglobin differ slightly in composition, and their dissociation curves are different so that oxygen can be transferred from maternal blood to fetal blood. If curve *c* in the graph below represents the dissociation curve of adult hemoglobin, which curve would represent fetal hemoglobin? (*Hint:* Must fetal hemoglobin have a higher or lower affinity for oxygen than maternal hemoglobin has?)
   a. a   c. c   e. e
   b. b   d. d

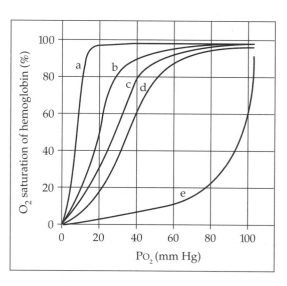

Use the following to answer questions 22–25.
   1. vena cava   4. right atrium
   2. left ventricle   5. aorta
   3. pulmonary vein   6. pulmonary capillaries

22. Which of the following sequences represents the flow of blood through the human body? (Obviously, many structures are not included.)
   a. 1-5-3-6-2-4
   b. 1-2-6-3-4-5
   c. 1-4-3-6-2-5
   d. 1-2-3-6-4-5
   e. 1-4-6-3-2-5

23. In which location would blood pressure be the greatest?
- **a.** 1
- **b.** 3
- **c.** 4
- **d.** 5
- **e.** 6

24. In which location would velocity of blood flow be the slowest?
- **a.** 1
- **b.** 3
- **c.** 4
- **d.** 5
- **e.** 6

25. Of these structures, which would have the thickest muscle layer?
- **a.** 1
- **b.** 2
- **c.** 3
- **d.** 4
- **e.** 5

26. The nurse tells you that your blood pressure is 112/70. What does the 70 refer to?
- **a.** your heart rate
- **b.** the velocity of blood during diastole
- **c.** the systolic pressure from ventricular contraction
- **d.** the diastolic pressure from the recoil of the arteries
- **e.** the venous pressure caused by the compression of the blood pressure cuff

27. What do the alveoli of mammalian lungs, the gill filaments of fish, and the tracheoles of insects all have in common?
- **a.** use of a circulatory system to transport absorbed oxygen
- **b.** respiratory surfaces that are invaginated
- **c.** countercurrent exchange
- **d.** a large, moist surface area for exchange
- **e.** all of the above

28. Which of the following would *not* be a possible function of a plasma protein in a mammal?
- **a.** pH buffer
- **b.** osmotic balance
- **c.** clotting agent
- **d.** oxygen transport
- **e.** antibody

29. What is the function of the cilia in the trachea and bronchi?
- **a.** movement of air into and out of the lungs
- **b.** increase the surface area for gas exchange
- **c.** vibrate when air rushes past them to produce sounds
- **d.** filter the air that rushes through them
- **e.** sweep mucus with its trapped particles up and out of the respiratory tract

30. Both birds and mammals have four-chambered hearts. Which of the following is the likely reason for this similarity?
- **a.** They shared a common ancestor that had a four-chambered heart.
- **b.** They are the only vertebrates with double circulation, and this double pumping requires four chambers.

- **c.** They are both endotherms, and the evolution of efficient circulatory systems was necessary to support the high metabolic rate of endotherms.
- **d.** This is an example of convergent evolution, because animals that obtain their oxygen from air require both a pulmonary and a systemic circuit.
- **e.** The more inefficient single atrium of amphibians and reptiles could not supply the higher oxygen needs of these endotherms.

31. What causes a heart murmur?
- **a.** a weak systole
- **b.** the snapping closed of the atrioventricular valves
- **c.** a leaking heart valve
- **d.** low blood pressure
- **e.** an interruption of the impulse from the SA to the AV node

32. How are arteries and veins different?
- **a.** An artery has a thicker muscular and connective tissue layer than does a vein.
- **b.** Veins do not have valves and arteries do.
- **c.** Blood is under lower pressure in arteries than in veins.
- **d.** Arteries carry blood to the heart; veins carry blood away from the heart.
- **e.** Blood in arteries is oxygen rich; blood in veins is always oxygen poor.

33. More O_2 is unloaded from hemoglobin in an actively metabolizing tissue than in a resting tissue. Why?
- **a.** The pH of an active tissue is higher as more CO_2 combines with water to produce H^+ and bicarbonate.
- **b.** The conformation of hemoglobin changes at the lower pH of an actively metabolizing tissue and O_2 dissociates more easily.
- **c.** Hemoglobin can bind four O_2 due to a change in conformation caused by the lower pH of the active tissue.
- **d.** More oxygen is bound to myoglobin at lower levels of oxygen concentration.
- **e.** More CO_2 is drawn out of tissue cells as it is converted to bicarbonate in red blood cells.

34. Which of the following does *not* contribute to the respiratory adaptations of deep-diving animals?
- **a.** an unusually large volume of blood
- **b.** large quantity of myoglobin in the muscles
- **c.** reduction in blood flow to noncritical body areas
- **d.** extra-large lungs
- **e.** large spleen that stores blood and releases it during a dive

CHAPTER 43

THE BODY'S DEFENSES

FRAMEWORK

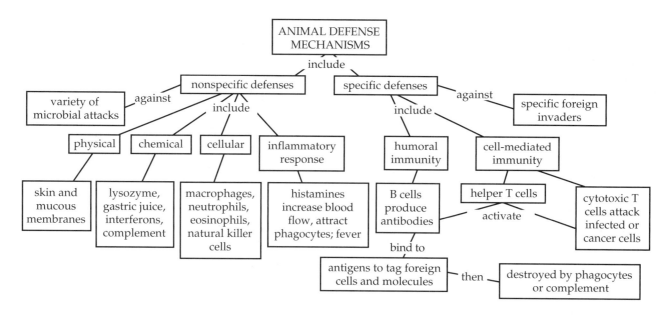

CHAPTER REVIEW

The defense system protects an animal from bacteria, viruses, other pathogens, and early-stage cancer cells. Nonspecific defense mechanisms include both the external physical barriers of the skin and mucous membranes and the internal antimicrobial proteins, phagocytic cells, and the inflammatory response. The third line of defense is the immune system, which reacts specifically to threat with the production of antibodies and the actions of a group of white blood cells called lymphocytes.

Nonspecific Defenses Against Infection

The skin and mucous membranes provide first-line barriers to infection (901)

Skin and the mucous membranes lining the digestive, respiratory, and genitourinary tracts are physical barriers to microbes. Skin secretions maintain a low pH, which discourages colonization by microbes. **Lysozyme,** an enzyme that attacks bacterial cell walls, is present in tears, saliva, and mucus. The ciliated, mucus-coated epithelial lining of the respiratory tract traps and removes microbes. The acidity of gastric juice kills most bacteria that reach the stomach.

Phagocytic cells, inflammation, and antimicrobial proteins function early in infection (901–904)

The second line of defense relies on **phagocytosis** by white blood cells.

Phagocytic and Natural Killer Cells   Short-lived **neutrophils** leave the blood and enter infected tissue in response to chemical signals (called chemotaxis) released from damaged cells and phagocytose microbes. **Monocytes** migrate into tissues and develop into **macrophages,** large, long-lived amoeboid cells that engulf microbes and destroy them with toxic forms of oxygen (superoxide anion and nitric oxide) and digestive enzymes. Macrophages may migrate through the body or become permanently attached in organs such as the lungs, liver, kidney, brain, connective tissue, lymph nodes, and spleen.

Eosinophils are leukocytes that attack larger parasitic invaders with destructive enzymes. **Natural killer (NK) cells** destroy the body's infected or abnormal cells by attacking their membranes.

The Inflammatory Response Physical injury or microorganisms can trigger an **inflammatory response,** characterized by redness, swelling, and heat. Chemical signals released from microbes or from body cells in response to injury initiate the inflammatory response. **Basophils** in the blood and **mast cells** in connective tissue release **histamine,** which triggers dilation and leakiness of blood vessels; leukocytes and damaged tissue cells also release prostaglandins and other chemicals that promote blood flow to the damaged area. Blood-clotting elements delivered to the area begin vessel repair and help to seal off infections. Vasodilation and chemotactic factors called **chemokines** result in the congregation of phagocytic cells. Macrophages phagocytose pathogens as well as damaged tissue cells and dead neutrophils. Accumulated pus is a combination of dead phagocytic cells and fluid from the capillaries.

A systemic nonspecific response to an infection may include an increase in the number of circulating leukocytes and a fever. Fever may be triggered by toxins produced by pathogens or by **pyrogens** released by certain leukocytes. Fevers stimulate phagocytosis and inhibit growth of microorganisms. *Septic shock* is a dangerous condition resulting from a widespread systemic inflammatory response.

Antimicrobial Proteins The **complement system** is a group of about 20 serum proteins that cooperate with both nonspecific and specific defense mechanisms, resulting in lysis of microbes or attraction of phagocytes. **Interferons** are proteins produced by virus-infected cells that diffuse to neighboring cells, stimulating production of proteins that inhibit viral reproduction in those cells. One type of interferon also activates phagocytes. Interferons produced by recombinant DNA technology are being tested for their effectiveness in treating viral infections and cancer.

How Specific Immunity Arises

Lymphocytes are the key cells of the immune system, the body's specific, third line of defense.

Lymphocytes provide the specificity and diversity of the immune system (904–905)

B lymphocytes (B cells) and **T lymphocytes (T cells)** circulate in blood and lymph and are concentrated in the spleen and lymph nodes. They both display *specificity* in their ability to recognize and respond to a particular **antigen** or foreign molecule. Certain B cells

■ **INTERACTIVE QUESTION 43.1**

Complete the following table that summarizes the functions of the cells and compounds of the nonspecific defense mechanisms.

Cells or Compounds	Functions
Neutrophils	a.
Monocytes	b.
Macrophages	c.
Eosinophils	d.
Natural killer (NK) cells	e.
Basophils and mast cells	f.
Histamine	g.
Interferons	h.
Complement system	i.
Lysozyme	j.
Prostaglandins	k.
Pyrogens	l.
Chemokines	m.

respond to the particular molecular shape of an antigen and secrete proteins called **antibodies** that interact with that antigen.

B and T cells both have membrane-bound **antigen receptors** that allow them to recognize specific antigens: transmembrane antibody molecules called *membrane antibodies* in B cells, and **T cell receptors,** which are structurally related to membrane antibodies but are never produced to be secreted. Each B or T lymphocyte carries about 100,000 receptors of exactly the same specificity. A lymphocyte's particular type of receptor is determined during early embryonic development by random genetic recombinations of segments of antibody or receptor genes. This process results in a huge *diversity* of lymphocytes that are able to respond to millions of different antigens.

Antigens interact with specific lymphocytes, inducing immune responses and immunological memory (905–906)

When one of the antigens from a microorganism interacts with receptors on specific B cells or T cells, those particular lymphocytes are activated to divide and differentiate into two clones—a large number of short-lived **effector cells,** which combat that antigen, and a clone of long-lived **memory cells,** all of which carry receptors for that antigen. By this **clonal selection,** a small number of cells is selected by their interaction with a specific antigen to produce thousands of cells keyed to that particular antigen.

The body mounts a **primary immune response** upon first exposure to an antigen. About 10 to 17 days are required for selected lymphocytes to proliferate and differentiate to yield the maximum response produced by effector T cells and the antibody-producing effector B cells, called **plasma cells.** Should the body reencounter the same antigen, the **secondary immune response** is more rapid, effective, and prolonged. The long-lived T and B memory cells are responsible for this *immunological memory.* This secondary immune response provides long-term protection against a previously encountered pathogen.

Lymphocyte development gives rise to an immune system that distinguishes self from nonself (906–908)

Blood cells, including lymphocytes, develop from pluripotent stem cells in the bone marrow or liver of a fetus. Lymphocytes either migrate to the thymus and differentiate into T cells or continue to develop in the bone marrow as B cells.

Immune Tolerance for Self As B cells and T cells mature, those that bear receptors for molecules already present in the body are either inactivated or self-destruct by programmed cell death (apoptosis), providing the *capacity to distinguish self from nonself.* This critical *self-tolerance* means that normally there are no mature lymphocytes that react against self components.

The Role of Cell Surface Markers in T Cell Function and Development T cells do interact with an important group of self molecules. A family of genes called the **major histocompatibility complex (MHC)** codes for a group of cell surface glycoproteins, known as the *HLA* (for human leukocyte antigens) in humans. **Class I MHC molecules** are found on all nucleated cells; **class II MHC molecules** are found primarily on macrophages and B cells. The hundreds of different alleles for each class I and class II MHC gene result in a unique biochemical fingerprint for each individual (except identical twins).

The role of MHC molecules is **antigen presentation** to T cells. Infected body cells display fragments of a pathogen's proteins in their class I MHC molecules, which are recognized by the antigen receptors of **cytotoxic T cells (T_C).** These T cells respond by destroying the infected cell. Macrophages and B cells are called **antigen-presenting cells (APCs)** when they engulf pathogens (either by phagocytosis or receptor-mediated endocytosis) and display their antigens in class II MHC molecules to **helper T cells (T_H).** These T cells activate other cells to fight that particular pathogen.

The differentiation of either cytotoxic T cells or helper T cells in the thymus depends on their affinity for class I MHC or class II MHC molecules, respectively, during development.

■ **INTERACTIVE QUESTION　43.2**

Describe the four attributes that characterize the immune system.

a.

b.

c.

d.

Immune Responses

Humoral immunity involves B cell activation and production of antibodies that circulate in the blood and lymph and defend against free bacteria, toxins, and viruses. **Cell-mediated immunity** involves T cells that react against body cells infected by bacteria or viruses and against fungi, protozoa, and worms. The cell-mediated branch also responds to tissue transplants and cancerous cells.

Helper T lymphocytes function in both humoral and cell-mediated immunity: *an overview (908–909)*

Some B cells and macrophages serve as antigen-presenting cells when they engulf foreign antigens and present fragments in their class II MHC molecules to helper T cells. A T cell surface protein called **CD4** enhances the interaction between an APC and a helper T cell, which results in the proliferation and differentiation of a clone of activated helper T cells and memory helper T cells.

Activated helper T cells secrete **cytokines,** such as **interleukin-2 (IL-2),** which help activate B cells and cytotoxic T cells. An antigen-presenting macrophage secretes **interleukin-1 (IL-1),** which activates the helper T cell to produce cytokines. The IL-2 secreted by the helper T cell also acts on the cell itself to enhance its reproduction and cytokine production.

In the cell-mediated response, cytotoxic T cells counter intracellular pathogens: *a closer look* (809–911)

All nucleated cells continuously produce class I MHC molecules, which capture fragments of cellular proteins and carry them to the surface. Should a cell be infected with a replicating virus, viral peptides are captured and exposed in its class I MHC molecules. **CD8** surface proteins on cytotoxic T cells bind to a part of class I MHC molecules and enhance the interaction between an infected cell and the cytotoxic T cell with a specific receptor for the exposed antigen. IL-2 released from a helper T cell also stimulates activation of the cytotoxic T cell into a killer cell, which then releases **perforin,** a protein that kills the target cell by forming holes in its membrane. Pathogens released from the destroyed cell are marked by circulating antibodies for destruction.

The class I MHC molecules on tumor cells present fragments of **tumor antigen** to cytotoxic T cells, which recognize them as foreign and destroy them.

■ **INTERACTIVE QUESTION   43.4**

a. What surface molecule of a helper T cell facilitates the interaction with a class II MHC of an APC and the helper T cell? _____

b. What surface molecule on a cytotoxic T cell assists in the interaction with class I MHC proteins displayed on infected cells? _____

c. What does an activated helper T cell release? _____

d. What does a cytotoxic T cell attached to an infected body cell release? _____

In the humoral response, B cells make antibodies against extracellular pathogens: *a closer look* (911–915)

Selective activation of a B cell occurs when an antigen binds to a membrane-bound antibody. The activation is aided by IL-2 and other cytokines released from helper T cells also activated by that antigen. The B cell proliferates into a clone of plasma cells and a clone of memory B cells. Most protein antigens are **T-dependent antigens** that require the aid of helper T cells to stimulate antibody production.

T-independent antigens trigger antibody production by B cells without the aid of IL-2. Found in bacterial capsules and flagella, these antigenic molecules

■ **INTERACTIVE QUESTION   43.3**

Label the components in this diagram that shows a helper T cell being activated by interaction with an APC and the central role of the helper T cell in activating both the humoral and cell-mediated branches of the immune response.

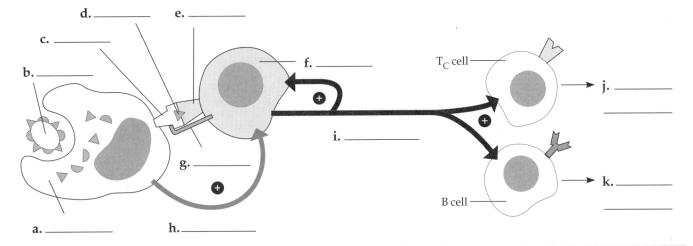

are usually polysaccharides or proteins with long chains of repeating subunits that apparently bind to sufficiently numerous antigen receptors on a B cell to activate the cell.

A variety of B cells will be stimulated in a given humoral response, each giving rise to thousands of plasma cells. And each plasma cell can secrete about 2,000 antibody molecules per second during its 4–5-day life span.

■ INTERACTIVE QUESTION 43.5

a. How do B cells serve as APCs?

b. Why is it thought that B cells function as important APCs in the secondary response to an antigen but not in the primary response to that antigen?

Antibody Structure and Function   Most antigens are proteins or polysaccharides that are surface components of various microbes or transplanted cells. Antibodies recognize a localized region, called an antigenic determinant, or **epitope,** of an antigen. Antigens may have many different epitopes that stimulate different B cells to produce antibodies.

Antibodies are a class of proteins called **immunoglobulins (Igs).** The Y-shaped antibody molecule consists of four polypeptide chains: two identical **light chains** and two identical **heavy chains,** linked together by disulfide bridges. Both heavy and light chains have variable regions at the ends of the two arms of the Y, which form two identical antigen-binding sites. The amino acid composition of the variable region creates the unique contours and specific antibody-antigen binding potential of these sites.

Researchers can use the specificity of antigen-antibody binding to create useful tools. Some antibody tools are *polyclonal* because they were formed by several different B cell clones, each specific for a different epitope. A technique for making **monoclonal antibodies** can supply quantities of identical antibodies for biological research, clinical testing, and medical applications.

There are five types of heavy-chain constant regions in the tail of the Y-shaped antibody, creating five classes of antibodies: IgM, IgG, IgA, IgD, and IgE. The constant region of each class determines its function.

Antibody-Mediated Disposal of Antigens   Antibodies label antigens for disposal by one of several effector mechanisms. In **neutralization,** antibodies may block the activity of an antigen, as when antibodies cover the binding sites of a virus. In **opsonization,** antibodies coat microbes and enhance phagocytosis by macrophages. Because each antibody molecule has at least two antigen-binding sites, the formation of antigen-antibody complexes may cause bacteria to **agglutinate** or soluble antigen molecules to precipitate. The resulting clumps are then engulfed by phagocytes.

In an important mechanism known as **complement fixation,** antigen-antibody complexes may activate the complement system. Many pathogens can be lysed by complement, either by the *classical pathway*, which is part of the humoral immune response and triggered by antibodies bound to antigens, or by the *alternative pathway*, which is a nonspecific defense triggered by substances found on many bacteria, yeasts, viruses, and protist parasites.

In the classical pathway, when antibody molecules target an invader such as a bacterial cell by attaching to it, a complement protein links two antibody molecules. This molecular association activates complement proteins to form a **membrane attack complex (MAC),** which produces a pore in the cell membrane, causing the cell to lyse.

Complement can amplify the inflammatory response by stimulating histamine release from basophils and mast cells. Several complement proteins attract phagocytes to infection sites and coat microorganisms, stimulating phagocytosis by opsonization. In **immune adherence,** microbes coated with antibodies and complement proteins adhere to the walls of blood vessels, facilitating their destruction by phagocytic cells.

■ INTERACTIVE QUESTION 43.6

List four ways in which antibodies mediate the disposal of antigens. Which of these enhance phagocytosis by macrophages?

a.

b.

c.

d.

Invertebrates have a rudimentary immune system (915)

The ability to distinguish self from nonself is well developed in invertebrates. Many invertebrates have amoeboid cells called *coelomocytes* that phagocytose foreign substances and produce interleukin-1 to attract more coelomocytes. Although not a specific defense, some invertebrates have lymphocyte-like cells that make antibody-like molecules. Tissue graft experiments in earthworms have established that their defense systems reject foreign tissue and develop a memory response to these grafts.

Immunity in Health and Disease

Immunity can be achieved naturally or artificially (916)

Active immunity can be acquired when the body produces antibodies and develops immunological memory from either exposure and recovery from an infectious disease or from **immunization**, also called **vaccination**. A vaccine may be an inactivated toxin, a killed or weakened microbe, or a portion of a microbe. In **passive immunity**, temporary immunity is provided by antibodies supplied through the placenta to a fetus, through milk to a nursing infant, or by an antibody injection.

The immune system's capacity to distinguish self from nonself limits blood transfusion and tissue transplantation (916–917)

Blood Groups and Blood Transfusion   The immune response to the chemical markers that determine **ABO blood groups** must be considered in blood transfusions. Antibodies to blood group antigens arise in response to normal bacterial flora and circulate in the blood plasma, where they will induce a devastating transfusion reaction to transfused blood cells with matching antigens. These anti-blood group antibodies are in the IgM class and do not cross the placenta.

An Rh-negative mother may develop antibodies against the **Rh factor,** another red blood cell antigen, if fetal blood from an Rh-positive child leaks across the placenta. Should she carry a second Rh-positive fetus, her immunological memory may result in the production of IgG antibodies that cross the placenta and destroy fetal red blood cells. Treatment of the mother with anti-Rh antibodies just after delivery destroys any Rh antigen that may have leaked into her circulation and prevents the mother's immunological response to the antigen.

Tissue Grafts and Organ Transplantation   Transplanted tissues and organs are rejected because the foreign MHC molecules are antigenic and trigger immune responses. The use of closely related donors, as

■ INTERACTIVE QUESTION   43.7

Fill in the following table to review your understanding of the antigens and antibodies of the ABO blood groups. Remember to compare the antigens on the donor cells with the antibodies in the recipient's plasma.

Blood Type	Antigens on RBCs	Antibodies in Plasma	Can Receive Blood from	Can Donate Blood to

well as drugs such as cyclosporin A and FK506 that suppress cell-mediated immunity, help to reduce the immune response after a transplant operation.

In bone marrow transplants, used to treat leukemia and blood cell diseases, the graft itself may be the source of immune rejection. The recipient's bone marrow cells are destroyed by irradiation, eliminating the recipient's immune system. The lymphocytes in the bone marrow transplant may produce a **graft-versus-host reaction** if the MHC molecules of donor and recipient are not well matched.

Abnormal immune function can lead to disease (917–919)

Allergies   Allergies are hypersensitivities to certain environmental antigens, or allergens. IgE antibodies that bind to mast cells in connective tissue can trigger allergic reactions. When antigens bind to these cell surface antibodies, the mast cells degranulate and release histamines, which create an inflammatory response that may include sneezing, a runny nose, and difficulty in breathing due to smooth muscle contractions. Antihistamines are drugs that combat these symptoms by blocking receptors for histamine. **Anaphylactic shock** is a severe allergic response in which the abrupt dilation of peripheral blood vessels caused by a rapid release of histamines leads to a life-threatening drop in blood pressure.

Autoimmune Diseases   Sometimes the immune system turns against self, leading to autoimmune diseases, such as lupus, rheumatoid arthritis, insulin-dependent diabetes, and multiple sclerosis (MS). These diseases may be caused by a failure in the regulation of self-reactive lymphocytes.

Immunodeficiency Diseases   A deficiency may occur in any of the components of the immune system. In the rare congenital disease known as *severe combined immunodeficiency (SCID),* both humoral and cell-mediated immune systems are nonfunctional. Gene therapy has been used to treat individuals with a SCID caused by a deficiency of the enzyme adenosine deaminase (ADA), although results thus far are equivocal. Certain cancers, such as Hodgkin's disease, and AIDS suppress the immune system.

There is growing evidence that general emotional health and immunity are related. Hormones secreted during stress affect the number of leukocytes; nerve fibers penetrate deep into the thymus, and receptors for chemical signals from nerve cells have been found on lymphocytes.

AIDS is an immunodeficiency disease caused by a virus (919–921)

Investigation of an increasing incidence of Kaposi's sarcoma and *Pneumocystis* pneumonia in the early 1980s led to the recognition of **acquired immunodeficiency syndrome (AIDS).** Individuals with AIDS are highly susceptible to *opportunistic diseases* that take advantage of a suppressed immune system. The infectious agent responsible for AIDS, known as **HIV (human immunodeficiency virus),** was identified in 1983. This lethal pathogen probably evolved from an HIV-like virus in chimpanzees in central Africa and first appeared in humans between 1915 and 1940.

Of the two major strains of the virus, HIV-1 is more widely distributed and virulent. Both strains infect cells with surface CD4 molecules, including helper T cells, macrophages, some B lymphocytes, and brain cells. CD4 and a protein *coreceptor* are required for viral entry. These coreceptors include fusin (CXCR4) on helper T cells and CCR5 on macrophages, both of which normally function in chemokine reception.

Following viral entry, HIV RNA is copied by reverse transcription into DNA, which is integrated as a provirus into the host cell genome, from where it directs production of new viral particles. Its existence as a provirus and the frequent mutational changes during replication make HIV difficult for both humoral and cell-mediated responses to irradicate. The immune system produces anti-HIV antibodies, which are detected in the blood within 1 to 12 months after infection and are the most common basis for identifying persons who are *HIV-positive.* This HIV antibody test is used to screen blood donations.

During the early drop in HIV level in the blood while the immune system mounts a defense, HIV continues to replicate in the lymph nodes. The subsequent rise in HIV blood levels is caused by the breakdown of lymphatic tissue function, the release of the virus from lymph tissue, and the depletion of helper T cells resulting from viral infection. It takes about 10 years for an HIV infection to progress to severe helper T cell depletion and AIDS. Although T cell levels are still monitored, it appears that measures of viral load are a better indicator of disease prognosis.

While not able to cure HIV, new drug combinations are slowing the progression to AIDS. Drugs that slow viral replication include DNA-synthesis inhibitors, reverse transcriptase inhibitors (such as AZT and ddI), and protease inhibitors that prevent the synthesis of HIV proteins.

HIV is transmitted by transfer of body fluids such as blood or semen. Nonsterile needles, unprotected sex, unscreened blood supplies, and transmission from mother to child during fetal development or nursing are all means by which HIV can spread. AIDS cases are expected to increase by nearly 20% per year, and education may be the best approach to slowing that spread.

■ INTERACTIVE QUESTION   43.8

a. Why is AIDS such a deadly disease?

b. Why has it proved so difficult to prevent and cure this disease?

WORD ROOTS

agglutinat- = glued together (*agglutination*: an antibody-mediated immune response in which bacteria or viruses are clumped together, effectively neutralized, and opsonized)

an- = without; **-aphy** = suck (*anaphylactic shock*: an acute, life threatening, allergic response)

anti- = against; **-gen** = produce (*antigen*: a foreign macromolecule that does not belong to the host organism and that elicits an immune response)

chemo- = chemistry; **-kine** = movement (*chemokine*: a group of about 50 different proteins secreted by blood vessel endothelial cells and monocytes; these molecules bind to receptors on many types of leukocytes and induce numerous changes central to inflammation)

cyto- = cell (*cytokines*: in the vertebrate immune system, protein factors secreted by macrophages and helper T cells as regulators of neighboring cells)

epi- = over; **-topo** = place (*epitope*: a localized region on the surface of an antigen that is chemically recognized by antibodies)

immuno- = safe, free; **-glob** = globe, sphere (*immunoglobulin:* one of the class of proteins comprising the antibodies)

inter- = between; **leuko-** = white (*interleukin-2:* a cytokine that helps B cells that have contacted an antigen to differentiate into antibody-secreting plasma cells)

macro- = large; **-phage** = eat (*macrophage:* an amoeboid cell that moves through tissue fibers, engulfing bacteria and dead cells by phagocytosis)

mono- = one (*monocyte:* an agranular leukocyte that is able to migrate into tissues and transform into a macrophage)

neutro- = neutral; **-phil** = loving (*neutrophil:* the most abundant type of leukocyte; neutrophils tend to self destruct as they destroy foreign invaders, limiting their lifespan to but a few days)

perfora- = bore through (*perforin:* a protein that forms pores in a target cell's membrane)

pyro- = fire (*pyrogen:* molecules that set the body's thermostat to a higher temperature. They are released by certain leukocytes)

STRUCTURE YOUR KNOWLEDGE

This chapter contains a wealth of information that is probably fairly new to you. If you take a little time and pull out the key players of the immune system and organize them first into very basic concept clusters and then develop more interrelated concept maps, you will find that this information is both understandable and fascinating.

1. Fill in the table below on some of the molecules involved in the immune system.

2. Create a concept map outlining specific defense mechanisms, showing the cells involved in humoral and cell-mediated immunities and their functions.

3. Describe the structure of an antibody molecule and relate this structure to its function.

4. Briefly explain clonal selection.

TEST YOUR KNOWLEDGE

MULTIPLE CHOICE: *Choose the one best answer.*

1. Which of the following is *incorrectly* paired with its effect?
   a. gastric juice—kills bacteria in the stomach
   b. fever—stimulates phagocytosis and inhibits microbial growth
   c. histamine—causes blood vessels to dilate
   d. vaccination—creates passive immunity
   e. lysozyme—attacks cell walls of bacteria

2. Which of the following would release interferon?
   a. a macrophage that has become an APC
   b. an injured epithelial cell of a blood vessel
   c. a cell infected by a virus
   d. a mast cell that has bound an antigen
   e. a helper T cell bound to an APC

3. Antibodies are
   a. proteins or polysaccharides usually found on the cell surface of invading bacteria or viruses.
   b. proteins that consist of two light and two heavy polypeptide chains.
   c. proteins circulating in the blood that tag foreign cells for complement fixation.
   d. proteins embedded in B cell membranes.
   e. b, c, and d are all correct.

Molecules	Where Produced or Found	Action
Complement		
Antibodies		
Interleukin-1		
Interleukin-2		
Perforin		
Class I MHC		
Class II MHC		

4. A secondary immune response is more rapid and greater in effect than a primary immune response because
   a. histamines and prostaglandins cause rapid vasodilation.
   b. the second response is an active immunity, whereas the primary one was a passive immunity.
   c. helper T cells are available to activate other blood cells.
   d. interleukins cause the rapid accumulation of phagocytic cells.
   e. memory cells respond to the pathogen and rapidly clone more effector cells.

5. Lymphocytes capable of reacting against "self" molecules
   a. are usually not a problem until a woman's second pregnancy.
   b. are usually inactivated or destroyed before birth.
   c. are usually kept separate from the immune system.
   d. contribute to immunodeficiency diseases.
   e. are characterized by class I MHC molecules.

6. The major histocompatibility complex
   a. is involved in the ability to distinguish self from nonself.
   b. is a collection of cell surface glycoproteins.
   c. may trigger T cell responses after transplant operations.
   d. presents antigen fragments on infected cells.
   e. All of the above are correct.

7. In opsonization,
   a. complement proteins and/or antibodies coat microorganisms and help phagocytes bind to and engulf the foreign cell.
   b. a set of complement proteins lyses a hole in a foreign cell's membrane.
   c. antibodies precipitate soluble antigens.
   d. a flood of histamines is released that may result in anaphylactic shock.
   e. microbes coated with antibodies and complement proteins adhere to vessel walls.

8. Severe combined immunodeficiency
   a. is an autoimmune disease.
   b. is a form of cancer in which the membrane surface of the cell has changed.
   c. is a disease in which both T and B cells are absent or inactive.
   d. is an immune disorder in which the number of helper T cells is greatly reduced.
   e. results from a few types of cancers, such as Hodgkin's disease.

9. A transfusion of type B blood given to a person who has type A blood would result in
   a. the recipient's anti-B antibodies reacting with the donated red blood cells.
   b. the recipient's B antigens reacting with the donated anti-B antibodies.
   c. the recipient forming both anti-A and anti-B antibodies.
   d. no reaction, because B is a universal donor type of blood.
   e. the introduced blood cells being destroyed by nonspecific defense mechanisms.

10. Which of the following are *incorrectly* paired?
    a. variable region—determines antibody specificity for an epitope
    b. immunoglobulins—glycoproteins that form epitopes
    c. constant region—determines class and function of antibody
    d. IgG—most abundant circulating antibodies, confer passive immunity to fetus
    e. IgE—receptor molecules attached to mast cells and basophils

11. Which of the following does *not* destroy a target cell by creating a hole in the membrane that causes the cell to lyse?
    a. neutrophil   d. complement proteins
    b. cytotoxic T cell   e. perforin
    c. natural killer cell

12. A T-independent antigen
    a. does not need the aid of T_H cells to bind to antibody molecules, whereas T-dependent antigens do.
    b. needs to bind to T_H cells to activate antibody production.
    c. is often a large molecule that simultaneously binds with several antigenic receptors on a B cell at one time.
    d. will result in the production of both plasma cells and memory cells.
    e. is responsible for the primary immune response.

13. What do antibodies, T cell receptors, and MHC molecules have in common?
    a. They are found exclusively in cells of the immune system.
    b. They are all part of the complement system.
    c. They are antigen-presenting molecules.
    d. They are or can be membrane-bound proteins.
    e. They are involved in the cell-mediated portion of the immune system.

14. Which of the following is an effective defense against bacteria but does *not* work against viral particles?
    a. secretion of interferon by an infected cell
    b. neutralization by antibodies

c. the enzyme lysozyme

d. a secondary immune response

e. humoral immunity

15. How are antibodies and complement related?
    a. They are both coded for by genes that have hundreds of alleles.
    b. They are both used in nonspecific defenses.
    c. They are both produced by plasma cells.
    d. Antibodies bound to antigens on a pathogen's membrane may combine with complement proteins to activate a membrane attack complex.
    e. Antibodies tag foreign cells for destruction; complement is attracted to the infection and destroys the invader by phagocytosis.

16. Which of the following describes the main difference between an inflammatory response and an immune response?
    a. The inflammatory response responds only to free pathogens in a localized area; the immune response responds only to pathogens that have entered body cells.
    b. The inflammatory response involves only leukocytes, whereas the immune response involves only lymphocytes.
    c. The inflammatory response relies on phagocytes to destroy pathogens, whereas the immune response relies on antibodies to destroy pathogens.
    d. The inflammatory response is nonspecific, whereas the immune response reacts to specific microbes on the basis of their different antigens.
    e. Complement proteins participate in the immune response but not in the inflammatory response.

17. Clonal selection is responsible for the
    a. proliferation of clones of effector and memory cells specific for an encountered antigen.
    b. recognition of class I MHC molecules by cytotoxic T cells.
    c. rearrangement of antibody genes for the light and heavy chains.
    d. formation of cell cultures in the commercial production of monoclonal antibodies.
    e. transformation of a clone of helper T cells into cytotoxic T cells keyed to a specific antigen.

18. What role does a macrophage play in the immune response?
    a. activates complement proteins by the classical pathway
    b. binds to the CD8 receptors on cytotoxic T cells to activate their production of perforin
    c. releases IL-2 to activate B cells to produce clones of plasma cells
    d. activates both humoral and cell-mediated immunity by releasing interferons after it has ingested a virus

e. presents antigens of an engulfed pathogen in its class II MHC molecules to helper T cells, and releases IL-1

19. All of the following are involved with nonspecific defense mechanisms *except*
    a. the inflammatory response.
    b. plasma cells.
    c. antimicrobial proteins such as lysozyme.
    d. chemokines that attract phagocytes.
    e. basophils, neutrophils, eosinophils, and natural killer cells.

20. Which of the following in *not* true of invertebrate defense systems?
    a. They are able to distinguish self from nonself.
    b. They may have phagocytic cells that engulf foreign matter and release cytokines.
    c. They produce specific antibodies against bacterial antigens, but not against viral antigens.
    d. Most do not exhibit immunological memory, although earthworms do exhibit memory against tissue grafts.
    e. Some have cells that produce antibody-like molecules, such as the protein hemolin found in insects.

21. Helper T cells play which of the following roles in the immune response?
    a. bind to class I MHC and activate complement proteins to attack and lyse cancer cells
    b. bind to the antigens presented in the CD8 receptors on cytotoxic T cells and release perforin
    c. produce interferons and histamines that help initiate a specialized inflammatory response
    d. present antigens of an engulfed pathogen in its class II MHC molecules to B cells, which are then stimulated to develop into a clone of plasma cells
    e. activate both the humoral and cell-mediated immunities by releasing IL-2 and other cytokines after recognizing class II MHC-antigen complexes on an APC

22. Which of the following statements about humoral immunity is correct?
    a. It is a form of passive immunity produced by vaccination.
    b. It defends against free pathogens with effector mechanisms such as neutralization, agglutination, precipitation, opsonization, or complement fixation.
    c. It protects the body against pathogens that have invaded body cells as well as against abnormal body cells.
    d. It is mounted by lymphocytes that have matured in the thymus.
    e. It depends on the recognition of class I MHC molecules that are bound to a specific antigen to activate its effector mechanism.

23. What accounts for the huge diversity of antigens to which B cells can respond?
    a. The antibody genes have millions of alleles.
    b. The rearrangement of the antibody genes during development results in millions of combinations of the light and heavy polypeptide chains.
    c. The antigen-binding sites at the arms of the molecule can assume a huge diversity of shapes in response to the specific antigen encountered.
    d. B cells have thousands of copies of antibodies bound to their plasma membranes.
    e. B cells can be APCs when they take in antigens by endocytosis and display fragments in their class II MHC molecules.

24. What is the function of CD4?
    a. a surface molecule on a cytotoxic T cell that enhances its binding to a class I MHC molecule displaying a foreign antigen
    b. a membrane protein on an APC that helps a helper T cell recognize the MHC-antigen complex
    c. a receptor that normally functions for cytokines, but which HIV uses as its receptor
    d. a surface molecule on a helper T cell that enhances its binding to a class II MHC molecule displaying a foreign antigen
    e. a portion of the class I MHC molecule found on all nucleated cells that identify cells as "self"

25. From which of the following would an AIDS patient be *least* likely to suffer?
    a. Kaposi's sarcoma or other cancers
    b. tuberculosis
    c. rheumatoid arthritis
    d. pneumonia
    e. yeast infections of mucous membranes

26. The newest successful AIDS therapy involves which of the following?
    a. gene therapy that alters the gene for CCR5 so that HIV cannot use this coreceptor to enter cells
    b. use of AZT and ddI that inhibit reverse transcriptase
    c. treatment of the opportunistic diseases of AIDS patients with more powerful antiviral and antibacterial drugs
    d. a vaccine made from a modified virus and HIV genes
    e. a cocktail of protease inhibitors (which interfere with a viral enzyme needed to produce viral proteins), AZT, and ddI

27. All of the following are considered diseases or malfunctions of the immune system *except*
    a. MHC-induced transplant rejection.
    b. SCID (severe combined immunodeficiency).
    c. lupus, multiple sclerosis, and insulin-dependent diabetes.
    d. AIDS.
    e. allergic anaphylactic shock.

28. Which of the following may induce a graft-versus-host reaction?
    a. organ transplant
    b. blood transfusion
    c. bone marrow transplant
    d. skin transplant
    e. gene therapy

29. How does the immune system recognize malignant tumor cells?
    a. They do not display class I MHC molecules.
    b. They display fragments of tumor antigen in their MHC molecules.
    c. They display cancerous viral fragments in their class II MHC molecules.
    d. They have abnormal amounts of polysaccharides in their extracellular matrix that trigger a T-independent immune response.
    e. They undergo opsonization by complement proteins so that they are recognized by phagocytes.

30. Place the following steps in the T_H cell activation of cell-mediated and humoral immunity in the correct order:
    1. T_H cell secretes interleukin-2 and cytokines.
    2. Macrophage engulfs pathogen and presents antigen in class II MHC.
    3. Plasma cells secrete antibodies and T_C cells attack cells with class I MHC-antigen complex.
    4. T cell receptor recognizes class II MHC-antigen complex.
    5. Macrophage secretes interleukin-1.
    6. Activated B cells form plasma and memory cells, activated T cells form T_C cells and memory cells.

    a. 1, 3, 5, 6, 2, 4
    b. 5, 1, 2, 6, 4, 3
    c. 2, 4, 5, 1, 6, 3
    d. 5, 2, 4, 1, 3, 6
    e. 2, 1, 4, 5, 6, 3

CHAPTER 44

REGULATING THE INTERNAL ENVIRONMENT

FRAMEWORK

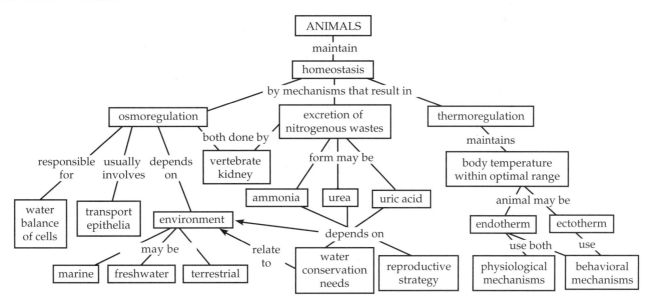

CHAPTER REVIEW

Animals are able to survive large fluctuations in their external environment by maintaining a relatively constant internal environment. This regulatory ability is called **homeostasis.** Animals are able to maintain a favorable range in their internal temperature **(thermoregulation),** regulate solute and water balance **(osmoregulation),** and dispose of nitrogenous wastes **(excretion).**

An Overview of Homeostasis

Regulating and conforming are the two extremes in how animals cope with environmental fluctuations (925–926)

For any given environmental variable, an animal may be a **regulator,** using homeostatic mechanisms to control internal fluctuations, or a **conformer,** allowing

some internal conditions to vary with certain environmental changes. Conformers often live in relatively stable environments.

Homeostasis balances an animal's gains versus losses for energy and materials (926–927)

As open systems, animals exchange energy and materials with the environment, and the rates of these gains and losses must be balanced over time. A net increase in inputs is required for growth and reproduction. Homeostasis requires an interconnected set of **budgets** of gains and losses in energy and materials.

Regulation of Body Temperature

Changes in temperature affect most biochemical and physiological processes. The two- to threefold increase in rates of enzyme-mediated reactions and metabolic processes for each 10°C temperature increase is known as the Q_{10} **effect.**

Each animal has an optimal temperature range, and many animals balance their heat budgets to maintain nearly constant internal temperatures.

Four physical processes account for heat gain or loss (927–928)

Heat is exchanged between an organism and the external environment by the physical processes of **conduction,** the direct transfer of thermal motion between objects in contact; **convection,** the transfer of heat by the flow of air or water past a surface; **radiation,** the emission of electromagnetic waves by all objects warmer than absolute zero; and **evaporation,** the loss of heat due to the conversion of the surface molecules of a liquid to a gas. Wind (convection) and the evaporation of sweat can greatly increase the loss of heat.

Ectotherms have body temperatures close to environmental temperature; endotherms can use metabolic heat to keep body temperature warmer than their surroundings (928–929)

The use of metabolic heat in determining body temperature differentiates ectotherms (most invertebrates, fishes, reptiles, and all amphibians) from endotherms (birds and mammals, and some fishes, reptiles, and insects). The environmental temperature determines the body temperature of an **ectotherm** because its low metabolic rate does not generate much heat. The high metabolic rate of an **endotherm** contributes enough heat to maintain a body temperature warmer than the environment.

Endotherms can sustain vigorous activity because their high body temperature produces a high aerobic metabolism. Endothermy also facilitates functioning within the more severe temperature fluctuations found on land. The advantages of endothermy, however, are quite energetically expensive.

■ **INTERACTIVE QUESTION   44.1**

Explain the following statement: Some ectotherms have body temperatures that are more constant, and some even have higher body temperatures, than those of endotherms.

Thermoregulation involves physiological and behavioral adjustments that balance heat gain and loss (929–930)

Four categories of adaptations enable endotherms and those ectotherms that thermoregulate to balance their heat budgets. (1) The rate of heat exchange with the environment can be adjusted by body insulation, **vasodilation, vasoconstriction,** and a **countercurrent heat exchanger,** which can minimize heat loss in body parts immersed in cold water. Dilation of superficial blood vessels increases heat loss, whereas constriction reduces blood flow and heat loss. (2) Cooling occurs by evaporative heat loss, which can be increased by panting, bathing, or sweating. (3) Behavioral responses, such as basking or moving to a warmer or cooler environment, can regulate heat loss. (4) Many mammals and birds can increase the rate of metabolic heat production when exposed to cold.

■ **INTERACTIVE QUESTION   44.2**

Describe how heat loss may be minimized from the legs of a Canada goose standing in cold water.

Most animals are ectothermic, but endothermy is widespread (930–935)

Mammals and Birds   Mammals and birds, which maintain high body temperatures within a narrow range, are able to regulate the rate of metabolic heat production as well as control the rate of heat gain or loss. Heat production can be increased by muscle contraction (by moving or shivering) and by **nonshivering thermogenesis (NST),** a rise in metabolic rate and the production of heat instead of ATP. **Brown fat** in some mammals is specialized for heat production by NST.

Vasodilation and vasoconstriction regulate heat exchange. Feathers and fur may be raised to trap more air and increase their insulation against heat loss. Marine mammals maintain their high body temperature by efficient heat-conserving mechanisms, including a thick layer of insulating blubber and countercurrent heat exchange in the flippers. When in warmer waters, these animals can dissipate metabolic heat by dilating superficial blood vessels. Terrestrial mammals and birds use evaporative cooling, which may be enhanced by panting, sweating, and spreading saliva on body surfaces. They may also allow a rise in body temperature to increase heat loss to the environment.

Amphibians and Reptiles Amphibians produce little heat and easily lose heat by evaporation from their moist skin. Behavioral adaptations usually enable them to maintain acceptable internal temperatures.

Behavioral adaptations, such as orienting the body to the sun or finding suitable microclimates, allow most reptiles to regulate their temperature within an optimal range. In some reptiles, vasoconstriction of surface vessels sends more blood to the body core to conserve heat. A few large reptiles use shivering to generate heat and become endotherms while incubating eggs.

Fishes The body temperature of fishes is determined mainly by the water temperature, as they lose metabolic heat readily when blood passes through the gills. Some large, active, endothermic fish are able to retain the heat produced by their swimming muscles with a countercurrent heat exchanger.

Invertebrates Some terrestrial invertebrates adjust their temperature by behavioral or physiological mechanisms. Endothermic flying insects may "warm up" before taking off by contracting their flight muscles. Many have a countercurrent heat exchanger that maintains a high thoracic temperature. The social organization of honeybees enables them to retain heat in cold temperatures by huddling together and to cool their hive during hot weather by bringing in water and fanning their wings.

Feedback Mechanisms in Thermoregulation The complex system of temperature regulation in mammals depends on feedback mechanisms. Both warm and cold temperature-sensing nerve cells are located in the skin, hypothalamus, and some other body regions. A group of neurons in the hypothalamus functions as the thermostat.

■ INTERACTIVE QUESTION 44.3

a. What cooling mechanisms are activated when the hypothalamus senses a body temperature above the set point?

b. What mechanisms are activated when body temperature falls below the set point?

Adjustments to Changing Temperatures Many animals are capable of **acclimatization,** a physiological adjustment to a different temperature range.

Ectotherms and endotherms acclimatize in different ways. Birds and mammals adjust the thickness of their insulating coats or vary their capacity for metabolic heat production to meet seasonal changes. Ectotherms must adjust their cellular physiology to acclimatize to changes in environmental and thus body temperature. They may produce more enzymes to compensate for reduced enzyme activity or enzymes with different temperature optima, or change the proportions of saturated and unsaturated lipids to help maintain membrane fluidity. Ectotherms in subzero environments may produce cryoprotectants to prevent ice formation in their cells.

Cells can make rapid adjustments to temperature increases and other stresses by producing **stress-induced proteins,** including **heat-shock proteins.** These molecules protect proteins that might otherwise be denatured and may prevent cell death when an organism faces severe environmental changes.

Torpor conserves energy during environmental extremes (935–936)

Torpor is a physiological state characterized by decreases in metabolism and activity. In **hibernation,** the body temperature is maintained at a lower level, greatly conserving energy and allowing the animal to withstand long periods of cold temperature and decreased food supply. A Belding's ground squirrel reduces its energy needs from 150 to an average of 5–7 kcal per day during its eight-month hibernation.

In **estivation,** animals survive long stretches of elevated temperature and diminished water supply by entering a period of inactivity and lowered metabolism. Hibernation and estivation may be triggered by seasonal changes in the length of daylight.

Many small mammals and birds that have very high metabolic rates enter a **daily torpor** controlled by the biological clock during the periods when they are not feeding.

Water Balance and Waste Disposal

Osmoregulation usually involves controlling the uptake and loss of solutes, with water following by osmosis. Homeostatic mechanisms temper changes in the internal body fluid that bathes the cells, either hemolymph in animals with open circulatory systems or interstitial fluid serviced by blood in those with closed circulatory systems.

Water balance and waste disposal depend on transport epithelia (936)

The transport of solutes across a **transport epithelium** is essential for both osmotic regulation and disposing of metabolic wastes. The cells of this epithelium are linked by impermeable tight junctions and regulate the passage of solutes between the internal fluid and the environment. Transport epithelia are usually arranged in tubular networks that provide large surface areas for exchange.

An animal's nitrogenous wastes are correlated with its phylogeny and habitat (936–938)

Ammonia, a small and toxic molecule, is produced when proteins and nucleic acids are metabolized for energy or converted to carbohydrates or fats. Some animals excrete ammonia; others expend energy to convert it to less toxic wastes such as urea or uric acid.

Ammonia   Aquatic animals can excrete nitrogenous wastes as ammonia because it is very soluble and easily permeates membranes. Many invertebrates lose ammonia across their whole body surface. In fishes, most of the ammonia is passed across the epithelia of the gills. In freshwater fishes, the gills exchange NH_4^+ ions for Na^+, helping to maintain Na^+ concentrations in body fluids.

Urea   Mammals, most adult amphibians, and many marine fishes and turtles produce **urea,** a much less toxic compound than ammonia. Urea can be tolerated in more concentrated form and excreted with less loss of water. Ammonia and carbon dioxide are combined in the liver to produce urea, which is then carried by the circulatory system to the kidneys. The production of urea requires energy.

Uric Acid   Land snails, insects, birds, and many reptiles produce **uric acid,** a compound of low solubility in water that can be excreted as a precipitate with very little water loss. Its synthesis, however, is energetically expensive.

The mode of reproduction of animal groups seems to have determined whether they excrete uric acid or urea as their nitrogenous waste product.

The form of nitrogenous wastes also relates to habitat. Terrestrial turtles excrete mainly uric acid, whereas aquatic turtles excrete both urea and ammonia. Some animals actually shift their nitrogenous waste product depending on environmental conditions.

■ **INTERACTIVE QUESTION 44.4**

Vertebrates that produce shelled eggs excrete _____ Mammals produce _____. What adaptive advantage do these types of nitrogenous wastes provide?

Cells require a balance between osmotic gain and loss of water (938–939)

Whatever an animal's habitat or nitrogenous waste product, its water gain must balance water loss.

Osmosis is the diffusion of water across a selectively permeable membrane that separates two solutions differing in **osmolarity** (moles of solute per liter). Osmolarity is expressed in units of milliosmoles per liter (mosm/L). Isoosmotic solutions are equal in osmolarity, and there is no net osmosis between them. There is a net flow of water from a hypoosmotic (more dilute) to a hyperosmotic (more concentrated) solution.

Osmoregulators expend energy to control their internal osmolarity; osmoconformers are isoosmotic with their surroundings (939–941)

Osmoconformers are isoosmotic with their surroundings and do not regulate their osmolarity. The body fluids of **osmoregulators** are not isoosmotic with their external environment. Osmoregulators must get rid of excess water if they live in a hypoosmotic medium or take in water to offset osmotic loss if they inhabit a hyperosmotic environment. Osmoregulation is energetically costly because animals must actively transport solutes in order to maintain osmotic gradients needed to gain or lose water.

Most animals are **stenohaline,** able to tolerate only small changes in external osmolarity. Animals that are **euryhaline** can survive large differences in osmotic environments by conforming or by maintaining a stable, internal osmolarity.

Maintaining Water Balance in the Sea   Most marine invertebrates are osmoconformers, whereas marine vertebrates, except for the hagfishes, are osmoregulators. Marine bony fishes are hypoosmotic to seawater and must drink large quantities of seawater to replace the water they lose by osmosis through their skin and gills. Excess salts are pumped out through the gills and other ions are excreted in the scanty urine.

Sharks have an internal salt concentration lower than that of seawater because their rectal glands and kidneys excrete salt out of the body. They maintain an osmolarity slightly higher than that of seawater, however, by retaining urea and trimethylamine oxide (which protects proteins from the damaging effects of urea) within their bodies. Water that enters a shark's body by osmosis is disposed of in urine produced in the kidneys.

Maintaining Osmotic Balance in Fresh Water Freshwater animals constantly take in water by osmosis and lose salts by diffusion. Protists use contractile vacuoles to pump out excess water. Freshwater fishes excrete large quantities of dilute urine. Salt supplies are replaced from their food or by active uptake of ions.

Special Problems of Living in Temporary Waters Some animals are capable of **anhydrobiosis,** or cryptobiosis—surviving dehydration in a dormant state. Anhydrobiotic roundworms produce large quantities of trehalose, a disaccharide that replaces water around membranes and proteins during dehydration.

Maintaining Osmotic Balance on Land Adaptations to prevent dehydration in terrestrial animals include water-impervious coverings and behavioral adaptations such as nocturnal lifestyles. Water is lost in urine and feces and through evaporation, and gained by drinking and eating moist food and by metabolic production during cellular respiration.

■ INTERACTIVE QUESTION 44.5

Indicate whether the following are osmoregulators or osmoconformers and whether they are isoosmotic, hyperosmotic, or hypoosmotic to their environment.

Animal	Osmoregulator or Osmoconformer?	Osmotic Relation to Environment
Marine invertebrate	a.	b.
Shark	c.	d.
Marine fish	e.	f.
Freshwater fish	g.	h.
Freshwater protozoan	i.	j.
Terrestrial animal	k.	l.

Excretory Systems

Most excretory systems produce urine by refining a filtrate derived from body fluids: *an overview* (941–942)

During **filtration,** water and small solutes are forced out of the blood or body fluids into the excretory system. Two mechanisms transform this filtrate to urine: In **secretion,** excess salts, toxins, and other solutes are added to the **filtrate,** and water and valuable solutes are removed from the filtrate in **selective reabsorption.**

Diverse excretory systems are variations on a tubular theme (942–944)

Protonephridia: Flame-Bulb Systems The **protonephridia** of flatworms are branched systems of closed tubules. Water and solutes from the interstitial fluid pass into the flame bulbs at the ends of the smallest tubules, propelled by the cilia of the flame bulbs. Urine empties into the external environment by way of nephridiopores. The flame-bulb system of freshwater flatworms is primarily osmoregulatory; metabolic wastes diffuse through the body surface or are excreted through the gastrovascular cavity. In parasitic flatworms, however, protonephridia function mainly in excretion of nitrogenous wastes.

Metanephridia Excretory tubules with internal openings that collect body fluids are found in most annelids. These **metanephridia** occur in pairs in each segment of an earthworm. An open ciliated funnel called a nephrostome collects coelomic fluid, which then moves through a folded tubule encased in capillaries. The transport epithelium of the tubule reabsorbs most solutes, which then reenter the blood. Hypoosmotic urine, carrying nitrogenous wastes and excess water absorbed by osmosis, exits by way of nephridiopores.

Malpighian Tubules **Malpighian tubules** are excretory organs in insects and other terrestrial arthropods. Transport epithelia lining these blind sacs, which open into the digestive tract, secrete solutes from the hemolymph into the tubule. The fluid passes through the hindgut and into the rectum, where a transport epithelium pumps most of the solutes back into the hemolymph, and water follows by osmosis. In this water-conserving system, nitrogenous wastes are eliminated as dry matter along with the feces.

Vertebrate Kidneys The numerous excretory tubules of most vertebrates are enclosed in a network of capillaries and arranged into compact kidneys. The kidneys, their associated blood vessels, and the structures that carry urine out of the body comprise the vertebrate excretory system.

■ INTERACTIVE QUESTION 44.6

Indicate whether these tubular systems function in osmoregulation, excretion, or both. If they function in osmoregulation, do they help conserve water or remove excess water that had entered by osmosis?

a. Protonephridia of freshwater planaria

b. Metanephridia of earthworms

c. Malpighian tubules of insects

d. Kidneys of terrestrial mammals

Nephrons and associated blood vessels are the functional units of the mammalian kidney (944–947)

Blood enters each of the pair of bean-shaped kidneys through a **renal artery** and leaves by way of a **renal vein.** Urine exits through a **ureter** and is temporarily stored in the **urinary bladder.** Urine exits the body through the **urethra.**

Structure and Function of the Nephron and Associated Structures The outer **renal cortex** and inner **renal medulla** of the kidney are packed with excretory tubules and blood vessels. The **nephron** consists of a long tubule with a cuplike **Bowman's capsule** at the blind end that encloses a ball of capillaries called the **glomerulus.**

Filtration of the blood occurs as blood pressure forces water, urea, salts, and small molecules through the porous capillary walls and specialized capsule cells called podocytes into Bowman's capsule, forming the filtrate. The filtrate passes through the **proximal tubule,** the **loop of Henle** with a descending limb and an ascending limb, and the **distal tubule. Collecting ducts** receive processed filtrate from many nephrons and pass the urine into the renal pelvis.

In the human kidney, 80% of the nephrons are **cortical nephrons** located entirely in the renal cortex. **Juxtamedullary nephrons,** found only in mammals and birds, have long loops of Henle that extend into the renal medulla and participate in the formation of hyperosmotic urine. The transport epithelium lining the nephrons and collecting ducts processes about 180 L of filtrate each day to produce about 1.5 L of urine.

An **afferent arteriole** supplies each nephron, subdividing to form the capillaries of the glomerulus, which then converge to form an **efferent arteriole.** This arteriole divides into a second capillary network—the **peritubular capillaries**—that surrounds the proximal and distal tubules. The **vasa recta** includes descending and ascending capillaries that parallel the loop of Henle. Exchange between the tubules and capillaries is via the interstitial fluid.

■ INTERACTIVE QUESTION 44.7

a. What is found in the filtrate that moves from Bowman's capsule into the tubule?

b. What remains in the capillaries?

From Blood Filtrate to Urine: a closer look (1) Reabsorption and secretion both take place in the proximal tubule. Its transport epithelium selectively secretes hydrogen ions, ammonia, and drugs and poisons (processed by the liver and delivered via peritubular capillaries) into the filtrate. Glucose, amino acids, potassium, and the buffer bicarbonate are actively or passively reabsorbed. Salt diffuses from the filtrate into the cells of the proximal tubule, which then actively transport Na^+ across the membrane into the interstitial fluid; Cl^- is transported passively out of the cells to balance the positive charge; and water follows by osmosis. Salt and water diffuse from the interstitial fluid into the peritubular capillaries.

(2) The descending limb of the loop of Henle is freely permeable to water but not to salt or other small solutes. The interstitial fluid is increasingly hyperosmotic toward the inner medulla, so water exits by osmosis, leaving behind a filtrate with a high solute concentration.

(3) The ascending limb of the loop of Henle is not permeable to water but is permeable to salt, which diffuses out of the thin lower segment of the loop, adding to the high osmolarity of the medulla. NaCl is actively transported out of the thick upper portion of the ascending limb. As salt leaves, the filtrate becomes less concentrated.

(4) The distal tubule is specialized for selective secretion and reabsorption. It regulates K^+ secretion and NaCl reabsorption, and it helps to regulate pH by secreting H^+ and reabsorbing bicarbonate.

(5) As the filtrate moves in the collecting duct back through the increasing osmotic gradient of the medulla, more and more water exits by osmosis. Urea diffuses out of the lower portion of the duct and helps to maintain the osmotic gradient. Salt excretion is controlled by the transport epithelium actively reabsorbing NaCl.

The mammalian kidney's ability to conserve water is a key terrestrial adaptation (947–951)

The osmolarity gradient of NaCl and urea in the renal medulla is produced by the loops of Henle and collecting ducts and enables the kidney to produce urine up to four times as concentrated as blood and normal interstitial fluid.

Conservation of Water by Two Solute Gradients   Filtrate leaving Bowman's capsule has an osmolarity of about 300 mosm/L, the same as blood. Both water and salt are reabsorbed in the proximal tubule. In the trip down the descending limb of the loop of Henle, water exits by osmosis, and the filtrate becomes more concentrated. Salt, which is now in high concentration in the filtrate, diffuses out as the filtrate moves up the salt-permeable but water-impermeable ascending limb—helping to create the osmolarity gradient.

The countercurrent flow of filtrate in the loop of Henle, along with the energy-consuming active transport of NaCl from the thick segment of the ascending limb, maintain the steep osmotic gradient between the cortex and medulla.

The vasa recta also has a countercurrent flow through the loop of Henle. As the blood in these vessels moves down into the inner medulla, water leaves

---

■ **INTERACTIVE QUESTION   44.8**

In this diagram of a nephron and collecting tubule, label the parts on the indicated lines. Label the arrows to indicate the movement of salt, water, nutrients, K^+, bicarbonate (HCO_3^-), H^+, and urea into or out of the tubule.

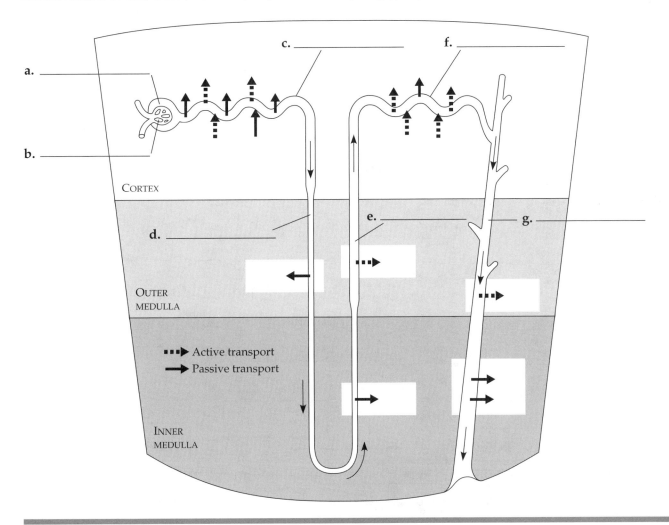

by osmosis and salt enters; as the blood moves back up toward the cortex, water moves back in and salt diffuses out. Thus the capillaries can carry nutrients and other supplies to the medulla without disrupting the osmolarity gradient.

The filtrate makes one final pass through the medulla, this time in the collecting duct, which is permeable to water and urea but not to salt. Water flows out by osmosis, and, as the filtrate becomes more concentrated, urea leaks into the interstitial fluid, adding to the high osmolarity of the inner medulla. The resulting concentrated urine may achieve an osmolarity as high as 1,200 mosm/L, isoosmotic to the interstitial fluid of the inner medulla.

How the Nervous System and Hormones Regulate Kidney Functions Osmolarity of urine in humans can vary from 70 to 1,200 mosm/L, depending on the body's water and salt balance and urea production.

Antidiuretic hormone (ADH) is produced in the hypothalamus and stored in the pituitary gland. When blood osmolarity rises above a set point of 300 mosm/L, osmoreceptor cells in the hypothalamus trigger the release of ADH. This hormone increases water permeability of the distal tubules and collecting ducts, increasing water reabsorption from the urine. After consuming water in food or drink, negative feedback decreases the release of ADH. When blood osmolarity is low, the absence of ADH results in the production of large volumes of dilute urine, called diuresis. Alcohol inhibits ADH release and can cause dehydration.

The **juxtaglomerular apparatus (JGA)**, located near the afferent arteriole leading to the glomerulus, responds to a drop in blood pressure or volume by releasing renin, an enzyme that converts the plasma protein angiotensinogen to **angiotensin II**. Angiotensin II functions as a hormone and constricts

■ INTERACTIVE QUESTION 44.9

Complete this concept map to help organize your understanding of the nervous and hormonal control of water and salt reabsorption in the kidneys.

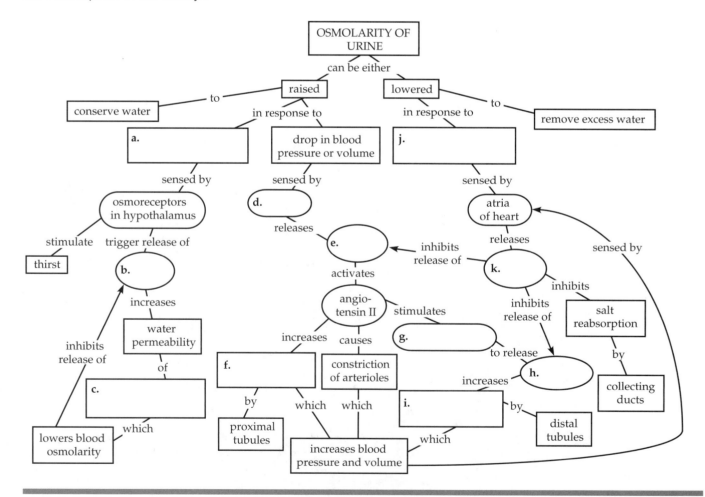

arterioles, stimulates the proximal tubules to reabsorb more NaCl and water, and stimulates the adrenal glands to release **aldosterone**. This hormone stimulates Na^+ and water reabsorption in the distal tubules. The **renin-angiotensin-aldosterone system (RAAS)** is a homeostatic feedback circuit that maintains adequate blood pressure and volume.

Atrial natriuretic factor (ANF) counters the RAAS. ANF, released by the atria of the heart in response to increased blood volume and pressure, inhibits the release of renin from the JGA, inhibits NaCl reabsorption by the collecting ducts, and reduces the release of aldosterone from the adrenals.

Diverse adaptations of the vertebrate kidney have evolved in different habitats (951)

Modifications of nephron structure and function correspond to osmoregulatory requirements in various habitats. Mammals in dry habitats, needing to conserve water and excrete hyperosmotic urine, have exceptionally long loops of Henle. Birds also have juxtamedullary nephrons, but their shorter loops of Henle prevent them from producing urine as hyperosmotic as that of mammals. The kidneys of reptiles have only cortical nephrons, and their urine is isoosmotic to body fluids; water is conserved, however, through reabsorption in the cloaca and the production of uric acid as their nitrogenous waste.

INTERACTIVE QUESTION 44.10

a. What type of nephron allows the production of hyperosmotic urine?

b. Which animal groups have this type of nephron?

Freshwater fishes are hyperosmotic to their environment and must excrete large quantities of very dilute urine. Salts are conserved by the efficient reabsorption of ions from the filtrate. When in fresh water, frogs and other amphibians actively absorb certain salts across their skin and excrete a dilute urine. When on land, water is reabsorbed from the urinary bladder.

The kidneys of marine bony fishes excrete very little urine and function mainly to get rid of divalent ions taken in by drinking seawater. Monovalent ions and nitrogenous wastes in the form of ammonium are excreted by the epithelia of the gills.

Interacting regulatory systems maintain homeostasis (951)

The vertebrate liver interacts with most organ systems and is important to homeostasis. Its functions include synthesis of plasma proteins; interconversion of glucose and glycogen; detoxification of drugs and toxins; and preparation of nitrogenous wastes.

WORD ROOTS

an- = without; **hydro-** = water; **-bios** = life (*anhydrobiosis:* the ability to survive in a dormant state when an organism's habitat dries up)

anti- = against; **-diure** = urinate (*antidiuretic hormone:* a hormone that helps regulate water balance)

con- = with; **-vect** = carried (*convection:* the mass movement of warmed air or liquid to or from the surface of a body or object)

counter- = opposite (*countercurrent heat exchanger:* a special arrangement of blood vessels that helps trap heat in the body core and is important in reducing heat loss in many endotherms)

-dilat = expanded (*vasodilation:* an increase in the diameter of superficial blood vessels triggered by nerve signals that relax the muscles of the vessel walls)

ecto- = outside; **-therm** = heat (*ectotherm:* an animal, such as a reptile, fish, or amphibian, that must use environmental energy and behavioral adaptations to regulate its body temperature)

endo- = inner (*endotherm:* an animal, such as a bird or mammal, that uses metabolic energy to maintain a constant body temperature)

eury- = broad, wide; **-halin** = salt (*euryhaline:* organisms that can tolerate substantial changes in external osmolarity)

glomer- = a ball (*glomerulus:* a ball of capillaries surrounded by Bowman's capsule in the nephron and serving as the site of filtration in the vertebrate kidney)

homeo- = like, same; **-stasis** = standing (*homeostasis:* the steady-state physiological condition of the body)

juxta- = near to (*juxtaglomerular apparatus:* a specialized tissue located near the afferent arteriole that supplies blood to the glomerulus)

meta- = with; **-nephri** = kidney (*metanephridium:* in annelid worms, a type of excretory tubule with internal openings called nephrostomes that collect body fluids and external openings called nephridiopores)

osmo- = pushing; **-regula** = regular (*osmoregulation:* adaptations to control the water balance in organisms living in hyperosmotic, hypoosmotic, or terrestrial environments)

peri- = around (*peritubular capillaries:* the network of tiny blood vessels that surrounds the proximal and distal tubules in the kidney)

podo- = foot; **-cyte** = cell (*podocytes:* specialized cells of Bowman's capsule that are permeable to water and small solutes but not to blood cells or large molecules such as plasma proteins)

proto- = first (*protonephridium:* an excretory system, such as the flame-cell system of flatworms, consisting of a network of closed tubules having external openings called nephridiopores and lacking internal openings)

reni- = a kidney; **-angio** = a vessel; **-tens** = stretched (*renin-angiotensin-aldosterone system:* a part of a complex feedback circuit that normally partners with antidiuretic hormone in osmoregulation)

steno- = narrow (*stenohaline:* organisms that cannot tolerate substantial changes in external osmolarity)

vasa- = a vessel; **-recta** = straight (*vasa recta:* the capillary system that serves the loop of Henle)

STRUCTURE YOUR KNOWLEDGE

1. Fill in the table below with a brief description of the general osmoregulatory, excretory, and thermoregulatory mechanisms for the animals listed.

2. List the substances that are filtered, secreted, and reabsorbed in the production of urine in mammals.
   a. Substances filtered:
   b. Substances secreted:
   c. Substances reabsorbed:

TEST YOUR KNOWLEDGE

MULTIPLE CHOICE: *Choose the one best answer.*

1. Contractile vacuoles most likely would be found in protists
   a. in a freshwater environment.
   b. in a marine environment.
   c. that are internal parasites.
   d. that are hypoosmotic to their environment.
   e. that are isoosmotic to their environment.

2. Transport epithelia are responsible for
   a. pumping water across a membrane.
   b. forming an impermeable boundary at an interface with the environment.
   c. the exchange of solutes for osmoregulation or excretion.
   d. transporting urine through the ureter and urethra.
   e. the passive transport of H^+ and HCO_3^- for the regulation of pH.

Animal	Osmoregulation	Excretory System	Thermoregulation
a. Marine invertebrate			
b. Marine bony fish			
c. Freshwater fish			
d. Flatworm			
e. Earthworm			
f. Insect			
g. Reptile			
h. Bird			
i. Mammal			

3. Which of the following is *incorrectly* paired with its excretory system?
   a. insect—Malpighian tubules
   b. flatworm—flame-bulb system
   c. earthworm—protonephridia
   d. amphibian—kidneys
   e. fish—kidneys

4. A freshwater fish would be expected to
   a. pump salt out through salt glands in the gills.
   b. produce copious quantities of dilute urine.
   c. diffuse urea out across the epithelium of the gills.
   d. have scales that reduce water loss to the environment.
   e. do all of the above.

5. Which of the following is *not* part of the filtrate entering Bowman's capsule?
   a. water, salt, and electrolytes
   b. glucose
   c. urea
   d. plasma proteins
   e. amino acids

6. Which is the correct pathway for the passage of urine in vertebrates?
   a. collecting tubule → ureter → bladder → urethra
   b. renal vein → renal ureter → bladder → urethra
   c. nephron → urethra → bladder → ureter
   d. cortex → medulla → bladder → ureter
   e. renal pelvis → medulla → bladder → urethra

7. Aldosterone
   a. is a hormone that stimulates thirst.
   b. is secreted by the adrenal glands in response to a high osmolarity of the blood.
   c. stimulates the active reabsorption of Na^+ in the distal tubules.
   d. causes diuresis.
   e. is converted from a blood protein by the action of renin.

8. Which of the following statements is *incorrect?*
   a. Long loops of Henle are associated with steep osmotic gradients and the production of hyperosmotic urine.
   b. Ammonia is a toxic nitrogenous waste molecule that passively diffuses out of the bodies of aquatic invertebrates.
   c. Uric acid is the form of nitrogenous waste that requires the least amount of water to excrete.
   d. In the mammalian kidney, urea diffuses out of the collecting duct and contributes to the osmotic gradient within the medulla.
   e. Uric acid is produced by a mammalian fetus and removed through the placenta to the mother's excretory system.

9. The process of reabsorption in the formation of urine insures that
   a. excess hydrogen ions are removed from the blood.
   b. drugs and other poisons are removed from the blood.
   c. urine is always hyperosmotic to interstitial fluid.
   d. glucose, salts, and water are returned to the blood.
   e. pH is maintained with a balance of hydrogen ions and bicarbonate.

10. The peritubular capillaries
    a. form the ball of capillaries inside Bowman's capsule from which filtrate is forced by blood pressure into the renal tubule.
    b. intertwine with the proximal and distal tubules and exchange solutes with the interstitial fluid.
    c. form a countercurrent flow of blood through the medulla to supply nutrients without interfering with the osmolarity gradient.
    d. surround the collecting ducts and reabsorb water, helping to create a hyperosmotic urine.
    e. rejoin to form the efferent arteriole.

11. ADH and the RAAS both increase water reabsorption, but they respond to different osmoregulatory problems. Which two of the following statements are true?
    1. ADH will be released in response to high alcohol consumption.
    2. ADH is released when osmoregulatory cells in the hypothalamus sense an increase in blood osmolarity.
    3. The RAAS will increase the osmolarity of urine due to the cooperative action of renin, aldosterone, and ANP.
    4. The RAAS is a response to a rise in blood pressure or volume.
    5. The RAAS is most likely to respond following an accident or severe case of diarrhea.

    a. 1 and 4
    b. 1 and 5
    c. 2 and 3
    d. 2 and 4
    e. 2 and 5

12. Which of the following is used by terrestrial animals as a mechanism to dissipate heat?
    a. hibernation
    b. countercurrent heat exchange between vessels that service the extremities
    c. raising fur or feathers
    d. vasodilation of surface vessels
    e. vasoconstriction of surface vessels

13. Which of the following sections of the mammalian nephron is *incorrectly* paired with its function?
    a. Bowman's capsule and glomerulus—filtration of blood
    b. proximal tubule—secretion of ammonia and H^+ into filtrate and transport of glucose and amino acids out of tubule
    c. descending limb of loop of Henle—diffusion of urea out of filtrate
    d. ascending limb of loop of Henle—diffusion and pumping of NaCl out of filtrate
    e. distal tubule—regulation of pH and potassium level

14. The rate of metabolic heat production can be increased by
    a. nonshivering thermogenesis.
    b. storage of brown fat.
    c. shivering and vasoconstriction.
    d. thick layers of blubber and countercurrent heat exchangers.
    e. all of the above.

15. The function of stress-induced proteins is to
    a. provide proteins that function at a lower temperature.
    b. change the composition of plasma membranes to maintain fluidity.
    c. regulate blood osmolarity and volume following an injury.
    d. regulate daily torpor in small mammals with high metabolic rates.
    e. protect cellular proteins during rapid temperature increases or pH changes.

16. All of the following are functions of the vertebrate liver *except*
    a. detoxification of drugs and toxins.
    b. synthesis of plasma proteins.
    c. interconversion of glucose and glycogen.
    d. synthesis of urine.
    e. synthesis of bile (which is then stored in the gallbladder).

17. One of the advantages of the production of uric acid by birds is that uric acid
    a. has low toxicity and can be safely stored in the egg.
    b. is very soluble in water and takes very little energy to produce.
    c. contributes to the production of the egg shell and can thus serve two purposes.
    d. takes less energy to produce than urea and is nontoxic.
    e. requires a moderate amount of water to excrete and is isoosmotic to body fluids.

18. What is the mechanism for the filtration of blood within the nephron?
    a. the active transport of Na^+ and glucose, followed by osmosis
    b. both active and passive secretion of ions, toxins, and NH_3 into the tubule
    c. high hydrostatic pressure of blood forcing water and small molecules out of the capillary
    d. the high osmolarity of the medulla that was created by active and passive transport of salt from the tubule and passive diffusion of urea from the collecting duct
    e. a lower osmotic pressure in Bowman's capsule compared to the osmotic pressure in the glomerulus

19. What stimulus causes the atria of the heart to release ANF (atrial natriuretic factor)?
    a. a rise in blood pressure
    b. a drop in blood pressure
    c. a drop in blood pH
    d. a rise in blood osmolarity
    e. a drop in blood osmolarity

20. What stimulus causes the juxtaglomerular apparatus to release renin?
    a. a drop in blood pH
    b. a rise in blood pH
    c. a drop in blood pressure
    d. a rise in blood osmolarity
    e. the release of ANF by the atria of the heart

21. Which of these animals would be likely to experience a daily torpor?
    a. bear
    b. hummingbird
    c. lizard
    d. groundhog
    e. crow

22. To produce urine that is hyperosmotic to blood and interstitial fluids requires
    a. juxtamedullary nephrons.
    b. an increasing osmotic concentration down through the medulla.
    c. the inhibition of ADH release.
    d. all three of the above.
    e. only a and b above.

23. Which of the following would be *least* true of an animal that is a regulator?
    a. It can live in a variable climate because of its homeostatic mechanisms.
    b. It may have a larger geographic range than a conformer.
    c. Much of its energy budget can be allocated to reproduction.

d. It may acclimatize to winter by increasing the thickness of its insulating coat.

e. It has behavioral as well as physiological mechanisms for responding to changing conditions.

24. Which of the following is an osmoregulator but not a thermoregulator?
    a. spider crab
    b. great white shark
    c. hawk moth
    d. vampire bat
    e. earthworm

25. Which of the following would *not* be used by an animal to balance its heat budget?
    a. countercurrent heat exchanger
    b. lowering its Q_{10} to 2.0
    c. behaviors such as basking or bathing
    d. raising fur or feathers
    e. increasing metabolic heat production

26. Which of the following is *not* descriptive of endotherms?
    a. ability to sustain vigorous activity
    b. efficient circulatory and gas exchange systems
    c. high metabolic rate and retained metabolic heat
    d. energetically efficient and relatively inexpensive
    e. ability to deal with terrestrial temperature fluctuations

27. Which of the following would be likely to produce cryoprotectants?
    a. polar bears and artic wolves
    b. euryhaline animals
    c. estivating amphibians
    d. ectotherm in subzero environment
    e. anhydrobiotic roundworms

CHAPTER 45

CHEMICAL SIGNALS IN ANIMALS

FRAMEWORK

This chapter introduces the intricate system of chemical control and communication within animals. Endocrine cells or neurosecretory cells produce hormones that regulate the activity of other organs. Steroid hormones bind with receptors within their target cells and move into the nucleus where they influence gene expression. The signal-transduction pathways of most other hormones involve binding with cell surface receptors and producing second messengers that trigger a cascade of metabolic reactions in target cells.

The hypothalamus and pituitary gland play coordinating roles by integrating the nervous and endocrine systems and producing many tropic hormones that control the synthesis and secretion of hormones in other endocrine glands and organs.

CHAPTER REVIEW

An animal **hormone** is a chemical signal usually transported through the circulatory system that elicits a specific response from **target cells.**

An Introduction to Regulatory Systems

Coordination and communication among the specialized parts of complex animals are achieved by the nervous system and the **endocrine system.** The nervous system conveys high-speed messages along neurons, whereas the endocrine system produces chemical messengers called hormones, which travel more slowly and regulate many biological processes. Endocrine cells are usually assembled into **endocrine glands,** which are ductless secretory organs that release hormones into body fluids.

The endocrine system and the nervous system are structurally, chemically, and functionally related (956)

The endocrine system and nervous system function together in maintaining homeostasis. They are structurally related in that many endocrine glands contain **neurosecretory cells,** specialized nerve cells that secrete hormones.

The endocrine and nervous systems are chemically related in that several chemicals are used by both systems, as hormones in one and neurotransmitters in the other. These systems are functionally related in that many physiological functions are coordinated by both nervous and hormonal communications. Both systems are usually regulated by feedback mechanisms.

Invertebrate regulatory systems clearly illustrate endocrine and nervous system interactions (956–957)

In invertebrates, reproduction, development, and behavior are usually controlled by an integration of endocrine and nervous systems.

Arthropods have extensive endocrine systems. Insects have three hormones that interact to control molting and development of adult characteristics: The hormone **ecdysone** functions to trigger molts and to promote development of adult characteristics. **Brain hormone (BH),** produced by neurosecretory cells in the brain, stimulates the prothoracic glands to secrete ecdysone. **Juvenile hormone (JH)** counters the action of ecdysone and promotes retention of larval characteristics during molting. When the JH level is high, ecdysone-induced molting produces larger larval stages; only after the JH level has decreased does molting result in a pupa.

Chemical Signals and Their Modes of Action

Hormones are chemical signals that travel throughout the body; local regulators act on neighboring cells; and pheromones communicate between different individuals.

A variety of local regulators affect neighboring target cells (958)

Local regulators are chemical signals that affect target cells near their points of secretion. The neurotransmitter released from a neuron to its target cell is a type of local regulator.

Growth factors are peptides and proteins that are required in the extracellular environment for many types of cells to divide and develop.

Nitric oxide (NO) is a local regulator that can serve various roles as a neurotransmitter, smooth muscle relaxant, and defense chemical.

Prostaglandins (PGs) are modified fatty acids that are released from most cells and have a wide range of effects on nearby target cells. In mammals, prostaglandins help to induce labor. Aspirin and ibuprofen inhibit the synthesis of prostaglandins and thus reduce their fever- and inflammation-inducing and pain-intensifying actions. The balance of antagonistic signals, such as the prostaglandins PGE and PGF that regulate blood flow to the lungs, is an important regulatory mechanism.

■ **INTERACTIVE QUESTION   45.1**

Briefly review the characteristics of the following:

a. hormone

b. local regulator

c. endocrine cell

d. neurosecretory cell

Most chemical signals bind to plasma-membrane proteins, initiating signal-transduction pathways (958–959)

The *reception* of a signal is its binding to a specific protein receptor on or in the target cell. *Signal transduction* leads to a *response* in the target cell.

The majority of chemical signals have plasma membrane receptors that are the first part of **signal-transduction pathways,** which convert and amplify extracellular signals to produce specific intracellular responses.

Steroid hormones, thyroid hormones, and some local regulators enter target cells and bind with intracellular receptors (960)

The protein receptors for most steroid and thyroid hormones and some local regulators are located within the target cells. These small nonpolar molecules cross the membrane and bind to an intracellular receptor; the hormone-receptor complex is usually

a transcription factor that binds to regulatory sites on DNA, either inducing or suppressing specific gene expression.

For signals that bind to cell surface proteins or intracellular receptors, a given signal can have different effects on different target cells and can produce different effects in different species.

The Vertebrate Endocrine System

Vertebrate hormones may target a few or many tissues in the body. **Tropic hormones** target other endocrine glands and are particularly important to chemical coordination. Many organs whose main functions are nonendocrine, such as the heart, digestive tract, kidney, and liver, also secrete important hormones. The thymus secretes thymosin and other messengers that stimulate development and differentiation of T lymphocytes.

The hypothalamus and pituitary integrate many functions of the vertebrate endocrine system (962–964)

The **hypothalamus,** situated in the lower brain, plays a key role in integrating the endocrine and nervous systems. The hypothalamus receives nerve signals from throughout the body, and its neurosecretory cells release hormones that are stored in or regulate the **pituitary gland** at the base of the hypothalamus.

The pituary gland has two discrete parts that develop separately and have different functions. The **anterior pituitary,** or **adenohypophysis,** develops from the roof of the embryonic mouth. It is composed of endocrine tissue and synthesizes several hormones. A set of hypothalamic neurosecretory cells produces **releasing hormones** and **inhibiting hormones** that regulate the anterior pituitary. These hormones are released into capillaries at the base of the hypothalamus and travel via a short portal vessel to capillary beds in the anterior pituitary.

The **posterior pituitary,** or **neurohypophysis,** is an extension of the brain that grows downward from a small bulge of the hypothalamus. It stores and secretes two hormones that are produced by the hypothalamus.

Posterior Pituitary Hormones   **Oxytocin** induces uterine contractions during birth and milk ejection during nursing. **Antidiuretic hormone (ADH)** functions in osmoregulation. Osmoreceptor cells in the hypothalamus, responding to an increase in blood osmolarity, signal neurosecretory cells of the hypothalamus to release ADH from their tips, located in the posterior pituitary. Binding of ADH to receptors on collecting ducts in the kidney sets off a signal-transduction pathway that results in increased reabsorption of water. The osmoreceptors also stimulate thirst. Drinking and the reabsorption of water from the urine lower blood osmolarity and cause the hypothalamus to stop the release of ADH, completing the negative feedback loop.

■ INTERACTIVE QUESTION 45.2

What hormones are produced by the hypothalamus? To where and how are they transported?

Anterior Pituitary Hormones   The anterior pituitary produces a number of hormones. **Growth hormone (GH)** affects a variety of target tissues, promoting growth directly and stimulating the release of growth factors, such as **insulinlike growth factors (IGFs)** that are produced by the liver and stimulate bone and cartilage growth. Gigantism and acromegaly are human growth disorders caused by excessive GH. Hypopituitary dwarfism can now be treated with genetically engineered GH.

The hormone **prolactin (PRL)** is a protein very similar to GH, with diverse effects in different vertebrate species, ranging from milk production and secretion in mammals, to delay of metamorphosis in amphibians, to osmoregulation in fish.

Three of the tropic hormones produced by the anterior pituitary are similar glycoproteins. **Thyroid-stimulating hormone (TSH)** regulates production of thyroid hormones. **Follicle-stimulating hormone (FSH)** and **luteinizing hormone (LH),** also called **gonadotropins,** stimulate gonad activity.

Several hormones come from fragments of pro-opiomelanocortin, a large protein cleaved into short pieces inside pituitary cells. **Adrenocorticotropic hormone (ACTH)** stimulates the adrenal cortex to produce and secrete its steroid hormones. **Melanocyte-stimulating hormone (MSH)** regulates the activity of pigment-containing cells in the skin of some vertebrates and appears to play a role in fat metabolism in mammals. **Endorphins** are hormones that inhibit pain perception. Endorphins are also produced by certain neurons in the brain. Heroin and other opiates bind to endorphin receptors in the brain.

The pineal gland is involved in biorhythms (964–965)

The **pineal gland,** located near the center of the mammalian brain, secretes **melatonin,** which regulates functions related to light and changes in day length.

The secretion of melatonin at night may function with a biological clock for daily or seasonal activities such as reproduction.

■ INTERACTIVE QUESTION 45.3

Fill in the following table to review the hormones stored and released by the posterior pituitary (**a** and **b**) and secreted by the anterior pituitary (**c–i**).

Hormone	Main Actions
Oxytocin	**a.**
ADH	**b.**
GH	**c.**
Prolactin	**d.**
TSH	**e.**
FSH and LH	**f.**
ACTH	**g.**
MSH	**h.**
Endorphins	**i.**

Thyroid hormones function in development, bioenergetics, and homeostasis (965–966)

The **thyroid gland** produces two hormones, **triiodothyronine (T_3)** and **thyroxine (T_4).** Thyroid hormones are critical to vertebrate development and maturation; an inherited deficiency in humans results in retarded skeletal and mental development, a condition known as cretinism.

The thyroid gland contributes to homeostasis in mammals, helping to maintain normal blood pressure, heart rate, muscle tone, digestion, and reproductive functions. T_3 and T_4 increase cellular metabolism. Excess thyroid hormone results in hyperthyroidism, with symptoms of weight loss, irritability, and high body temperature and blood pressure. Hypothyroidism can cause cretinism or weight gain and lethargy in adults. A goiter is an enlarged thyroid gland and may be associated with a lack of iodine in the diet.

A negative feedback loop controls secretion of thyroid hormones. Secretion of TSH-releasing hormone, or TRH, by the hypothalamus stimulates the anterior pituitary to secrete thyroid-stimulating hormone. TSH binds to receptors in the thyroid gland, triggering the synthesis and release of thyroid hormones. High levels of T_3, T_4, and TSH inhibit the secretion of TRH.

The mammalian thyroid gland also secretes **calcitonin,** a hormone that lowers calcium levels in the blood.

Parathyroid hormone and calcitonin balance blood calcium (966)

The four **parathyroid glands** secrete **parathyroid hormone (PTH),** which stimulates Ca^{2+} reabsorption in the kidney, and its release from bone to raise blood calcium levels. PTH also activates **Vitamin D** in the kidney, which then acts as a hormone to increase Ca^{2+} uptake from food in the intestines. A balance between the antagonistic PTH and calcitonin maintains blood calcium homeostasis.

Endocrine tissues of the pancreas secrete insulin and glucagon, antagonistic hormones that regulate blood glucose (966–969)

Scattered within the exocrine tissue of the **pancreas** are clusters of endocrine cells known as the **islets of Langerhans.** Within each islet are **alpha cells** that secrete the hormone **glucagon** and **beta cells** that secrete the hormone **insulin.** These antagonistic hormones regulate glucose concentration in the blood, and negative feedback controls their secretion.

Insulin lowers blood glucose levels by promoting the movement of glucose from the blood into body cells, by slowing the breakdown of glycogen in the liver, and by inhibiting the conversion of amino acids and glycerol (from fats) to sugar. Glucagon raises glucose concentrations by stimulating the liver to increase glycogen hydrolysis, convert amino acids and glycerol to glucose, and release glucose to the blood.

In diabetes mellitus, the absence of insulin in the bloodstream or the loss of response to insulin in target tissues reduces glucose uptake by cells. Glucose accumulates in the blood and is excreted in the urine, with an accompanying loss of water. The body must use fats for fuel, and acids from fat breakdown may lower blood pH.

Type I diabetes mellitus, also known as insulin-dependent diabetes, is an autoimmune disorder in which pancreatic cells are destroyed. This type of diabetes is treated by regular injections of genetically engineered human insulin. More than 90 percent of diabetics have **type II diabetes mellitus,** or noninsulin-dependent diabetes, characterized either by insulin deficiency or reduced responsiveness of target cells. Exercise and dietary control are often sufficient to manage this disease.

The adrenal medulla and adrenal cortex help the body manage stress (969–971)

In mammals, the **adrenal glands** consist of two different glands: the outer **adrenal cortex** and the central **adrenal medulla.** The adrenal medulla produces **epinephrine** (adrenaline) and **norepinephrine** (noradrenaline)—both of which are **catecholamines,** a class of compounds synthesized from the amino acid tyrosine.

■ INTERACTIVE QUESTION 45.4

Complete this concept map on the regulation of blood glucose levels.

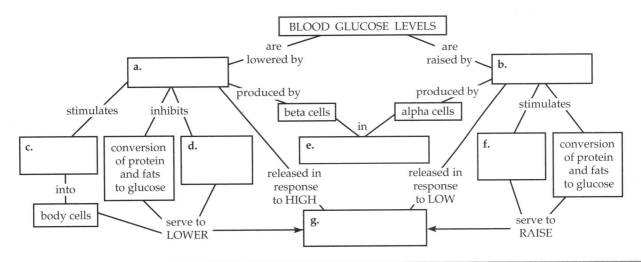

Epinephrine and norepinephrine, released in response to positive or negative stress, increase the availability of energy sources by stimulating glycogen hydrolysis in skeletal muscle and the liver, glucose release from the liver, and fatty acid release from fat cells. These hormones increase metabolic rate and the rate and volume of the heartbeat, dilate bronchioles in the lungs, and influence the contraction or relaxation of smooth muscles to increase blood supply to the heart, brain, and skeletal muscles, while reducing the supply to other organs. The sympathetic division of the autonomic nervous system stimulates the release of epinephrine from the adrenal medulla. Norepinephrine's primary role is in sustaining blood pressure.

The adrenal cortex responds to endocrine signals released in response to stress. A releasing hormone from the hypothalamus causes the anterior pituitary to release ACTH. This tropic hormone stimulates the adrenal cortex to synthesize and secrete **corticosteroids,** a group of steroid hormones.

The two main human corticosteroids are the **glucocorticoids,** such as cortisol, and the **mineralocorticoids,** such as aldosterone. Glucocorticoids promote synthesis of glucose from noncarbohydrates (such as muscle proteins) and thus increase energy supplies during stress. Cortisone has been used to treat serious inflammatory conditions such as arthritis, but its immunosuppressive effects can be dangerous.

Mineralocorticoids affect salt and water balance. Aldosterone stimulates kidney cells to reabsorb sodium ions and water from the filtrate. Aldosterone secretion is regulated by the RAAS system, although ACTH secreted by the anterior pituitary in response to severe stress increases aldosterone secretion. Both glucocorticoids and mineralocorticoids appear to help maintain homeostasis during extended periods of stress.

A third group of corticosteroids produced by the adrenal cortex are sex hormones, in particular androgens, which appear to influence female sex drive.

■ INTERACTIVE QUESTION 45.5

Describe how the adrenal gland responds to short-term and long-term stress.

a. Short-term stress:

b. Long-term stress:

Gonadal steroids regulate growth, development, reproductive cycles, and sexual behavior (972)

The testes of males and ovaries of females produce steroids that affect growth, development, and reproductive cycles and behaviors. The three major categories of gonadal steroids—androgens, estrogens, and progestins—are found in different proportions in males and females.

The testes primarily synthesize **androgens,** such as **testosterone,** which determine the gender of the developing embryo and stimulate development of the male reproductive system and secondary sex characteristics. **Estrogens** regulate the development and maintenance of the female reproductive system and secondary sex characteristics. In mammals, **progestins** help prepare and maintain the uterus for the growth of an embryo.

A hypothalamic releasing hormone, GnRH, controls secretion of FSH and LH, gonadotropins from the anterior pituitary gland that control the synthesis of estrogens and androgens.

WORD ROOTS

adeno- = gland; **-hypo** = below (*adenohypophysis:* also called the anterior pituitary, a gland positioned at the base of the hypothalamus)

andro- = male; **-gen** = produce (*androgens:* the principal male steroid hormones, such as testosterone, which stimulate the development and maintenance of the male reproductive system and secondary sex characteristics)

anti- = against; **-diure** = urinate (*antidiuretic hormone:* a hormone that helps regulate water balance)

cata- = down; **-chol** = anger (*catecholamines:* a class of compounds, including epinephrine and norepinephrine, that are synthesized from the amino acid tyrosine)

-cortico = the shell; **-tropic** = to turn or change (*adrenocorticotropic hormone:* a peptide hormone released from the anterior pituitary, it stimulates the production and secretion of steroid hormones by the adrenal cortex)

ecdys- = an escape (*ecdysone:* a steroid hormone that triggers molting in arthropods)

endo- = inside (*endorphin:* a hormone produced in the brain and anterior pituitary that inhibits pain perception)

epi- = above, over (*epinephrine:* a hormone produced as a response to stress; also called adrenaline)

gluco- = sweet (*glucagon:* a peptide hormone secreted by pancreatic endocrine cells that raises blood glucose levels; an antagonistic hormone to insulin)

lut- = yellow (*luteinizing hormone:* a gonadotropin secreted by the anterior pituitary)

melan- = black (*melatonin:* a modified amino acid hormone secreted by the pineal gland)

neuro- = nerve (*neurohypophysis:* also called the posterior pituitary, it is an extension of the brain)

oxy- = sharp, acid (*oxytocin:* a hormone that induces contractions of the uterine muscles and causes the mammary glands to eject milk during nursing)

para- = beside, near (*parathyroid glands:* four endocrine glands, embedded in the surface of the thyroid gland, that secrete parathyroid hormone and raise blood calcium levels)

pro- = before; **-lact** = milk (*prolactin:* a hormone produced by the anterior pituitary gland, it stimulates milk synthesis in mammals)

tri- = three; **-iodo** = violet (*triiodothyrodine:* one of two very similar hormones produced by the thyroid gland and derived from the amino acid tyrosine)

STRUCTURE YOUR KNOWLEDGE

1. Briefly describe the two general mechanisms by which chemical signals trigger responses in target cells, depending on the location of the receptor.

2. Most homeostatic functions are maintained by a balance between hormones with antagonistic effects. Describe how the thyroid and parathyroid glands regulate calcium levels in the blood.

TEST YOUR KNOWLEDGE

MATCHING: *Match the hormone and gland or organ that produces it to the descriptions. Choices may be used more than once, and not all choices are used.*

Hormones	Gland or Organ
A. ACTH	a. adrenal cortex
B. androgens	b. adrenal medulla
C. ADH	c. hypothalamus
D. calcitonin	d. pancreas
E. epinephrine	e. parathyroid
F. glucagon	f. pineal
G. glucocorticoids	g. pituitary
H. insulin	h. testis
I. melatonin	i. thymus
J. oxytocin	j. thyroid
K. PTH	
L. thyroxine	

Hormone	Gland	Hormone Action
_____	_____	1. involved in biological clock and seasonal activities
_____	_____	2. break down muscle protein for conversion to glucose
_____	_____	3. increase blood sugar, glycogen breakdown in liver
_____	_____	4. stimulate development of male reproductive system
_____	_____	5. stimulate adrenal cortex to synthesize corticosteroids
_____	_____	6. increase available energy, heart rate, metabolism
_____	_____	7. regulate metabolism, growth, and development
_____	_____	8. lower blood calcium levels
_____	_____	9. increase reabsorption of water by kidney
_____	_____	10. stimulate contraction of uterus, milk secretion

MULTIPLE CHOICE: *Choose the one best answer.*

1. Which of the following is *not* an accurate statement about hormones?
   a. Not all hormones are secreted by endocrine glands.
   b. Most hormones move through the circulatory system to their destination.
   c. Target cells have specific protein receptors for hormones.
   d. Hormones are essential to homeostasis.
   e. Steroid hormones often function as neurotransmitters.

2. The best description of the difference between pheromones and hormones is that
   a. pheromones are small, volatile molecules, whereas hormones are steroids.
   b. pheromones are involved in reproduction, whereas hormones are not.
   c. pheromones are a form of neural communication; hormones are a form of chemical communication.
   d. pheromones are signals that function between animals, whereas hormones communicate among the parts within an animal.
   e. pheromones are local regulators, whereas hormones travel greater distances.

3. Which one of the following hormones is *incorrectly* paired with its origin?
   a. releasing hormones—hypothalamus
   b. growth hormone—anterior pituitary
   c. progestins—ovary
   d. TSH—thyroid
   e. mineralocorticoids—adrenal cortex

4. Which of the following is an example of a positive feedback mechanism?
   a. the liver's production of insulin-like growth factors in response to growth hormone, which promotes skeletal growth
   b. the ability of the neurotransmitter acetylcholine to cause skeletal muscle to contract, heart muscle to relax, and cells of the adrenal medulla to secrete epinephrine
   c. prostaglandins released from placental cells promoting muscle contraction during childbirth, with muscle contractions stimulating more prostaglandin release
   d. the interplay between prostaglandin E and prostaglandin F on blood vessels servicing the lungs
   e. elevated levels of stress resulting in neural stimulation of the adrenal medulla and hormonal stimulation of the adrenal cortex

5. Ecdysone
   a. is a steroid hormone produced in insects that promotes retention of larval characteristics.
   b. is responsible for the color changes in amphibians.
   c. is a hormone secreted from specialized neurons that stimulates egg laying in the mollusk *Aplysia*.
   d. is secreted by prothoracic glands in insects and triggers molts and development of adult characteristics.
   e. is involved in metamorphosis in amphibians.

6. Which of the following local regulators amplifies the sensation of pain?
   a. prostaglandins
   b. PGE and PGF
   c. growth factors
   d. interleukins
   e. nitric oxide

7. Antidiuretic hormone (ADH)
   a. is produced by cells in the kidney and liver in response to low blood volume or pressure.
   b. stimulates the reabsorption of Na^+ from the urine.
   c. is released from the posterior pituitary and increases the permeability of kidney collecting tubules to water.
   d. is a steroid hormone produced by the adrenal cortex.
   e. is part of the RAAS that regulates salt and water balance.

8. The anterior pituitary
   a. stores oxytocin and ADH produced by the hypothalamus.
   b. receives releasing and inhibiting hormones from the hypothalamus through portal vessels connecting capillary beds.
   c. produces several releasing and inhibiting hormones.
   d. is responsible for nervous and hormonal stimulation of the adrenal glands.
   e. produces only tropic hormones.

9. Acromegaly, the abnormal growth of bones of the hands, feet, and head, is caused by
   a. an autoimmune disorder.
   b. an excess of thyroxine.
   c. an excess of glucocorticoids.
   d. an abnormally high androgen-to-estrogen ratio.
   e. an excess of growth hormone.

10. Which of the following hormones is *not* involved with increasing the blood glucose concentration?
    a. glucagon
    b. epinephrine
    c. glucocorticoids
    d. ACTH (adrenocorticotropic hormone)
    e. insulin

11. Pro-opiomelanocortin
    a. is produced by the hypothalamus and is transported through a portal vessel.
    b. is cleaved into several hormones, including ACTH, MSH, and endorphins.
    c. produces several releasing hormones.
    d. mimics the effects of heroin and opiates.
    e. is a precursor to the hormones of the adrenal cortex.

12. Which of the following is *not* true of norepinephrine?
    a. It is secreted by the adrenal medulla.
    b. It maintains blood pressure and, along with epinephrine, it increases the rate and volume of heartbeat and constricts selected blood vessels.
    c. Its release is stimulated by ACTH.
    d. It serves as a neurotransmitter in the nervous system.
    e. It is part of the flight-or-fight response to stress.

Choose from the following hormones to answer questions 13–15.
    a. ACTH
    b. parathyroid hormone
    c. epinephrine
    d. estrogen
    e. insulin

13. Which of the above is a tropic hormone?

14. Which hormone is a steroid hormone?

15. Which hormone is released in response to a nervous impulse?

16. A lack of iodine in the diet can lead to formation of a goiter because
    a. excess thyroxine production causes the thyroid gland to enlarge.
    b. with limited iodine available, triiodothyronine (T_3) rather than thyroxine (T_4) is produced.
    c. iodine is a key ingredient of the growth hormones that control growth of the thyroid.
    d. little thyroxine is produced, TRH and TSH production is not inhibited, and thyroid stimulation continues.
    e. the hypothalamus is stimulated to increase its production of TRH.

17. A tropic hormone is a hormone
    a. whose target tissue is another endocrine gland.
    b. that is produced by the hypothalamus but stored and released from the posterior pituitary.
    c. that acts by negative feedback to regulate its own level in the body.
    d. from the hypothalamus that regulates the synthesis and secretion of hormones of the posterior pituitary.
    e. that is released in response to nervous stimulation.

18. What is the best description of the mechanism of action of steroid hormones?
    a. transported by neurosecretory cells directly to target tissues
    b. form a hormone-receptor complex inside the cell that regulates gene expression
    c. amplified response using second messengers
    d. bind to membrane-bound receptors and initiate a signal-transduction pathway
    e. secreted into the interstitial fluid and act as a local regulator

CHAPTER 46

ANIMAL REPRODUCTION

FRAMEWORK

This chapter covers the patterns and mechanisms of animal reproduction. In sexual reproduction, fertilization may occur externally or internally. Development of the zygote may take place internally within the female, externally in a moist environment, or in a protective, resistant egg. Eutherian (placental) mammals provide nourishment as well as shelter for the embryo, and they nurse their young.

The human reproductive system is described, including its organs, glands, hormones, gamete formation, and sexual response. The chapter also covers pregnancy and birth, contraception, and recently developed reproductive technologies.

CHAPTER REVIEW

Overview of Animal Reproduction

Both asexual and sexual reproduction occur in the animal kingdom (975)

In **asexual reproduction,** a single individual produces offspring, usually using only mitotic cell division. In **sexual reproduction,** two meiotically formed haploid **gametes** fuse to form a diploid **zygote,** which develops into an offspring. Gametes are usually a relatively large, nonmotile **ovum** and a small, flagellated **spermatozoon.** Sexual reproduction combines genes from two parents and produces offspring with varying phenotypes.

Diverse mechanisms of asexual reproduction enable animals to produce identical offspring rapidly (976)

Many invertebrates can reproduce asexually by **fission,** in which a parent is separated into two or more equal-sized individuals; by **budding,** in which a new individual grows out from the parent's body; or by **fragmentation,** in which the body is broken into several pieces, each of which develops into a complete animal. **Regeneration** is necessary for a fragment to develop into a new organism. Some invertebrates release specialized groups of cells that grow into new individuals, like the **gemmules** produced by sponges.

Reproductive cycles and patterns vary extensively among animals (976–978)

Periodic reproductive cycles may be linked to favorable conditions or energy supplies. A combination of environmental and hormonal cues control the timing of these cycles.

The freshwater crustacean *Daphnia,* like aphids and rotifers, produces eggs that develop by **parthenogenesis,** as well as eggs that are fertilized. In bees, wasps, and ants, males are produced parthenogenetically, whereas sterile worker females and reproductive females are produced from fertilized eggs. In a few parthenogenetic fishes, amphibians, and lizards, doubling of chromosomes creates diploid "zygotes."

In **hermaphroditism,** found in some sessile, burrowing, and parasitic animals, each individual has functioning male and female reproductive systems. Mating results in fertilization of both individuals. In some fishes and oysters, individuals reverse their sex during their lifetime, a pattern known as **sequential hermaphroditism.** Sex reversal may be related to age or to the relative advantage conferred by size. In a **protandrous** species, an organism is first a male; in a **protogynous** species, it is first a female.

■ INTERACTIVE QUESTION 46.1

a. What adaptive advantages would asexual reproduction provide?

b. What adaptive advantage would sexual reproduction provide?

Mechanisms of Sexual Reproduction

In **external fertilization,** eggs and sperm are shed, and fertilization occurs in the environment. **Internal fertilization** involves the placement of sperm in or near the female reproductive tract so that the egg and sperm unite internally.

Internal and external fertilization both depend on mechanisms ensuring that mature sperm encounter fertile eggs of the same species (978)

Behavioral cooperation, as well as copulatory organs and sperm receptacles, are required for internal fertilization.

External fertilization occurs almost exclusively in moist habitats, where the gametes and developing zygote are not in danger of desiccation.

Pheromones are small, volatile chemical signals that may function as mate attractants.

■ INTERACTIVE QUESTION 46.2

List three mechanisms that may help to ensure that gamete release is synchronized when fertilization is external.

a.

b.

c.

Species with internal fertilization usually produce fewer zygotes but provide more parental protection than species with external fertilization (978–979)

Developing embryos may receive some type of protection. The amniote eggs of birds, reptiles, and monotremes protect the embryo in a terrestrial environment. The embryos of eutherian mammals develop within the uterus, nourished from the mother's blood supply through the placenta. Parental care of young is widespread among vertebrates and even among many invertebrates.

Complex reproductive systems have evolved in many animal phyla (979–980)

The simplest reproductive systems do not even have **gonads** to produce gametes. In polychaete annelids, eggs or sperm develop from cells lining the coelom. Gametes may be shed through excretory openings or released by the splitting of the parent.

Parasitic flatworms are hermaphroditic and have complex reproductive systems.

Most insects have separate sexes and complex reproductive systems. Sperm develop in the testes, are stored in the seminal vesicles, and are ejaculated into the female. Eggs are produced in the ovaries and fertilized in the vagina. Females may have a **spermatheca,** or sperm-storing sac.

With the exception of most mammals, vertebrates have a common opening, a **cloaca,** for the digestive, excretory, and reproductive systems. Many nonmammalian vertebrates evert the cloaca to ejaculate.

Most mammals have a separate opening for the digestive tract. The urethra is used by both the excretory and reproductive systems in males, whereas most female mammals have a separate vagina and urethra.

Mammalian Reproduction

Human reproduction involves intricate anatomy and complex behavior (980–984)

Reproductive Anatomy of the Human Male The external male reproductive organs include the scrotum and penis. Sperm are produced in the highly coiled **seminiferous tubules** of the **testes. Leydig cells** produce testosterone and other androgens. Sperm production requires a cooler temperature than the internal body temperature of most mammals, so the **scrotum** suspends the testes below the abdominal cavity.

Sperm pass from a testis into the coiled **epididymis,** in which they mature and gain motility. During **ejaculation,** sperm are propelled through the **vas deferens,** into a short **ejaculatory duct,** and out through the **urethra** which runs through the penis.

Three sets of accessory glands add secretions to the **semen.** The **seminal vesicles** contribute an alkaline fluid containing mucus, fructose (as an energy source for the sperm), a coagulating enzyme, ascorbic acid, and prostaglandins (which stimulate uterine contractions). The **prostate gland** produces a secretion that contains anticoagulant enzymes and citrate. Benign enlargement of the prostate and prostate cancer are common medical problems in older men. The **bulbourethral glands** produce a mucus that neutralizes urine remaining in the urethra.

■ INTERACTIVE QUESTION 46.3

Label the indicated structures in these diagrams of the human male and female reproductive system.

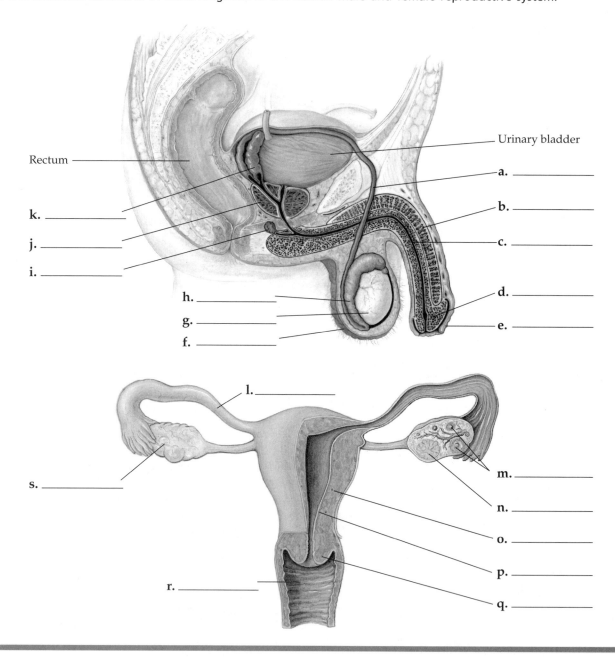

The 2–5 ml of ejaculated semen may contain over 500 million sperm. The slightly alkaline semen neutralizes the vagina. Semen initially coagulates, then liquifies as the sperm start swimming through the female tract.

The **penis** is composed of spongy tissue that engorges with blood during sexual arousal, producing an erection that facilitates insertion of the penis into the vagina. Some mammals have a **baculum,** a bone that helps stiffen the penis. The head of the penis,

called the **glans penis,** is covered by a fold of skin called the **prepuce,** or foreskin.

Reproductive Anatomy of the Human Female   The external reproductive structures include the clitoris and two sets of labia. The female gonads, the **ovaries,** contain many **follicles,** which are sacs of cells that nourish and protect the egg cell contained within each of them. During each menstrual cycle, a maturing follicle produces estrogens, the primary female sex hormones. Following **ovulation,** the follicle forms a solid mass called the **corpus luteum,** which secretes progesterone and estrogens.

The egg cell is expelled into the abdominal cavity and swept by cilia into the **oviduct,** or fallopian tube, through which it is transported to the **uterus.** The **endometrium,** or lining of the uterus, is highly vascularized. The neck of the uterus, the **cervix,** opens into the **vagina,** the thin-walled birth canal and repository for sperm during copulation.

The vaginal opening is initially covered by a vascularized membrane called the **hymen.** The separate openings of the vagina and urethra are in a region called the **vestibule,** enclosed by two pairs of skin folds, the inner **labia minora** and the outer **labia majora.** The **clitoris,** composed of erectile tissue, is at the top of the vestibule. **Bartholin's glands** secrete mucus into the vestibule during sexual arousal.

A **mammary gland,** located within a breast, is composed of fatty tissue and a series of small milk-secreting sacs that drain into ducts that open at the nipple. The lack of estrogen in males prevents their mammary glands from enlarging and developing secretory functions.

Human Sexual Response   The human sexual response cycle includes two types of physiological reactions: **vasocongestion,** increased blood flow to a tissue, and **myotonia,** increased muscle tension. The excitement phase involves vasocongestion and vaginal lubrication, preparing the vagina and penis for **coitus,** or sexual intercourse. The plateau phase continues vasocongestion and myotonia, and breathing rate and heart rate increase. In both sexes, **orgasm** is characterized by rhythmic, involuntary contractions of reproductive structures. In males, emission deposits semen in the urethra, and ejaculation occurs when the urethra contracts and semen is expelled. In the resolution phase, vasocongested organs return to normal size, and muscles relax.

Spermatogenesis and oogenesis both involve meiosis but differ in three significant ways (984–986)

Spermatogenesis, the production of mature sperm cells, occurs continuously in the seminiferous tubules of the testes as **spermatogonia,** the stem cells that give rise to sperm, differentiate into spermatocytes that undergo meiosis. The haploid nucleus of a sperm is contained in a head, tipped with an **acrosome,** which contains enzymes that help the sperm penetrate the egg. Mitochondria provide ATP for movement of the flagellum, or tail.

Oogenesis begins in the female embryo. **Oogonia,** the stem cells that give rise to ova, divide and differentiate into **primary oocytes,** which are arrested in prophase I of meiosis. At birth, an ovary already contains all of its primary oocytes. Following puberty, FSH periodically stimulates a primary oocyte to finish meiosis I and develop into a **secondary oocyte,** which is arrested in metaphase II. In humans, meiosis is completed if a sperm penetrates the ovum. Meiotic cytokinesis is unequal, producing one large ovum and up to three small haploid polar bodies that disintegrate.

■ INTERACTIVE QUESTION   46.4

List the three important ways in which oogenesis differs from spermatogenesis.

a.

b.

c.

A complex interplay of hormones regulates reproduction (986–989)

The Male Pattern   Androgens are responsible for the male primary (associated with reproduction) and secondary (associated with voice deepening, hair distribution, and muscle growth) sex characteristics. The most important androgen is testosterone, produced mainly by Leydig cells of the testes. Androgens are also determinants of sexual and other behaviors.

■ INTERACTIVE QUESTION   46.5

Fill in the blanks in the following description of the control of male reproductive hormones.

a. _____ produced by the hypothalamus regulates the release of gonadotropic hormones from the anterior pituitary. **b.** _____ stimulates the seminiferous tubules to increase spermatogenesis. **c.** _____ stimulates **d.** _____ production by Leydig cells. Androgens contribute to spermatogenesis and development of **e.** _____.
f. _____ control the production of GnRH, LH, and FSH, maintaining fairly constant hormone levels in human males.

The Female Pattern   Female humans and many primates have **menstrual cycles,** during which the endometrium lining thickens to prepare for the implantation of the embryo and then is shed if fertilization does not occur. This bleeding, called **menstruation,** occurs on a cycle of approximately 28 days in humans. Other mammals have **estrous cycles,** during which the endometrium thickens, but, if fertilization does not occur, it is reabsorbed. Estrous cycles are often coordinated with season, and females are receptive to sexual activity only during **estrus,** or heat, the period surrounding ovulation.

The human menstrual cycle begins with the **menstrual flow phase,** the few days of menstrual bleeding; followed by the **proliferative phase,** the week or two during which the endometrium begins to thicken; and the **secretory phase,** about two weeks during which the endometrium becomes more vascularized and develops glands that secrete a glycogen-rich fluid.

The **ovarian cycle** begins with the **follicular phase,** during which an egg cell enlarges, its follicular cells become multilayered and enclose a fluid-filled cavity, and the large, mature follicle forms a bulge near the ovary surface. **Ovulation** occurs with the rupture of the follicle and adjacent ovary wall. The remaining follicular tissue develops into the hormone-secreting corpus luteum during the **luteal phase.**

Five hormones coordinate the menstrual and ovarian cycles using positive and negative feedback. During the follicular phase, the hypothalamus secretes GnRH (gonadotropin-releasing hormone), which stimulates the anterior pituitary to secrete small amounts of FSH (follicle-stimulating hormone) and LH (luteinizing hormone). FSH stimulates follicular growth, and the cells of the follicle secrete estrogens. The slow rise in estrogens inhibits the release of pituitary gonadotropins (FSH and LH). When the secretion of estrogens rises sharply, the hypothalamus is stimulated to increase GnRH output, which results in a rise in LH and FSH release.

By positive feedback, the increase in LH, caused by increased secretion of estrogens from the follicle, induces maturation of the follicle, and ovulation occurs about a day after the LH surge.

In the luteal phase, LH stimulates the transformation of the ruptured follicle and maintains the corpus luteum, which secretes estrogens and progesterone. The rising level of these hormones exerts negative feedback on the hypothalamus and pituitary, inhibiting secretion of LH and FSH. The corpus luteum disintegrates, thereby dropping the levels of estrogens and progesterone. This drop releases the inhibition of the hypothalamus and pituitary, and FSH and LH secretion begins again, stimulating growth of new follicles and the start of the next follicular phase.

The hormones of the ovarian cycle synchronize the menstrual cycle and the preparation of the uterus for possible implantation of an embryo. Estrogens cause the endometrium to begin to thicken in the proliferative phase. After ovulation, estrogens and progesterone stimulate increased vascularization and gland development of the endometrium. Thus, the luteal phase of the ovarian cycle corresponds with the secretory phase of the menstrual cycle. The rapid drop of ovarian hormones caused by the disintegration of the corpus luteum reduces blood supply to the endometrium and begins its disintegration, leading to the menstrual flow phase of the next cycle.

Estrogens are also responsible for female secondary sex characteristics and influence sexual behavior.

Menopause, the cessation of ovulation and menstruation, results from a decline in the production of estrogens as the ovaries become less responsive to FSH and LH.

Embryonic and fetal development occur during pregnancy in humans and other eutherian (placental) mammals (989–995)

From Conception to Birth   The development of one or more **embryos** in the uterus, preceded by **conception** and ending with birth, is called **pregnancy** or **gestation.** The gestation period correlates with body size and the degree of development of the young at birth. Human pregnancy averages 266 days (38 weeks).

Human gestation can be divided into three **trimesters.** About 24 hours after fertilization in the oviduct, the zygote begins to divide; it travels to the uterus in about 3 to 4 days. After about a week of **cleavage,** the zygote develops into a ball of cells called a **blastocyst** and implants in the endometrium in about 5 more days. Tissues grow out of the developing embryo to mingle with the endometrium and form the **placenta,** a disk-shaped organ in which gas and nutrient exchange and waste removal take place between the maternal and embryonic circulations. Development proceeds to **organogenesis,** and, by the eighth week, the embryo has all the rudimentary structures of the adult and is called a **fetus.**

The embryo secretes **human chorionic gonadotropin (HCG),** which maintains the corpus luteum's secretion of progesterone and estrogens through the first trimester. High progesterone levels initiate growth of the maternal part of the placenta, enlargement of the uterus, cessation of ovulation and menstrual cycling, and breast enlargement.

During the second trimester, HCG declines, the corpus luteum degenerates, and the placenta secretes its own progesterone, which maintains the pregnancy. The fetus grows rapidly and is quite active. The third trimester is a period of rapid fetal growth.

■ INTERACTIVE QUESTION 46.6

In the diagram of the human female reproductive cycle, label the lines indicating the levels of the gonadotropic and the ovarian hormones and the phases of the ovarian and menstrual cycles.

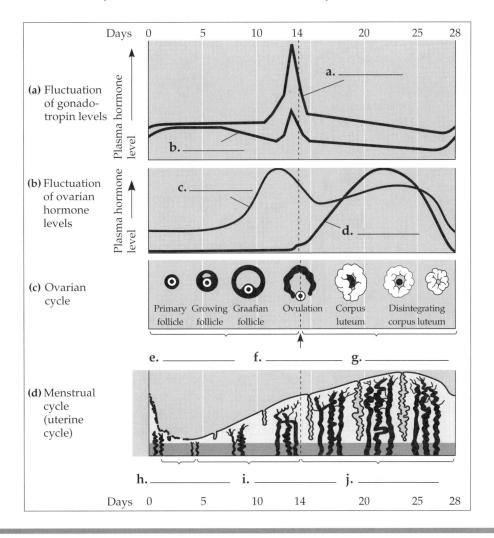

Through a positive feedback system, hormones and local regulators induce labor. High levels of estrogen stimulate the development of oxytocin receptors on the uterus. Oxytocin produced by the fetus and the mother's posterior pituitary stimulates contractions and uterine secretion of prostaglandins, which further enhance uterine contractions.

Labor consists of a series of strong contractions of the uterus and results in birth, or **parturition.** During the first stage of labor, the cervix dilates. The second stage consists of the contractions that force the fetus out of the uterus and through the vagina. The placenta is delivered in the final stage of labor.

Lactation is unique to mammals. Prolactin secretion by the anterior pituitary stimulates milk production, and oxytocin controls the release of milk during nursing.

■ INTERACTIVE QUESTION 46.7

List the functions of the following hormones and where they are secreted:

a. human chorionic gonadotropin

b. progesterone

c. oxytocin

Reproductive Immunology   Several hypotheses attempt to explain why a mother does not reject an embryo, which has paternal as well as maternal chemical markers. There is evidence that the trophoblast, a protective layer that develops from the blastocyst and surrounds the embryo, induces the development of white blood cells that suppress other white cells from mounting an immune attack. This suppression may occur only after an initial immune response to the trophoblast. Some researchers suggest that if the initial response is too weak, then suppression may not occur, and the continued immunological attack results in spontaneous abortion of the embryo.

Contraception   **Contraception,** the deliberate prevention of pregnancy, can be accomplished by several methods: preventing release of egg or sperm, preventing fertilization, preventing implantation, or aborting the embryo. Complete abstinence from sexual intercourse is the most reliable form of birth control. The **rhythm method,** or **natural family planning,** is based on refraining from intercourse during the period in which conception is most likely, the few days before and after ovulation.

Barrier methods of contraception include **condoms** and **diaphragms,** which, when used in conjunction with spermicidal foam or jelly, present a physical and chemical barrier to fertilization.

The intrauterine device (IUD) is a small, plastic or metal device that prevents implantation of the blastocyst. IUDs have a low failure rate but have been associated with harmful side effects.

The release of gametes may be prevented by chemical contraception, as in **birth control pills,** and by sterilization, as in **tubal ligation** in women or **vasectomy** in men. Birth control pills are combinations of synthetic estrogens and progestin, which act by negative feedback to stop the release of GnRH by the hypothalamus and FSH and LH by the pituitary, resulting in a cessation of ovulation and follicle development. A progestin-only minipill alters the cervical mucus so that it blocks sperm entry to the uterus. Norplant and Depo-Provera are longer lasting progestin treatments. Cardiovascular problems are a potential effect of estrogen-containing birth control pills.

Abortion is the termination of a pregnancy. Spontaneous abortion, or miscarriage, occurs in as many as one-third of all pregnancies. The drug mifepristone (RU-486) is an analog of progesterone that blocks progesterone receptors in the uterus and can be used to terminate a pregnancy within the first 7 weeks.

Condoms are the only form of birth control that offer some protection against sexually transmitted diseases.

■ INTERACTIVE QUESTION   46.8

List the three general types, along with examples, of birth control methods. Which examples are most likely to prevent pregnancy? Which are least likely to do so?

a.

b.

c.

Modern technology offers solutions for some reproductive problems (995)

Some genetic diseases and congenital defects can be detected while the fetus is in the uterus. The use of ultrasound imaging produces an image of the fetus from the echoes of high-frequency sound waves. In amniocentesis and chorionic villus sampling, fetal cells can be removed and tested for genetic defects. In a new noninvasive technique, fetal cells can be identified in the mother's blood sample and then tested.

For couples unable to conceive because of male infertility, sperm banks can provide sperm from anonymous donors. *In vitro* **fertilization** involves the removal of ova from a woman whose oviducts are blocked, fertilization within a culture dish, and implantation of the developing embryo in the uterus.

The development of male chemical contraception is currently being researched.

WORD ROOTS

a- = not, without (*asexual reproduction:* a type of reproduction involving only one parent that produces genetically identical offspring)

acro- = tip; **-soma** = body (*acrosome:* an organelle at the tip of a sperm cell that helps the sperm penetrate the egg)

bacul- = a rod (*baculum:* a bone that is contained in, and helps stiffen, the penis of rodents, raccoons, walruses, and several other mammals)

blasto- = produce; **-cyst** = sac, bladder (*blastocyst:* a hollow ball of cells produced one week after fertilization in humans)

coit- = a coming together (*coitus:* the insertion of a penis into a vagina, also called sexual intercourse)

contra- = against (*contraception:* the prevention of pregnancy)

-ectomy = cut out (*vasectomy:* the cutting of each vas deferens to prevent sperm from entering the urethra)

endo- = inside (*endometrium:* the inner lining of the uterus, which is richly supplied with blood vessels)

epi- = above, over (*epididymis:* a coiled tubule located adjacent to the testes where sperm are stored)

gyno- = female (*protogynous:* a form of sequential hermaphroditism in which the female sex occurs first)

labi- = lip; **major-** = larger (*labia majora:* a pair of thick, fatty ridges that enclose and protect the labia minora and vestibule)

lact- = milk (*lactation:* the continued production of milk)

menstru- = month (*menstruation:* the shedding of portions of the endometrium during a menstrual cycle)

minor- = smaller (*labia minora:* a pair of slender skin folds that enclose and protect the vestibule)

myo- = muscle (*myotonia:* increased muscle tension)

oo- = egg; **-genesis** = producing (*oogenesis:* the process in the ovary that results in the production of female gametes)

partheno- = a virgin (*parthenogenesis:* a type of reproduction in which females produce offspring from unfertilized eggs)

partur- = giving birth (*parturition:* the expulsion of a baby from the mother, also called birth)

proto- = first; **andro-** = male (*protandrous:* a form of sequential hermaphroditism in which the male sex occurs first)

-theca = a cup, case (*spermatheca:* a sac in the female reproductive system where sperm are stored)

tri- = three (*trimester:* a three month period)

vasa- = a vessel (*vasocongestion:* the filling of a tissue with blood caused by increased blood flow through the arteries of that tissue)

STRUCTURE YOUR KNOWLEDGE

1. Trace the path of a human sperm from the point of production to the point of fertilization, briefly commenting on the functions of both the structures it passes through and the associated glands.

2. Answer the following questions concerning the human menstrual cycle.
   a. What does GnRH do?

   b. What does FSH stimulate?

   c. What causes the spike in LH level?

   d. What does this LH surge induce?

   e. What does LH maintain during the luteal phase?

   f. What inhibits secretion of LH and FSH?

   g. What allows LH and FSH secretion to begin again?

3. Describe how birth control pills work. How does the French drug RU-486 (mifepristone) function?

TEST YOUR KNOWLEDGE

FILL IN THE BLANKS

_____ 1. type of asexual reproduction in which a new individual grows while attached to the parent's body

_____ 2. sequential hermaphroditism in which an organism is first a male and then becomes a female

_____ 3. development of egg without fertilization

_____ 4. individual with functioning male and female reproductive systems

_____ 5. common opening of digestive, excretory, and reproductive systems in nonmammalian vertebrates

_____ **6.** type of reproductive cycle in which thickened endometrium is reabsorbed

_____ **7.** hormone that maintains the uterine lining during pregnancy

_____ **8.** common duct for urine and semen in mammalian males

_____ **9.** filling of a tissue with blood due to increased blood flow

_____ **10.** period when ovulation and menstruation cease in humans

MULTIPLE CHOICE: *Choose the one best answer.*

1. Which of the following is an explanation for the periodicity of reproductive cycles in animals?
   a. Reproduction may correspond with periods of increased food supply, during which energy can be invested in gamete formation.
   b. Seasonal cycles may allow offspring to be produced during favorable environmental conditions when chances of survival are highest.
   c. Hormonal control of reproduction may be tied to biological clocks and seasonal cues.
   d. Synchronicity in release of gametes increases probability of fertilization.
   e. All of these may contribute to periodic reproductive activity.

2. Which of the following is *least* likely to be hermaphroditic?
   a. earthworm
   b. barnacle
   c. tapeworm
   d. grasshopper
   e. liver fluke

3. Which of the following is *incorrectly* paired with its function?
   a. seminiferous tubules—add fluid containing mucus, fructose, and prostaglandins to semen
   b. scrotum—encases testes and suspends them below abdominal cavity
   c. epididymis—tubules in which sperm gain motility
   d. prostate gland—adds fluid to semen
   e. vas deferens—transports sperm from epididymis to ejaculatory duct

4. The function of the corpus luteum is to
   a. nourish and protect the egg cell.
   b. produce prolactin in the milk sacs of the mammary gland.
   c. produce progesterone and estrogens.
   d. produce estrogens and disintegrate following ovulation.
   e. maintain pregnancy by production of human chorionic gonadotropin.

5. Which of the following hormones is *incorrectly* paired with its function?
   a. androgens—responsible for primary and secondary male sex characteristics
   b. oxytocin—stimulates uterine contractions during parturition
   c. estrogens—responsible for primary and secondary female sex characteristics
   d. LH—stimulates production of androgens by Leydig cells of testes
   e. prolactin—stimulates breast development at puberty

6. Myotonia is
   a. a congenital birth defect.
   b. the hormone responsible for breast development.
   c. muscle tension.
   d. the filling of a tissue with blood.
   e. responsible for delivery of the placenta.

7. The secretory phase of the menstrual cycle
   a. is associated with dropping levels of estrogens and progesterone.
   b. is when the endometrium begins to degenerate and menstrual flow occurs.
   c. involves the initial proliferation of the endometrium.
   d. corresponds with the follicular phase of the ovarian cycle.
   e. corresponds with the luteal phase of the ovarian cycle.

8. Examples of birth control methods that prevent release of gametes from the gonads are
   a. sterilization and chemical contraception.
   b. birth control pills and IUDs.
   c. condoms and diaphragms.
   d. abstinence and chemical contraception.
   e. the progestin minipill and RU-486.

9. The ability of a pregnant woman not to reject her "foreign" fetus may be due to
   a. the fact that fetal and maternal blood never mix.
   b. the suppression of her immune response to the paternal chemical markers on fetal tissue.

c. the protection of the fetus within the trophoblast that is made of maternal tissue.

d. the production of human chorionic gonadotropin that maintains the pregnancy.

e. the masking of paternal markers by specialized white blood cells.

10. Certain maternal diseases, drugs, alcohol, and radiation are most dangerous to embryonic development
    a. during the first 2 weeks when the embryo has not yet implanted and spontaneous abortion may occur.
    b. during the first 2 months when organogenesis is occurring.
    c. during the first and second trimesters when the embryonic liver is not yet filtering toxins.
    d. during the second trimester when the corpus luteum no longer secretes estrogens and progesterone.
    e. during the third trimester when the most rapid growth is occurring.

Use the following to answer questions 11–15.
    a. estrogen
    b. progesterone
    c. LH (luteinizing hormone)
    d. FSH (follicle-stimulating hormone)
    e. HCG (human chorionic gonadotropin)

11. Which hormone stimulates ovulation and the development of the corpus luteum?

12. Which hormone is produced by the developing follicle and initiates thickening of the endometrium?

13. Which hormone is produced by the embryo during the first trimester and is necessary for maintaining a pregnancy?

14. Which hormone stimulates the production of sperm in the seminiferous tubules of the testes?

15. Which hormone is produced by the corpus luteum and later by the placenta and is responsible for maintaining a pregnancy?

16. In the birth process, which of the following is/are involved in triggering and maintaining labor?
    a. HCG produced by the fetus
    b. oxytocin produced by fetus and mother, and prostaglandins produced by the placenta
    c. a drop in progesterone caused by the disintegration of the corpus luteum
    d. prolactin produced by the fetus and mother
    e. a surge in the production of LH

17. In which location does fertilization usually take place in a human female?
    a. ovary
    b. oviduct
    c. uterus
    d. cervix
    e. vagina

18. The major advantage of sexual reproduction is that
    a. it is easier than asexual reproduction.
    b. offspring have a better start when produced from a union of egg and sperm than when simply budded off a parent.
    c. it produces diploid offspring, whereas asexually produced offspring are haploid.
    d. it produces variation in the offspring, and new genetic combinations may be better adapted to a changing environment.
    e. it requires both male and female members of a species to be present in a population and have behavioral interactions.

19. Which of the following are the stem cells that give rise to sperm?
    a. spermatogonia
    b. spermatozoa
    c. Sertoli cells
    d. spermatocytes
    e. spermatids

20. How does meiosis differ in the production of human sperm and ova?
    a. Meiosis occurs while a female is an embryo, but does not begin until puberty in males.
    b. Each meiotic division produces four sperm but only two ova.
    c. Meiosis occurs in the testes of males but in the oviducts of females.
    d. Female stem cells, the primary oocytes, are produced before birth, whereas male stem cells continue to divide throughout life.
    e. Meiosis is an uninterrupted process in males, whereas it begins again in females when a follicle matures and is only completed when a sperm penetrates the egg cell.

CHAPTER 47

ANIMAL DEVELOPMENT

FRAMEWORK

Fertilization initiates physical and molecular changes in the egg cell. Early embryonic development includes cleavage of the fertilized egg to form a blastula, gastrulation, and organogenesis. The amount of yolk in the egg and the evolutionary history of the animal determine how these processes occur in any particular animal group.

Development and the differentiation of cells result from the control of gene expression by both cytoplasmic determinants and cell-cell induction. Pattern formation depends on the positional information a cell receives within a developing structure. Developmental biologists, using transplant experiments and techniques from molecular biology, are gradually unraveling some of the mechanisms underlying the complex processes of animal development.

CHAPTER REVIEW

Animal development charts the course from a single fertilized egg to an organism made of many differentiated cells organized into specialized tissues and organs. The combination of molecular genetics, classical studies, and the use of well-known research organisms is helping developmental biologists to study the physical, cellular, and molecular events of embryonic development.

The Stages of Early Embryonic Development

From egg to organism, an animal's form develops gradually: *the concept of epigenesis* **(999)**

Preformation, or the belief that the egg or sperm contains a miniature embryo, was once the favored explanation of animal development. **Epigenesis** is the now accepted theory that the form of an embryo gradually develops from an egg. The developmental plan of an embryo is predetermined by the zygote genome and by the distribution of maternal mRNA and proteins in the egg cytoplasm. As cell division separates cytoplasmic components, nuclei are exposed to different environments that affect which genes are expressed by different cells. Cell signaling conveys developmental messages between cells that influence gene expression and differentiation. Morphogenesis produces the body shape.

Fertilization activates the egg and brings together the nuclei of sperm and egg (999–1002)

Fertilization, the union of egg and sperm, combines the haploid sets of chromosomes of these specialized cells and activates the egg by initiating metabolic reactions that trigger embryonic development.

The Acrosomal Reaction   In sea urchin fertilization, the acrosome at the tip of the sperm discharges hydrolytic enzymes when it comes in contact with the jelly coat of an egg. This **acrosomal reaction** allows the acrosomal process to elongate through the jelly coat. Fertilization within the same species is assured when proteins on the surface of the acrosomal process attach to specific receptor molecules on the vitelline layer of the egg, which is external to the plasma membrane.

In response to the fusion of sperm and egg plasma membranes, ion channels open in the egg membrane, and sodium ions flow into the egg. The resulting depolarization of the membrane prevents other sperm cells from fusing with the egg, providing a **fast block to polyspermy.**

The Cortical Reaction   Membrane fusion also initiates the **cortical reaction,** involving a signal-transduction pathway that triggers release of calcium ions from the egg's endoplasmic reticulum. In response to the Ca^{2+} increase in the cytoplasm, **cortical granules** located in the outer cortex of the egg release their contents into the perivitelline space. The released enzymes and macromolecules cause the vitelline layer to elevate and harden to form the **fertilization envelope,** which functions as a **slow block to polyspermy.**

Activation of the Egg   The rise in Ca^{2+} concentration activates the egg by increasing the rates of cellular respiration and protein synthesis. Hydrogen ions are transported out of the cell, and the increase in cellular pH may be involved in activation.

Parthenogenetic development can be initiated by injecting calcium into an egg. Even an enucleated egg can be activated to begin protein synthesis, showing that inactive mRNA had been stockpiled in the egg.

After the sperm nucleus fuses with the egg nucleus, DNA replication begins in preparation for the cleavage division that begins the development of the embryo.

■ INTERACTIVE QUESTION   47.1

a. What assures that only sperm of the right species fertilize sea urchin eggs?

b. What assures that only one sperm will fertilize an egg?

Fertilization in Mammals   Following internal fertilization, secretions of the mammalian female reproductive tract enhance motility and alter surface molecules of the sperm. A thus capacitated sperm migrates through the follicle cells released with the egg to the **zona pellucida,** a three-dimensional filament network made of three types of glycoproteins. Complementary binding of a molecule on the sperm head to the glycoprotein ZP3 induces an acrosomal reaction. The hydrolytic enzymes released from the acrosome enable the sperm cell to penetrate the zona pellucida.

Binding of a sperm membrane protein with the egg membrane triggers a depolarization and fast block to polyspermy. Release of granules from the egg cortex during the cortical reaction causes alterations of the zona pellucida and a slow block to polyspermy.

Microvilli of the egg enclose the entire sperm. The basal body of the sperm flagellum divides to form two centrosomes (with centrioles) that function in cell division. Nuclear envelopes disperse and the first mitotic division occurs.

Cleavage partitions the zygote into many smaller cells (1002–1005)

Cleavage is a succession of rapid cell divisions during which the embryo becomes partitioned into many small cells, called **blastomeres.** Cleavage parcels different regions of the cytoplasm, which contain different cytoplasmic components, into cells and sets the stage for later development.

The polarity of the egg of many frogs and other animals is defined by the **vegetal pole,** where the stored nutrients in **yolk** are most concentrated, and the opposite end called the **animal pole,** where the polar bodies budded from the egg during meiosis and the anterior part of the embryo often forms.

In the eggs of many frogs, the animal hemisphere has melanin granules in the outer cytoplasm (the cortex), whereas the vegetal hemisphere contains the yellow yolk. Fertilization results in a rotation of the cortex toward the point of sperm entry, creating an opposite narrow **gray crescent,** which the first cleavage division will bisect and which marks the future dorsal side of the embryo.

Yolk impedes cell division, and smaller cells are produced in the animal hemisphere in the frog. In species in which eggs have little yolk, blastomeres are of equal size, although an animal-vegetal axis results from unequal distribution of other substances.

The first two cleavage divisions are vertical, or meridional; the third division is equatorial. Deuterostomes—echinoderms and chordates—share similarities in embryonic development that differ from most protostomes—annelids, arthropods, and mollusks.

Further cleavage produces a **morula,** a solid ball of cells. A fluid-filled cavity called the **blastocoel** develops, creating the stage called the **blastula.** The frog blastula has its blastocoel restricted to the animal hemisphere due to unequal cell division.

The large amount of yolk in eggs of birds results in **meroblastic cleavage** restricted to the small disk of cytoplasm on top of the yolk. **Holoblastic cleavage** is the complete division of eggs with small or moderate amounts of yolk. In the yolk-rich eggs of insects, the zygote nucleus undergoes repeated divisions and the nuclei migrate to the egg surface. Plasma membranes eventually form around the nuclei, creating a blastula made of cells surrounding a mass of yolk.

■ INTERACTIVE QUESTION   47.2

Identify the early embryonic stages diagrammed below and label the indicated regions or structures.

a. _____

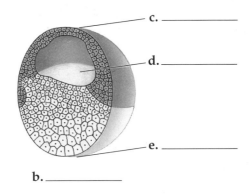

c. _____

d. _____

e. _____

b. _____

Gastrulation rearranges the blastula to form a three-layered embryo with a primitive gut (1005–1007)

Changes in cell motility, shape, and adhesion are part of the morphogenetic rearrangements of cells in **gastrulation** that result in a three-layered embryo called the **gastrula.** The gastrula consists of the embryonic germ layers—the outer **ectoderm,** the middle **mesoderm,** and the inner **endoderm** that lines the embryonic digestive tract. This triploblastic body plan is characteristic of most animal phyla.

In a sea urchin, gastrulation occurs as cells from the vegetal pole detach and move into the blastocoel as migratory *mesenchyme cells.* The remaining cells at the vegetal pole flatten into a plate that then buckles inward by a process known as **invagination.** The invagination deepens to form a narrow pouch called the **archenteron,** a cavity that will become the digestive tract. The archenteron opening is the **blastopore,** which develops into the anus. Cytoplasmic extensions (filopodia) from mesenchyme cells at the tip of the archenteron pull the archenteron across the blastocoel, where the endoderm fuses with the ectoderm to form the mouth.

Gastrulation in frog development is more complicated because of the multilayered blastula wall and the large, yolk-filled cells of the vegetal hemisphere. A group of invaginating cells begins gastrulation in the region of the gray crescent, forming a tuck that will become the **dorsal lip** of the blastopore. In a process called **involution,** surface cells roll over the dorsal lip and migrate along the roof of the blastocoel. The blastopore eventually forms a circle surrounding a **yolk plug** of large, yolk-filled cells. Gastrulation produces an endoderm-lined archenteron surrounded by mesoderm, with the outer layer of the gastrula composed of ectoderm.

■ INTERACTIVE QUESTION   47.3

Label the parts in these diagrams of gastrulation in a sea urchin and a frog gastrula.

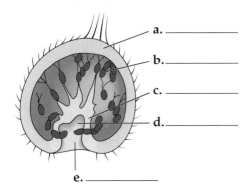

a. _____
b. _____
c. _____
d. _____
e. _____

Gastrulation in sea urchin

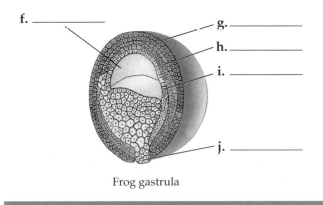

f. _____
g. _____
h. _____
i. _____
j. _____

Frog gastrula

In organogenesis, the organs of the animal body form from the three embryonic germ layers (1007–1011)

Organs begin to develop from the embryonic germ layers in a process known as **organogenesis.** Morphogenetic changes include folds, splits, and dense clustering (condensation) of cells. In a frog embryo, the **notochord** forms from condensation of dorsal mesoderm along the roof of the archenteron. Ectoderm above the developing notochord forms a neural plate, which folds inward to form a hollow **neural tube,** from which will develop the central nervous system.

Mesoderm along the sides of the notochord condenses and separates into blocks, called **somites,** which give rise to vertebrae and muscles associated with the axial skeleton, arranged in segmented fashion. Lateral to the somites, the mesoderm splits to form the lining of the coelom.

In vertebrates, a band of cells called the *neural crest* separates from the meeting margins of the neural tube when it forms. These cells later migrate to form pigment cells in the skin, bones and muscles of the skull, teeth, the adrenal medulla, and components of the peripheral nervous system.

■ INTERACTIVE QUESTION 47.4

Label the indicated structures in this diagram of organogenesis in a frog embryo.

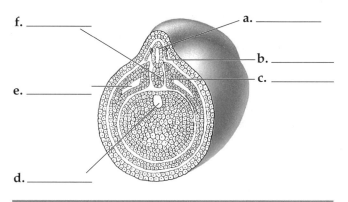

f. _____   a. _____

b. _____

c. _____

e. _____

d. _____

Amniote embryos develop in a fluid-filled sac within a shell or uterus (1007)

Reptiles, birds, and mammals are **amniotes,** meaning that they create the aqueous environment necessary for embryonic development with a fluid-filled sac surrounded by a membrane called the amnion.

Avian Development Meroblastic cleavage of the fertilized egg produces a cap of cells called the **blastodisc** on top of the large, undivided yolk.

The blastomeres separate into an upper epiblast and a lower hypoblast layer, forming a cavity between them comparable to a blastocoel. A linear invagination, called the **primitive streak,** marks the anterior-posterior axis of the embryo and is the site of gastrulation. Cells of the epiblast move through the primitive streak, forming the mesoderm and the endoderm. Cells remaining in the epiblast become ectoderm. The borders of the embryonic disc fold down

and join, forming a three-layered tube attached by a stalk to the yolk. The hypoblast cells form part of the yolk sac and connecting stalk. Organogenesis proceeds in a fashion similar to that of the frog embryo.

The primary germ layers also give rise to four **extraembryonic membranes,** essential to development within the avian egg shell.

■ INTERACTIVE QUESTION 47.5

List the four extraembryonic membranes found in an avian egg and briefly describe their functions.

a.

b.

c.

d.

Mammalian Development Fertilization occurs in the oviduct, and embryonic development begins on the journey to the uterus. The mammalian egg and zygote show no obvious polarity. There is little yolk, thus cleavage is holoblastic. Gastrulation and early organogenesis, however, are similar in pattern to those of birds and reptiles. *Compaction* occurs at the eight-cell stage, when surface proteins are formed that make the cells adhere tightly to each other.

The **blastocyst** consists of an outer epithelium, the **trophoblast,** surrounding a cavity into which protrudes a cluster of cells, called the **inner cell mass.** These inner cells develop into the embryo and some of the extraembryonic membranes. The **trophoblast** secretes enzymes that enable the blastocyst to embed in the uterine lining (endometrium) and extends projections that will develop, along with mesodermal tissue, into the fetal portion of the placenta.

The inner cell mass forms a flat embryonic disk with an upper epiblast and lower hypoblast, homologous to the two-layered blastodisc of birds. In gastrulation, cells from the epiblast move through a primitive streak to form the mesoderm and endoderm.

The four extraembryonic membranes are homologous to those of reptiles and birds. The chorion, which develops from the trophoblast, surrounds the embryo and the other membranes. The epiblast froms the amnion, enclosing the embryo in a fluid-filled amniotic cavity. The yolk-sac membrane, enclosing a small fluid-filled cavity, is the site of early formation of blood cells. The allantois forms blood vessels that connect the embryo with the placenta through the umbilical cord.

Organogenesis begins with the formation of the notochord, neural tube, and somites. In the human embryo, all major organs have begun development by the end of the first trimester.

■ INTERACTIVE QUESTION 47.6

Label the indicated cell layers in this diagram of a human blastocyst implanting into the endometrium. List the embryonic tissues and membranes that will develop from each cell layer.

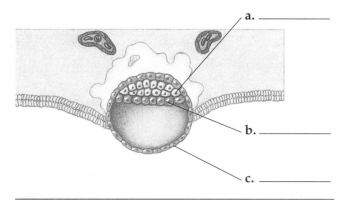

a. _____

b. _____

c. _____

The Cellular and Molecular Basis of Morphogenesis and Differentiation in Animals

Morphogenesis in animals involves specific changes in cell shape, position, and adhesion (1012–1014)

Changes in cell shape usually involve reorganization of the cytoskeleton. Extension of microtubules to elongate a cell and then contraction of microfilaments at the apical end create a wedge shape, initiating invaginations and evaginations.

Migrating embryonic cells use their cytoskeleton to extend and retract cellular protrusions. Cells at the leading edge of migrating tissue may extend filopodia and then drag along sheets of cells. By **convergent extension,** cells of a tissue layer crawl between each other, causing the sheet of cells to elongate and become narrower.

Glycoproteins such as fibronectins in the extracellular matrix (ECM) provide anchorage for crawling cells and tracts that direct the movement of migrating cells. The orientation of contractile microfilaments in the cytoskeleton of migrating cells corresponds to the fibronectin fibers secreted by underlying nonmigratory cells. Glycoproteins called **cell adhesion molecules (CAMs)** hold the cells of specific tissues together and help regulate morphogenesis. **Cadherins** are important adhesion molecules whose genes are differentially expressed in specific locations during development.

The developmental fate of cells depends on cytoplasmic determinants and cell-cell induction: *a review* (1014)

The mechanisms underlying the differentiation of cells during embryonic development are organized into two basic concepts: heterogeneous cytoplasm in the egg is partitioned through cleavage into different blastomeres; and the interactions between cells (induction) affect a cell's developmental fate. Both of these phenomena influence gene expression, which leads to the differentiation of cells.

Fate mapping can reveal cell genealogies in chordate embryos (1014–1015)

In the 1920s, Vogt developed a **fate map** for amphibian embryos by labeling regions of the blastula with dye to determine where specific cells showed up in later developmental stages. Cell lineage analysis follows the mitotic descendants of individual blastomeres. The developmental history of every cell has been determined for the nematode *Caenorhabditis elegans.*

In most animals, early "founder cells" can be identified that will generate specific tissues. Experiments have also shown that a cell's *developmental potential* becomes restricted as development proceeds.

The eggs of most vertebrates have cytoplasmic determinants that help establish the body axes and differences among cells of the early embryo (1015–1016)

Polarity and the Basic Body Plan   In mammals, polarity does not seem to be established until after cleavage has begun. In most other species, the major axes are established much earlier, even in the unfertilized egg.

Restriction of Cellular Potency   Early blastomeres that can be separated and still produce complete embryos are said to be totipotent. Even though cytoplasmic determinants are asymmetrically distributed in the frog egg, the first cleavage division bisects the gray crescent and equally separates these determinants.

The zygote is the only totipotent cell in many species. Mammalian blastomeres, however, remain totipotent until the inner cell mass forms. The progressive restriction of a cell's developmental potency is characteristic of animal development. Experimental manipulation shows that the tissue-specific fates of cells of late gastrulae have been fixed.

■ INTERACTIVE QUESTION 47.7

If an eight-cell stage of a sea urchin embryo is split vertically into two four-cell groups, both develop into normal larvae. If the division is made horizontally, abnormal larvae develop. Explain this experimental result.

Inductive signals drive differentiation and pattern formation in vertebrates (1016–1019)

In induction, one group of cells influences the development of another through physical contact or chemical signals that switch on sets of genes.

The "Organizer" of Spemann and Mangold Using transplant experiments with amphibian embryos, Mangold and Spemann established that the dorsal lip is a "primary organizer" of the embryo because of its influence in early organogenesis. The molecular basis of induction by Spemann's organizer appears to be tied to the inactivation of bone morphogenic protein 4 (BMP-4) on the dorsal side of the embryo. Proteins similar to BMP-4 and its inhibitors appear to play a role in the development of many different organisms.

Development of the vertebrate eye involves a series of inductions from cells in all three germ layers, progressively determining the fate of epidermal cells that become lens cells.

Pattern Formation in the Vertebrate Limb **Pattern formation** is the ordering of cells and tissues into the characteristic structures in the proper locations in the animal. Cells and their offspring differentiate based on **positional information,** molecular cues that indicate their position with respect to body axes.

A limb bud of a chick consists of a core of mesoderm and a layer of ectoderm. Researchers have identified two organizer regions that secrete proteins that provide positional information to cells.

Cells of the **apical ectodermal ridge (AER)** at the tip of the limb bud secrete proteins of the fibroblast growth factor (FGF) family. These growth signals appear to promote limb bud outgrowth and patterning along the proximal-distal axis. The AER and other ectoderm also appear to signal the limb's dorsal-ventral axis.

The **zone of polarizing activity (ZPA),** located at the posterior attachment of the bud, appears to communicate position along the anterior-posterior axis in a developing limb. Cells of the ZPA secrete a protein growth factor called Sonic hedgehog. Similar proteins

have been identified as important positional cues in a number of development processes in other organisms, such as fruit flies and mice.

Cells respond to positional information relating to their location in a developing organ according to their developmental histories. Thus, patterns of gene expression first determine whether cells will become part of a wing or of a leg; then positional cues indicate what part of the wing or leg the cells will help form. A hierarchy of gene activations leads to the expression of homeobox-containing *(Hox)* genes that are involved in specifying various regions of the developing limb.

■ INTERACTIVE QUESTION 47.8

What would be the effect of implanting cells genetically engineered to produce and secrete Sonic hedgehog to the anterior attachment of a limb bud?

WORD ROOTS

acro- = the tip *(acrosomal reaction:* the discharge of a sperm's acrosome when the sperm approaches an egg)

arch- = ancient, beginning *(archenteron:* the endoderm-lined cavity, formed during the gastrulation process, that develops into the digestive tract of an animal)

blast- = bud, sprout; **-pore** = a passage *(blastopore:* the opening of the archenteron in the gastrula that develops into the mouth in protostomes and the anus in deuterostomes)

blasto- = produce; **-mere** = a part *(blastomere:* small cells of an early embryo)

cortex- = shell *(cortical reaction:* a series of changes in the cortex of the egg cytoplasm during fertilization)

ecto- = outside; **-derm** = skin *(ectoderm:* the outermost of the three primary germ layers in animal embryos)

endo- = within *(endoderm:* the innermost of the three primary germ layers in animal embryos)

epi- = above; **-genesis** = origin, birth *(epigenesis:* the progressive development of form in an embryo)

extra- = beyond *(extraembryonic membrane:* four membranes that support the developing embryo in reptiles, birds, and mammals)

fertil- = fruitful *(fertilization:* the union of haploid gametes to produce a diploid zygote)

gastro- = stomach, belly (*gastrulation:* the formation of a gastrula from a blastula)

holo- = whole (*holoblastic cleavage:* a type of cleavage in which there is complete division of the egg)

in- = into; **vagin-** = a sheath (*invagination:* the infolding of cells)

involut- = wrapped up (*involution:* cells rolling over the edge of a lip into the interior)

mero- = a part (*meroblastic cleavage:* a type of cleavage in which there is incomplete division of yolk-rich egg, characteristic of avian development)

meso- = middle (*mesoderm:* the middle primary germ layer of an early embryo)

morul- = a little mulberry (*morula:* a solid ball of blastomeres formed by early cleavage)

noto- = the back; **-chord** = a string (*notochord:* a long flexible rod that runs along the dorsal axis of the body in the future position of the vertebral column)

poly- = many (*polyspermy:* fertilization by more than one sperm)

soma- = a body (*somites:* paired blocks of mesoderm just lateral to the notochord of a vertebrate embryo)

tropho- = nourish (*trophoblast:* the outer epithelium of the blastocyst, which forms the fetal part of the placenta)

zona = a belt; **pellucid-** = transparent (*zona pellucida:* the extracellular matrix of a mammalian egg)

STRUCTURE YOUR KNOWLEDGE

1. Create a flowchart that shows the sequence of key events in the fertilization of a sea urchin egg and indicate the functions of these events.

2. Fill in the table below, briefly describing the early stages of development for sea urchin, frog, bird, and mammalian embryos.

Animal	Cleavage	Blastula	Gastrula
Sea urchin			
Frog			
Bird			
Mammal			

TEST YOUR KNOWLEDGE

MULTIPLE CHOICE: *Choose the one best answer.*

1. The vitelline layer of a sea urchin egg
   a. is inside the fertilization envelope.
   b. releases calcium, which initiates the cortical reaction.
   c. has receptor molecules that are specific for proteins on the acrosomal process of sperm.
   d. is a thick jelly coat that is hydrolyzed by the release of enzymes from the acrosome.
   e. fuses with the sperm plasma membrane, initiating the depolarization of the membrane.

2. The slow block to polyspermy
   a. prevents sperm from other species from fertilizing the egg.
   b. is directly produced by the depolarization of the membrane.
   c. is a result of the formation of the fertilization envelope.
   d. is caused by the expulsion of hydrogen ions.
   e. involves all of the above.

3. The blastocoel
   a. develops into the archenteron or embryonic gut.
   b. is a fluid-filled cavity in the blastula.
   c. opens to the exterior through a blastopore.
   d. forms a hollow chamber during gastrulation.
   e. is lined with mesoderm.

4. Which of the following groups does *not* have eggs with definite polarity?
   a. sea urchin
   b. frog
   c. fruit fly
   d. bird
   e. mammal

5. Cytoplasmic determinants
   a. are unevenly distributed cytoplasmic components that influence the developmental fates of cells.
   b. are involved in the regulation of gene expression.
   c. usually include maternal mRNA.
   d. are often separated in the first few cleavage divisions.
   e. are all of the above.

6. Which of the following is *incorrectly* paired with its embryonic germ layer?
   a. muscles—mesoderm
   b. central nervous system—ectoderm
   c. lens of the eye—mesoderm
   d. liver—endoderm
   e. notochord—mesoderm

7. In a frog embryo, gastrulation
   a. is impossible because of the large amount of yolk.
   b. proceeds by involution as cells roll over the dorsal lip of the blastopore.
   c. produces a blastocoel displaced into the animal hemisphere.
   d. occurs along the primitive streak.
   e. involves the formation of the notochord and neural tube.

8. The primitive streak of mammalian embryos is analogous to
   a. the dorsal lip of the frog embryo.
   b. the blastodisc of a bird embryo.
   c. the multinucleated stage of a fruit fly embryo.
   d. the archenteron of a sea urchin embryo.
   e. none of the above.

9. A function of the allantois in birds is to
   a. provide for nutrient exchange between the embryo and yolk sac.
   b. store nitrogenous wastes in the form of urea.
   c. form a respiratory organ in conjunction with the chorion.
   d. provide an aqueous environment for the developing embryo.
   e. produce the blood vessels of the umbilical cord.

10. Somites are
    a. blocks of mesoderm circling the archenteron.
    b. condensations of cells from which the notochord arises.
    c. serially arranged mesoderm blocks that develop into vertebrae and skeletal muscles.
    d. mesodermal derivatives of the notochord that line the coelom.
    e. produced during organogenesis in sea urchins, frogs, birds, and mammals, but not in fruit flies.

11. What forms the fetal portion of the placenta?
    a. the epiblast
    b. the trophoblast and some mesoderm
    c. the allantois and yolk sac
    d. the endometrium
    e. the amnion

12. In her transplant experiments with frog embryos, Hilde Mangold found that a part of the dorsal lip of the blastopore moved to another location would result in a second gastrulation and even the development of a doubled, face-to-face tadpole. Which of the following is the best explanation for these results?
    a. morphogenetic movements
    b. separation of cytoplasmic determinants
    c. homeobox regulation of gene expression
    d. fate map determination
    e. induction by dorsal lip tissue

13. Pattern formation appears to be determined by
    a. positional information a cell receives from gradients of chemical signals called morphogens.
    b. differentiation of cells, which then migrate into developing organs.
    c. the movement of cells along fibronectin fibrils.
    d. the induction of cells by mesoderm cells found in the center of organs.
    e. gastrulation and the formation of the three tissue layers.

Use this chick embryo diagram for questions 14–16.

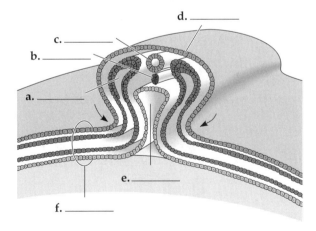

14. Which line points to the notochord?

15. Which line points to the archenteron?

16. To what structure does line **f** point?
    a. the primitive streak through which gastrulation occurs
    b. the hypoblast
    c. the mesodermal lining of the coelom
    d. the blastocoel between the epiblast and the hypoblast
    e. the developing extraembryonic membranes

17. The early cleavage divisions of a human zygote most closely resemble
    a. those of a sea urchin because neither egg has a large store of yolk and cleavage is holoblastic.
    b. those of a chick because both birds and mammals evolved from a common reptilian ancestor.
    c. those of a frog because the cells of the animal hemisphere are smaller than those of the vegetal hemisphere in both types of embryos.
    d. those of a *Drosophila* because both pass through an early multinucleated stage before cell membranes form.
    e. all of the above animals because they are all deuterostomes and have similar embryonic patterns.

18. Gastrulation and organogenesis in mammals most closely resemble
    a. those of a sea urchin because both types of eggs have little yolk and cleavage is holoblastic.
    b. those of a chick because both birds and mammals evolved from a common reptilian ancestor.
    c. those of a frog because gastrulation takes place through the dorsal lip of the blastopore in both types of eggs.

    d. neither a chick nor a frog because mammals have some embryonic cells that develop into extraembryonic membranes.
    e. those of a *Drosophila* because homeotic genes control pattern development in both organisms.

19. Which of the following groups of structures is formed from endoderm?
    a. notochord, nerve cord, medulla of adrenal gland
    b. skeletal muscles, vertebrae, circulatory system
    c. lining of coelom and archenteron
    d. lining of digestive tract, liver, pancreas
    e. lens of eye, teeth, reproductive system

20. What is secreted by cells of the apical ectodermal ridge?
    a. Sonic hedgehog that communicates location along the anterior-posterior axis
    b. cadherins that help to hold migrating cells together as they form tissues of the leg
    c. ZPA, which forms fibronectin fibers that guide migrating cells
    d. fibroblast growth factors that promote limb bud outgrowth
    e. CAMs that cause the limb bud to elongate by convergent extension

CHAPTER 48

NERVOUS SYSTEMS

FRAMEWORK

This chapter describes the structural components of nervous systems and how they functionally integrate, coordinate, and transmit information from the internal and external environment.

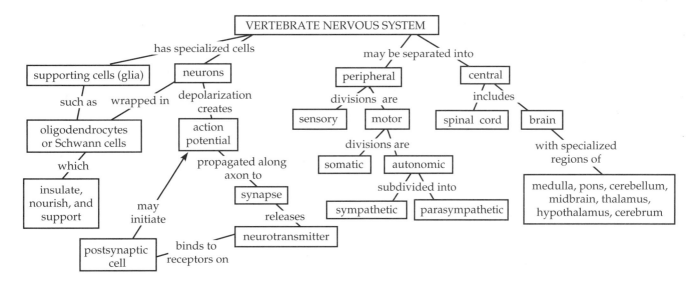

CHAPTER REVIEW

The maintenance of homeostasis involves the interaction and cooperation of the nervous, endocrine, and immune systems. The incredible structural complexity of the nervous system enables an animal to integrate vast amounts of information and respond rapidly to environmental stimuli. Diverse nervous systems have evolved in various animal phyla, but they are remarkably similar at the cellular level.

An Overview of Nervous Systems

Nervous systems perform the three overlapping functions of sensory input, integration, and motor output (1023)

Sensory input gathered by **sensory receptors** is conducted to integration centers in the brain and spinal cord (**central nervous system** or **CNS**). **Motor output**

is conveyed to **effector cells,** muscle or gland cells. Integration of sensory input and motor output is a continuous background activity within the CNS. The **peripheral nervous system (PNS)** carries sensory and motor information between the body and the central nervous system in **nerves,** bundles of neuron extensions wrapped in connective tissue. Neurons relay information through a combination of electrical and chemical signals.

Networks of neurons with intricate connections form nervous systems (1023–1026)

Neuron Structure and Synapses   A **neuron,** or **nerve cell,** consists of a **cell body,** which contains the nucleus and organelles, and the fiberlike processes, **dendrites** and **axons,** which conduct signals. The highly branched, short dendrites carry messages toward the cell body; the longer axons transmit signals to other cells.

Axons originate from a region of the cell body called the **axon hillock.** Many axons are wrapped in an insulating **myelin sheath.** Terminal branches of axons have specialized endings called **synaptic terminals,** which release **neurotransmitters** that relay signals across the **synapse** to a **postsynaptic cell,** another neuron or effector cell. The transmitting cell is called the **presynaptic cell.**

A Simple Nerve Circuit—the Reflex Arc   A **reflex arc,** the simplest type of nerve circuit, regulates an automatic response, called a **reflex.** A **sensory neuron** transmits information from a sensory receptor to a **motor neuron,** which signals an **effector cell** to carry out the response. Even the simple knee-jerk reflex, however, includes a second nerve circuit in which the sensory neuron relays the information from the stretch receptor in the thigh muscle to **interneurons** in the spinal cord, which then inhibit motor neurons to the flexor muscles. Most nerve circuits involve large numbers of interneurons, which are constantly communicating with each other.

Cell bodies of motor neurons and interneurons are located in the gray matter of the CNS. The white matter surrounding the butterfly-shaped gray matter in the spinal cord contains axons. A sensory neuron's cell body is located outside the spinal cord in a dorsal root ganglion. Nerve cell bodies, often of similar function, are clustered into **ganglia** in the PNS. Clusters of cell bodies in the brain are called **nuclei.**

Types of Nerve Circuits   Three types of nerve circuits include circuits in which information is spread from one presynaptic neuron to several postsynaptic neurons; convergent circuits in which several presynaptic neurons feed information into a single postsynaptic neuron; and circular paths in which information flows from one neuron to others and then back to its source.

Supporting Cells (Glia)   The very numerous **supporting cells,** or **glia,** give structural integrity and physiological support to the nervous system. Recent studies indicate that glia may have synaptic communication with each other and with neurons. Radial glia guide the embryonic growth of neurons. **Astrocytes** are supporting glia in the CNS that induce the formation of the **blood-brain barrier,** which restricts the passage of most substances into the brain. Both **oligodendrocytes** (in the CNS) and **Schwann cell**s (in the PNS) insulate axons in a myelin sheath by wrapping around them and forming concentric membrane layers. Gaps between adjacent Schwann cells are called nodes of Ranvier.

■ **INTERACTIVE QUESTION   48.1**

Label the indicated structures on this diagram of a vertebrate motor neuron. Indicate the direction of impulse transmission.

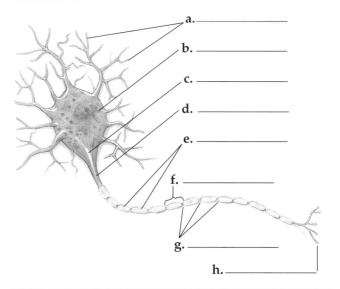

a. _____

b. _____

c. _____

d. _____

e. _____

f. _____

g. _____

h. _____

The Nature of Nerve Signals

Every cell has a voltage, or membrane potential, across its plasma membrane (1026–1028)

An electrical charge difference, or **membrane potential,** exists across the plasma membrane of all cells. Due to the difference in ion concentrations between the cytoplasm and the extracellular fluid, the plasma membrane is electrically polarized.

Measuring Membrane Potentials   Electrophysiologists can measure the magnitude of the charge separation by placing microelectrodes connected to a voltmeter inside and outside a cell. The membrane potential, or **resting potential,** of a typical nontransmitting neuron is about −70 mV.

How a Cell Maintains a Membrane Potential   A membrane potential is a result of differences in the concentrations of ions on either side of the membrane and of the membrane's selective permeability to these ions. The principal cation outside cells is sodium (Na^+) and the principal anion is chloride (Cl^-). Inside cells, potassium (K^+) is the principal cation, and negatively charged proteins, amino acids, phosphate, sulfate, and other ions are grouped together and symbolized as the principal anion, A^-.

The number and type of ion channels determine a membrane's permeability to different ions. Cells are usually much more permeable to K^+ than to Na^+. The internal anions (A^-) contribute a constant internal negative charge because most are too large to cross the membrane.

The potassium concentration gradient drives K^+ out of the cell until the electrical gradient, created by the nonmoving A^- and the exiting of positive charge, starts to drive K^+ back across the membrane. The resulting equilibrium potential for potassium is about -85 mV. Despite the low permeability of the membrane to Na^+, the concentration gradient and inner negative charge will move some Na^+ into the cell, increasing the internal positive charge and raising the membrane potential of a resting neuron to about -70 mV. Sodium-potassium pumps maintain these resting concentration gradients using energy from ATP to move sodium ions back out of the cell and potassium ions into the cell.

■ INTERACTIVE QUESTION 48.2

a. What is the principle cation inside the cell?_____ outside the cell?_____

b. What is the principle anion inside the cell? _____ outside the cell?_____

c. A plasma membrane is polarized. Which side of the membrane has a negative charge?

d. What causes this charge separation, or membrane potential?

Changes in the membrane potential of a neuron give rise to nerve impulses (1028–1031)

Neurons and muscle cells are **excitable cells;** they are able to generate changes in their membrane potentials, called their **resting potentials.** A change in the polarization of the plasma membrane may generate an electrical impulse.

Neurons and some other cells have **gated ion channels** that open or close in response to stimuli, causing a change in membrane potential. **Chemically-gated ion channels** respond to a chemical stimulus, such as a neurotransmitter. **Voltage-gated ion channels** respond to a change in membrane potential.

Graded Potentials: Hyperpolarization and Depolarization Stimuli may increase or decrease the electrical gradient across the membrane. For example, should the stimulus open potassium channels, the efflux of K^+ will increase the polarization of the membrane, resulting in **hyperpolarization.** When sodium channels open and Na^+ is allowed to flow in, the inside of the cell becomes less negative and the electrical gradient is reduced, **depolarizing** the membrane. With this type of **graded potential,** the magnitude of a voltage change is proportional to the strength of the stimulus: the stronger the stimulus, the more gated ion channels that open.

The Action Potential: All-or-Nothing Depolarization Once depolarization of a typical neuron reaches a **threshold potential** of about -55 to -50 mV, an **action potential,** or nerve impulse, is triggered. A graded depolarization from a dendrite or cell body spreads along the membrane to an axon, where an action potential can be generated. The action potential is an all-or-none event, always creating the same voltage spike once the threshold potential is reached, regardless of the intensity of the stimulus. An action potential causes the membrane potential first to reverse polarity, and then return rapidly to the resting potential.

The voltage-gated ion channels in an axon's plasma membrane are activated when a graded depolarization reaches the threshold potential. Voltage-gated potassium channels open slowly in response to depolarization. Sodium activation gates open rapidly in response to depolarization. The sodium inactivation gates, which are open in the resting state, close slowly in response to depolarization. The initial massive influx of Na^+ through the sodium activation gates, while K^+ is just starting to exit through the potassium channels and the sodium inactivation gates are just starting to close, creates the rapid spike of depolarization.

A rapid repolarization follows the action potential spike. The sodium channel inactivation gates close, greatly reducing sodium permeability, and the potassium gates completely open, allowing K^+ to flow rapidly from the cell and helping return the cell to its resting-stage negativity. During the *undershoot,* the potassium channels have not yet closed, and the continued outflow of K^+ temporarily hyperpolarizes the membrane. During the **refractory period,** which occurs before the sodium inactivation gates have reopened, the neuron cannot respond to another stimulus.

Though the amplitude of an action potential is always the same, the frequency of action potentials increases with the intensity of a stimulus.

■ INTERACTIVE QUESTION 48.3

This diagram shows the changes in voltage-gated ion channels during an action potential. Label the channels and gates, ions, and five phases of the action potential. Label the axes of the graph and show where each phase occurs. Describe the ion movements associated with each phase.

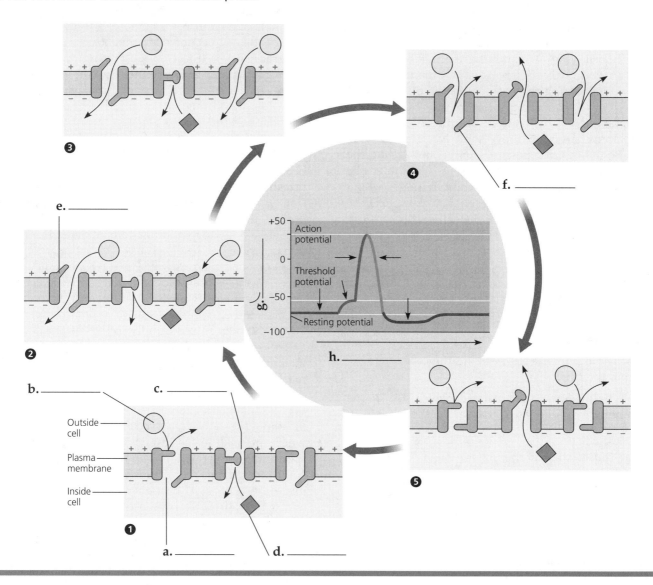

Nerve impulses propagate themselves along an axon (1031–1032)

As sodium ions move into the cell during the action potential, they depolarize adjacent sections of the membrane, bringing them to the threshold potential. Local depolarizations and action potentials across the membrane result in the propagation of serial action potentials along the length of the neuron. Because of the brief refractory period, the action potential is propagated in only one direction.

Resistance to current flow is inversely proportional to the cross-sectional area of the conducting "wire."

The greater the axon diameter, the faster action potentials are propagated. Some invertebrates, such as squid and lobsters, have giant axons that conduct impulses very rapidly.

In vertebrates, voltage-gated ion channels are concentrated in the nodes of Ranvier, small gaps between successive Schwann cells where the axon membrane has contact with extracellular fluid. Action potentials can be generated only at these nodes, and a nerve impulse "jumps" from node to node, resulting in a faster mode of transmission known as **saltatory conduction.**

Chemical or electrical communication between cells occurs at synapses (1033–1034)

Synapses can occur between two neurons, between sensory receptors and neurons, or between neurons and muscle cells or gland cells.

Electrical Synapses   Electrical synapses allow action potentials to flow directly from presynaptic to postsynaptic cells via gap junctions. Electrical synapses are found in the giant axons of some crustaceans but are less common than chemical synapses in vertebrates and most invertebrates.

Chemical Synapses   At a chemical synapse, the electrical message is converted to a chemical message that travels from the presynaptic cell across a **synaptic cleft** to the postsynaptic cell, where it is converted back into an electrical signal. A synaptic terminal contains numerous **synaptic vesicles,** in which thousands of molecules of **neurotransmitter** are stored. The depolarization of the **presynaptic membrane** opens voltage-gated calcium channels in the membrane. The influx of Ca^{2+} causes the synaptic vesicles to fuse with the presynaptic membrane and release neurotransmitter into the cleft.

The **postsynaptic membrane** contains receptor proteins associated with particular ion channels. Binding of neurotransmitter to the receptors on these chemically gated channels allows ions to cross the membrane, either depolarizing or hyperpolarizing the postsynaptic membrane. Enzymes rapidly break down the neurotransmitter, or it is taken up into adjacent cells.

■ INTERACTIVE QUESTION   48.4

Identify the components of this chemical synapse following the depolarization of the synaptic terminal.

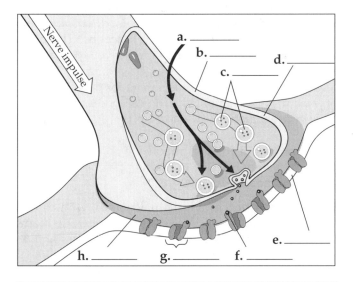

Neural integration occurs at the cellular level (1034–1036)

A neuron may receive information from many neighboring neurons at thousands of synapses, some of which are excitatory and others inhibitory. At an excitatory synapse, binding of neurotransmitter to receptors opens a chemically gated channel that allows Na^+ to flow into, and K^+ to flow out of, the cell. The electrochemical gradient drives more Na^+ into the cell. This net inflow of positive charge depolarizes the membrane, creating an **excitatory postsynaptic potential (EPSP)** and bringing the membrane potential closer to the threshold potential. At an inhibitory synapse, binding of neurotransmitter opens ion gates that allow K^+ to flow out and/or Cl^- to move into the cell, hyperpolarizing the membrane and producing an **inhibitory postsynaptic potential (IPSP).**

EPSPs and IPSPs are graded potentials. Their magnitude depends on the number of neurotransmitter molecules that bind to receptors, and they spread their voltage changes along the membrane of the postsynaptic cell. **Summation** of several EPSPs is usually necessary to bring the axon hillock to threshold potential. The membrane potential of the axon hillock at any given time is determined by the sum of all EPSPs and IPSPs.

■ INTERACTIVE QUESTION   48.5

a. _____summation occurs with repeated release of neurotransmitter from one or more synaptic terminals before the postsynaptic potential returns to its resting potential.

b. _____summation occurs when several different presynaptic terminals, usually from different neurons, release neurotransmitter simultaneously.

The same neurotransmitter can produce different effects on different types of cells (1036–1038)

Dozens of molecules have been identified as neurotransmitters. Many neurotransmitters alter the membrane permeability of the postsynaptic cell by binding to receptors on ion channels; others trigger signal-transduction pathways in the postsynaptic cell.

Acetylcholine   **Acetylcholine** is a very common neurotransmitter in invertebrates and vertebrates. Depending on the type of receptor, it can be inhibitory or excitatory in the vertebrate CNS. In neuromuscular junctions, acetylcholine released from a motor axon depolarizes the postsynaptic muscle cell.

Biogenic Amines   **Biogenic amines,** neurotransmitters derived from amino acids, often trigger signal-transduction pathways in postsynaptic cells. **Epinephrine, norepinephrine,** and **dopamine** are derived from tyrosine. **Serotonin** (synthesized from tryptophan) and dopamine affect sleep, mood, attention, and learning. Imbalances of these neurotransmitters have been associated with several disorders.

Other Chemical Neurotransmitters   The amino acids **glycine, glutamate, aspartate,** and **gamma aminobutyric acid (GABA)** function as neurotransmitters in the CNS. GABA is the most common inhibitory transmitter in the brain. **Neuropeptides** are short chains of amino acids that function as neurotransmitters, often through signal-transduction pathways. **Substance P** is an excitatory neurotransmitter that functions in pain perception. **Endorphins** are neuropeptides produced in the brain during physical or emotional stress that have painkilling and other functions. Opiates bind to endorphin receptors in the brain.

Gaseous Signals of the Nervous System   Neurons of the vertebrate CNS and PNS use nitric oxide (NO) and carbon monoxide as local regulators. During sexual arousal in human males, NO released by neurons in the penis cause blood vessels to dilate, filling spongy erectile tissue with blood. Some neurons signal endothelial cells in blood vessels to synthesize and release NO, triggering relaxation of smooth muscle cells and vessel dilation.

■ **INTERACTIVE QUESTION   48.6**

Acetylcholine stimulates skeletal muscle contraction but inhibits or slows cardiac muscle contraction. How can this neurotransmitter have such opposite effects?

Evolution and Diversity of Nervous Systems

The ability of cells to respond to the environment has evolved over billions of years (1038)

Mechanisms to sense and respond to changes in the environment evolved in prokaryotes. Modification of this ability to respond to environmental chemicals allowed for the development of chemical communication between the cells of multicellular organisms. The cells and signaling mechanisms of nervous systems were present over 600 million years ago.

Nervous systems show diverse patterns of organization (1038–1039)

A simple diffuse **nerve net** controls the radially symmetrical body cavity in cnidarians such as hydras. Echinoderms have a central nerve ring with radial nerves connected to a nerve net in each arm.

Cephalization, the concentration of sense organs, feeding structures, and neural structures in the head, evolved in bilaterally symmetrical animals. Flatworms have a central nervous system consisting of a simple brain and two longitudinal **nerve cords.** Mollusks and arthropods have a complicated brain and a ventral nerve cord with segmentally arranged ganglia. Vertebrates have a dorsal nerve cord.

■ **INTERACTIVE QUESTION   48.7**

The organization of an organism's nervous system correlates with how the animal interacts with the environment. How do the different nervous systems of clams and squid illustrate this generalization?

Vertebrate Nervous Systems

Vertebrate nervous systems have central and peripheral components (1040)

All vertebrate nervous systems have a high degree of cephalization. The CNS consists of the spinal cord and brain. The PNS carries information to and from the CNS and regulates homeostasis.

The CNS develops from the embryonic dorsal hollow nerve cord. Spaces in the brain called **ventricles** are continuous with the narrow **central canal** of the spinal cord, and all are filled with **cerebrospinal fluid.** This fluid, formed by filtration of the blood, cushions the brain and carries out circulatory functions. The brain and spinal cord are covered by protective layers of connective tissue called the meninges. Axons are in bundles or tracts; **white matter** is named for the white color of their myelin sheaths. Neuron cell bodies, dendrites, and unmyelinated axons make up the **gray matter.**

The divisions of the peripheral nervous system interact in maintaining homeostasis (1040–1042)

Paired **cranial nerves** and **spinal nerves,** with associated ganglia, make up the vertebrate PNS. The mammalian PNS contains 12 pairs of cranial nerves and 31 pairs of spinal nerves. Most cranial and all spinal nerves contain both sensory and motor neurons.

Functionally, the peripheral nervous system consists of the **sensory division** with sensory, or afferent, neurons that bring information to the CNS, and the **motor division** with efferent neurons that carry signals away from the CNS to effector cells. The motor division has two parts: The **somatic nervous system** carries signals to skeletal muscles, the movement of which is largely under conscious control. The **autonomic nervous system** maintains the internal environment by its involuntary control over smooth and cardiac muscles and various organs.

The autonomic nervous system is also subdivided: The **sympathetic division** accelerates the heart and metabolic rate, arousing an organism for action and generating energy. The **parasympathetic division** carries signals that enhance self-maintenance activities that conserve energy, such as digestion and slowing the heart rate.

Embryonic development of the vertebrate brain reflects its evolution from three anterior bulges of the neural tube (1042–1043)

As the neural tube differentiates during embryonic development, the **forebrain, midbrain,** and **hindbrain** are evident as three bilaterally symmetrical, anterior bulges. In birds and mammals, the forebrain becomes quite large. The most sophisticated integrative center,

the **cerebrum,** is an outgrowth of the forebrain. In humans, five brain regions develop from these initial bulges: the telencephalon and diencephalon from the forebrain, the mesencephalon from the midbrain, and the metencephalon and myelencephalon from the hindbrain.

During the second and third months of human development, the cerebrum expands to form the right and left cerebral hemispheres, extending over many of the other brain regions. The **cerebral cortex** is the highly convoluted outer gray matter. The cerebral hemispheres also contain white matter and internal gray matter (nuclei). The thalamus, epithalamus, and hypothalamus develop from the diencephalon portion of the forebrain.

The adult brainstem includes the midbrain (from the mesencephalon), the pons (from the metencephalon), and the medulla oblongata (from the myelencephalon). The cerebellum, which is not part of the brainstem, develops from the metencephalon.

Evolutionarily older structures of the vertebrate brain regulate essential automatic and integrative functions (1043–1046)

The Brainstem The brainstem is a stalk with a cap-like swelling at the end of the spinal cord that extends deep within the brain. The **medulla oblongata,** or

---

■ INTERACTIVE QUESTION 48.8

Complete this concept map to help you learn the functional organization of the vertebrate nervous system.

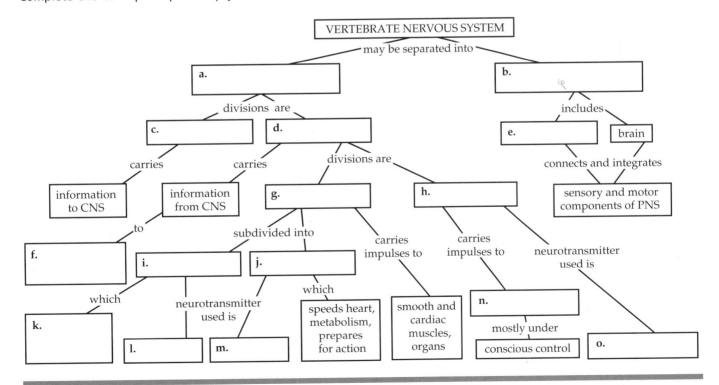

medulla, contains control centers for such homeostatic functions as breathing, swallowing, heart and blood vessel actions, and digestion. The **pons** functions with the medulla in some of these activities and in conducting information between the rest of the brain and the spinal cord. The tracts of motor neurons from the mid- and forebrain cross in the medulla, so that the right side of the brain controls much of the movement of the left side of the body, and vice versa.

The **midbrain** receives and integrates sensory information and sends this information to specific regions of the forebrain. Fibers involved in hearing pass through or terminate in the inferior colliculi. The superior colliculi form prominent optic lobes in nonmammalian vertebrates but only coordinate visual reflexes in mammals, in which vision is integrated in the cerebrum.

Arousal is a state in which an individual is aware of the external world, whereas sleep is a state in which the individual is not conscious of external stimuli. The **reticular formation** is a system of over 90 nuclei that extends through the brain stem and filters the sensory information reaching the cerebral cortex. The reticular activating system (RAS) is a part of the filtering system and regulates sleep and arousal. The level of arousal relates to the amount of input the cortex receives. Sleep-producing nuclei are located in the pons and medulla, and serotonin may be the neurotransmitter involved in these areas. A center that causes arousal is found in the midbrain.

Patterns in the electrical activity of the brain can be recorded by an **electroencephalogram,** or **EEG.** Slow, synchronous alpha waves are produced by a person lying quietly with eyes closed. Rapid, irregular beta waves are associated with opened eyes or thinking about a complex problem. Quite slow and highly synchronized delta waves occur during deep sleep. Periods of delta waves alternate with periods of a desynchronized EEG and rapid eye movements, called REM sleep, when most dreaming occurs.

All birds and mammals have a characteristic sleep-wake cycle. Sleep may be important in learning and memory.

The Cerebellum   The **cerebellum,** which develops from part of the metencephalon, may be involved in learning motor responses. The cerebellum integrates information from the auditory and visual systems with sensory input from the joints and muscles as well as motor pathways from the cerebrum to provide automatic coordination of movements and balance.

The Thalamus and Hypothalamus   The epithalamus, thalamus, and hypothalamus all develop from the embryonic diencephalon. The **epithalamus** includes the projecting pineal gland and a choroid plexus, one of several clusters of capillaries producing cerebrospinal fluid. The **thalamus** is a major input center for sensory

information going to the cerebrum and an output center for motor information from the cerebrum. It also receives input from the cerebrum and other parts of the brain regulating emotion and arousal.

The **hypothalamus** is the major brain region for homeostatic regulation. It produces the posterior pituitary hormones and the releasing hormones that control the anterior pituitary. The hypothalamus contains the regulating centers for many autonomic functions and also plays a role in sexual and mating behaviors, the alarm response, and pleasure.

Studies of daily, regular behaviors show that an internal mechanism, called the **biological clock,** is important for maintaining these circadian rhythms. In mammals, the **suprachiasmatic nuclei (SCN),** located in the hypothalamus, function as the biological clock. Sensory neurons provide visual information to the SCN to keep the mammalian clock synchronized with natural cycles of light and dark. Humans kept in free-running conditions have about a 25-hour cycle.

The cerebrum is the most highly evolved structure of the mammalian brain (1046–1047)

The cerebrum develops from the embryonic telencephalon. It is divided into right and left **cerebral hemispheres,** each with an outer covering of gray matter, called the cerebral cortex, and inner white matter. The **basal nuclei** (also called basal ganglia), located deep in the white matter, are important in planning and learning movements.

As the largest and most complex part of the mammalian brain, the cerebral cortex has changed the most during vertebrate evolution. The **neocortex,** an outer six layers of neurons running along the brain surface, is unique to mammals. Convolutions increase its surface area, and the size of the neocortex correlates with sophisticated behavior and greater cognitive ability.

The right side of the cortex receives information from and controls the movement of the left side of the body, and vice versa. Communication between the two hemispheres travels through the **corpus callosum,** a thick band of fibers.

Cognitive functions of the cerebrum include learning and cognition—the process of knowing, consciousness, and decision making.

Regions of the cerebrum are specialized for different functions (1047–1051)

The surface of each hemisphere is divided into four lobes with specialized functions localized in each lobe. Found at the boundary between the frontal and parietal lobe of the cerebral cortex, the primary motor cortex sends signals to skeletal muscles and the primary somatosensory cortex receives and partially integrates input from touch, pain, pressure, and temperature receptors. The proportion of somatosensory and motor cortex devoted to controlling each part of the body is correlated with the importance of that area.

■ INTERACTIVE QUESTION 48.9

Identify the structures of the human brain. Match the functions listed below to these structures.

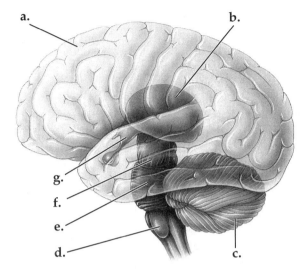

Structure	Function
a. _____	_____
b. _____	_____
c. _____	_____
d. _____	_____
e. _____	_____
f. _____	_____

Functions

1. Coordination, balance, movement
2. Aids medulla in some functions, conducts information between brain and spinal cord
3. Screens and relays information to cerebrum
4. Regulates breathing, heart rate, digestion
5. Integrates sensory and motor information, thinking
6. Produces hormones, homeostatic regulation
7. Sends sensory information to forebrain, contains nuclei involved in hearing and vision

---

Integrative Function of the Association Areas Sensory information (visual, auditory, somatosensory) first goes to primary sensory areas located in different lobes of the cortex. Olfactory information is routed to "primitive" cortical regions, then, via the thalamus, to the interior frontal lobe. Association areas adjacent to the primary sensory areas integrate ("associate") the different sensory inputs and send signals to association areas in the frontal lobes. These areas determine appropriate motor response plans, which the primary motor cortex directs to the skeletal muscles.

During mammalian evolution, the neocortex has increased in size, with the expansion and regional specialization of association areas facilitating complex behavior and learning.

Lateralization of Brain Function As the brain of a child develops, functions become segregated in opposite cerebral hemispheres. The left hemisphere becomes specialized for language, math, and logic, whereas the right hemisphere is involved in pattern recognition, spatial perception, and emotional processing. The left hemisphere is better at processing fine visual and auditory data and directing detailed movement. The right hemisphere seems to specialize in perceiving images within the whole context.

Language and Speech Studies of brain lesions and imaging of brain activity using PET (positron emission tomography) show that several association areas in different lobes are involved in understanding and generating language. Broca's area in the frontal lobe is involved in speech generation; Wernicke's area in the posterior temporal lobe is involved in speech comprehension. Researchers studying the activation of brain areas during different tasks have found that novel situations cause the greatest mobilization of brain areas and that practice allows activities to be accomplished with a lower level of brain activity.

Emotions The hippocampus and olfactory cortex (both common to the reptilian and mammalian cortex), along with sections of the thalamus, hypothalamus, and inner portions of the cortex, form a ring around the brain stem called the **limbic system.** This system interacts with sensory and other areas of the neocortex in generating emotions, both primary emotions, such as laughing and crying, and emotional feelings associated with the survival-related functions of the brain stem, such as feeding and sexuality.

Emotional brain circuits of the limbic system form early in development; thus infants can bond to a caretaker, recognize elements of a human face, and express fear, distress, and anger. The amygdala, a nucleus in the temporal lobe, functions in laying down emotional memories. This emotional memory system appears to develop earlier than the system that allows recall of events, which involves the hippocampus. The association of primary emotions with different situations involves portions of the neocortex, particularly the frontal lobes.

Memory and Learning   Human memory consists of **short-term memory,** the immediate perception of an object or idea, and **long-term memory,** the retention of this information for later recall. Transferring information from short-term to long-term memory involves the hippocampus and is facilitated by rehearsal, a positive or negative emotional state, and associations with previously learned and stored information. Various association areas of the cortex are involved with storing and retrieving words and images.

Remembering facts and numbers may involve rapid changes in the strength of existing nerve connections. Learning and remembering skills and procedures seems to involve nerve cells making new connections. Skill memories are formed by repetition of motor activities and do not require conscious recall of specific steps.

The storage of memories and emotions seems to be related to functional changes in synapses in the hippocampus and amygdala. Repeated, weak stimulation may lead to a decrease in responsiveness to an action potential by a postsynaptic cell, a change called **long-term depression (LTD). In long-term potentiation (LTP),** a postsynaptic cell has an enhanced response to a single action potential as a result of repeated bursts of action potentials from a presynaptic cell that cause a strong depolarization of the postsynaptic membrane. The presynaptic cell releases the excitatory neurotransmitter glutamate, which binds with special receptors in the postsynaptic membrane, opening gated channels for Ca^{2+}. The influx of Ca^{2+} triggers a cascade of changes that contribute to LTP.

Human Consciousness   The study of human consciousness has been facilitated by brain-imaging techniques but remains a challenging endeavor. A central hypothesis is that consciousness is an emergent property of the brain that perhaps involves a repetitive scanning mechanism of multiple areas of the brain to produce a unified conscious moment.

▪ INTERACTIVE QUESTION 48.10

How would you define human consciousness?

Research on neuron development and neural stem cells may lead to new approaches for treating CNS injuries and diseases (1051–1053)

The mammalian central nervous system cannot repair itself when brain cells are destroyed by injury or disease, although surviving brain cells can make new connections and sometimes compensate for damage.

Nerve Cell Development   New research in neurobiology centers on the differentiation and migration of neurons, the growth of axons, and the formation of synapses with the proper target cells. The **growth cone** at the leading edge of an axon appears to respond to signal molecules released from cells along the growth path. These signals may act as attractants or repellents to direct axonal growth. Cell adhesion molecules (CAMs) on the growth cone also interact with CAMs on surrounding cells that provide tracks for the axon to follow. Growth is stimulated by nerve growth factor released by astrocytes and by growth-promoting proteins produced by the axon. Because this process takes place early in development, it may be difficult to replicate in damaged nervous systems. The genes and basic mechanism of this process are quite similar in nematodes, insects, and vertebrates and have apparently been conserved during evolution.

Neural Stem Cells   In 1998, F. Gage and P. Eriksson reported finding newly formed nerve cells in the hippocampus of cancer victims who had been treated with a chemical marker that labels the DNA of dividing cells. In 2001, Gage and his colleagues reported that they had cultured neural progenitor cells removed from brain tissue samples. These cells were not as developmentally flexible as embryonic stem cells, but were able to differentiate into neurons and astrocytes. A goal of future research is to find ways to induce neural stem cells to differentiate into glia or neurons when needed, and to culture neural stem cells for transplantation into damaged neurological tissue.

WORD ROOTS

astro- = a star; **-cyte** = cell (*astrocytes:* glial cells that provide structural and metabolic support for neurons)

auto- = self (*autonomic nervous system:* a subdivision of the motor nervous system of vertebrates that regulates the internal environment)

bio- = life; -genic = producing (*biogenic amines:* neurotransmitters derived from amino acids)

cephalo- = head (*cephalization:* the clustering of sensory neurons and other nerve cells to form a small brain near the anterior end and mouth of animals with elongated, bilaterally symmetrical bodies)

dendro- = tree (*dendrite:* one of usually numerous, short, highly branched processes of a neuron that conveys nerve impulses toward the cell body)

de- = down, out (*depolarization:* an electrical state in an excitable cell whereby the inside of the cell is made less negative relative to the outside)

electro- = electricity; **-encephalo** = brain (*electroencephalogram:* a medical test that measures different patterns in the electrical activity of the brain)

endo- = within (*endorphin:* a hormone produced in the brain and anterior pituitary that inhibits pain perception)

epi- = above, over (*epithalamus:* a brain region, derived from the diencephalon, that contains several clusters of capillaries that produce cerebrospinal fluid; it is located above the thalamus)

glia = glue (*glia:* supporting cells that are essential for the structural integrity of the nervous system and for the normal functioning of neurons)

hyper- = over, above, excessive (*hyperpolarization:* an electrical state whereby the inside of the cell is made more negative relative to the outside than at the resting membrane potential)

hypo- = below (*hypothalamus:* the ventral part of the vertebrate forebrain that functions in maintaining homeostasis, especially in coordinating the endocrine and nervous systems; it is located below the thalamus)

inter- = between (*interneurons:* an association neuron; a nerve cell within the central nervous system that forms synapses with sensory and motor neurons and integrates sensory input and motor output)

neuro- = nerve; **trans-** = across (*neurotransmitter:* a chemical messenger released from the synaptic terminal of a neuron at a chemical synapse that diffuses across the synaptic cleft and binds to and stimulates the postsynaptic cell)

oligo- = few, small (*oligodendrocytes:* glial cells that form insulating myelin sheaths around the axons of neurons in the central nervous system)

para- = near (*parasympathetic division:* one of two divisions of the autonomic nervous system)

post- = after (*postsynaptic cell:* the target cell at a synapse)

pre- = before (*presynaptic cell:* the transmitting cell at a synapse)

salta- = leap (*saltatory conduction:* rapid transmission of a nerve impulse along an axon resulting from the action potential jumping from one node of Ranvier to another, skipping the myelin-sheathed regions of membrane)

soma- = body (*somatic nervous system:* the branch of the motor division of the vertebrate peripheral nervous system composed of motor neurons that carry signals to skeletal muscles in response to external stimuli)

supra- = above, over (*suprachiasmatic nuclei:* a pair of structures in the hypothalamus of mammals that functions as a biological clock)

syn- = together (*synapse:* the locus where a neuron communicates with a postsynaptic cell in a neural pathway)

STRUCTURE YOUR KNOWLEDGE

1. Develop a flow chart, diagram, or description of the sequence of events in the creation and propagation of an action potential and in the transmission of this potential across a chemical synapse.

2. List the location and functions of these important brain nuclei or functional systems.
   a. reticular formation
   b. suprachiasmatic nucleus
   c. basal nuclei
   d. limbic system
   e. amygdala and hippocampus

TEST YOUR KNOWLEDGE

MULTIPLE CHOICE: *Choose the one best answer.*

1. Which of the following does *not* distinguish the nervous system from the endocrine system?
   a. Transmission of nervous impulses is more rapid.
   b. The nervous system uses chemical communication.
   c. Nervous system messages are delivered directly to target cells or organs.
   d. The structural complexity of the nervous system allows for integration of more information and responses.
   e. The response of the body to nervous communication is more rapid.

2. Interneurons
   a. may connect sensory and motor neurons.
   b. are confined to the PNS.
   c. are confined to the spinal cord.
   d. are electrical synapses between neurons.
   e. do not have cell bodies.

3. Nodes of Ranvier are
   a. gaps where Schwann cells abut and at which action potentials are generated.
   b. neurotransmitter-containing vesicles located in the synaptic terminals.
   c. the major components of the blood-brain barrier that restrict the passage of substances into the brain.
   d. clusters of receptor proteins located on the postsynaptic membrane.
   e. ganglia adjacent to the spinal cord.

4. Which of the following is *not* true of the resting potential of a typical neuron?
   a. The inside of the cell is more negative than is the outside.
   b. There are concentration gradients with more sodium outside the cell and a higher potassium concentration inside the cell.
   c. It is about -70 mV and can be measured by using microelectrodes placed inside and outside the cell.
   d. It is formed by the sequential opening of voltage-gated channels.
   e. It results from the combined equilibrium potentials of potassium and sodium.

5. After the depolarization of an action potential, repolarization occurs due to the
   a. closing of sodium activation and inactivation gates.
   b. opening of sodium activation gates.
   c. refractory period in which the membrane is hyperpolarized.
   d. delay in the action of the sodium-potassium pump.
   e. opening of voltage-gated potassium channels and the closing of sodium channels.

6. The threshold potential of a membrane
   a. is a positive value, often equal to about 35 mV.
   b. opens voltage-gated channels and permits the rapid outflow of sodium ions.
   c. is the depolarization that is needed to generate an action potential.
   d. is a graded potential that is proportional to the strength of a stimulus.
   e. is an all-or-none event.

7. Which of the following is *not* true of chemical synapses?
   a. Synaptic terminals at the ends of branching axons contain synaptic vesicles, which enclose the neurotransmitter.
   b. The influx of sodium when an action potential reaches the presynaptic membrane causes synaptic vesicles to release their neurotransmitter into the cleft.
   c. The binding of neurotransmitter to receptors on the postsynaptic membrane changes the membrane's permeability to certain ions.
   d. An excitatory postsynaptic potential forms when sodium channels open and the membrane potential moves closer to an action potential threshold.
   e. Neurotransmitter is rapidly degraded in the synaptic cleft.

8. An inhibitory postsynaptic potential occurs when
   a. sodium flows into the postsynaptic cell.
   b. enzymes do not break down the neurotransmitter in the synaptic cleft.
   c. binding of the neurotransmitter opens ion gates that result in the membrane becoming hyperpolarized.
   d. acetylcholine is the neurotransmitter.
   e. norepinephrine is the neurotransmitter.

9. Which of the following do (does) *not* function as a neurotransmitter or local regulator released by neurons?
   a. amino acids such as glutamate
   b. neuropeptides such as endorphin
   c. biogenic amines such as dopamine
   d. steroids
   e. NO

10. Which of the following animals is mismatched with its nervous system?
    a. sea star (echinoderm)—modified nerve net, central nerve ring with radial nerves
    b. hydra (cnidarian)—ring of ganglia, paired ventral nerve cords
    c. annelid worm—brain, ventral nerve cord with segmental ganglia
    d. vertebrate—dorsal central nervous system of brain and spinal cord and peripheral nervous system
    e. squid (mollusk)—large complex brain, ventral nerve cord, some giant axons

11. Which of the following is *not* true of the autonomic nervous system?
    a. It is a subdivision of the somatic nervous system.
    b. It consists of the sympathetic and parasympathetic divisions.
    c. It is part of the peripheral nervous system.
    d. It controls smooth and cardiac muscles.
    e. It is part of the motor division.

12. If you needed to obtain nuclei from the cell bodies of neurons for an experiment, you would want to make preparations of
    a. astrocytes of the brain.
    b. the gray matter of the brain.
    c. the inner portion of the spinal cord.
    d. sympathetic ganglia near the spinal cord.
    e. either b, c, or d.

13. The superior and inferior colliculi
    a. control biorhythms and are found in the thalamus.
    b. are the part of the limbic system found in the midbrain.

c. are located in the parietal and frontal lobe, respectively, and involved in language and speech.

d. are nuclei in the midbrain involved in hearing and vision.

e. are found in the inner cortex and are involved in memory storage.

14. Which of the following structures is *incorrectly* paired with its function?
   a. pons—conducts information between spinal cord and brain
   b. cerebellum—contains the tracts that cross motor neurons from one side of the brain to the other side of the body
   c. thalamus—screens and relays incoming impulses to the cerebrum
   d. corpus callosum—band of fibers connecting left and right hemispheres
   e. hypothalamus—homeostatic regulation, pleasure centers

15. REM sleep
   a. is characterized by a highly synchronized EEG.
   b. occurs when the reticular activating system is waking us up.
   c. is deep sleep from which one cannot easily be awakened.
   d. is when dreaming occurs.
   e. is characterized by rapid eye movements and slow and synchronized delta waves.

16. Long-term potentiation seems to be involved in
   a. memory storage and learning.
   b. enhanced response of a presynaptic cell resulting from strong depolarization from bursts of action potentials.
   c. the attachment of emotion to information to be learned.
   d. the release of the excitatory neurotransmitter glutamate that opens sodium-gated channels.
   e. all of the above.

17. What is the main effect of the neurotransmitter GABA in the CNS?
   a. stimulate instinctual behavior
   b. create inhibitory postsynaptic potentials
   c. create excitatory postsynaptic potentials
   d. induce sleep
   e. decrease pain and induce euphoria

18. The function of motor neurons is to carry impulses
   a. from the spinal cord to the brain.
   b. from the cerebellum to the cerebrum.
   c. from sensory receptors to the spinal cord.
   d. from the CNS to effectors (muscles and organs).
   e. between interneurons and the PNS.

19. Which of the following is *incorrectly* paired with its function?
   a. axon hillock—region of neuron where action potential originates
   b. Schwann cells—create myelin sheath around axon in PNS
   c. synapse—space between presynaptic and postsynaptic cell into which neurotransmitter is released
   d. synaptic terminal—receptor that is part of ion channel that is keyed to specific neurotransmitter
   e. dendrite—carry nerve impulse toward neuron cell body

20. If the binding of a neurotransmitter to its receptor opens Cl^- channels, what would be the effect on the postsynaptic cell?
   a. It would hyperpolarize, receiving an IPSP.
   b. It would reach threshold and an action potential would fire.
   c. It would depolarize and form an EPSP, but probably not fire an action potential.
   d. It would initiate a signal-transduction pathway.
   e. Its membrane potential would not change because neither sodium nor potassium channels opened.

21. Movement of an action potential in only one direction along a neuron is a function of
   a. saltatory conduction.
   b. the pathway from dendrite to axon.
   c. the refractory period when sodium inactivation gates are still closed.
   d. the localized depolarization of the surrounding membrane.
   e. the formation of a threshold potential, which creates an all-or-none firing.

22. What makes up the white matter of the spinal cord?
   a. the meninges
   b. myelinated sheaths of axons
   c. cerebrospinal fluid
   d. motor and interneuron cell bodies
   e. nodes of Ranvier

23. Why is signal transmission faster in myelinated axons?
   a. These axons are thinner, and there is less resistance to the voltage flow.
   b. These axons use electrical synapses rather than chemical synapses.
   c. The action potential can jump from node to node along the insulating myelin sheath.
   d. These axons are thicker and provide less resistance to voltage flow.
   e. These axons have higher depolarization values than do unmyelinated axons.

24. How is an increase in the strength of a stimulus communicated by a neuron?
    a. The spike of the action potential reaches a higher voltage.
    b. The frequency of action potentials generated along the neuron increases.
    c. The length of an action potential (the duration of the depolarization phase) increases.
    d. The action potential travels along the neuron faster.
    e. All action potentials are the same; the nervous system cannot discriminate between different strengths of stimuli.

25. Lateralization of the cerebrum refers to the fact that
    a. the left hemisphere usually controls verbal and analytic ability, while the right hemisphere is specialized for emotional processing and spatial relations.
    b. the right side of the brain controls the left side of the body, and vice versa.
    c. the primary somatosensory cortex receives sensory information that is mapped onto one region of the cerebral cortex, while the primary motor cortex localized in another region controls motor output to the body.
    d. visceral functions are controlled by the hindbrain, whereas thinking and language are centered in association areas of the cortex.
    e. memory formation involves circuits that include the limbic system, hippocampus, and association areas of the cortex.

26. Which of the following neurotransmitters is released by neurons from the sympathetic division of the autonomic nervous system at synapses with smooth muscle in the stomach?
    a. GABA
    b. Ca^{2+}
    c. acetylcholine
    d. substance P
    e. norepinephrine

27. The release of the neurotransmitter glutamate opens Ca^{2+} channels in neurons, activating enzymes that change the postsynaptic cell so that it is more sensitive to future stimulation. What phenomenon does this describe?
    a. an emotion
    b. a spinal reflex
    c. short-term memory
    d. sleep
    e. long-term potentiation

28. A function of the reticular formation is to
    a. keep you awake during lecture.
    b. coordinate movement and balance.
    c. produce emotions.
    d. control visceral functions such as breathing and heart rate.
    e. create consciousness.

29. When did the ability of cells to sense and respond to chemical signals evolve?
    a. during the Cambrian explosion within the various animal phyla
    b. with the development of chemical signals that control the path of axon growth
    c. in prokaryotes who were able to sense and respond to environmental chemicals
    d. in the ancestor of the first animal
    e. in the first multicellular organism

30. Which of these structures is found only in the mammalian brain?
    a. telencephalon and diencephalon
    b. optic lobes
    c. olfactory cortex
    d. neocortex
    e. hippocampus

31. Which of the following is part of the mechanism by which axons grow along a path to their proper target cell?
    a. production of nerve growth factor by astrocytes
    b. reception of signal molecules by the growth cone that either attract or repel the growing axon
    c. interaction of cell adhesion molecules on the growth cone with those of cells along the growth path
    d. activation of genes in the axon that produce growth-promoting proteins
    e. all of the above are involved in this process

32. Neural progenitor cells
    a. were first cultured in 1998 from the hippocampus of some cancer victims.
    b. retain the ability to divide and differentiate into both neurons and astrocytes.
    c. are similar to embryonic stem cells in that they can be coaxed to differentiate into neurons, blood stem cells, and astrocytes.
    d. have been transplanted into individuals with spinal cord injuries and were able to grow and make connections with the proper target cells.
    e. are the first cells formed as the neural tube rolls up, and they give rise to both glia and neurons.

CHAPTER 49

SENSORY AND MOTOR MECHANISMS

FRAMEWORK

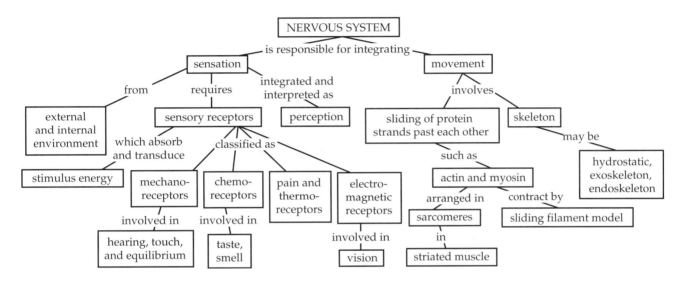

CHAPTER REVIEW

Sensing, Acting, and Brains

The brain's processing of sensory input and motor output is cyclical rather than linear (1058)

In a continuous cycle, animals are constantly moving, exploring and sensing their environment, and processing that information in the brain, which then directs their next movements. The processing of sensations received from sensory receptors is a very complex process, with signals first going to the thalamus, which selectively sends them on to the many parts of the brain that contribute to the formation of perceptions. The brain also undergoes a complex decision-making process in response to those perceptions to generate the next motor actions.

Introduction to Sensory Reception

Information is transmitted in the nervous system as action potentials. **Sensations,** or impulses arriving

from sensory neurons, are routed to different parts of the brain that interpret them, producing **perceptions** of various stimuli.

Sensory receptors transduce stimulus energy and transmit signals to the nervous system (1059–1060)

Sensory reception is the detection of energy from a particular stimulus by sensory cells. **Sensory receptors** are usually modified neurons or epithelial cells that occur singly or in groups within sensory organs. Sensory receptors may be **exteroreceptors** or **interoreceptors,** which detect stimuli from the external or internal environment, respectively.

Sensory Transduction   **Sensory transduction** is the conversion of stimulus energy into a **receptor potential,** a graded change in the membrane potential of a receptor cell that is proportional to the strength of the stimulus.

Amplification   The stimulus energy may need to undergo **amplification,** either by accessory structures of sense organs or as part of the transduction process.

Transmission   **Transmission** of the receptor potential to the central nervous system may occur either as a result of the generation of an action potential when the sensory receptor is a sensory neuron, or by the release of neurotransmitter from a receptor cell into a synapse with a sensory neuron, which then may transmit an action potential. The strength of the stimulus and the receptor potential correlates with the frequency of action potentials or the quantity of neurotransmitter released, which then translates into the frequency of action potentials in the sensory neuron.

Integration   **Integration,** or processing, of sensory information begins with the summation of graded potentials from receptors. Ongoing stimulation may lead to **sensory adaptation** in which continuous stimulation results in a decline in sensitivity of the receptor cell. A receptor cell's threshold for firing may vary with conditions. Sensory information is further integrated by complex receptors and the CNS.

■ INTERACTIVE QUESTION   49.1

List the four functions common to all receptor cells as they absorb stimulus energy and transmit it to the nervous system.

a.

b.

c.

d.

Sensory receptors are categorized by the type of energy they transduce (1060–1063)

Mechanoreceptors respond to the mechanical energy of pressure, touch, stretch, motion, and sound. Bending or stretching of the mechanoreceptor cell membrane increases its permeability to sodium and potassium ions, creating a receptor potential.

In humans, modified dendrites of sensory neurons function as touch receptors. The length of skeletal muscles is monitored by **muscle spindles,** stretch receptors that are stimulated by stretching of the muscles.

Hair cells, mechanoreceptors that detect motion, are found in the vertebrate ear and lateral line organs of fishes. When motion produces bending in the cilia or microvilli projecting from a hair cell, ion permeabilities

either increase or decrease, and the rate of action potential firing changes.

Pain receptors in humans are naked dendrites in the epidermis called **nociceptors.** Different groups of receptors respond to excess heat, pressure, or chemicals released by injured cells. Histamines and acids stimulate pain receptors, and prostaglandins increase pain by sensitizing receptors.

Thermoreceptors in the skin and hypothalamus respond to heat or cold and help to regulate body temperature.

Chemoreceptors include both general receptors that monitor the total solute concentration and specific receptors that respond to individual kinds of important molecules. **Gustatory** (taste) and **olfactory** (smell) **receptors** respond to groups of related chemicals.

Electromagnetic receptors respond to various forms of electromagnetic energy. **Photoreceptors** detect visible light and are often organized into eyes. Some animals have receptors that detect infrared rays, electric currents, and magnetic fields.

Photoreceptors and Vision

Despite the diversity of photoreceptors that have evolved in the animal kingdom, all contain light-absorbing pigment molecules, and may be homologous with a common genetic basis.

A diversity of photoreceptors has evolved among invertebrates (1063–1064)

The **eye cup** of planarians, which detects light intensity and direction, consists of receptor cells within a cup formed from darkly pigmented cells. The brain compares impulses from the left and right eye cups to help the animal navigate a direct path away from a light source.

The **compound eye** of insects and crustaceans contains up to thousands of light detectors called **ommatidia,** each with its own lens. The different intensities of light entering the many ommatidia produce a mosaic image. The compound eyes of insects are adept at detecting movement and color, and some detect ultraviolet radiation.

Some jellies, polychaetes, spiders, and many mollusks have a **single-lens eye,** in which light is focused through the single lens onto the retina containing light-transducing receptor cells.

Vertebrates have single-lens eyes (1064–1065)

The eyeball consists of a tough, outer connective tissue layer called the **sclera** and a thin, pigmented inner layer called the **choroid.** The **conjunctiva** is a mucous membrane covering the sclera. At the front of the eye, the sclera becomes the transparent **cornea,** which functions as a fixed lens, and the choroid forms the colored **iris,** which regulates the amount of light entering through the **pupil.** The **retina,** the innermost layer, contains the photoreceptor cells. The optic nerve attaches to the eye at the optic disk, forming a blind spot on the retina.

The transparent **lens** focuses an image onto the retina. The **ciliary body** produces the **aqueous humor** that fills the anterior eye cavity; jellylike **vitreous humor** fills the posterior cavity. Many fishes focus by moving the lens backward or forward. Mammals focus on close objects by **accommodation,** in which ciliary muscles contract, causing suspensory ligaments to slacken and the elastic lens to become rounder.

Rod cells and **cone cells** are the photoreceptors in the retina. The relative proportion of each of these receptors correlates with the activity pattern of the animal: Rods are more light sensitive and enable night vision, whereas cones distinguish colors. In the human eye, rods are most concentrated toward the edge of the retina, whereas the center of the visual field, the **fovea,** is filled with cones.

The light-absorbing pigment rhodopsin triggers a signal-transduction pathway (1065–1067)

Both rods and cones have visual pigments embedded in a stack of folded membranes or disks in the outer segment of each cell. **Retinal** is the light-absorbing molecule and is bonded to a membrane protein called an **opsin.**

Rods contain the visual pigment **rhodopsin.** When retinal absorbs light, it changes shape and separates from opsin. In the dark, retinal is converted back to its original shape and it recombines with opsin. In bright light, the rhodopsin remains "bleached," and cones are responsible for vision.

The light-induced dissociation of retinal changes the conformation of opsin. The altered opsin activates a membrane-bound G protein called transducin, which then activates an enzyme that alters the second messenger, cyclic guanosine monophosphate. In the dark, cGMP is bound to sodium ion channels, keeping the rod-cell membrane depolarized and releasing glutamate into its synapses with **bipolar cells.** Depending on the postsynaptic receptor molecule on the bipolar cells, this neurotransmitter excites some cells and inhibits others. The light-triggered rhodopsin signal-transduction pathway converts cGMP to GMP, closing sodium channels, hyperpolarizing the membrane, and decreasing its release of glutamate, which releases some bipolar cells from inhibition or suppresses others.

The three subclasses of red, green, and blue cones, each with its own type of opsin that binds with retinal to form a **photopsin,** are named for the color of light they are best at absorbing.

■ **INTERACTIVE QUESTION 49.2**

Label the parts of the vertebrate eye.

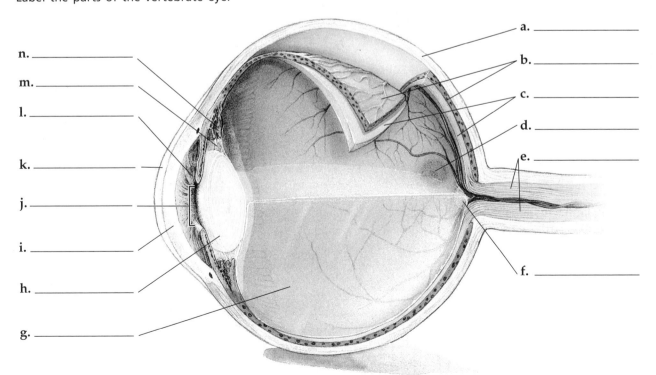

n. _____

m. _____

l. _____

k. _____

j. _____

i. _____

h. _____

g. _____

a. _____

b. _____

c. _____

d. _____

e. _____

f. _____

■ INTERACTIVE QUESTION 49.3

Summarize the effects of light on rod cells and rhodopsin by filling in the following table.

	In the Dark	In the Light
Rhodopsin	a.	b.
cGMP	c.	d.
Sodium channels	e.	f.
Rod-cell membrane	g.	h.
Neuro-transmitter	i.	j.
Bipolar cells	k.	l.

■ INTERACTIVE QUESTION 49.4

The receptive field of a ganglion cell includes the rods or cones that supply information to the bipolar cells connected to it.

a. Would a large or small receptive field produce the sharpest image?

b. Where would the smallest receptive fields be found?

The retina assists the cerebral cortex in processing visual information (1067–1069)

Rods and cones synapse with **bipolar cells** in the retina, which in turn synapse with **ganglion cells,** whose axons form the optic nerve and transmit action potentials to the brain. **Horizontal cells** and **amacrine cells** in the retina help to integrate visual information.

Signals from rods and cones may follow the vertical pathway directly from receptor cells to bipolar cells to ganglion cells. A lateral pathway involves lateral integration by horizontal cells, which carry signals from one receptor cell to others and to several bipolar cells, and by amacrine cells, which relay information from one bipolar cell to several ganglion cells. **Lateral inhibition** of nonilluminated receptors enhances contrast.

The left and right optic nerves meet at the **optic chiasma** at the base of the cerebral cortex. Information from the left visual field of both eyes travels to the right side of the brain, whereas what is sensed in the right field of view goes to the left side. Most axons of the ganglion cells go to the **lateral geniculate nuclei** of the thalamus, where neurons lead to the **primary visual cortex** in the occipital lobe of the cerebrum. Interneurons also carry information to other visual processing and integrating centers in the cortex.

Hearing and Equilibrium

In mammals and most terrestrial vertebrates, the mechanoreceptors for hearing and balance are hair cells in fluid-filled canals in the ear.

The mammalian hearing organ is within the inner ear (1069–1070)

The ear consists of three regions: The external pinna and the auditory canal make up the **outer ear.** The **tympanic membrane** (eardrum), separating the outer ear from the **middle ear,** transmits sound waves to three ossicles—the **malleus** (hammer), **incus** (anvil), and **stapes** (stirrup)—which conduct the waves to the inner ear by way of a membrane called the **oval window.** The **Eustachian tube,** connecting the pharynx and the middle ear, equalizes pressure within the middle ear. The **inner ear** is a fluid-filled labyrinth of channels within the temporal bone.

The coiled **cochlea** of the inner ear has two large fluid-filled chambers—an upper vestibular canal and a lower tympanic canal—separated by the smaller cochlear duct. The **organ of Corti,** located on the basilar membrane that forms the floor of the cochlear duct, contains receptor hair cells. Their hairs extend into the cochlear duct, and some attach to the tectorial membrane, which overhangs the organ of Corti.

Pressure waves in air—transmitted by the tympanic membrane, the three bones of the middle ear, and the oval window—produce pressure waves in the cochlear fluid, which travel from the vestibular canal through the tympanic canal and dissipate when they strike the **round window.** These pressure waves vibrate the basilar membrane, bending the hairs of its receptor cells against the tectorial membrane, which opens ion channels and allows K^+ to enter the cells. The resulting depolarization increases neurotransmitter release from the hair cells and increases the frequency of action potentials generated in their associated sensory neuron and carried through the auditory nerve to the brain.

Volume is a result of the amplitude, or height, of the sound wave; a stronger wave bends the hair cells more and results in more action potentials. **Pitch** is related to the frequency of sound waves, usually expressed in hertz (Hz). Different regions of the basilar membrane vibrate in response to different frequencies, and neurons associated with the vibrating region transmit action potentials to specific auditory regions of the cortex where the sensation is perceived as a particular pitch.

■ **INTERACTIVE QUESTION** 49.5

Label the parts in this diagram of the human ear.

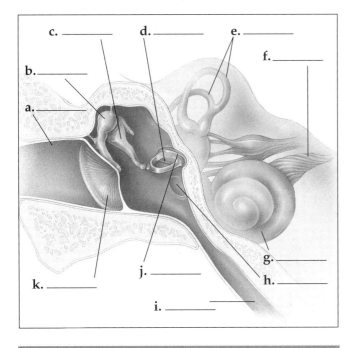

The inner ear also contains the organs of equilibrium (1070–1072)

Within the inner ear are two chambers, the **utricle** and **saccule,** and three **semicircular canals,** responsible for detection of body position and balance. Hair cells in the utricle and saccule project into a gelatinous material containing calcium carbonate particles called otoliths. These heavy particles exert a pull on the hairs in the direction of gravity; different head angles stimulate different hair cells, thereby altering the output of neurotransmitters and the resulting action potentials of sensory neurons of the vestibular nerve. Other hair cells in the swellings at the base of the semicircular canals (the ampullae) project into a gelatinous mass called the cupula and are able to sense rotational movements by the pressure of the endolymph in the semicircular canals against the cupula.

A lateral line system and inner ear detect pressure waves in most fishes and aquatic amphibians (1072–1073)

Fishes and aquatic amphibians have inner ears, consisting of a saccule, utricle, and semicircular canals, within which sensory hairs are stimulated by the movement of otoliths. Sound-wave vibrations pass through the skeleton of the head to the inner ear. Some fishes have a Weberian apparatus, a series of bones that transmits vibrations from the swim bladder to the inner ear.

Mechanoreceptor units called neuromasts—clusters of hair cells with sensory hairs embedded in a gelatinous cap or cupula—are contained in the **lateral line system** of fishes and aquatic amphibians. Water moving through a tube along both sides of the body bends the cupula and stimulates the hair cells, enabling the fish to perceive its own movement, water currents, sounds of low frequency, and the pressure waves generated by other moving objects.

The inner ear is the organ of hearing and equilibrium in terrestrial vertebrates. In amphibians, sound vibrations are conducted by a tympanic membrane on the body surface and a single middle ear bone to the inner ear. A cochlea has evolved in birds and mammals.

Many invertebrates have gravity sensors and are sound-sensitive (1073)

Most invertebrates have **statocysts,** which often consist of a layer of hair cells around a chamber containing **statoliths,** dense granules or grains of sand. The stimulation of the hair cells under the statoliths provides positional information to the animal.

Many insects sense sounds with body hairs that vibrate in response to sound waves of specific frequencies. Localized "ears" are found on many insects, consisting of a tympanic membrane with attached receptor cells stretched over an internal air chamber.

■ **INTERACTIVE QUESTION** 49.6

What organs in mammals, fishes, and invertebrates are used to sense movement or equilibrium? How are these organs similar?

a. Mammals:

b. Fishes:

c. Invertebrates:

d. Similarities:

Chemoreception—Taste and Smell

Perceptions of taste and smell are usually interrelated (1074–1075)

For terrestrial animals, the sense of taste (gustation) detects chemicals in solution, whereas smell (olfaction) detects airborne chemicals. Both rely on chemoreceptors.

Sensillae, or sensory hairs containing chemoreceptor taste cells, are found on the feet and mouthparts of insects. Insects also have olfactory sensillae, usually located on their antennae.

In mammals, the sensations of taste and smell are produced when a molecule dissolved in liquid binds to a protein in a receptor cell membrane, triggering a membrane depolarization and the release of neurotransmitter. **Taste buds,** which contain groups of modified epithelial cells, are scattered on the tongue and mouth. The four primary taste perceptions are sweet, sour, salty, and bitter. The brain integrates the input from the chemoreceptors in the taste buds as they differentially respond to various chemicals, and it creates the perception of a complex flavor.

Olfactory receptor cells are neurons that line the upper part of the nasal cavity; binding of solubilized chemicals to specific receptor molecules on their cilia triggers a signal-transduction pathway. The second messenger cAMP opens Na^+ channels, depolarizing the membrane and sending action potentials to the olfactory bulb of the brain.

■ INTERACTIVE QUESTION   49.7

For what purposes do animals use their senses of gustation and olfaction?

Movement and Locomotion

Locomotion is the active movement from place to place. Animals move to obtain food, escape from danger, and find mates.

Locomotion requires energy to overcome friction and gravity (1075–1077)

All forms of locomotion require energy to overcome the forces of friction and gravity in a water, land, or air environment.

Swimming   Overcoming gravity is less of a problem for swimming animals due to the buoyancy of water. The density of water, however, makes resistance a major problem. Most fast swimmers have a fusiform shape that reduces friction. Animals may swim by using their legs as oars, using water to jet-propel themselves, or by undulating their bodies and tails from side to side or up and down.

Locomotion on Land   Walking, running, hopping, or crawling on land requires the animal to support itself and move against gravity; but air offers little resistance to movement. Powerful leg muscles and strong skeletal support are important for propelling an animal forward and supporting it. A running or hopping animal may store some energy in its tendons or legs to facilitate the next step. Maintaining balance is another requirement for walking, running, or hopping. A crawling animal must overcome the friction produced by its contact with the ground.

Flying   A flying animal uses its wings to provide enough lift to overcome the force of gravity.

Cellular and Skeletal Underpinnings of Locomotion   On a cellular level, all animal movement depends on energy expended to move protein strands past each other, either in microtubules, which are responsible for beating of cilia and flagella, or in microfilaments, which are involved in amoeboid movement and muscle contraction.

■ INTERACTIVE QUESTION   49.8

a. Which mode of locomotion is the most energy efficient?

b. Which mode is the most energetically expensive per distance traveled?

c. Which mode consumes the most energy per unit of time?

d. In animals specialized for the same mode of transport, does a larger animal or a smaller one travel more efficiently?

Skeletons support and protect the animal body and are essential to movement (1077–1078)

Hydrostatic Skeletons   Fluid under pressure in a closed body compartment creates a **hydrostatic skeleton.** As muscles change the shape of the fluid-filled compartment, the animal moves. Earthworms and other annelids move by **peristalsis** using rhythmic waves of contractions of circular and longitudinal muscles.

Exoskeletons   Typical of mollusks and arthropods, **exoskeletons** are hard coverings deposited on the surface of animals. The cuticle of an arthropod contains fibrils of the polysaccharide **chitin** embedded in a protein matrix. Where protection is most needed, this flexible support is hardened by cross-linking proteins and addition of calcium salts.

Endoskeletons   The supporting elements of an **endoskeleton** are embedded in the soft tissues of the animal. Sponges have endoskeletons composed of spicules or fibers, and echinoderms have hard plates beneath the skin. The endoskeletons of chordates are composed of cartilage and/or bone. The vertebrate axial skeleton consists of skull, vertebral column, and rib cage; the appendicular skeleton contains the pectoral and pelvic girdles and the limb bones. Joints allow for flexibility in body movement.

■ INTERACTIVE QUESTION   49.9

List an advantage and a disadvantage for each of the three types of skeletons.

a. hydrostatic skeleton

b. exoskeleton

c. endoskeleton

Physical support on land depends on adaptations of body proportions and posture (1078–1080)

Body proportions of small and large animals vary, partly as a result of the physical relationship between the diameter of a support and the strain produced by increasing size and weight. Posture is an important structural factor in supporting body weight. Most of the weight of large land mammals is supported by muscles and tendons, which hold the legs relatively straight and directly under the body.

Muscles move skeletal parts by contracting (1080–1081)

Because muscles can only contract, antagonistic muscle pairs are required to move body parts in opposite directions.

Structure and Function of Vertebrate Skeletal Muscle

Vertebrate **skeletal muscle** consists of a bundle of multinucleated muscle cells, called fibers, running the length of the muscle. Each fiber is a bundle of **myofibrils,** each composed of two kinds of **myofilaments: Thin filaments** consist of two strands of actin coiled with two strands of regulatory protein, and **thick filaments** are made of myosin molecules.

The regular arrangement of myofilaments produces repeating light and dark bands; thus skeletal muscle is also called striated muscle. The repeating units are called **sarcomeres** and are delineated by **Z lines.** Thin filaments attach to the Z line and project toward the center of the sarcomere. Thick filaments lie in the center, forming the **A band,** with an H zone in the center where the thin filaments do not reach. At the edges of the sarcomere, where thick filaments do not extend when the muscle is at rest, is an **I band** of only thin filaments.

Interactions between myosin and actin generate force during muscle contractions (1081–1083)

According to the **sliding-filament model** of muscle contraction, the thick and thin filaments do not change length but simply slide past each other, increasing their area of overlap.

The mechanism for the sliding of filaments is the hydrolyzing of ATP by the globular head of a myosin molecule, which then changes to a high-energy configuration that binds to actin and forms a cross-bridge. When relaxing to its low-energy configuration, the myosin head bends and pulls the attached thin filament toward the center of the sarcomere. When a new molecule of ATP binds to the myosin head, it breaks its bond to actin, and the cycle repeats with the high-energy configuration attaching to an actin molecule farther along on the thin filament. Most of the energy for muscle contraction comes from phosphagens, such as creatine phosphate in vertebrates, which regenerate ATP.

Calcium ions and regulatory proteins control muscle contraction (1083–1084)

When the muscle is at rest, the myosin-binding sites of the actin molecules are blocked by a strand of the regulatory protein **tropomyosin,** whose position is controlled by another set of regulatory proteins, the **troponin complex.** When troponin binds to calcium ions, the tropomyosin-troponin complex changes shape, and myosin-binding sites on the actin are exposed.

Calcium ions are actively transported into the **sarcoplasmic reticulum** of a muscle cell. When an action potential of a motor neuron causes the release of acetylcholine into the neuromuscular junction, the muscle cell depolarizes and an action potential spreads into the **T (transverse) tubules,** infoldings of the muscle cell plasma membrane. The action potential changes the permeability of the sarcoplasmic reticulum, and calcium ions are released into the cytosol. Calcium binding with troponin exposes myosin-binding sites and contraction begins. Contraction ends when the sarcoplasmic reticulum pumps Ca^{2+} back out of the cytosol, and the tropomyosin-troponin complex again blocks the binding sites.

■ INTERACTIVE QUESTION   49.10

Identify the components in this diagram of a section of a skeletal muscle fiber.

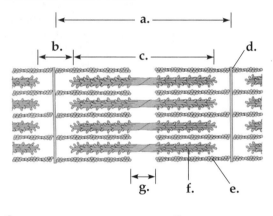

a. _____   e. _____
b. _____   f. _____
c. _____   g. _____
d. _____

Diverse body movements require variation in muscle activity (1084–1086)

The response of a single muscle fiber to an action potential of a motor neuron is an all-or-none contraction, producing a twitch. More graded contraction of a muscle fiber occurs by temporal summation when action potentials arrive rapidly, producing a greater level of tension and a state of smooth and sustained contraction called **tetanus.**

Graded contraction of whole muscles also results from involving additional fibers. A single motor neuron and all the muscle fibers it innervates make up a **motor unit.** The strength of a muscle contraction depends on the size and number of motor units involved. Muscle tension can be increased by the activation of additional motor units, called **recruitment.**

The muscle fatigue of extended tetanus is caused by depletion of ATP, loss of ion gradients required for depolarization, and accumulation of lactate. Some postural support muscles are always in a state of partial contraction, in which different motor units take turns contracting.

Fast and Slow Muscle Fibers   **Slow muscle fibers** have less sarcoplasmic reticulum; thus calcium remains in the cytosol longer and twitches last longer. Slow fibers, common in muscles that must sustain long contractions, have many mitochondria, a good blood supply, and **myoglobin,** which extracts oxygen from the blood and stores it. **Fast muscle fibers** produce rapid and powerful contractions.

Other Types of Muscles   Vertebrate **cardiac muscle** cells are connected by **intercalated disks** through which action potentials spread to all the cells of the heart. The plasma membrane of a cardiac muscle cell can generate action potentials by rhythmic depolarizations. The action potentials of cardiac muscle cells last quite long and have a role in controlling the duration of contraction.

The spiral arrangement of actin and myosin filaments in **smooth muscle** accounts for its nonstriated appearance. Smooth muscle lacks a T tubule system and well-developed sarcoplasmic reticulum. Its relatively slow contractions do not generate as much tension as striated muscle, but can occur over a greater range of lengths.

Invertebrates have muscle cells similar to the skeletal and smooth muscle cells of vertebrates. The flight muscles of insects are capable of independent and rapid contraction.

WORD ROOTS

ama- = together (*amacrine cell:* neurons of the retina that help integrate information before it is sent to the brain)

aqua- = water (*aqueous humor:* the clear, watery solution that fills the anterior cavity of the eye)

bi- = two (*bipolar cell:* neurons that synapse with the axons of rods and cones in the retina of the eye)

chemo- = chemical (*chemoreceptor:* a receptor that transmits information about the total solute concentration in a solution or about individual kinds of molecules)

coch- = a snail (*cochlea:* the complex, coiled organ of hearing that contains the organ of Corti)

electro- = electricity (*electromagnetic receptor:* receptors of electromagnetic energy, such as visible light, electricity, and magnetism)

endo- = within (*endoskeleton:* a hard skeleton buried within the soft tissues of an animal, such as the spicules of sponges, the plates of echinoderms, and the bony skeletons of vertebrates)

exo- = outside (*exoskeleton:* a hard encasement on the surface of an animal, such as the shells of mollusks or the cuticles of arthropods, that provides protection and points of attachment for muscles)

externo- = outside (*exteroreceptor:* sensory receptors that detect stimuli outside the body, such as heat, light, pressure, and chemicals)

fovea- = a pit (*fovea:* the center of the visual field of the eye)

gusta- = taste (*gustatory receptors:* taste receptors)

hydro- = water (*hydrostatic skeleton:* a skeletal system composed of fluid held under pressure in a closed body compartment; the main skeleton of most cnidarians, flatworms, nematodes, and annelids)

inter- = between; **-cala** = insert (*intercalated disks:* specialized junctions between cardiac muscle cells which provide direct electrical coupling among cells)

interno- = inside (*interoreceptor:* sensory receptors that detect stimuli within the body, such as blood pressure and body position)

mechano- = an instrument (*mechanoreceptor:* a sensory receptor that detects physical deformations in the body's environment associated with pressure, touch, stretch, motion, and sound)

myo- = muscle; **-fibro** = fiber (*myofibril:* a fibril collectively arranged in longitudinal bundles in muscle cells; composed of thin filaments of actin and a regulatory protein and thick filaments of myosin)

noci- = harm (*nociceptor:* a class of naked dendrites in the epidermis of the skin)

olfact- = smell (*olfactory receptor:* smell receptors)

omma- = the eye (*ommatidia:* the facets of the compound eye of arthropods and some polychaete worms)

peri- = around; **-stalsis** = a constriction (*peristalsis:* rhythmic waves of contraction of smooth muscle that push food along the digestive tract)

photo- = light (*photoreceptor:* receptors of light)

rhodo- = red (*rhodopsin:* a visual pigment consisting of retinal and opsin)

sacc- = a sack (*saccule:* a chamber in the vestibule behind the oval window that participates in the sense of balance)

sarco- = flesh; **-mere** = a part (*sarcomere:* the fundamental, repeating unit of striated muscle, delimited by the Z lines)

sclero- = hard (*sclera:* a tough, white outer layer of connective tissue that forms the globe of the vertebrate eye)

semi- = half (*semicircular canals:* a three-part chamber of the inner ear that functions in maintaining equilibrium)

stato- = standing; **-lith** = a stone (*statolith:* sensory organs that contain mechanoreceptors and function in the sense of equilibrium)

tetan- = rigid, tense (*tetanus:* the maximal, sustained contraction of a skeletal muscle, caused by a very fast frequency of action potentials elicited by continual stimulation)

thermo- = heat (*thermoreceptor:* an interoreceptor stimulated by either heat or cold)

trans- = across; **-missi** = send (*transmission:* the conduction of impulses to the central nervous system)

tropo- = turn, change (*tropomyosin:* the regulatory protein that blocks the myosin binding sites on the actin molecules)

tympan- = a drum (*tympanic membrane:* another name for the eardrum)

utric- = a leather bag (*utricle:* a chamber behind the oval window that opens into the three semicircular canals)

vitre- = glass (*vitreous humor:* the jellylike material that fills the posterior cavity of the vertebrate eye)

STRUCTURE YOUR KNOWLEDGE

1. Describe the path of a light stimulus from where it enters the vertebrate eye to its transmission as an action potential through the optic nerve.

2. Arrange the following structures in an order that enables you to describe the passage of sound waves from the external environment to the perception of sound in the brain.

round window	basilar membrane
oval window	cerebral cortex
pinna	vestibular and tympanic canals
auditory nerve	tectorial membrane
cochlea	tympanic membrane
auditory canal	malleus, incus, stapes
cochlear duct	organ of Corti with hair cells

3. Trace the sequence of events in muscle contraction from an action potential in a motor neuron to the relaxation of the muscle.

TEST YOUR KNOWLEDGE

MATCHING: *Match the term with its description.*

_____ 1. oxygen-storing compound in muscles

_____ 2. stretch receptor in muscle, monitors position

_____ 3. stimulus energy converted into receptor potential

_____ 4. light-absorbing pigment

_____ 5. naked dendrites that detect pain

_____ 6. bones that transmit vibrations to inner ear of some fish

_____ 7. invertebrate mechanoreceptors for position

_____ 8. energy-storing compound in muscles

_____ 9. changing shape of lens to focus

_____ 10. decrease in sensitivity of receptor during constant stimulation

A. accommodation

B. adaptation

C. ampulla

D. creatine phosphate

E. cupula

F. ganglion cells

G. muscle spindle

H. myofibrils

I. myoglobin

J. nociceptors

K. ommatidia

L. retinal

M. statocyst

N. transduction

O. transmission

P. tropomyosin

Q. Weberian apparatus

MULTIPLE CHOICE: *Choose the one best answer.*

1. What is a sensory perception?
   a. the opening of ion channels in response to mechanical stimulation
   b. the conversion of stimulus energy into a receptor potential by a sensory receptor or sensory neuron
   c. the interpretation of sensation according to the region of the brain that receives the nerve impulse
   d. the amplification and transmission of sensory data to the CNS
   e. the emotional response to a sensation

2. Which of the following is an interoreceptor?
   a. temperature receptor in hypothalamus
   b. pain receptor in fingertip
   c. taste receptor on tongue
   d. chemoreceptor in nasal passages
   e. rod or cone cell in eye

3. The compound eye found in insects and crustaceans
   a. is composed of thousands of prisms that focus light onto a few photoreceptor cells.
   b. cannot sense colors.
   c. contains many ommatidia, each of which contains a lens that focuses light from a tiny portion of the field of view.
   d. may detect infrared and ultraviolet radiation.
   e. does all of the above.

4. Which of the following statements is *not* true?
   a. The iris regulates the amount of light entering the pupil.
   b. The choroid is the thin, pigmented inner layer of the eye that contains the photoreceptor cells.
   c. The ciliary muscle functions in changing the shape of the lens.
   d. Thin aqueous humor fills the anterior eye cavity.
   e. The conjunctiva is a mucous membrane surrounding the sclera.

5. Which of the following is *incorrectly* paired with its function?
   a. cones—respond to different light wavelengths producing color vision
   b. horizontal cells—produce lateral inhibition of receptor and bipolar cells
   c. bipolar cells—relay impulses between rods or cones and ganglion cells
   d. ganglion cells—synapse with the optic nerve at the optic disk or blind spot
   e. fovea—center of the visual field, containing only cones in humans

6. The molecule rhodopsin
   a. dissociates when struck by light and is functional only in complete dark.
   b. is recombined by enzymes from retinal and opsin in the light.
   c. is more sensitive to light than are photopsins and is involved in black and white vision in dim light.
   d. activates a signal-transduction pathway that increases a rod cell's permeability to sodium.
   e. is or does all of the above.

7. Rotation of the head of a vertebrate is sensed by
   a. the organ of Corti located in the utricle and saccule.
   b. changes in the action potentials of hair cells in response to the pull on them by calcium carbonate particles.
   c. bending of hair cells in the ampulla of a semicircular canal in response to the inertia of the endolymph.
   d. hair cells in response to movement of fluid in the coiled cochlea.
   e. statocysts when statoliths stimulate hair cells.

8. Fish can perceive pressure waves from moving objects with their
   a. lateral line system containing mechanoreceptor units called neuromasts.
   b. tectorial membrane as it is stimulated by the bending of sensory hairs embedded in a gelatinous cap.
   c. inner ears, within which sensory hairs are stimulated by the movement of otoliths.
   d. Weberian apparatus, a series of bones that transmits vibrations from the swim bladder to the inner ear.
   e. sensillae collected into ampullae in the lateral line system.

9. The volume of a sound is determined by
   a. the area in the brain that receives the signal.
   b. the section of the basilar membrane that vibrates.
   c. the number of hair cells that are stimulated.
   d. the frequency of the sound wave.
   e. the degree to which hair cells are bent and the resulting increase in the number of action potentials produced in the sensory neuron.

10. Hydrostatic skeletons are used for movement by all of the following *except*
    a. cnidarians.
    b. echinoderms.
    c. nematodes.
    d. annelids.
    e. flatworms.

11. When striated muscle fibers contract,
    a. the Z lines are pulled closer together.
    b. the I band remains the same.
    c. the A band becomes shorter.
    d. the H zone widens slightly.
    e. all of the above occur.

12. Tetanus
    a. is muscle fatigue resulting from a buildup of lactate during anaerobic respiration.
    b. is the all-or-none contraction of a single muscle fiber.
    c. is the result of stimulation of additional motor neurons.
    d. is the result of a volley of action potentials, whose summation produces an increased and sustained contraction.
    e. is the rigidity of muscles following death.

13. What is the role of ATP in muscle contraction?
    a. to form cross-bridges between thick and thin filaments
    b. to break the cross-bridge when it binds to myosin and provide energy to myosin to form its high-energy configuration

    c. to remove the tropomyosin-troponin complex from blocking the binding sites on actin
    d. to bend the cross-bridge and pull the thin filaments toward the center of the sarcomere
    e. to replace the supply of creatine phosphate required for movement of actin past myosin

14. How does calcium affect muscle contraction?
    a. It is released from the T tubules in response to an action potential to initiate contraction.
    b. The binding of acetylcholine opens calcium channels in the plasma membrane, creating an action potential that travels down the T tubules.
    c. It binds to tropomyosin and helps to stabilize cross-bridge formation.
    d. Its binding to troponin causes tropomyosin to move away from the myosin-binding sites on the actin filament.
    e. Its release from the sarcoplasmic reticulum changes the membrane potential of the muscle cell so that contraction can occur.

15. A motor unit is
    a. a sarcomere that extends from one Z line to the next Z line.
    b. a motor neuron and the muscle fibers it innervates.
    c. the myofibrils of a fiber and the transverse tubules that connect them.
    d. the muscle fibers that make up a muscle.
    e. the antagonistic set of muscles that flex and extend a body part.

16. Which of the following is *not* a characteristic of cardiac muscle?
    a. spiral arrangement of actin and myosin filaments
    b. intercalated disks that spread action potentials between cells
    c. action potentials that last a long time
    d. ability to generate action potentials without nervous input
    e. striations

17. Smooth muscle contracts relatively slowly because
    a. the only ATP available is supplied by fermentation.
    b. its contraction is stimulated by hormones, not motor neurons.
    c. it does not have a well-developed sarcoplasmic reticulum, and calcium enters the cell through the plasma membrane during an action potential.
    d. it is not striated.
    e. it is composed exclusively of slow muscle fibers.

18. The absorption of light by rhodopsin in a rod cell leads to
    a. an increase in the amount of neurotransmitter released.
    b. the opening of sodium channels caused by the binding of cGMP.
    c. a hyperpolarized membrane and a decrease in the release of neurotransmitter.
    d. the bleaching of rhodopsin and the inhibition of the postsynaptic cell.
    e. the closing of sodium channels and the release of neurotransmitter.

19. What is the function of the cornea?
    a. regulate the amount of light that enters the eye
    b. accommodation to focus on close objects
    c. tough, protective connective tissue layer surrounding the eye
    d. transparent covering that acts as a fixed lens to bend light rays
    e. produce aqueous humor that helps to focus images on the optic disk

20. A receptive field in the retina would involve a *single*
    a. bipolar cell.
    b. photoreceptor (rod or cone) cell.
    c. amacrine cell.
    d. horizontal cell.
    e. ganglion cell.

21. What is the function of the three bones of the middle ear?
    a. respond to different frequencies of sound waves
    b. transmit pressure waves to the oval window
    c. equalize pressure within the inner ear by communicating through the Eustachian tube
    d. sense rotation in three different planes (*x*, *y*, and *z* axes)
    e. support the tympanic membrane (eardrum)

22. The sensory taste receptors on the feet of flies are
    a. mechanoreceptors.
    b. photoreceptors.
    c. pheromone receptors.
    d. chemoreceptors.
    e. naked dendrites.

23. What do the lateral line of fish, invertebrate statocysts, and the cochlea of your ear all have in common?
    a. They are used to sense sound or pressure waves.
    b. They are organs of equilibrium.
    c. They use hair cells as their receptor cells.
    d. They use a second-messenger pathway of signal transduction.
    e. They use granules to stimulate their receptor cells.

24. In a contracting muscle fiber, creatine phosphate is used to
    a. supply extra O_2 for aerobic respiration to generate ATP.
    b. phosphorylate ADP to replenish ATP supplies.
    c. provide the energy for the sliding of the myosin filament.
    d. recycle lactate to glucose.
    e. transduce the action potential of the T tubules into stimulation of the sarcoplasmic reticulum.

25. When you pull up your lower arm and "make a muscle" in your biceps, you are
    a. contracting a flexor.
    b. relaxing a flexor.
    c. contracting an extensor.
    d. contracting a tendon.
    e. producing a simple muscle twitch.

26. Different regions of the basilar membrane vibrate in response to different
    a. volumes of sounds.
    b. amplitudes of sound waves.
    c. pitches.
    d. action potential frequencies.
    e. directions of rotation of the head.

27. How does an animal with a jointed exoskeleton move?
    a. alternate contractions of circular and longitudinal muscles
    b. expansion of the exoskeleton at joints, followed by contraction of the skeleton
    c. alternate contraction and extension of muscles that attach to opposite sides of a joint
    d. contraction of muscles that cause the exoskeleton to work against the hydrostatic skeleton, creating movement of body parts
    e. alternate contraction of antagonistic muscles (extensors and flexors) that span a joint

28. Which of the following modes of locomotion is the most energy efficient for an animal that is specialized for moving in that fashion?
    a. swimming
    b. running
    c. flying
    d. hopping
    e. crawling

ECOLOGY

CHAPTER 50

AN INTRODUCTION TO ECOLOGY AND THE BIOSPHERE

FRAMEWORK

This chapter describes the organizational levels at which ecological questions are asked, the abiotic factors to which organisms have adapted in both an ecological and an evolutionary time frame, and the major world communities, or biomes, in which adaptations to climate and abiotic factors have produced similar and characteristic life forms.

CHAPTER REVIEW

Ecology studies the distribution and abundance of organisms and their interactions with the environment.

The Scope of Ecology

The interactions between organism and their environments determine the distribution and abundance of organisms (1093)

Ecology had its foundation in natural history, but experimental approaches in both the laboratory and the field are increasingly used to investigate ecological questions. Many ecologists develop mathematical models and computer simulations to study the interactions of variables and gain insight into complex ecological questions.

Organisms are affected by and, in turn, affect both the **abiotic** and **biotic components** of their environments.

Ecology and evolutionary biology are closely related sciences (1093)

Interactions between organisms and their environments occur within **ecological time.** The cumulative effects of these interactions are realized on the scale of **evolutionary time.** The distribution and abundance of organisms are the result of both evolutionary history and present interactions with the environment.

Ecological research ranges from the adaptations of individual organisms to the dynamics of the biosphere (1093–1095)

Organismal ecology considers behavioral, morphological, and physiological responses of an organism to its biotic and abiotic environment. **Population ecology** is concerned with the factors that control the size of **populations,** which are groups of individuals of the same species in an area. The **community** includes all the populations of organisms in an area; **community ecology** looks at interactions such as predation and competition.

Ecosystem ecology considers abiotic factors as well as the biological community, and it addresses such topics as the flow of energy and chemical cycling. A **landscape** or **seascape** includes several linked ecosystems; **landscape ecology** looks at the impact of the juxtapositions of different ecosystems.

Ecology is a multidisciplinary science, incorporating many fields of biology and the physical sciences. Ecology is a challenging science because of the complexity and scale of ecological questions.

The **biosphere** is the global ecosystem, the layer of Earth inhabited by life.

■ INTERACTIVE QUESTION   50.1

List the five levels of ecological study and give examples of the focus of inquiry at each level.

a.

b.

c.

d.

e.

Ecology provides a scientific context for evaluating environmental issues (1095)

The science of ecology is not synonymous with the growing ecological consciousness about current environmental problems, but it helps us to understand these complicated problems and their possible solutions. Ecological information is always increasing but never complete. The **precautionary principle** should guide decisions in which not all answers are known.

Factors Affecting Distributions of Organisms

Biogeography studies the past and present distribution of species.

Species dispersal contributes to the distribution of organisms (1096–1098)

In studying the **dispersal** of organisms, ecologists work through a series of logical steps to determine what limits geographic distributions.

Species Transplants   Transplant experiments can indicate whether dispersal or biotic or abiotic factors limit the distribution of a species. A successful transplant shows that the potential range of a species is larger than its actual range. A species may not occupy its potential range because of dispersal (area inaccessible) or behavior (habitat selection). If the transplant is unsuccessful, the limiting factors may be biotic (predation, competition, lack of interdependent species) or abiotic (physical or chemical factors). Transplant experiments should include a control to test whether simply moving organisms affects their viability. Ecologists rarely conduct such experiments, however, but they study the results of accidental and other intentional transplants.

Problems with Introduced Species   Increasingly over the past 200 years, humans have purposefully or accidentally introduced species to areas to which they had not dispersed, often with dramatic and disastrous results. Generally, the *tens rule* applies to introduced species: One in ten introduced species is successful, and of these, one in ten becomes a pest.

■ INTERACTIVE QUESTION   50.2

Give an example of a purposefully introduced species and an accidentally introduced species that have become pests in North America.

Behavior and habitat selection contribute to the distribution of organisms (1098–1099)

Sometimes the behavior of organisms in habitat selection keeps them from occupying all their potential range. Habitat selection by ovipositing insects, which often choose only certain host plants, may limit their distribution. When environmental conditions change rapidly, as they often do as a result of human interventions, habitat selection behaviors that evolved over a long time frame may no longer be adaptive.

Biotic factors affect the distribution of organisms (1099)

The inability of transplanted organisms to survive and reproduce may be due to predation, disease, competition, or lack of mutual symbiosis. "Removal and addition" experiments test whether predators limit the distribution of prey species. Sea urchins were shown to limit the abundance of seaweeds in subtidal zones.

Abiotic factors affect the distribution of organisms (1099)

Global patterns of geographic distributions reflect abiotic factors such as temperature, rainfall, salinity, and light.

Temperature   Temperature is an important environmental factor because of its effects on metabolism and enzyme activity. Most organisms cannot maintain body temperatures that differ much from ambient temperature.

Water   The availability of water in different habitats can vary greatly. Organisms must maintain their water balance, compensating for different osmolarities in aquatic environments and avoiding desiccation in terrestrial habitats.

Sunlight Light energy drives almost all ecosystems. The intensity and quality of light are limiting factors in aquatic environments. Many plants and animals are sensitive to photoperiod, which serves as a reliable indicator of seasonal changes.

Wind Wind increases the rate of heat and water loss in organisms and can affect plant morphology.

Rocks and Soil Soils, which vary in their physical structure, pH, and mineral composition, affect the distribution of plants and, in turn, the distribution of animals. Substrate composition in aquatic environments influences water chemistry and the types of organisms that can inhabit intertidal and benthic zones.

Temperature and moisture are the major climatic factors determining the distribution of organisms (1100–1106)

The **climate,** or prevailing weather conditions of a locality, is determined by temperature, water, light, and wind.

Climate and Biomes **Biomes** are major types of ecosystems found in broad geographic regions. A climograph plots annual mean temperature and rainfall for a region; generally these values correlate with the distribution of various biogeographic regions. Overlaps of biomes on a climograph indicate the importance of seasonal patterns of variation in rainfall and temperatures.

Global Climate Patterns The absorption of solar radiation heats the atmosphere, land, and water, setting patterns for temperature variations, air circulation, and water evaporation that cause latitudinal variations in climate. The shape of the Earth and the tilting of its axis create seasonal variations in day length and temperature that increase with latitude. The **tropics** receive the greatest amount of and least variation in solar radiation.

The global circulation of air begins as intense solar radiation near the equator causes warm, moist air to rise, producing the characteristic wet tropical climate, the arid conditions around 30° north and south as dry air descends, the fairly wet though cool climate about 60° latitude as air rises again, and the cold and rainless climates of the arctic and antarctic regions. Air flowing in the lower levels of the three major air circulation cells on either side of the equator produces predictable global wind patterns.

Local and Seasonal Effects on Climate Regional climatic patchiness is influenced by proximity to water and topographical features. Coastal areas are generally more moist, and large bodies of water moderate the climate.

Seasonal changes affect local climate. Seasonal changes in wind patterns can produce wet and dry seasons and affect ocean currents, sometimes causing upwellings of cold, nutrient-rich water. Seasonal temperature changes produce the biannual **turnover** of waters in lakes that brings oxygenated water to the bottom and nutrient-rich water to the surface.

■ **INTERACTIVE QUESTION 50.3**

Mountains affect local climate. Describe their influence in the following three areas:

a. solar radiation:

b. temperature:

c. rainfall:

Microclimate Climate also varies on a very small scale. **Microclimates** within an area have differences in abiotic features that affect the distributions of organisms.

Long-term Climate Change Global warming will have a great effect on the distributions of plants and animals. Fossil pollen deposits have documented the rates of northward expansion of various tree species following the last continental glacier. As the geographic climatic limits change with global warming, seed dispersal may not be rapid enough for some species of plants to migrate into their new range.

Aquatic and Terrestrial Biomes

Aquatic biomes occupy the largest part of the biosphere (1106–1112)

One of the chemical differences in aquatic habitats is salt concentration—less than 1% for freshwater biomes versus an average of 3% for marine biomes. Freshwater biomes are closely linked with and shaped by the surrounding terrestrial biomes. Three-fourths of Earth is covered by oceans, which influence global rainfall, climate, and wind patterns. Marine algae and photosynthetic bacteria produce a large portion of the world's oxygen and consume enormous amounts of carbon dioxide.

Vertical Stratification of Aquatic Biomes Many aquatic biomes are stratified in availability of light and temperature. The **photic zone** receives sufficient light for photosynthesis, whereas little light penetrates into the lower **aphotic zone**. A narrow **thermocline** separates warmer surface waters from the cold bottom layer. The bottom substrate, called the **benthic zone,** is home to organisms collectively called **benthos.** Settling **detritus** provides food for the benthos.

Freshwater Biomes Small bodies of standing fresh water are called ponds; larger ones are lakes. The shallow waters close to shore, called the **littoral zone,** support a diverse community of rooted and floating aquatic plants, algae, mollusks, arthropods, fishes, and amphibians. The open surface waters of the **limnetic zone** are occupied by a variety of phytoplankton and zooplankton, supporting a food chain of larger organisms. Detritus sinks into the deep, aphotic **profundal zone** and the benthic zone, where it is decomposed by microbes.

   Oligotrophic lakes are deep, clear, nutrient poor, and fairly nonproductive. The shallower, nutrient-rich waters of **eutrophic** lakes support large, productive phytoplankton communities. **Mesotrophic** lakes have a moderate amount of nutrients and phytoplankton. Runoff carrying nutrients and sediment may gradually convert oligotrophic lakes into eutrophic lakes. Dumping of municipal wastes and runoff from lawns and farms leads to "cultural eutrophication."

   Streams and rivers are freshwater, flowing habitats, whose physical and chemical characteristics vary from their source to their point of entry into oceans or lakes. Overhanging vegetation contributes to nutrient content. Oxygen levels are high in turbulently flowing water and low in murky, warm waters. Organisms show evolutionary adaptations that allow them to live in moving waters. Human impact on streams and rivers has included pollution, stream channelization, and dams.

Wetlands Defined as areas covered with water and supporting hydrophytes ("water plants"), **wetlands** range from marshes to swamps to bogs. Topography results in basin wetlands in shallow basins, riverine wetlands along shallow and periodically flooded rivers and streams, and fringe wetlands with rising and falling water levels along coasts of large lakes and seas. These richly diverse biomes are important to flood control and water quality. Wetlands are now being protected and regulated.

Estuaries Where a freshwater river or stream meets the ocean, an **estuary** is formed. These highly productive areas are often bordered by wetlands called mudflats and saltmarshes. Salinity varies both spatially and daily with the rise and fall of tides. Salt-marsh grasses, algae, and phytoplankton are the major producers. Estuaries serve as feeding and breeding areas for marine invertebrates, fish, and waterfowl. Little undisturbed estuary habitat remains.

■ **INTERACTIVE QUESTION   50.4**

Indicate with a + or − whether the following are relatively high or low in oxygen level, nutrient content, and productivity.

Biome		Oxygen Level	Nutrient Content	Productivity
Oligotrophic lake	**a.**			
Eutrophic lake	**b.**			
Headwater of stream	**c.**			
Turbid river	**d.**			
Estuary	**e.**			

Zonation in Marine Communities The marine environment has a photic zone, in which phytoplankton, zooplankton, and many fishes are found, and an aphotic zone, into which light does not penetrate. The **intertidal zone** is along the shoreline. The **neritic zone** is the shallow region over the continental shelf. The **oceanic zone** reaches great depths. Open water of any depth is the **pelagic zone,** the bottom surface of which is the **benthic zone.**

Intertidal Zones The daily cycle of tides exposes the shoreline to variations in water, nutrients, and temperature, and to the mechanical force of wave action. Rocky intertidal communities are vertically stratified, with organisms adapted to firmly attach to the hard substrate. Sandy or mudflat intertidal zones are home to burrowing worms, clams, and crustaceans. Recreational use and intertidal pollutants such as oil have severely reduced species diversity and population numbers.

Coral Reefs Found in tropical waters in the neritic zone, **coral reefs** are highly diverse and productive biomes. The structure of the reef is produced by the calcium carbonate skeletons of the coral (various cnidarians) and serves as a substrate for other corals, sponges, and algae. The coral animals feed on microscopic organisms and organic debris and are also nourished by symbiotic photosynthetic dinoflagellates. Coral reefs are easily damaged by pollution, development, native and introduced predators, and high water temperatures.

The Oceanic Pelagic Biome　The water of the **oceanic pelagic biome** is typically nutrient poor because detritus sinks to the benthic zone. Seasonal mixing of temperate oceans stimulates phytoplankton growth. Phytoplankton flourish in the photic region and are grazed on by numerous types of zooplankton, invertebrate larvae, and fishes. Morphological adaptations of plankton help them maintain buoyancy. Free-swimming animals, called nekton, include squid, fishes, turtles, and marine mammals. Many pelagic birds feed from the surface waters.

Benthos　Nutrients reach the benthic zone as detritus falling from the waters above. Neritic benthic communities receive sunlight and are very diverse and productive. Various invertebrates and fishes inhabit the **abyssal zone,** the deep benthic region where light does not penetrate, and are adapted to darkness, cold, and high water pressure. Chemoautotrophic prokaryotes form the basis of a collection of organisms adapted to the hot, low-oxygen environment surrounding **deep-sea hydrothermal vents.**

■ **INTERACTIVE QUESTION   50.5**

Different marine environments can be classified on the basis of light penetration, distance from shore, and open water or bottom. Match the following zones to their corresponding numbers on the diagram below:

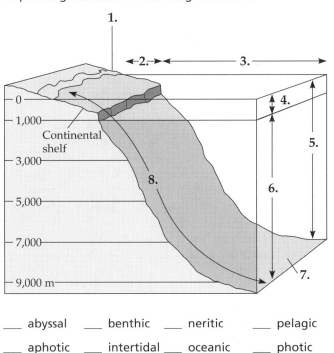

___ abyssal   ___ benthic   ___ neritic   ___ pelagic

___ aphotic   ___ intertidal   ___ oceanic   ___ photic

The geographic distribution of terrestrial biomes is based mainly on regional variations in climate (1112–1117)

The geographic distribution of the world's major terrestrial biomes is related to abiotic factors—in particular, the prevailing climate. Biomes are usually named for their predominant vegetation and major climatic features. Each biome also has characteristic microorganisms, fungi, and animals.

Terrestrial biomes may have vertical stratification, such as the layers in a forest from **canopy,** low-tree stratum, shrub understory, herbaceous plant ground layer, and forest floor (litter) to root layer. The root layer in arctic tundra is shallow due to the underlying **permafrost.** Vertical stratification of vegetation provides diverse habitats for animals.

The area where biomes grade into each other is called an ecotone. Species composition of any one biome varies locally. Often, convergent evolution has produced a superficial resemblance of unrelated "ecological equivalents."

The extensive patchiness typical of most biomes results from natural or human disturbances. Grasslands, savannas, chaparral, and many coniferous forests are maintained by the periodic disturbance of fire. Urban and agricultural biomes cover a large portion of the Earth's land mass.

Tropical forests occur within 23° latitude of the equator. Variations in rainfall result in tropical dry forests, tropical deciduous forests, and tropical rain forests, where rainfall is abundant. The tropical rain forest has pronounced vertical stratification.

Savannas are tropical and subtropical grasslands with scattered trees and rainy and dry seasons. Fires and large grazing mammals restrict vegetation to grasses and forbs, small broad-leaved plants.

Characterized by low and unpredictable precipitation, **deserts** may be hot or cold, depending on location. Desert animals have physiological and behavioral adaptations to dry conditions. Plants may use CAM photosynthesis, and have water storage adaptation and protective spines and poisons.

Chaparral is common along coastlines in midlatitudes that have mild, rainy winters and hot, dry summers. The dominant vegetation—dense, spiny evergreen shrubs—is maintained by and adapted to periodic fires.

Temperate grasslands are maintained by fire, seasonal drought, and grazing by large mammals. Soils are deep and rich in nutrients.

Characterized by broad-leaved deciduous trees, **temperate deciduous forests** grow in midlatitude regions that have adequate moisture to support the growth of large trees. Trees drop their leaves before winter.

A large biome found in northern latitudes, the **coniferous forest,** or taiga, is characterized by harsh winters with heavy snowfall. Coniferous trees grow in dense, uniform stands. Coastal temperate rain forests are also coniferous forests.

The northernmost limit of plant growth is in the **tundra** and is characterized by dwarfed or matlike vegetation. The alpine tundra, found at all latitudes on high mountains above the tree line, has similar flora and fauna.

■ INTERACTIVE QUESTION   50.6

Temperature and precipitation are two of the key factors that influence the vegetation found in a biome. On the climograph shown below, label the North American biomes (arctic and alpine tundra, coniferous forest, desert, grassland, temperate forest, and tropical forest) represented by each area of temperature and precipitation.

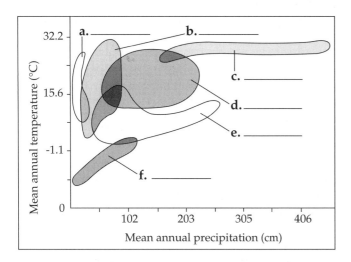

The Spatial Scale of Distributions

Different factors may determine the distribution of a species on different scales (1117)

The range of a species when drawn in terms of worldwide occurrence is different from the actual local areas it occupies throughout its range. Abiotic factors such as climate may limit species on the global scale, but more subtle biotic interactions may explain its location in any one area.

Most species have small geographic ranges (1117–1118)

Only a minority of species are common and widespread; most species in all taxonomic groups are relatively rare.

WORD ROOTS

a- = without; **bio-** = life (*abiotic components:* nonliving chemical and physical factors in the environment)

abyss- = deep, bottomless (*abyssal zone:* the very deep benthic communities near the bottom of the ocean; this region is characterized by continuous cold, extremely high water pressure, low nutrients, and near or total absence of light)

bentho- = the depths of the sea (*benthic zone:* the bottom surfaces of aquatic environments)

estuar- = the sea (*estuary:* the area where a freshwater stream or river merges with the ocean)

eu- = good, well; **troph-** = food, nourishment (*eutrophic:* shallow lakes with high nutrient content in the water)

geo- = the Earth (*biogeography:* the study of the past and present distribution of species)

hydro- = water; **therm-** = heat (*deep-sea hydrothermal vents:* a dark, hot, oxygen-deficient environment associated with volcanic activity; the food producers are chemoautotrophic prokaryotes)

inter- = between (*intertidal zone:* the shallow zone of the ocean where land meets water)

limn- = a lake (*limnetic zone:* the well-lit, open surface waters of a lake farther from shore)

littor- = the seashore (*littoral zone:* the shallow, well-lit waters of a lake close to shore)

oligo- = small, scant (*oligotrophic lake:* a nutrient-poor, clear, deep lake with minimum phytoplankton)

meso- = middle (*mesotrophic:* lakes with moderate amounts of nutrients and phytoplankton productivity intermediate to oligotrophic and eutrophic systems)

micro- = small (*microclimate:* very fine scale variations of climate, such as the specific climatic conditions underneath a log)

pelag- = the sea (*oceanic pelagic biome:* most of the ocean's waters far from shore, constantly mixed by ocean currents)

perman- = remaining (*permafrost:* a permanently frozen stratum below the arctic tundra)

-photo = light (*aphotic zone:* the part of the ocean beneath the photic zone, where light does not penetrate sufficiently for photosynthesis to occur)

profund- = deep (*profundal zone:* the deep aphotic region of a lake)

thermo- = heat; **-clin** = slope (*thermocline:* a narrow stratum of rapid temperature change in the ocean and in many temperate-zone lakes)

STRUCTURE YOUR KNOWLEDGE

1. a. Define *ecology*.
   b. What methods are used to answer ecological questions?
2. a. What are biomes?
   b. What accounts for the similarities in life forms found in the same type of biome in geographically separated areas?

TEST YOUR KNOWLEDGE

MULTIPLE CHOICE: *Choose the one best answer.*

1. Which level of ecology considers energy flow and chemical cycling?
   a. community
   b. ecosystem
   c. organismal
   d. population
   e. abiotic

2. Which level of ecological research would consider how a community is affected by neighboring ecosystems?
   a. ecosystem ecology
   b. landscape ecology
   c. community ecology
   d. population ecology
   e. biosphere ecology

3. Ecologists use mathematical models and computer simulations because
   a. ecological experiments are too broad in scope to be performed.
   b. most of them are mathematicians.
   c. ecology is becoming a more descriptive science.
   d. these approaches allow them to study the interactions of multiple variables and simulate large-scale experiments.
   e. variables can be manipulated with computers but not in field experiments.

4. Which of the following would affect the distribution of a species?
   a. dispersal ability
   b. interactions with mutualistic symbionts
   c. climate and physical factors of the environment
   d. predators, parasites, competitors
   e. All of the above influence where species are found.

5. According to the precautionary principle,
   a. ecological research should provide all the answers before policy decisions are made.
   b. environmental decisions must ignore political, economic, and social concerns and be based strictly on science.
   c. ecological research should not try to manipulate variables in natural settings but only in the laboratory or computer simulations.
   d. environmental decisions should be made carefully, taking into account the complexity of ecosystems and the potential effects of such decisions.
   e. ecologists, not legislators, should make environmental policy and funding decisions.

6. What is the "tens rule"?
   a. One in every ten species has a widespread geographic distribution.
   b. One of ten introduced species is successful; and of these, one in ten becomes common enough to be a pest.
   c. It takes most tree species ten years to disperse ten kilometers.
   d. For every 10° increase in mean temperature, there is a tenfold increase in species diversity.
   e. The experimental addition of a predator species decreases the prey abundance tenfold; the removal of a predator increases prey abundance tenfold.

7. In which of the following biomes is light most likely to be a limiting factor?
   a. desert
   b. estuary
   c. coral reef
   d. grassland
   e. ocean pelagic zone

8. Which of the following is *incorrectly* paired with its description?
   a. neritic zone—shallow area over continental shelf
   b. abyssal zone—benthic region where light does not penetrate
   c. littoral zone—area of open water
   d. intertidal zone—shallow area at edge of water
   e. profundal zone—deep, aphotic region of lakes

9. Phytoplankton are the basis of the food chain in
   a. streams.
   b. wetlands.
   c. the oceanic photic zone.
   d. rocky intertidal zones.
   e. deep-sea hydrothermal vents.

10. The ample rainfall of the tropics and the arid areas around 30° north and south latitudes are caused by
    a. ocean currents that flow clockwise in the northern hemisphere and counterclockwise in the southern hemisphere.
    b. the global circulation of air initiated by intense solar radiation near the equator producing wet and warm air.
    c. the tilting of the earth on its axis and the resulting seasonal changes in climate.
    d. the heavier rain on the windward side of mountain ranges and the drier climate on the leeward side.
    e. the location of tropical rain forests and deserts.

11. The permafrost of the arctic tundra
    a. prevents plants from getting established and growing.
    b. protects small animals during the long winters.
    c. prevents plant roots from penetrating very far into the soil.
    d. helps to keep the soil from getting wet since water cannot soak in.
    e. Both c and d are correct.

12. Many plant species have adaptations for dealing with the periodic fires typical of a
    a. savanna.
    b. chaparral.
    c. temperate grassland.
    d. temperate deciduous forest.
    e. a, b, or c.

13. Two communities have the same mean temperature and rainfall but very different compositions and characteristics. The best explanation for this phenomenon is that the two
    a. are found at different altitudes.
    b. are composed of species that have very low dispersal rates.
    c. are found on different continents.
    d. receive different amounts of sunlight.
    e. have different seasonal temperatures and patterns of rainfall throughout the year.

14. Upwellings in the ocean
    a. are locations of reef communities.
    b. occur over deep-sea hydrothermal vents.
    c. are responsible for ocean currents.
    d. bring nutrient-rich water to the surface.
    e. are most common in tropical waters, where they bring oxygen-rich water to the surface.

15. Why do the tropics and the windward side of mountains receive more rainfall than areas a-round 30° latitude or the leeward side of mountains?
    a. Rising air expands, cools, and drops its moisture.
    b. Descending air condenses and drops its moisture.
    c. The tropics and the windward side of mountains are closer to the ocean.
    d. There is more solar radiation in the tropics and on the windward side of mountains.
    e. The rotation of the earth determines global wind patterns.

MATCHING: *Match the biotic description with its biome.*

Biome	Biotic Description
_____ 1. chaparral	A. broad-leaved deciduous trees
_____ 2. desert	B. lush growth, vertical layers
_____ 3. savanna	C. evergreen shrubs, fire-adapted vegetation
_____ 4. taiga	D. tropical grasses and forbs
_____ 5. temperate forest	E. coniferous forests
_____ 6. temperate grassland	F. low shrubby or matlike vegetation
_____ 7. tropical rain forest	G. grasses in relatively cool regions
_____ 8. tundra	H. widely scattered shrubs, cacti, succulents

CHAPTER 51

BEHAVIORAL BIOLOGY

FRAMEWORK

This chapter introduces the complex and fascinating subject of animal behavior.

Behaviors can range from simple fixed-action patterns in response to specific stimuli to insight learning in novel situations. Behaviors result from interactions among environmental stimuli, experience, and individual genetic makeup. The parameters of behavior are controlled by genetics and thus are acted upon by natural selection. Behavioral ecology focuses on the ultimate cause of reproductive fitness, which can be used to interpret foraging behavior, agonistic interactions, mating patterns, and altruistic behavior. Sociobiology extends evolutionary interpretations to social behavior.

CHAPTER REVIEW

Introduction to Behavior and Behavioral Ecology

What is Behavior? (1122)

Behavior is what an animal does and how it does it. Animal behaviors are observable movements or actions as well as nonmuscular physiological or neural changes.

Behaviors have both proximate and ultimate causes (1122)

The ultimate causation of a particular animal behavior relates to the evolutionary basis of the behavior—why it has been favored by natural selection. Proximate causation explains the immediate cause of a behavior in terms of the cues or stimuli that trigger it and the internal mechanisms that produce it. Proximate causes produce behaviors that evolve because they result in increased Darwinian fitness. Behavioral biologists use phylogenetic comparisons to study the evolutionary histories of behaviors in a particular lineage.

■ INTERACTIVE QUESTION   51.1

Many animals breed in the spring or early summer.

a. What is a probable proximate cause of this behavior?

b. What is the probable ultimate cause of this behavior?

Behavior results from both genes and environmental factors (1122–1123)

All of an animal's anatomical, physiological, and behavioral traits have genetic and environmental components. All these phenotypic traits exhibit a norm of reaction (a range of variation) that relates to the environment in which they are expressed. Most behavioral traits have broad norms of reaction and are determined by multiple genes. The environmental factors affecting behavior include all the physical and chemical conditions within cells and during development, the interactions within the animal, and all the interactions with its physical environment and other organisms.

Innate behavior is developmentally fixed (1123–1124)

Innate behavior is performed virtually the same by all individuals, regardless of environmental differences. These automatic behaviors must have maximized fitness to the extent that genes for variant behavior were lost.

Classical ethology presaged an evolutionary approach to behavioral biology (1124–1126)

Ethology, the study of how animals behave in their natural environment, originated in the 1930s. The work of Lorenz, Tinbergen, and von Frisch provided evidence for the innate components of behavior.

A **fixed-action pattern (FAP)** is a highly stereotyped sequence of behaviors that, once begun, is usually carried through to completion. A FAP is triggered by a **sign stimulus**—some external stimulus that is often a limited subset of available sensory information.

■ **INTERACTIVE QUESTION 51.2**

a. What is the sign stimulus for attack behavior in male stickleback fish?

b. Give an example of a FAP in a human infant and the sign stimulus that illicits it.

Behavioral ecology emphasizes evolutionary hypotheses (1126–1128)

Behavioral ecology looks to evolution for explanations of animal behavior, with the underlying expectation that natural selection favors behaviors that enhance reproductive success.

Songbird Repertoires Many songbirds have a repertoire of song types, which can be distinguished with a sound spectrograph. A behavioral ecology approach to studying this phenomenon may produce testable predictions that older, more experienced males will have larger song repertoires and that females will mate more often or earlier in the season with males having large repertoires. Experimental studies have supported the hypothesis that, at least in some species, larger repertoires have resulted in females pairing earlier with experienced males, thus giving offspring a better chance to survive and increasing the reproductive fitness of the parents.

Cost-Benefit Analysis of Foraging Behavior Behavioral ecologists use cost-benefit analysis to study the proximate and ultimate causes of feeding behaviors or **foraging**. According to the **optimal foraging theory,** feeding behaviors will maximize energy intake over expenditure. Factors involved in an optimal foraging strategy include the distance to a food item; the time and energy required to pursue, catch, and handle the item; the amount of usable energy and nutrients contained in it; and the risk of being preyed upon during feeding. Studies such as those on prey selection by bluegill sunfish have indicated that animals can modify their foraging behavior in ways that tend to maximize overall energy intake. This ability appears to be innate, although experience and physical maturation are thought to increase foraging efficiency.

■ **INTERACTIVE QUESTION 51.3**

Explain how Zack's study of whelk-eating crows supports the optimal foraging theory.

Learning

Learning is experience-based modification of behavior (1128)

Learning is the modification of behavior as a result of experience and can affect even innate behaviors.

Different human languages and the specific alarm calls of vervet monkeys are examples of learned behavioral variations of an innate ability to learn a language or give calls in response to threatening objects.

Learning versus Maturation Improved performance of innate behaviors may result from neuromuscular development, a process called **maturation.** Birds prevented from flying until older are able, upon release, to fly without practice or learning. In other cases, such as cross-fostered herring and laughing gull chicks, learning can be shown to modify a developmentally fixed behavior.

Habituation A simple type of learning called **habituation** is the loss of sensitivity to unimportant stimuli or to stimuli not associated with appropriate feedback. Habituation may increase fitness by allowing an animal's nervous system to focus on important stimuli.

Imprinting is learning limited to a sensitive period (1129–1130)

Learning that occurs during a specific time period and is generally irreversible is called **imprinting.** Mother-offspring bonding is critical for survival of offspring and for the reproductive fitness of the parent. The *imprinting stimulus* for newly hatched geese is the movement of an object, whether it be their mother or Konrad Lorenz, away from them.

Imprinting is characterized by a limited **sensitive period** during which it may occur. The timing and duration of the sensitive period of sexual imprinting may vary.

Bird song provides a model system for understanding the development of behavior (1130–1132)

Studies of bird songs show that the techniques for learning these calls vary for different songbird species. White-crowned sparrows appear to have a 50-day sensitive period in which they memorize the song of their species, forming a template. If raised in isolation during this sensitive period, a bird does not develop a typical adult song, but it can form a template if it listens to tape recordings of its species. In a second learning phase, a juvenile bird sings a subsong, which gradually improves as the bird practices, apparently comparing its own singing with the memorized template. The adult sparrow then sings this crystallized song for the rest of its life.

Experiments also show that the innate tendency to learn only a white-crowned sparrow song can be changed by experience. When sparrows older than 50 days can interact with singing adult birds of another species, the length of their sensitive period for learning songs is longer, and they are able to learn the song of the other species.

People also have a sensitive period for learning vocalizations; learning foreign languages is easier up until the teen years.

Different species of birds show a great deal of diversity in their ability to learn songs. Some can develop nearly normal songs even when raised in isolation or deafened early in life.

Canaries do not crystallize an adult song, but learn a new, more elaborate song each breeding season. Nottebohm's research has identified the forebrain region responsible for song learning in canaries. Changes in the size of this area correlate with the plastic phase and then the learning of new songs each breeding season, as well as with the size of a song repertoire.

Many animals can learn to associate one stimulus with another (1132)

In **associative learning,** animals learn to associate one stimulus with another. **Classical conditioning,** described through the work of Pavlov in the 1900s with salivating dogs, illustrates the association of an irrelevant stimulus with a fixed physiological response.

Operant conditioning refers to trial-and-error learning through which an animal associates a behavior with a reward or punishment. Skinner's work with rats is the best-known laboratory study of operant conditioning.

Practice and exercise may explain the ultimate bases of play (1132–1133)

Species that engage in **play** often have a highly social lifestyle. Despite its use of energy and the risk of injury, play behavior may confer adaptive advantages. Play that simulates stalking and attacking may allow animals to practice these behaviors. Play also promotes muscular and cardiovascular fitness.

■ INTERACTIVE QUESTION 51.4

Indicate the type of learning illustrated by the following examples:

a. Ewes will adopt and nurse a lamb shortly after they give birth to their own lamb but will butt and reject a lamb introduced a day or two later.

b. A dog, whose early "accidents" were cleaned up with paper towels accompanied with harsh discipline, hides under the bed any time a paper towel is used in the household.

c. Ducklings eventually ignore a cardboard silhouette of a hawk that is repeatedly flown over them.

d. Kittens stalk and pounce on each other, biting and kicking as they roll around together.

e. In Pavlov's experiments, the ringing of a bell caused a dog to salivate.

Animal Cognition

Problem-solving behavior is most often observed in mammals, especially in primates and dolphins, although such behavior has also been documented in some bird species.

The study of cognition connects nervous system function with behavior (1133–1136)

Cognition refers to an animal's ability to perceive, store, process, and use information from its sensory receptors. **Cognitive ethology** considers animal awareness or consciousness and the connection between nervous system function and animal behavior.

Animals use various cognitive mechanisms during movement through space (1134–1136)

Kinesis and Taxis A **kinesis** is a simple change in activity or turning rate in response to a stimulus. Although kinetic movements are randomly directed, they tend to maintain organisms in favorable environments. A **taxis** is an oriented movement toward or away from a stimulus, usually performed automatically.

Use of Landmarks Within a Familiar Area Animals may learn and use a particular set of **landmarks** to find their way within their area. Experiments with honeybees fitted with transponders mapped the exploratory, nonforaging flights of young bees and the direct flights of older bees to nectar-rich flowers.

Cognitive Maps More complicated than a set of learned landmarks, **cognitive maps** are internal representations of the spatial relationships of objects in an animal's surroundings. Evidence for cognitive maps comes from research with jays that are able to retrieve stored food from thousands of caches.

Migration Behavior **Migrations** are long-distance, regular movements, often involving a round trip each year. Three mechanisms may be used in migration: Traveling from one familiar landmark to another is called *piloting. Orientation* occurs when an animal detects compass directions and moves in a straight path toward its destination. *Navigation* involves the ability to detect compass direction and to determine present location relative to destination (a "map sense").

Some species of birds and other animals orient and navigate by using a combination of compass cues: the sun, stars, and Earth's magnetic field. Many migrating animals use internal timing mechanisms to adjust to the changing positions of the sun and stars, both throughout the day and along the migratory route.

■ **INTERACTIVE QUESTION   51.5**

Sow bugs are placed in experimental chambers that are either humid or dry and have both light and dark areas. In the humid chamber, the sow bugs move into the dark area and stop moving. In the dry chamber, they move into the dark area and continue to move about in that area. Explain these experimental results.

The study of consciousness poses a unique challenge for scientists (1136)

It is difficult to study scientifically whether or not nonhuman animals are consciously aware of themselves and their surroundings. Griffin and others have argued that conscious thinking is an inherent part of behavior in many animals and that consciousness has adaptive advantages and has arisen by natural selection, forming an evolutionary continuum through many animal groups.

Social Behavior and Sociobiology

Sociobiology places social behavior in an evolutionary context (1137)

Social behavior involves interaction between two or more animals, usually of the same species. Growing from the work of Hamilton and the publication of Wilson's *Sociobiology,* the discipline of **sociobiology** uses evolutionary theory as the basis for studies of social behavior.

Competitive social behaviors often represent contests for resources (1137–1140)

Agonistic Behavior **Agonistic behavior** involves a contest to determine which competitor gains access to a resource, such as food or a mate. The encounter may include a test of strength or, more commonly, symbolic behavior or **ritual,** which serves to avoid actual physical conflicts. Threat displays often lead to a display of submission or appeasement by one of the competitors. Scarcity of resources may result in more intense combat.

Within more permanent social groups, individuals may engage in **reconciliation behaviors** following a conflict.

Dominance Hierarchies Many social groups are structured by agonistic behavior. **Dominance hierarchies,** as illustrated by the "pecking order" of hens, prevent continual combat by establishing which animals get first access to resources.

Territoriality A **territory** is an area established for feeding, mating, and/or rearing young from which conspecifics are excluded. Territory size varies with species, function, and resource abundance. A home range is the area in which an animal may roam, but a territory is the area that the animal defends. Agonistic behavior is used to establish and defend territories. Vocal displays, scent marks, and patrolling may be used to continually proclaim ownership.

■ **INTERACTIVE QUESTION   51.6**

a. Why are many interactions between members of the same species agonistic?

b. What mechanisms reduce violent encounters between conspecifics?

Dominance systems and territoriality help to stabilize population density by ensuring that at least some individuals have a sufficient supply of a limited resource in order to reproduce.

Natural selection favors mating behavior that maximizes the quantity or quality of mating partners (1140–1142)

Courtship Courtship, the behavior patterns that lead up to copulation (or gamete release), often consist of a series of displays by one or both partners. Complex courtship interactions, unique to each species, assure that individuals are of the proper species, sex, and physiological mating condition.

Courtship behaviors may have evolved because of sexual selection. Females, who usually have more **parental investment** in each offspring due to greater time and resource allocation, usually show more discrimination in choosing mates than do males. Males, whose reproduction success may be maximized by fertilizing the eggs of many females, often compete with each other for access to females. Female choice, however, seems to have been a strong selection factor in the evolution of vigorous courtship displays and extreme secondary sexual characteristics, which may indicate good health.

Mating Systems Many species have **promiscuous** mating, with no strong pair bonds forming. Longer lasting relationships may be **monogamous** or **polygamous.** Polygamous relationships are most often **polygynous** (one male and many females), although a few are **polyandrous.**

The needs of offspring are an ultimate factor in the reproductive pattern of the parents. If young require more food than one parent can supply, a male may increase his reproductive fitness by helping to care for offspring rather than going off in search of more mates. With mammals, the female often provides all the food, and males are often polygynous.

Certainty of paternity also influences mating systems and parental care. With internal fertilization, the acts of mating and egg laying or birth are separated, and paternity is less certain than when eggs are fertilized externally.

Social interactions depend on diverse modes of communication (1142–1144)

Defining Animal Signals and Communication Communication is the intentional transmission of information between individuals using special behaviors called **signals.** A change in behavior of the "receiver" is an indication that **communication** has occurred.

Communication may involve visual, auditory, chemical, tactile, and electrical signals, depending on the lifestyle and sensory specializations of a species.

■ **INTERACTIVE QUESTION 51.7**

a. Explain the basis for the distinction between male competition and female choice in courtship behavior

b. Natural selection has resulted in exclusive male parental care being much more frequent in species with external fertilization, where the male's genetic contribution to the offspring is more certain. Explain how such behavior could evolve.

Pheromones **Pheromones** are chemical signals commonly used by mammals and insects in reproductive behavior to attract mates and to trigger specific courtship behaviors. The trailing behavior of ants is based on pheromones. Social or hive bees use pheromones to maintain the social order of the colony.

The Dance Language of the Honeybee Bees use ritualized dances to communicate the location of food sources. Round dances are used when the food source is relatively close to the hive. Regurgitated nectar provides a scent to direct other bees to the food. The location of more distant food is communicated by waggle dances, in which the duration and vertical orientation of the dance on the comb seem to indicate the distance from the hive and the direction to the food relative to the horizontal angle to the sun. Sounds and odors coming from a dancing bee may also communicate information about a food resource.

■ **INTERACTIVE QUESTION 51.8**

Why is most communication among mammals olfactory and auditory, whereas communication among birds is visual and auditory?

The concept of inclusive fitness can account for most altruistic behavior (1144–1147)

Many social behaviors are selfish, benefiting an individual at the expense of others. **Altruism** is behavior that reduces an individual's fitness while increasing the fitness of the recipient.

Inclusive Fitness   Hamilton was the first behavioral ecologist to explain altruistic behavior in terms of **inclusive fitness,** the ability of an individual to pass on its genes either by producing its own offspring or by helping close relatives produce their offspring.

Hamilton's Rule and Kin Selection   Hamilton developed a quantitative measure, called **Hamilton's rule,** that predicts that natural selection would favor altruistic acts among related individuals if $rB > C$. B and C are the benefit to the recipient and the cost to the altruist, measured by the change in the average number of offspring produced as a result of the altruistic act. The term r refers to the **coefficient of relatedness,** a measure of the probability of a gene being inherited by two individuals from a commom ancestor. **Kin selection** is the term for the natural selection of altruistic behavior that enhances the reproductive success of related individuals.

Studies show that most cases of altruistic behavior involve close relatives and thus improve the individual's inclusive fitness.

DNA analysis of naked mole rats shows that all members of a colony are closely related to the reproducing queen and kings. Cooperative behavior in common mole rat colonies, where genetic relatedness is much less, may relate to the harsh, arid environment in which these colonies are found. Resource limitations may have favored the evolution of cooperative colonies.

When altruistic behavior involves nonrelated animals, the explanation offered is **reciprocal altruism;** there is no immediate benefit for the altruistic individual, but some future benefit may occur when the helped animal may "return the favor." Reciprocal altruism often is used to explain altruism in humans.

■ INTERACTIVE QUESTION   51.9

a. According to kin selection, would an individual be more likely to exhibit altruistic behavior toward a parent, a sibling, or a first (full) cousin?

b. Explain your answer in terms of the coefficient of relatedness and Hamilton's rule.

Sociobiology connects evolutionary theory to human culture (1147–1148)

In *Sociobiology,* Wilson speculated on the evolutionary basis of certain social behaviors of humans.

The parameters of human social behavior may be set by genetics, but the environment undoubtedly shapes behavioral traits just as it influences the expression of physical traits. Due to our capacity for learning and integration, human behavior appears to be quite plastic. Our structured societies, with their regulations on behaviors, including behaviors that would enhance an individual's fitness, may be the one unique characteristic separating humans and other animals.

WORD ROOTS

agon- = a contest (*agonistic behavior:* a type of behavior involving a contest of some kind that determines which competitor gains access to some resource, such as food or mates)

andro- = a man (*polyandry:* a polygamous mating system involving one female and many males)

etho- = custom, habit (*ethology:* the study of animal behavior in natural conditions)

gyno- = a woman (*polygyny:* a polygamous mating system involving one male and many females)

kine- = move (*kinesis:* a change in activity rate in response to a stimulus)

mono- = one; **-gamy** = reproduction (*monogamous:* a type of relationship in which one male mates with just one female)

poly- = many (*polygamous:* a type of relationship in which an individual of one sex mates with several of the other sex)

socio- = a companion (*sociobiology:* the study of social behavior based on evolutionary theory)

STRUCTURE YOUR KNOWLEDGE

1. How does the nature-versus-nurture controversy apply to behavior?
2. How does the concept of Darwinian fitness apply to behavior?

TEST YOUR KNOWLEDGE

MULTIPLE CHOICE: *Choose the one best answer.*

1. Behavioral ecology is the
   a. mechanistic study of the behavior of animals, focusing on stimulus and response.
   b. application of human emotions and thoughts to other animals.
   c. study of animal cognition.
   d. study of animal behavior from an evolutionary perspective of Darwinian fitness.
   e. study of the ecological basis of behavior.

2. Proximate causes
   a. explain the evolutionary significance of a behavior.
   b. are immediate causes of behavior such as environmental stimuli.
   c. are environmental, whereas ultimate causes are genetic.
   d. are endogenous, although they may be set by exogenous cues.
   e. show that nature is more important than nurture.

3. Animals appear to maximize their energy intake-to-expenditure ratio. What is this behavior called?
   a. optimal foraging
   b. territoriality
   c. a fixed-action pattern
   d. maturation
   e. learning

4. Which of the following is an example of a fixed-action pattern?
   a. a bluegill sunfish feeding on larger *Daphnia* when prey are abundant
   b. a chick pecking at the red spot on a parent's moving beak
   c. a whale migrating long distances to its feeding territory
   d. a songbird learning its song after listening to a taped song of its species
   e. a bird learning to avoid monarch butterflies

5. A change in behavior as a result of experience is called
   a. habituation.
   b. imprinting.
   c. insight.
   d. learning.
   e. maturation.

6. A sensitive period
   a. is the time right after birth when sexual identity is developed.
   b. usually follows the receiving of a sign stimulus.
   c. is a limited time during which imprinting can occur.
   d. is the period during which birds can learn to fly.
   e. is the time during which social animals play.

7. In operant conditioning,
   a. an animal improves its performance of a fixed-action pattern (FAP).
   b. an animal learns as a result of trial and error.
   c. sensitivity to unimportant or repetitive stimuli decreases.
   d. a bird can learn the song of a related species if it hears only that song.
   e. an irrelevant stimulus can elicit a response because of its association with a normal stimulus.

8. Which modality of intraspecies communication signal would be best suited to a nocturnal species such as an owl?
   a. auditory
   b. visual
   c. chemical
   d. tactile
   e. electrical

9. A kinesis
   a. is a randomly directed movement that is not caused by external stimuli.
   b. is a movement that is directed toward or away from a stimulus.
   c. is a change in activity rate in response to a stimulus.
   d. is illustrated by trout swimming upstream.
   e. often involves piloting but not orientation or navigation.

10. A dominance hierarchy
    a. may be established by agonistic behavior.
    b. determines which animals get first access to resources.
    c. helps to avoid potential injury of competitors.
    d. may help to stabilize population density.
    e. applies to all of the above.

11. An animal's territory
    a. may be larger than its home range.
    b. may decrease in size if resources dwindle and expand if resources become more plentiful.
    c. excludes both conspecifics and members of other species.
    d. may be proclaimed by scent marks, vocal displays, and patrolling.
    e. applies to all of the above.

12. In a species in which females provide all the needed food and protection for the young,
    a. males are likely to be promiscuous.
    b. mating systems are likely to be monogamous.
    c. mating systems are likely to be polyandrous.
    d. males most likely will show sexual selection.
    e. females will have a higher Darwinian fitness than males.

13. The ability of honeybees to fly directly to a food source, after having to wait several hours from the time of the waggle dance, indicates that
    a. the waggle dance provided directions relative only to the position of the hive.
    b. bees have an internal clock that compensates for the movement of the sun during the elapsed time.
    c. the bees must have been to that food source before.
    d. bees are directed more by olfactory cues than by directional cues.
    e. the intensity of the dance provided directional cues.

14. A crow that aids its parents in raising siblings is increasing its
    a. reproductive success.
    b. altruistic behavior.
    c. inclusive fitness.
    d. coefficient of relatedness.
    e. certainty of paternity.

15. Sociobiology
    a. explains the evolutionary basis of behavioral characteristics within animal societies.
    b. applies evolutionary explanations to human social behaviors.
    c. studies the roles of culture and genetics in human social behavior.
    d. considers communication, mating systems, and altruism from the viewpoint of Darwinian fitness.
    e. does all of the above.

16. According to the concept of kin selection,
    a. an animal would be more likely to aid a stranger if the "kindness" could be reciprocated.
    b. an animal would aid its parent before it would help its sibling.
    c. animals are more likely to choose close relatives as mates.
    d. examples of altruism usually involve close relatives and increase an animal's inclusive fitness.
    e. evolution is the proximate cause of animal behavior.

17. A female bird would most likely increase her Darwinian fitness by
    a. mating with as many males as possible.
    b. choosing a mate based on evidence that he is a "good provider" or has "good genes."
    c. reproducing many times in her lifetime.
    d. being polyandrous.
    e. foraging in a large flock.

18. Which of the following would *not* contribute to the migratory behavior of a bird that travels over long distances through all types of weather to return to its breeding grounds?
    a. ability to sense Earth's magnetic field
    b. navigation
    c. orientation
    d. piloting
    e. kinesis and taxis

19. When a white-crowned sparrow sings a subsong,
    a. it is practicing the songs of other bird species in its vicinity.
    b. it has developed the crystallized song that it will sing for the rest of its life.
    c. it is in the sensitive period during which the song template is formed.
    d. it apparently compares the template it formed during its sensitive period to its own singing.
    e. it is in the plastic phase when it is creating a new, more elaborate song for the next breeding season.

20. According to Hamilton's rule, natural selection would favor altruistic acts when
    a. the probability that the altruist will lose its life is less than 0.5 and the coefficient of relatedness is greater than 0.25 *(rC=B)*.
    b. the cost to the altruist times the coefficient of relatedness is less than the benefit to the receiver *(rC < B)*.
    c. the benefit to the receiver times the coefficient of relatedness is greater than the cost to the altruist *(rB > C)*.
    d. the cost to the receiver times the coefficient of relatedness is greater than the cost to the altruist *(rC > B)*.
    e. the benefit to the altruist times the coefficient of relatedness is less than the cost to the receiver *(rB < C)*.

21. In which species would you be most likely to find reconciliation behaviors following a conflict?
    a. canaries
    b. chicken
    c. chimpanzees
    d. wolves
    e. caged ibex (a type of wild goat)

22. Which of the following examples of behavior provide evidence of animal cognition?
    a. chimpanzee stacking up boxes to reach a banana
    b. ravens pulling up string to obtain attached food item
    c. injury-feigning behavior of ground-nesting bird
    d. the cognitive map of a jay that enables it to retrieve food from its many caches
    e. All of the above show evidence of information processing and animal cognition.

CHAPTER 52

POPULATION ECOLOGY

FRAMEWORK

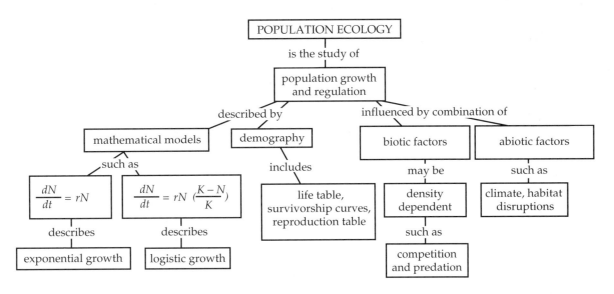

CHAPTER REVIEW

The continuing growth of the human population in the face of limited resources is a critical biological phenomenon. Population ecology is the study of the fluctuations in population size and composition and their ecological causes.

Characteristics of Populations

A **population** is a group of individuals of the same species that occupy the same area, use the same resources, and respond to similar environmental factors. The characteristics of a population are determined by interactions with the environment on both ecological and evolutionary time scales.

Two important characteristics of any population are density and the spacing of individuals (1152–1153)

Every population has geographic boundaries; ecologists use boundaries based upon the type of organism and the research question being asked. The number of individuals per unit area or volume is the population

density; the spacing of those individuals within the population is referred to as **dispersion.**

Measuring Density Population density is often measured by using one of a variety of sampling techniques to count and estimate population size. Indirect indicators, such as burrows or nests, also may be used. In the **mark-recapture method,** animals are trapped, marked, released, and trapped again. The proportion of the total second trapping that are marked indicates the proportion of the population that was marked; dividing the initial number marked by this proportion estimates population size.

■ **INTERACTIVE QUESTION 52.1**

In a mark–recapture study, an ecologist traps, marks, and releases 25 voles in a small wooded area. A week later she resets her traps and captures 30 voles, 10 of which were marked. What is her estimate of the population of voles in that area?

Patterns of Dispersion Individuals may be dispersed in the population's geographic range in several patterns. **Clumping** may indicate a heterogeneous environment, with organisms congregating in suitable microenvironments; clumping also may be related to social interactions between individuals.

Uniform distribution is usually related to competition for resources, resulting in interactions that produce territories or spaces between individuals. **Random** spacing, indicating the absence of strong attractions or repulsions between individuals, is not very common.

Populations of a species may also show dispersion patterns within their geographic range.

Demography is the study of factors that affect the growth and decline of populations (1153–1156)

Population size changes in response to the relative rates of birth and immigration versus mortality (death) and emigration. **Demography** is the study of the vital statistics that affect population size.

Life Tables and Survivorship Curves A **life table** presents age-specific survival data for a population. It can be constructed by following a **cohort** of organisms from birth to death.

A **survivorship curve** shows the number or proportion of members of a cohort still alive at each age. There are three general types of survivorship curves. Type I, with low mortality during early and middle age and a rapid increase with old age, is typical of populations that produce relatively few offspring and provide parental care. In a Type II curve, death rate is relatively constant throughout the life span. A Type III curve is typical of populations that produce many offspring, most of which die off rapidly. The few that survive are likely to reach adulthood. Many species show intermediate or more complex survivorship patterns.

Reproductive Rates In sexually reproducing species, demographers usually follow only female reproduction, often only of female offspring. A **reproductive table** gives the age-specific reproductive rates in a population. Such a fertility schedule can be constructed by following the reproductive output of a cohort, measuring the number of female offspring by age group. The net output of daughters per individual in an age class is the product of the proportion of females breeding and the number of female offspring they produce.

■ **INTERACTIVE QUESTION** **52.2**

Identify the types of survivorship curves shown below and give examples of groups that exhibit each curve.

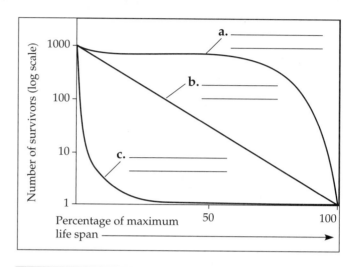

Life Histories

The **life history** of an organism—its schedule of reproduction and survival—affects population growth in ecological time but evolved as a result of natural selection operating in evolutionary time.

Life histories are highly diverse, but they exhibit patterns in their variability (1156)

Some species put all their reproductive resources into a single reproductive effort, often called **big-bang reproduction.** Other species follow the strategy of **repeated reproduction,** making smaller reproductive efforts over a span of time. A key factor in the selection for and evolution of **semelparity** (big-bang) versus **iteroparity** (repeated reproduction) is the survival rate of young offspring. Repeated reproduction may be favored when chances of offspring survival are low.

Limited resources mandate trade-offs between investments in reproduction and survival (1157–1158)

Darwinian fitness is measured by how many offspring survive to produce their own offspring. Because organisms have a finite energy budget, they cannot maximize all life history factors simultaneously. Reproductive costs often include a reduction in survival or in future reproductions. Evolution has fashioned how individuals in a population "choose" how often to breed, at what age to begin reproduction, and how many offspring to produce at a time.

The production of large numbers of offspring is related to the selective pressures of high mortality rates of young in uncertain environments or from intense predation. Parental investment in the size of offspring, incubation, and/or parental care increases survival chances of the young.

■ INTERACTIVE QUESTION 52.3

Mortality, number of offspring per reproduction, and parental investment are usually interrelated. On the following graphs, sketch the relationship you would predict between the variables.

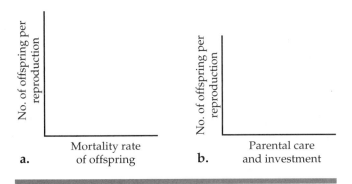

a. Mortality rate of offspring

b. Parental care and investment

Population Growth

Observations, experiments, and mathematical modeling are used to determine rates of population growth, to study variables affecting growth, and to predict population sizes.

The exponential model of population growth describes an idealized population in an unlimited environment (1159–1160)

Growth of a small population in a very favorable environment will be restricted only by the physiological ability of that species to reproduce. Ignoring immigration and emigration, the change in population size during a specific time period is equal to the number of births minus deaths. Births and deaths can be expressed in terms of a per capita birth rate and death rate; the change in population size per unit of time can be signified as $\Delta N/\Delta t = bN - dN$. The difference in per capita birth rate and death rate is the per capita population growth rate, symbolized by r; $r = b - d$. **Zero population growth (ZPG)** occurs when $r = 0$. The formula describing change in the population at

any one instant uses the notation of differential calculus and is written as $dN/dt = rN$.

The **intrinsic rate of increase** (r) is the fastest growth rate possible for a population reproducing under ideal conditions. This **exponential population growth,** expressed as $dN/dt = rN$, produces a J-shaped growth curve when graphed. The larger the population (N) becomes, the faster the population grows. A population's maximum growth rate (r_{max}) is determined by its life history features. Periods of geometric (exponential) growth may occur in some populations that exploit an unfilled environment or rebound from a catastrophic event.

The logistic model of population growth incorporates the concept of carrying capacity (1160–1163)

A population may grow exponentially for only a short time before its increased density limits the resources available for continued growth and reproduction of its members. Crowding and resource limitation may lead to decreased per capita birth rates and increased per capita death rates. The **carrying capacity** (K) is the maximum stable population size that a particular environment can support without harm.

The Logistic Growth Equation The mathematical model of **logistic population growth** ($dN/dt = rN(K-N)/K$) includes the expression $(K-N)/K$ to reflect the impact of the increasing population size on r as the population approaches the carrying capacity.

When N is small, $(K-N)/K$ is close to 1, and growth is approximately exponential (rN). As population size approaches the carrying capacity, the $(K-N)/K$ term becomes a small fraction, and rN is reduced by that fraction. When N reaches K, the term $(K-N)/K$ is 0, and the growth rate is 0. At this point, birth rate equals death rate, and the size of the population does not increase. According to the logistic model, population growth is density dependent. The logistic model produces an S-shaped growth curve, and maximum increase in population numbers occurs when N is intermediate.

How Well Does the Logistic Model Fit the Growth of Real Populations? Some laboratory populations of small animals and microorganisms show logistic growth. Natural populations may grow logistically but few reach a stable carrying capacity.

The logistic model makes the assumption that any increase in population numbers will have a negative effect on population growth. The *Allee effect* is seen in some populations, however, when individuals benefit as the population grows, either from physical support, as in plants, or from social interactions important to reproduction.

The logistic model also assumes that populations approach their carrying capacity smoothly, but many populations overshoot and then oscillate above and below a general carrying capacity.

There are populations for which population density is not an important factor. These populations are often reduced by environmental conditions before resources have a chance to become limiting.

■ INTERACTIVE QUESTION 52.4

Label the exponential and logistic growth curves, and show the equation associated with each curve. What is *K* for the population shown with curve **b**?

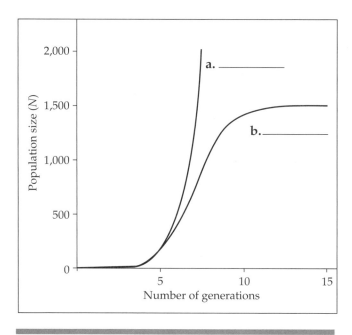

The Logistic Population Growth Model and Life Histories Natural selection will favor different life history traits depending on population densities and environmental conditions. Populations at high density, close to their carrying capacity, may experience ***K*-selection** or density-dependent selection for traits such as competitive ability and efficient resource utilization. In variable environments in which population density fluctuates or in open habitats with low population density, ***r*-selection** or density-independent selection would favor traits that maximize population growth, such as increased fecundity and early maturity. Laboratory studies that have varied population densities and conditions have produced different proportions of *K*-selected and *r*-selected traits in populations of the same species.

Population-Limiting Factors

The ecological questions of what stops population growth and what determines population density in different habitats have practical applications for conservation and agriculture.

Density-dependent decreases in birth rate and increases in death rate may regulate populations through **negative feedback.** If the rates of birth and death do not change with population density, they are called **density independent** and do not regulate growth by feedback. An equilibrium density may be reached in a population as long as birth rate or death rate or both are density dependent.

Negative feedback prevents unlimited population growth (1164-1165)

A variety of factors may produce the negative feedback that changes birth and death rates. Numerous field studies have attempted to identify these factors. Problems that make clear generalizations on limiting mechanisms difficult include the lack of studies in tropical and polar regions, the focus on bird and mammal populations without representative insect studies, and the need for very long-term experimental studies on population dynamics.

Limited food supply often limits reproductive output. The availability of territorial space may be the limiting resource for some animals. Increased population densities may affect health and survivorship in both plant and animal populations. The accumulation of toxic metabolic wastes and the transmission rate of disease may also be limiting factors.

Predation may be a density-dependent factor when a predator shows switching behavior and feeds preferentially on a prey population that has reached a high density.

Intrinsic factors may also regulate population size. Studies of mice have shown that even when food or shelter is not limiting, population size stabilizes when high densities induce a stress syndrome of hormonal changes that inhibit reproduction and increase mortality.

Population dynamics reflect a complex interaction of biotic and abiotic influences (1165–1167)

Most populations show long-term fluctuations in numbers. The decline in the population of northern pintail ducks has been linked to loss of prairie ponds, cultivation of nesting fields, and increased predation due to the concentration of nests in the remaining natural vegetation. Dramatic fluctuations in populations of large grazers and browsers may be linked to the severity of winter.

The erratic fluctuations of some populations have been linked to a combination of biotic and abiotic factors—in the case of the Dungeness crab, to density-dependent cannibalism and minor changes in ocean currents.

Some populations have regular boom-and- bust cycles (1167–1168)

The ten-year cycles in density of snowshoe hares and lynx in boreal forests have been studied to determine whether food shortages, predator overexploitation, or a combination of both causes the cyclic collapses in hare populations. Experimentally increasing the food supply raised the carrying capacity for hares, but the density cycles continued to occur in both experimental and control areas. Field ecologists determined that 90% of hare deaths are due to predators. Experiments that excluded predators from one area and both excluded predators and added food to another area support the hypothesis that excessive predation mainly drives hare cycles, but available winter food supply is a contributing influence.

Cyclical population patterns have been observed in small herbivorous mammals, birds, and insects, and the causes of these fluctuations may vary. One current idea is that crowding increases social aggression and causes endocrine changes that may result in population cycles. Cyclical patterns may also result from a time lag in predator response to increased prey density, as predators reproduce more slowly than prey. Predator cycles most likely follow the population cycles of their prey, and these may be accentuated when predators turn on one another as prey become scarce.

■ **INTERACTIVE QUESTION 52.5**

a. List some density-dependent factors that may limit population growth.

b. List some abiotic factors that may cause population fluctuaFtions.

Human Population Growth

The human population has been growing almost exponentially for three centuries but cannot do so indefinitely (1168–1169)

From 1650, it took 200 years for the global population to double to 1 billion. In the next 80 years, it doubled to 2 billion; it doubled again in the next 45 years and

is projected to reach 7.8 billion by the year 2025 if the present growth rate is maintained.

Demographic Transition Population stability can be reached in one of two ways: Zero population growth = high birth rates − high death rates; or zero population growth = low birth rates − low death rates. The movement from the first configuration to the second is called the **demographic transition.** Death rates declined rapidly in most third world countries after 1950, but birth rate decline has been more uneven—from rapid in China to just beginning in most of Africa. The world's population growth is regional; it is near equilibrium (0.1%) in developed nations and 1.7% in less developed countries, where 80% of the world's population lives.

Age Structure The age structure of a population influences present and future growth. A large proportion of individuals of reproductive age or younger results in more rapid growth. Age structure also predicts future social conditions and needs.

Estimating Earth's carrying capacity for humans is a complex problem (1169–1171)

Wide Range of Estimates for Carrying Capacity Estimates of Earth's carrying capacity have varied greatly and average around 10–15 billion humans. These estimates may be based on the logistic equation or the amount of inhabitable land. Estimates based on food as the limiting factor must take into account the amount of agricultural land, average yield of crops, the type of diet, and the number of calories allocated to each person.

Ecological Footprint The concept of an **ecological footprint** takes into account the multiple constraints involved in estimating carrying capacity. Six types of ecologically productive areas are used to calculate each country's ecological footprint: arable land, pasture, forest, ocean, built-up land, and fossil energy land (vegetative area required to absorb CO_2 from burning fossil fuels). Each type of ecologically productive area is converted to land area per person and totaled for the planet, adding up to about 2 hectares (ha) per person. Land for parks and conservation reduces this estimate to 1.7 ha per person of land available for human use. Ecological footprints vary greatly by country, as do the available ecological capacities of each nation. In a 1997 study, the United States had an ecological footprint of 8.4 ha per person and an available ecological capacity of only 6.2 ha. Thus, the United States has already exceeded its carrying capacity and is overutilizing the world's resources. This study also suggests that the world human population as a whole is already slightly above its carrying capacity.

The ultimate carrying capacity of Earth may be determined by food supplies, space, nonrenewable resources, degradation of the environment, or several interacting factors.

Human population growth is unique in that it can be consciously controlled by voluntary contraception or government-supported birth control programs. The key to the demographic transition is reduced family size. In many cultures, women are delaying marriage and reproduction, thus slowing population growth.

When and how human population growth will level off is an issue of great ecological consequence.

WORD ROOTS

co- = together (*cohort:* a group of individuals of the same age, from birth until all are dead)

demo- = people; **-graphy** = writing (*demography:* the study of statistics relating to births and deaths in populations)

itero- = to repeat (*iteroparity:* a life history in which adults produce large numbers of offspring over many years; also known as repeated reproduction)

semel- = once; **-parity** = to beget (*semelparity:* a life history in which adults have but a single reproductive opportunity to produce large numbers of offspring, such as the life history of the Pacific salmon; also known as "big-bang reproduction")

STRUCTURE YOUR KNOWLEDGE

1. Create a concept map to organize your understanding of the exponential and logistic equations—the mathematical models of population growth.

2. What is the best collection of life history traits that would maximize reproductive success?

TEST YOUR KNOWLEDGE

MULTIPLE CHOICE: *Choose the one best answer.*

1. In a range with a heterogeneous distribution of suitable habitats, the dispersion pattern of a population probably would be
   a. clumped.
   b. uniform.
   c. random.
   d. unpredictable.
   e. dense.

2. Which of the following is *not* true of life tables?
   a. They were first used by life insurance companies to estimate survival patterns.
   b. They show the age-specific mortality or death rate for a species.
   c. Ecologists have collected them for many natural populations.
   d. They can be used to construct survivorship curves.
   e. They are often constructed by following a cohort from birth to death.

3. In a species in which offspring survival is quite inconsistent or low, one would predict
   a. the production of a small number of large offspring.
   b. the production of a large number of large offspring.
   c. iteroparity or repeated reproduction.
   d. semelparity or big-bang reproduction.
   e. more *r*-selected traits such as fast population growth.

4. A Type I survivorship curve is level at first, with a rapid increase in mortality in old age. This type of curve is
   a. typical of many invertebrates that produce large numbers of offspring.
   b. typical of humans and other large mammals.
   c. found most often in *r*-selected populations.
   d. almost never found in nature.
   e. typical of all species of birds.

5. The middle of the S growth curve in the logistic growth model
   a. shows that at middle densities, individuals of a population do not affect each other.
   b. is best described by the term *rN*.
   c. shows that reproduction will occur only until the population size reaches *K* and *dN/dt* becomes 0.
   d. is the period when competition for resources is highest.
   e. is the period when the population is increasing the fastest.

6. A few members of a population have reached a favorable habitat with few predators and unlimited resources, but their population growth rate is slower than that of the parent population. What is a possible explanation for this situation?
   a. The genetic makeup of these founders may be less favorable than that of the parent population.
   b. The parent population may still be in the exponential part of its growth curve and not yet limited by density-dependent factors.
   c. The Allee effect may be operating; there are not enough population members present for successful reproduction.

d. a, b, and c may apply.

e. This scenario would not happen.

7. The term $(K-N)/K$

    a. is the carrying capacity for a population.

    b. is greatest when K is very large.

    c. is zero when population size equals carrying capacity.

    d. increases in value as N approaches K.

    e. accounts for the overshoot of carrying capacity.

8. Which of the following would *not* be a density-dependent factor limiting a population's growth?

    a. switching behavior of a predator

    b. a limited number of available nesting sites

    c. a stress syndrome that alters hormone levels

    d. a very early fall frost

    e. intraspecific competition

9. The carrying capacity for a population is estimated at 500; the population size is currently 400; and r is 0.01. What is dN/dt?

    a. 0.01

    b. 0.8

    c. 8

    d. 40

    e. 50

10. In order to maintain the largest sustainable fish harvest, fishing efforts should

    a. take only postreproductive fish.

    b. maintain the population close to its carrying capacity.

    c. reduce the population to a very low number to take advantage of exponential growth.

    d. maintain the population density close to ½ K.

    e. be prohibited.

11. The human population is growing at such an alarmingly fast rate because

    a. technology has increased our carrying capacity and, thus, density-dependent factors have not slowed reproduction.

    b. the death rate has greatly decreased since the Industrial Revolution.

    c. the age structure of many countries is highly skewed toward younger ages.

    d. infant mortality has decreased.

    e. all of the above are true.

12. As a population approaches its carrying capacity in a particular environment, which of the following would be expected to occur?

    a. N moves toward and becomes 0.

    b. r becomes negative.

    c. Birth rate decreases and death rate increases.

    d. Births cease and zero population growth is achieved.

    e. Exponential growth begins to slow.

Questions 13–15. Use the following choices to indicate how these life history characteristics would be affected by the described changes.

    a. increase

    b. decrease

    c. stay the same

    d. no relationship or unable to predict

13. For a population regulated by density-dependent factors, how might clutch or seed crop size change with increased population density?

14. For a population in a particular environment, how would K be expected to change with an increase in N?

15. In a population showing exponential growth, how would dN/dt be expected to change with an increase in N?

16. Experimental studies of the population cycles of the snowshoe hare and the lynx have shown that

    a. the hare population is regulated by its food resources because adding food increased the carrying capacity of experimental areas.

    b. hares are as likely to die of starvation as of predation.

    c. lynx are the only predators of hares and increases in the lynx population cause the cycles in the hare populations.

    d. the stress of overcrowding causes the population cycles in both hare and lynx.

    e. the hare population is regulated by a combination of food and predators (not just the lynx); the lynx population appears to cycle in response to its prey availability.

17. The demographic transition is the gradual shift from

    a. a Type I survivorship curve to a Type II curve.

    b. semelparity to iteroparity.

    c. an age structure skewed toward the younger ages to an even age distribution.

    d. high birth rates and high death rates to low birth rates and low death rates.

    e. exponential growth to logistic growth.

18. An ecological footprint is an estimate of

    a. the carrying capacity of each nation.

    b. the available ecological capacity of each nation.

    c. the amount of land needed per person to meet the current demand on resources.

    d. the size of a population in relationship to the resources it uses.

    e. how much land is needed to produce food for a vegetarian versus a meateater.

CHAPTER 53

COMMUNITY ECOLOGY

FRAMEWORK

Communities are composed of populations of various species that may interact through competition, predation, parasitism, or mutualism. The structure of a community—its species composition and relative abundance—is determined by these interactions and the trophic structure of the community. Disturbances keep most communities in a state of nonequilibrium. Species richness and biodiversity relate to a community's size and geographic location.

CHAPTER REVIEW

The aggregate of different species living close enough to allow for interaction is called a biological **community.** Community ecology studies the factors involved in determining species composition and the relative abundance of species in a community.

What Is a Community?

Species richness, the number of different species found in a community, and **relative abundance,** the relative numbers of individuals in each species, are both components of a community's structure.

Contrasting views of communities are rooted in the individualistic and interactive hypotheses (1175–1176)

Ecologists in the early 1900s developed two differing views on community composition based upon plant distributions. Gleason advanced the **individualistic hypothesis** that saw communities as chance groupings of species found in the same area because of similar abiotic requirements. Clements advocated the **interactive hypothesis** of a community functioning as an integrated unit, with species linked together by their interrelationships.

The individualistic hypothesis predicts that species have independent distributions along environmental gradients and that boundaries between communities are indistinct. The interactive hypothesis, in contrast, predicts that species are clustered into discrete communities.

■ INTERACTIVE QUESTION 53.1

Species composition and distribution in most plant communities appear to be individualistic. What may explain the occasional occurrence of sharp delineations in species composition between communities?

The debate continues with the rivet and redundancy models (1176)

The **rivet model** of communities, first proposed by P. and A. Ehrlich in 1981, suggests that changing the composition or abundance of species in a community would affect many of the members of that interwoven community. The **redundancy model,** proposed by Walker in 1992, views communities as loosely connected assemblages, in which species are redundant—disappearing members will be replaced by other species. The relationships among members of most communities probably fall between these two polar models.

Interspecific Interactions and Community Structure

Interspecific interactions occur between the different species living in a community.

Populations may be linked by competition, predation, mutualism, and commensalism (1176–1181)

The effect of interspecific interactions on population densities can be signified by $+$ and $-$ signs. For example, in a $+/-$ interaction such as predation, the interaction is beneficial to one species and detrimental to the other.

Competition If populations of two species use the same limited resource, **interspecific competition** may affect the densities of both populations.

Gause's laboratory experiments with *Paramecium* showed that two species of protists that rely on the same limited resource could not coexist in the same community. This **competitive exclusion principle** predicts that the less efficient competitor will be locally eliminated.

An organism's **ecological niche** is described as its place in an ecosystem—its habitat and use of biotic and abiotic resources. The competitive exclusion principle holds that two species with identical niches cannot coexist in a community.

Resource partitioning, slight variations in niche that allow ecologically similar species to coexist, provides circumstantial evidence that competition was a selection factor in evolution. **Character displacement** of some morphological trait or resource use allows closely related sympatric species to avoid competition. When these species are allopatric, their differences may be much less.

Predation **Predation** involves a predator eating prey, but also includes **herbivory,** in which herbivores eat plants, and **parasitism,** in which parasites live and feed on or in their hosts.

Adaptations to increase success in predation may include acute senses, speed and agility, camouflage coloration, and physical structures such as claws, fangs, teeth, and stingers.

Plants may defend themselves with mechanical devices, such as thorns, or chemical compounds. Distasteful or toxic chemicals include such well-known compounds as strychnine, morphine, nicotine, tannins, and various spices.

Animals can defend against predation by hiding, fleeing, or defending. Potential prey may use camouflage in the form of **cryptic coloration** to blend in with the background. Mechanical and chemical defenses discourage predation. From the food they eat, some animals passively accumulate compounds that are toxic to their predators; others may synthesize their own toxins. Bright, conspicuous, **aposematic coloration** warns predators not to eat animals with chemical defenses.

Mimicry may be used by prey to exploit the warning coloration of other species. Predators may use mimicry to "bait" their prey.

Parasites that live within a **host** are **endoparasites;** those that feed on the surface of a host are **ectoparasites.** In **parasitoidism,** insects lay eggs on hosts, on which their larvae then feed. Pathogens, such as bacteria,

viruses, protists, or even fungi or prions, are like microscopic parasites. Pathogens may kill their hosts; parasites do not usually cause lethal harm to the host on which they feed.

■ INTERACTIVE QUESTION 53.2

Name the following two types of mimicry:

a. harmless species resembling a poisonous or distasteful species

b. mutual imitation by two or more distasteful species

Mutualism In **mutualism,** interactions between species benefit both participants. Mutualistic interactions require the coevolution of adaptations in both species. Many cases of mutual symbiosis may have evolved when organisms became able to derive some benefit from their predator or parasite.

Commensalism In **commensalism,** only one member appears to benefit from the interaction. Examples include "hitchhiking" species and species that feed on food incidentally exposed by another.

■ INTERACTIVE QUESTION 53.3

Name and give examples of the interspecific interactions symbolized in the table.

	Interaction	Examples
+ / +	**a.**	
+ / 0	**b.**	
+ / −	**c.**	
− / −	**d.**	

Trophic structure is a key factor in community dynamics (1181–1183)

The **trophic structure** of a community is its feeding relationships. A **food chain** shows the transfer of food energy from one **trophic level** to the next, from producers to primary consumers (herbivores) to secondary, tertiary, or quaternary consumers (carnivores) and eventually to decomposers.

Food Webs A food web diagrams the complex trophic relationships within a community. The complicated connections of a food web arise because most consumers are omnivores, feeding at various trophic levels. Food webs can be simplified by grouping species into functional groups such as primary consumers, or by isolating partial food webs that interact little with the more complex web.

What Limits the Length of a Food Chain? Within a food web, each food chain consists of only a few links. Two hypotheses are proposed to explain why most food chains are limited to five or fewer links. According to the **energetic hypothesis,** food chains are limited by the inefficiency of energy transfer (only about 10%) from one trophic level to the next. The **dynamic stability hypothesis** suggests that short food chains are more stable than long ones because an environmental disruption that reduces production at lower levels will be magnified at higher trophic levels as food supply is reduced all the way up the chain.

■ INTERACTIVE QUESTION 53.4

Experimental data from tree hole communities showed that food chains were longest when food supply at the producer level was greatest. Which hypothesis about what limits food chain length do these results support?

Dominant species and keystone species exert strong controls on community structure (1183–1185)

Dominant Species Species in a community that have the highest abundance or largest **biomass** are a major influence on the occurrence and distribution of other species. A species may become a **dominant species** due to its competitive use of resources or success at avoiding predation. The removal of a dominant species from a community may adversely affect any species that relied exclusively on that species, but its role will probably quickly be filled by other species, as predicted by the redundancy model.

Keystone Species A **keystone species** has a large impact on community structure as a result of its ecological role. Paine's study of a predatory sea star demonstrated its role in maintaining species richness in an intertidal community by reducing the density of mussels, a highly competitive prey species.

The structure of a community may be controlled bottom-up by nutrients or top-down by predators (1185–1186)

Arrows can be used to indicate the effect of an increase in the biomass of one trophic level on another trophic level. According to the **bottom-up model** of community organization, $N \rightarrow V \rightarrow H \rightarrow P$, an increase in mineral nutrients yields an increase in biomass at each succeeding trophic level: vegetation, herbivores, and predators. The **top-down model,** $N \leftarrow V \leftarrow H \leftarrow P$, assumes that predation controls community organization, with a series of $+/-$ effects cascading down the trophic levels. According to this model, also called the trophic cascade model, increasing predators will decrease herbivores, which results in increased vegetation and lowered levels of nutrients.

■ INTERACTIVE QUESTION 53.5

Many freshwater lake communities appear to be organized along the top-down model. What actions might ecologists take if they wanted to use *biomanipulation* to control excessive algal blooms in a lake with four trophic levels (algae, zooplankton, primary predator fish, and top predator fish)?

Disturbance and Community Structure

Traditionally, biological communities were viewed as existing in a state of equilibrium, held there by interspecific interactions. The ability of a community to reach and maintain this relatively constant species composition and to return to this equilibrium following a disturbance is known as **stability.** The **nonequilibrium model** emphasizes the nonstable, changing structure and composition of communities as a result of disturbances.

Most communities are in a state of nonequilibrium owing to disturbances (1186–1188)

Disturbances such as fire, drought, storms, overgrazing, or human activities change resource availability, reduce or eliminate some populations, and may create opportunities for new species. Communities may often be in a state of recovery from disturbance. Small-scale disturbances may enhance environmental patchiness and help maintain species diversity. Human prevention of some natural disturbances may lead to large-scale disturbances, such as the destructive fires in Yellowstone National Park in 1988.

Humans are the most widespread agents of disturbance (1188)

Human activities have altered the structure of communities all over the world. The destruction of tropical rain forests and the creation of vast barren areas in Africa are the results of human disturbances. A common human impact on communities is to reduce species diversity. Humans use about 60% of Earth's land, often as monocultures of crops or pulpwood and lumber. Intensive grazing often results in the replacement of several native plants with only a few introduced species.

Ecological succession is the sequence of community changes after a disturbance (1189–1191)

Transition in species composition in a community, usually following some disturbance, is known as **ecological succession**. If no soil was originally present, as on a new volcanic island or on the moraine left by a retreating glacier, the process is called **primary succession**. A series of colonizers usually begins with autotrophic bacteria and moves through lichens, mosses, grasses, shrubs, and trees until the community reaches its prevalent form of vegetation. **Secondary succession** occurs when an existing community is disrupted by fire, logging, or farming, but the soil remains intact. Herbaceous species may colonize first, followed by woody shrubs and eventually forest trees.

Early colonizers may *facilitate* the arrival of other species by improving the environment. Or the actions of early species may *inhibit* the establishment of later species. Early and later species may be independent in their colonization and *tolerate* each other's presence.

Ecologists have studied moraine succession over the 200-year retreat of glaciers at Glacier Bay in Alaska. Alder is the dominant plant 30–80 years after

deglaciation, followed by Sitka spruce for the next 120 years. The community becomes a spruce-hemlock forest by the third century after deglaciation, except in flat, poorly drained areas, where *Sphagnum* invades. These mosses make the soil waterlogged and acid, killing the trees and creating *Sphagnum* bogs.

During early successional stages following glacial retreat, inhibition seems to restrict the colonization of new vegetation. But then *Dryas* (an herbaceous angiosperm) and alder enrich the soil and facilitate the arrival of spruce.

■ **INTERACTIVE QUESTION 53.6**

Describe the effects of the alder stage of succession on soil pH and fertility.

Biogeographic Factors Affecting the Biodiversity of Communities

A community's **biodiversity** correlates with its size and geographic location. Biodiversity is greatest in the tropics and on larger versus smaller islands.

Community biodiversity measures the number of species and their relative abundance (1191–1192)

When ecologists describe biodiversity, they measure **heterogeneity,** which considers both **species richness** (number) and **relative abundance.** Obtaining such data may be difficult due to sampling errors and the rarity of most species in a community. Measuring biodiversity is critical for conservation biology. Large-scale patterns are often studied by lumping species into broader categories.

Species richness generally declines along an equatorial-polar gradient (1191–1193)

Biodiversity studies of tree, ant, snake, bird, and fish species have documented much greater numbers of species in tropical habitats than in temperate and polar regions. Tropical communities are older, partly because of their longer growing season and partly because they have not had to "start over" after glaciation, as has been the case several times for many polar and temperate communities. This longer span of evolutionary time would allow for more speciation events to have occurred.

The higher solar energy input and water availability of the tropics are major climatic explanations for their higher biodiversity. Evapotranspiration is the amount of water evaporated from soil and transpired by plants, and is a measure of solar energy, temperature, and water availability. Evapotranspiration rates have been shown to correlate with species richness of trees and vertebrates in North America.

Species richness is related to a community's geographic size (1193–1194)

A **species-area curve** illustrates the correlation between the size of a community and the number of species found there. The slope of the curve may vary for different taxa and types of communities. Use of such curves in conservation biology can allow predictions on how a loss of habitat may affect biodiversity.

Species richness on islands depends on island size and distance from the mainland (1194–1195)

Any habitat surrounded by a significantly different habitat is considered an island and allows ecologists to study factors that affect species diversity. In the 1960s, MacArthur and Wilson developed a general hypothesis of island biogeography, stating that the size of the island and its closeness to the mainland (or source of dispersing species) are important variables directly correlated with species diversity. Larger islands closer to the mainland will have a higher species diversity than smaller or more distant islands. The eventual number of species on the island depends on the immigration rate of new species and the extinction rate of island species. These rates change as the number of species on the island increases, and when the rates become equal, an equilibrium in species diversity develops, although species composition may continue to change.

While the island biogeography hypothesis may apply over relatively short time periods in cases where colonization determines species composition, over long periods adaptive evolutionary changes and abiotic disturbances on the island may determine community structure and composition.

■ **INTERACTIVE QUESTION   53.7**

Many biogeographic studies have found that large islands have greater species richness than small islands. Label the lines on the following graph that show how immigration rate and extinction rate vary with the number of species on large and small islands. Indicate the location of the equilibrium number on the *x* axis for a small and a large island.

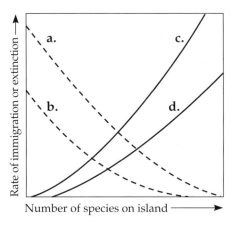

Word Roots

crypto- = hidden, concealed (*cryptic coloration:* a type of camouflage that makes potential prey difficult to spot against its background)

ecto- = outer (*ectoparasites:* parasites that feed on the external surface of a host)

endo- = inner (*endoparasites:* parasites that live within a host)

herb- = grass; **-vora** = eat (*herbivory:* the consumption of plant material by an herbivore)

hetero- = other, different (*heterogeneity:* a measurement of biological diversity considering richness and relative abundance)

inter- = between (*interspecific competition:* competition for resources between plants, between animals, or between decomposers when resources are in short supply)

mutu- = reciprocal (*mutualism:* a symbiotic relationship in which both the host and the symbiont benefit)

STRUCTURE YOUR KNOWLEDGE

1. Complete this concept map to organize your understanding of the important factors that structure a community.

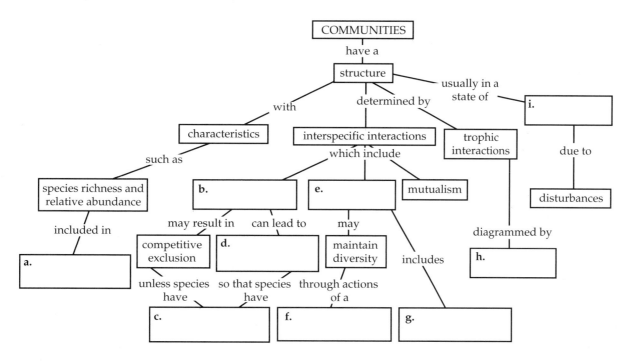

2. Community ecologists develop models or hypotheses to describe community structure and the factors that contribute to such structure. Briefly explain the following models that were described in this chapter.

   a. Individualistic hypothesis
   b. Interactive hypothesis
   c. Rivet model
   d. Redundancy model
   e. Competitive exclusion principle
   f. Energetic hypothesis
   g. Dynamic stability hypothesis
   h. Bottom-up model
   i. Top-down (trophic cascade) model
   j. Nonequilibrium model

TEST YOUR KNOWLEDGE

MULTIPLE CHOICE: *Choose the one best answer.*

1. Which of the following is *not* part of Gleason's individualistic concept of communities?
   a. Communities are chance collections of species that are in the same area because of similar environmental requirements.
   b. There should be no distinct boundaries between communities.
   c. The consistent composition of a community is based on interactions that cause it to function as an integrated unit.
   d. Species are distributed independently along environmental gradients.
   e. Most plant communities studied meet the predictions made by this concept.

2. Two species, A and B, occupy adjoining environmental patches that differ in several abiotic factors. When species A is experimentally removed from a portion of its patch, species B colonizes the vacated area and thrives. When species B is experimentally removed from a portion of its patch, species A does not successfully colonize the area. What might you conclude from these results?
   a. Both species A and species B are limited to their range by abiotic factors.
   b. Species A is limited to its range by competition, and species B is limited by abiotic factors.
   c. Both species are limited to their range by competition.
   d. Species A is limited to its range by abiotic factors, and species B is limited to its range because it cannot compete with species A.
   e. Species A is a predator of species B.

3. The species richness of a community refers to
   a. the relative numbers of individuals in each species.
   b. the number of different species found in a community.
   c. the feeding relationships or trophic structure within the community.
   d. the heterogeneity of that community.
   e. its stability or ability to persist through disturbances.

4. The rivet model of communities is most similar to
   a. the dynamic stability hypothesis.
   b. the top-down model.
   c. the redundancy model.
   d. the individualistic hypothesis.
   e. the interactive hypothesis.

5. Through resource partitioning,
   a. two species can compete for the same prey item.
   b. slight variations in niche allow closely related species to coexist in the same habitat.
   c. two species can share the same niche in a habitat.
   d. competitive exclusion results in the success of the superior species.
   e. two species with identical niches do not share the same habitat and thus avoid competition.

6. Which of the following organisms and trophic levels is mismatched?
   a. algae—producer
   b. phytoplankton—primary consumer
   c. fungi—decomposer
   d. carnivorous fish larvae—secondary consumer
   e. eagle—tertiary or quaternary consumer

7. Aposematic coloring is most commonly found in
   a. prey whose body morphology is cryptic.
   b. predators who are able to sequester toxic plant compounds in their bodies.
   c. prey species that have chemical defenses.
   d. good-tasting prey that evolve to look like each other.
   e. prey species that are camouflaged to match their environment.

8. A palatable (good-tasting) prey species may defend against predation by
   a. Müllerian mimicry.
   b. Batesian mimicry.
   c. secondary compounds.
   d. aposematic coloration.
   e. either a or b.

9. When one species was removed from a tidepool, the species richness became significantly reduced. The removed species was probably
   a. a strong competitor.
   b. a potent parasite.
   c. a resource partitioner.
   d. a keystone species.
   e. the species with the highest relative abundance.

10. A highly successful parasite
    a. will not harm its host.
    b. may benefit its host.
    c. will be able to feed without killing its host.
    d. will kill its host fairly rapidly.
    e. will have coevolved into a commensalistic interaction with its host.

11. Why do most food chains consist of only three to five links?
    a. There are only five trophic levels: producers; primary, secondary, and tertiary consumers; and decomposers.
    b. Most communities are controlled bottom-up by mineral nutrient supply, and few communities have enough nutrients to support more links.
    c. The dominant species in most communities consumes the majority of prey; thus, not enough food is left to support higher predators.
    d. According to the energetic hypothesis, the inefficiency of energy transfer from one trophic level to the next limits the number of links that can exist.
    e. According to the trophic cascade model, increasing the biomass of top trophic levels causes a decrease in the biomass of lower levels, so that the top levels can no longer be supported.

12. During succession, inhibition
    a. may prevent the achievement of a stable community.
    b. may slow down both the rate of colonization and the rate of extinction, depending on the size of the area and distance from the source of dispersing species.
    c. results from the frequent disturbances that often eliminate early colonizers.
    d. may slow down the successful colonization by other species.
    e. may involve changes in soil pH or accelerated accumulation of humus.

13. According to the nonequilibrium model,
    a. chance events such as disturbances play major roles in the structure and composition of communities.
    b. species composition in a community is always in flux as a result of human interventions.
    c. food chains are limited to a few links because long chains are more unstable in the face of environmental disturbances.
    d. the communities with the most diversity have the least stability or resistance to change.
    e. early colonizers inhibit other species, whereas later colonizers facilitate the arrival of new species.

14. An island that is small and far from the mainland, in contrast to a large island close to the mainland, would be expected to
    a. have lower biodiversity.
    b. be in an earlier successional stage.
    c. have higher biodiversity but a much lower abundance of organisms.
    d. have a higher rate of colonization but a higher rate of extinction.
    e. have a lower rate of colonization and a lower rate of extinction.

15. According to the top-down (trophic cascade) model of community control, which trophic level would you *decrease* if you wanted to *increase* the vegetation level in a community?
    a. nutrients
    b. vegetation
    c. secondary consumers (carnivores)
    d. tertiary consumers
    e. omnivores

16. A major explanation for the decline in species richness along an equatorial-polar gradient is the correlation of high levels of solar radiation and water availability with biodiversity. Which of the following is also suggested as a factor in the high species richness of tropical communities?
    a. the inverse relationship between biodiversity and evapotranspiration
    b. the greater age of these communities (longer growing season and fewer climatic setbacks), providing more time for speciation events
    c. the large area of the tropics and corresponding richness predicted by the species-area curve
    d. the lack of disturbances in tropical areas
    e. the greater immigration rate and lower extinction rate found on large tropical islands

17. Ecologists survey the tree species in two forest plots of different ages. Plot 1 has seven different species and 95% of all trees belong to just one species. Plot 2 has five different species, each of which is represented by approximately 20% of the trees. How would you describe plot 2 as compared with plot 1?
    a. higher species richness
    b. most heterogeneous
    c. lower relative abundance
    d. lower species richness
    e. Both b and d are correct.

18. Which of the following interspecific interactions is *not* an example of predation?
    a. ectoparasite and host
    b. herbivore and plant
    c. honeybee and flower
    d. pathogen and host
    e. carnivore and prey

CHAPTER 54

ECOSYSTEMS

FRAMEWORK

This chapter describes energy flow and chemical cycling through ecosystems. Producers convert light energy into chemical energy, which is then passed, with a loss of energy at each level, through the food web and ultimately to detritivores. Energy makes a one-way trip through ecosystems. Chemical elements are cycled in the ecosystem from abiotic reservoirs through producers, consumers, and detritivores, and back to the reservoirs.

An ecosystem's primary production may be limited by nutrients, temperature, or moisture. The low trophic efficiency in the transfer of energy from one level to the next is reflected in pyramids of production, biomass, and numbers.

Human activities are altering chemical cycles, causing climate change, depleting atmospheric ozone, and changing ecosystems.

CHAPTER REVIEW

An **ecosystem** is a community and its physical environment; it includes all the biotic and abiotic components of an area. Most ecosystems are powered by energy from sunlight, which is transformed to chemical energy by autotrophs, passed to a series of heterotrophs in the organic compounds of food, and continually dissipated in the form of heat. Chemical elements are cycled between the abiotic and biotic components of the ecosystem as autotrophs incorporate them into organic compounds and the processes of metabolism and decomposition return them to the soil, air, and water.

The Ecosystem Approach to Ecology

Ecosystem ecologists group species populations into **trophic levels** based on their main source of nutrition.

Trophic relationships determine the routes of energy flow and chemical cycling in an ecosystem (1199)

Most **primary producers,** or **autotrophs,** use light energy to photosynthesize sugars for use as fuel in respiration and as building materials for other organic compounds. **Heterotrophs** depend on autotrophs for their organic compounds. The **primary consumers** are herbivores; **secondary consumers** are carnivores. **Detritivores,** or **decomposers,** consume **detritus,** which is organic wastes, fallen leaves, and dead organisms.

Decomposition connects all trophic levels (1199–1200)

Detritivores that feed on plant remains often form a link between producers and consumers.

Fungi and prokaryotes are the most important decomposers in most ecosystems, converting organic materials from all trophic levels to inorganic compounds that can be recycled by autotrophs.

The laws of physics and chemistry apply to ecosystems (1200)

According to the first law of thermodynamics, energy cannot be created or destroyed, only transformed. Energy flows through ecosystems from its input as solar radiation to its conversion into chemical energy to its dissipation as heat. The second law of thermodynamics states that, in each energy conversion, some energy is lost as heat. Ecologists trace the energy flow in ecosystems and the efficiency of ecological energy conversions.

Whereas energy makes a one-way trip through ecosystems, chemical elements move through the trophic levels and are then recycled to abiotic reservoirs and back into producers.

Primary Production in Ecosystems

Primary production is a measure of the amount of light energy converted to chemical energy—the photosynthetic output of an ecosystem's autotrophs.

An ecosystem's energy budget depends on primary production (1200–1201)

The Global Energy Budget The intensity of solar energy striking the Earth varies by latitude and, depending on cloud cover and dust in the air, by region. Only a small portion of incoming solar radiation strikes photosynthetic organisms, and, of that, only 1–2% is converted to chemical energy. Nevertheless, worldwide photosynthetic production is about 170 billion tons of organic material per year.

Gross and Net Primary Production Net primary production (NPP) is the **gross primary production (GPP)** minus the energy used by plants in their own cellular respiration (R). NPP = GPP − R.

Primary production can be expressed as energy per unit area per unit time ($J/m^2/yr$) or as **biomass** measured in terms of dry weight of organic material added ($g/m^2/yr$). **Standing crop** is the total biomass of plants in an ecosystem. Primary production and the contribution to Earth's total productivity vary by ecosystem.

In aquatic ecosystems, light and nutrients limit primary production (1201–1205)

Production in Marine Ecosystems The depth to which light penetrates affects primary production in oceans. Nutrients, however, limit marine production more than light. Nitrogen and phosphorus levels are very low in the photic zone of the open ocean, limiting the growth of phytoplankton. Nutrient-addition experiments in polluted coastal waters indicate that nitrogen limits algal growth more than does phosphorus. Studies of the Sargasso Sea have shown that the micronutrient iron is the **limiting nutrient** in these unproductive waters. The addition of iron to test regions in tropical oceans resulted in an increase in cyanobacteria that fix nitrogen, which then stimulated the growth of eukaryotic phytoplankton.

Production in Freshwater Ecosystems In freshwater ecosystems, light intensity, temperature, and availability of minerals affect production. **Eutrophication,** the shift in composition of phytoplankton communities from domination by green algae and diatoms to blooms of cyanobacteria, has been linked to phosphorus pollution from sewage and fertilizer runoff.

In terrestrial ecosystems, temperature, moisture, and nutrients limit primary production (1205)

Both water supply and temperature vary more in terrestrial ecosystems. Production in terrestrial ecosystems is related to precipitation and temperature. Nutrients may limit production on a local scale. A limiting nutrient is one whose inadequate supply restricts productivity. Nitrogen or phosphorus is the limiting nutrient in many ecosystems.

a. List some ecosystems with high rates of production.

b. List some ecosystems with low rates of production.

c. The open ocean has low net primary production yet contributes the greatest percentage of Earth's net primary production. Explain.

d. Antarctic seas are often more productive than most tropical seas, even though they are colder and receive lower light intensity. Explain.

Secondary Production in Ecosystems

The rate at which consumers in an ecosystem produce new biomass from their food is called **secondary production.**

The efficiency of energy transfer between trophic levels is usually less than 20% (1206–1208)

Production Efficiency Herbivores consume only a fraction of the plant material produced; they cannot digest all they eat; and much of the energy they do absorb is used for cellular respiration. Only the chemical energy stored as growth or in offspring is available as food to higher trophic levels. The proportion of assimilated food energy that is used for net secondary production is a measure of the efficiency of energy transformation: **production efficiency** = net secondary production/assimilation of primary production. Production efficiencies vary from 1–3% for "warm-blooded" birds and mammals, to 10% for fishes, to 40% for insects.

Trophic Efficiency and Ecological Pyramids **Trophic efficiency** is the percentage of the energy of one trophic level that makes it to the next level, usually ranging from 5–20%. Trophic efficiencies are always less than production efficiencies due to the loss of energy via respiration, feces, and the organic material not consumed by the next trophic level. A **pyramid of production** shows this multiplicative loss of energy.

A **biomass pyramid** illustrates the standing crop biomass of organisms at each trophic level. This pyramid usually narrows rapidly from producers to the top trophic level. Some aquatic ecosystems have inverted biomass pyramids in which zooplankton (consumers) outlive and outweigh the highly productive, but heavily consumed, phytoplankton. Phytoplankton have a short **turnover time,** determined by dividing standing crop biomass by production. The production pyramid for this ecosystem, however, is normal in shape.

The **pyramid of numbers** illustrates that higher trophic levels contain small numbers of individuals, resulting from the larger size of these animals and the greatly decreased energy availability illustrated by the pyramid of production.

■ INTERACTIVE QUESTION   54.2

a. Why is production efficiency higher for fishes than for birds and mammals?

b. Assuming a 10% trophic efficiency (transfer of energy to the next trophic level), approximately what proportion of the chemical energy produced in photosynthesis makes it to a tertiary consumer?

Herbivores consume a small percentage of vegetation: the green world hypothesis (1208)

Most ecosystems are green. The **green world hypothesis** proposes six factors that keep herbivore populations from stripping Earth's vegetation: plant defenses; limited essential nutrients that restrict herbivore growth and reproduction; abiotic fluctuations; intraspecific competition; and interspecific interactions such as predation, parasitism, and disease. This final, "top-down" explanation is proposed as the most important factor limiting herbivore populations.

Cycling of Chemical Elements in Ecosystems

Chemical elements are passed between abiotic and biotic components of ecosystems through **biogeochemical cycles.** Plants and other autotrophs use inorganic nutrients to build organic matter, which is passed through the food chain. Chemicals are returned to the atmosphere, water, or soil through respiration and the action of decomposers.

Biological and geologic processes move nutrients between organic and inorganic compartments (1209–1212)

The route of a biogeochemical cycle depends on the element and the trophic structure of an ecosystem. Gaseous carbon, oxygen, sulfur, and nitrogen have global cycles involving atmospheric reservoirs. Less mobile elements, such as phosphorus, potassium, calcium, and the trace elements, have a more localized cycle in which soil is the main abiotic reservoir.

A General Model of Chemical Cycling   Most nutrients are found in four types of reservoirs or compartments: organic material in living organisms or detritus, available to other organisms; unavailable organic material in "fossilized" deposits; available inorganic elements and compounds in water, soil, or air; and unavailable elements in rocks. Nutrients may leave the unavailable reservoirs through weathering of rock, erosion, or burning of fossil fuels.

The actual movement of elements through biogeochemical cycles is quite complex, with influx and loss of nutrients from ecosystems occurring in many ways. Ecologists have studied this movement by adding radioactive tracers to chemical elements in several ecosystems.

The water cycle is basically a physical process. Except in photosynthesis, water is not chemically altered as it cycles through ecosystems. The water cycle involves evaporation, precipitation, and transpiration, with a net flow of water evaporating by solar energy from the oceans, moving as water vapor to the land where it precipitates, and returning to the oceans through runoff and groundwater.

In the carbon cycle, plants take CO_2 from the atmosphere for photosynthesis and organisms release it in cellular respiration. Fossil fuel combustion is increasing the amount of atmospheric CO_2.

The Nitrogen Cycle   Plants require nitrogen in the form of NH_4^+ or NO_3^-; they cannot assimilate atmospheric nitrogen. A small amount of nitrogen enters the ecosystem through atmospheric deposition, in rain or the settling of particulates. Most nitrogen enters through **nitrogen fixation:** Soil bacteria and symbiotic bacteria in root nodules fix nitrogen in terrestrial ecosystems, and some cyanobacteria do so in aquatic ecosystems. Fertilizers also add a significant amount of nitrogen to ecosystems.

The excess ammonia released by nitrogen-fixing bacteria is protonated to ammonium in slightly acidic soil. Evaporation of ammonia back to the atmosphere from neutral soils then raises NH_4^+ concentrations in rainfall.

In a process called **nitrification,** some soil bacteria oxidize ammonium to NO_2^- (nitrite) and NO_3^- (nitrate). Plants primarily absorb nitrate. Animals obtain nitrogen in organic form from plants or other animals. Some bacteria convert nitrate to N_2 to obtain their oxygen. This **denitrification** returns N_2 to the atmosphere.

Bacterial and fungal decomposers, in a process called **ammonification,** break down organic nitrogenous compounds and return ammonium to the soil. Most local nitrogen cycling involves decomposition and reassimilation.

The Phosphorus Cycle Weathering of rock adds phosphorus to the soil in the form of PO_4^{3-}, which is absorbed by plants. Organic phosphate is transferred from plants to consumers and returned to the soil through the action of decomposers or by animal excretion. Humus and soil particles usually bind phosphate, keeping it available locally for recycling. The phosphorus that leaches into water eventually travels to the sea, where sediments become incorporated into rocks. These rocks may eventually return to terrestrial ecosystems as a result of geologic processes occurring within geologic time.

Decomposition rates largely determine the rates of nutrient cycling (1212–1213)

Temperature and the availability of water and O_2 influence decomposition rates; thus, nutrient cycling times vary in different ecosystems. Nutrients cycle rapidly in a tropical rain forest; the soil contains only about 10% of the ecosystem's nutrients. In temperate forests where decomposition is slower, 50% of organic nutrients are stored in the detritus and soil. Decomposition rates are slow in aquatic ecosystems, and sediments constitute a nutrient sink.

Nutrient cycling is strongly regulated by vegetation (1213–1214)

Research groups are conducting **long-term ecological research (LTER)** to follow ecosystem dynamics over relatively long time periods.

A team of scientists has looked at nutrient cycling in the Hubbard Brook forest ecosystem since 1963. The mineral budget for each of six valleys was determined by measuring the influx of key nutrients in rainfall and their outflow through the creek that drained each watershed. About 60% of the precipitation exited through the stream; the rest was lost by transpiration and evaporation. Most minerals were recycled within the forest ecosystem.

■ **INTERACTIVE QUESTION 54.3**

Label the organisms and compounds that are illustrated in this nitrogen cycle.

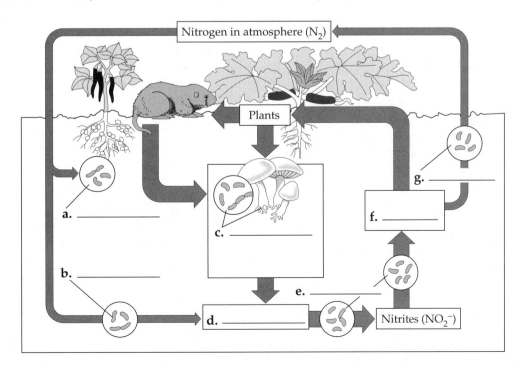

The effect of deforestation on nutrient cycling was measured for three years in a valley that was completely logged and sprayed with herbicides. Compared with a control, water runoff from the deforested valley increased 30–40%; net loss of minerals, such as Ca^{2+}, K^+, and nitrate, was large. Nitrate increased in concentration in the creek 60-fold, removing this critical soil nutrient and contaminating drinking water.

Long-term data from Hubbard Brook indicate that acid precipitation has removed most of the Ca^{2+} from the soils. This lack of Ca^{2+} appears to have halted forest growth in the last decade.

■ INTERACTIVE QUESTION   54.4

a. In which natural ecosystem do nutrients cycle the fastest? Why?

b. In which natural ecosystem do nutrients cycle the slowest? Why?

c. What is the effect of loss of vegetation on nutrient cycling?

Human Impact on the Ecosystems of the Biosphere

The human population is disrupting chemical cycles throughout the biosphere (1214–1216)

Agricultural Effects on Nutrient Cycling   The harvesting of crops removes nutrients that would otherwise recycle into the soil. After depleting the organic and inorganic reserves of nutrients, crops require the addition of expensive synthetic fertilizers. The addition of nitrogen fertilizers, increased legume cultivation, and burning have doubled the supply of fixed nitrogen. Excess soil nitrogen can contaminate groundwater and, when released as nitrogen oxides by denitrifiers, contribute to global warming, ozone thinning, and acid precipitation.

Critical Load and Nutrient Cycles   Nitrogen that exceeds the **critical load,** the amount of added nutrient that can be absorbed by plants without damaging the ecosystem, can contaminate groundwater and enter lakes and rivers and drain into the ocean.

Accelerated Eutrophication of Lakes   The lack of mineral nutrients in an oligotrophic lake keeps primary production low. The additional nutrients in mesotrophic and eutrophic lakes increase primary and overall production. Sewage, factory wastes, and runoff of animal wastes and fertilizers from agricultural lands can lead to **cultural eutrophication** of lakes. The rapid increase in nutrients can cause an explosive increase in algae and cyanobacteria. Oxygen shortages, due to algal and bacterial respiration at night and the metabolism of decomposers that work on the accumulating organic material, kill off many fish and other lake organisms.

Combustion of fossil fuels is the main cause of acid precipitation (1216–1217)

The burning of wood, and coal and other fossil fuels releases oxides of sulfur and nitrogen, which form sulfuric and nitric acid in the atmosphere. These acids return to the earth as **acid precipitation,** defined as rain, snow, or fog with a pH less than 5.6. Emissions from ore smelters and electrical plants drift downwind and create acid precipitation over vast and distant areas. Nutrients leach from soils as acid precipitation changes soil chemistry, and forests have been damaged. Fish populations in lakes across North America and Europe have declined and community compositions have changed.

Toxins can become concentrated in successive trophic levels of food webs (1217–1218)

Many toxic chemicals dumped into ecosystems are nonbiodegradable; some may become more harmful as they react with other environmental factors. Organisms absorb these toxins from food and water and may retain them within their tissues. Chlorinated hydrocarbons, such as DDT, and polychlorinated biphenyls, or PCBs, have been implicated in endocrine system problems in many animal species. In a process known as **biological magnification,** the concentration of such compounds increases in each successive link of the food chain.

Human activities may be causing climate change by increasing carbon dioxide concentration in the atmosphere (1218–1220)

Rising Atmospheric CO_2   The concentration of CO_2 in the atmosphere has been increasing since the Industrial Revolution as a result of the combustion of fossil fuel and the burning of wood removed by deforestation. If C_3 plants become able to outcompete C_4 plants with the increase in CO_2, species composition in natural and agricultural communities may be significantly altered.

The Greenhouse Effect Through a phenomenon known as the **greenhouse effect,** CO_2 and water vapor in the atmosphere absorb infrared radiation reflected from Earth and rereflect it back to Earth, causing an increase in temperature.

Global Warming Scientists use mathematical models to try to estimate the extent and consequences of increasing CO_2 levels. Climatologists predict a temperature rise of 2°C in the next 100 years if CO_2 levels continue to rise. The effects on vegetation of previous global warming trends can be used to predict the effects of increasing temperatures. Controlling the level of CO_2 emissions in increasingly industrialized societies is a huge international challenge.

■ INTERACTIVE QUESTION 54.5

List some of the potential consequences of global warming.

Human activities are depleting atmospheric ozone (1220)

A layer of ozone molecules (O_3) in the lower stratosphere absorbs damaging ultraviolet radiation. This layer has been gradually thinning since 1975, largely as a result of the accumulation of breakdown products of chlorofluorocarbons in the atmosphere. The dangers of ozone depletion may include increased incidence of skin cancer and cataracts and unpredictable effects on phytoplankton, crops, and natural ecosystems.

WORD ROOTS

auto- = self; **troph-** = food, nourishment (*autotroph:* an organism that obtains organic food molecules without eating other organisms)

bio- = life; **geo-** = the Earth (*biogeochemical cycles:* the various nutrient circuits which involve both biotic and abiotic components of ecosystems)

de- = from, down, out (*denitrification:* the process of converting nitrate back to nitrogen)

detrit- = wear off; **-vora** = eat (*detritivore:* a consumer that derives its energy from nonliving organic material)

hetero- = other, different (*heterotroph:* an organism that obtains organic food molecules by eating other organisms or their by-products)

STRUCTURE YOUR KNOWLEDGE

1. Two processes that emerge at the ecosystem level of organization are energy flow and chemical cycling. Develop a concept map that explains, compares, and contrasts these two processes.

2. Describe four or five human intrusions in ecosystem dynamics that have detrimental effects.

TEST YOUR KNOWLEDGE

MULTIPLE CHOICE: *Choose the one best answer.*

1. Which of the following groups is absolutely essential to the functioning of an ecosystem?
   a. producers
   b. producers and herbivores
   c. producers, herbivores, and carnivores
   d. detritivores
   e. producers and detritivores

2. Primary production
   a. is equal to the standing crop of an ecosystem.
   b. is greatest in freshwater ecosystems.
   c. is the rate of conversion of light to chemical energy in an ecosystem.
   d. is inverted in some aquatic ecosystems.
   e. is all of the above.

3. Which of the following is an accurate statement about ecosystems?
   a. Energy is recycled through the trophic structure.
   b. Energy is usually captured from sunlight by primary producers, passed to secondary producers in the form of organic compounds, and lost to detritivores in the form of heat.
   c. Chemicals are recycled between the biotic and abiotic sectors, whereas energy makes a one-way trip through the food web and is eventually dissipated as heat.
   d. There is a continuous process by which energy is lost as heat, and chemical elements leave the ecosystem through runoff.
   e. A food web shows that all trophic levels may feed off each other.

4. In the experiment in which iron was added to the Sargasso Sea, the growth of eukaryotic phytoplankton was stimulated because
   a. iron was the limiting nutrient for eukaryotic phytoplankton growth.
   b. the iron interacted with bottom sediments, releasing nitrogen and phosphorus into the water.
   c. iron interacted with phosphorus, making that nutrient available to the phytoplankton.
   d. the iron reached the critical load necessary to promote photosynthesis.
   e. iron stimulated the growth of nitrogen-fixing cyanobacteria, which then made nitrogen available for phytoplankton growth.

5. The open ocean and tropical rain forest are the two largest contributors to Earth's net primary production because
   a. both have high rates of net primary production.
   b. both cover huge surface areas of the Earth.
   c. nutrients cycle fastest in these two ecosystems.
   d. the ocean covers a huge surface area and the tropical rain forest has a high rate of production.
   e. both a and b are correct.

6. Production in terrestrial ecosystems is affected by
   a. temperature.
   b. light intensity.
   c. availability of nutrients.
   d. availability of water.
   e. all of the above.

7. Secondary production
   a. is measured by the standing crop.
   b. is the rate of biomass production in consumers.
   c. is greater than primary production.
   d. is 10% less than primary production.
   e. is the gross primary production minus the energy used for respiration.

8. Which of the following is *not* true of a pyramid of production?
   a. Only about 10% of the energy in one trophic level is passed into the next level.
   b. Because of the loss of energy at each trophic level, most food chains are limited to three to five links.
   c. The pyramid of production of some aquatic ecosystems is inverted because of the large zooplankton primary consumer level.
   d. Eating grain-fed beef is an inefficient means of obtaining the energy trapped by photosynthesis.
   e. A pyramid of numbers is usually the same shape as a pyramid of production.

9. In which of the following would you expect production efficiency to be the greatest?
   a. plants
   b. mammals
   c. fish
   d. insects
   e. birds

10. Biogeochemical cycles are global for elements
    a. that are found in the atmosphere.
    b. that are found mainly in the soil.
    c. such as carbon, nitrogen, and phosphorus.
    d. that are dissolved in water.
    e. in the nonavailable reservoirs.

11. Which of these processes is *incorrectly* paired with its description?
    a. nitrification—oxidation of ammonium in the soil to nitrite and nitrate
    b. nitrogen fixation—reduction of atmospheric nitrogen into ammonia
    c. denitrification—removal of nitrogen from organic compounds
    d. ammonification—decomposition of organic compounds into ammonium
    e. atmospheric deposition—nitrogen added to soil in rain or dust particles

12. Clear-cutting tropical forests yields agricultural land with limited productivity because
    a. it is too hot in the tropics for most food crops.
    b. the tropical forest regrows rapidly and chokes out agricultural crops.
    c. few of the ecosystem's nutrients are stored in the soil; most are in the forest trees.
    d. phosphorus, not nitrogen, is the limiting nutrient in those soils.
    e. decomposition rates are high but primary production is low in the tropics.

13. Which of the following was *not* shown by the Hubbard Brook Experimental Forest study?
    a. Most minerals recycle within a forest ecosystem.
    b. Deforestation results in a large increase in water runoff.
    c. Mineral losses from a valley were great following deforestation.
    d. Nitrate was the mineral that showed the greatest loss.
    e. Acid rain increased as a result of deforestation.

14. The finding of harmful levels of DDT in grebes (fish-eating birds) following years of trying to eliminate bothersome gnat populations in a lakeshore town is an example of
    a. eutrophication.
    b. biological magnification.

c. the biomass pyramid.
d. chemical cycling.
e. increasing resistance to pesticides.

15. The greenhouse effect
    a. could change global climate and lead to the flooding of coastal areas.
    b. could result in more C_4 plants in plant communities that were previously dominated by C_3 plants.
    c. causes an increase in temperature when CO_2 absorbs more sunlight entering the atmosphere.
    d. could increase precipitation in central continental areas.
    e. could do all of the above.

16. According to the green world hypothesis, herbivores eat only a small portion of an ecosystem's vegetation because
    a. primary production is much greater than secondary production.
    b. plants have a very short turnover time.

c. herbivores cannot digest most of what they eat.
d. predators, parasites, and disease keep herbivore populations in check.
e. the production efficiency of herbivores is very low.

17. Which of the following trophic levels would have the largest numbers of individuals?
    a. primary producers
    b. omnivores
    c. primary consumers
    d. herbivores
    e. tertiary consumers

18. A serious effect of the thinning of the ozone layer is
    a. a reduction in species diversity.
    b. global warming.
    c. acid precipitation.
    d. an increase in harmful UV radiation reaching Earth.
    e. cultural eutrophication.

CHAPTER 55

CONSERVATION BIOLOGY

FRAMEWORK

Biodiversity at the genetic, species, and ecosystem levels is crucial to human welfare. This chapter explores the causes of the current biodiversity crisis and several of the approaches to preserving the diversity of species on Earth. Conservation biologists focus on determining the habitat needs of endangered species, establishing and managing nature reserves that are often in human-dominated landscapes, and restoring degraded areas. Ecological research and our biophilia may help to achieve the goal of sustainable development—the long-term perpetuation of human societies *and* the ecosystems that support them.

CHAPTER REVIEW

The goal of **conservation biology** is to remedy the **biodiversity crisis,** the current alarming decline in the diversity of organisms on Earth. About 1.5 million species have been formally identified; estimates of total numbers of species on Earth range from 10 million to 80 million. Many of these species reside in tropical forests that are being rapidly destroyed. Human activities are altering all ecosystem processes, and extinction rates resulting from these disruptions may be 1,000 times higher than at any time in the past 100,000 years.

The Biodiversity Crisis

The three levels of biodiversity are genetic diversity, species diversity, and ecosystem diversity (1225–1226)

Loss of Genetic Diversity   Loss of the genetic diversity within and between populations lessens a species' adaptive potential.

Loss of Species Diversity   An **endangered species,** according to the U.S. Endangered Species Act (ESA), is one that is "in danger of extinction throughout all or a significant portion of its range." A **threatened species** is defined as one that is likely to become endangered. There are many well-documented examples of recent extinctions and endangered species of birds, plants, and fishes. Because millions of the world's species are unknown, it is difficult to assess local and global extinctions and their effects on the structure and function of ecosystems.

Loss of Ecosystem Diversity   Biodiversity also includes ecosystem diversity; loss of an ecosystem may affect the whole biosphere.

Biodiversity at all three levels is vital to human welfare (1226–1228)

There are aesthetic and ethical bases for preserving biodiversity.

Benefits of Species Diversity and Genetic Diversity A loss of biodiversity is a loss of the genetic potential held in the genomes of species. Biodiversity is a natural resource that can provide medicines, fibers, industrial chemicals, and food.

Ecosystem Services   Humans depend on Earth's ecosystems; their loss risks our own survival. **Ecosystem services** include such things as purification of air and water, detoxification and decomposition of wastes, nutrient cycling, flood control, pollination of crops, and access to aesthetic beauty. Some ecologists value these ecosystem services at twice the gross national product of all countries combined. The failure of Biosphere II to produce a sustainable biosphere illustrates how difficult it is to engineer a system that provides the services of our natural ecosystems.

The four major threats to biodiversity are habitat destruction, introduced species, overexploitation, and food chain disruptions (1228–1232)

Habitat Destruction   The greatest threat to biodiversity is habitat destruction, caused by agriculture, urban development, forestry, mining, and environmental pollution. Both terrestrial ecosystems and marine habitats have been damaged. Habitat destruction is cited for 73% of the species listed as extinct, endangered, vulnerable, or rare by the International Union for Conservation of Nature and Natural Resources (IUCN). Fragmentation of natural habitats is a common occurrence and almost always leads to species loss.

Introduced Species **Introduced species,** sometimes called exotic species, compete with or prey upon native species and have probably been responsible for about 40% of extinctions in the past 250 years. Humans have transplanted thousands of species, intentionally and unintentionally, with huge economic costs in damage and control efforts in addition to the loss of native species.

Overexploitation Overexploitation involves harvesting plants or animals at rates higher than their populations' ability to reproduce. Species of large animals with low intrinsic reproductive rates and species on small islands are particularly vulnerable to extinction. Overfishing, particularly using new harvesting techniques, has drastically reduced populations of many commercially important fish species.

Disruption of Food Chains When a predator feeds exclusively on one species, extinction of that species also dooms its predator.

■ INTERACTIVE QUESTION 55.1

Give an example of how each of the following causes of the biodiversity crisis has reduced population numbers or caused extinctions.

a. habitat destruction

b. introduced species

c. overexploitation

d. disruptions of food chains

Conservation at the Population and Species Levels

According to the small population approach, a population's small size can draw it into an extinction vortex (1232–1236)

Very small populations are considered endangered. According to the **small population approach,** the

inbreeding and genetic drift characteristic of a small population may draw it into an **extinction vortex,** in which the loss of genetic variation leads by positive feedback loops to smaller and smaller population size.

How Small Is Too Small for a Population? Conservation biologists often use computer models to estimate a species's **minimum viable population size (MVP),** the smallest number of individuals necessary to sustain a population. **Population viability analysis (PVA)** makes use of MVP to predict the long-term viability of a population.

The **effective population size (N_e)** is based on a population's breeding potential and is determined by a formula that includes data on the number of individuals that breed and the sex ratio of the population: $N_e = (4N_f N_m)/(N_f + N_m)$. Other formulas take into account other life history or genetic factors. Conservation efforts should be based on the minimum number of reproductively active individuals needed to prevent extinction.

■ INTERACTIVE QUESTION 55.2

Is the effective population size usually larger or smaller than the actual number of individuals in the population?

Case Study: The Greater Prairie Chicken and the Extinction Vortex As agriculture fragmented their populations, the number of prairie chickens in Illinois declined from millions in the 19th century to 50 in 1993. A comparison of DNA from museum specimens and from the endangered Illinois population showed a decline in genetic variation, supporting the extinction vortex hypothesis of low genetic diversity as linked to this population's decline. After transplanting birds from larger populations in other states, researchers noted an increase in egg viability and the Illinois population rebounded.

Case Study: Population Viability Analysis for Two Popular Herbs Nantel's population viability analysis of the declining American ginseng and wild leek populations in Canada predict that these populations will not persist if harvested at all.

Case Study: Analysis of Grizzly Bear Populations
Shaffer performed a PVA as part of a long-term study of grizzly bears in Yellowstone National Park. He estimated the population size targets for threatened grizzly populations of at least 100 bears. Due to habitat limitations, the recovery targets (the goals mandated by the Endangered Species Act) have been set at fewer than 100 bears. Allendorf's computer model using life history and kinship data indicates that the effective population size (N_e) is only about 25% of the total population, and that even Yellowstone's population of 200 bears may experience a loss of genetic variability and perhaps fitness. He has modeled the effect of introducing unrelated bears in order to preserve viability.

■ **INTERACTIVE QUESTION   55.3**

Explain the basic premise of the small population approach. What conservation strategy is recommended for preserving small populations?

The declining-population approach is a proactive conservation strategy of detecting, diagnosing, and halting population declines (1236–1237)

The emphasis of the **declining-population approach** is to identify populations that may be declining, identify the environmental factors that caused that decline, and then recommend corrective measures.

Steps in the Diagnosis and Treatment of Declining Populations   The following logical steps are part of the declining-population approach: assess population trends and distribution to establish that a species is in decline; determine its environmental requirements; list all possible causes of the decline and the predictions that arise from each of these hypotheses; test the most likely hypothesis to see if the population rebounds if this suspected factor is altered; apply the results to the management of the threatened species.

Case Study: Diagnosing and Treating the Decline of the Red-Cockaded Woodpecker   Logging and agriculture have fragmented the mature pine forest habitats of the red-cockaded woodpecker, driving this species toward extinction. Historically, periodic fires kept the understory around the pines low, another habitat requirement. Recognition of the social organization of

this species and the factors that slow its dispersal to new territories has aided in its recovery. Management strategies now include protection of some longleaf pine forests, controlled fires, and the excavation of breeding cavities in unoccupied habitat to encourage establishment of new breeding groups.

■ **INTERACTIVE QUESTION   55.4**

Describe the declining-population approach to the conservation of endangered species.

Conserving species involves weighing conflicting demands (1237–1238)

Preserving habitat for endangered species often conflicts with human economic and recreational desires. Keystone species exert more influence on community structure and ecosystem processes, and prioritizing the species to be saved on the basis of their ecological role may be key to the survival of whole communities.

Conservation at the Community, Ecosystem, and Landscape Levels

Conservation efforts increasingly are directed at sustaining the biodiversity of whole communities and ecosystems. Many species use more than one kind of ecosystem or live on borders between ecosystems. A **landscape** is a collection of interacting ecosystems, and **landscape ecology** applies ecological principles to large land-use patterns.

Edges and corridors can strongly influence landscape biodiversity (1238–1239)

Landscapes include ecosystems separated by boundaries or *edges*, which have their own sets of physical conditions and communities of organisms. Edge communities may be important sites of speciation, but their proliferation due to human fragmentation of habitats may serve to reduce biodiversity as edge species become predominant.

Movement corridors are narrow strips or clumps of habitat that connect isolated patches. Artificial corridors are sometimes constructed when habitat patches have been separated by major human disruptions.

■ **INTERACTIVE QUESTION 55.5**

What are some potential benefits of corridors? How may they be harmful?

Conservation biologists face many challenges in setting up protected areas (1239–1240)

Currently, about 7% of Earth's land has been set aside as reserves. Current ecological research is applied to the designation and management of these protected areas. **Biodiversity hot spots,** small areas with very high concentrations of endemic, threatened, and endangered species, are good choices for nature reserves. Hot spots vary, however, by taxonomic group, and their protection would in no way conserve all of biodiversity.

New information on the requirements for minimum viable population sizes indicates that most national parks and reserves are much too small—the *biotic boundary* needed to sustain a population is usually much larger than the *legal boundary* set aside in a reserve.

■ **INTERACTIVE QUESTION 55.6**

What factors would favor the creation of larger, extensive preserves? What factors favor smaller, unconnected preserves?

Nature reserves must be functional parts of landscapes (1241–1242)

Even though nature reserves provide islands of protected habitat, the concept of nonequilibrium ecology with natural disturbances applies to them as well as to their surrounding disturbed and fragmented landscapes. Patch dynamics, edges, and corridor effects must be considered in the design and management of preserves.

The approach to landscape management called zoned reserve systems attempts to surround **zoned reserves** with buffer zones in which the human social and economic climate is stable and activities are regulated to promote the long-term viability of the protected zones.

It is likely that less than 10% of the biosphere will ever be protected in nature reserves, so the preservation of biodiversity involves working to create reserves in landscapes that are human dominated.

Restoring degraded areas is an increasingly important conservation effort (1242–1244)

Areas degraded by farming, mining, or environmental pollution are often abandoned. **Restoration ecology** uses ecological principles to return degraded areas to close to their natural state. The natural time frame for recovery relates to the size of the area disturbed. Restoration ecologists attempt to identify and manipulate the factors that most limit recovery time in order to speed the successional processes involved in a community's recovery from human disturbances. An example of this type of augmentation of ecosystem processes is using plant species that thrive in nutrient-poor soils to facilitate recolonization of native species.

Bioremediation uses prokaryotes, fungi, or plants to detoxify polluted ecosystems. Some plants may be able not only to extract metals from contaminated soils, but to concentrate them for commercial use. Prokaryotes are being used to metabolize toxins in dump sites and clean up oil spills.

Restoration ecologists often apply **adaptive management,** in which they experiment with promising management approaches and learn as they work in each unique and complex disturbed ecosystem.

The goal of sustainable development is reorienting ecological research and challenging all of us to reassess our values (1244)

Sustainable development is a concept that emphasizes the long-term prosperity of human societies and the ecosystems that support them. The Ecological Society of America endorses a research agenda, the **Sustainable Biosphere Initiative,** to encourage studies of global change, biodiversity, and maintenance of the productivity of natural and artificial ecosystems. An important goal is developing the ecological knowledge necessary to make intelligent and responsible decisions concerning Earth's abiotic and biotic resources.

Declining biodiversity, an expanding human population, and continuing environmental degradation require individuals and nations to address their values that affect the long-term maintenance and health of the biosphere.

The future of the biosphere may depend on our biophilia (1245)

E. O. Wilson calls our attraction to Earth's diversity of life and our affinity for natural environments *biophilia*. Perhaps this connection is innate and will provide the ethical resolve to protect species from extinction and ecosystems from destruction. By coming to know and understand nature through the study of biology, we may be more able to appreciate and value the processes and diversity of the biosphere.

WORD ROOTS

bio- = life (*biodiversity hot spot*: a relatively small area with an exceptional concentration of species)

STRUCTURE YOUR KNOWLEDGE

1. What are the major threats to biodiversity, listed in order of importance?

2. How does the loss of biodiversity threaten human welfare?

3. What do edges and movement corridors have to do with habitat fragmentation?

TEST YOUR KNOWLEDGE

MULTIPLE CHOICE: *Choose the one best answer.*

1. According to the Endangered Species Act, what is the definition of a threatened species?
   a. an exotic species that cannot successfully compete with indigenous organisms
   b. an endemic species that is found nowhere else in the world
   c. a species that is found in disturbed habitats
   d. a species that is in danger of extinction in all or a large part of its range
   e. a species that is likely to become endangered

2. Ecosystem services include all of the following *except*
   a. pollination of crops.
   b. production of antibiotics and drugs.
   c. access to aesthetic beauty.
   d. decomposition of wastes.
   e. moderation of weather extremes.

3. Which of the following is the most serious threat to biodiversity?
   a. competition from introduced species
   b. commercial harvesting
   c. habitat destruction
   d. overexploitation
   e. disruptions of food chains

4. Some grassland and conifer forest preserves have effective fire prevention programs. What is the most likely result of such programs?
   a. an increase in species diversity because fires are prevented
   b. a change in community composition because fires are natural disturbances that maintain the community structure
   c. the preservation of endangered species in the area
   d. no change in the species composition of the preserved community
   e. succession to a deciduous forest

5. According to the small population approach, what is the most important remedy for preserving an endangered species?
   a. establish a large nature reserve around its habitat
   b. control the populations of its natural predators
   c. determine the reason for its decline
   d. encourage dispersal and increase in genetic variability
   e. set up artificial breeding programs

6. Which of the following is typical of biodiversity hot spots?
   a. a large number of endemic species
   b. a high rate of habitat degradation
   c. little species diversity
   d. a large land or aquatic area
   e. very large populations of migratory birds

7. Which of the following may occur when a population drops below its minimum viable population size?
   a. genetic drift
   b. a further reduction in population size
   c. inbreeding
   d. a loss of genetic variability
   e. All of the above are characteristics of an extinction vortex that the population may enter.

8. What are movement corridors?
   a. strips or clumps of habitat that connect isolated habitats
   b. the routes taken by migratory animals
   c. a landscape that includes several different ecosystems

d. the areas forming the boundary or edge between two ecosystems

e. buffer zones that promote the long-term viability of protected areas

9. What does it mean if a population's effective population size (N_e) is the same as its actual population size?

a. The population is not in danger of becoming extinct.

b. The population has high genetic variability.

c. All the members of the population breed.

d. The population's minimum viable population will not sustain the population.

e. The population is being drawn into an extinction vortex.

10. The focus of the declining-population approach to conservation is to

a. predict a species' minimum viable population size.

b. transplant members from other populations to increase genetic variation.

c. perform a population viability analysis to predict the long-term viability of a population in a particular habitat.

d. determine the cause of a species' decline and take remedial action.

e. establish zoned reserves that ensure that human landscapes surrounding reserves support the protected habitats.

11. With limited resources, conservation biologists need to prioritize their efforts. Of the following choices, which should receive the greatest conservation attention?

a. the northern spotted owl

b. declining keystone species in a community

c. a commercially important species

d. endangered and threatened vertebrate species

e. all declining species

12. Restoration ecology

a. uses the zoned reserve system to buffer nature reserves.

b. identifies biodiversity hot spots for protection.

c. may use bioremediation and augmentation of successional processes to return degraded areas to their natural state.

d. uses the research agenda of the Sustainable Biosphere Initiative to study biodiversity and preserve Earth's ecosystems.

e. uses adaptive management to restore and maintain the productivity of artificial ecosystems.

ANSWER SECTION

CHAPTER 1: INTRODUCTION: TEN THEMES IN THE STUDY OF LIFE

**SUGGESTED ANSWERS TO STRUCTURE
YOUR KNOWLEDGE**

1. See Table 1.1, page 22 in main text.

CHAPTER 2: THE CHEMICAL CONTEXT OF LIFE

■ INTERACTIVE QUESTIONS

2.1 **a.** Mass is a measure of the amount of matter within an object.
b. Weight is how strongly that mass is pulled upon by gravity.

2.2 calcium, phosphorus, potassium, sulfur, sodium, chlorine, magnesium

2.3 neutrons; 15, 15, 16, 31

2.4 absorb; released

2.5 **a.** Nitrogen
$_7$N

b. Oxygen
$_8$O

c. Magnesium
$_{12}$Mg

d. Chlorine
$_{17}$Cl

2.6 **a.** protons
b. atomic number
c. element

d. neutrons
e. mass number or atomic weight
f. isotopes
g. electrons
h. electron shells or energy levels
i. valence shell

2.7 **a.** 1 **b.** 2 **c.** 3 **d.** 4

2.8 **a.** Nonpolar; even though N has a high electronegativity, the three pairs of electrons are shared equally between the two N atoms because each atom has an equally strong attraction for the electrons.
b. Nonpolar; C and H have similar electronegativities and equally share electrons between them.
c. Polar; N is more electronegative than H and pulls the shared electrons in each covalent bond closer to itself.
d. The C=O bond is polar because O is more electronegative than C; the C–H bonds are nonpolar.

2.9 **a.** $CaCl_2$ **b.** Ca^{2+} is the cation.

2.10

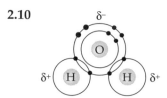

2.11 6, 6, 6

SUGGESTED ANSWERS TO STRUCTURE YOUR KNOWLEDGE

1.

Particle	Charge	Mass	Location
Proton	+1	1 dalton	nucleus
Neutron	0	1 dalton	nucleus
Electron	−1	negligible	orbitals in electron shells

2. a. The atoms of each element have a characteristic number of protons in their nuclei, referred to as the **atomic number**. In a neutral atom, the atomic number also indicates the number of electrons.

The **mass number** is an indication of the approximate mass of an atom and is equal to the number of protons and neutrons in the nucleus.

The **atomic weight** is equal to the mass number and is measured in the atomic mass unit of daltons. Protons and neutrons both have a mass of approximately 1 dalton.

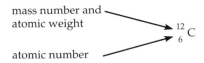

mass number and atomic weight

atomic number

$^{12}_{6}C$

b. The **valence**, an indication of bonding capacity, is the number of unpaired electrons that an atom has in its valence shell.

c. The valence of an atom is most related to the chemical behavior of an atom because it is an indication of the number of bonds the atom will make, or the number of electrons the atom must share in order to reach a filled valence shell.

3. Ionic and nonpolar covalent bonds represent the two ends of a continuum of electron sharing between atoms in a molecule. In ionic bonds the electrons are completely pulled away from one atom by the other, creating negatively and positively charged ions (anions and cations). In nonpolar covalent bonds the electrons are equally shared between two atoms. Polar covalent bonds form when a more electronegative atom pulls the shared electrons closer to it, producing a partial negative charge associated with that portion of the molecule and a partial positive charge associated with the atom from which the electrons are pulled.

ANSWERS TO TEST YOUR KNOWLEDGE

Multiple Choice:

1. b	6. e	11. a	16. c	21. b
2. d	7. d	12. c	17. d	22. c
3. a	8. b	13. c	18. b	
4. e	9. e	14. b	19. a	
5. a	10. c	15. e	20. c	

CHAPTER 3: WATER AND THE FITNESS OF THE ENVIRONMENT

■ INTERACTIVE QUESTIONS

3.1

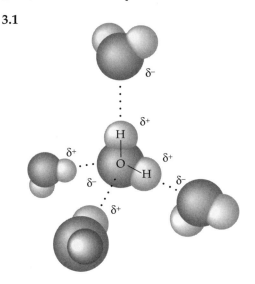

3.2 a. polar water molecules
b. absorbed
c. released
d. specific heat
e. heat of vaporization
f. evaporative cooling
g. solar heat
h. rain
i. ice forms

3.3 a. olive oil — hydrophobic — nonpolar
b. sugar — hydrophilic — polar
c. salt — hydrophilic — ionic
d. candle wax — hydrophobic — nonpolar

3.4 **a.** The molecular weight of $C_3H_6O_3$ is 90 d, the combined atomic weights of its atoms. A mole of lactic acid = 90 g. A 0.5 M solution would require ½ mol, or 45 g.
b. 228 grams of NaCl

3.5

[H^+]	[OH^-]	pH	Acidic, Basic, or Neutral?
10^{-3}	10^{-11}	3	acidic
10^{-8}	10^{-6}	8	basic
10^{-7}	10^{-7}	7	neutral
10^{-1}	10^{-13}	1	acidic

3.6 carbonic acid   bicarbonate   hydrogen ion

$$H_2CO_3 \rightleftharpoons HCO_3^- + H^+$$

H^+ donor H^+ acceptor

a. Bicarbonate acts as a base to accept excess H^+ ions when the pH starts to fall; the reaction moves to the left.
b. When the pH rises, H^+ ions are donated by carbonic acid, and the reaction shifts to the right.

SUGGESTED ANSWERS TO STRUCTURE YOUR KNOWLEDGE

1. **a.** Cohesion, adhesion
    b. A water column is pulled up through plant vessels.
    c. Heat is absorbed or released when hydrogen bonds break or form. Water absorbs or releases a large quantity of heat for each degree of temperature change.
    d. High heat of vaporization
    e. Solar heat is absorbed by tropical seas.
    f. Evaporative cooling
    g. Evaporation of water cools surfaces of plants and animals.
    h. Hydrogen bonds in ice space water molecules apart, making ice less dense.
    i. Floating ice insulates bodies of water so they don't freeze solid.
    j. Versatile solvent
    k. Polar water molecules surround and dissolve ionic and polar solutes.

2. See concept map below.

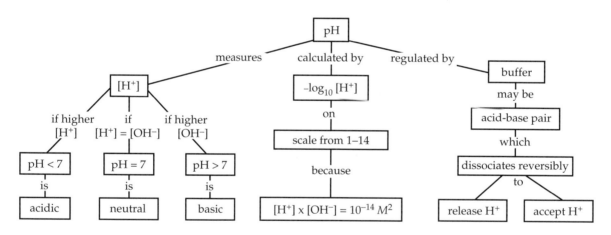

ANSWERS TO TEST YOUR KNOWLEDGE

Multiple Choice:

1. c	5. b	9. e	13. c	17. d
2. a	6. d	10. d	14. c	18. e
3. e	7. e	11. b	15. e	19. d
4. e	8. b	12. d	16. a	20. e

CHAPTER 4: CARBON AND THE MOLECULAR DIVERSITY OF LIFE

■ INTERACTIVE QUESTIONS

4.1

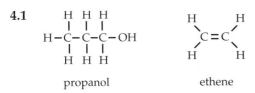

propanol   ethene

4.2 Ethanol and dimethyl ether, structural isomers, have the same number and kinds of atoms but a different bonding sequence and very different properties. Maleic acid and fumaric acid are geometric isomers whose double bonds fix the spatial arrangement of the molecule. Although the enantiomers *l*- and *d*-lactic acid look similar in a flat representation of their structures, they are not superimposable.

SUGGESTED ANSWERS TO STRUCTURE YOUR KNOWLEDGE

1.

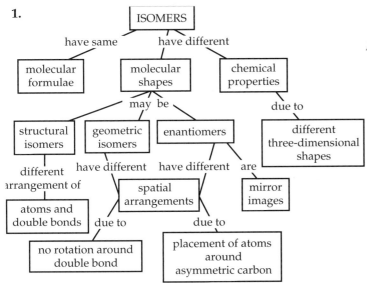

2.

Functional Group	Molecular Formula	Names and Characteristics of Organic Compounds Containing Functional Group
Hydroxyl	–OH	Alcohols; polar group
Carbonyl	$>C=O$	Aldehyde or ketone; polar group
Carboxyl	–COOH	Carboxylic acid; release H^+
Amino	$-NH_2$	Amines; basic, accept H^+
Sulfhydryl	–SH	Thiols; cross-links stabilize protein structure
Phosphate	$-OPO_3^{-2}$	Used in energy transfers

ANSWERS TO TEST YOUR KNOWLEDGE

Multiple Choice:

1. c   4. a   7. d
2. b   5. d   8. c
3. e   6. a

Matching:

1. a, c   4. e   7. b, f   10. a, d
2. b, f   5. e   8. e   11. e
3. a, d, e 6. a, c, d, e 9. a, c   12. c

CHAPTER 5: THE STRUCTURE AND FUNCTION OF MACROMOLECULES

■ INTERACTIVE QUESTIONS

5.1 **a.** hydroxyl
  b. carbonyl
  c. aldose
  d. ketose
  e. rings

5.2

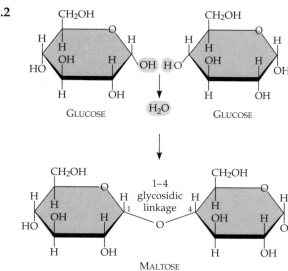

5.3 **a.** monosaccharides
  b. $(CH_2O)_n$
  c. energy compounds
  d. carbon skeletons, monomers
  e. glycosidic linkages
  f. disaccharides
  g. polysaccharides
  h. glycogen
  i. animals
  j. starch
  k. cellulose
  l. chitin

5.4 **a.** fats, triacylglycerides
  b. phospholipids
  c. glycerol
  d. fatty acids
  e. unsaturated: C=C bonds
  f. saturated: no C=C, all possible C—H bonds
  g. phosphate group
  h. cell membrane
  i. steroids
  j. hormones, cell membrane component (cholesterol)

5.5

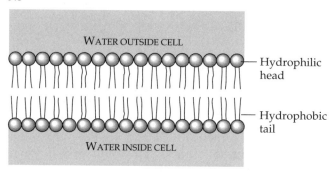

5.6

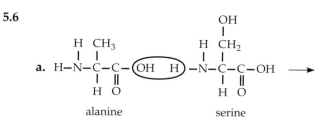

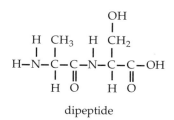

b. serine's R group is polar; alanine's R group is nonpolar
c. a polypeptide backbone

5.7 **a.** hydrogen bond
  b. hydrophobic and van der Waals interactions
  c. disulfide bridge
  d. ionic bond
These interactions between R groups produce tertiary structure.

5.8 **a.** A change in pH alters the availability of H^+, OH^-, or other ions, thereby disrupting the hydrogen bonding and ionic bonds that maintain protein shape.
b. A protein in an organic solvent would turn inside out as the hydrophilic regions became clustered on the inside of the molecule and the hydrophobic regions interacted with the nonpolar solvent.
c. The return to its functional shape indicates that a protein's conformation is intrinsically determined by its primary structure—the sequence of its amino acids.

5.9

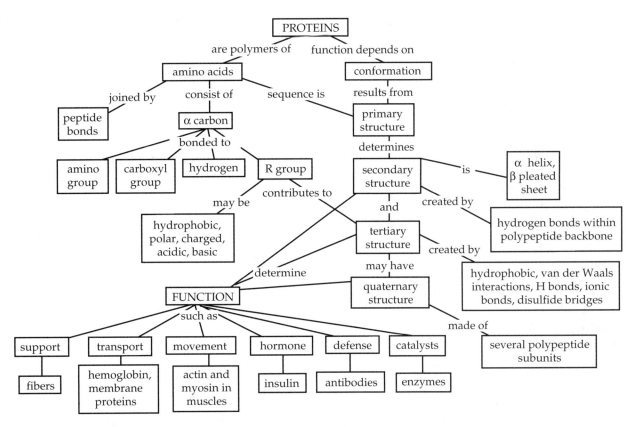

5.10 DNA → RNA → Protein

5.11 a.

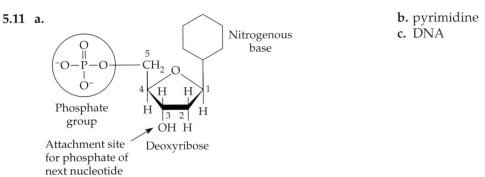

b. pyrimidine
c. DNA

5.12

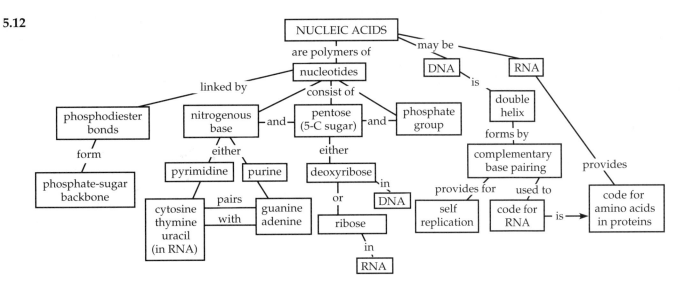

SUGGESTED ANSWERS TO STRUCTURE YOUR KNOWLEDGE

1. The primary structure of a protein is the specific, genetically coded sequence of amino acids in a polypeptide chain. The secondary structure involves the coiling (α helix) or folding (β pleated sheet) of the protein, stabilized by hydrogen bonds along the polypeptide backbone. The tertiary structure involves interactions between the side chains of amino acids and produces a characteristic three-dimensional shape for a protein. Quaternary structure occurs in proteins composed of more than one polypeptide chain.

2. **a.** amino acid (glycine)
    b. fatty acid
    c. nitrogenous base, purine (adenine)
    d. glycerol
    e. phosphate group
    f. pentose (ribose)
    g. sugar (triose)

    1. b, d   3. c, e, f   5. c   7. e, f
    2. a   4. f, g   6. a   8. b

ANSWERS TO TEST YOUR KNOWLEDGE

Matching:

1. A   4. C   7. B   10. A
2. B   5. C   8. C
3. D   6. D   9. A

Multiple Choice:

1. e   5. c   9. d   13. b   17. c
2. c   6. b   10. c   14. d   18. b
3. a   7. c   11. c   15. d   19. c
4. e   8. a   12. d   16. a   20. b

Fill in the Blanks:

1. F. Sanger   6. quaternary
2. guanine   7. glycogen
3. purines   8. phospholipids
4. nucleotide   9. cellulose
5. primary structure   10. chaperonins
   (amino acid sequence)

CHAPTER 6: AN INTRODUCTION TO METABOLISM

■ INTERACTIVE QUESTIONS

6.1 **a.** capacity to do work
     b. kinetic
     c. motion
     d. potential
     e. position
     f. conserved
     g. created nor destroyed
     h. first
     i. transformed or transferred
     j. entropy
     k. second

6.2

	System with high free energy	System with low free energy
Stability	low	high
Spontaneous	change will be	change will not be
Equilibrium	moves toward	is at
Work capacity	high	low

6.3

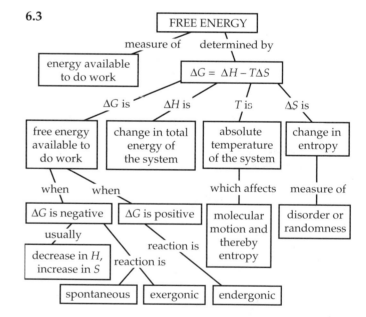

6.4 **a.** adenine
     b. ribose
     c. three phosphate groups
     d. The phosphate portion of the molecule is most unstable, and the terminal phosphate bond is most likely to break.

e. The negatively charged phosphate groups are crowded together, and their mutual repulsion makes the bonds unstable. A hydrolysis reaction breaks the terminal phosphate bond and releases a molecule of inorganic phosphate.

f. The products of hydrolysis (ADP and P_i) are more stable than ATP. The chemical change to a more stable state releases energy.

6.5 **a.** free energy
b. transition state
c. E_A (free energy of activation) with enzyme
d. E_A without enzyme
e. ΔG of reaction

6.6

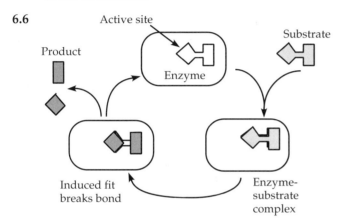

6.7 A competitive inhibitor would mimic the shape of the substrate and compete with the substrate for the active site. A noncompetitive inhibitor would be a shape that could bind to another site on the enzyme molecule and would change the conformation of the active site such that the substrate could no longer fit.

SUGGESTED ANSWERS TO STRUCTURE YOUR KNOWLEDGE

1. Metabolism is the totality of chemical reactions that take place in living organisms. To create and maintain the structural order required for life requires an input of free energy—from sunlight for photosynthetic organisms and from energy-rich food molecules for other organisms. A cell couples catabolic, exergonic reactions ($-\Delta G$) with anabolic, endergonic reactions ($+\Delta G$), using ATP as the primary energy shuttle between the two.

2. Enzymes are essential for metabolism because they lower the activation energy of the specific reactions they catalyze and allow those reactions to occur extremely rapidly at a temperature conducive to life. By regulating the enzymes it produces, a cell can regulate which of the myriad of possible chemical reactions take place at any given time. Metabolic control also occurs through allosteric regulation and feedback inhibition.

ANSWERS TO TEST YOUR KNOWLEDGE

Multiple Choice:

1. c	5. c	9. e	13. b	17. b
2. b	6. c	10. c	14. b	18. d
3. a	7. a	11. b	15. e	19. c
4. e	8. e	12. c	16. d	20. a

Fill in the Blanks:

1. metabolism	6. free energy of activation
2. anabolic	7. competitive inhibitors
3. kinetic	8. coenzymes
4. allosteric	9. feedback inhibition
5. entropy	10. phosphorylated intermediates

CHAPTER 7: A TOUR OF THE CELL

■ INTERACTIVE QUESTIONS

7.1 **a.** the study of cell structure
b. the internal ultrastructure of cells
c. the three-dimensional surface topography of a specimen
d. Light microscopy enables study of living cells and does not introduce the artifacts that may be introduced by TEM and SEM.

7.2 **a.** 10^2, or 100 times the surface area
b. 10^3, or 1000 times the volume

7.3 The genetic instructions for specific proteins are transcribed from DNA into messenger RNA (mRNA), which then passes into the cytoplasm to complex with ribosomes where it is translated into the primary structure of proteins.

7.4 **a.** smooth ER—in different cells may house enzymes that synthesize lipids, metabolize carbohydrates, detoxify drugs and alcohol; stores and releases calcium ions in muscle cells
b. nuclear envelope—double membrane that encloses nucleus; pores regulate passage of materials

c. rough ER—attached ribosomes produce proteins that enter cisternae; produces secretory proteins and membranes

d. transport vesicle—carries products of ER and Golgi apparatus to various locations

e. Golgi apparatus—processes products of ER; makes polysaccharides, packages products in vesicles targeted to specific locations

f. plasma membrane—selective barrier that regulates passage of materials into and out of cell

g. vacuole—membrane-enclosed sac that may perform diverse functions

h. lysosome—houses hydrolytic enzymes to digest macromolecules

7.5

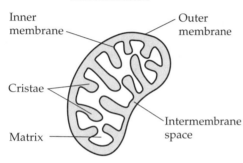

Mitochondrion

Inner membrane · Outer membrane · Cristae · Intermembrane space · Matrix

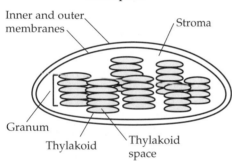

Chloroplast

Inner and outer membranes · Stroma · Granum · Thylakoid · Thylakoid space

7.6 Peroxisomes do not bud from the endomembrane system, but grow by incorporating proteins and lipids from the cytosol; they increase in number by dividing.

7.7 **a.** hollow tube, formed from columns of tubulin dimers; 25-nm diameter

b. cell shape and support (compression resistant), tracks for moving organelles, chromosome movement, beating of cilia and flagella

c. two twisted chains of actin molecules; 7-nm diameter

d. muscle contraction, maintain (tension bearing) and change cell shape, pseudopod movement, cytoplasmic streaming, sol-gel transformations

e. supercoiled fibrous proteins of keratin family; 8–12 nm

f. reinforce cell shape, anchor nucleus

7.8   **Two adjacent plant cells**

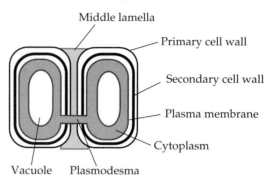

Middle lamella · Primary cell wall · Secondary cell wall · Plasma membrane · Cytoplasm · Vacuole · Plasmodesma

SUGGESTED ANSWERS TO STRUCTURE YOUR KNOWLEDGE

1. **a.** nucleus, chromosomes, centrioles, microtubules (spindle), microfilaments (actin-myosin aggregates pinch apart cell)

    b. nucleus, chromosomes, DNA $\longrightarrow$ mRNA $\longrightarrow$ ribosomes $\longrightarrow$ enzymes and other proteins

    c. mitochondria

    d. ribosomes, rough and smooth ER, Golgi apparatus, vesicles

    e. smooth ER (peroxisomes also detoxify substances)

    f. lysosomes, food vacuoles

    g. peroxisomes

    h. cytoskeleton: microtubules, microfilaments, intermediate filaments, extracellular matrix

    i. cilia and flagella (microtubules), microfilaments (actin) in muscles and pseudopodia

    j. plasma membrane, vesicles

    k. desmosomes, tight and gap junctions, ECM

2. **a.** structural support, middle lamella glues cells together

    b. storage, waste disposal, protection, growth

    c. photosynthesis, production of carbohydrates

    d. starch storage

    e. cytoplasmic connections between cells

3. **a.** rough endoplasmic reticulum

    b. smooth endoplasmic reticulum

    c. chromatin

    d. nucleolus

    e. nuclear envelope

    f. nucleus

    g. ribosomes

    h. Golgi apparatus

    i. plasma membrane

    j. mitochondrion

    k. lysosome

    l. cytoskeleton

    m. microtubules

    n. intermediate filaments

o. microfilaments
p. microvilli
q. peroxisome
r. centrosome (contains a pair of centrioles)
s. flagellum

4.

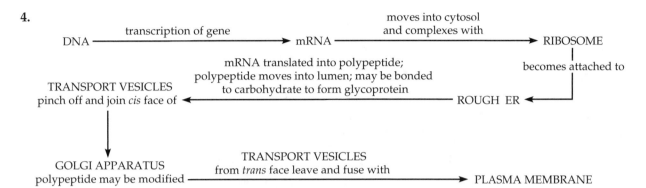

ANSWERS TO TEST YOUR KNOWLEDGE

Multiple Choice:

1. c	5. a	9. b	13. b	17. c
2. b	6. d	10. d	14. c	18. e
3. d	7. a	11. e	15. e	19. b
4. e	8. c	12. b	16. d	20. a

Fill in the Blanks:

1. transport vesicle
2. cristae
3. extracellular matrix
4. peroxisomes
5. grana
6. basal body
7. cytosol
8. cytoskeleton
9. tight junction
10. tonoplast

CHAPTER 8: MEMBRANE STRUCTURE AND FUNCTION

■ INTERACTIVE QUESTIONS

8.1 **a.** phospholipid bilayer
b. hydrocarbon tail—hydrophobic
c. phosphate head—hydrophilic
d. hydrophobic region of protein
e. hydrophilic region of protein

8.2 **a.** In hybrid human/mouse cells, membrane proteins rapidly intermingle.
b. The cell may increase the proportion of unsaturated phospholipids in its membrane.

8.3 transport, enzymatic activity, signal transduction, intercellular attachment, cell-cell recognition, attachment to cytoskeleton and ECM

8.4 Ions and larger polar molecules, such as glucose, are impeded by the hydrophobic center of the plasma membrane's lipid bilayer.

8.5 Side A is initially hypertonic; side B is initially hypotonic; water will move from B to A.

8.6 **a.** The hypertonic protists will gain water from their hypotonic environment.

b. They may have membranes that are less permeable to water and contractile vacuoles that expel excess water.

8.7 **a.** isotonic
b. hypotonic

8.8 Although it may speed diffusion, facilitated diffusion is still passive transport because the solute is moving down its concentration gradient; the process is driven by the concentration gradient and not energy expended by the cell.

8.9 Three Na^+ ions are pumped out of the cell for every two K^+ ions pumped in, resulting in a net movement of positive charge from the cytoplasm to the extracellular fluid.

8.10 **a.** Human cells use receptor-mediated endocytosis to take in cholesterol.
b. LDL receptor proteins in the plasma membrane are defective and low-density lipoproteins cannot bind and be transported into the cell.

SUGGESTED ANSWERS TO STRUCTURE YOUR KNOWLEDGE

1.

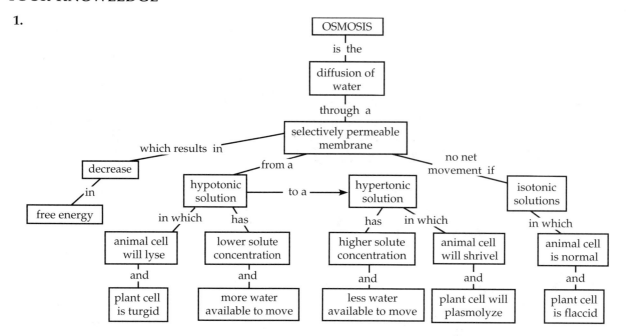

2. **a.** II represents facilitated diffusion because the solute is moving through a transport protein and down its concentration gradient. The cell does not expend energy in this transport. Polar molecules and ions may be moved by facilitated diffusion.

    b. III represents active transport because the solute is clearly moving against its concentration gradient. Energy from the cell must be expended to drive this transport against the gradient.

    c. I illustrates diffusion through the lipid bilayer. The solute molecules must be nonpolar or very small polar molecules.

    d. Both diffusion and facilitated diffusion are considered passive transport because the solute moves down its concentration gradient and the cell does not expend energy in the transport.

ANSWERS TO TEST YOUR KNOWLEDGE

Multiple Choice:

1. c	6. a	11. e	16. b	21. d
2. b	7. e	12. a	17. e	22. b
3. a	8. c	13. b	18. c	23. d
4. e	9. b	14. c	19. a*	24. c
5. d	10. a	15. b	20. e	25. d

*Explanation for answer to question 19: This problem involves both osmosis and diffusion. Although the solution is initially isotonic, glucose will diffuse down its concentration gradient until it reaches dynamic equilibrium with a 1.5 *M* concentration on both sides. The increasing solute concentration on side A will cause water to move into this hypertonic side, and the water level will rise.

CHAPTER 9: CELLULAR RESPIRATION: HARVESTING CHEMICAL ENERGY

■ INTERACTIVE QUESTIONS

9.1 $C_6H_{12}O_6$, 6 CO_2, energy (ATP + heat)

9.2 **a.** oxidized
   b. oxidizing agent
   c. reduced

9.3 **a.** oxygen
   b. glucose
   c. Some is stored in ATP and some is released as heat.

9.4 **a.** electron acceptor or oxidizing agent
   b. NADH

9.5 **a.** glycolysis: glucose $\longrightarrow$ pyruvate (not technically considered part of cellular respiration because it does not require O_2)
   b. Krebs cycle
   c. electron transport chain and oxidative phosphorylation
   d. substrate-level phosphorylation
   e. substrate-level phosphorylation
   f. oxidative phosphorylation
   The top two arrows show electrons carried by NADH to the electron transport chain.

9.6 **a.** 2ATP
   b. 2 glyceraldehyde phosphate
   c. 2 NAD^+
   d. 2 NADH
   e. 4 ATP
   f. 2 pyruvate

9.7 **a.** pyruvate (from glycolysis)
   b. CO_2
   c. NADH
   d. CoA
   e. acetyl CoA
   f. oxaloacetate
   g. citrate
   h. CO_2

   i. NADH
   j. CO_2
   k. NADH
   l. ATP
   m. $FADH_2$
   n. NADH

9.8 **a.** intermembrane space
   b. inner mitochondrial membrane
   c. mitochondrial matrix
   d. electron transport chain
   e. NADH + H^+
   f. NAD^+
   g. H^+
   h. 2 H^+ + ½ O_2
   i. H_2O
   j. ATP synthase
   k. ADP + ⓟ_i
   l. ATP

9.9 **a.** -2
   b. 4
   c. Krebs cycle
   d. 32 or 34
   e. 38
   f. 2
   g. 6
   h. 2
   i. 2
   j. 2

9.10 Respiration yields up to 19 times more ATP than does fermentation. By oxidizing pyruvate to CO_2 and passing electrons from NADH (and $FADH_2$) through the electron transport chain, respiration can produce a maximum of 38 ATP compared to the 2 net ATP that are produced by fermentation.

SUGGESTED ANSWERS TO STRUCTURE YOUR KNOWLEDGE

1. See Interactive Questions 9.5, 9.6, 9.7, and 9.8.

2.

Process	Main Function	Inputs	Outputs
Glycolysis	Oxidation of glucose to 2 pyruvate, 2 ATP net	glucose 2 ATP 2 NAD$^+$ 4 ADP + $\circledP_i$	2 pyruvate 4 ATP (2 net) 2 NADH
Pyruvate to acetyl CoA	Oxidation of pyruvate to acetyl CoA, which then enters Krebs cycle	2 pyruvate 2 CoA 2 NAD$^+$	2 acetyl CoA 2 CO$_2$ 2 NADH
Krebs cycle	Acetyl COA is combined with oxaloacetate to produce citrate, which is cycled back to oxaloacetate as redox reactions produce NADH and FADH$_2$, ATP is formed by substrate-level phosphorylation, and CO$_2$ is released.	2 acetyl CoA 2 oxaloacetate 2 ADP + $\circledP_i$ 6 NAD$^+$ 2 FAD	2 CoA 4 CO$_2$ 2 ATP 6 NADH 2 FADH$_2$
Electron transport chain and chemiosmosis	NADH (from glycolysis and Krebs) and FADH$_2$ (from Krebs) transfer electrons to carrier molecules in mitochondrial membrane. In a series of redox reactions, H$^+$ is pumped into intermembrane space, and electrons are delivered to 1/2 O$_2$. Proton-motive force drives H$^+$ through ATP synthase to make ATP.	10 NADH 2 FADH$_2$ 24 H$^+$ 6 O$_2$ 34 ADP + $\circledP_i$	10 NAD$^+$ 2 FAD 12 H$_2$O 34 ATP
Fermentation	Anaerobic catabolism: glycosis followed by regeneration of NAD$^+$ so glycolysis can continue. Pyruvate is either reduced to ethyl alcohol and CO$_2$ or to lactate.	See glycosis above 2 pyruvate 2 NADH	2 ATP 2 NAD$^+$ 2 ethanol and 2 CO$_2$ or 2 lactate

3.

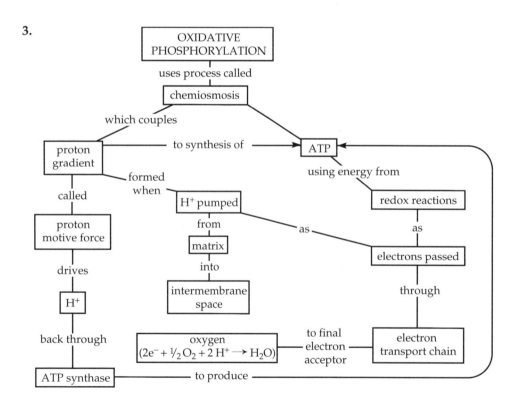

ANSWERS TO TEST YOUR KNOWLEDGE

1. a	7. c	13. c	19. c	25. a
2. a	8. e	14. e	20. e	26. e
3. c	9. b	15. c	21. c	27. c
4. d	10. b	16. c	22. b	28. d
5. e	11. a	17. b	23. d	
6. d	12. e	18. d	24. e	

CHAPTER 10: PHOTOSYNTHESIS

■ INTERACTIVE QUESTIONS

10.1 **a.** outer membrane
b. granum
c. inner membrane
d. thylakoid space
e. thylakoid
f. stroma

10.2 **a.** light
b. H_2O
c. light reactions taking place in thylakoid menbranes
d. O_2
e. ATP
f. NADPH
g. CO_2
h. Calvin cycle taking place in stroma
i. CH_2O (sugar)

10.3 blue light

10.4 The solid line is the absorption spectrum; the dotted line is the action spectrum. Some wavelengths of light, particularly in the yellow and orange range, result in a higher rate of photosynthesis than would be indicated by the absorption of those wavelengths by chlorophyll *a*. These differences are partially accounted for by accessory pigments, such as chlorophyll *b* and the carotenoids, that absorb light energy from different wavelengths and pass that energy on to chlorophyll *a*.

10.5 A photosystem contains an antenna complex of chlorophyll *a*, chlorophyll *b*, and carotenoid molecules; a reaction-center chlorophyll *a* (P700 or P680); and a primary electron acceptor.

10.6 **a.** photosystem II
b. P680, reaction-center chlorophyll *a*
c. water (H_2O)
d. oxygen ($\frac{1}{2} O_2$)
e. electrons ($2e^-$)
f. primary electron acceptor
g. electron transport chain
h. photophosphorylation by chemiosmosis
i. ATP
j. photosystem I
k. P700
l. primary electron acceptor
m. $NADP^+$ reductase
n. NADPH

ATP and NADPH provide the chemical energy and reducing power for the Calvin cycle.

10.7 **a.** Ferredoxin passes the electrons to the cytochrome complex in the electron transport chain, from which they return to P700.
b. Electrons from P680 are not passed to P700. Without the oxidizing agent P680, water is not split. Fd does not pass electrons to $NADP^+$ reductase to form NADPH.
c. Electrons do pass down the electron transport chain, and the energy released by their "fall" drives cyclic photophosphorylation.

10.8 **a.** in the thylakoid space (pH 5)
b. (1) transport of protons into the thylakoid space as Pq transfers electrons to the cytochrome complex; (2) protons from the splitting of water remain in the thylakoid space; (3) removal of H^+ in the stroma during the reduction of $NADP^+$.

10.9 Photorespiration may be an evolutionary relic from the time when there was little O_2 in the atmosphere and the ability of rubisco to distinguish between O_2 and CO_2 was not critical. Now, in our oxygen-rich atmosphere, photorespiration seems to be an agricultural liability.

10.10 **a.** in the bundle-sheath cells
b. Carbon is initially fixed into a four-carbon compound in the mesophyll cells by PEP carboxylase. When this compound is broken down in the bundle-sheath cells, CO_2 is maintained at a high enough concentration that rubisco does not accept O_2 and cause photorespiration.

10.11 a. carbon fixation
   b. reduction
   c. regeneration of CO_2 acceptor (RuBP)
   d. 3 CO_2
   e. ribulose bisphosphate (RuBP)
   f. rubisco
   g. 3-phosphoglycerate
   h. 6 ATP $\longrightarrow$ 6 ADP

   i. 1,3-bisphosphoglycerate
   j. 6 NADPH $\longrightarrow$ 6 $NADP^+$
   k. 6 (P)_i
   l. glyceraldehyde 3-phosphate (G3P)
   m. G3P
   n. glucose and other organic molecules
   o. 3 ATP $\longrightarrow$ 3 ADP + 3 (P)_i

SUGGESTED ANSWERS TO STRUCTURE YOUR KNOWLEDGE

1.

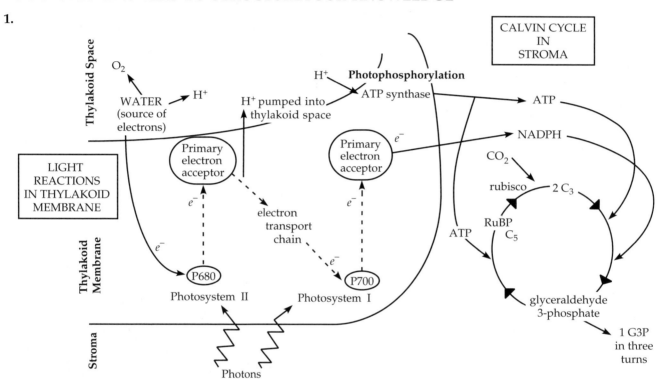

2.

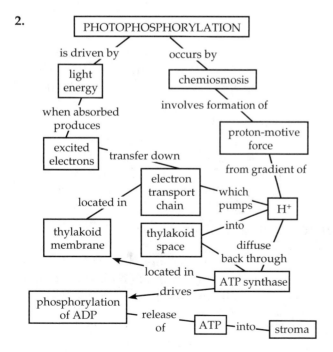

ANSWERS TO TEST YOUR KNOWLEDGE

Multiple Choice:

1. d	7. e	13. d	19. e	25. b
2. a	8. c	14. d	20. d	26. a
3. b	9. a	15. c	21. d	27. d
4. a	10. e	16. b	22. e	28. a
5. b	11. a	17. c	23. c	
6. b	12. b	18. d	24. a	

CHAPTER 11: CELL COMMUNICATION

■ INTERACTIVE QUESTIONS

11.1 a. Yes

b. In plants, hormones may reach their target cells by traveling through vessels, through cells via plasmodesmata, or even through the air as a gas.

11.2 The G protein ia also a GTPase enzyme that hydrolyzes its bound GTP to GDP and inactivates itself. It then dissociates from the enzyme it had activated, and that enzyme returns to its original state.

11.3 a. signal molecules
b. α helix in the membrane
c. tyrosine kinase part of receptor
d. phosphates
e. ATP $\longrightarrow$ ADP
f. tyrosines
g. activated relay proteins
h. cellular responses

11.4 a. A target cell has a protein receptor that is specific for that signal.
b. Signals that bind to surface receptors are large, polar, and/or ionic. These hydrophilic molecules cannot pass through the membrane. Intracellular signals are small or hydrophobic and can move through the lipid membrane.

11.5 a. A protein kinase transfers a phosphate group from ATP to a protein; the resulting conformation change usually activates the protein.
b. A protein phosphatase removes a phosphate group from a protein, usually inactivating the protein.
c. A phosphorylation cascade is a series of protein kinase relay molecules that are sequentially phosphorylated.

11.6 a. signal molecule (first messenger)
b. G-protein-linked receptor
c. activated G protein (GTP bound)
d. adenylyl cyclase
e. ATP
f. cAMP (second messenger)
g. protein kinase A
h. phosphorylation cascade to cellular response

11.7 a. signal molecule
b. dimer
c. phosphorylated tyrosines
d. phospholipase C
e. IP$_3$
f. Ca^{2+}
g. endoplasmic reticulum
h. calmodulin

11.8 a. Signal molecules reversibly bind to receptors, and when they leave a receptor, the receptor reverts to its inactive form.
b. Activated G proteins are inactivated when the GTPase portion of the protein converts GTP to GDP.
c. Phosphodiesterase converts cAMP to AMP, thus damping out this second messenger.
d. Protein phosphatases remove phosphate groups from activated proteins. The balance of active protein kineses and active phosphatases regulates the activity of many proteins.

SUGGESTED ANSWERS TO STRUCTURE YOUR KNOWLEDGE

1. Cell-signaling pathways are essential for communication between cells. Unicellular organisms use signals to relay environmental or reproductive information. Communication in multicellular organisms allows for the development and coordination of specialized cells. Extracellular signals control the crucial activities of cells, such as cell division, differentiation, metabolism, and gene expression.

2. Cell signaling occurs through signal-transduction pathways that include reception, transduction, and response. First a signal molecule binds to a specific receptor. The message is transduced as the activated receptor activates a protein that may relay the message through a sequence of activations, finally leading to the activation of proteins that produce the specific cellular response.

3. The sequence described in the previous answer results in the activation of cellular proteins. When the signal is transduced to activate a transcription factor, however, the cellular response is a change in gene expression and the production of new proteins.

4. In an enzyme cascade, each step in the pathway activates multiple substrates of the next step, thus amplifying the original message to produce potentially millions of activated proteins and thus a large cellular response to a few signals.

ANSWERS TO TEST YOUR KNOWLEDGE

1. c	4. b	7. c	10. a	13. b
2. d	5. e	8. e	11. c	14. d
3. a	6. d	9. b	12. e	15. c

CHAPTER 12: THE CELL CYCLE

■ INTERACTIVE QUESTIONS

12.1 genome

12.2 a. 46
b. 23

12.3 a. Growth—most organelles and cell components are produced continuously throughout these subphases.
b. DNA synthesis

12.4 Refer to main text Fig. 12.5 for chromosome diagrams.
a. G$_2$ of interphase
b. prophase
c. prometaphase
d. metaphase
e. anaphase
f. telophase and cytokinesis
g. centrosomes (with centrioles)
h. aster
i. chromatin (duplicated)
j. nuclear envelope
k. nucleolus
l. early mitotic spindle
m. pair of centrioles
n. nonkinetochore microtubules
o. kinetochore microtubules
p. metaphase plate
q. spindle
r. cleavage furrow
s. nuclear envelope forming

12.5 a. MPF is a complex of cyclin and Cdk that initiates mitosis by phosphorylating proteins and other kinases.
b. MPF concentration is high as it triggers the onset of mitosis but is reduced at the end of mitosis because it depends on the concentration of cyclin in the cell. The Cdk level is constant through the cell cycle, but the level of cyclin varies because active MPF starts a process that degrades cyclin. Thus MPF regulates its own level and can only become active when sufficient cyclin accumulates again during the next interphase.

SUGGESTED ANSWERS TO STRUCTURE YOUR KNOWLEDGE

1. **Interphase:** 90% of cell cycle; growth and DNA replication.

- G$_1$ phase: The chromosome consists of long, thin chromatin fibers made of DNA and associated proteins. RNA molecules are being transcribed from genes that are switched on.
- S phase—synthesis of DNA: The chromosome is replicated; two exact copies, called sister chromatids, are produced and held together by proteins along their length. Growth continues.
- G$_2$ phase: Growth continues.

Mitosis phase: cell division
- Prophase: The chromatids become tightly coiled and folded.
- Prometaphase: Kinetochore fibers from opposite ends of the mitotic spindle attach to the kinetochores of the sister chromatids; the chromosome moves toward midline.
- Metaphase: The centromere of the chromosome is aligned at the metaphase plate along with the centromeres of the other chromosomes.
- Anaphase: The sister chromatids separate (now considered to be individual chromosomes) and move to opposite poles.
- Telophase: Chromatin fiber of chromosome uncoils and is surrounded by re-forming nuclear membrane.

2.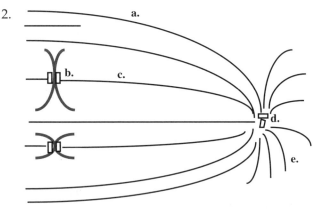

a. nonkinetochore microtubules: push poles apart by sliding past microtubules from the opposite pole
b. kinetochore: protein and DNA structure in region of centromere where microtubules attach; contains motor proteins
c. kinetochore microtubules: move chromosomes to metaphase plate and separate chromosomes as motor proteins of kinetochores "walk" toward the pole and the microtubules disassemble

d. centrosome and centrioles: region of mitotic spindle formation

e. aster: radiating spindle fibers (in animal cells)

3. **a.** anaphase
   b. interphase
   c. late telophase
   d. metaphase

ANSWERS TO TEST YOUR KNOWLEDGE

Fill in the Blank:

1. G_0
2. anaphase
3. prophase
4. cytokinesis
5. S phase
6. metaphase
7. telophase
8. prophase
9. prometaphase
10. G_1 phase

Multiple Choice:

1. d	5. b	9. b	13. a
2. c	6. c	10. a	14. e
3. d	7. c	11. c	15. a
4. e	8. c	12. e	16. d

CHAPTER 13: MEIOSIS AND SEXUAL LIFE CYCLES

■ INTERACTIVE QUESTIONS

13.1 **a.** 14
  b. 7
  c. 28
  d. 1

13.2 Area of each life cycle from fertilization to meiosis should be shaded in as diploid.
  a. meiosis
  b. fertilization
  c. zygote
  d. gametes
  e. fertilization
  f. $2n$
  g. zygote
  h. meiosis
  i. meiosis
  j. spores
  k. mitosis
  l. gametes
  m. fertilization
  n. zygote

13.3 2^{23}, approximately 8 million

13.4 **a.** metaphase II
  b. prophase I

  c. anaphase I
  d. interphase
  e. metaphase I
  f. anaphase II
  Proper sequence: **d. b. e. c. a. f.**

SUGGESTED ANSWERS TO STRUCTURE YOUR KNOWLEDGE

1. **a.** Chromosome replication, sister chromatids attached at centromere
   b. Synapsis of homologous (tetrads) pairs, crossing over at chiasmata, spindle forms
   c. Homologous pairs line up at metaphase plate, with kinetochores attached to spindle fibers from opposite poles
   d. Homologous pairs of chromosomes separate and homologues move toward opposite poles, sister chromatids remain attached
   e. Haploid set of chromosomes, each consisting of two sister chromatids, aligns at metaphase plate
   f. Centromeres separate and nonreplicated chromosomes move to opposite poles

2.

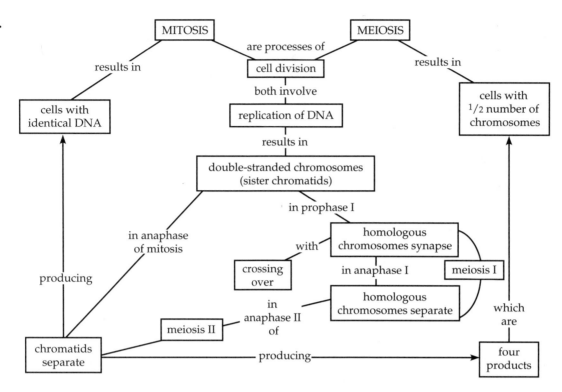

ANSWERS TO TEST YOUR KNOWLEDGE

Multiple Choice:

1. b	4. a	7. c	10. b	13. c	16. b	19. e
2. d	5. e	8. c	11. a	14. d	17. d	20. b
3. e	6. b	9. b	12. b	15. e	18. b	21. a

CHAPTER 14: MENDEL AND THE GENE IDEA

■ INTERACTIVE QUESTIONS

14.1 **a.** Ⓡ

b. ⓡ

c. F₁ Generation

d. *Rr*

e. F₂ Generation

f. Ⓡ

g. ⓡ

h. ◯ *Rr*

i. ◯ *Rr*

j. ⬭ *rr*

k. 3 round : 1 wrinkled

l. 1 *RR* : 2 *Rr* : 1 *rr*

14.2 **a.** all *Tt* tall plants
b. 1 : 1 tall (*Tt*) to dwarf (*tt*)

14.3 **a.** tall purple plants
b. *TtPp*
c. *TP, Tp, tP, tp*
d.

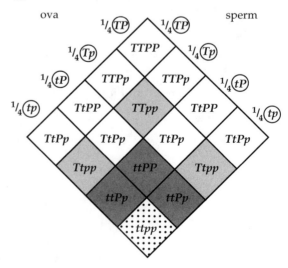

e. 9 tall purple : 3 tall white : 3 dwarf purple : 1 dwarf white
f. 12 : 4 or 3 : 1 tall to dwarf; 12 : 4 or 3 : 1 purple to white

14.4 You could determine this in two ways. The probability of getting both recessive alleles in a gamete is ¼, and for two such gametes to join is ¼ × ¼, or 1/16. An easier way is to treat each gene involved as a monohybrid cross. The probability of obtaining a homozygous recessive offspring in a monohybrid cross is ¼. The probability of the offspring being homozygous recessive for both genes is ¼ × ¼, or 1/16.

14.5 **a.** Consider the outcome for each gene as a monohybrid cross. The probability that a cross of *Aa* × *Aa* will produce an *A_* offspring is ¾. The probability that a cross of *Bb* × *bb* will produce a *B_* offspring is ½. The probability that a cross of *cc* × *CC* will produce a *C_* offspring is 1. To have all of these events occur simultaneously, multiply their probabilities: ¾ × ½ × 1 = ⅜.
b. Offspring could be *A_bbC_*, *aaB_C_*, or *A_B_C_*. The genotype *A_B_cc* is not possible.
Probability of *A_bbC_* = ¾ × ½ × 1 = ⅜
Probability of *aaB_C_* = ¼ × ½ × 1 = ⅛
Probability of *A_B_C_* = ¾ × ½ × 1 = ⅜
Probability of offspring showing at least two dominant traits is the total of these independent probabilities, or 7/8. The probability of only one dominant trait (*aabbCc*) is ¼ × ½ × 1 = ⅛.

14.6 **a.** A : I^AI^A and I^Ai
b. B : I^BI^B and I^Bi
c. AB : I^AI^B
d. O : *ii*

14.7 The ratio of offspring from this *Mm Bb* × *Mm Bb* cross would be 9 : 3 : 4, a common ratio when one gene is epistatic to another.

Phenotype	Genotype	Ratio
Black	*M_B_*	¾ × ¾ = 9/16
Brown	*M_bb*	¾ × ¼ = 3/16
White	*mm__*	¼ × 1 = ¼ or 4/16

14.8 **a.** The parental cross produced 25-cm tall F_1 plants, all *AaBbCc* plants with 3 units of 5 cm added to the base height of 10 cm.
b. As a general rule, the number of phenotypic classes in a quantitative cross of heterozygotes equals the number of alleles involved plus one. In this case, 6 alleles (AaBbCc) + 1 = 7. Or, of the 64 possible combinations of the 8 types of F_1 gametes, there will be 7 different phenotypic classes in the F_2, varying from 6 dominant alleles (40 cm), 5 dominant (35 cm), 4 dominant (30 cm), and so on, to all 6 recessive alleles (10 cm).

14.9 **a.** This trait is recessive. If it were dominant, albinism would be present in every generation, and it would be impossible to have albino children with nonalbino (homozygous recessive) parents.
b. father = *Aa*; mother = *Aa*, because neither parent is albino and they have albino offspring (*aa*)
c. mate 1 = *AA* (probably); mate 2 = *Aa*; grandson 4 = *Aa*
d. The genotype of son 3 could be *AA* or *Aa*. If his wife is *AA*, then he could be *Aa* (both his parents are carriers) and the recessive allele never would be expressed in his offspring. Even if he and his wife were both carriers (heterozygotes), there would be a 243/1024 (¾ × ¾ × ¾ × ¾ × ¾) or 24% chance that all five children would be normally pigmented.

14.10 **a.** ¼
b. ⅔. There is a probability that ¾ of the offspring will have a normal phenotype. Of these, ⅔ would be predicted to be heterozygotes and, thus, carriers of the recessive allele.

14.11 Both sets of prospective grandparents must have been carriers. The parents do not have the disorder so they are not homozygous recessive. Thus each has a ⅔ chance of being a heterozygote carrier. The probability that both parents are carriers is ⅔ × ⅔ = 4/9; the chance that two heterozygotes will have a recessive homozygous child is ¼. The overall chance that a child will inherit the disease is 4/9 × 4/9 = 1/9. Should this couple have a baby that has the disease, this would establish that they are both carriers, and the chance that a subsequent child would have the disease is ¼.

SUGGESTED ANSWERS TO STRUCTURE YOUR KNOWLEDGE

1. Mendel's law of segregation occurs in anaphase I, when alleles segregate as homologous chromosomes move to opposite poles of the cell. The two cells formed from this division have one-half the number of chromosomes and one copy of each gene. Mendel's law of independent assortment relates to the lining up of synapsed chromosomes at the equatorial plate in a random fashion during metaphase I. Genes on different chromosomes will assort independently into gametes.

2. One allele does not "dominate" another. All alleles operate independently within a cell, coding for their gene products, usually enzymes, as specified by their particular sequence of DNA nucleotides. If both alleles code for functional enzymes or products, then the alleles may be codominant with both traits expressed in the heterozygote, as illustrated by MN or AB blood types. If the recessive allele codes for a nonfunctional enzyme, then the resulting recessive trait may be obvious only when homozygous (such as O blood type in which there are no carbohydrate molecules on the cell membrane). If one copy of an allele produces sufficient product that the dominant trait is expressed, then we have complete dominance where the heterozygote phenotype is indistinguishable from the homozygote dominant (for example, smooth peas in which the enzyme converts sugar to starch). With incomplete dominance, the heterozygote is distinguishable from homozygous dominants and recessives. The functional allele is insufficient to produce the full trait (as with the pink color of a heterozygote in carnations). Genotypes translate into phenotypes by way of biochemical pathways, which are controlled by enzymes coded for by genes. Thus molecular processes produce physical, physiological, and behavioral results.

ANSWERS TO GENETICS PROBLEMS

1. White alleles are dominant to yellow alleles. If yellow were dominant, then you should be able to get white squash from a cross of two yellow heterozygotes.

2. **a.** ¼   [½ (to get AA) × ½ (bb)]
   b. ⅛   [¼ (aa) × ½ (BB)]
   c. ½   [1 (Aa) × ½ (Bb) × 1 (Cc)]
   d. 1/32   [¼ (aa) × ¼ (bb) × ½ (cc)]

3. Since flower color shows incomplete dominance, there will be six phenotypic classes in the F_2 instead of the normal four classes found in a 9:3:3:1 ratio. You could find the answer with a Punnett square, but multiplying the probabilities of the monohybrid crosses is more efficient.

Tall red	T_RR = ¾ × ¼	3/16
Tall pink	T_Rr = ¾ × ½	⅜ or 6/16
Tall white	T_rr = ¾ × ¼	3/16
Dwarf red	$ttRR$ = ¼ × ¼	1/16
Dwarf pink	$ttRr$ = ¼ × ½	⅛ or 2/16
Dwarf white	$ttrr$ = ¼ × ¼	1/16

4. Determine the possible genotypes of the mother and child. Then find the blood groups for the father that could not have resulted in a child with the indicated blood group.
   a. no groups exonerated
   b. A or O
   c. A or O
   d. AB only
   e. B or O

5. The parents are *CcBb* and *Ccbb*. Right away you know that all *cc* offspring will die and no *BB* black offspring are possible because one parent is *bb*. Only four phenotypic classes are possible. Determine the proportion of each type by applying the law of multiplication.

   Lethal (*cc_ _*) = ¼ of embryos do not develop
   Normal brown (*CCBb*) = ¼ × ½ = ⅛
   Normal white (*CCbb*) = ¼ × ½ = ⅛
   Deformed brown (*CcBb*) = ½ × ½ = ¼
   Deformed white (*Ccbb*) = ½ × ½ = ¼

   Ratio of viable offspring: 1 : 1 : 2 : 2

6. Father's genotype must be *Pp* since polydactyly is dominant and he has had one normal child. Mother's genotype is *pp*. The chance of the next child having normal digits is ½ or 50% because the mother can only donate a *p* allele and there is a 50% chance that the father will donate a *p* allele.

7. The genotypes of the puppies were ⅜ *B_S_*, ⅜ *B_ss*, ⅛ *bbS_*, and ⅛ *bbss*. Because recessive traits show up in the offspring, both parents had to have had at least one recessive allele for both genes. Black:chestnut occurs in a 6 : 2 or 3 : 1 ratio, indicating a heterozygous cross. Solid:spotted occurs in a 4 : 4 or 1 : 1 ratio, indicating a cross between a heterozygote and a homozygous recessive. Parental genotypes were *BbSs* × *Bbss*.

8. The 1 : 1 ratio of the first cross indicates a cross between a heterozygote and a homozygote. The second cross indicates that the hairless hamsters cannot have a homozygous genotype, because then only hairless hamsters would be produced. Let us say that hairless hamsters are *Hh* and normal-haired are *HH*. The second cross of heterozygotes should yield a 1 : 2 : 1 genotype ratio. The 1 : 2 ratio indicates that the homozygous recessive genotype may be lethal, with embryos that are *hh* never developing.

9. **a.** Tail length appears to be a quantitative trait. A general rule for the number of alleles in a cross involving a quantitative trait is one less than the number of phenotypic classes. Here there are five

phenotypic classes, thus four alleles or two gene pairs. In this cross, the phenotypic ratio is approximately 1 : 4 : 6 : 4 : 1. The sum of 16 also indicates a dihybrid cross. Each dominant gene appears to add 5 cm in tail length. The 15-cm pigs are double recessives; the 35-cm pigs are double dominants.

b. A cross of a 15-cm and a 30-cm pig would be equivalent to *aabb* × *AABb*. Offspring would have either one or two dominant alleles and would have tail lengths of 20 or 25 cm.

10. a. The black parent would be *CC^ch*. The Himalayan parent could be either *C^h c^h* or *C^h c*. First list the alleles in order of dominance: *C*, *C^ch*, *C^h*, *c*. To approach this problem you should write down as much of the genotype as you can be sure about for each phenotype involved. The parents were black (*C_*) and Himalayan (*C^h_*). The offspring genotypes would be black (*C_*) and Chinchilla (*C^ch_*). Right away you can tell that the black parent must have been *CC^ch* because half of the offspring are Chinchilla; *C^ch* is dominant over *C^h* and *c* and could not have been present in the Himalayan parent. The genotype of this parent could be either *C^h C^h* or *C^h c*. A testcross would be necessary to determine this genotype.

b. You cannot definitely determine the genotype of the parents in the second cross. You can, however, eliminate some genotypes. The black parent could not be *CC* or *CC^ch*, because both the *C* and *C^ch* alleles are dominant to *C^h*, and Himalayan offspring were produced. The Chinchilla parent could not be *C^ch C^ch* for the same reason. One or both parents had to have *C^h* as their second allele; one, but not both, might have *c* as their second allele.

11. a. First figure out possible genotypes: _ _*E_* = golden (any *B* combination with at least 1 *E*), *B_ee* = black, *bbee* = brown. All you know of the first parents at the start is _ _*E_* × _ _*E_*. Since you see black and brown offspring, you know that both had to be heterozygous for *Ee*, and at least one was heterozygous for *Bb* (to get a black dog). Since black and brown are in a 1:1 ratio (like a testcross ratio), then one dog was *Bb* and the other was *bb*. You need to consider the results from the second cross to know which dog was *Bb*.

b. For the second cross, you now know that Dog 2 is *?bEe* and Dog 3 is *B?ee*. A ratio of approximately 3:1 black to brown looks like a

cross of heterozygotes, so both parents must be *Bb*. So Dog 3 is *Bbee*, and Dog 2 must be *BbEe*. That means that Dog 1 must be *bbEe*.

12.

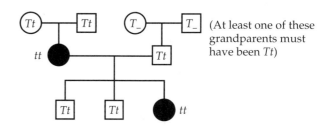

(At least one of these grandparents must have been *Tt*)

13. Since the parents were true-breeding for two characters, the F₁ would be doubly heterozygous. Since all F₁ had red, terminal flowers, those two traits must be dominant and their genotypes could be represented as *RrTt*. One would predict an F₂ phenotypic ratio of 9 red, terminal : 3 red, axial : 3 white, terminal: 1 white, axial. If 100 offspring were counted, one would expect to get 19 (³⁄₁₆ × 100) plants with red, axial flowers.

14. Because the F₂ generation is a ratio based on 16, one can conclude that two genes are involved in the inheritance of this flower color character. The parent plants were probably double homozygotes (*AABB* for red and *aabb* for white). The all-red F₁ must be *AaBb*, and the red color must require at least one dominant allele from both genes. The 9 : 6 : 1 ratio of the F₂ indicates that 9 plants had at least one dominant allele for both genes to produce the red color; 1 plant was doubly homozygous recessive and thus white; and the 6 pale purple were caused by having at least one dominant allele of only one of the two genes involved in flower color (*A_bb* or *aaB_*).

ANSWERS TO TEST YOUR KNOWLEDGE

Matching:

1. I		**3.** F		**5.** L		**7.** H		**9.** C
2. A		**4.** E		**6.** J		**8.** D		**10.** K

Multiple Choice:

1. c		**5.** b		**9.** b		**13.** d
2. a		**6.** d		**10.** c		**14.** d
3. e		**7.** c		**11.** d		**15.** c
4. c		**8.** d		**12.** a		**16.** b

CHAPTER 15: THE CHROMOSOMAL BASIS OF INHERITANCE

■ INTERACTIVE QUESTIONS

15.1

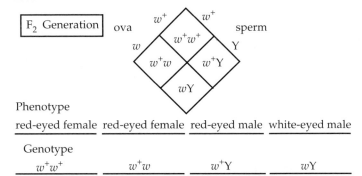

15.2 a. tall, purple-flowered and dwarf, white-flowered

b. tall, white-flowered and dwarf, purple-flowered

15.3 Since linked genes have their loci on the same chromosome, they travel together during meiosis and *more* parental offspring are produced. Recombinants are the result of crossing over between chromatids of homologous chromosomes.

15.4 Solving a linkage problem is often a matter of trial and error. Sometimes it helps to lay out the loci with the greatest distance between them and fit the other genes between or on either side by adding or substracting map unit distances.

$$
\overline{15}
$$

m —— 9 —— **j** —— 6 —— **l** —— 6 —— **k**

$$
\overline{12}
$$

15.5 The gene is sex-linked, so a good notation is X^C, X^c, and Y so that you will remember that the Y does not carry the gene. Capital C indicates normal sight. Genotypes are:

1. $X^C Y$
2. $X^C X^c$
3. $X^C X^C$ or $X^C X^c$ (probably $X^C X^C$ since 4 sons are $X^C Y$)
4. $X^C X^c$
5. $X^C Y$
6. $X^c Y$
7. $X^C X^c$

15.6 a. A trisomic organism ($2n + 1$) has an extra copy of one chromosome, usually caused by a nondisjunction during meiosis. A triploid organism ($3n$) has an extra set of chromosomes, possibly caused by a total nondisjunction in gamete formation.

b. The genetic balance of a trisomic organism would be more disrupted than that of an organism with a complete extra set of chromosomes.

15.7 translocation, deletion, and inversion

15.8 Aneuploidies of sex chromosomes appear to upset genetic balance less, perhaps because relatively few genes are located on the Y chromosome and extra X chromosomes are inactivated as Barr bodies.

SUGGESTED ANSWERS TO STRUCTURE YOUR KNOWLEDGE

1. Genes that are not linked assort independently, and the ratio of offspring from a testcross with a dihybrid heterozygote should be $1:1:1:1$ (*AaBb* × *aabb* gives *AaBb*, *Aabb*, *aaBb*, *aabb* offspring and a recombination frequency of 50%). Genes that are linked and do not cross over should produce a $1:1$ ratio in this testcross (*AaBb* and *aabb* because a heterozygote derived from a parental cross of *AABB* × *aabb* produces only *AB* and *ab* gametes). If crossovers between distant genes always occur, the heterozygote will produce equal quantities of *AB*, *Ab*, *aB*, and *ab* gametes, and the genotype ratio of offspring will be $1:1:1:1$, the same as it is for unlinked genes. Because each 1% recombination frequency is equal to 1 map unit, this measurement ceases to be meaningful at relative distances of 50 or more map units. However, crosses with intermediate genes on the chromosome could establish both that the genes A and B are on the same chromosome and that they are a certain map unit distance apart.

2. A cross between a mutant female fly and a normal male should produce all normal females (who get a wild-type allele from their father) and all mutant male flies (who get the mutant allele on the X chromosome from their mother). $X^m X^m \times X^+ Y$ produces $X^+ X^m$ and $X^m Y$ offspring (normal females and mutant males).

3. The serious phenotypic effects that are associated with these chromosomal alterations indicate that normal development and functioning is dependent on genetic balance. Most genes appear to be vital to an organism's existence, and extra copies of genes upset genetic balance. Inversions and translocations, which do not disrupt the balance of genes, can alter phenotype because of effects on gene functioning from neighboring genes.

ANSWERS TO GENETICS PROBLEMS

1. **a.** The trait is recessive and probably sex-linked. Two sets of unaffected parents in the second generation have offspring with the trait, indicating that it must be recessive. More males than females display the trait. Females 3 and 5 must be carriers since their father has the trait, and they each pass it on to a son.
b. Using the symbols X^T for the dominant allele and X^t for the recessive:
1. X^tY
2. X^TX^t
3. X^TX^t
4. X^TY
5. X^TX^t
6. X^TX^t or X^TX^T
7. X^TX^t
c. Individual #6 has a brother who has the trait and her mother's father had the trait, so her mother must be a carrier of the trait, meaning there is a ½ probability that #6 is a carrier. Since she is mated to a phenotypically normal male, none of her daughters will show the trait (0 probability). Her sons have a ½ chance of having the trait if she is a carrier. ½ × ½ = ¼ probability that her sons will have the trait. There is a ½ chance that a child would be male, so the probability of an affected child is ½ × ¼, or ⅛.

2. c a b d

3. The genes appear to be linked because the parental types appear most frequently in the offspring. Recombinant offspring represent 10 out of 40 total offspring for a recombination frequency of 25 percent, indicating that the genes are 25 map units apart.

4. One of the mother's X chromosomes carries the recessive lethal allele. One-half of male fetuses would be expected to inherit that chromosome and spontaneously abort. Assuming an equal sex ratio at conception, the ratio of girl to boy children would be 2 : 1, or 6 girls and 3 boys.

5. **a.** For female chicks to be black, they must have received a recessive allele from the male parent. If all female chicks are black, the male parent must have been Z^bZ^b. If the male parent was homozygous recessive, then all male offspring will receive a recessive allele, and the female parent would have to be Z^BW to produce all barred males.
b. For female chicks to be both black and barred, the male parent must have been Z^BZ^b. If the female parent were Z^BW, only barred male chicks would be produced. To get an equal number of black and barred male chicks, the female parent must have been Z^bW.

ANSWERS TO TEST YOUR KNOWLEDGE

Multiple Choice:

1. e	5. a	9. a	13. c	17. e
2. a	6. e	10. e	14. a	18. d
3. b	7. b	11. d	15. e	19. c
4. d	8. c	12. b	16. d	20. d

CHAPTER 16: THE MOLECULAR BASIS OF INHERITANCE

■ INTERACTIVE QUESTIONS

16.1 **a.** They grew T2 with *E. coli* with radioactive sulfur to tag phage proteins.
b. T2 was grown with *E. coli* in the presence of radioactive phosphorus to tag phage DNA.
c. Radioactivity was found in the supernatant, indicating that the phage protein did not enter the bacterial cells.
d. In the samples with the labeled DNA, most of the radioactivity was found in the bacterial cell pellet.
e. They concluded that viral DNA is injected into the bacterial cells and serves as the hereditary material for viruses.

16.2 **a.** sugar-phosphate backbone
b. complementary base pair
c. pyrimidine base
d. purine base
e. hydrogen bonds
f. cytosine
g. guanine
h. adenine
i. thymine
j. nucleotide
k. deoxyribose
l. phosphate group
m. 3.4 nm
n. 0.34 nm
o. 2 nm

16.3

DNA strands   Density bands

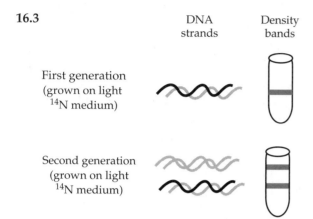

First generation (grown on light ^{14}N medium)

Second generation (grown on light ^{14}N medium)

16.4 The phosphate end of each strand is the 5' end, and the hydroxyl group extending from the 3' carbon of the sugar marks the 3' end. See Figure 16.2 in the text.

16.5 **a.** helicase
b. single-strand binding protein
c. DNA polymerase
d. leading strand
e. lagging strand
f. DNA ligase
g. Okazaki fragment
h. DNA Polymerase (replacing primer)
i. RNA primer
j. primase
k. replication fork
l. 3' end of parental strand
m. 5' end

SUGGESTED ANSWERS TO STRUCTURE YOUR KNOWLEDGE

1. Watson and Crick used the X-ray crystallography data from Franklin to deduce that DNA was a helix 2 nm wide, with nitrogenous bases stacked 0.34 nm apart, and making a full turn every 3.4 nm. Using molecular models of wire, they experimented with various arrangements and finally placed the sugar-phosphate backbones on the outside of the helix with the bases extending inside. Pairing of a purine base with a pyrimidine base produced the proper diameter. Specificity of base pairing (A with T and C with G) is assured by hydrogen bonds.

2. Replication bubbles form where proteins recognize specific base sequences and open up the two strands. **Helicase,** an enzyme that works at the replication fork, untwists the helix and separates the strands. **Single-strand binding proteins** support the separated strands while replication takes place. **Primase** lays down about 10 RNA bases to start the new strand. After a proper base pairs up on the exposed template, **DNA polymerase** joins the nucleotide to the 3' end of the new strand. On the lagging strand, short Okazaki fragments are formed by primase and polymerase (again moving 5' ⟶ 3'). A DNA polymerase replaces the primer. Ligase joins the 3' end of one fragment to the 5' end of its neighbor. Proofreading enzymes check for mispaired bases, and nucleases, DNA polymerase, and other enzymes repair damage or mismatches.

ANSWERS TO TEST YOUR KNOWLEDGE

Multiple Choice:

1. c	6. a	10. a	14. a	18. d
2. a	7. b	11. e	15. d	19. e
3. b	8. e	12. d	16. b	20. c
4. d	9. c	13. e	17. c	21. a
5. b				

CHAPTER 17: FROM GENE TO PROTEIN

■ INTERACTIVE QUESTIONS

17.1   transcription   translation
DNA   →   RNA   →   protein

17.2 Met Pro Asp Phe Lys stop

17.3 **a.** Initiation: Transcription factors bind to promoter and facilitate the binding of polymerase II, forming a transcription initiation complex; polymerase separates DNA strands at initiation site.
b. Elongation: Polymerase II moves along DNA strand, connecting RNA nucleotides that have paired to the DNA template to the 3' end of the growing RNA strand.
c. Termination: After transcribing past a termination sequence, polymerase II cuts and releases the pre-mRNA.

17.4 A 5' cap consisting of a modified guanosine triphosphate is added to the leader segment. A poly(A) tail consisting of up to 250 adenine nucleotides is attached to the 3' trailer segment. Spliceosomes have excised the introns and spliced the exons together.

17.5

DNA Triplet $3' \rightarrow 5'$	mRNA Codon $5' \rightarrow 3'$	Anticodon $3' \rightarrow 5'$	Amino Acid
TAC	AUG	UAC	methionine
GGA	CCU	GGA	proline
TTC	AAG	UUC	lysine
ATC	UAG	AUC	stop

17.6 **1.** Codon recognition: An elongation factor (not shown) helps an aminoacyl tRNA into the A site where its anticodon hydrogen-bonds to the mRNA codon; 2 GTP are required.
2. Peptide bond formation: Ribosome catalyzes peptide bond formation between new amino acid and polypeptide held in P site.
3. Translocation: The tRNA in the P site is moved to the E site and released; the tRNA now holding the polypeptide is moved from the A to the P site, taking the mRNA with it; one GTP is required.
4. Termination: Release factor binds to stop codon in the A site. Free polypeptide and tRNA are released from the P site. Ribosomal subunits and other assembly components separate.

a. amino end of growing polypeptide
b. aminoacyl tRNA
c. large subunit of ribosome
d. A site
e. small subunit of ribosome
f. 5' end of mRNA
g. peptide bond formation
h. E site
i. release factor
j. termination (stop) codon
k. P site
l. free polypeptide

17.7 A ribosome that is translating an mRNA that codes for a secretory or membrane protein will become bound to the ER when an SRP binds to the initial signal peptide and then to an ER receptor protein.

17.8 **a.** Messenger RNA carries the code from DNA that specifies the amino acid sequence of proteins to ribosomes.
b. Transfer RNA carries a specific amino acid to its position in a polypeptide based on matching its anticodon to an mRNA codon.
c. Ribosomal RNA makes up 60% of ribosomes and has structural and catalytic functions.
d. Small nuclear RNA is part of spliceosome and plays a catalytic role in splicing pre-mRNA.
e. SRP RNA is part of signal-recognition particle that binds to signal peptides of polypeptides targeted to the ER.

17.9 **a.** Silent: a base-pair substitution in the third nucleotide of a codon that still codes for the same amino acid.
b. Missense: a base-pair substitution or frameshift mutation that results in a codon for a different amino acid.
c. Nonsense: a base-pair substitution or frameshift mutation that creates a stop codon and prematurely terminates translation.
d. Frameshift: an insertion or deletion of one, two, or more than three nucleotides that disrupts the reading frame and creates extensive missense and nonsense mutations.

SUGGESTED ANSWERS TO STRUCTURE YOUR KNOWLEDGE

1.

	Transcription	Translation
Template	DNA	RNA
Location	nucleus (cytoplasm in prokaryotes)	cytoplasm; ribosomes can be free or attached to ER
Molecules involved	RNA nucleotides, DNA template strand, RNA polymerase, transcription factors	amino acids, tRNA, mRNA, ribosomes, ATP, GTP, enzymes, initiation, elongation, and release factors
Enzymes involved	RNA polymerases, RNA processing enzymes, ribozymes	aminoacyl-tRNA synthetase, ribosomal enzymes (ribozymes)
Control—start and stop	transcription factors locate promoter region with TATA box, terminator sequence	initiation factors, initiation sequence (AUG), stop codons, release factor
Product	primary transcript (pre-mRNA)	protein
Product processing	RNA processing: 5' cap and poly(A) tail, splicing of pre-mRNA—introns removed by snRNPs in spliceosomes	spontaneous folding, disulfide bridges, signal peptide removed, cleaving, quaternary structure, modification with sugars, etc.
Energy source	ribonucleoside triphosphate	ATP and GTP

2. The genetic code includes the sequence of nucleotides on DNA that is transcribed into the codons found on mRNA and translated into their corresponding amino acids. There are 64 possible mRNA codons created from the four nucleotides used in the triplet code (4^3).

Redundancy of the code refers to the fact that several triplets may code for the same amino acid. Often these triplets differ only in the third nucleotide. The wobble phenomenon explains the fact that there are only about 45 different tRNA molecules that pair with the 61 possible codons (three codons are stop codons). The third nucleotide of many tRNAs can pair with more than one base. Because of the redundancy of the genetic code, these wobble tRNAs still place the correct amino acid in position.

The genetic code is nearly universal; each codon codes for exactly the same amino acid in almost all organisms. (Some exceptions to this universality have been found in a few unicellular eukaryotes and in certain mitochondria and chloroplasts.) This universality points to an early evolution of the code in the history of life and the evolutionary relationship of all life on Earth.

3.

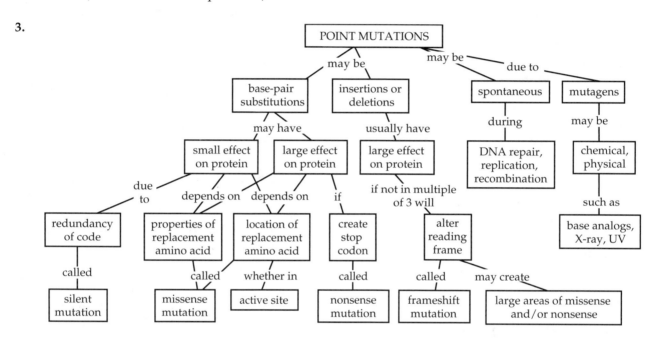

ANSWERS TO TEST YOUR KNOWLEDGE

Multiple Choice:

1. d	4. e	7. c	10. e	13. d	16. e	19. b	22. d
2. a	5. c	8. c	11. d	14. c	17. e	20. a	23. e
3. b	6. b	9. b	12. d	15. b	18. b	21. a	24. b

CHAPTER 18: MICROBIAL MODELS: THE GENETICS OF VIRUSES AND BACTERIA

■ INTERACTIVE QUESTIONS

18.1 1. Phage attaches to host cell and injects DNA.
2. Phage DNA forms circle.
3. New phage DNA and proteins are synthesized and self-assemble into phages.
4. Bacterium lyses, releasing phages.
5. Phage DNA integrates into bacterial chromosome.
6. Bacterium reproduces, passing prophage to daughter cells.

7. Colony of infected bacteria forms.
8. Occasionally, prophage exits bacterial chromosome and begins lytic cycle.
a. phage DNA
b. bacterial chromosome
c. new phages
d. prophage
e. replicated bacterial chromosome with prophage

18.2 RNA → DNA → RNA; viral reverse transcriptase, host RNA polymerase

18.3 Proto-oncogenes generally code for proteins that regulate the cell cycle, such as cellular growth factors or the receptor proteins for growth factors.

18.4 **a.** DNA
b. RNA
c. protein capsid
d. host
e. bacterium
f. lytic or lysogenic cycles
g. animal
h. oncogenes
i. viral envelope
j. retrovirus
k. reverse transcriptase
l. plant
m. viroids
n. plasmids
o. transposons

18.5 A temperate phage that leaves the lysogenic cycle and enters the lytic cycle may carry bacterial genes that were adjacent to the insertion site of the prophage.

18.6 **a.** circular chromosome
b. F plasmid, R plasmid
c. mutation
d. transformation
e. naked DNA
f. transduction
g. phage
h. conjugation
i. F⁺ or Hfr, and F⁻ cells
j. transposons
k. antibiotic-resistant genes
l. adaptation to environment/evolution

18.7 **a.** regulatory gene
b. promoter
c. operator
d. genes
e. operon
f. RNA polymerase
g. active repressor
h. inducer (allolactose)
i. mRNA for enzymes of pathway

18.8 **a.** anabolic; corepressor; on; inactive
b. catabolic; inducers; off; active

SUGGESTED ANSWERS TO STRUCTURE YOUR KNOWLEDGE

1.

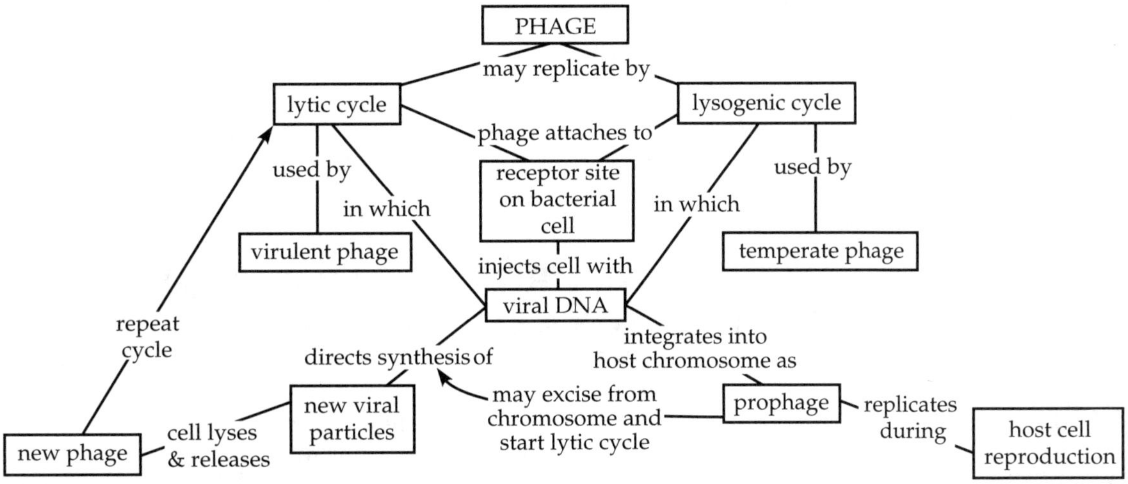

2.

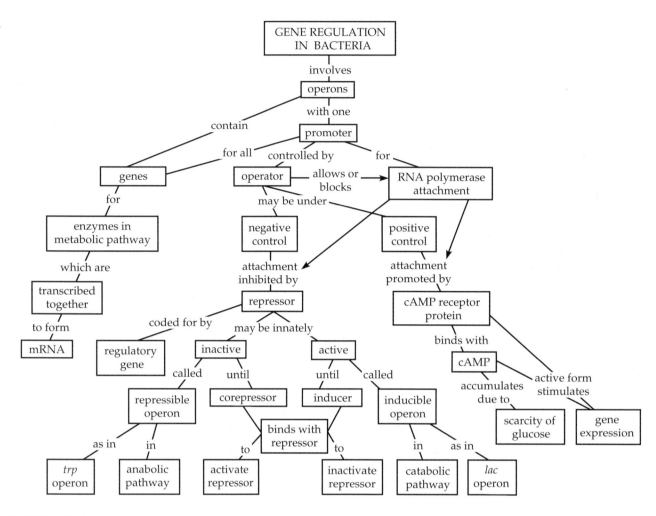

ANSWERS TO TEST YOUR KNOWLEDGE

Multiple Choice:

1. b	4. e	7. b	10. e	13. a	16. d	19. b	22. e	25. a
2. d	5. a	8. b	11. a	14. e	17. c	20. a	23. b	26. a
3. c	6. d	9. c	12. b	15. c	18. e	21. d	24. c	

Matching:

1. C	3. G	5. F	7. A
2. D	4. E	6. B	8. H

CHAPTER 19: THE ORGANIZATION AND CONTROL OF EUKARYOTIC GENOMES

■ INTERACTIVE QUESTIONS

19.1 nucleosomes; 30-nm chromatin fiber; looped domains; coiling and folding of looped domains into highly condensed metaphase chromosome

19.2 fragile X syndrome and Huntington's disease

19.3 Different combinations of the multiple variable and constant regions come together during cellular differentiation, creating millions of different B lymphocytes, which produce millions of different antibodies.

19.4 a. Barr body—compacted X chromosome in cells of female

b. Histone deacetylation would decrease transcription because it would make genes in the nucleosome less accessible.

19.5 a. enhancers
b. activators
c. transcription factors
d. promoter
e. TATA box
f. RNA polymerase

19.6 a. express different genes
b. available or unavailable for transcription
c. Regulation of gene expression
d. binding of transcription factors to enhancer sequences and promoters
e. posttranscriptional mechanisms

SUGGESTED ANSWERS TO STRUCTURE YOUR KNOWLEDGE

1. a. DNA packing into nucleosomes; methylation decreases and histone acetylation increases transcription; gene amplification or loss (multiple copies of rRNA genes in ovum); gene rearrangements (immunoglobulin genes)

b. Transcription factors that bind with enhancer and promoter regions to form transcription initiation complex; steroid hormones or other chemical messages may bind with receptor proteins, producing transcription factors

c. RNA processing and export (alternative splicing, 5′ cap and poly(A) tail added); mRNA degradation

d. Repressor proteins may prevent ribosome binding so mRNA can be stockpiled (awaiting fertilization in ovum); initiation factors necessary for mRNA and ribosomes to join

e. Protein processing by cleavage or modification; transport to target location; selective degradation of proteins

2. a. Proto-oncogenes are key genes that control cellular growth and division. When such genes mutate to form a more active product, become amplified, or have changes in their normal control mechanisms, they may become oncogenes and produce the uncontrolled cell division that leads to formation of a tumor.

b. Tumor-suppressor genes code for proteins that regulate cell division. Loss or mutation of both alleles of a tumor-suppressor gene may allow tumors to develop. Usually mutations or other changes must occur both in oncogenes and in several suppressor genes for cancer to develop.

ANSWERS TO TEST YOUR KNOWLEDGE

Multiple Choice:

1. b	6. e	11. c	16. d	19. b
2. a	7. c	12. e	17. c	20. d
3. e	8. d	13. a	18. e	21. a
4. b	9. c	14. b		
5. c	10. e	15. b		

CHAPTER 20: DNA TECHNOLOGY AND GENOMICS

■ INTERACTIVE QUESTIONS

20.1 The third sequence, because it has the same sequence running in opposite directions. The enzyme would probably cut between G and A, producing AATT and TTAA sticky ends..

20.2 Use cDNA that has no introns; attach near promoters and proper transcription sequences in vector DNA.

20.3 a. bacterial plasmid
b. *amp*^{R} (ampicillin resistance) gene
c. *lacZ* gene
d. same restriction sites
e. human DNA containing gene of interest
f. complementary sticky ends

g. human DNA fragment
h. nonfunctional *lacZ* gene
i. recombinant plasmid
j. plate with ampicillin and X-gal

1. Plasmid and gene of interest are isolated.
2. Both DNAs are cut with same restriction enzyme (disrupts *lacZ* gene of plasmid). Fragments are mixed and some foreign fragments base-pair with plasmid. DNA ligase seals ends.
3. Recombinant plasmid transforms bacteria.
4. Clone cells. Cells containing recombinant plasmid are identified by their ability to grow in presence of the antibiotic and by their white color. Blue colonies contain plasmids that resealed and thus have a functioning *lacZ* gene.

5. Identify clone carrying gene of interest with nucleic acid probe or by protein product.

20.4. a. By amplifying the gene prior to cloning, the later task of identifying clones carrying the desired gene is simplified.

b. There is a limit to the number of accurate copies that can be made by PCR due to the accumulation of relatively rare copying errors. Large quantities of a gene are better prepared by DNA cloning in cells.

20.5 Southern blotting procedure:
a. restriction enzyme treatment of samples
b. gel electrophoresis
c. blotting
d. hybridization with labeled probe
e. autoradiography
Crime samples contain blood from the victim and, presumably, the perpetrator of the crime. After identifying the bands from the victim, the remaining fragments match those of Suspect 2's fragment profile.

20.6 The sequence of nucleotides in the original fragment would be complementary to the sequence shown by the order of fragments. It would read, from the bottom up, A G C T T C A G T C.

20.7 The public consortium followed a hierarchy of three stages: (1) genetic (linkage) mapping that established about 200 markers/chromosome; (2) physical mapping that cloned and ordered smaller and smaller overlapping fragments (using YAC or BAC vectors for cloning the large fragments); and (3) DNA sequencing of each small fragment, followed by assembly of the overall sequence. The Celera approach omitted the first two stages. Each chromosome was cut into small fragments, which were cloned in plasmid or phage vectors. The sequence of each fragment was determined, and powerful computers assembled the overlapping fragments to determine the overall sequence.

20.8 a. Before, members of the extended family had to be tested to determine the variant of the RFLP marker to which the disease-causing allele was linked. Then RFLP analysis was done with the patient's blood to determine which marker, and thus which allele, he or she had inherited. Now the cloned gene can be used as a probe, and hybridization analysis can detect the abnormal allele.
b. The major difficulty is to assure that proper control mechanisms are present so that the gene is expressed at the proper time, in the proper place, and to the proper degree. If genes can be inserted in cells that will continue to divide, the therapy should last longer, although such results

have not yet been achieved. Ethical considerations include how available such expensive treatments will be and whether genetic engineering should be done on germ cells, thus influencing the genetic makeup of future generations.

SUGGESTED ANSWERS TO STRUCTURE YOUR KNOWLEDGE

1. a. Bacterial enzymes that cut DNA at restriction sites, creating "sticky ends" that can base-pair with other fragments. Use: make recombinant DNA, form restriction fragments used for many other techniques

b. Mixture of molecules applied to gel in electric field; molecules separate, moving at different rates due to charge and size. Use: separate restriction fragments into pattern of distinct bands, fragments can be removed from gel and retain activity or can be identified with probes

c. mRNA isolated from cell is treated with reverse transcriptase to produce a complementary DNA strand, and then a double-stranded DNA gene, minus introns and control regions. Use: creates genes that are easier to clone in bacteria, produces library of genes that are active in cell

d. Radioactively or fluorescently labeled single-stranded DNA or mRNA used to base-pair with complementary sequence of DNA or RNA. Use: locate gene in clone of bacteria, identify bands on gels, diagnose infectious diseases

e. DNA fragments separated by gel electrophoresis, transferred by blotting onto filter or membrane, labeled probe added, rinsed, autoradiography. Use: analyze DNA for homologous sequences, DNA fingerprinting

f. Single-stranded DNA fragments are incubated with four nucleotides, DNA polymerase, and one of four dideoxy nucleotides that interrupt synthesis, samples separated by high-resolution gel electrophoresis, sequence of nucleotides read from the four sets of bands (or from sequence of fluorescent tags). Use: determine nucleotide sequence

g. Polymerase chain reaction: DNA is mixed with heat-resistant DNA polymerase, nucleotides, and primers having complementary sequences for targeted DNA section and repeatedly heated to separate, cooled to pair with primers and replicate. Use: rapidly produce multiple copies of a gene or section of DNA *in vitro*

h. Restriction fragments analyzed by Southern blotting to compare different band patterns caused by DNA differences in restriction sites. Use: DNA fingerprints for forensic use, map chromosomes using RFLP markers, diagnose genetic diseases

2. Agricultural applications: (1) production of vaccines and hormones, which will improve the health or productivity of livestock, (2) improvement of the genomes of agricultural plants and animals, (3) improvements in the nitrogen-fixing capacity of bacteria or plants, (4) development of plant varieties that have genes for resistance to diseases, herbicides, and insects

Medical applications: (1) development of vaccines, (2) diagnosis of genetic diseases, (3) treatment of genetic disorders, (4) production of insulin, human growth factor, TPA, and other useful products

ANSWERS TO TEST YOUR KNOWLEDGE

Multiple Choice:

1. c	7. c	13. d	19. c	25. d
2. b	8. a	14. a	20. b	26. d
3. e	9. a	15. c	21. a	
4. e	10. b	16. d	22. a	
5. c	11. c	17. b	23. e	
6. c	12. b	18. d	24. e	

CHAPTER 21: THE GENETIC BASIS OF DEVELOPMENT

■ INTERACTIVE QUESTIONS

21.1 Model organisms ideally should have easily observable embryos, short generation times, high reproductive rates, and relatively simple genomes, and there should be a preexisting knowledge base of their genetics and biology.

21.2 DNA methylation helps to regulate gene expression, essential to embryonic development. The DNA of many cloned embryos has been found to be improperly methylated.

21.3 The action of MyoD must depend on a combination of regulatory proteins, some of which may be lacking in the cells that it is not able to transform into muscle cells.

21.4 In the screening, a mutation-carrying parent is crossed with a wild-type parent, and the F_1 includes some heterozygous offspring who carry the mutation. When these heterozygotes are bred, ¼ of their offspring are homozygous recessive for the mutation. These flies are only recognized when F_2 females breed. The inviable embryos had mothers who were homozygous recessive and thus did not produce and supply the necessary mRNAs to the unfertilized egg for the morphogens that determine the body axes.

21.5 **a.** a mutation in both *bicoid* alleles of the mother fly so that no *bicoid* mRNA was secreted into the anterior of the egg and thus no bicoid protein gradient was available to establish the anterior end
b. a mutation in a pair-rule gene so that either the odd or even segments failed to develop
c. a mutation in a homeotic gene so that a head segment was identified as a thoracic segment and genes for leg development were activated

21.6 All of these genes are master regulatory genes that control the expression of batteries of other genes. The homeobox codes for the DNA-binding homeodomain of a transcription factor.

21.7 **a.** sequential inductions
b. concentration
c. signal-transduction pathways; transcriptional regulation

21.8 **a.** The basic mechanism for programmed cell death must have evolved early in animal evolution.
b. Programmed cell death in humans is important in the normal development of the nervous system, fingers, and toes; for normal functioning of the immune system; and to prevent the development of cancerous cells (by the internal triggering of apoptosis in cells with DNA damage).

21.9 **a.**

Whorl	Genes Active	Organs in Normal	Genes Active Mutant A	Organs in Mutant A
1	*A*	sepals	*C*	carpels
2	*AB*	petals	*BC*	stamens
3	*BC*	stamens	*BC*	stamens
4	*C*	carpels	*C*	carpels

b. Gene *A* would be expressed alone in each whorl, producing a flower with four whorls of sepals.

SUGGESTED ANSWERS TO STRUCTURE YOUR KNOWLEDGE

1. Proteins translated from maternal mRNA deposited in the egg cell (transcripts of egg-polarity genes) diffuse through the embryo's multinucleated cytoplasm and regulate the activity of zygotic genes. These morphogen gradients activate gap genes, whose products activate pair-rule genes, whose products activate the segment polarity genes. Most of the products of these segmentation genes (as well as the maternal effect genes) are transcription factors that control this hierarchy of gene activation responsible for the segmentation pattern of the embryo. The appropriate homeotic genes are activated in each segment, and they regulate the transcription of the genes that control development of segmental structures.

2. Most cytoplasmic determinants are mRNA for transcription factors that are divided by the first few mitotic divisions. They are present in the cells and can enter the nucleus and regulate transcription. Inducers must communicate between cells. They are often proteins that bind to cell surface receptors and initiate a signal-transduction pathway involving a cascade of enzyme activations, usually leading to the activation of transcription factors within the target cell.

ANSWERS TO TEST YOUR KNOWLEDGE

Multiple Choice:

1. e	5. d	9. c	13. a
2. b	6. c	10. c	14. c
3. c	7. e	11. a	15. d
4. e	8. d	12. b	16. e

CHAPTER 22: DESCENT WITH MODIFICATION: A DARWINIAN VIEW OF LIFE

■ INTERACTIVE QUESTIONS

22.1 a. 1. E f   3. B e   5. F a   7. G g
      2. A b   4. C c, d, g   6. D c
  b. 5, 1, 2, 4, 3, 7, 6

22.2 The excessive production of offspring sets up the struggle for existence; only a small proportion can live to leave offspring of their own. Natural selection is the differential reproductive success of individuals within a population that are best suited to the environment, which leads over generations to greater adaptation of populations of organisms to their environment.

22.3 Treatment with the drug 3TC prevents most HIV from reproducing when their enzyme reverse transcriptase inserts this cytosine mimic into its DNA, halting replication and production of new HIV. The 3TC-resistant HIV have a version of reverse transcriptase that discriminates between cytosine and 3TC, and they are still able to reproduce. With HIV's generation time of two days, it takes very little time for 3TC-resistant HIV to be strongly selected for by this drug and 100% of a patient's HIV population to be 3TC resistant.

22.4 a. biogeography
    b. fossil record
    c. homologies

    d. island species and mainland species or neighboring species in different habitats
    e. ancestral and transitional forms
    f. homologous structures
    g. embryological development
    h. molecular biology
    i. DNA and proteins
    j. descent from a common ancestor

SUGGESTED ANSWERS TO STRUCTURE YOUR KNOWLEDGE

1. The two major components of Darwin's evolutionary theory are that all life has descended from a common ancestral form and that this evolution has been the result of natural selection. The theory of natural selection is based on several key observations and inferences. The overproduction of offspring in conjunction with limited resources leads to a struggle for existence and the differential reproductive success of those organisms best suited to the local environment. The unequal survival and reproduction of the most fit individuals in a population leads to the gradual accumulation of adaptive characteristics in a population.

ANSWERS TO TEST YOUR KNOWLEDGE

Multiple Choice:

1. b	3. e	5. c	7. e	9. d	11. c	13. b	15. c
2. c	4. a	6. a	8. d	10. d	12. e	14. d	16. c

CHAPTER 23: THE EVOLUTION OF POPULATIONS

■ INTERACTIVE QUESTIONS

23.1 a. The 98 *BB* mice contribute 196 *B* alleles, and the 84 *Bb* mice contribute 84 *B* alleles to the gene pool. These 84 *Bb* mice also contribute 84 *b* alleles, and the 18 *bb* mice contribute 36 *b* alleles. Of a total of 400 alleles, 280 are *B* and 120 are *b*. Allele frequencies are 0.7 *B* and 0.3 *b*.
b. The frequencies of genotypes are 0.49 *BB*, 0.42 *Bb*, and 0.09 *bb* ($^{98}/_{200}$, $^{84}/_{200}$, $^{18}/_{200}$).

23.2 0.7; 0.3; 0.49; 0.42; 0.09

23.3 a. 0.36; 0.48; 0.16. Plug *p* (0.6) and *q* (0.4) into the expanded binomial: $p^2 + 2pq + q^2$.
b. 0.6; 0.4. Add the frequency of the homozygous dominant genotype to ½ the frequency of the heterozygote for the frequency of *p*. For *q*, add the homozygous recessive frequency and ½ the heterozygote frequency. Alternatively, to determine *q*, take the square root of the homozygous recessive frequency if you are sure the population is in Hardy-Weinberg equilibrium. The frequency of *p* is then 1 − *q*.
c. 0.9; 0.1. Determine the genotype frequencies by dividing the number of each genotype by the total number counted; $^{65}/_{80} = 0.81$, $^{14}/_{80} = 0.18$, $^{1}/_{80} = 0.01$. Then determine allele frequencies as above. Another approach is to determine the number of alleles in the population (160) and the number of *S* and *s* alleles contributed by each genotype; *S* is $^{144}/_{160} = 0.9$ and *s* is $^{15}/_{160} = 0.1$.

23.4 a. genetic drift
b. small population
c. bottleneck effect
d. founder effect
e. natural selection
f. better reproductive success
g. gene flow
h. migration between populations
i. mutations

23.5 a. mutation
b. sexual recombination
c. Bacteria and microorganisms have very short generation times and a new beneficial mutation can increase in frequency rapidly in an asexually reproducing bacterial population. Although mutations are the source of new alleles, they are so infrequent that their contribution to genetic variation in a large, diploid population is minimal. However, sexual recombination in the production and union of gametes produces zygotes with unique combinations of alleles.

23.6 a. Diploidy—The sickle-cell allele is hidden from selection in heterozygotes.
b. Heterozygote advantage—Heterozygotes are resistant to malaria and have a selective advantage in areas where malaria is a major cause of death.

23.7 a. 1. The two given fitness values are less than 1, so *Bb* must produce the most offspring and have a relative fitness of 1.
b. The relative fitness for *Bb* of 1 is twice the fitness of 0.5. *Bb* would produce twice the number of offspring, or 200. *bb* with a fitness of 0.25 would produce ¼ as many offspring as *Bb*, or 50.
c. The higher relative fitness of *Bb* may be due to heterozygote advantage.

SUGGESTED ANSWERS TO STRUCTURE YOUR KNOWLEDGE

1. a. The Hardy-Weinberg theorem states that allele frequencies within a population will remain constant from one generation to the next as long as the population is large, mutation and migration are negligible, mating is random, and no selective pressure operates. This Hardy-Weinberg equilibrium provides a null hypothesis to allow population geneticists to test for microevolution.
b. $p^2 + 2pq + q^2 = 1$. In the Hardy-Weinberg equation, *p* and *q* refer to the frequencies of two alleles in the gene pool. The frequency of homozygous offspring is ($p \times p$) or p^2 and ($q \times q$) or q^2. Heterozygous individuals can be formed in two ways, depending on whether the ovum or sperm carries the *p* or *q* allele, and their frequency is equal to 2*pq*.

2. Genetic variation is retained within a population by diploidy and balanced polymorphism. Diploidy masks recessive alleles from selection when they occur in the heterozygote. Thus, less adaptive or even harmful alleles are maintained in the gene pool and are available should selection pressures change. Balanced polymorphism results in several alleles at a gene locus being maintained in a population. In situations in which there is heterozygote advantage, the two alleles will be retained in roughly equal frequencies within the gene pool. Frequency-dependent selection, in which morphs present in higher numbers are selected against by predators or other factors, is another cause of balanced polymorphism.

3. The allele frequencies estimated from this sample are $B = 0.6$ and $b = 0.4$. Assuming this sample truly represents the individuals in the larger population, then we must say that the population is not in Hardy-Weinberg equilibrium. If it were, we would expect to have sampled 36 *BB*, 48 *Bb*, and 16 *bb* individuals. The larger numbers of both homozygous genotypes may indicate the action of some selection agent on speckled butterflies that reduces the numbers of heterozygotes in the population.

ANSWERS TO TEST YOUR KNOWLEDGE

1. e	6. d	11. e	16. b	21. a
2. a	7. e	12. c	17. c	22. e
3. c	8. d	13. d	18. b	23. d
4. c	9. b	14. a	19. b	24. c
5. d	10. e	15. b	20. c	

CHAPTER 24: THE ORIGIN OF SPECIES

■ INTERACTIVE QUESTIONS

24.1 **a.** reduced hybrid fertility   **b.** post-
c. gametic isolation   **d.** pre-
e. mechanical isolation   **f.** pre-
g. temporal isolation   **h.** pre-
i. hybrid breakdown   **j.** post-
k. behavioral isolation   **l.** pre-
m. reduced hybrid viability   **n.** post-
o. habitat isolation   **p.** pre-

24.2 **a.** reproductive isolation
b. morphological
c. ecological
d. pluralistic
e. evolutionary history resulting in genetic closeness and unique genetic markers

24.3 The Hawaiian Archipelago is a series of relatively young, isolated, and physically diverse islands whose thousands of endemic species are examples of adaptive radiation resulting from multiple colonizations and allopatric speciations.

24.4 **a.** 20 chromosomes
b. 24 chromosomes

24.5 In the gradual tempo model of evolution, small changes accumulate within populations as a result of chance events and natural selection, leading to the gradual evolution of new life forms. The punctuated equilibrium model holds that evolution occurs in spurts of relatively rapid change interspersed with long periods of stasis.

24.6 Feathers and wings may have first developed as structures for social displays; light, honeycombed bones may have increased the agility of bipedal dinosaurs.

24.7 **a.** allometric growth
b. paedomorphosis
c. heterochrony

SUGGESTED ANSWERS TO STRUCTURE YOUR KNOWLEDGE

1. Speciation, by which a new species evolves from a parent species, is part of macroevolution and the increase in biological diversity. Microevolution is the process by which changes occur within the gene pool of a population as a result of either chance events or natural selection. If the makeup of the gene pool changes enough, microevolution may lead to speciation.

2. Modification of existing structures: either the gradual development of a more complex structure from one that served the same function, or exaptation, in which a structure that evolved in

one context is gradually changed to serve a novel function. "Evo-devo": changes in genes that control aspects of development, such as growth rates, timing, and spatial arrangements of body parts, may result in novel designs.

1. b	5. c	9. b	13. b	17. c
2. c	6. a	10. d	14. c	18. b
3. d	7. b	11. b	15. c	19. e
4. e	8. c	12. a	16. e	20. d

CHAPTER 25: PHYLOGENY AND SYSTEMATICS

■ INTERACTIVE QUESTIONS

25.1 a. 17,190 years. Carbon-14 has a half-life of 5,730 years. In 5,730 years the ratio of C-14 to C-12 would be reduced by ½; in 11,460 years it would be reduced by ¼; and in 17,190 years it would be reduced by ⅛.
b. 2.6 billion years. The fossil has ¼ the amount of potassium-40. It would take two half-lives to decay that amount.

25.2 Organisms that were abundant, widespread, existed over a long period of time, had hard shells or skeletons, and lived in regions in which the geology favored fossil formation are most likely overrepresented in the fossil record.

25.3 Marsupials probably migrated to Australia through South America and Antarctica while the continents were still joined. When Pangaea broke up, marsupials adaptively radiated on isolated Australia. Eutherian (placental) mammals evolved and diversified on the other continents.

25.4 Mass extinctions empty many biological niches, which may then be exploited by species that survived extinction.

25.5 Adaptations to different environments may lead to large morphologic differences within related groups, and convergent evolution can produce similar structures in unrelated organisms in response to similar selective pressures. Comparisons of embryonic development can help sort homology from analogy.

25.6

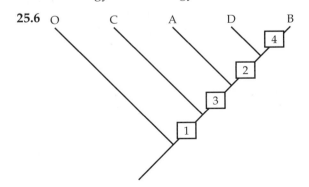

25.7 A C G T G C A C G
A   G T G   A G G
The second DNA segment shows two deletions and one base substitution.

25.8 The four-chambered heart of birds and mammals is analogous, not homologous. It evolved independently in the two groups. An abundance of evidence supports the hypothesis that birds and reptiles are closer relatives, and that birds and mammals evolved from different reptilian ancestors.

SUGGESTED ANSWERS TO STRUCTURE YOUR KNOWLEDGE

1. The four eras include the Precambrian, Paleozoic, Mesozoic, and Cenozoic. The remarkable adaptive radiation of most modern animal phyla in the Cambrian explosion marks the boundary between the Precambrian and Paleozoic eras (543 mya). At the boundary between the Paleozoic and Mesozoic eras (245 mya), continental drift caused the formation of the supercontinent called Pangaea. The major climatic and habitat changes associated with that event may have caused the Permian extinctions. The Cretaceous extinctions, perhaps caused by Earth's collision with an asteroid, mark the boundary between the Mesozoic and Cenozoic eras (65 mya). The extinction of most dinosaur lineages may have opened adaptive zones, leading to the adaptive radiation of mammals.

2.

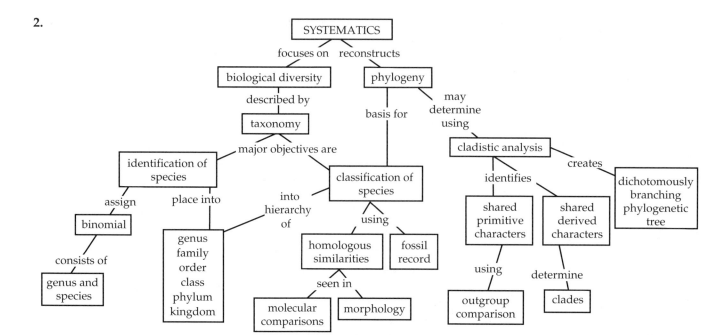

ANSWERS TO TEST YOUR KNOWLEDGE

Multiple Choice:

1. c	5. a	9. e	13. a
2. a	6. b	10. c	14. e
3. b	7. d	11. a	15. d
4. e	8. e	12. d	16. b

True or False:

1. True
2. False—change the first *more* to the *less* they vary.

3. True
4. False—change *A cladogram* to *An evolutionary tree that combines cladistic analysis with fossil and morphological data.*
5. True
6. False—change to: because *the half-life of carbon-14 is too short to date fossils that are that old.*
7. False—change to: indicative of *an asteroid having crashed into the Earth.*
8. True

CHAPTER 26: EARLY EARTH AND THE ORIGIN OF LIFE

■ INTERACTIVE QUESTIONS

26.1 The evolution of cellular respiration allowed these surviving prokaryotes to use oxygen to more efficiently obtain energy from organic molecules.

26.2 Refer to Figure 26.1 in the textbook.

26.3 **a.** The abiotic synthesis and accumulation of organic molecules, such as amino acids, simple sugars, and nucleotides
b. The linking of organic monomers into polymers, such as proteins, lipids, and nucleic acids
c. RNA molecules serve as catalysts and templates for self-replication and also for sequences of amino acids and polypeptide synthesis

d. The aggregation of abiotically produced molecules into protobionts, separated from the surroundings by selectively permeable membranes that permitted the concentration of critical molecules, and the emerging properties of metabolism, excitability, and primitive reproduction. Natural selection could act on protobionts containing genetic material.

SUGGESTED ANSWERS TO STRUCTURE YOUR KNOWLEDGE

1. The primitive Earth is believed to have had seas, volcanoes, and large amounts of ultraviolet radiation passing through a thin atmosphere,

which probably consisted of H_2O, CO, CO_2, N_2, CH_4, and NH_3. Life could evolve in this environment because the reducing (electron-adding) atmosphere and the availability of energy from lightning and UV radiation facilitated the abiotic synthesis of organic molecules, and hot lava rocks, clay, or pyrite may have aided the polymerization of these monomers.

An alternative hypothesis is that deep-sea vents may have provided the energy and chemical precursors needed for the origin of life.

2.

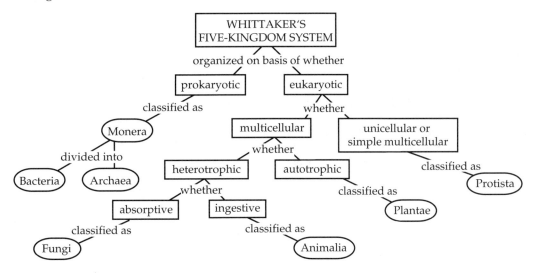

3. Molecular systematics and cladistic analysis show that prokaryotes diverged early into two distinct lineages, each of which can be assigned to a separate domain. Molecular systematics also shows that the traditional kingdom Protista is not monophyletic, but can be divided into five or more kingdoms. The goal of a classification system is to reconstruct phylogeny. New data allow systematists to revise and refine their hypotheses of the evolutionary history of the diversity of life.

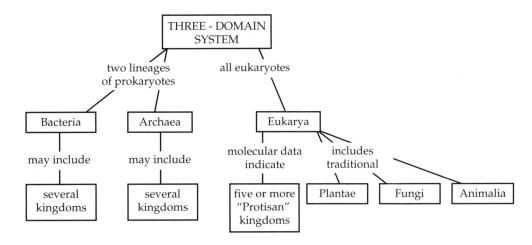

ANSWERS TO TEST YOUR KNOWLEDGE

Multiple Choice:

1. b	4. a	7. c	10. b
2. d	5. c	8. a	11. d
3. e	6. b	9. e	12. a

Matching:

1. C	3. E
2. A	4. D

CHAPTER 27: PROKARYOTES AND
THE ORIGINS OF METABOLIC DIVERSITY

■ INTERACTIVE QUESTIONS

27.1 a. Bacteria
b. Archaea
c. Eukarya

27.2 a. spheres (cocci), rods (bacilli), or helices (spirilla)
b. 1–5 μm in diameter, one-tenth the size of eukaryotic cells
c. cell wall made of peptidoglycan; gram-negative bacteria also have outer lipopolysaccharide membrane; archaea lack peptidoglycan; sticky capsule for adherence; pili for attachment or conjugation
d. flagella; filaments that spiral around cell and cause cell to corkscrew; or glide in secreted slime; may show taxis to stimuli
e. infoldings of plasma membrane may be used in metabolic functions
f. circular DNA molecule (genophore) with little associated protein found in nucleoid region; may have plasmids with antibiotic resistance and other genes
g. binary fission; rapid geometric growth into colony; some genetic recombination in transformation, transduction, or conjugation; most variation from mutation

27.3 a. light
b. phototroph
c. chemicals
d. CO_2
e. organic nutrients
f. heterotroph
g. photoautotroph
h. photoheterotroph
i. chemoheterotroph
j. cyanobacteria and other prokaryotes; algae, plants
k. certain prokaryotes
l. certain prokaryotes
m. many prokaryotes and protists; fungi; animals

27.4 Photosynthetic prokaryotes are found in diverse lineages of prokaryotes. If heterotrophic nutrition was the primitive condition in the ancestral prokaryote of these lines, then the complex process of photosynthesis must have evolved independently several different times. A more parsimonious hypothesis assumes that photosynthesis evolved very early, and the heterotrophic groups that are on the same branches as photosynthetic groups evolved by the loss of photosynthetic ability.

27.5 a. endospore; thick-walled, resistant cell that can withstand harsh conditions for long periods of time, breaking dormancy when conditions are favorable
b. cyanobacteria; the cell is specialized for nitrogen fixation

SUGGESTED ANSWERS TO STRUCTURE YOUR KNOWLEDGE

1. Molecular evidence, especially comparison of rRNA signature sequences, indicates that bacteria and archaea diverged very early in evolutionary history and that archaea may share a more recent common ancestor with eukaryotes.

2. Decomposers—recycle nutrients; bioremediation—sewage treatment, clean up oil spills, pesticide breakdown.
Nitrogen fixers—provide nitrogen to the soil in nodules on legume roots.
Symbionts—enteric bacteria, some produce vitamins.
Biotechnology—production of antibiotics, hormones, and many useful products.

ANSWERS TO TEST YOUR KNOWLEDGE

Multiple Choice:

1. c	4. c	7. a	10. d	13. e
2. e	5. b	8. c	11. e	14. c
3. b	6. d	9. a	12. b	15. e

Fill in the Blanks:

1. cocci
2. nucleoid region
3. Gram stain
4. pili
5. taxis
6. endospore
7. saprobes, decomposers
8. exotoxins
9. bioremediation
10. cyanobacteria

CHAPTER 28: THE ORIGINS OF EUKARYOTIC DIVERSITY

▪ INTERACTIVE QUESTIONS

28.1 **a.** ruptured coats of cysts, 2.1 billion years old; "chemical signatures," 2.7 billion years old
b. freshwater, marine, moist terrestrial habitats, symbionts
c. photoautotrophs, heterotrophs, or mixotrophs
d. most have flagella or cilia at some point in life cycle
e. asexual by mitosis (process may differ); some have sexual reproduction and various life cycle patterns; resistant cysts

28.2 **a.** from invaginations of the plasma membrane
b. similarities between bacteria and mitochondria and chloroplasts: size, membrane enzymes and transport systems, division process, circular DNA molecule with little associated protein, and ribosomes; also the existence of endosymbiotic relationships in modern organisms
c. Mitochondria or genetic remnants of them are found in all eukaryotes, indicating an earlier evolution than chloroplasts, which are present in fewer eukaryotic lineages.

28.3 **a.** Plastids with more than two membranes were probably acquired by secondary endosymbiosis when a eukaryotic alga (with its two-membraned plastid) became an endosymbiont within a heterotrophic protist. In the process of invagination into the host cell, another membrane would have been added around the alga. The algal endosymbiont may then have lost most of its components (including its membrane), but its plastid remained, encased in the new host cell's membrane.
b. Cyanobacteria were the original source of all plastids.

28.4 *Paramecium* is a ciliate in the protistan clade Alveolata.
a. cilia
b. oral groove
c. food vacuoles
d. contractile vacuole
e. anal pore
f. macronucleus
g. micronucleus

28.5 **a.** Stramenopila
b. "Hetero" refers to the two different flagella—one smooth and one with the hairlike projections characteristic of the Stramenopila
c. diatoms, golden algae, brown algae
d. Their plastids have four membranes and probably originated by secondary endosymbiosis of a red alga.

28.6 These three groups represent distinct eukaryotic lineages, but their phylogenetic relationship to other eukaryotes is still uncertain.

28.7 **a.** diploid
b. multinucleate mass, plasmodium
c. amoeboid or flagellated cells
d. haploid
e. solitary amoeboid cells until aggregate to fruiting body
f. two amoebas fuse to form resistant zygote

SUGGESTED ANSWERS TO STRUCTURE YOUR KNOWLEDGE

1. Unicellular—and a few colonial and multicellular—eukaryotic organisms, most of which use aerobic respiration, have flagella or cilia at some point in their life, reproduce asexually (many have sexual reproduction), require water or moist habitats (which may be inside the bodies of hosts), form cysts to withstand harsh conditions, may be photoautotrophic, heterotrophic, or mixotrophic

2. Originally, Whittaker assigned only unicellular eukaryotes to the kingdom Protista. The kingdom was expanded to include some colonial and multicellular eukaryotes. However, accumulating evidence indicates that there are many different lineages of protists, and thus, the kingdom Protista is phylogenetically untenable. Molecular studies are helping to unravel protistan phylogeny and eight clades that may be considered kingdoms are described in the text: Diplomonadida, Parabasala, Euglenozoa, Alveolata, Stramenopila, Rhodophyta, Chlorophyta, and Mycetozoa. The phylogenies of the groups that move using pseudopods remain unclear.

ANSWERS TO TEST YOUR KNOWLEDGE

Multiple Choice:

1. d	4. c	7. c	10. e
2. a	5. a	8. d	11. b
3. c	6. d	9. b	12. a

Matching:

1. C	5. H
2. E	6. B
3. F	7. A
4. G	8. D

CHAPTER 29: PLANT DIVERSITY I: HOW PLANTS COLONIZED LAND

■ INTERACTIVE QUESTIONS

29.1 a. origin of plants from algal ancestors
b. origin and diversification of vascular plants
c. origin of seed plants
d. radiation of flowering plants

29.2 a. gametes
b. mitosis
c. zygote
d. mitosis
e. sporophyte
f. spores
g. meiosis
h. gametophyte

29.3 a. Hollow columns of dead xylem cells carry water and minerals up from the roots.
b. The living cells of phloem form tubes that distribute organic nutrients throughout the plant.

29.4 In an ancestral charophycean, a zygote may have first divided by mitosis into a group of diploid cells. This delay in meiosis increased the number of spores that could be produced from a zygote.

29.5 a. gametophyte
b. archegonia
c. antheridia
d. swim
e. sporophyte
f. meiosis
g. sporangium (capsule)
h. gametophytes

29.6 a. meiosis
b. spores
c. gametophyte
d. antheridium
e. sperm

f. archegonium
g. egg
h. fertilization
i. zygote
j. gametophyte
k. young sporophyte
l. mature sporophyte
m. sporangia (in sori)
The gametophyte (upper portion of the diagram) is haploid; the sporophyte (lower portion) is diploid.

SUGGESTED ANSWERS TO STRUCTURE YOUR KNOWLEDGE

1. protected and nourished embryo, sporopollenin-walled spores, apical meristems, alternation of generations, multicellular gametangia, vascular tissue strengthened with lignin, cuticle, stomata

2. **a.** origin of plants
   b. early vascular plants appear
   c. early seed plants appear
   d. radiation of flowering plants
   e. ancestral green algae (charophyceans)
   f. bryophytes (liverworts, hornworts, mosses)
   g. pteridophytes—seedless vascular plants (ferns)
   h. seed plants
   i. gymnosperms (conifers)
   j. angiosperms (flowering plants)

ANSWERS TO TEST YOUR KNOWLEDGE

Multiple Choice:

1. c	5. d	9. d	13. b
2. e	6. c	10. e	14. a
3. d	7. d	11. d	15. c
4. a	8. a	12. e	

CHAPTER 30: PLANT DIVERSITY II: THE EVOLUTION OF SEED PLANTS

■ INTERACTIVE QUESTIONS

30.1 a. integuments ($2n$)
b. megasporangium ($2n$)
c. megaspore (n)
d. seed coat ($2n$)

e. food supply (from female gametophyte tissue) (n)
f. embryo ($2n$)

30.2 Resistant, airborne pollen grains transport sperm to the female gametophyte, eliminating the need for a moist environment for sperm to swim to reach egg cells.

30.3 a. ovulate cone
 b. ovule
 c. megasporangium
 d. pollen cone
 e. microsporangium
 f. pollen grain
 g. surviving megaspore
 h. female gametophyte
 i. egg in archegonium
 j. sporophyte embryo
 k. food supply (female gametophyte)
 l. seed coat
 m. seed
Meiosis occurs within the microsporangia to produce microspores that develop into pollen grains, and within the megasporangium when the megaspore mother cell gives rise to a megaspore that develops into the female gametophyte. *Pollination* occurs when pollen is drawn into the ovule through the micropyle. *Fertilization* occurs, usually more than a year later, when a sperm nucleus discharged from the pollen tube fertilizes an egg contained in an archegonium. *Gametophyte structures* in the pine life cycle include pollen grains and the multicellular female gametophyte contained within the megasporangium.

30.4 a. sepals, petals, stamens, carpels
 b. mature ovary that may be modified to help disperse seeds

30.5 a. male gametophyte—contained within pollen grain, consists of two haploid cells, one of which divides to form two sperm; female gametophyte—embryo sac contained in the ovule that often has eight haploid nuclei in seven cells.
 b. embryo and endosperm within a seed coat derived from the integuments of the ovule
 c. Double fertilization may coordinate development of food storage in the seed with the successful fertilization of an egg.

SUGGESTED ANSWERS TO STRUCTURE YOUR KNOWLEDGE

1. **a.** progymnosperms
 b. Phyla Ginkgophyta, Cycadophyta, Gnetophyta, and Coniferophyta
 c. In addition to the monocots, phylum Anthophyta now includes the clades *Amborella,* water lilies, other early angiosperm lines, and dicots

2. gametophytes develop within the resistant walls of spores; the protected female gametophyte retained on sporophyte plant; nonswimming pollen; seed for protection, dispersal, and nourishment of embryo

3. xylem vessels for more efficient transport; animal pollinators; fruit for protection and dispersal of seeds

ANSWERS TO TEST YOUR KNOWLEDGE

Multiple Choice:

1. e	**3.** a	**5.** c	**7.** b	**9.** c
2. b	**4.** d	**6.** d	**8.** b	**10.** e

True or False:

1. False—add *or algae.*
2. True
3. False—change *gymnosperms* to *angiosperms.*
4. True
5. True
6. False—change *fruit* to *seed,* or change *an embryo . . .* to *a mature ovary.*
7. False—change *tracheids* to *vessel elements.*
8. True
9. False—change *angiosperms* to *gymnosperms,* or change *haploid cells . . .* to *an embryo sac with a few haploid cells.*
10. True

CHAPTER 31: FUNGI

■ INTERACTIVE QUESTIONS

31.1 a. network of hyphae, typical body form
 b. cross-walls between cells, with pores
 c. multinucleated hyphae, aseptate
 d. two nuclei from different sources exist together in hyphae
 e. fusion of hyphal cytoplasm
 f. fusion of nuclei

31.2 a. conidia (asexual spores)
 b. ascocarp

 c. ascospores
 d. asci
 e. ascomycete
 f. sporangia
 g. zygosporangium
 h. zygomycete
 i. basidiocarp
 j. basidium
 k. basidiospores
 l. basidiomycete

31.3 a. Their growth on bare rock prepares the way for the succession of plants. Lichens are an important food source for grazing animals in the arctic tundra.

b. Almost all vascular plants have mycorrhizae, and these association greatly increase mineral absorption by plants.

SUGGESTED ANSWERS TO STRUCTURE YOUR KNOWLEDGE

1. **a.** chytrids, aquatic saprobes, or parasites
    b. coenocytic hyphae
    c. uniflagellated zoospores
    d. zygote fungi, black bread mold, mycorrhizae
    e. coenocytic hyphae
    f. spores in sporangium on aerial hyphae
    g. hyphae fuse, resistant zygosporangia, karyogamy, meiosis
    h. sac fungi, yeasts, truffles, mycorrhizae
    i. septate hyphae, some unicellular
    j. spores called conidia in chains or clusters
    k. dikaryotic hyphae in ascocarp, karyogamy in asci, ascospores
    l. club fungi, mushrooms, rusts, mycorrhizae
    m. extensive mycelium, dikaryotic hyphae
    n. less frequent than sac fungi
    o. basidiocarps, karyogamy in basidia, basidiospores

2. The role of fungi as saprobes and decomposers of organic material is central to the recycling of chemicals between living organisms and their physical environment. Parasitic fungi have economic and health effects on humans. Plant pathogens cause extensive loss of food crops and trees. Mycorrhizae are found on the roots of most plants. These mutualistic associations greatly benefit the plant.

3. Protein and ribosomal RNA comparisons indicate that fungi are more closely related to animals than to plants. Molecular systematics indicates that both fungi and animals evolved from a protistan ancestor.

ANSWERS TO TEST YOUR KNOWLEDGE

Fill in the Blanks:

1. septum	5. basidium	9. coenocytic
2. heterokaryon	6. asci	10. Chytrids
3. chitin	7. mycorrhizae	
4. exoenzymes	8. zygosporangia	

Multiple Choice:

1. a	3. c	5. e	7. a	9. c
2. c	4. d	6. b	8. a	

CHAPTER 32: INTRODUCTION TO ANIMAL EVOLUTION

■ INTERACTIVE QUESTIONS

32.1 a. Parazoa: Porifera (sponges)
b. Eumetazoa
c. Radiata: Cnidaria (hydras, jellies, anemones) and Ctenophora (comb jellies)
d. Bilateria
e. Acoelomates: Platyhelminthes (flatworms)
f. Pseudocoelomates: Rotifera (rotifers) and Nematoda (nematodes or round worms)
g. Coelomates
h. Protostomes: Mollusca (mollusks), Annelida (segmented worms), and Arthropoda (arthropods)
i. Deuterostomes: Echinodermata (echinoderms) and Chordata (chordates)

32.2 The Ecdysozoa include the phyla Arthropoda and Nematoda (a pseudocoelomate). Ecdysozoans get their name from the shedding of their exoskeleton, called ecdysis, as they grow.

32.3 The Lophotrochozoa include the phyla Annelida and Mollusca (both of which have a trochophore larva), the acoelomate Platyhelminthes, the pseudocoelomate Rotifera, and the lophophorate phyla Bryozoa, Phoronida, and Brachiopoda. The name comes from the combination of lophophore-bearing phyla and the phyla with a trochophore larva.

SUGGESTED ANSWERS TO STRUCTURE YOUR KNOWLEDGE

1. **a.** spiral and determinate
    b. radial and indeterminate
    c. schizocoelous, split in solid mass of mesoderm
    d. enterocoelous, buds from archenterons
    e. mouth, second opening for anus
    f. anus, second opening for mouth

2. They agree on the two deepest branches (the parazoa–eumetazoa and radiata–bilateria dichotomies) and on the deuterostome clade as a monophyletic branch of the coelomates.

3. Molecular systematics establishes two protostome clades (Ecdysozoa and Lophotrochozoa). The arthropods and the pseudocoelomate nematodes are grouped into the Ecdysozoa, and the annelids, mollusks, acoelomate flatworms, pseudocoelomate rotifers, and three lophophorate phyla are placed in the Lophotrochozoa.

ANSWERS TO TEST YOUR KNOWLEDGE

Multiple Choice:

1. c	5. c	9. c	13. b
2. d	6. a	10. d	14. d
3. e	7. e	11. a	15. e
4. e	8. d	12. e	

CHAPTER 33: INVERTEBRATES

■ INTERACTIVE QUESTIONS

33.1 a. line body cavity (spongocoel); create water current, trap and phagocytose food particles, may produce gametes
b. wander through mesohyl; digest food and distribute nutrients, produce skeletal fibers, may produce gametes

33.2 a. polyp
b. medusa
c. mouth/anus
d. tentacle
e. mesoglea
f. gastrovascular cavity

33.3 a. Planarians have an extensively branched gastrovascular cavity; food is drawn in and undigested wastes are expelled through the mouth at the tip of a muscular pharynx.
b. Tapeworms are bathed in predigested food in their host's intestine.

33.4 a. 1. muscular foot for movement
2. visceral mass containing internal organs
3. mantle that may secrete shell and enclose visceral mass to form mantle cavity that houses gills
b. 1. Snails creep slowly, using their radula to rasp up algae.
2. Clams are sedentary suspension feeders that use siphons to draw water over the gills, where food is trapped in mucus.
3. Squid are active predators that capture and crush prey with jaws.

33.5 a. septum
b. intestine
c. metanephridium
d. setae
e. ventral nerve cord
f. dorsal and ventral blood vessels
g. longitudinal and circular muscles
h. cuticle

33.6 a. anterior feeding appendages modified as pincers or fangs
b. Many spiders trap their prey in webs they have spun from abdominal gland secretions. Spiders kill prey with poison-equipped, fang-like chelicerae. After masticating and adding digestive juices, they suck up liquid, partially digested food.

33.7 a. The head has fused segments and has one pair of antennae, a pair of compound eyes, and several pairs of mouthparts modified for various types of ingestion.
b. The segmented thorax has three pairs of walking legs and may have one or two pairs of wings.
c. The segmented abdomen has no appendages.

33.8 a. mostly terrestrial
b. mostly aquatic
c. fly, walk, burrow
d. swim, float (planktonic)
e. tracheal system
f. gills or body surface
g. Malpighian tubules
h. diffusion across cuticle, glands regulate salt balance
i. one pair
j. two pairs
k. unbranched; mandibles and modified mouthparts for chewing, piercing, or sucking; thorax has three pairs of walking legs and may have wings (extensions of cuticle)
l. branched; mandibles and multiple mouthparts, walking legs on thorax, abdominal appendages

33.9 **a.** radial anatomy as adults, usually five spokes; sessile or slow moving
b. endoskeleton of calcareous plates
c. water vascular system with tube feet for locomotion, feeding, respiration

SUGGESTED ANSWERS TO STRUCTURE YOUR KNOWLEDGE

1. **a.** Parazoa
   b. Eumetazoa
   c. Radiata
   d. Bilateria
   e. Protostomia
   f. Deuterostomia
   g. Lophotrochozoa
   h. Ecdysozoa
   i. Porifera (sponges)
   j. Cnidaria (hydras, jellies, sea anemones, corals)
   k. Ctenophora (comb jellies)
   l. Platyhelminthes (flatworms: planaria, flukes, tapeworms)
   m. Rotifera (rotifers)
   n. Nemertea (proboscis worms)
   o. Bryozoa (bryozoans)
   p. Phoronida (phoronids)
   q. Brachiopoda (brachiopods)
   r. Mollusca (clams, snails, squids)
   s. Annelida (segmented worms: earthworms, polychaetes, leeches)
   t. Nematoda (roundworms)
   u. Anthropoda (curstaceans, insects, spiders)
   v. Echinodermata (sea stars, sea urchins)
   w. Chordata (vertebrates, lancelets, tunicates)

ANSWERS TO TEST YOUR KNOWLEDGE

Matching:

1. C l	5. H d	9. D f
2. B e	6. B j	10. C i
3. E g	7. E b	11. H m
4. A h	8. B a	12. E c

Multiple Choice:

1. b	4. e	7. c	10. b	13. d	16. e
2. c	5. a	8. b	11. a	14. a	17. b
3. c	6. c	9. a	12. c	15. d	18. c

CHAPTER 34: VERTEBRATE EVOLUTION AND DIVERSITY

■ INTERACTIVE QUESTIONS

34.1 lancelet, subphylum Cephalochordata
   a. mouth
   b. *pharyngeal slits
   c. intestine
   d. segmental muscles
   e. anus
   f. *postanal tail
   g. *dorsal, hollow nerve cord
   h. *notochord
   * Chordate characteristics

34.2 **a.** cephalization with sensory organs and enlarged brain to aid navigation
   b. cranium and vertebral column to protect brain and spinal cord and provide strong attachments for muscles
   c. a closed circulatory system to support an active metabolism

34.3 **a.** vertebral column
   b. jaws; two sets of paired appendages; ossified skeletons and teeth
   c. lungs or lung derivatives
   d. legs

e. amniotic egg
f. milk
g. feathers
h. gnathostomes
i. tetrapods
j. amniotes
k. hagfishes (class Myxini)
l. lampreys (class Cephalaspidomorphi)
m. sharks and rays (class Chondrichthyes)
n. bony fishes (former class Osteichthyes)
o. ray-finned fishes (class Actinopterygii)
p. lobe-finned fishes (class Actinistia)
q. lungfishes (class Dipnoi)
r. frogs and salamanders (class Amphibia)
s. mammals (class Mammalia)
t. turtles ("class" Testudines)
u. lizards and snakes; tuatara ("class" Lepidosauria or "classes" Squamata and Sphenodontia)
v. crocodiles and alligators (class Crocodilia)
w. birds (class Aves)

34.4 **a.** increase the area for absorption and slow the passage of food through the short shark intestine
   b. common opening for the reproductive tract, excretory system, and digestive tract

34.5 a. The common fusiform body shape of fast-swimming sharks, bony fish, and aquatic mammals is an example of convergent evolution, an adaptation to reduce drag while swimming.
b. See answer to Interactive Question 34.3.

34.6 a. The fossil record shows a series of intermediates from lobe-finned fishes through lungfishes that would qualify as tetrapods.
b. the Carboniferous period, from 363 to 290 mya

34.7 a. amnion—encases embryo in fluid, prevents dehydration, and cushions shocks
b. allantois—stores metabolic wastes, functions with chorion in gas exchange
c. chorion—functions in gas exchange
d. yolk sac—expands over yolk; blood vessels transport stored nutrients into embryo

34.8 a. amniotic egg, skin covered with waterproof scales, lungs, ectothermic
b. turtles (Testudines), tuataras (Sphenodontia), lizards and snakes (Squamata), alligators and crocodiles (Crocodilia)
c. The class Reptilia is not monophyletic because it does not include the birds. Birds share the most recent ancestor with the dinosaurs.

34.9 light, strong skeleton, no teeth, and absence of some organs to reduce weight; feathers that shape wings into airfoil; strong keel on sternum to which flight muscles attach; high metabolic rate supported by efficient circulatory system, respiratory system, and endothermy; well-developed visual and motor areas of brain

34.10 a. hair; milk produced by mammary glands; endothermic with active metabolism; diaphragm to ventilate lungs; four-chambered heart; larger brains, increased learning capacity; differentiated teeth and remodeled jaw
b. See answer to Interactive Question 34.3.

34.11 a. 5 to 7 million years ago
b. increased brain size; change in jaw shape and dentition; bipedal posture; reduced size difference between the sexes; changes in family structure (monogamy, extended parental care)

SUGGESTED ANSWERS TO STRUCTURE YOUR KNOWLEDGE

1. See Interactive Question 34.3 on page 260.

2. (1) Pharyngeal slits used for suspension feeding became adapted for gas exchange; (2) skeletal supports for gills became adapted for use as hinged jaws; (3) lungs of bony fish were transformed to air bladders, aiding in buoyancy; (4) fleshy fins supported by skeletal elements developed into limbs for terrestrial animals; (5) feathers, probably originally used for insulation, came to be used for flight; (6) dexterous hands important for an arboreal life evolved to manipulate tools.

3. Amphibians remained in damp habitats, burrowing in mud during droughts; some secrete foamy protection for eggs laid on land. Reptiles developed scaly, waterproof skin, an amniote egg that provided an aquatic environment for the developing embryo, and behavioral adaptations to modulate changing temperatures. All terrestrial groups are tetrapods, using limbs for locomotion on land.

4. According to the multiregional hypothesis, modern humans evolved in several areas from Neanderthals and other regional archaic *Homo sapiens,* between 1 and 2 million years ago. According to the replacement hypothesis, a relatively recent dispersal (100,000 years ago) from Africa replaced the regional descendants of *Homo erectus,* and gave rise to human geographic diversity.
   Some paleoanthropologists claim that the fossil records best support the multiregional model. Molecular data from mtDNA, nuclear DNA, and the Y chromosome indicate a more recent divergence of human groups, supporting the replacement hypothesis.

ANSWERS TO TEST YOUR KNOWLEDGE

Fill in the Blanks:

1.	Urochordata	7.	diapsids
2.	Cephalochordata	8.	dinosaurs (theropods)
3.	somites	9.	monotremes
4.	gnathostomes	10.	diaphragm
5.	operculum	11.	Anthropoidea
6.	amniotic egg	12.	*Australopithecus*

Multiple Choice:

1.	a	5.	a	9.	b	13.	d
2.	b	6.	d	10.	e	14.	d
3.	e	7.	a	11.	a	15.	c
4.	d	8.	c	12.	b		

CHAPTER 35: PLANT STRUCTURE AND GROWTH

■ INTERACTIVE QUESTIONS

35.1 a. point of stem at which leaf is attached
b. at the node in the angle between leaf and stem
c.

Monocots	Dicots
one cotyledon	two cotyledons
fibrous root system	taproot
leaf veins parallel	leaf veins netlike
may lack petioles	have petioles
flower parts in multiples of 3	flower parts in multiples of 4 or 5
pith inside stele in root	xylem spokes in stele with phloem in between
vascular bundles in stem scattered	vascular bundles in stem in ring, pith inside
no secondary growth	vascular and cork cambiums may produce secondary growth

35.2 a. sclerenchyma (fibers and sclereids), tracheids, and vessel elements
b. the cells listed in **a.** and sieve-tube members

35.3 Primary growth continues to add to the tips of roots and shoots, while secondary growth thickens and strengthens older regions of a woody plant.

35.4 a. epidermis (dermal)—root hairs function in absorption
b. cortex (ground)—uptake of minerals and water, food storage
c. stele (vascular)—central vascular cylinder
d. endodermis—regulates water and mineral movement into stele
e. pericycle—origin of lateral roots
f. pith—central parenchyma cells
g. xylem—water and mineral transport
h. phloem—nutrient transport

35.5 a. cuticle
b. upper epidermis
c. palisade parenchyma
d. spongy parenchyma
e. guard cells
f. xylem
g. phloem
h. stomata

35.6 See table in Interative Question 35.1c.

35.7 H, F, A, B, G, D, C, E

35.8 Microtubules of the preprophase band leave behind ordered actin microfilaments that determine the orientation of the nucleus and the location of the cell plate during cytokinesis. Microtubules may constrain the movement of cellulose-producing enzymes and thus orient the cellulose microfibrils in the cell walls that will determine the direction in which cells can expand.

SUGGESTED ANSWERS TO STRUCTURE YOUR KNOWLEDGE

1. **a.** protoderm
   b. ground meristem
   c. cortex
   d. cork cells
   e. procambium
   f. primary phloem
   g. vascular cambium
   h. secondary phloem
   i. secondary xylem

2. This drawing is a dicot stem, indicated by the ring of vascular bundles and the central pith.
   a. sclerenchyma
   b. phloem
   c. xylem
   d. ray of ground tissue connecting pith to cortex
   e. cortex
   f. pith
   g. vascular bundle
   h. epidermis

ANSWERS TO TEST YOUR KNOWLEDGE

Matching:

1. H	**5.** I	**9.** J
2. F	**6.** A	**10.** G
3. B	**7.** C	
4. E	**8.** D	

Multiple Choice:

1. a	**5.** b	**9.** e	**13.** c
2. e	**6.** a	**10.** d	**14.** c
3. b	**7.** e	**11.** e	**15.** e
4. c	**8.** c	**12.** b	**16.** d

CHAPTER 36: TRANSPORT IN PLANTS

■ INTERACTIVE QUESTIONS

36.1 a. $\Psi_P = 0$, $\Psi_S = -0.6$, $\Psi = -0.6$
b. $\Psi_P = 0.6$, $\Psi_S = -0.6$, $\Psi = 0$. The turgor pressure of the cell that develops with movement of water into the cell finally offsets the solute potential of the hypertonic cell. No net osmosis occurs, and the water potentials of both cell and solution are 0.
c. $\Psi_P = 0$, $\Psi_S = -0.8$, $\Psi = -0.8$. The cell would **plasmolyze,** losing water until the solute potential of the cell would equal that of the bathing solution. There would be no turgor pressure.

36.2 a. symplastic   **f.** endodermis
b. apoplastic   **g.** cortex
c. Casparian strip   **h.** epidermis
d. xylem vessels   **i.** root hair
e. stele

36.3 a. The loss of water vapor from the air spaces through the stomata causes evaporation from the water film coating the mesophyll cells, which creates a tension on the remaining water film.
b. Water molecules hold together due to hydrogen bonding, and a pull on one water molecule is transmitted throughout the column of water.
c. Water molecules adhere to the hydrophilic walls of narrow xylem cells, which helps to support the column of water against gravity.
d. The transpiration of water from the leaf creates a tension which produces a gradient of water potentials that extends from the leaf to the root.

36.4 The illumination of blue-light receptors activates proton pumps that transport H^+ out of the guard cell. The resulting membrane potential facilitates the movement of K^+ through specific membrane channels into the cell. This increase in solute concentration lowers the water potential of the cell, and water enters by osmosis. Photosynthesis in the guard cell produces the ATP needed for the proton pumps.

36.5 A storage organ, such as a root or tuber, stores carbohydrates during the summer and is a sugar sink, but when its starch is broken down to sugar in the spring to supply growing buds, it becomes a sugar source.

36.6 a. cotransport protein
b. sucrose

c. H^+
d. proton pump (ATPase)
e. ATP

SUGGESTED ANSWERS TO STRUCTURE YOUR KNOWLEDGE

1. Solutes may move across membranes by passive transport when they move down their concentration gradient. Transport proteins, which speed this passive transport, may be specific carrier proteins or selective channels. Active transport usually involves a proton pump that creates a membrane potential as positively charged hydrogen ions are moved out of the cell. Cations may now move down their electrochemical gradient. In cotransport, a solute is moved by a transport protein driven by the inward diffusion of H^+.

2. The transpiration-cohesion-tension mechanism explains the ascent of xylem sap. The cohesion of water molecules transmits the pull resulting from transpiration throughout the column, and the adhesion of water to the hydrophilic walls of the xylem vessels aids the flow. Tension created by the evaporation of water from the menisci that form in the leaf's air spaces sets up this bulk flow of water.

   A pressure flow mechanism also explains the movement of phloem sap, but the plant must expend energy to create the differences in water potential that drive translocation. The active accumulation of sugars in sieve-tube members greatly decreases water potential at the source end, resulting in an inflow of water. The removal of sugar from the sink end of a phloem tube results in the osmotic loss of water. The resulting difference in hydrostatic pressure between the source and sink end of the phloem tube causes the bulk flow of phloem sap.

ANSWERS TO TEST YOUR KNOWLEDGE

1. b	6. b	11. a	16. d
2. e	7. d	12. c	17. d
3. d	8. d	13. e	18. b
4. a	9. a	14. b	19. d
5. c	10. c	15. c	20. c

CHAPTER 37: PLANT NUTRITION

■ INTERACTIVE QUESTIONS

37.1 a. component of chlorophyll, activates many enzymes
b. component of nucleic acids, phospholipids, ATP, some coenzymes
c. component of cytochromes, cofactor for chlorophyll synthesis

37.2 in the younger, growing parts of a plant

37.3 A loam, which is a mixture of sand, silt, and clay, has enough large particles to provide air spaces and enough small particles to retain water and minerals. Cations bind to clay particles and are made available to roots by cation exchange. Humus in a soil helps to retain water and provide a steady supply of mineral nutrients, such as nitrogen, phosphorous, and potassium.

37.4 a variety of conservation-minded, environmentally safe, and profitable farming methods that permit the use of soil as a renewable resource

37.5 a. nitrogen-fixing bacteria
b. ammonifying bacteria
c. nitrifying bacteria
d. denitrifying bacteria
e. nitrate and nitrogenous organic compounds

SUGGESTED ANSWERS TO STRUCTURE YOUR KNOWLEDGE

1.

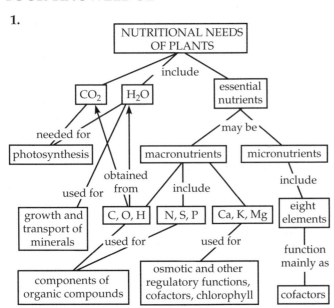

2. Root nodules are mutualistic associations between legumes and nitrogen-fixing bacteria of the genus *Rhizobium*. These bacteria receive nourishment from the plant and provide the plant with ammonium. Mycorrhizae are mutualistic associations between the roots of most plants and various species of fungi. The fungi stimulate root growth, increase surface area for absorption, and secrete acids that increase the solubility of some minerals. In return, the fungi receive nourishment from the plant.

3. Both these types of symbiotic relationships involve chemical recognition between specific plant species and bacterial or fungal species. And both appear to involve similar molecular mechanisms. The signal molecules from the fungi and bacteria are either chitins or chitinlike molecules. Early "infection" in both cases increases the concentration of plant hormones known as cytokinins in roots. And in both cases, the plant's early nodulin genes are activated and probably initiate growth responses in the root tissues.

ANSWERS TO TEST YOUR KNOWLEDGE

1. b	4. b	7. e	10. b	13. c
2. a	5. a	8. d	11. c	14. e
3. c	6. e	9. c	12. d	15. b

CHAPTER 38: PLANT REPRODUCTION AND BIOTECHNOLOGY

■ INTERACTIVE QUESTIONS

38.1
a. stamen
b. anther
c. filament
d. petal
e. carpel
f. stigma
g. style
h. ovary
i. sepal
j. ovule

Pollen is formed in the anther. Pollination occurs when pollen lands on the stigma. Fertilization occurs within the embryo sac in the ovule.

38.2
a. a pollen grain with a thick coat surrounding a generative cell (which will divide to form two sperm) and a tube cell
b. the embryo sac, often containing seven cells and eight nuclei

38.3 Double fertilization conserves resources by allowing for nutrient development only when fertilization has occured.

38.4
a. seed coat
b. epicotyl
c. hypocotyl
d. cotyledons
e. radicle
f. plumule
g. cotyledon (scutellum)
h. endosperm
i. radicle
j. coleorhiza
k. plumule
l. coleoptile

38.5 Hormonal interactions induce softening of the pulp, a change in color, and an increase in sugar content.

38.6
a. The hypocotyl hook continues to elongate but does not straighten. Light is the cue that causes the hook to straighten and thus pull the cotyledons out of the ground.
b. In peas, a hook forms in the epicotyl, lifting the shoot tip out of the soil while the pea cotyledons remain in the ground. In maize and other grasses, the coleoptile pushes through the soil, and the shoot tip is protected as it grows up through the tubular sheath.

38.7 Apomixis provides for the dispersal and dormancy advantages of seeds without the need for sexual reproduction and genetic recombination.

SUGGESTED ANSWERS TO STRUCTURE YOUR KNOWLEDGE

1.

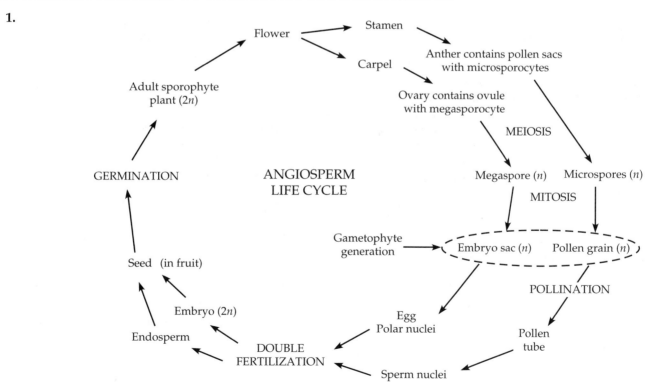

2. *Sexual:* Advantages include increased genetic variability, which provides the potential to adapt to changing conditions, and the dispersal and dormancy capabilities provided by seeds. Disadvantages are that the seedling stage is very vulnerable, sexual reproduction is very energy intensive (since most seeds and seedlings don't survive), and genetic recombination may separate adaptive traits.

Asexual: Advantages include the hardiness of vegetative propagation and the maintenance of genetically well-adapted plants in a given environment. Disadvantages relate to the advantages of sexual reproduction: There is no genetic variability from which to choose should conditions change, and there are no seeds for dormancy or dispersal.

3. *Benefits:* reduce use of chemical pesticides; weed crops with herbicides instead of erosion-producing tillage; increase nutritional value of crops; produce disease-resistant crops; increase drought resistance or saline tolerance in crops

Dangers: introduce allergens into human foods; harm nontarget species with pesticides; create hybrid "superweeds" due to crop-to-weed transgene escape; produce other unanticipated harmful results that cannot be stopped once GM organisms are released into the environment

ANSWERS TO TEST YOUR KNOWLEDGE

Fill in the Blanks:

1. ovary	6. epicotyl
2. sporophyte	7. hypocotyl or epicotyl hook
3. monoecious	8. scion
4. embryo sac	9. protoplast
5. radicle	10. callus

Multiple Choice:

1. c	6. a	11. a	16. e
2. a	7. c	12. e	17. c
3. d	8. e	13. e	18. e
4. d	9. c	14. a	
5. e	10. d	15. d	

CHAPTER 39: PLANT RESPONSES TO INTERNAL AND EXTERNAL SIGNALS

■ INTERACTIVE QUESTIONS

39.1 a. The signal is light; the receptor is a phytochrome located in the cytosol.
b. Light-activated phytochrome activates G proteins, which in turn activate at least two pathways, one producing the second messenger cGMP that sets off a phosphorylation cascade, and one leading to a Ca^{2+}-calmodulin complex that activates specific kinases. Both of these pathways activate various transcription factors (and deactivate negative transcriptional factors) that change gene expression.
c. The plant response involves growth changes (slowing of stem elongation, leaf expansion, elongation of roots) and the production of chlorophyll and enzymes involved in photosynthesis.

39.2 a. ethylene
b. cytokinins
c. abscisic acid
d. auxin
e. brassinosteroids
f. gibberellins

39.3 When the chromophore component of a phytochrome absorbs light, a change in its conformation may activate the kinase domain, which phosphorylates cellular proteins and triggers the cell's response to light.

39.4 Free-running periods are circadian rhythms that vary from exactly 24 hours when an organism is kept in a constant environment without environmental cues.

39.5 a. not flower   **f.** flower
b. flower   **g.** not flower
c. not flower   **h.** flower
d. flower   **i.** not flower
e. not flower   **j.** flower

In the last two light regimens, a flash of far-red light cancels the effect of the red flash, and the red flash in the last one cancels the effect of the far-red flash.

39.6 a. positive phototropism, negative gravitropism
b. thigmotropism; mechanical stimulation causing unequal growth rates of cells on opposite sides of the tendril

39.7 a. reduce transpirational water loss by stomatal closing and wilting; deep roots continue to grow in moister soil; heat-shock proteins produced to stabilize plant proteins

b. change membrane lipid composition by increasing unsaturated fatty acids; air tubes may develop in cortex of root

SUGGESTED ANSWERS TO STRUCTURE YOUR KNOWLEDGE

1. cytokinin spray used to keep flowers fresh; ethylene used to ripen stored fruit; synthetic auxins to induce seedless fruit set; gibberellins sprayed on Thompson seedless grapes; 2,4-D (auxin) used as herbicide

2.

3. **1.** binding of pathogen ligand to specific receptor based on *R-Avr* recognition
2. STP, signal-transduction pathway that triggers HR and production of alarm signal
3. HR, hypersensitive response consisting of production of phytoalexins, PR proteins, and reinforcement of cell walls to seal off infection
4. production of signal hormone, probably salicylic acid
5. distribution of signal to rest of plant
6. STP, signal triggers signal-transduction pathway in distant cells
7. SAR, activation of systemic acquired resistance with production of antimicrobial molecules

ANSWERS TO TEST YOUR KNOWLEDGE

True or False:

1. False—change *abscisic acid* to *gibberellin*, or say *The removal of abscisic acid.*

2. True

3. False—change *gibberellins* to *cytokinins.*

4. False—change *thigmomorphogenesis* to *triple response*; or say *Thigmomorphogenesis is a change in growth form in response to mechanical stress.*

5. True

6. True

7. False—change *circadian rhythm* to *photoperiodism*; or change *day or night length* to *on a 24-hour cycle.*

8. True

9. False—change *A virulent* to *An avirulent*

10. False—change *cryptochrome* to *chromophore*; or say *Cryptochrome is a blue-light photoreceptor.*

Multiple Choice:

1. b	4. b	7. b	10. b	13. d
2. c	5. a	8. e	11. b	14. d
3. e	6. b	9. c	12. e	15. c

CHAPTER 40: AN INTRODUCTION TO ANIMAL STRUCTURE AND FUNCTION

■ INTERACTIVE QUESTIONS

40.1 a. Stratified squamous; outer skin; thick layer is protective, and new cells produced near basement membrane replace those sloughed off.
b. Simple columnar; lining of digestive tract; cells with large cytoplasmic volumes are specialized for secretion or absorption.

40.2 a. osteon (Haversian system)
b. central canal
c. osteocyte
d. bone
e. macrophage
f. fibroblast
g. collagenous fiber
h. elastic fiber
i. reticular fiber
j. loose connective tissue
k. white blood cells
l. red blood cells
m. platelet
n. blood

40.3 a. Cardiac; dark bands are intercalated discs that relay electrical signals from cell to cell during heartbeat.
b. smooth muscle

40.4 digestive, circulatory, respiratory, immune and lymphatic, excretory, endocrine, reproductive, nervous, integumentary, skeletal, muscular

40.5 The animal can live on land, and its organ systems can control the composition of the fluid bathing its cells.

40.6 a. sweat glands
b. thermostat in the brain
c. set point

40.7 Short bursts of activity rely on ATP present in muscle cells and generated anaerobically by glycolysis. Sustained activity depends on the generation of ATP by aerobic respiration, and an endotherm's cellular respiration rate is about 10 times greater than an ectotherm's rate.

SUGGESTED ANSWERS TO STRUCTURE YOUR KNOWLEDGE

1.

Tissue	Structural Characteristics	General Functions	Specific Examples
Epithelial	Tightly packed cells; basement membrane; cuboidal, columnar, squamous shapes; simple or stratified	Protection, absorption, secretion; lines body surfaces	Mucous membrane, may be ciliated; skin; lining of blood vessels
Connective	Few cells that secrete extracellular matrix of protein fibers in liquid, gel, or solid substance	Connect and support other tissues	Loose connective; adipose; fibrous connective; bone; cartilage; blood
Muscle	Long cells (fibers) with myofibrils of actin and myosin	Contraction, movement	Skeletal (voluntary); smooth (involuntary); cardiac
Nervous	Neurons with cell bodies, axons, and dendrites	Sense stimuli, conduct impulses	Nerves, brain

2. All organs are covered with epithelial tissue, and internal lumens, ducts, and blood vessels are lined with epithelium. Most organs contain some types of connective tissue. Muscle tissue would be present if the organ must contract. Nervous tissue innervates most organs. The stomach is an example of an organ composed of many tissues. An epithelial mucosa lines the lumen; the submucosa contains blood vessels and loose connective tissue; the muscularis, a layer of circular and longitudinal smooth muscle, comes next; and the outer serosa is an epithelium.

ANSWERS TO TEST YOUR KNOWLEDGE

Multiple Choice:

1. b	**3. d**	**5. b**	**7. b**	**9. b**	**11. c**
2. a	**4. e**	**6. d**	**8. a**	**10. e**	**12. d**

CHAPTER 41: ANIMAL NUTRITION

■ INTERACTIVE QUESTIONS

41.1 Vegetarians must eat a combination of plant foods, complementary in amino acids and consumed in the same day, to prevent protein deficiencies. Because the body cannot store amino acids, a deficiency of a single essential amino acid may prevent protein synthesis and limit the use of other amino acids.

41.2 The proteins, fats, and carbohydrates that make up food are too large to pass through cell membranes. Monomers also must be reorganized to form the macromolecules specific to each animal.

41.3 **a.** pharynx—sucks food through mouth
b. esophagus—passes food to crop
c. crop—stores and moistens food
d. gizzard—grinds food
e. intestine—digests and absorbs
The typhlosole increases the surface area for absorption.

41.4 **a.** Polysaccharides (starch and glycogen) broken down in the mouth by salivary amylase to smaller polysaccharides and maltose; hydrolysis continues until salivary amylase is inactivated by the low pH in the stomach.
b. Proteins in the stomach are broken down by pepsin to smaller polypeptides.

41.5 **a.** pancreatic amylases, disaccharidases*
b. trypsin, chymotrypsin, aminopeptidase, carboxypeptidase, and dipeptidases*
c. nucleases, nucleotidases*, nucleosidases*
d. lipase (aided by bile salts)

* attached to epithelial cells

41.6 Carnivores are characterized by sharp incisors, fanglike canines, and jagged premolars and molars; herbivores have broad molars for grinding plant material. The intestine of an herbivore is longer to facilitate digestion of plant material, and, unlike carnivores, its cecum may contain cellulose-digesting microorganisms.

SUGGESTED ANSWERS TO STRUCTURE YOUR KNOWLEDGE

1. **a.** energy/fuel
   b. carbon skeletons/organic raw materials
   c. essential nutrients
   d. undernourished
   e. biosynthesis
   f. malnourishment
   g. essential amino acids
   h. vitamins
   i. coenzymes or parts of coenzymes
   j. minerals

2. **a.** salivary glands
   b. oral cavity
   c. pharynx
   d. esophagus
   e. stomach
   f. small intestine
   g. large intestine (colon)
   h. anus
   i. rectum
   j. pancreas
   k. pyloric sphincter
   l. gallbladder
   m. liver

ANSWERS TO TEST YOUR KNOWLEDGE

Matching:

1. M	5. F	9. N
2. I	6. G	10. E
3. D	7. J	
4. B	8. H	

Multiple Choice:

1. d	6. d	11. b	16. e	21. b
2. c	7. a	12. a	17. a	22. e
3. b	8. c	13. c	18. d	
4. b	9. b	14. c	19. c	
5. d	10. e	15. b	20. d	

CHAPTER 42: CIRCULATION AND GAS EXCHANGE

■ INTERACTIVE QUESTIONS

42.1 a. The heart pumps hemolymph through vessels into sinuses, and body movements squeeze the sinuses, forcing hemolymph through the body. When the heart relaxes, hemolymph is drawn back in through pores.
b. The heart pumps blood through large vessels that narrow for exchange within organs and then regroup to return blood to the heart.

42.2 a. capillaries of head and forelimbs
b. left pulmonary artery
c. aorta
d. capillaries of left lung
e. left pulmonary vein
f. left atrium
g. aorta
h. capillaries of abdomen and hind limbs
i. left ventricle
j. right ventricle
k. posterior vena cava
l. right atrium
m. right pulmonary vein
n. capillaries of right lung
o. right pulmonary artery
p. anterior vena cava

The pulmonary veins, left side of the heart, and aorta and arteries leading to the systemic capillary beds carry oxygen-rich blood. See text figure 42.4 for numbers. Note that oxygen-rich blood in the aorta splits and travels either to the head region or lower body, returning to the heart through the anterior or posterior vena cava, respectively.

42.3 a. atrioventricular
b. semilunar
c. SA (sinoatrial) node
d. AV (atrioventricular) node

42.4 a. artery walls
b. contraction of ventricle
c. diastole
d. systole
e. lower number (diastolic pressure)
f. higher number (systolic pressure)
g. peripheral resistance

h. contract
i. relax
j. nerves, hormones

42.5 a. plasma
b. electrolytes
c. osmotic balance, buffering, muscle and nerve functioning
d. proteins
e. buffers, osmotic factors, lipid escorts, antibodies, clotting
f. nutrients, wastes, respiratory gases, hormones
g. erythrocytes (red blood cells)
h. transport O_2 and NO
i. leukocytes (white blood cells)
j. phagocytes
k. defense and immunity
l. platelets
m. blood clotting

42.6 smoking, not exercising, obesity, and eating a fat-rich diet

42.7 a. water
b. gill
c. pump water by moving jaws and operculum
d. air
e. tracheoles of tracheal system
f. body movements compress and expand air tubes
g. air (and water)
h. lungs and moist skin
i. positive pressure breathing: fill mouth, close nostrils, raise jaw, and blow up lungs
j. air
k. alveoli in lungs
l. negative pressure breathing: lower diaphragm, raise ribs, thus increasing volume and decreasing pressure of lungs; air flows in

42.8 This graph shows an effect called the **Bohr shift.** The dissociation curve for hemoglobin is shifted to the right at a lower pH, meaning that at any given partial pressure of O_2, hemoglobin is less saturated with oxygen. A rapidly metabolizing tissue produces more CO_2, which lowers pH, and hemoglobin will unload more of its O_2 to that tissue.

SUGGESTED ANSWERS TO STRUCTURE YOUR KNOWLEDGE

1. **a.** aorta
    b. pulmonary artery
    c. pulmonary veins
    d. left atrium
    e. left ventricle
    f. right ventricle
    g. posterior vena cava
    h. atrioventricular valve
    i. semilunar valve
    j. right atrium
    k. anterior vena cava

    Blood flow from venae cavae → right atrium → right ventricle → pulmonary artery → lung → pulmonary vein → left atrium → left ventricle → aorta

2. Nasal cavity → pharynx → through glottis to larynx → trachea → bronchus → bronchiole → alveoli → interstitial fluid → alveolar capillary → hemoglobin molecule → venule → pulmonary vein → left atrium → left ventricle → aorta → renal artery → arteriole → capillary in kidney

ANSWERS TO TEST YOUR KNOWLEDGE

Multiple Choice:

1. c	8. b	15. a	22. e	29. e
2. c	9. a	16. d	23. d	30. c
3. d	10. e	17. b	24. e	31. c
4. e	11. d	18. c	25. b	32. a
5. a	12. d	19. c	26. d	33. b
6. b	13. c	20. b	27. d	34. d
7. a	14. a	21. b	28. d	

CHAPTER 43: THE BODY'S DEFENSES

■ INTERACTIVE QUESTIONS

43.1 **a.** short-lived leukocytes that phagocytose microbes; about 70% of leukocytes
b. leukocytes that migrate from blood to tissues, where they develop into macrophages
c. long-lived, large amoeboid phagocytes that engulf microbes and dead tissue cells and neutrophils
d. leukocytes that attack large parasites with enzymes
e. leukocytes that attack membranes of the body's infected or abnormal cells
f. release histamine to initiate inflammatory response
g. causes vasodilation and increased permeability of blood vessels
h. proteins released by virus-infected cells that stimulate neighboring cells to produce proteins that inhibit viral replication
i. set of serum proteins that cause lysis of microbes and are involved with nonspecific and specific defenses
j. enzyme in tears, saliva, and mucus that attacks bacterial cell walls
k. released by damaged tissue cells and leukocytes; promote blood flow
l. chemicals released by some leukocytes that raise body temperature, producing a fever
m. chemical attractants that guide phagocyte migration

43.2 **a.** The immune system is able to react *specifically* to pathogens and foreign molecules by recognizing antigens.
b. The *diversity* of the immune system results from the huge variety of lymphocytes, each of which has receptors for a specific antigen to which it responds.
c. The immune system creates a *memory* of the antigens it encounters and can react against them more promptly upon reexposure.
d. The immune system is able to distinguish *self from nonself*. The recognition by specific T cells of the combination of self (MHC molecule) and nonself (antigen) displayed by body cells is central to the immune response.

43.3 **a.** APC (macrophage)
b. bacterium
c. class II MHC molecule
d. antigen fragment
e. T cell receptor
f. helper T cell
g. CD4
h. IL-1 activates T_H cell
i. IL-2 and other cytokines activate T_H, B, and T_C cells
j. cell-mediated immunity (attack on infected cells)
k. humoral immunity (secretion of antibodies by plasma cells)

43.4 a. CD4
b. CD8
c. interleukin-2 and other cytokines that activate T_H, T_C, and B cells
d. perforin that lyses the target cell

43.5 a. B cells that bind specifically to an antigen also take in a few antigen molecules by receptor-mediated endocytosis and then present pieces of this antigen in their class II MHC molecules to helper T cells.
b. There are not many B cells specific for a particular antigen present to serve as APCs in the primary response. In the secondary response to an antigen, however, the large number of memory B cells probably serve as important APCs to activate helper T cells.

43.6 a. neutralization—blocks viral binding sites; coats microbes in opsonization
b. agglutination of antigen-bearing microbes
c. precipitation of soluble antigens
d. complement fixation and formation of a membrane attack complex

The first three mechanisms tag antigens for phagocytosis.

43.7

Blood Type	Antigens on RBCs	Antibodies in Plasma	Can Receive Blood from	Can Donate Blood to
A	A	anti-B	A, O	A, AB
B	B	anti-A	B, O	B, AB
AB	AB	none	A, B, AB, O	AB
O	none	anti-A + B	O	A, B, AB, O

43.8 a. HIV destroys helper T cells, thus crippling the humoral and cell-mediated immune systems and leaving the body unable to fight the HIV virus and other opportunistic diseases such as Kaposi's sarcoma and *Pneumocystis* pneumonia.
b. HIV can remain hidden as a provirus for years, and infected individuals can unknowingly pass the virus on through unscreened blood donations, unprotected sex, and needle sharing. The antigenic changes of HIV during replication impair the ability of the immune system to mount an effective defense and greatly complicate the development of a vaccine against HIV. Newly developed drugs may slow the replication of the virus or treat the opportunistic infections, but no cure has been found.

SUGGESTED ANSWERS TO STRUCTURE YOUR KNOWLEDGE

1. Complement: a set of proteins that circulate in blood. They interact with antibodies in classical pathway to form membrane attack complex that lyses cells; can also lyse pathogens in nonspecific defense in alternative pathway; also functions in opsonization, histamine release, immune adherence, and chemotaxis.
    Antibodies: immunoglobulins made by plasma cells derived from B cells. They circulate in blood and lymph, form antigen receptors on B cells, and attach as receptors to mast cells and basophils. They bind to specific antigen, mark foreign cells and molecules for destruction, neutralize microbes, agglutinate soluble antigens, and produce opsonization.
    Interleukin-1: a cytokine secreted by antigen-presenting macrophage (APC) after binding with helper T cell. It stimulates helper T cell to release interleukin-2.
    Interleukin-2: a cytokine released by activated T_H cell. It activates T_H, T_C, and B cells.
    Perforin: a protein released by T_C cells when attached to cells displaying class I MHC-antigen complexes. It forms pore in target cell's membrane, causing it to lyse.
    Class I MHC: major histocompatibility complex, a set of glycoproteins on plasma membrane of all nucleated body cells. They present fragments of antigens, recognized by T_C cells; CD8 helps interaction.
    Class II MHC: MHC molecules found on macrophages and B cells. They present fragments of antigens, recognized by T_H-cell receptor; CD4 helps interaction.

2.

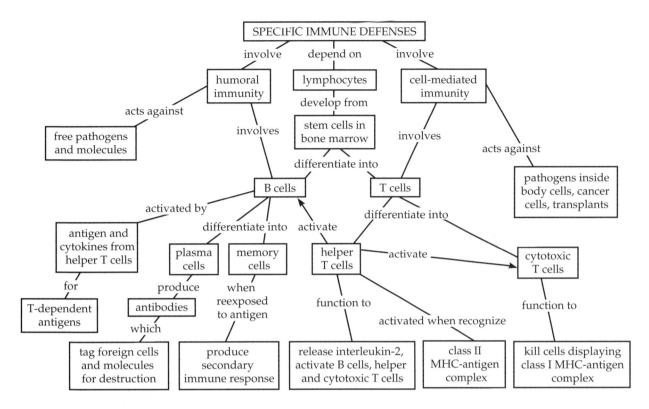

3. An antibody is a protein that typically consists of two identical light polypeptide chains and two identical heavy chains, held together by disulfide bonds in a Y-shaped molecule. The amino acid sequences in the variable sections of the light and heavy chains in the arms of the Y account for the specificity in binding between antibodies and antigenic determinants. The constant region of the antibody tail determines its effector function: five types of these constant regions correspond to five classes of antibodies. As an example, IgE immunoglobulins are attached to mast cells, which create allergic responses, and IgA is found in body secretions and is the major antibody in colostrum, the first breast milk produced.

4. According to the concept of clonal selection, T and B cells become fixed early in their development to produce a specific antigen receptor or antibody. The body's ability to respond to a great variety of antigens depends on a lymphocyte population with a huge diversity of receptor specificities. When a T or B cell encounters its antigen, it is selectively activated to proliferate and produce a clone of effector cells and a clone of memory cells, all having the same antigenic specificity.

ANSWERS TO TEST YOUR KNOWLEDGE

Multiple Choice:

1. d	7. a	13. d	19. b	25. c
2. c	8. c	14. c	20. c	26. e
3. e	9. a	15. d	21. e	27. a
4. e	10. b	16. d	22. b	28. c
5. b	11. a	17. a	23. b	29. b
6. e	12. c	18. e	24. d	30. c

CHAPTER 44: REGULATING THE INTERNAL ENVIRONMENT

■ INTERACTIVE QUESTIONS

44.1 Because the environmental temperature determines its body temperature, an ectotherm that lives where temperatures fluctuate very little or where temperatures are very high can have a body temperature more stable than or higher than that of an endotherm.

44.2 The arteries and veins in the leg are arranged as a countercurrent heat exchanger. As arterial blood flows down the leg, heat is transferred to the cooler venous blood returning to the body.

The thermal gradient along the length of the legs maintains maximal heat exchange.

44.3 **a.** vasodilation of surface arterioles, activation of sweat glands
b. vasoconstriction of surface vessels, shivering by contraction of skeletal muscles

44.4 *Uric acid* can be stored safely within the egg while the embryo develops. *Urea* is soluble and less toxic than ammonia and can be removed by the mother's blood. Urea is less expensive to produce than uric acid.

44.5 **a.** conformer   **b.** isoosmotic
c. regulator   **d.** slightly hyperosmotic
e. regulator   **f.** hypoosmotic
g. regulator   **h.** hyperosmotic
i. regulator   **j.** hyperosmotic
k. regulator   **l.** hyperosmotic

44.6 **a.** osmoregulation; remove excess water
b. both; remove excess water
c. both; conserve water
d. both; usually conserve water

44.7 **a.** water, salts, nitrogenous wastes, glucose, vitamins, and other small molecules
b. blood cells and large molecules such as plasma proteins

44.8 **a.** Bowman's capsule
b. glomerulus
c. proximal tubule
d. descending limb, loop of Henle
e. ascending limb, loop of Henle
f. distal tubule
g. collecting duct

• HCO_3^-, NaCl*, H_2O, nutrients*, and K^+ are reabsorbed from the proximal tubule. H^+* and NH_3 are secreted into the proximal tubule.

• Water is reabsorbed from the descending limb of the loop of Henle.

• NaCl diffuses from the thin segment of the ascending limb and is pumped out* of the thick segment.

• NaCl*, H_2O, and HCO_3^-* are reabsorbed from the distal tubule. K^+* and H^+* are secreted into the tubule.

• NaCl* is reabsorbed; urea and H_2O diffuse out of the collecting tubule.

(* indicates active transport.)

44.9 **a.** high blood osmolarity
b. ADH, antidiuretic hormone
c. distal tubules and collecting ducts
d. JGA, juxtaglomerular apparatus
e. renin
f. NaCl and water reabsorption
g. adrenal glands
h. aldosterone
i. Na^+ and water reabsorption
j. increase in blood pressure or volume
k. ANP, atrial natriuretic protein

44.10 **a.** Juxtamedullary nephrons; the longer the loop of Henle, the more hypertonic the urine.
b. Mammals and birds; loops of Henle are especially long in those taxa inhabiting dry habitats.

SUGGESTED ANSWERS TO STRUCTURE YOUR KNOWLEDGE

1. **a.** Isoosmotic to seawater, osmoconformer; diffusion of ammonia through body wall; ectothermic
    b. Drink water to compensate for loss to hyperosmotic seawater, gills pump out salt; ammonia diffuses through gills, little urine excreted; ectothermic, may have countercurrent heat exchanger
    c. Copious dilute urine to compensate for osmotic gain, may pump salts in through gills; ammonia diffuses across epithelium of gills; ectothermic
    d. Protonephridia, flame-bulb system, dilute fluid excreted; metabolic wastes diffuse through body wall or excreted into gastrovascular cavity; ectothermic
    e. Metanephridia, excrete dilute urine to offset osmosis; metanephridia remove wastes from coelomic fluid; ectothermic
    f. Malpighian tubules, salt and most water reabsorbed in rectum; Malpighian tubules remove wastes from hemolymph, uric acid conserves water; ectothermic, behavioral responses, some endothermic
    g. Nephron in kidney adjusts salts, no loop of Henle, cloaca reabsorbs water, thick skin reduces water loss; kidney removes nitrogenous wastes, uric acid conserves water, related to shelled egg; ectothermic, behavioral responses
    h. Nephron in kidney, juxtamedullary nephrons allow concentration of urine, nasal salt glands in sea birds, uric acid conserves water, related to shelled egg; nephrons remove nitrogenous wastes; endothermic
    i. Nephron in kidney, concentration of urine related to habitat and hydration needs, juxtamedullary nephrons allow concentration of urine; nephrons remove wastes; endothermic

2. **a.** water, salts, glucose, amino acids, vitamins, urea, other small molecules
   b. ammonia, drugs and poisons, H^+, K^+
   c. water, glucose, amino acids, vitamins, K^+, NaCl, bicarbonate

ANSWERS TO TEST YOUR KNOWLEDGE

Multiple Choice:

1. a	6. a	11. e	16. d	21. b	25. b
2. c	7. c	12. d	17. a	22. e	26. d
3. c	8. e	13. c	18. c	23. c	27. d
4. b	9. d	14. a	19. a	24. e	
5. d	10. b	15. e	20. c		

CHAPTER 45: CHEMICAL SIGNALS IN ANIMALS

■ INTERACTIVE QUESTIONS

45.1 a. chemical messages that usually travel through the bloodstream to target cells
b. chemical messages, such as interleukins, neurotransmitters, NO, growth factors, and prostaglandins, that communicate locally between cells
c. cell that produces hormones
d. modified nerve cell that produces hormones

45.2 Oxytocin and ADH are synthesized by neurosecretory cells and transported down the axons to the posterior pituitary. Releasing and inhibiting hormones are secreted by neurosecretory cells into capillaries that drain through portal vessels to a capillary bed in the anterior pituitary.

45.3 a. stimulates contraction of uterus and milk release
b. increases the permeability of kidney collecting ducts to water, increasing water reabsorption
c. promotes growth and the release of growth factors
d. various effects, depending on species, including stimulation of milk production
e. stimulates thyroid gland to produce hormones
f. stimulates activity of gonads
g. stimulates adrenal cortex to produce and secrete hormones
h. regulates skin pigment cells in amphibians
i. reduce perception of pain

45.4 a. insulin
b. glucagon
c. movement of glucose
d. glycogen hydrolysis in liver
e. islets of Langerhans in pancreas
f. glycogen hydrolysis in liver and glucose release
g. blood glucose level

45.5 a. Nervous stimulation from the hypothalamus to the adrenal medulla causes the secretion of the catecholamine hormone epinephrine, which increases blood pressure, rate and volume of heart beat, breathing rate, and metabolic rate, stimulates glycogen hydrolysis and fatty acid release, and changes blood flow patterns. Norepinephrine release maintains blood pressure, as well as contributing to the effects of epinephrine.
b. A releasing hormone from the hypothalamus stimulates ACTH release from the pituitary which stimulates the adrenal cortex to release corticosteroids. Glucocorticoids increase blood glucose through conversion of proteins and fats. Mineralocorticoids increase blood volume and pressure by stimulating the kidney to reabsorb sodium ions and water.

SUGGESTED ANSWERS TO STRUCTURE YOUR KNOWLEDGE

1. Chemical signals that bind to plasma membrane receptors initiate signal-transduction pathways that may activate cellular enzymes or affect gene expression. Steroid hormones bind with protein receptors inside a cell, and the hormone-receptor complex acts as a transcription factor to turn on (or off) specific genes.

2. Calcitonin is secreted by the thyroid in response to a rise in Ca^{2+} levels above a set point. Its effects are to stimulate Ca^{2+} deposit in bones and reduce Ca^{2+} uptake in the intestines and kidneys. As Ca^{2+} levels in the blood fall, PTH (parathyroid hormone) is secreted by the parathyroid glands and reverses the effects of calcitonin.

ANSWERS TO TEST YOUR KNOWLEDGE

Matching:

1. I f	5. A g	9. C c, released from g
2. G a	6. E b	10. J c, released from g
3. F d	7. L j	
4. B h	8. D j	

Multiple Choice:

1. e	5. d	9. e	13. a	17. a
2. d	6. a	10. e	14. d	18. b
3. d	7. c	11. b	15. c	
4. c	8. b	12. c	16. d	

CHAPTER 46: ANIMAL REPRODUCTION

■ INTERACTIVE QUESTIONS

46.1 a. Rapid colonization of new areas and perpetuation of successful genotypes in stable habitats; eliminates need to locate mate.
b. Varying genotypes and phenotypes of offspring may enhance reproductive success of parents in fluctuating environments.

46.2 a. courtship behaviors that trigger the release of gametes and may provide a means for mate selection
b. chemical signals
c. environmental signals (temperature, day length)

46.3 a. vas deferens
b. erectile tissue of penis
c. urethra
d. glans penis
e. prepuce
f. scrotum
g. testis
h. epididymis
i. bulbourethral gland
j. prostate gland
k. seminal vesicle
l. oviduct
m. eggs within follicles
n. corpus luteum
o. uterine wall (muscular layer)
p. endometrium (lining of uterus)
q. cervix
r. vagina
s. ovary

46.4 a. Spermatogonia continue to divide by mitosis and spermatogenesis occurs continuously, whereas a female's supply of primary oocytes is present at birth.
b. Each meiotic division produces four sperm, but only one ovum (and polar bodies that disintegrate).
c. Spermatogenesis is an uninterrupted process. Oogenesis occurs in stages: prophase I before birth; after puberty, meiosis I and II up to metaphase II in a maturing follicle just before ovulation; and completion of meiosis II (in humans) when a sperm cell penetrates the egg.

46.5 a. GnRH (gonadotropin-releasing hormone)
b. FSH (follicle-stimulating hormone)
c. LH (luteinizing hormone)
d. androgen
e. primary and secondary sex characteristics
f. Negative feedback loops

46.6 a. LH
b. FSH
c. estrogens
d. progesterone
e. follicular phase
f. ovulation
g. luteal phase
h. menstrual flow phase
i. proliferative phase
j. secretory phase

46.7 a. secreted by embryo, maintains corpus luteum (and thus progesterone) in first trimester
b. secreted by corpus luteum (estrogen and progesterone secretion) and later by placenta, regulates growth and maintenance of placenta, formation of mucus plug in cervix, cessation of ovulation, growth of uterus
c. secreted by fetus and posterior pituitary; stimulates uterine contractions and uterine secretion of prostaglandins during birth; controls release of milk during nursing

46.8 a. prevent fertilization—abstinence, rhythm method, condom, diaphragm, progestin minipill, coitus interruptus, cervical cap, spermicides, morning after pills (MAP)
b. prevent implantation of embryo—IUD, RU-486, MAP

c. prevent release of gametes from gonads—sterilization, combination birth control pills, subcutaneous progestin implants

The most effective methods are abstinence, sterilization, and chemical contraception. Least effective are the rhythm method and coitus interruptus.

SUGGESTED ANSWERS TO STRUCTURE YOUR KNOWLEDGE

1. Path of sperm from formation to fertilization

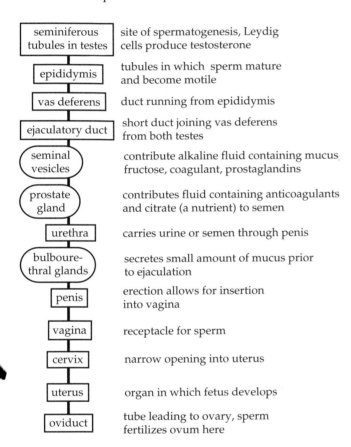

seminiferous tubules in testes	site of spermatogenesis, Leydig cells produce testosterone
epididymis	tubules in which sperm mature and become motile
vas deferens	duct running from epididymis
ejaculatory duct	short duct joining vas deferens from both testes
seminal vesicles	contribute alkaline fluid containing mucus, fructose, coagulant, prostaglandins
prostate gland	contributes fluid containing anticoagulants and citrate (a nutrient) to semen
urethra	carries urine or semen through penis
bulbourethral glands	secretes small amount of mucus prior to ejaculation
penis	erection allows for insertion into vagina
vagina	receptacle for sperm
cervix	narrow opening into uterus
uterus	organ in which fetus develops
oviduct	tube leading to ovary, sperm fertilizes ovum here

2. **a.** GnRH is a releasing hormone of the hypothalamus that stimulates the anterior pituitary to secrete FSH and LH.

b. FSH stimulates growth of follicles.

c. The LH surge is caused by an increase in GnRH production that resulted from increasing levels of estrogens produced by the developing follicle.

d. The LH surge induces the maturation of the follicle and ovulation, and it transforms the ruptured follicle to the corpus luteum.

e. LH maintains the corpus luteum, which secretes estrogens and progesterone.

f. High levels of estrogens and progesterone act on the hypothalamus and pituitary to inhibit the secretion of LH and FSH.

g. When the corpus luteum disintegrates, the production of estrogens and progesterone ceases, which allows LH and FSH secretion to begin again.

3. Birth control pills, which contain a combination of synthetic estrogens and progestin, act by negative feedback to prevent the release of GnRH by the hypothalamus and FSH and LH by the pituitary, thus preventing the development of follicles and ovulation. RU-486 is an analog of progesterone that blocks progesterone receptors in the uterus, thus preventing progesterone from maintaining the endometrium.

ANSWERS TO TEST YOUR KNOWLEDGE

Fill in the Blanks:

1. budding
2. protandrous
3. parthenogenesis
4. hermaphrodite
5. cloaca
6. estrous cycle
7. progesterone
8. urethra
9. vasocongestion
10. menopause

Multiple Choice:

1. e	5. e	9. b	13. e	17. b
2. d	6. c	10. b	14. d	18. d
3. a	7. e	11. c	15. b	19. a
4. c	8. a	12. a	16. b	20. e

CHAPTER 47: ANIMAL DEVELOPMENT

■ INTERACTIVE QUESTIONS

47.1 a. specific binding between receptors on vitelline layer and protein on acrosomal process
b. fast block to polyspermy caused by membrane depolarization during acrosomal reaction and slow block to polyspermy caused by formation of the fertilization envelope during cortical reaction

47.2 **a.** morula
b. blastula of frog embryo
c. animal pole
d. blastocoel
e. vegetal pole

47.3 **a.** ectoderm
b. mesenchyme cells
c. endoderm
d. archenteron
e. blastopore
f. archenteron
g. ectoderm
h. mesoderm
i. endoderm
j. yolk plug in blastopore

47.4 **a.** neural tube
b. neural crest
c. somite
d. archenteron
e. coelom
f. notochord

47.5 **a.** The **yolk sac** grows to enclose the yolk and develops blood vessels to carry nutrients to the embryo.

b. The **amnion** encloses a fluid-filled sac, which provides an aqueous environment for development and acts as a shock absorber.
c. The **allantois,** growing out of the hindgut, serves as a receptacle for the nitrogenous waste, uric acid.
d. The vascularized allantois, in conjunction with the **chorion** (pressed against the inside of the egg shell) provides gas exchange for the developing embryo.

47.6 **a.** epiblast—three primary germ layers, amnion, some mesoderm becomes part of placenta, allantois from outpocketing of gut
b. hypoblast—yolk sac
c. trophoblast—chorion, fetal portion of placenta

47.7 Cytoplasmic determinants in the sea urchin egg must have an unequal polar distribution. The developmental fates of cells in the animal and vegetal halves of the embryo are determined by the first horizontal division.

47.8 The developing limb would be a mirror image with posterior digits facing both forward and backward. This is the same effect as transplanting a donor ZPA to that location.

SUGGESTED ANSWERS TO STRUCTURE YOUR KNOWLEDGE

1.

	Event	Function
Acrosomal Reaction	Sperm contacts egg; hydrolytic enzymes released from acrosome	Allows acrosomal process to penetrate jelly coat
	Protein on acrosomal process attaches to receptor on vitelline layer	Assures that egg is fertilized by sperm of same species
	Enzymes digest through vitelline layer	Brings plasma membranes in contact
	Sperm and egg plasma membranes fuse; sperm nucleus enters egg cytoplasm	Two haploid nuclei in same egg; membrane fusion causes ion channels to open
	Ion channels open, Na$^+$ flows into egg, and membrane depolarizes	Depolarization creates **fast block to polyspermy**
Cortical Reaction	Calcium released in egg cytoplasm	Triggered by depolarization, calcium release causes cortical reaction
	Exocytosis by cortical granules; fertilization envelope forms	Enzymes loosen vitelline layer, swells by osmotic uptake of water, forms **slow block to polyspermy**
Activation of Egg	pH rises, rates of cellular respiration and protein synthesis increase	Activation of egg, synthesis of proteins from mRNA stockpiled in egg
	Sperm and egg nucleus fuse; DNA replication begins	Prepares for first cell division, ready for development

2.

	Cleavage	Blastula	Gastrula
Sea urchin	Holoblastic; equal-sized blastomeres	Hollow ball of cells, one cell thick, surrounding blastocoel	Invagination through blastopore to form endoderm of archenteron, migrating mesenchyme forms mesoderm
Frog	Holoblastic, but yolk impedes divisions; unequal-sized blastomeres	Blastocoel in animal hemisphere; walls several cells thick	Involution of cells at dorsal lip of blastopore, migrating cells form mesoderm and endoderm around archenteron
Bird	Meroblastic, only in cytoplasmic disk on top of yolk	Blastodisc; blastocoel cavity between epiblast and hypoblast	Migration of epiblast cells through primitive streak forms mesoderm and endoderm, lateral folds join to form archenteron
Mammal	Holoblastic, no polarity to egg, blastomeres equal	Embryonic disc forms from innercell mass with epiblast and hypoblast; trophoblast surrounds embryo	Migration of epiblast cells through primitive streak forms mesoderm and endoderm

ANSWERS TO TEST YOUR KNOWLEDGE

Multiple Choice:

1. c	4. e	7. b	10. c	13. a	16. e	19. d
2. c	5. e	8. a	11. b	14. b	17. a	20. d
3. b	6. c	9. c	12. e	15. e	18. b	

CHAPTER 48: NERVOUS SYSTEMS

■ INTERACTIVE QUESTIONS

48.1 a. dendrites
  b. cell body
  c. axon hillock
  d. axon
  e. myelin sheath
  f. Schwann cell
  g. nodes of Ranvier
  h. synaptic terminal
The impulse moves from the dendrites to the cell body and out the axon.

48.2. a. K^+ inside the cell; Na^+ outside the cell
  b. Negatively charged amino acids, proteins, phosphate, sulfate, and other ions, collectively called A^-, inside the cell; Cl^- outside the cell
  c. The inside of the cell is more negative compared to outside the cell.
  d. The membrane is more permeable to K^+ than to Na^+. K^+ leaks out of the cell; the internal anions remain inside the cell; a small amount of Na^+ leaks into the cell. There is an overall net loss of positive charge, creating a region right inside the membrane that is more negative than the other side of the membrane.

48.3 a. sodium channel
  b. Na^+
  c. potassium channel
  d. K^+
  e. sodium activation gate
  f. sodium inactivation gate
  g. membrane potential (mV)
  h. time
  1. Resting state: sodium activation gates and potassium channels are closed, sodium inactivation gates are open.
  2. Threshold: some Na^+ channels open; if Na^+ influx reaches threshold potential, additional Na^+ gates open and action potential is triggered.
  3. Depolarizing phase: sodium activation gates open and Na^+ enters cell, membrane is depolarized; voltage-gated K^+ channels are slowly beginning to open.
  4. Repolarizing phase: sodium inactivation gates close, potassium gates completely open, potassium leaves cell and membrane repolarizes.
  5. Undershoot: both gates of the sodium channels are closed, potassium channels still open, and membrane becomes hyperpolarized.

48.4 **a.** Ca²⁺ flowing into presynaptic cell
    b. synaptic terminal
    c. synaptic vesicles containing neurotransmitter
    d. presynaptic membrane
    e. postsynaptic membrane
    f. receptor with bound neurotransmitter
    g. open ion channel
    h. synaptic cleft

48.5 **a.** Temporal
    b. Spatial

48.6 The type of postsynaptic receptor and its mode of action determine neurotransmitter function. Binding of acetylcholine to receptors in heart muscle activates a signal transduction pathway that makes it harder to generate an action potential, thus slowing the strength and rate of contraction. The acetylcholine receptor in skeletal muscle cells opens ion channels and has a stimulatory effect, depolarizing the muscle cell membrane.

48.7 Sessile mollusks such as clams have little cephalization and only simple sense organs. Cephalopod mollusks such as squid have large brains, image-forming eyes, and giant axons, all of which contribute to their active predatory life and their ability to learn.

48.8 **a.** peripheral nervous system
    b. central nervous system
    c. sensory/afferent
    d. motor/efferent
    e. spinal cord
    f. effector muscles, glands, and organs
    g. autonomic
    h. somatic
    i. parasympathetic
    j. sympathetic
    k. slows heart, conserves energy
    l. acetylcholine
    m. norepinephrine at target organ (acetylcholine by preganglionic neuron)
    n. skeletal muscles
    o. acetylcholine

48.9 **a.** cerebral cortex   5
    b. thalamus   3
    c. cerebellum   1
    d. medulla   4
    e. pons   2
    f. midbrain   7
    g. hypothalamus   6

48.10 the awareness of self and surroundings in the present moment, along with the ability to access memories and think about the future

SUGGESTED ANSWERS TO STRUCTURE YOUR KNOWLEDGE

1.

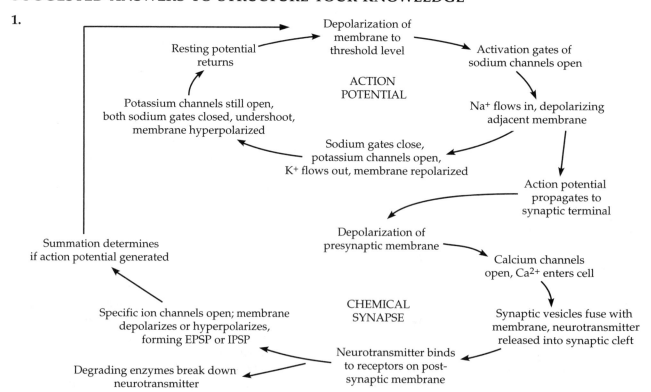

2. a. system of over 90 nuclei extending through the brainstem; regulates state of arousal; filters all sensory information going to cortex
b. in hypothalamus; functions as biological clock for circadian rhythms
c. cluster of nuclei deep in white matter of cerebrum; centers for motor coordination and planning and learning movements
d. functional center in parts of thalamus, hypothalamus, and inner cerebral cortex (olfactory cortex, hippocampus, and amygdala); centers of human emotions and memories
e. two nuclei in inner cerebral cortex; involved in limbic system and memory; amygdala is in- volved in forming emotional memories; hippocampus functions in learning and memory storage

ANSWERS TO TEST YOUR KNOWLEDGE

Multiple Choice:

1. b	7. b	13. d	19. d	25. a	31. e
2. a	8. c	14. b	20. a	26. e	32. b
3. a	9. d	15. d	21. c	27. e	
4. d	10. b	16. a	22. b	28. a	
5. e	11. a	17. b	23. c	29. c	
6. c	12. e	18. d	24. b	30. d	

CHAPTER 49: SENSORY AND MOTOR MECHANISMS

■ INTERACTIVE QUESTIONS

49.1 a. sensory transduction
b. amplification
c. transmission
d. integration

49.2 a. sclera
b. choroid
c. retina
d. fovea (center of visual field)
e. optic nerve
f. optic disk (blind spot)
g. vitreous humor
h. lens
i. aqueous humor
j. pupil
k. cornea
l. iris
m. suspensory ligament
n. ciliary body

49.3 a. inactive
b. active; retinal separates from opsin
c. bound to sodium channels
d. converted by signal transduction to GMP, dissociates from sodium channels
e. open
f. closed
g. depolarized
h. hyperpolarized
i. glutamate released
j. glutamate release slowed
k. some inhibited, some activated
l. one class released from inhibition, the other class is suppressed

49.4 a. small
b. fovea

49.5 a. auditory canal
b. malleus (hammer)
c. incus (anvil)
d. stapes (stirrup)
e. semicircular canals
f. auditory nerve
g. cochlea
h. round window
i. Eustachian tube
j. oval window
k. tympanic membrane (ear drum)

49.6 a. utricle and saccule for position, semicircular canals for movement
b. lateral line system with neuromasts
c. statocysts containing statoliths
d. The balance organs of mammals have otoliths and those of invertebrates contain statoliths. These granules settle by gravity and stimulate hair cells to signal position and movement. Sensory hairs on the hair cells in the neuromasts of fishes are embedded in a gelatinous cap which is bent by moving water and stimulates the hair cells. Hair cells in the ampullae of the semicircular canals project into a gelatinous cap which is bent by moving endolymph. All of these are mechanoreceptors.

49.7 find mates, recognize territory, navigate, communicate, and locate and taste food

49.8 a. swimming
b. running
c. flying
d. larger

49.9 a. no extra materials needed, good shock absorber; not much protection or support for lifting animal off the ground
b. quite protective; in arthropods, must be molted in order to grow, restricts size
c. various types of joints allow for flexible movement in vertebrates, good structural support; not very protective

49.10 a. sarcomere
b. I band
c. A band
d. Z line
e. thin filament (actin)
f. thick filament (myosin)
g. H zone

SUGGESTED ANSWERS TO STRUCTURE YOUR KNOWLEDGE

1. Light enters the eye through the pupil and is focused by the lens onto the retina. When a rhodopsin (photopigment in rods) absorbs light energy, it sets off a signal-transduction pathway that decreases the receptor cell's permeability to sodium and hyperpolarizes the membrane. The reduction in the release of neurotransmitter by a rod cell releases some connected bipolar cells from inhibition, which in turn generates action potentials in ganglion cells—the sensory neurons that form the optic nerve. Horizontal and amacrine cells provide lateral integration of visual information.

2. Sound waves are collected by the *pinna* and travel down the *auditory canal* to the *tympanic membrane*, where they are transmitted by the *malleus, incus,* and *stapes.* Vibration of the stapes against the *oval window* sets up pressure waves in the fluid in the *vestibular canal* within the *cochlea,* from which they are transmitted to the *tympanic canal* and then dissipated when they strike the *round window.* The pressure waves vibrate the *basilar membrane,* on which the *organ of Corti* is located within the *cochlear duct.* Tips of the hair cells arising from the organ of Corti are

embedded in the *tectorial membrane.* When vibrated, they bend, triggering a depolarization, release of neurotransmitter, and initiation of action potentials in the sensory neurons of the *auditory nerve,* which leads to the *cerebral cortex.*

3. A motor neuron releases acetylcholine into the neuromuscular junction, initiating an action potential within the muscle fiber, which spreads into the cell through the transverse tubules. The action potential changes the permeability of the sarcoplasmic reticulum membrane, which releases calcium ions into the cytosol. The calcium binds with troponin, changes the shape of the tropomyosin-troponin complex, and exposes the myosin-binding sites of the actin molecules. Heads of myosin molecules in their high-energy configuration bind to these sites, forming cross-bridges. Myosin relaxes to its low-energy state and its head bends, pulling the thin filament toward the center of the sarcomere. When myosin binds to an ATP, the cross-bridge breaks. Hydrolysis of the ATP returns the myosin to its high-energy configuration, and it binds further along the actin molecule. This sequence continues as long as there is ATP and until calcium is pumped back into the sarcoplasmic reticulum, when the tropomyosin-troponin complex again blocks the actin sites.

ANSWERS TO TEST YOUR KNOWLEDGE

Matching:

1. I	4. L	7. M	10. B
2. G	5. J	8. D	
3. N	6. Q	9. A	

Multiple Choice:

1. c	8. a	15. b	22. d
2. a	9. e	16. a	23. c
3. c	10. b	17. c	24. b
4. b	11. a	18. c	25. a
5. d	12. d	19. d	26. c
6. c	13. b	20. e	27. e
7. c	14. d	21. b	28. a

CHAPTER 50: AN INTRODUCTION TO ECOLOGY AND THE BIOSPHERE

■ INTERACTIVE QUESTIONS

50.1 a. organism: physiological, behavioral, and morphological adaptations to the biotic and abiotic environment

b. population: growth and regulation of population size

c. community: interactions such as predator-prey and competition that affect community structure

d. ecosystem: energy flow and chemical cycling
e. landscape: interactions among linked ecosystems

50.2 The aggressive African honeybee was brought to Brazil to breed a better species of bee. It escaped and has spread through the Americas, stinging animals and humans and competing with established honeybee colonies. The zebra mussel has spread from its accidental introduction in the Great Lakes throughout lakes and river systems, clogging water systems and changing native communities.

50.3 a. South-facing slopes in the northern hemisphere receive more sunlight and are warmer and drier than north-facing slopes.
b. Air temperature drops with an increase in elevation, and high-altitude communities may be similar to communities in higher latitudes.
c. The windward side of a mountain range receives much more rainfall than the leeward side. The warm, moist air rising over the mountain releases moisture, and the drier, cooler air absorbs moisture as it descends the other side.

50.4 a. + − −
b. − + +
c. + − −
d. − + −
e. + + +

50.5
6	abyssal	2	neritic
7	aphotic	3	oceanic
8	benthic	5	pelagic
1	intertidal	4	photic

50.6 a. desert
b. grassland
c. tropical forest
d. temperate forest
e. coniferous forest
f. arctic and alpine tundra

SUGGESTED ANSWERS TO STRUCTURE YOUR KNOWLEDGE

1. **a.** Ecology is the study of the relationships of organisms to their abiotic and biotic environments.
   b. Experimental manipulation of variables is done both in the laboratory and in the field, along with observational measurements of various physical factors and kinds and numbers of organisms. Mathematical models and computer simulations are useful for predicting the effects of variables, especially in situations where experimentation is impractical or impossible.

2. **a.** Biomes are characteristic communities, usually identified by the predominant vegetation and climate that range over broad geographic areas.
   b. Convergent evolution, common adaptations of different organisms to similar environments, accounts for similarities in life forms within geographically separated biomes.

ANSWERS TO TEST YOUR KNOWLEDGE

Multiple Choice:

1. b	5. d	9. c	13. e
2. b	6. b	10. b	14. d
3. d	7. e	11. c	15. a
4. e	8. c	12. e	

Matching:

1. C	4. E	7. B
2. H	5. A	8. F
3. D	6. G	

CHAPTER 51: BEHAVIORAL BIOLOGY

INTERACTIVE QUESTIONS

51.1 a. The stimulus of an increase in day length may result in reproductive behavior.
b. Breeding is most successful at this time due to warm temperature and abundant food supply

51.2 a. the red belly of an intruder into i
b. An infant's smile and grasp a sponse to visual or tactile stim

51.3 Zack measured the average quired to break whelk she and calculated the avera

by computing the total flight drops × height per drop). H erage flight height for cr breaking behavior (5.2 very close to the 5-m an optimal

51.5 The sow bugs were showing negative phototaxis and a kinesis for relative humidity.

51.6 a. Members of the same species occupy the same niche and are thus competitors for resources such as food, territory, and mates. In agonistic encounters, one animal may establish its claim to these resources.

b. ritual behavior (threat displays, appeasement gestures), reconciliation behavior, dominance hierarchies, and territories

51.7 a. Females usually have more parental investment in offspring due to the larger resource allocation in eggs and maternal care. Their reproductive fitness is improved by selecting mates that provide offspring with good genes, which may be reflected in showy displays or secondary sex characteristics such as the long eye stalks in stalk-eyed flies. A male has little resource commitment in sperm and may potentially improve his fitness by mating as often and with as many partners as possible; thus the ability to compete for mates would be selected for.

b. If a male's parental care resulted in his greater reproductive success, then the genes for this behavior would increase in frequency in a population. But if the male were parenting offspring that were not his own, then he would not pass these "parental care" genes on to the next generation in a higher proportion because he was also helping to perpetuate the genes of other males.

51.8 Mammals are mostly nocturnal; birds are diurnal.

51.9 a. a parent or a sibling

b. The individual shares more genes in common with a parent or a sibling. The coefficient of relatedness is 0.5, whereas it is only 0.125 for a cousin. Thus, with Hamilton's rule $rB > C$, the benefit to the recipient is discounted less by a larger value of r.

SUGGESTED ANSWERS TO STRUCTURE YOUR KNOWLEDGE

1. There is controversy over how much of animal behavior is innate and genetically programmed and how much is a product of experience and learning. Fixed-action patterns are clearly developmentally fixed, although experience may improve the performance of such behaviors. In other cases, genetics may set the parameters for an organism's behavior; however, experience can modify behavior, and learning is clearly evident.

2. According to the concept of Darwinian fitness, an animal's behavior should help to increase its chance of survival and reproduction of viable offspring. Survival behaviors that are genetically programmed would be most likely to be passed on, and species would evolve many innate behaviors for foraging, migrating, mating, and care of offspring. The evolution of cognitive ability would help individuals deal with novel situations. Social behaviors and communication may help reduce potentially harmful competition. Parental investment and certainty of paternity may result in differences in mating systems and parental care that influence an individual's reproductive success. Altruistic behavior may be explained on the basis of kin selection; the inclusive fitness of an animal increases if its altruistic behavior benefits related animals who may be carrying a high proportion of the same genes.

ANSWERS TO TEST YOUR KNOWLEDGE

Multiple Choice:

1. d	6. c	11. d	16. d	21. c
2. b	7. b	12. a	17. b	22. e
3. a	8. a	13. b	18. e	
4. b	9. c	14. c	19. d	
5. d	10. e	15. e	20. c	

CHAPTER 52: POPULATION ECOLOGY

STIONS

ptured in the   d trap-

52.2 a. Type I, humans and many large mammals

b. Type II, some annual plants, *Hydra* and various invertebrates, some lizards and rodents

III, many fishes and marine inverte-